SYMBOLS

n	An integer (1, 2, 3, . . .)
n	Principal quantum number
n	Turns per unit length
N	Number of turns
N	Normal (perpendicular) force
p	Object distance
p	Pressure
P	Power
PE	Potential Energy
q	Image distance
Q	Electric charge
Q	Heat
r	Angle of reflection, refraction
r	Internal resistance
r, R	Radius
R	Range
R	Rate of flow
R	Electrical resistance
s	Displacement from equilibrium position
s	Distance; displacement
S	Shear modulus
t	Time
T	Period
T	Temperature
T	Tension
v	Speed
$\mathbf{v}$	Velocity
V	Potential difference

V	Volume
w	Energy density
w	Weight
W	Work
X_C	Capacitive reactance
X_L	Inductive reactance
Y	Young's Modulus
Z	Atomic number
Z	Impedance
α (alpha)	Angular acceleration
α (alpha)	Temperature coefficient of resistivity
β (beta)	Sound intensity level
γ (gamma)	Electromagnetic photon
γ (gamma)	Surface tension
Δ (delta)	"Change in"
η (eta)	Viscosity
θ (theta)	An angle
λ (lambda)	Wavelength
μ (mu)	Coefficient of friction
μ (mu)	Magnetic permeability
ρ (rho)	Resistivity
σ (sigma)	Stefan-Boltzmann constant
Σ (sigma)	"Sum of"
τ (tau)	Torque
ϕ (phi)	Phase angle
Φ (phi)	Magnetic flux
ω (omega)	Angular velocity

POWERS OF TEN

10^{-10}	$= 0.000,000,000,1$	10^0	$= 1$
10^{-9}	$= 0.000,000,001$	10^1	$= 10$
10^{-8}	$= 0.000,000,01$	10^2	$= 100$
10^{-7}	$= 0.000,000,1$	10^3	$= 1000$
10^{-6}	$= 0.000,001$	10^4	$= 10,000$
10^{-5}	$= 0.000,01$	10^5	$= 100,000$
10^{-4}	$= 0.000,1$	10^6	$= 1,000,000$
10^{-3}	$= 0.001$	10^7	$= 10,000,000$
10^{-2}	$= 0.01$	10^8	$= 100,000,000$
10^{-1}	$= 0.1$	10^9	$= 1,000,000,000$
10^0	$= 1$	10^{10}	$= 10,000,000,000$

MULTIPLIERS FOR SI UNITS

a	atto-	10^{-18}		da	deka-	10^1
f	femto-	10^{-15}		h	hecto-	10^2
p	pico-	10^{-12}		k	kilo-	10^3
n	nano-	10^{-9}		M	mega-	10^6
μ	micro-	10^{-6}		G	giga-	10^9
m	milli-	10^{-3}		T	tera-	10^{12}
c	centi-	10^{-2}		P	peta-	10^{15}
d	deci-	10^{-1}		E	exa-	10^{18}

PHYSICS

THIRD EDITION

PHYSICS

ARTHUR BEISER

THIRD EDITION

THE BENJAMIN/CUMMINGS PUBLISHING COMPANY, INC.

Menlo Park, California • Reading, Massachusetts • London • Amsterdam • Don Mills, Ontario • Sydney

Sponsoring Editor: Philip Hagopian
Production Editor: Karen Bierstedt
Copy Editor: Robert E. Whitlock
Book and Cover Designer: Janet Bollow
Artist: Georg Klatt
Computer Illustrations: Chris Morgan
Cover Photograph: Ben F. Laposky

The "oscillons," or electronic abstractions, that appear on the cover and open each chapter were created by Ben F. Laposky by manipulating various waveforms on a cathode ray oscilloscope.

Library of Congress Cataloging in Publication Data

Beiser, Arthur.
  Physics.

  Includes index.
  1. Physics.   I. Title.
QC23.B4144   1982   530   81-17020
ISBN 0-8053-0381-2   AACR2

ABCDEFGHIJ–DO–898765432

The Benjamin/Cummings Publishing Company, Inc.
2727 Sand Hill Road
Menlo Park, California 94025

PREFACE

SCOPE OF THE BOOK

Physics is the science of matter and is concerned with its fundamental structure, properties, and behavior. Like other scientists, physicists are in search of ideas that unify their vast subject—ideas that permit them to understand the workings of atoms and molecules, of stars and galaxies, and of everything in between. Physics is thus the master science, the concepts and principles of which are drawn upon by such other sciences as chemistry, biology, geology, and astronomy in their explorations of narrower aspects of the universe around us and of which we are part. These same concepts and principles constitute the foundations of engineering and have made possible the technologically advanced world of today.

This book is intended primarily for students who will require some competence in physics in their subsequent work. The organization is straightforward, from the laws of motion to the properties of matter in bulk, vibrations and waves, heat and thermodynamics, electromagnetism, optics, and finally modern physics. Wherever possible, the origin of a significant result is described, since the essence of science is that nothing is ever pulled out of a hat but must have a firm basis in experiment, observation, or established theory. The mathematical level has been kept as low as possible. Only elementary algebra and the simplest trigonometry are employed, and these are reviewed in an appendix to the extent required by the book. The use of electronic calculators for such purposes as finding powers and roots is also described.

More than 700 illustrations (virtually all in color) are provided to help make the reader's task easier. All of the illustrations for this edition have been redrawn and many new ones have been added. Additional aids are marginal notes, worked examples, and lists of important terms and formulas at the end of each chapter. A comprehensive glossary is included at the end of the book. The nearly two thousand exercises (a quarter of them new) are divided into three categories: multiple-choice questions for quick review, straightforward exercises for practice in manipulating numbers and formulas, and problems that are moderately challenging. A few more elaborate problems serve to extend the text as well as the reader, but none is really difficult. Answers to all the multiple-choice questions and outline solutions for the odd-numbered exercises and problems are provided. Each outline solution shows how the problem can be attacked and gives the answer (usually to one more significant figure than is justified), but some of the intermediate steps are omitted. In this way, the reader who is puzzled by a certain type of problem can find out how to proceed and still have something left to do on his or her own. The outline solutions thus supplement the worked examples in the body of the text to provide about a thousand model answers to typical questions and problems. Mastering the solved exercises should bring the unsolved even-numbered ones within the competence of the reader.

A unique supplementary paperback—*Solving Physics Problems,* by John Ward of the University of Michigan—provides students with a step-by-step method for the detailed solutions to more than three hundred of the odd-numbered exercises and problems in this edition.

NEW IN THE THIRD EDITION

Although the Third Edition of *Physics* differs little in organization from its predecessor, much of the text was rewritten to sharpen the exposition and to broaden its coverage. The former two chapters on kinematics were combined into a single one, and the discussion of temperature, heat, and thermodynamics was rearranged in a more logical pattern. Among topics given expanded treatments are satellite and planetary motion, equilibrium, surface tension, the Doppler effect, the decibel scale, relative humidity, the kinetic theory of gases, heat transfer, electromagnetic waves, the scattering of light, the camera, the wave-particle duality, semiconductor devices, radioactive decay and radiometric dating, and elementary particles and their connection with the fundamental interactions. Fission and fusion energy and energy sources in general are considered in more detail than before to help the reader thread his or her way through the jungle of conflicting arguments about the most appropriate technologies to pursue in a time of increasing pollution and dwindling reserves of fossil fuels.

Some examples of new topics in this edition are pumps, the damped harmonic oscillator, musical sounds, shock waves and supersonic motion, the heat pump, the cathode ray tube and its applications, magnetic hysteresis, ac filters, the rainbow, relativisitic velocity addition, the electron microscope, atomic sizes, optical properties of solids, and radiation detection. Applications of basic physics to geology and astronomy are now included, as well as applications to biology and technology; for instance, the nature of the earth's interior, the origin of the universe and of the elements, supernovas, neutron stars, and black holes. Nevertheless, the emphasis of the book remains concentrated on the fundamentals of physics.

A major change in the third edition is to the exclusive use of SI units. The relationships between SI and British units are described because the latter are familiar to the reader, and the British equivalents of some SI results are given to help in appreciating their magnitudes, but all the exercises and problems are in SI units.

In preparing this edition of *Physics,* I have had the benefit of comments by Albert Altman, University of Lowell; Bennet B. Brabson, Indiana University; Sheldon Brown, California State University, Fresno; David J. Ernst, Texas A & M University; Ed Graff, Lake Michigan College; H. Kimball Hansen, Brigham Young University; Sanford Kern, Colorado State University; Paul L. Lee, California State University, Northridge; Bernard F. Long, Foothill College; Frederick P. Montana, Middlesex County College; Stanton Truxillo, University of Tampa; and Loren Weaver, Elgin Community College. Their help was of great value and is much appreciated.

Readers who have suggestions for improvement of the book are invited to send them to me in care of the publisher.

Arthur Beiser

INTRODUCTION

Each of us is a member of the human community, and each of us participates in some way in the continuing evolution of this community. Problems of politics and economics involve us all, and few people would wish to be ignorant of them.

Each of us is also part of the physical universe, whose evolution we can no more escape than we can escape the actions of our fellow human beings. We are made of atoms linked into molecules, liquids, and solids, and we live on a planet bound to a star that is a member of one of the galaxies that occupy the reaches of space. Atoms, galaxies, and everything in between have certain regularities of behavior in common, which are the "laws of nature." These same laws of nature underlie the technology that today dominates the lives of people and nations. An acquaintance with science is as vital to understanding our place in modern civilization as it is to understanding our place in the universe.

The scientist explores the natural world directly by experiment and observation and indirectly by abstract reasoning. Experiment and observation yield numbers, the raw material of science. Abstract reasoning yields theories—relationships that not only generalize the results of many measurements but also permit inferences to be drawn about phenomena either as yet undiscovered or impossible to examine directly. We will never visit the sun's interior or the interior of an atom, yet we know a great deal about what goes on in both places. The evidence is indirect but persuasive.

What has made science such a powerful tool for investigating nature is the constant testing of its findings, both experimental and theoretical. Nothing is accepted on anybody's personal authority, on the basis of "common sense," or because it is part of a religious or political doctrine. Because every generalization in science is subject to modification in the light of fresh evidence, science is a living body of information and not a collection of dogmas. Challengers of currently accepted ideas are not scorned— if their views are well-founded, anyway—nor is disgrace attached to a scientist whose contributions do not survive indefinitely. To rock the boat is part of the game; to overturn it is one way to win.

Broadly speaking, the science of physics is concerned with the elementary particles of which all matter is composed, the ways in which these particles interact with one another, and the behavior of the composite bodies (atoms, molecules, liquids, solids) formed by elementary particles.

Consider a sample of matter of any sort—a drop of water, a grain of sand, a blade of grass—and imagine that it is somehow chopped up into finer and finer pieces. How far can this go on? Eventually each sample is reduced to its constituent atoms, of which nearly a hundred are found in nature. If the atoms themselves are broken up, we are left with just three kinds of elementary particles: protons, neutrons, and electrons. Other kinds of elementary particles occur as well, but these are the ones primarily involved in ordinary matter. (In fact, protons and neutrons seem to be composed of still more elementary particles called quarks. Quarks apparently cannot exist independently, but the evidence for them is nevertheless quite convincing.)

Protons, neutrons, and electrons interact with one another in only four different ways, which are responsible for the structure and behavior of the entire universe. These four fundamental interactions are the strong nuclear, the weak nuclear, the electromagnetic, and the gravitational. Protons and neutrons join together by virtue of the strong nuclear interaction to form atomic nuclei, the exact compositions of which are partly determined by the weak nuclear interaction as well. Electromagnetic forces hold electrons to nuclei to form atoms. Electromagnetic interactions between nearby atoms yield stable clumps of them called molecules and larger aggregates that constitute liquids and solids. On a larger scale, gravitation pulls matter together into the planets, stars, and galaxies that populate space. Recent evidence suggests that the weak nuclear interaction is actually an aspect of the electromagnetic interaction, and that there is an intimate link between them and the strong nuclear interaction. The final step in understanding how nature operates would be a single theory that ties together all the interactions, including gravitation, and all the elementary particles. This goal, though just barely visible in the distance today, does not seem at all impossible of achievement.

Three particles and four interactions—these hardly seem enough to account for the span of things and events in everyday life, let alone for the evolution of the universe. Yet nothing else is needed—apparently—to explain phenomena ranging from the falling of a dropped stone to the development of plants and animals from inanimate matter.

To perceive such exquisite unity in the diversity around us took four hundred years of effort. The time scale is not arbitrary, for modern science began with Galileo (1564–1642). Before Galileo, explanations of the natural world were sought in terms of self-evident principles, ideas supposed to be so clearly correct that experiments were unnecessary. Such a supposedly self-evident principle was Aristotle's assertion that a heavy object falls faster than a light one; another was the belief that all motion must be sustained by an applied force. The trouble with these principles, and many others like them, is that they are wrong—in the real world, moving bodies behave quite differently. Galileo's greatest contribution, overshadowing his specific discoveries in mechanics and astronomy, was the notion that statements about nature must be tested by observation.

It is not easy to get a firm intellectual grip on nature in the rough. Nothing is as simple as it seems. We think of the earth as being round, but in fact it is shaped more like a grapefruit with a bumpy skin than like a perfect sphere. We say that the earth moves in an elliptical orbit about the sun, but in fact the orbit has wiggles no ellipse ever had. In order to extract the essence of a phenomenon, the physicist must idealize reality with the help of models. By choosing a sphere as a model for the actual earth and an ellipse as a model for its actual orbit, the most important features of the earth and its motion are isolated for analysis. If the starting point were a squashed, corrugated earth moving in an irregular path, it would be much more difficult to make any progress.

Another kind of model arises when the physicist tries to understand the microscopic world of atoms and molecules. A useful model of a gas, for example, pictures it as a collection of myriad tiny particles like billiard balls that fly about in all directions. This model is quite successful in accounting for many aspects of the behavior of gases. But it is still a model and not the whole story, since if we could somehow look directly at the structure of a gas, we would find that the particles of which it is composed are not at all like miniature billiard balls and in certain respects do not even act like particles in the usual sense.

There is no clear line between a model and a theory. Usually the term theory is reserved for a large-scale generalization that not only accounts for existing data but from which predictions about the outcome of new experiments can be made. A theory may or may not be based upon a particular model. Thus the kinetic theory of gases is a logical structure based on the model of a gas as a collection of particles in rapid, random motion. On the other hand, there is no model directly associated with Newton's theory of gravitation, which deals with concepts that do not need to be pictured in order to be related to one another. Nevertheless, the key to formulating this theory was Newton's choice of spherical models for the sun, moon, and planets and of ellipses as models for the planetary orbits.

More than one model may be convenient or necessary in some cases. Light is a notable example. In many practical applications it is quite sufficient to imagine that light consists of "rays," thin pencils of something-or-other, which are perfectly straight unless reflected or refracted, when they are bent through definite angles. A more sophisticated picture of light capable of explaining many more aspects of its behavior is the wave model. Without even specifying what kind of waves are involved, the wave model accounts for the interference, diffraction, and polarization of light, and interprets reflection and refraction in a straightforward way.

The next level of model-building brings even more rewards of understanding, but there is an unexpected (and at first glance disturbing) problem: *two* models seem to be needed, of equal validity but different areas of application. One of them is a further development of the simple wave model, with the waves now identified as electromagnetic in character. The other model employs no waves at all but regards light as a stream of tiny particles. Certain phenomena can only be interpreted on the basis of the electromagnetic wave model, others only on the basis of the particle model.

Then what is light? The answer is that light is what it is, and no model that we can visualize in terms of our everyday experience can encompass its ultimate nature. But a single, comprehensive theory of light has been devised, the conclusions of which are in superb agreement with experiment, and which makes comprehensible the need for two models. These models are not wrong, but each is limited in scope, which is true of all models in physics. Models are useful devices, but they are seldom the last word.

Nearly always in this book we shall be considering models rather than actual things. It is a good habit to try to pick out those aspects of the phenomenon under study that are incorporated in its model and those that are left out, and to ask ourselves what is gained and what is lost in each case.

CONTENTS

PHYSICS

THIRD EDITION

SCALARS AND VECTORS

1

The raw material of the physicist consists of numbers obtained by observation, and each conclusion is a quantitative statement that can be tested by measuring something and comparing the result with the predicted value. In turn, these conclusions are used by other scientists in their own investigations of nature and by engineers to design such things as bridges, aircraft, and computers. Before we actually take up the study of physics, then, we must review how physical quantities are expressed and treated mathematically. A quantity is called *scalar* when it possesses a magnitude only (time is an example), and *vector* when a certain direction is associated with it as well (force is an example). Many of the most important quantities in physics are vector in character.

1–1 UNITS

The process of measurement is essentially one of comparison. A certain standard quantity of some kind, called a *unit,* is first established, and other

TABLE 1-1
Basic SI units

Quantity	Symbol	Unit	Unit Symbol
Length	L	meter	m
Mass	m	kilogram	kg
Time	t	second	s
Electric current	I	ampere	A
Temperature	T	kelvin	K
Luminous intensity	I_L	candela	cd

Measurement requires units

quantities of the same kind are compared with it. When we say a ladder is 2.6 meters long, we mean that its length is 2.6 times a certain distance called the meter whose magnitude is fixed by international agreement. The result of every measurement must therefore have two parts, a number to answer the question "How many?" and a unit to answer the question "Of what?"

Four basic units are sufficient for most quantities

Nearly all quantities in the physical world can be expressed in terms of only four fundamental units, those of length, time, mass, and electric current. Thus every unit of area (the square meter, for instance) is the product of a length unit and a length unit; every unit of speed (the kilometer per hour, for instance) is a length unit divided by a time unit; every unit of force (the newton, for instance) is the product of a mass unit and a length unit divided by the square of a time unit; and every unit of electric charge (the coulomb, for instance) is the product of a unit of electric current and a time unit.

SI units are used in science

A *system of units* is a set of specified units of length, time, mass, and electric current from which all other units are to be derived. All units are arbitrary. Thousands of different ones have been adopted at various times and places in the course of history, a number of which survive. The most widely used is the *metric system,* which was introduced in France nearly two centuries ago. The virtues of the metric system soon led to its universal adoption by scientists, more slowly to its spread throughout the world for engineering and everyday life. The Système Internationale (SI) is the current version of the metric system and is the one emphasized in this book. The fundamental SI units are listed in Table 1–1.

The British system is rarely used outside the U.S.

Although the *British system* is on the way out everywhere else (including Great Britain), it is still widely used in commerce and industry in the United States. In the British system the units of length and mass are respectively the foot and the slug (the pound is a unit of force). Units of time and electrical units are the same in both systems.

Calculations are easy with SI units

Because a basic unit in a system may not always be convenient in size for a given measurement, other units have come into use within each system. Thus long distances are usually given in miles rather than in feet, or in kilometers rather than in meters. The great advantage of the SI system is that it is wholly decimalized, which facilitates calculation, whereas the British system is quite irregular in this respect. (1 kilometer = 1000 meters, for example, but 1 mile = 5280 feet.) The chief units in each system, together with their equivalents in the other system, are given inside the back cover of the book. The conventional prefixes used with SI units of all kinds are listed in Table 1–2; powers-of-ten notation is reviewed in Appendix B.

It is essential that standard units be absolutely constant in magnitude. Whenever possible in recent years they have been redefined in terms of quantities in nature that are regarded as invariant under all circumstances and not subject to change in time.

Prefix	Power of Ten	Abbre- viation	Pronunciation	Example
atto-	10^{-18}	a	at' toe	1 aC $= 1$ attocoulomb $= 10^{-18}$ C
femto-	10^{-15}	f	fem' toe	1 fm $= 1$ femtometer $= 10^{-15}$ m
pico-	10^{-12}	p	pee' koe	1 pf $= 1$ picofarad $= 10^{-12}$ f
nano-	10^{-9}	n	nan' oe	1 ns $= 1$ nanosecond $= 10^{-9}$ s
micro-	10^{-6}	μ	my' kroe	1 μA $= 1$ microampere $= 10^{-6}$ A
milli-	10^{-3}	m	mil' i	1 mg $= 1$ milligram $= 10^{-3}$ g
centi-	10^{-2}	c	sen' ti	1 cl $= 1$ centiliter $= 10^{-2}$ 1
kilo-	10^{3}	k	kil' oe	1 kN $= 1$ kilonewton $= 10^{3}$ N
mega-	10^{6}	M	meg' a	1 MW $= 1$ megawatt $= 10^{6}$ W
giga-	10^{9}	G	ji' ga	1 GeV $= 1$ gigaelectronvolt $= 10^{9}$ eV
tera-	10^{12}	T	ter' a	1 Tm $= 1$ terameter $= 10^{12}$ m
peta-	10^{15}	P	pe' ta	1 Ps $= 1$ petasecond $= 10^{15}$ s
exa-	10^{18}	E	ex' a	1 EJ $= 1$ exajoule $= 10^{18}$ J

TABLE 1-2
Subdivisions and multiples of SI units are widely used, and each is designated by a prefix according to the corresponding power of ten

The standard meter, for instance, is now defined as exactly 1,650,763.73 wavelengths of one of the spectral lines of the krypton isotope ^{86}Kr. The inch is in turn defined in terms of the centimeter—one inch represents exactly 2.54 cm. Measuring rods can be directly compared with the new "standard meter" in any well-equipped laboratory, and no longer must the entire world rely upon the measurement between scratches on a certain platinum-iridium bar kept at Sèvres, France.

Standard units

Often a quantity expressed in terms of a certain unit must instead be expressed in terms of another unit of the same kind. For instance, we might find from a European map that Amsterdam is 648 km from Berlin and want to know what this distance is in miles. To carry out such a conversion we must keep in mind two rules:

How to convert units

1. Units are treated in an equation in exactly the same way as any algebraic quantity, and may be multiplied and divided by one another;

2. Multiplying or dividing a quantity by 1 does not affect its value.

To convert 648 km to its equivalent in miles we note from the table inside the back cover that

$$1 \text{ km} = 0.621 \text{ mi}$$

Therefore

$$0.621 \frac{\text{mi}}{\text{km}} = 1$$

and multiplying or dividing any quantity by 0.621 mi/km does not affect its value but only changes the units in which it is given. Hence we have

$$d = (648 \text{ km}) \left(0.621 \frac{\text{mi}}{\text{km}}\right) = 402 \text{ mi}$$

since km/km $= 1$ and so drops out.

Example The piston displacement of the Volvo MD21A diesel engine is 2.11 liters, where 1 liter $= 1000$ cm^3. Express this volume in cubic inches.

Solution The displacement of the engine is $V = 2110$ cm^3. Since 1 cm $= 0.394$ in., 1 cm$^3 = (0.394$ in.$)^3 = 0.0612$ in.^{3}, and

$$V = (2110 \, \cancel{cm^3}) \left(0.0612 \, \frac{in.^3}{\cancel{cm^3}} \right) = 129 \, in.^3$$

■

Example Express a speed of 80 km/h in meters per second.

Solution Here two units are to be converted, which we can do in a single step:

$$v = \left(80 \, \frac{\cancel{km}}{\cancel{h}} \right) \left(1000 \, \frac{m}{\cancel{km}} \right) \left(\frac{1}{3600 \, s/\cancel{h}} \right) = 22 \, m/s$$

■

1–2 SCALAR AND VECTOR QUANTITIES

A scalar quantity has magnitude only

Some physical quantities require only a number and a unit to be completely specified. It is quite sufficient to say that the mass of a man is 85 kilograms, that the area of a farm is 160 acres, that the frequency of a sound wave is 660 cycles per second, or that a light bulb consumes electrical energy at the rate of 100 watts. These are examples of *scalar quantities*.

A vector quantity has magnitude and direction

Quantitative statements cannot always be confined to numbers and units alone. A man's mass is the same whether he stands on his feet or on his head, but the motion of a car traveling north at 60 km/h is hardly the same as the motion of a car traveling in a circle at 60 km/h. A quantity whose direction is significant is called a *vector quantity*. Vector quantities occur frequently in physics, and their arithmetic is different from the arithmetic of scalar quantities.

Displacement is a vector quantity

The simplest example of a vector quantity is *displacement*, which is a change in position. If we are told that a certain airplane leaves New York City and flies for 640 km (a scalar statement), we know nothing about its actual path or final destination; it could land anywhere within a circle of radius 640 km whose center is New York. If we are told instead that the airplane flies for 640 km to the west (a vector statement), we know where it is headed and can identify its destination as Cleveland (Fig. 1–1).

Velocity and force

Two other vector quantities of special significance are *velocity* and *force*. The velocity of a body incorporates both its speed (a scalar quantity) and its direction: the speed of a car might be 60 km/h, whereas its velocity is 60 km/h to the north. This may seem a minor distinction, but it is not. Given its speed and direction, we can readily calculate where a moving body will be after a certain time interval; but if we do not know the direction, we cannot determine the path of the body at all.

The newton is the SI unit of force and weight

A force is often spoken of as a "push" or a "pull." Although there is more to the concept of force than this description indicates, it is adequate for the time being, and we shall postpone a more elaborate discussion until later. Forces are important because they are responsible for all changes in motion; a force is needed to start a stationary object in motion and to deviate or stop a moving one. The direction as well as the magnitude of a force must be known to determine its effects, and vector methods must

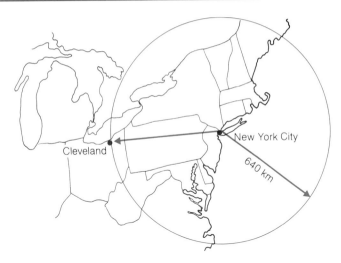

FIG. 1–1 An airplane that leaves New York and flies for a distance of 640 km may land anywhere within the circle shown. If the airplane undergoes a displacement of 640 km to the west, it lands in Cleveland. Distance is a scalar quantity; displacement is a vector quantity.

be used to treat forces as well as to treat displacements and velocities. In the SI system, the unit of force (and of weight, since weight is a force) is the *newton* (N), where 1 N = 0.225 lb.

1–3 VECTORS

The most straightforward way to represent a vector quantity is to draw a straight line with an arrowhead at one end to indicate the direction of the quantity. The length of the line is proportional to the magnitude of the quantity. A line of this kind is called a *vector*.

Figure 1–2 shows how a 640-km westward displacement might be represented by a vector. The compass rose establishes the orientation of the displacement, and the distance scale establishes the relationship between length in the diagram and the corresponding actual length.

Vector quantities other than displacements can also be represented by vectors. The length of the vector in each case is proportional to the magnitude of the quantity it represents, and its direction is the direction of the quantity. Thus a velocity whose magnitude is 25 m/s is represented on a scale of 1 mm = 1 m/s by an arrow 25 mm long, and a force whose magnitude is 1500 N is represented on a scale of 1 mm = 100 N by an arrow 15 mm long (Fig. 1–3). The directions of the arrows indicate the directions of the quantities.

The symbols of vector quantities are customarily printed in boldface type (**F** for force), while italics are used both for scalar quantities (*m* for mass) and for the magnitudes of vectors. Thus we might denote a 240-km westward displacement by the symbol **A** and its magnitude of 240 km by the symbol *A*. In handwriting, vector quantities are indicated by placing arrows over their symbols—for instance, $\vec{A}$.

FIG. 1–2 The arrow is a vector that represents a 640-km westward displacement.

A vector is a directed line segment that represents a vector quantity

Symbols for vector and scalar quantities

1–4 VECTOR ADDITION

Ordinary arithmetic is used to add two or more scalar quantities of the same kind together. Ten kg of potatoes plus 4 kg of potatoes equals 14 kg of potatoes. The same

Adding scalar quantities

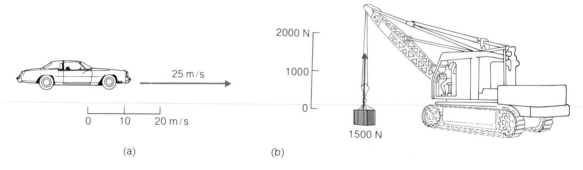

(a)   (b)

FIG. 1–3   (a) A vector that represents a velocity of 25 m/s. (b) A vector that represents a force of 1500 N.

kind of arithmetic is used for vector quantities of the same kind when the directions are the same. Thus a total upward force of 100 N is required to lift a 100-N weight, regardless of whether the force is applied by one, two, or ten people. It is the total force on a body, not the individual forces, that affects the motion of the body.

What do we do when the directions are different? A man who walks 10 km north and then 4 km west does *not* undergo a net displacement of 14 km from his starting point, even though he has walked a total of 14 km. A vector diagram provides a convenient method of determining his actual displacement. The procedure is to make a scale drawing of the successive displacements **A** and **B,** as in Fig. 1–4, and to join the starting point and terminal point with a single vector **R.** The required net displacement is **R,** whose length corresponds to 10.8 km and whose direction corresponds to 22° west of north.

Adding vector quantities

The general rule for adding vectors is illustrated in Fig. 1–5. To add **B** to **A,** shift **B** parallel to itself until its tail is at the head of **A.** In its new position **B** must still have its original length and direction. The vector sum **A + B** is a vector **R** (often called the *resultant*) drawn from the tail of **A** to the head of **B.** To find the magnitude R of the resultant **R,** measure the length of **R** on the diagram and compare this length with the scale. The direction of **R** with respect to **A** or **B** may be determined with a protractor.

The sum of two or more vectors is called their resultant

The same procedure may be used with any number of vectors: place the tail of each vector at the head of the previous one, keeping their lengths and original directions unchanged, and draw a vector **R** from the tail of the first vector to the head of the last.

FIG. 1–4   The successive displacements of 10 km north, **A,** and 4 km west, **B,** result in a net displacement **R** of 10.8 km in a direction 22° west of north.

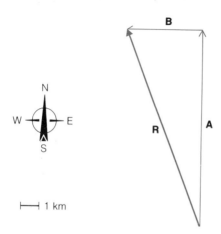

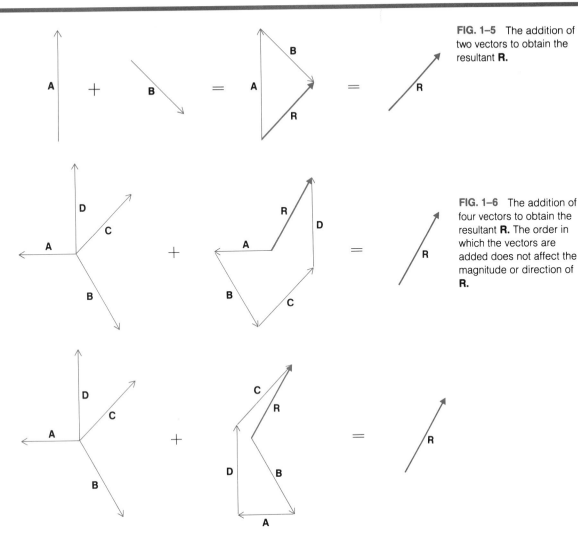

FIG. 1–5 The addition of two vectors to obtain the resultant **R.**

FIG. 1–6 The addition of four vectors to obtain the resultant **R.** The order in which the vectors are added does not affect the magnitude or direction of **R.**

R is the sum required. Figure 1–6 shows how four vectors, whose initial magnitudes and directions differ, are added together. We note that the order in which the vectors are added does not affect the result; that is,

$$\mathbf{A} + \mathbf{B} = \mathbf{B} + \mathbf{A} \qquad \qquad \textit{Vector addition} \quad (1\text{–}1)$$

Vectors can be added in any order

Example   A boat heads east at 10.0 km/h in a river that flows south at 3.0 km/h. What is the boat's velocity relative to the earth?

Solution   We add the vector that represents the river current to the vector that represents the boat's velocity relative to the river, as in Fig. 1–7. This procedure leads to a single vector, which indicates that the boat's velocity relative to the earth is 10.4 km/h in a direction about 17° south of east.

For rough calculations it is sufficient to determine the magnitude and direction of a resultant directly from a vector diagram with ruler and protractor. More accurate

FIG. 1–7   A boat headed east at 10.0 km/h in a river that flows south at 3.0 km/h moves relative to the earth at 10.4 km/h in a direction 17° south of east.

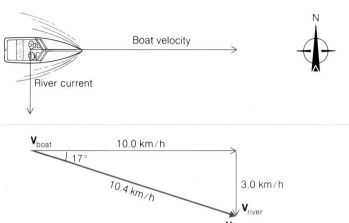

results may be obtained with the help of trigonometry. In the case of this problem, the fact that $\mathbf{v}_{boat}$ is perpendicular to $\mathbf{v}_{river}$ means that we can use the Pythagorean theorem to find the magnitude v of the resultant velocity $\mathbf{v}$. According to this theorem, the square of the hypotenuse (the long side) of a right triangle equals the sum of the squares of the other two sides. Hence

$$v^2 = v_{boat}^2 + v_{river}^2$$
$$v = \sqrt{v_{boat}^2 + v_{river}^2}$$
$$= \sqrt{(10.0\,\text{km/h})^2 + (3.0\,\text{km/h})^2} = \sqrt{109\,(\text{km/h})^2} = 10.4\,\text{km/h}$$

The angle θ between $\mathbf{v}$ and $\mathbf{v}_{boat}$ is specified by

$$\tan\theta = \frac{v_{river}}{v_{boat}} = \frac{3.0\,\text{km/h}}{10.0\,\text{km/h}} = 0.30$$

When we examine a table of trigonometric functions, we find that the angle whose tangent is closest to 0.30 is 17°, since tan 17° = 0.306. Hence θ is slightly less than 17°. To the nearest degree,

$$\theta = 17°$$

Alternatively, we can use an electronic calculator to find θ by entering 0.30 and pressing the $\tan^{-1}$ (sometimes arctan or inv tan) key. Elementary trigonometry is reviewed in Appendix A-7 and summarized in Fig. 1–8. ∎

FIG. 1–8   Basic trigonometric formulas.

$$\sin\theta = \frac{\text{opposite side}}{\text{hypotenuse}} = \frac{A}{C}$$

$$\cos\theta = \frac{\text{adjacent side}}{\text{hypotenuse}} = \frac{B}{C}$$

$$\tan\theta = \frac{\text{opposite side}}{\text{adjacent side}} = \frac{A}{B}$$

Pythagorean theorem: $A^2 + B^2 = C^2$

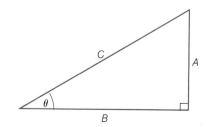

1–5   VECTOR SUBTRACTION

It is sometimes necessary to subtract one vector from another. To subtract **A** from **B**, for example, we first form the negative of **A**, denoted −**A**, a vector that has the same length as **A** but points in the opposite direction (Fig. 1–9). We then add −**A** to **B** in the usual manner. This procedure may be summarized as

The negative of a vector points in the opposite direction

$$\mathbf{B} - \mathbf{A} = \mathbf{B} + (-\mathbf{A}) \qquad\qquad \textit{Vector subtraction} \quad (1\text{–}2)$$

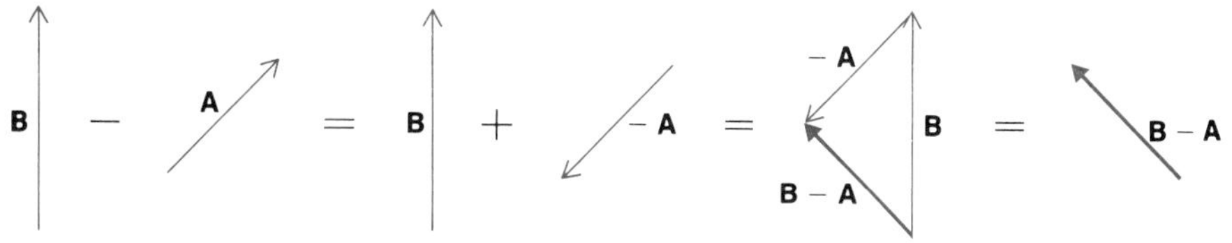

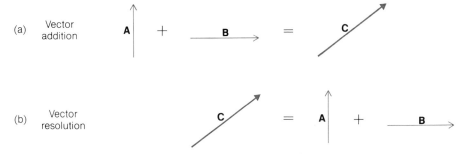

FIG. 1–9   Vector subtraction.

1–6   RESOLVING A VECTOR

Just as we can add together two or more vectors to give a single resultant vector, so we can break up a vector into two or more others. The process of replacing one vector by two or more others is called *resolving* the vector, and the new vectors are called the *components* of the original one (Fig. 1–10). Often the best way to analyze a physical problem is to resolve a vector into components. Almost invariably the components of a vector are chosen to be perpendicular to one another.

Components of a vector

Figure 1–11 shows a boy pulling a wagon with a rope at the angle θ above the ground. Only part of the force he exerts affects the horizontal motion, since the wagon moves horizontally while the force **F** is not a horizontal one. We can resolve **F** into two components, $\mathbf{F}_x$ and $\mathbf{F}_y$, where

Horizontal and vertical components

$\mathbf{F}_x$ = horizontal component of **F**

$\mathbf{F}_y$ = vertical component of **F**

It is a good habit to distinguish between the original vector and its components when all are on the same diagram by drawing two short lines across the original vector, as in Fig. 1–11. This is a reminder that the original vector is no longer to be considered since it has been replaced by its components.

FIG. 1–10   (a) In vector addition, two or more vectors are combined to form a single vector. (b) In vector resolution, a single vector is replaced by two or more others whose sum is the same as the original vector. Here **A** and **B** are the components of the vector **C**.

FIG. 1–11   The resolution of a force vector into horizontal and vertical components. The original vector is indicated by two short lines across it.

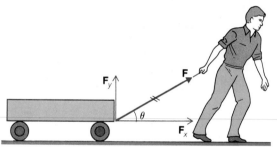

$$F_x = F \cos \theta,$$
$$F_y = F \sin \theta$$

The magnitudes of $\mathbf{F}_x$ and $\mathbf{F}_y$ are

$$F_x = F \cos \theta \tag{1-3}$$

$$F_y = F \sin \theta \tag{1-4}$$

The horizontal component $\mathbf{F}_x$ is responsible for the wagon's motion, while the vertical component $\mathbf{F}_y$ merely pulls upward on it. $\mathbf{F}_x$ is the projection of $\mathbf{F}$ in the horizontal direction, and $\mathbf{F}_y$ is the projection of $\mathbf{F}$ in the vertical direction.

Example   If the force the boy exerts on the wagon in Fig. 1–11 is 60 N and $\theta = 24°$, find F_x and F_y. Does their sum equal 60 N? Should it?

Solution   We have (Fig. 1–12)

$$F_x = F \cos \theta = (60\,\text{N})(\cos 24°) = 54.8\,\text{N}$$

$$F_y = F \sin \theta = (60\,\text{N})(\sin 24°) = 24.4\,\text{N}$$

The algebraic sum of the magnitudes F_x and F_y is 54.8 N + 24.4 N = 79.2 N, although the boy has exerted a force of only 60 N. Where is the mistake? The answer is that there is no mistake: $\mathbf{F}_x$ and $\mathbf{F}_y$ are vectors whose directions are different, so they can *only* be added vectorially. The algebraic sum of the magnitudes F_x and F_y has no meaning. If we add $\mathbf{F}_x$ and $\mathbf{F}_y$ vectorially, with the help of the Pythagorean theorem we find that

$$F^2 = F_x^2 + F_y^2$$
$$F = \sqrt{F_x^2 + F_y^2} = \sqrt{(54.8\,\text{N})^2 + (24.4\,\text{N})^2} = \sqrt{3598\,\text{N}^2} = 60\,\text{N}$$

which is equal to the force the boy exerts.   ■

Example   A car weighing 12,000 N (2700 lb) is parked in a driveway that is at a 15° angle with the horizontal. Find the components of the car's weight parallel and perpendicular to the driveway.

Solution   The weight $\mathbf{w}$ of anything is the gravitational force the earth exerts on it, a force that acts vertically downward (Fig. 1–13). Because $\mathbf{w}$ is vertical and $\mathbf{F}_\perp$ is per-

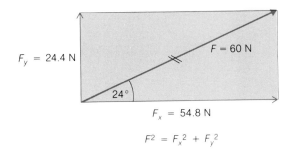

FIG. 1–12

$F_y = 24.4\,N$

$F = 60\,N$

$24°$

$F_x = 54.8\,N$

$F^2 = F_x{}^2 + F_y{}^2$

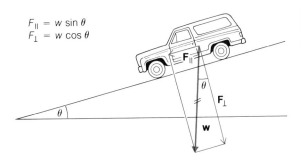

$F_\parallel = w\sin\theta$
$F_\perp = w\cos\theta$

FIG. 1–13 Weight is a force that acts vertically downward, but it can be resolved into components parallel and perpendicular to a surface of arbitrary orientation.

pendicular to the road, the angle θ between **w** and $F_\perp$ is equal to the angle $\theta = 15°$ between the road and the horizontal. Hence

$$F_\parallel = w\sin\theta = (12{,}000\,N)(\sin 15°) = 3106\,N$$
$$F_\perp = w\cos\theta = (12{,}000\,N)(\cos 15°) = 11{,}591\,N \qquad \blacksquare$$

1–7 VECTOR ADDITION BY COMPONENTS

As we have seen, it is easy to add two or more vectors together graphically by drawing them head-to-tail and joining the tail of the first vector to the head of the last to form the resultant **R**. There are two ways to solve a problem of this kind by trigonometry. One is the direct procedure of adding two of the initial vectors together, then adding a third to the sum of the first two, a fourth to the sum of the first three, and so on. This is usually a laborious process that presents many chances for error.

A better method for adding several vectors is to work in terms of their components. **Component method of** The procedure is as follows for vectors that lie in the same plane: **vector addition**

1. Resolve the initial vectors into their components in the x and y directions.
2. Add the components in the x direction to give $\mathbf{R}_x$ and add the components in the y direction to give $\mathbf{R}_y$. That is,

$$\mathbf{R}_x = x\text{-component of }\mathbf{R}$$
$$= \mathbf{A}_x + \mathbf{B}_x + \mathbf{C}_x + \ldots = \text{sum of }x\text{-components} \qquad (1\text{–}5)$$

FIG. 1–14 Vector addition by components.

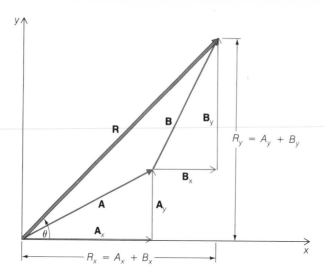

$$\mathbf{R}_y = y\text{-component of } \mathbf{R}$$

$$= \mathbf{A}_y + \mathbf{B}_y + \mathbf{C}_y + \ldots = \text{sum of } y\text{-components} \qquad (1\text{–}6)$$

3. Find the magnitude and direction of the resultant **R** from the components $\mathbf{R}_x$ and $\mathbf{R}_y$. From the Pythagorean theorem,

$$R = \sqrt{R_x^2 + R_y^2} \qquad (1\text{–}7)$$

The direction of **R** can be found from the values of the components by trigonometry; the best way to do this depends upon the problem at hand. Figure 1–14 shows the connection between head-to-tail vector addition and the component method for the case of two vectors.

Example A sailboat is headed due north at a forward speed of 6.0 knots (kn). The pressure of the wind on its sails causes the boat to move sideways to the east at 0.5 kn. A tidal current is flowing to the southwest at 3.0 kn. What is the velocity of the sailboat relative to the earth's surface? [A *knot* is a unit of speed equal to one nautical mile per hour. The nautical mile is widely used in air and sea navigation because it is the same in length as one minute (1′) of latitude, where 60′ = 1°. Since 1 nautical mile = 1.852 km = 6076 ft, 1 kn = 1.852 km/h = 1.151 mi/h.]

Solution For convenience, we shall call north the $+y$-direction, south the $-y$-direction, east the $+x$-direction, and west the $-x$-direction. With the help of Fig. 1–15 we see that the magnitudes of the components of the three velocity vectors are

$$A_x = 0 \qquad\qquad C_x = -(3.0\,\text{kn})(\cos 45°)$$

$$A_y = 6.0\,\text{kn} \qquad\qquad = -2.1\,\text{kn}$$

$$B_x = 0.5\,\text{kn} \qquad\qquad C_y = -(3.0\,\text{kn})(\sin 45°)$$

$$B_y = 0 \qquad\qquad = -2.1\,\text{kn}$$

FIG. 1–15

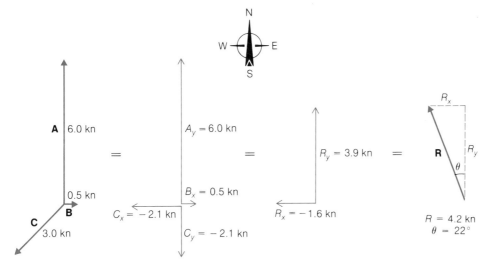

We then add together the components in each direction:

$$R_x = A_x + B_x + C_x \qquad\qquad R_y = A_y + B_y + C_y$$
$$= 0 + 0.5\,\text{kn} - 2.1\,\text{kn} \qquad = 6.0\,\text{kn} + 0 - 2.1\,\text{kn}$$
$$= -1.6\,\text{kn} \qquad\qquad\qquad = 3.9\,\text{kn}$$

The magnitude of the resultant velocity is

$$R = \sqrt{R_x^2 + R_y^2} = \sqrt{1.6^2 + 3.9^2}\,\text{kn} = \sqrt{17.8}\,\text{kn} = 4.2\,\text{kn}$$

Thus the speed of the boat relative to the earth is 4.2 kn. Its direction, as shown in the last part of the diagram, is west of north at an angle θ with north. We can find the value of θ by first finding the value of $\tan\theta$:

$$\tan\theta = \frac{R_x}{R_y} = \frac{1.6}{3.9} = 0.410 \qquad \text{so} \qquad \theta = 22° \qquad \blacksquare$$

IMPORTANT TERMS

To measure a quantity means to compare it with a standard quantity of the same kind called a **unit.** Almost all physical quantities can be expressed in terms of four fundamental measurements: length, mass, time, and electric current. In a **system of units**, a set of units of length, mass, time, and electric current is specified from which all other units in the system are derived. The **SI** (metric) system is used in everyday life in much of the world and universally by scientists.

A **scalar quantity** is one that has magnitude only. A **vector quantity** is one that has both magnitude and direction. Thus time is a scalar quantity, and force is a vector quantity.

A **vector** is an arrowed line whose length is proportional to the magnitude of some vector quantity and whose direction is that of the quantity. A **vector diagram** is a scale drawing of the various forces, velocities, or other vector quantities involved in the motion of a body.

In the graphical method of **vector addition,** the tail of each successive vector is placed at the head of the previous one, with their lengths and original directions kept unchanged. The **resultant** is a vector drawn from the tail of the first vector to the head of the last.

A vector can be **resolved** into two or more other vectors called the **components** of the original vector. Usually the components of a vector are chosen to be in mutually perpendicular directions.

IMPORTANT FORMULAS

Vector addition: $\mathbf{A} + \mathbf{B} = \mathbf{B} + \mathbf{A}$

Vector subtraction: $\mathbf{B} - \mathbf{A} = \mathbf{B} + (-\mathbf{A})$

Components of a vector: $\mathbf{A} = \mathbf{A}_x + \mathbf{A}_y$

$$A_x = A \cos \theta$$
$$A_y = A \sin \theta$$

Vector addition by components:
$$R = \sqrt{R_x^2 + R_y^2}$$
$$\text{where } R_x = A_x + B_x + C_x + \dots$$
$$R_y = A_y + B_y + C_y + \dots$$

Angle θ between $\mathbf{R}$ and $\mathbf{R}_x$: $\tan \theta = \dfrac{R_y}{R_x}$

MULTIPLE CHOICE

1. Of the following, the shortest is
   (a)   1 mm.   (b) 0.01 in.
   (c)   0.001 ft.   (d) 0.00001 km.

2. Of the following, the longest is
   (a) 10^4 in.   (b) 10^4 m.
   (c) 10^3 ft.   (d) 0.1 mi.

3. The number of cubic centimeters in a cubic foot is approximately
   (a) 1.7×10^3.   (b) 1.7×10^4.
   (c) 2.8×10^4.   (d) 1.7×10^5.

4. The number of seconds in a day is closest to
   (a) 10^4.   (b) 10^5.
   (c) 10^6.   (d) 10^7.

5. The number of seconds in a month is approximately
   (a) 2.6×10^6.   (b) 2.6×10^7.
   (c) 2.6×10^8.   (d) 2.6×10^9.

6. Of the following units, which could be associated with a vector quantity?
   (a)   meters/minute   (b) quarts/second
   (c)   hours   (d) cubic feet

7. Which of the following statements is incorrect?
   (a) All vector quantities have directions.
   (b) All vector quantities have magnitudes.
   (c) All scalar quantities have directions.
   (d) All scalar quantities have magnitudes.

8. The minimum number of unequal forces whose vector sum can equal zero is
   (a)   1.   (b) 2.
   (c)   3.   (d) 4.

9. An engine block is supported by a rope hoist attached to an overhead beam. When the block is pulled to one side by a horizontal force exerted by another rope, the tension in the rope hoist
   (a) is less than before.
   (b) is unchanged.
   (c) is greater than before.
   (d) may be any of the above, depending on the magnitude of the horizontal force.

10. Which of the following pairs of displacements cannot be added to give a resultant displacement of 2 m?
   (a) 1 m and 1 m   (b) 1 m and 2 m
   (c) 1 m and 3 m   (d) 1 m and 4 m

11. Which of the following sets of forces cannot have a vector sum of zero?
   (a) 10, 10, and 10 N   (b) 10, 10, and 20 N
   (c) 10, 20, and 20 N   (d) 10, 20, and 40 N

12. Which of the following sets of displacements might be able to return a car to its starting point?
   (a) 2, 8, 10, and 25 km
   (b) 5, 20, 35, and 65 km
   (c) 60, 120, 180, and 240 km
   (d) 100, 100, 100, and 400 km

13. A displacement of 9 m and another 6 m can be added to give a resultant displacement of
   (a) 0 m.   (b) 1.5 m.
   (c) 4 m.   (d) 16 m.

14. A 10-N force and a 5-N force act on a body. The resultant force on the body must be
   (a) between 5 and 10 N.   (b) between 5 and 15 N.
   (c) more than 5 N.   (d) more than 10 N.

15. A man walks 8 km north and then 5 km in a direction 60° east of north. His resultant displacement from his starting point is
   (a) 11 km.   (b) 12 km.
   (c) 13 km.   (d) 14 km.

16. The resultant of a 4-N force acting upward and a 3-N force acting horizontally is
   (a) 1 N.   (b) 5 N.
   (c) 7 N.   (d) 12 N.

17. The angle between the resultant of Question 16 and the vertical is approximately

    (a) 37°.   (b) 45°.

    (c) 53°.   (d) 60°.

18. An airplane travels 100 km to the north and then 200 km to the east. The displacement of the airplane from its starting point is approximately

    (a) 100 km.   (b) 200 km.

    (c) 220 km.   (d) 300 km.

19. At what angle east of north should the airplane of Question 18 have headed in order to reach its destination in a straight flight?

    (a) 22°   (b) 45°

    (c) 50°   (d) 63°

20. A vector **A** lies in a plane and has the components A_x and A_y. The magnitude A_x of $\mathbf{A}_x$ is equal to

    (a) $A - A_y$.   (b) $\sqrt{A} - \sqrt{A_y}$.

    (c) $\sqrt{A - A_y}$.   (d) $\sqrt{A^2 - A_y^2}$.

21. An escalator has a velocity of 3.0 m/s at an angle of 60° above the horizontal. The vertical component of its velocity is

    (a) 1.5 m/s.   (b) 1.8 m/s.

    (c) 2.6 m/s.   (d) 3.5 m/s.

EXERCISES

1. Is it correct to say that scalar quantities are abstract, idealized quantities with no precise counterparts in the physical world, whereas vector quantities can be said to properly represent reality?

2. What kind of quantity is the magnitude of a vector quantity? What kind of quantity is the resultant of two vector quantities of the same kind?

3. The resultant of three vectors is zero. Must they all lie in a plane?

4. A bird lands on an overhead power cable, which sags slightly under its weight. Is it possible to prevent any such sagging by applying enough tension to the cable?

5. The tallest tree in the world is a Sequoia in California that is 368 ft high. How high is this in meters? In kilometers?

6. The density of the element osmium is 22.6 g/cm³. Express this density in kilograms per cubic meter.

7. An acre contains 4840 yd², where 1 yd = 3 ft. How many square meters is this? How many acres are there in a square kilometer?

8. Find the volume in cubic meters of a swimming pool whose dimensions are 50 ft × 20 ft × 6 ft.

9. A "board foot" is a unit of lumber measure that corresponds to the volume of a piece of wood 1 ft square and 1 in. thick. How many cubic inches are there in a board foot? How many cubic feet? How many cubic centimeters?

10. Water emerges from the nozzle of a fountain in Arizona at 75 km/h and reaches a height of 170 m. Express the water speed in meters per second and in miles per hour and the height in feet.

11. The speedometer of a European car is calibrated in kilometers per hour. What is the car's speed in miles per hour when the speedometer reads 50?

12. In 1968 the horse Dr. Fager ran a mile in 1 min 32.2 s. Find his average speed in meters per second.

13. A person walks 70 m to an elevator and then ascends 40 m. Find the magnitude and direction of the person's displacement from the starting place.

14. A driver becomes lost and travels 12 km west, 5 km south, and then 8 km east. Find the magnitude and direction of the car's displacement from the starting place.

15. A woman pushes a 50-N lawn mower with a force of 25 N. If the handle of the lawn mower is 45° above the horizontal, how much downward force is being exerted on the ground by the lawn mower?

16. A weight of 80 N is at rest on an inclined plane that makes an angle of 40° with the horizontal. Find the components of this weight parallel and perpendicular to the plane.

17. A sailboat cannot sail directly to windward but must "tack" back and forth at a certain angle with respect to the direction from which the wind is blowing. Which sailboat has the greater component of velocity to windward, the *Alpha* whose velocity is 5 km/h at an angle of 40° off the wind, or the *Beta* whose velocity is 6 km/h at an angle of 50° off the wind?

18. The ship *Salmonella* is heading due east at 18 km/h in the presence of a north wind of 12 km/h. What is the horizontal component of velocity of the smoke relative to the ship as it leaves the funnel? Relative to the earth?

19. A 100-N sack of potatoes is suspended by a rope. A man pushes sideways on the sack with a force of 40 N. What is the tension in the rope?

PROBLEMS

1. Two billiard balls are rolling on a flat table. One has the velocity components $v_x = 1$ m/s, $v_y = 2$ m/s. The other has the velocity components $v_x = 2$ m/s, $v_y = 3$ m/s. If both balls started from the same point, what is the angle between their paths?

2. Two cars leave a crossroads at the same time, one headed north at 50 km/h and the other headed east at 70 km/h. How far apart are they after 0.5 h? After 2 h?

3. A man on the ground observes an airplane climbing at an angle of 37° above the horizontal. He gets in his car and by driving at 70 km/h is able to stay directly below the airplane. What is the airplane's speed?

4. An airplane whose speed is 150 km/h climbs from a runway at an angle of 20° above the horizontal. What is its altitude 1 min after takeoff? How many kilometers does it travel in a horizontal direction in this period of time?

5. An airplane is heading southeast when it takes off at an angle of 25° above the horizontal at 200 km/h. (a) What is the vertical component of its velocity? (b) What is the horizontal component of its velocity? (c) What is the component of the velocity of the plane toward the south?

6. On a windless day, raindrops that fall on the side windows of a car moving at 10 m/s are found to make an angle of 50° with the vertical. Find the speed of the raindrops relative to the ground.

7. A horizontal and a vertical force combine to give a resultant force of 10 N that acts in a direction 40° above the horizontal. Find the magnitudes of the horizontal and vertical forces.

8. The resultant of two perpendicular forces has a magnitude of 40 N. If the magnitude of one of the forces is 25 N, what is the magnitude of the other force?

9. Two tugboats are towing a ship. Each exerts a horizontal force of 5.0 tons, and the angle between the two ropes is 30°. What is the resultant force exerted on the ship?

10. The following forces act on an object resting on a level, frictionless surface: 10 N to the north, 20 N to the east, 10 N at an angle 40° south of east, and 20 N at an angle 50° west of south. Find the magnitude and direction of the resultant force acting on the object.

11. An airplane flies 200 km east from city *A* to city *B*, then 200 km south from city *B* to city *C*, and finally 100 km northwest to city *D*. How far is it from city *A* to city *D*? In

what direction must the airplane head to return directly to city *A* from city *D*?

12. A boat moving at 15 km/h is crossing a river 2 km wide in which the current is flowing at 5 km/h. (a) If the boat heads directly for the other shore, how long will the trip take? (b) In what direction should the boat head if it is to reach a point on the other shore directly opposite the starting point? (c) How long will the crossing take in case (b)?

13. The ketch *Minots Light* is heading northwest at 7 knots through a tidal stream which is flowing southwest at 3 knots. Find the magnitude and direction of its velocity relative to the earth's surface.

14. A car is headed east at 50 km/h and a truck is headed northwest at 70 km/h. Find (a) the velocity of the car relative to the truck, and (b) the velocity of the truck relative to the car.

15. The yacht *Quicksilver* is headed north at 7.0 knots and the yacht *Lianda* is headed east at 8.0 knots. Find (a) the velocity of *Quicksilver* relative to *Lianda,* and (b) the velocity of *Lianda* relative to *Quicksilver.*

16. The vector **A** has a magnitude of 20 cm and points 20° clockwise from the +*y*-direction. The vector **B** has a magnitude of 10 cm and points 60° clockwise from the +*y*-direction. Find the magnitude and direction of **A** + **B**, **A** − **B**, and **B** − **A**.

17. The vector **A** has a magnitude of 10 cm and points 37° clockwise from the +*y*-direction. The vector **B** has a magnitude of 10 cm and points 37° clockwise from the +*x*-direction. Find the magnitude and direction of **A** + **B**, **A** − **B**, and **B** − **A**.

ANSWERS TO MULTIPLE CHOICE

1. b	**6.** a	**11.** d	**16.** b
2. b	**7.** c	**12.** c	**17.** a
3. c	**8.** c	**13.** c	**18.** c
4. b	**9.** c	**14.** b	**19.** d
5. a	**10.** d	**15.** a	**20.** d
			21. c

ANALYZING MOTION

2

Everything in the physical world is in motion, from the elementary particles within atoms to the largest galaxies of stars. Since the goal of physics is an understanding of the nature and behavior of this world, physicists are directly concerned with things in motion in nearly all their work. We shall find in this chapter that introducing the quantities speed, velocity, and acceleration to supplement the fundamental ones of time and distance permits us to deal with many kinds of motion, including that characteristic of falling bodies, in a simple, straightforward way. The use of vector methods extends the scope of the analysis to include such complex motions as that of a projectile.

2–1 FRAME OF REFERENCE

When something has changed its position with respect to its surroundings, we say it has *moved*.

All motion is relative to a frame of reference

There are two separate ideas here. One is that of *change*: when something has moved, the world is not exactly the same as it was before. The other idea is that of *frame of reference*: if we are to notice that something is moving, we must be able to check its position relative to something else. The choice of a frame of reference for reckoning motion depends upon the situation. For a car, the most convenient frame of reference is the earth's surface; for a sailor, it is the deck of the ship; for a planet, it is the sun; for an electron in an atom, it is the atom's nucleus.

Choice of frame of reference may be important

The key to solving many problems in physics is the proper selection of a frame of reference. Newton was able to interpret the motions of the earth and planets in terms of the gravitational pull of the sun only because he recognized the sun as the center of the solar system. If instead he had considered the earth as the center, with the sun and the other planets moving around it, their paths would have appeared to be very irregular. It is unlikely that he would have been able to discover the law of gravitation by analyzing these seemingly complicated motions.

How shall we describe the motion of something once an appropriate frame of reference has been chosen? All sorts of motions are possible, even in the simple case of a car on a road. The car may be headed in either of two directions; it may travel at any speed up to a certain maximum; the speed does not have to be constant but can increase or decrease from time to time; and the car can even come to a stop and then reverse its direction. What we need is a procedure for specifying exactly what is happening as the car (or anything else) changes its position as time goes on.

2–2   SPEED AND VELOCITY

As we all know, the *speed* of a moving object is the rate at which it covers distance. It is important to distinguish between average speed and instantaneous speed. The *average speed* $\bar{v}$ of something that travels the distance s in the time interval t is

Average speed

$$\bar{v} = \frac{s}{t} \qquad\qquad Average\ speed \quad (2\text{–}1)$$

$$Average\ speed = \frac{distance\ traveled}{time\ interval}$$

Thus a car that has gone 180 km in 3 h had an average speed of

$$\bar{v} = \frac{s}{t} = \frac{180\,\text{km}}{3\,\text{h}} = 60\,\text{km/h}$$

The average speed of the car is only part of the story of its journey, however, because knowing $\bar{v}$ does not tell us whether the car had the same speed for the entire 3 h or sometimes went faster than 60 km/h and sometimes slower.

Let us suppose we are in a car starting from rest and that we note the times at which its odometer indicates it has covered 100 m, 200 m, 300 m, and so on. The data might appear as in Table 2–1. When we plot these data on a graph, as in Fig. 2–1, we find that the line joining the various points is not a straight line but shows a definite upward curve. In each successive equal time interval (as marked off at the bottom of the graph), the car covers a greater distance than before—it is going faster and faster.

Instantaneous speed

Even though the car's speed is changing, at every moment it has a certain definite value (which is what is indicated by its speedometer). To find this *instantaneous speed*

Total distance, m	0	100	200	300	400	500	TABLE 2–1
Elapsed time, s	0	28	40	49	57	63	

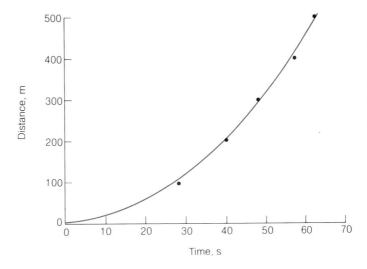

FIG. 2–1 A graph of the data in Table 2–1. Each point represents the result of one measurement. A smooth curve does not quite go through all the points, but we may reasonably attribute the discrepancies to experimental error.

v at a particular time t, we draw a straight line tangent to the distance-time curve at that value of t. The length of the line does not matter. Then we determine v from the tangent line from the formula

$$v = \frac{\Delta s}{\Delta t} \qquad\qquad \textit{Instantaneous speed} \quad (2\text{–}2)$$

where Δs is the distance interval between the ends of the tangent and Δt is the time interval between them. (Δ is the Greek capital letter *delta*.) The instantaneous speed of the car at $t = 40$ s is, from Fig. 2–2,

$$v = \frac{\Delta s}{\Delta t} = \frac{100\,\text{m}}{10\,\text{s}} = 10\,\text{m/s}$$

Table 2–2 shows the instantaneous speeds of the car at 10-s intervals as determined from the graph.

When the instantaneous speed of an object does not change, it is moving at *constant speed*. Figure 2–3 is a distance-time graph of a car that has a constant speed of 7.5 m/s; the curve is, of course, a straight line. For the case of constant speed, Eq. (2–2) gives us these useful formulas: **Constant speed**

$$s = vt \qquad (v = \text{constant}) \qquad\qquad (2\text{–}3)$$

Distance = speed × time

$$t = \frac{s}{v} \qquad (v = \text{constant}) \qquad\qquad (2\text{–}4)$$

Time = $\dfrac{\text{distance}}{\text{speed}}$

TABLE 2–2	Elapsed time, s	0	10	20	30	40	50	60
	Instantaneous speed, m/s	0	2.5	5.0	7.5	10	12.5	15

FIG. 2–2 The procedure for finding the instantaneous speed at t = 40 s from the data of Table 2–1. Because the graph of distance versus time is not a straight line, the instantaneous speed is changing continuously.

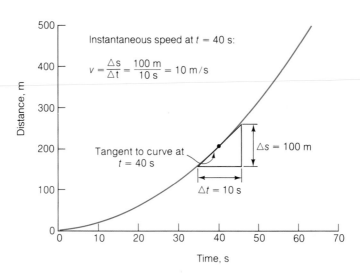

Instantaneous speed at t = 40 s:

$$v = \frac{\Delta s}{\Delta t} = \frac{100 \text{ m}}{10 \text{ s}} = 10 \text{ m/s}$$

Tangent to curve at t = 40 s

Δs = 100 m

Δt = 10 s

FIG. 2–3 The distance-time graph of a car traveling at a constant speed of 7.5 m/s is a straight line.

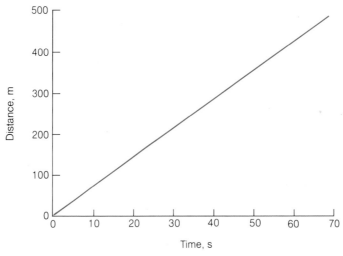

Thus if the speed of something is constant, we can predict exactly how far it will go in a given period of time; or, given the distance, we can determine the time required.

Example   Echoes return in 2.5 s to a person standing in front of a cliff. How far away is the cliff? The speed of sound in air at sea level is 343 m/s when the temperature is 20°C.

Solution The total distance a sound travels in 2.5 s is

$$s = vt = (343 \text{ m/s})(2.5 \text{ s}) = 858 \text{ m}$$

The sound must travel to the cliff and back, and so the cliff is half this distance away, which is $\frac{1}{2} \times 858 \text{ m} = 429 \text{ m}$. ∎

Example How many seconds does a car traveling at the constant speed of 50 km/h take to cover 100 m?

Solution Since 100 m = 0.1 km,

$$t = \frac{s}{v} = \frac{0.1 \text{ km}}{50 \text{ km/h}} = 0.002 \text{ h}$$

There are 3600 s in an hour, so

$$t = (0.002 \text{ h})(3600 \text{ s/h}) = 7.2 \text{ s}$$

With practice, it will be found easier to perform a calculation of this kind in one step:

$$t = \frac{s}{v} = \left(\frac{0.1 \text{ km}}{50 \text{ km/h}}\right)\left(3600 \frac{\text{s}}{\text{h}}\right) = 7.2 \text{ s}$$ ∎

The speed of a moving object is a scalar quantity. The object's *velocity,* however, includes the direction in which it is moving, and is a vector quantity. If the object undergoes a displacement **s** in the time interval t, its average velocity $\bar{\mathbf{v}}$ during this interval is

Velocity is a vector quantity

$$\bar{\mathbf{v}} = \frac{\mathbf{s}}{t} \qquad\qquad \textit{Average velocity} \quad (2–5)$$

Instantaneous velocity **v** is the value of $\Delta\mathbf{s}/\Delta t$ at a particular moment, and for straight-line motion is found by the same procedure as that used for instantaneous speed v but with the direction specified as well. In general, we must find $\Delta\mathbf{s}$ by vector subtraction; in Chapter 5 this is done for motion along a circular path.

Evidently "speed" has two meanings: the rate of covering distance and the magnitude of velocity. In many situations both meanings give the same result. A woman who runs north for 400 m on a straight path in 64 s has an average speed of $\bar{v} = 6.25$ m/s whether we consider $\bar{v}$ as the rate at which she covers distance or as the magnitude of her northward average velocity $\bar{\mathbf{v}}$. But if the woman runs one lap around an oval 400-m track in 64 s, her displacement from her starting point is $\mathbf{s} = 0$ and so her average velocity is $\bar{\mathbf{v}} = 0$. If we consider the woman's average speed here as the magnitude of her velocity, $\bar{v} = 0$. However, if we are using average speed as the rate of covering distance, $\bar{v} = 6.25$ m/s as before. We must therefore be careful as to which meaning "speed" has in a given case. The instantaneous speed of an object and the magnitude of its instantaneous velocity are always the same.

Speed has two meanings

2–3 ACCELERATION

In the real world few objects move at constant velocity for very long. Something whose velocity changes is said to be *accelerated*. This term is applied regardless of whether **v** is increasing, decreasing, or changing in direction. Figure 2–4 shows three examples of accelerated motion.

Force and acceleration

What is it that causes something to be accelerated? The answer is a force, which for the moment we can think of as a push or a pull. In the absence of a force, an object at rest remains at rest and an object in motion continues in motion at constant velocity. These matters are discussed in detail in Chapter 3.

Acceleration is rate of change of velocity

Just as velocity is the rate of change of displacement with time, *acceleration* is the rate of change of velocity with time. The symbol for acceleration is **a,** since it is a vector quantity. For the time being we are considering straight-line (or *linear*) motion only, so we shall consider acceleration as a scalar quantity and refer to acceleration as the rate of change of speed.

If an object's speed is v_0 to begin with and changes to v_f during a time interval t, its acceleration is given by the formula

$$a = \frac{v_f - v_0}{t} \qquad \qquad \textit{Acceleration} \quad (2\text{–}6)$$

$$\text{Acceleration} = \frac{\text{change in speed}}{\text{time interval}}$$

FIG. 2–4 The successive positions of three accelerated cars after equal periods of time. (a) The car is going faster and faster, so it travels a longer distance in each period of time. (b) The car is going slower and slower, so it travels a shorter distance in each period of time. (c) The magnitude of the car's velocity is constant, but its direction changes.

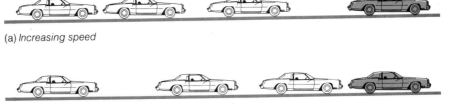

(a) *Increasing speed*

(b) *Decreasing speed*

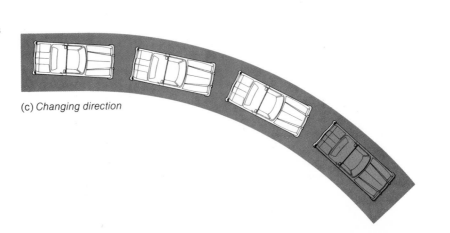

(c) *Changing direction*

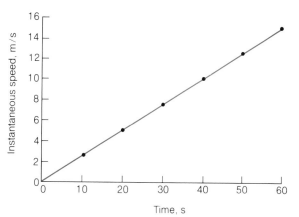

FIG. 2-5 A graph of instantaneous speed versus time for the data of Table 2–2. Although the car's speed is not constant, it varies in a uniform way with time. This is an example of constant acceleration.

When the final speed v_f is greater than the initial speed v_0, the acceleration is positive, which signifies that the object is going faster and faster. When the final speed is less than the initial speed, the acceleration is negative, which signifies that the object is going slower and slower.

Positive and negative accelerations

Let us return to the data on the car of Tables 2–1 and 2–2 and plot a graph of its instantaneous speed v versus time t, as in Fig. 2–5. All the points lie on a straight line, which means that v is directly proportional to t. Although the car's speed is not constant, it varies in a uniform way with time: as time goes on, the speed increases exactly in proportion. Therefore the car's acceleration is constant.

Constant acceleration

From its definition, acceleration is expressed in terms of

$$\frac{\text{Speed}}{\text{time}} = \frac{\text{distance/time}}{\text{time}} = \frac{\text{distance}}{\text{time} \times \text{time}} = \frac{\text{distance}}{\text{time}^2}$$

Units of acceleration

An object whose speed increases by 10 m/s in each second would accordingly have its acceleration expressed as

$$a = 10\frac{\text{m/s}}{s} = 10\,\text{m/s}^2$$

The initial speed of the car whose motion we have been considering is $v_0 = 0$ and after, say, $t = 20$ s it is $v_f = 5.0$ m/s. Hence the car's acceleration is

$$a = \frac{v_f - v_0}{t} = \frac{(5.0 - 0)\,\text{m/s}}{20\,\text{s}} = 0.25\,\text{m/s}^2$$

If we make the same calculation at the later time $t = 40$ s, we have, since v_f is now 10 m/s according to Table 2–2,

$$a = \frac{(10 - 0)\,\text{m/s}}{40\,\text{s}} = 0.25\,\text{m/s}^2$$

The value of a is the same because the acceleration is constant. If the acceleration were not constant, different values of a would be obtained at different times.

Not all accelerations are constant, of course, but a great many real motions are best understood by idealizing them in terms of constant accelerations.

FIG. 2–6 Instantaneous speed versus time for a certain car.

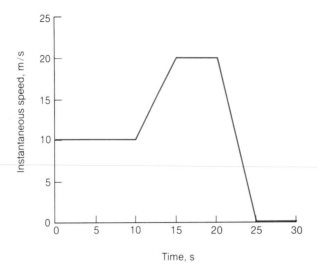

Time, s

Example   Discuss the motion of the car whose speed-time graph is shown in Fig. 2–6.

Solution   A horizontal line on a v-t graph means that v does not change, so the acceleration is 0 at first. A line sloping upward means a positive acceleration (v increasing), and a line sloping downward means a negative acceleration (v decreasing). Hence the car begins to move at the constant speed of 10 m/s, then is accelerated at 2 m/s^2 to the speed of 20 m/s, travels at 20 m/s for 5 s, and finally undergoes a negative acceleration of -4 m/s^2 until it comes to rest at $t = 25$ s.   ∎

2–4   SPEED AND ACCELERATION

Final speed of accelerated object

In the event an object starts to accelerate from some initial speed v_0, its change in speed at during the time interval t in which the acceleration a (assumed constant) occurs is added to v_0 (Fig. 2–7). Hence the final speed v_f at a time t after the acceleration begins is the initial speed v_0 plus the change in speed at:

$$v_f = v_0 + at \qquad (2\text{–}7)$$

Final speed = initial speed + speed change

The same formula can be obtained by solving Eq. (2–6) for v_f.

Equation (2–7) can be used regardless of the algebraic signs of v_0 and a. In the case of a car, a negative speed customarily implies motion opposite to a specified direction, which we might refer to as backward motion (though the car may actually be facing either way). A negative acceleration means that the car is slowing if v_0 is positive or increasing in backward speed if v_0 is negative.

If v_0 is negative and a is positive, the car is traveling backward at a diminishing speed. The importance of keeping track of the algebraic signs of the various quantities is evident.

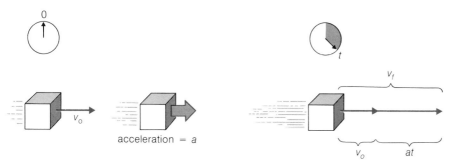

FIG. 2–7 The final speed v_f of an accelerated body is equal to its initial speed v_0 plus the change at that occurred during the interval t.

Example A car has an initial speed of 20 m/s and an acceleration of -1 m/s^2. Find its speed after 10 s and after 50 s.

Solution The speed after 10 s is

$$v_f = v_0 + at = 20\,\text{m/s} - (1\,\text{m/s}^2)(10\text{s}) = (20 - 10)\,\text{m/s} = 10\,\text{m/s}$$

which is half what it was initially. After 50s, the speed of the car, if a stays the same, is

$$v_f = v_0 + at = 20\,\text{m/s} - (1\,\text{m/s}^2)(50\,\text{s}) = (20 - 50)\,\text{m/s} = -30\,\text{m/s}$$

which is in the opposite direction and greater than the original speed v_0. ◼

2–5 DISTANCE AND ACCELERATION

As we have seen, the final speed of an object that has been accelerated for the time t is given by $v_f = v_0 + at$. The next question to ask is, how far does the object go during the time interval t?

We know that, in general, $s = \bar{v}t$, so if we can determine the average speed during the time interval t we can also find s, the distance through which the body moves. Because the acceleration a is constant, v is changing at a uniform rate, and

Average speed during constant acceleration

$$\bar{v} = \frac{\text{initial speed} + \text{final speed}}{2}$$

(If a is not constant, this formula does not hold.)

Here the initial speed is v_0 and the final speed is $v_0 + at$; hence

Distance covered by accelerated object

$$\bar{v} = \frac{v_0 + v_0 + at}{2} = v_0 + \tfrac{1}{2}at$$

The distance traveled is accordingly

$$s = \bar{v}t = (v_0 + \tfrac{1}{2}at) \times t$$

which yields the very useful formula

$$s = v_0t + \tfrac{1}{2}at^2 \qquad \textit{Distance under constant acceleration} \quad (2\text{–}8)$$

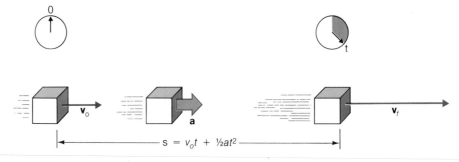

Figure 2–8 illustrates this result. When the initial speed is $v_0 = 0$, we have simply

$$s = \tfrac{1}{2}at^2 \qquad (v_0 = 0) \tag{2–9}$$

Example A car has an initial speed of 20 m/s and an acceleration of -1 m/s^2. Find its displacement after the first 10 s and after the first 50 s from the moment the acceleration begins.

Solution In the first 10 s the car travels

$$\begin{aligned} s &= v_0 t + \tfrac{1}{2}at^2 = (20\,\text{m/s})(10\text{s}) - \tfrac{1}{2}(1\,\text{m/s}^2)(10\,\text{s})^2 \\ &= (200 - 50)\,\text{m} = 150\,\text{m} \end{aligned}$$

and its displacement after 50 s is

$$\begin{aligned} s &= v_0 t + \tfrac{1}{2}at^2 = (20\,\text{m/s})(50\,\text{s}) - \tfrac{1}{2}(1\,\text{m/s}^2)(50\,\text{s})^2 \\ &= (1000 - 1250)\,\text{m} = -250\,\text{m} \end{aligned}$$

This result means that the car is 250 m *behind* its starting point after 50 s have elapsed (Fig. 2–9). ∎

2-6 DISTANCE, SPEED, AND ACCELERATION

Additional formulas for accelerated motion

It is not hard to find relationships among s, v_0, v_f, and a that do not directly involve the time t. The first step is to rewrite the defining formula for acceleration, which is

$$a = \frac{v_f - v_0}{t}$$

in the equivalent form

$$t = \frac{v_f - v_0}{a}$$

This formula gives the time during which the speed of the body changed from v_0 to its

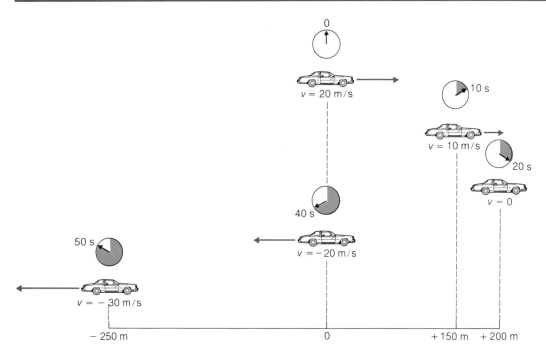

final value of v_f. Now we substitute this expression for t into the formula for the distance traveled:

$$s = v_0 t + \tfrac{1}{2} a t^2$$

The result is

$$
\begin{aligned}
s &= v_0 \frac{(v_f - v_0)}{a} + \tfrac{1}{2} a \frac{(v_f - v_0)^2}{a^2} \\
&= \frac{v_0 v_f}{a} - \frac{v_0^2}{a} + \frac{v_f^2}{2a} - \frac{v_0 v_f}{a} + \frac{v_0^2}{2a} \\
&= \frac{v_f^2 - v_0^2}{2a}
\end{aligned}
$$

Distance under constant acceleration   (2–10)

Also, by multiplying both sides of the last equation by $2a$ and rearranging the terms, we arrive at the formula

$$v_f^2 = v_0^2 + 2as$$

Speed under constant acceleration   (2–11)

FIG. 2–9 Position and speed, at various times, of a car that has a constant acceleration of -1 m/s^2 and whose initial speed is 20 m/s. At $t = 20$ s the car has come to a stop, and then it begins to move in the negative (backward) direction. At $t = 40$ s the car is back where it started, but moving in the opposite direction to its initial one. At $t = 50$ s the car is 250 m behind its starting point.

Example   How far will the car of the previous example have gone when it comes to a stop?

Solution   When the car is at rest, $v_f = 0$. Since $v_0 = 20$ m/s and $a = -1$ m/s^2, we have from Eq. (2–10)

$$s = \frac{v_f^2 - v_0^2}{2a} = \frac{0 - (20 \text{ m/s})^2}{(2)(-1 \text{ m/s}^2)} = \frac{400 \text{ m}^2/\text{s}^2}{2 \text{ m/s}^2} = 200 \text{ m}$$

The car comes to a stop after it has gone 200 m. Then, since the acceleration is negative and is assumed here to continue, it begins to move in the negative (backward) direction, as in Fig. 2–9. ∎

2–7 ACCELERATION OF GRAVITY

Experiment and observation are the foundations of science

Drop a stone, and it falls. Does the stone fall at a constant speed, or is it accelerated? Does the motion of the stone depend upon its weight, or its size, or its color? More than two thousand years ago questions such as these were answered by Greek philosophers, notably Aristotle, on the basis of "logical reasoning" only. To them it seemed reasonable that heavy things should fall faster than light things, for example. Almost nobody felt it necessary to perform experiments to seek information on the physical universe until Galileo (1564–1642) revolutionized science by doing just that: performing experiments. Modern science owes its success in understanding and utilizing natural phenomena to its reliance upon experiment and observation.

The acceleration of gravity near the earth's surface is approximately the same everywhere

What Galileo found, as the result of careful measurements made on balls rolling down inclined planes, was that *all freely falling objects have the same acceleration* at the same place near the earth's surface. This acceleration, which is called the acceleration of gravity (symbol g), has the value

$$g = 9.81 \text{ m/s}^2 \qquad \textit{Acceleration of gravity} \quad (2\text{–}12)$$

FIG. 2–10 All freely falling objects near the earth's surface have the same acceleration g. Hence, if air resistance can be neglected, the farther an object falls, the greater its final speed.

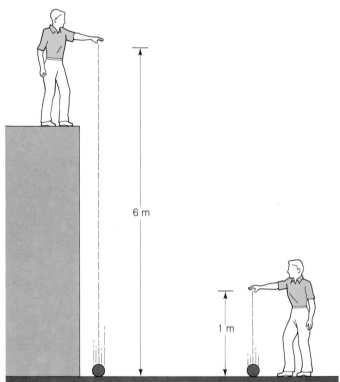

6 m

1 m

11 m/s   4.4 m/s

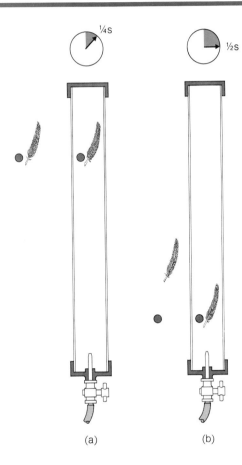

FIG. 2–11 All objects that fall in a vacuum near the earth's surface have the same downward acceleration.

(a)   (b)

to three significant figures (see Appendix B-4 for a discussion of significant figures). In most problems in elementary physics it is sufficient to let $g = 9.8$ m/s^2. In British units, $g = 32.2$ ft/s^2.

An object falling from rest thus has a speed of 9.8 m/s (32 ft/s) after the first second, a speed of 19.6 m/s (64 ft/s) after the next second, and so on. The greater the distance through which a stone falls after being dropped, the greater its speed when it hits the ground (Fig. 2–10). But the stone's *acceleration* is always the same.

Another aspect of Galileo's work deserves comment. His conclusion that all things fall with the same constant acceleration is an *idealization* of reality. The actual accelerations with which objects fall depend upon many factors: the location on the earth, the size and shape of the object, and the density and state of the atmosphere. For example, a bullet falls faster than a feather does in air because of the effects of buoyancy and air resistance. Galileo perceived that the essence of the phenomenon was a constant acceleration downward, with other factors acting merely to cause deviations from the constant value. In a vacuum the bullet and feather fall with exactly the same acceleration (Fig. 2–11).

The drag force due to air resistance on an object of given size and shape depends upon the speed of the object—the faster it goes, the more the drag. In the case of a falling object, the drag force increases as the speed increases until finally it cannot go

Terminal speed of falling object

FIG. 2–12 This graph compares how the speed of a falling body varies with time when it is in a vacuum and when it is in air. In a vacuum the speed is given by $v = gt$ and increases without limit until the body strikes the ground. In air the terminal speed of v_t is eventually reached and the body continues to fall at this speed.

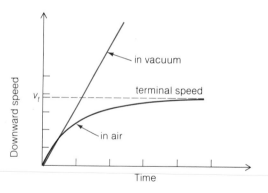

any faster. The object then continues to fall at a constant *terminal speed* (Fig. 2–12). The terminal speed of a person in free fall is about 54 m/s (120 mi/h), whereas it is only about 6.3 m/s (14 mi/h) with an open parachute. In the absence of air resistance, raindrops and hailstones would reach the ground at speeds high enough to be dangerous.

2–8 FREE FALL

We can apply the formulas we derived earlier for motion under constant acceleration to objects in free fall. It must be kept in mind, of course, that the direction of **g** is always downward, no matter whether we are dealing with a dropped object or with one that is initially thrown upward.

Example A stone is dropped from the top of New York's Empire State Buidling, which is 450 m high. Neglecting air resistance, how long does it take the stone to reach the ground? What is its speed when it strikes the ground? (See Fig. 2–13).

Solution The general formula for the distance traveled in the time t by an accelerated object is

$$s = v_0 t + \tfrac{1}{2}at^2$$

Here $v_0 = 0$, since the stone is simply dropped with no initial velocity, and the acceleration is $a = g$. Hence we have

$$h = \tfrac{1}{2}gt^2$$

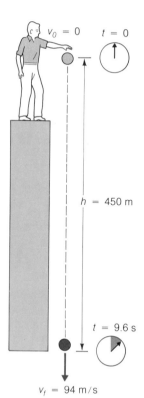

FIG. 2–13

where h represents vertical distance from the starting point. First we solve this formula for t, which yields

$$t = \sqrt{\frac{2h}{g}}$$

and then we substitute $h = 450$ m and $g = 9.8$ m/s^2 to obtain

$$t = \sqrt{\frac{2h}{g}} = \sqrt{\frac{2 \times 450\,\text{m}}{9.8\,\text{m/s}^2}} = \sqrt{92}\,\text{s} = 9.6\,\text{s}$$

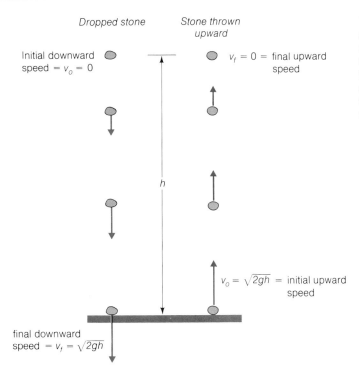

Dropped stone   *Stone thrown upward*

Initial downward speed = $v_0 = 0$

$v_f = 0$ = final upward speed

h

$v_0 = \sqrt{2gh}$ = initial upward speed

final downward speed = $v_f = \sqrt{2gh}$

FIG. 2–14 A stone dropped from a height h reaches the ground with the speed $\sqrt{2gh}$. In order to reach the height h, a stone thrown upward from the ground must have the minimum speed $\sqrt{2gh}$.

Knowing the duration of the fall makes it simple to compute the stone's final speed:

$$v_f = v_0 + at = 0 + gt = (9.8 \text{ m/s}^2)(9.6 \text{ s}) = 94 \text{ m/s}$$

The speed that a dropped object has when it reaches the ground is the same as the speed with which it must be thrown upward from the ground to rise to the same height (Fig. 2–14). To prove this statement, we refer to the formula $v_f^2 = v_0^2 + 2as$ and replace the s with h to give

The speed needed to reach a certain height equals the speed after falling from that height

$$v_f^2 = v_0^2 + 2ah$$

When a stone is dropped, its acceleration a equals the acceleration of gravity g, the initial velocity v_0 is zero, and so

$$v_f^2 = 0 + 2gh \qquad v_f = \sqrt{2gh}$$

When the stone is thrown upward, on the other hand, $a = -g$ (since the downward acceleration is opposite in direction to the upward initial velocity), and at the top of its path $v = 0$. Hence

$$0 = v_0^2 - 2gh \qquad v_0 = \sqrt{2gh}$$

The speed is the same in both cases.

FIG. 2–15 The path of a stone thrown upward with an initial speed of 16 m/s. Air resistance is neglected.

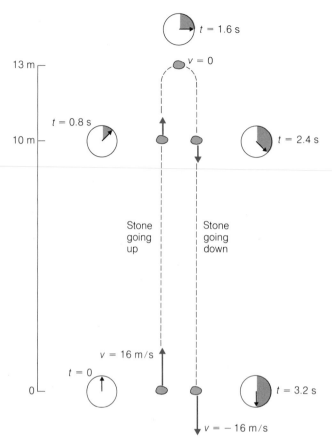

Example   A stone is thrown upward with an initial speed of 16 m/s (Fig. 2–15). (a) What will its maximum height be? (b) When will it return to the ground? (c) Where will it be in 0.8 s? (d) Where will it be in 2.4 s?

Solution   (a) We make use of Eq. (2–10) to find the highest point the stone will reach, reckoning up as positive $(+)$ and down as negative $(-)$. Here $v_0 = 16$ m/s and $a = -9.8$ m/s^2, and at the top of the path $s = h$ and $v_f = 0$. Hence

$$h = \frac{v_f^2 - v_0^2}{2a} = \frac{0 - (16 \,\text{m/s})^2}{2(-9.8 \,\text{m/s}^2)} = \frac{256 \,(\text{m/s})^2}{19.6 \,\text{m/s}^2} = 13 \,\text{m}$$

(b) When will the stone strike the ground? An object takes precisely as long to fall from a certain height h as it does to rise that high (provided that h is its maximum height, as it is here), just as an object's final speed when dropped from a height h is the same as the initial speed required for it to get that high. From Eq. (2–9),

$$h = \tfrac{1}{2}gt^2$$

from which we find that

$$t = \sqrt{\frac{2h}{g}} = \sqrt{\frac{2 \times 13\,\text{m}}{9.8\,\text{m/s}^2}} = 1.6\,\text{s}$$

Because the stone takes as long to rise as to fall, the total time it is in the air is twice 1.6 s, or 3.2 s.

(c) To find the height of the stone a given time after it was thrown upward, we make use of Eq. (2–8),

$$s = v_0 t + \tfrac{1}{2}at^2$$

with $s = h$, $v_0 = 16$ m/s, and $a = -9.8$ m/s^2. For $t = 0.8$ s,

$$h = (16\,\text{m/s})\,(0.8\,\text{s}) - \tfrac{1}{2}(9.8\ \text{m/s}^2)(0.8\ \text{s})^2 = 10\ \text{m}.$$

(d) When we substitute $t = 2.4$ s in the above formula, the result is again

$$h = (16\,\text{m/s})\,(2.4\,\text{s}) - \tfrac{1}{2}(9.8\ \text{m/s}^2)(2.4\ \text{s})^2 = 10\ \text{m}$$

All this apparently paradoxical result means is that at 0.8 s the stone is at a height of 10m on its way up, then it goes on further to its maximum height of 13 m, and at 2.4 s it is once more at a height of 10 m but now on the way down. ■

2–9 MOTION IN A VERTICAL PLANE

An object that moves through space usually has a curved path rather than a perfectly straight one. Our strategy in attacking problems of this kind is to resolve the object's acceleration **a** (assumed constant) and initial velocity $\mathbf{v}_0$, whose vector nature we must now take into account, into their horizontal components $\mathbf{a}_x$ and $\mathbf{v}_{0x}$ and vertical components $\mathbf{a}_y$ and $\mathbf{v}_{0y}$. With the help of Eqs. (2–7), (2–8), and (2–11), we then examine separately the object's motion in each of these directions. Finally we can combine $\mathbf{v}_x$ and $\mathbf{v}_y$ to find **v,** and $\mathbf{s}_x$ and $\mathbf{s}_y$ to find **s,** at any time t after the start of the motion by the vector addition procedure of Chapter 1.

Resolving velocity into vertical and horizontal components

Suppose that we drop a ball A from the edge of a table while rolling an identical ball B off to the side (Fig. 2–16). At the moment the balls leave the table, A has zero velocity while B has the horizontal velocity $\mathbf{v}_0$. The velocity components of the balls therefore have the magnitudes

$$v_{Ax} = 0 \qquad v_{Bx} = v_0 \qquad v_{Ay} = 0 \qquad v_{By} = 0$$

Both balls reach the floor at the same time, even though B has traveled some distance s away from the table. The reason for this behavior is that the acceleration of

FIG. 2–16   Ball *A* is
dropped from the edge of
a table while ball *B* is
simultaneously rolled off
the edge with the initial
horizontal speed v_0.

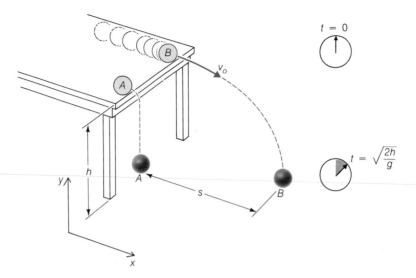

**Vertical motion is
independent of
horizontal motion**

gravity is the same for all bodies near the earth regardless of their state of motion; both *A* and *B* started out with no vertical velocity, both underwent the same downward acceleration, and so both took the same period of time to fall.

A body in free fall descends the distance

$$h = \tfrac{1}{2}gt^2$$

in the time *t* when it starts with no vertical component of velocity. Hence both balls require the time

$$t = \sqrt{\frac{2h}{g}}$$

to reach the floor. If the table is 1.0 m high, then

$$t = \sqrt{\frac{2 \times 1.0\,\text{m}}{9.8\,\text{m/s}^2}} = 0.45\,\text{s}$$

While it is falling, ball *B* is also moving horizontally with the speed v_0. When it strikes the floor it will have traveled the horizontal distance

$$s = v_0 t$$

Let us say that $v_0 = 5.0$ m/s. Therefore

$$s = (5.0\ \text{m/s})\,(0.45\ \text{s}) = 2.3\ \text{m}$$

Example   Find the speeds with which balls *A* and *B* strike the floor.

Solution   The final velocity of *A* has only the single component

$$v_{Ay} = gt$$

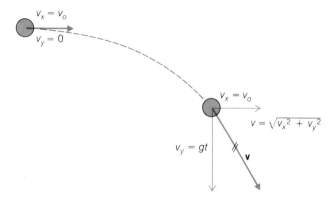

FIG. 2–17 The horizontal and vertical components of the velocity of ball B must be added vectorially to determine the magnitude of its velocity.

and so

$$v_A = v_{Ay} = (9.8 \text{ m/s}^2)(0.45 \text{ s}) = 4.4 \text{ m/s}$$

The final velocity of B, however, has both horizontal and vertical components, namely

$$v_{Bx} = v_0 \qquad v_{By} = gt$$

From Fig. 2–17 we see that since $\mathbf{v}_{Bx}$ is perpendicular to $\mathbf{v}_{By}$, their vector sum $\mathbf{v}_B$ has the magnitude

$$v_B = \sqrt{v_{Bx}^2 + v_{By}^2} = \sqrt{v_0^2 + (gt)^2}$$

Since $v_0 = 5.0 \text{ m/s}$ and $t = 0.45 \text{ s}$,

$$v_B = \sqrt{(5.0 \text{ m/s})^2 + (9.8 \text{ m/s}^2 \times 0.45 \text{ s})^2} = 6.7 \text{ m/s}$$

It is worth nothing that the *vector sum* of $\mathbf{v}_{Bx}$ and $\mathbf{v}_{By}$ has the magnitude 6.7 m/s, whereas the *algebraic sum* of v_{Bx} and v_{By} is 9.4 m/s. The latter figure is, of course, completely meaningless, since velocity is a vector quantity and velocity addition must obey the rules of vector addition. ∎

Velocity addition must be done by vector methods

2–10 PROJECTILE FLIGHT

A more general case of motion in a vertical plane is exemplified by the flight of a projectile, for instance a rocket. Let us ignore the curvature of the earth and the frictional resistance of the atmosphere to the passage of the rocket, and assume that the rocket uses up its fuel at a distance from its launching point that is small compared with the total distance it travels. If the initial velocity $\mathbf{v}_0$ of the rocket makes an angle of θ with level ground, we can resolve $\mathbf{v}_0$ into components whose magnitudes are

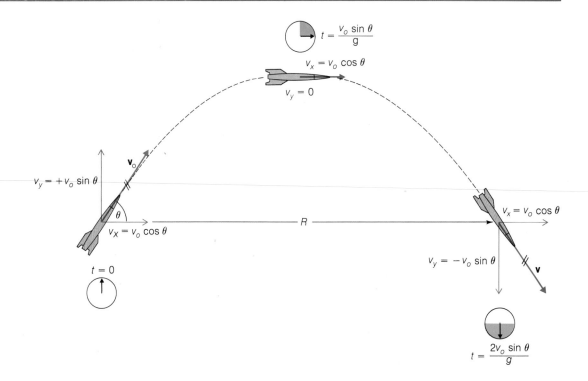

FIG. 2–18 In rocket flight, the horizonal component of a rocket's velocity is constant in the absence of air resistance.

$$v_x = v_0 \cos \theta \qquad v_y = v_0 \sin \theta$$

The horizonal velocity component v_x remains constant during the rocket's flight. The vertical component v_y, however, gradually drops to zero due to the downward acceleration of gravity, and then becomes more and more negative (meaning that there is faster and faster motion downward) until the rocket hits the ground (Fig. 2–18).

We can once again use the formulas for straight-line motion to discuss the horizontal and vertical aspects of the rocket's motion separately, since these are independent of each other. Let us first calculate the time of flight of the rocket. The rocket will continue to rise until the vertical component of its velocity, given by

Vertical component of projectile's velocity

$$v_y = v_0 \sin \theta - gt$$

is zero. At this time t,

$$0 = v_0 \sin \theta - gt \qquad \text{and} \qquad t = \frac{v_0 \sin \theta}{g}$$

The rocket requires the same period of time to return to the ground, and so its total time of flight T is

$$T = 2t = \frac{2v_0 \sin \theta}{g} \qquad\qquad \textit{Time of flight} \quad (2\text{–}13)$$

Range of projectile

We can now find the range R of the rocket; that is, we can see how far from its launching point it will strike the ground. Since the horizontal component v_x of the rocket's velocity is constant,

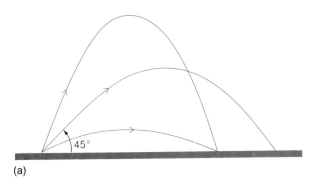

(a)

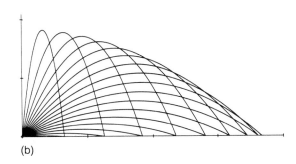

(b)

FIG. 2–19 (a) In the absence of air resistance, the maximum range of a projectile occurs when it is fired at an angle of 45°. (b) Equal-range parts of projectile paths.

$$R = v_x T = (v_0 \cos \theta)\left(\frac{2v_0 \sin \theta}{g}\right) = \frac{2v_0{}^2}{g} \sin \theta \cos \theta$$

This formula can be simplified by making use of the trigonometric identity

$$\sin \theta \cos \theta = \tfrac{1}{2} \sin 2\theta$$

The rocket's range may therefore be written

$$R = \frac{v_0{}^2}{g} \sin 2\theta \qquad\qquad \textit{Range of projectile} \quad (2\text{–}14)$$

This formula gives the range of a rocket (or any other projectile, for that matter, provided it obeys the restrictions given earlier) in terms of its initial velocity v_0 and the angle θ at which it is launched. We note that R is a maximum when $\sin 2\theta = 1$, since 1 is the highest value the sine function can have. Since $\sin 90° = 1$, the maximum range occurs when the initial angle θ is 45°. Any other angle, greater or smaller, will result in a shorter range (Fig. 2–19).

Maximum range occurs for $\theta = 45°$.

Example An arrow leaves a certain bow with a speed of 30 m/s. (a) What is its maximum range? (b) At what two angles could the archer point the arrow if it is to reach a target 70 m away?

Solution (a) The maximum range occurs when $\sin 2\theta = 1$ and is

$$R_{max} = \frac{v_0{}^2}{g} = \frac{(30 \text{ m/s})^2}{9.8 \text{ m/s}^2} = 92 \text{ m}$$

(b) From the general formula $R = (v_0{}^2/g) \sin 2\theta$ we obtain

$$\sin 2\theta = \frac{Rg}{v_0{}^2} = \frac{(70 \text{ m}) (9.8 \text{ m/s}^2)}{(30 \text{ m/s})^2} = 0.762$$

Thus

$$2\theta = \sin^{-1} 0.762 = 50° \qquad \text{and} \qquad \theta = 25°$$

There is also an angle greater than 45° that will give the same range. To find this angle, we require the trigonometric identity

$$\sin \phi = \sin (180° - \phi)$$

FIG. 2–20

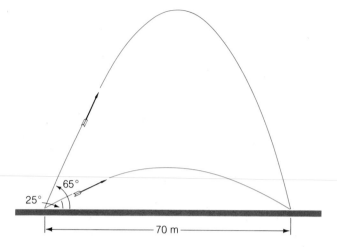

which holds for any angle ϕ. If we let $\phi = 2\theta$, we have

$$\sin 2\theta = \sin (180° - 2\theta)$$

from which we conclude that the two possible angles are θ and $90° - \theta$. The second angle is therefore $90° - 25° = 65°$ (Fig. 2–20). ∎

The examples of this chapter have been worked out not because they are in themselves especially significant, but because they illustrate the power of the mathematical approach to physical phenomena. By defining certain quantities and relating them to each other and to events that actually occur in the real world, a whole theoretical structure of equations may be built up. This structure is an instrument enabling us to solve problems that otherwise would each require a separate, perhaps difficult or impossible, experiment. We must remember that the validity of the theoretical structure depends upon its experimental basis; but once this is established, we may proceed to work out its consequences with pencil and paper.

IMPORTANT TERMS

The **average speed** of a moving object is the distance it covers in a time interval divided by the time interval. The object's **instantaneous speed** at a certain moment is the rate at which it is covering distance at that moment. Speed is a scalar quantity.

The **velocity** of a moving object is a specification of both its speed and its direction of motion. Velocity is a vector quantity, equal to the rate of change of displacement with respect to time.

The **acceleration** of a body is the rate at which its velocity changes with time; the change in velocity may be a change in magnitude or a change in direction or both.

The **acceleration of gravity** is the acceleration of a freely falling body near the earth's surface. The symbol of the acceleration of gravity is g, and its value is 9.8 m/s^2 (32 ft/s^2).

IMPORTANT FORMULAS

Average speed and average velocity: $v = \dfrac{s}{t}$

$$\mathbf{v} = \dfrac{\mathbf{s}}{t}$$

Constant speed: $s = vt$

$$t = \dfrac{s}{v}$$

Acceleration: $a = \dfrac{v_f - v_0}{t}$

Final speed under constant acceleration:

$v_f = v_0 + at$

Distance under constant acceleration:

$s = v_0 t + \frac{1}{2}at^2$

Speed under constant acceleration:

$v_f^{\,2} = v_0^{\,2} + 2as$

Free fall from rest: $h = \frac{1}{2}gt^2$

$v_f = \sqrt{2gh}$

Projectile range: $R = \dfrac{v_0^{\,2}}{g}\sin 2\theta$

[Air resistance is assumed negligible in the following exercises and problems]

MULTIPLE CHOICE

1. An example of an object whose motion is *not* accelerated is a car that
(a) turns a corner at the constant speed of 10 km/h.
(b) descends a hill at the constant speed of 30 km/h.
(c) descends a hill at a speed that increases from 20 km/h to 40 km/h uniformly.
(d) climbs a hill, goes over its crest, and descends on the other side, all at the constant speed of 30 km/h.

2. On a distance-time graph, a horizontal straight line corresponds to motion at
(a) zero speed.
(b) constant speed.
(c) increasing speed.
(d) decreasing speed.

3. On a distance-time graph, a straight line sloping upward to the right corresponds to motion at
(a) zero speed.
(b) constant speed.
(c) increasing speed.
(d) decreasing speed.

4. On a speed-time graph, the motion of a car traveling along a straight road with the uniform acceleration of 2 m/s^2 would appear as a
(a) horizontal straight line.
(b) straight line sloping upward to the right.
(c) straight line sloping downward to the right.
(d) curved line whose downward slope to the right increases with time.

5. The acceleration of a stone thrown upward is
(a) greater than that of a stone thrown downward.
(b) the same as that of a stone thrown downward.
(c) smaller than that of a stone thrown downward.
(d) zero until it reaches the highest point in its motion.

6. Ball *A* is thrown horizontally and ball *B* is dropped from the same height at the same moment.
(a) Ball *A* reaches the ground first.
(b) Ball *B* reaches the ground first.
(c) Ball *A* has the greater speed when it reaches the ground.
(d) Ball *B* has the greater speed when it reaches the ground.

7. A ball is thrown horizontally from a moving car. While it is in flight it is *not* true that
(a) its speed changes.
(b) its acceleration changes.
(c) its direction of motion relative to the car changes.
(d) its direction of motion relative to the road changes.

8. A bicycle travels 12 km in 40 min. Its average speed is
(a) 0.3 km/h.   (b) 8 km/h.
(c) 18 km/h.   (d) 48 km/h.

9. A car that travels at 40 km/h for 2 h, at 50 km/h for 1 h, and at 20 km/h for $\frac{1}{2}$ h has an average speed of
(a) 31.4 km/h.   (b) 40 km/h.
(c) 45 km/h.   (d) 55 km/h.

10. A pitcher takes 0.1 s to throw a baseball, which leaves his hand with a speed of 30 m/s. The ball's acceleration was
(a) 3 m/s^2.   (b) 30 m/s^2.
(c) 300 m/s^2.   (d) 3000 m/s^2.

11. How long does a car with an acceleration of 2 m/s^2 take to go from 10 m/s to 30 m/s?
(a) 10 s   (b) 20 s
(c) 40 s   (d) 400 s

12. A car undergoes a constant acceleration of 6 m/s^2 starting from rest. In the first second it travels
(a) 3 m.   (b) 6 m.
(c) 18 m.   (d) 36 m.

13. An airplane requires 20 s and 400 m of runway to become airborne, starting from rest. Its speed when it leaves the ground is
(a) 20 m/s.   (b) 32 m/s.
(c) 40 m/s.   (d) 80 m/s.

14. A car has an initial speed of 15 m/s and an acceleration of 1 m/s^2. In the first 10 s after the acceleration begins, the car travels
(a) 50 m.   (b) 150 m.
(c) 155 m.   (d) 200 m.

15. A car has an initial speed of 15 m/s and an acceleration of -1 m/s^2. In the first 10 s after the acceleration begins, the car travels

(a) 25 m.   (b) 50 m.
(c) 100 m.   (d) 145 m.

16. How far does the car of question 15 go before coming to a stop?

(a) 112.5 m   (b) 150 m
(c) 225 m   (d) 450 m

17. Two balls are thrown vertically upward, one with an initial speed twice that of the other. The ball with the greater initial speed will reach a height

(a) $\sqrt{2}$ that of the other.
(b) twice that of the other.
(c) 4 times that of the other.
(d) 8 times that of the other.

18. A wheel falls from an airplane flying horizontally at an altitude of 490 m. If there were no air resistance, the wheel would strike the ground in

(a) 10 s.   (b) 50 s.
(c) 80 s.   (d) 100 s.

19. The wheel of question 18 will strike the ground with a speed of

(a) 49 m/s.   (b) 98 m/s.
(c) 490 m/s.   (d) 9604 m/s.

20. A stone is dropped from a cliff. After it has fallen 30 m its speed is

(a) 17 m/s.   (b) 24 m/s.
(c) 44 m/s   (d) 588 m/s.

21. A ball thrown vertically upward at 25 m/s continues to rise for approximately

(a) 2.5 s.   (b) 5 s.
(c) 7.5 s.   (d) 10 s.

22. In question 21, how much time will elapse before the ball strikes the ground?

(a) 2.5 s   (b) 5 s
(c) 7.5 s   (d) 10 s

23. A ball is rolled off the edge of a table at 1.2 m/s. After 0.1 s the ball's speed is

(a) 1.48 m/s.   (b) 1.55 m/s.
(c) 2.18 m/s.   (d) 2.4 m/s.

24. A ball is thrown at a 30° angle above the horizontal with a speed of 3 m/s. After 0.5 s the horizontal component of its velocity will be

(a) 1.5 m/s.
(b) 2.6 m/s.
(c) 4.9 m/s.
(d) 5.5 m/s.

25. In the absence of wind and air resistance, a projectile has its maximum range when fired at an angle with the ground

(a) of 30°.
(b) of 45°.
(c) of 60°.
(d) that depends upon the initial speed of the projectile.

EXERCISES

1. Can a rapidly moving object have the same acceleration as a slowly moving one?

2. The acceleration of a certain moving object is constant in magnitude and direction. Must the path of the object be a straight line? If not, give an example.

3. A hunter aims a rifle directly at a squirrel on a branch of a tree. The squirrel sees the flash of the rifle's firing. Should the squirrel stay where it is or drop from the branch in free fall at the instant the rifle is fired?

4. A person at the masthead of a sailboat moving at constant velocity drops a wrench. The person is 20 m above the boat's deck at the time, and the stern of the boat is 20 m aft of the mast. Is there a minimum speed the sailboat can have such that the wrench does not land on the deck? If so, what is this speed?

5. Is it true that an object dropped from rest falls three times farther in the second second after being released than it does in the first second?

6. A movie is shown that appears to be of a ball falling through the air. Is there any way to determine from what appears on the screen if the movie is actually of a ball being thrown upward but the film is being run backward in the projector?

7. An airplane takes off at 9:00 A.M. and flies at 300 km/h until 1:00 P.M. At 1:00 P.M. its speed is increased to 400 km/h and it maintains this speed until it lands at 3:30 P.M. What is the airplane's average speed for the entire flight?

8. A car travels at 100 km/h for 2 h, at 60 km/h for the next 2 h, and finally at 80 km/h for 1 hr. Find its average speed for the entire 5 h.

9. (a) A car's speed increases from 8 m/s to 20 m/s in 10 s. Find its acceleration. (b) The car's speed then decreases from 20 m/s to 10 m/s in 5 s. Find its acceleration now.

10. The tires of a certain car begin to lose their grip on the pavement at an acceleration of 5 m/s^2. If the car has this acceleration, how many seconds does it require to reach a speed of 25 m/s starting from 10 m/s?

11. The figure shows distance-time graphs for nine cars.

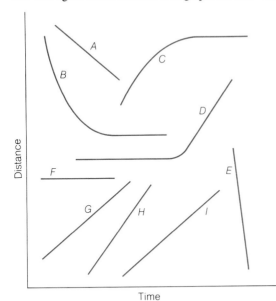

(a) Which cars are or have been moving in the forward direction?

(b) Which cars are or have been moving in the backward direction?

(c) Which car has the highest constant speed?

(d) Which car has the highest constant speed in the forward direction?

(e) Which car has the highest constant speed in the backward direction?

(f) Which cars have the same speed?

(g) Which car has not moved at all?

(h) Which car has been accelerated from rest to a constant speed?

(i) Which car has been brought to a stop from an initial speed in the forward direction?

(j) Which car has been brought to a stop from an initial speed in the backward direction?

12. The brakes of a car moving at 14 m/s are suddenly applied and the car comes to a stop in 4 s. (a) What was its acceleration? (b) How long would the car take to come to a stop starting from 20 m/s with the same acceleration? (c) How long would the car take to slow down from 20 m/s to 10 m/s with the same acceleration?

13. A car starts from rest and reaches a speed of 22 m/s in 20 s. (a) What was its acceleration? (b) How long would the car take to go from 22 m/s to 30 m/s with the same acceleration?

14. A spacecraft has an acceleration of magnitude $5g$. What distance is needed for it to attain a speed of 10 km/s?

15. A Porsche reaches a speed of 42 km/h from a standing start in 15.5 s. What distance does it cover while doing so?

16. A Ferrari covers 100 m from a standing start in 6 s at constant acceleration. Find its final speed.

17. A DC-8 airplane has a takeoff speed of 80 m/s, which it reaches 35 s after starting from rest. (a) How much time does the airplane spend in going from 0 to 20 m/s? What distance does it cover in doing so? (b) How much time does the airplane spend in going from 60 to 80 m/s? What distance does it cover in doing so? (c) What is the minimum length of the runway?

18. The brakes of a certain car produce an acceleration of -5 m/s^2. (a) If the car is moving at 20 m/s when the brakes are applied, how far does it go in the first second afterward? (b) How far does the car go in the course of being slowed down from 20 m/s to 10 m/s?

19. Divers in Acapulco, Mexico, leap from a point 36 m above the sea. What is their speed when they enter the water?

20. A stone is dropped from a cliff 490 m above its base. How long does the stone take to fall?

21. A ball dropped from the roof of a building takes 4 s to reach the street. How high is the building?

22. A stone is thrown vertically upward at 9.8 m/s. When will it reach the ground?

23. A ball is thrown vertically downward at 10 m/s. What is its speed 1 s later? 2 s later?

24. A ball is thrown vertically upward at 10 m/s. What is its speed and direction 1 s later? 2 s later?

25. A lead pellet is propelled vertically upward by an air rifle with an initial speed of 16 m/s. Find its maximum height.

26. The acceleration of gravity at the surface of Mars is 3.7 m/s^2. If a stone thrown upward on Mars reaches a height of 15 m, find its initial speed and the total time of flight.

27. Find the initial and final speeds of a ball thrown vertically upward that returns to the thrower 3 s later.

28. A bullet is fired vertically upward and returns to the ground in 20 s. Find the height it reaches.

29. Does the speed of a projectile sent off at a 45° angle of elevation vary in its path? If so, where is the speed greatest and where is it least?

30. What effect does doubling the initial speed of a projectile have on its range?

31. Find the minimum initial speed of a champagne cork that travels a horizontal distance of 11 m.

32. In April 1959 Miss Victoria Zacchini was fired 47 m from a cannon in Madison Square Garden, New York City. What was the minimum muzzle speed of the cannon?

33. A football leaves the toe of a punter at an angle of 40° above the horizontal. What is its minimum initial speed if it travels 40 m?

34. Find the range of an arrow that leaves a bow at 50 m/s at an angle of 50° above the horizontal. *Note:* sin (90° + θ) = cos θ.

PROBLEMS

1. A passenger in an airplane flying from New York to Los Angeles notes the time at which he passes over various cities and towns. With the help of a map he determines the distances between these landmarks, and compiles the table shown below. Plot the distance covered by the airplane versus time from these data, and describe the airplane's motion with the help of the graph.

Time (P.M.):	4:00	5:12	5:41	6:14
Distance (km):	0	660	926	1267

Time (P.M.):	6:39	7:54	9:18	10:00
Distance (km):	1525	2300	3028	3392

2. A European train passes successive kilometer posts at the times given below. Plot the data on a graph and determine whether the train's speed is constant over the entire distance or not. If it is not constant, plot the train's speed in each time interval versus time and find the acceleration. What are the train's initial and final speeds?

Distance (km):	0	1	2	3	4
Time (s):	0	64	114	156	193

Distance (km):		5	6	7	8
Time (s):		227	259	292	324

3. The odometer of a car is checked at 1-min intervals and the readings below are obtained. Calculate the speed of the car in each time interval and plot the results on a graph. Describe the motion of the car with the help of this graph.

Time (min):	0	1	2	3
Distance (km):	42.20	42.74	43.64	44.90

Time (min):	4	5	6	7
Distance (km):	46.34	47.78	49.22	50.66

Time (min):	8	9	10
Distance (km):	51.74	52.10	52.10

4. The next figure is a graph that shows the speed of an object plotted against time. Find the total distance the object travels during the period covered by the graph.

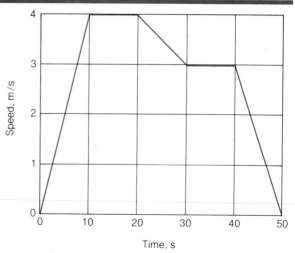

5. A woman jogs halfway to her destination at 8 km/h and walks the rest of the way at 6 km/h. What is her average speed for the entire trip?

6. An express train passes a certain station at 20 m/s. The next station is 2 km away and the train reaches it 1 min later. (a) Did the train's speed change? (b) If it did, what was its speed at the second station, assuming a constant acceleration?

7. The engineer of a train traveling at 80 km/h applies the brakes when he passes an amber signal. The next signal is one km down the track and the train reaches it 75 s later. Assuming a uniform deceleration, find the speed of the train at the second signal.

8. A bus travels 400 m between two stops. It starts from rest and accelerates at 1.5 m/s² until it reaches a speed of 9 m/s. The bus continues at this speed and then decelerates at 2 m/s² until it comes to a halt. Find the total time required for the journey.

9. A car is approaching a traffic light at 20 m/s when it turns red. The driver takes 1 s to react and put his foot on the brake pedal, after which the car decelerates at 2 m/s². Find the distance needed for the car to come to a stop after the light turned red.

10. A sprinter accelerates from rest until he reaches a speed of 12 m/s and then continues running at this speed. If he takes 11 s to cover 100 m, what was his acceleration and how long did it last?

11. A truck accelerates from rest at 0.3 m/s² for 30 s, continues at constant speed for 2 min, and then comes to a stop in 15 s. What distance did it cover?

12. A car is stationary in front of a red traffic light. As the light turns green, a truck goes past at a constant speed of 15 m/s. At the same moment, the car begins to accelerate at 1.25 m/s²; when it reaches 25 m/s, the car continues at that speed. When does the car pass the truck? How far will they have gone from the traffic light at that time?

13. An elevator has a maximum acceleration of ± 1.5 m/s^2 and a maximum speed of 6 m/s. Find the shortest period of time required for it to take a passenger to the tenth floor of a building from street level, a height of 50 m, with the elevator coming to a stop at this floor.

14. When a flea jumps, it accelerates through about 0.8 mm (a little less than the length of its legs) and is able to reach a height of as much as 10 cm. (a) Find the flea's acceleration (assumed constant) and its speed at takeoff when this occurs. (b) When a person jumps, the acceleration distance is about 50 cm. If the acceleration of a person were the same as that of a flea, find the speed at takeoff and the height that would be reached.

15. An orangutan throws a coconut vertically upward at the foot of a cliff 40 m high while his mate simultaneously drops another coconut from the top of the cliff. The two coconuts collide at an altitude of 20 m. What was the initial speed of the coconut that was thrown upward?

16. A helicopter is climbing at 8 m/s when it drops a pump near a leaking boat. The pump reaches the water 4 s afterward. How high was the helicopter when the pump was dropped? When the pump reached the water?

17. A girl throws a ball vertically upward at 10 m/s from the roof of a building 20 m high. (a) How long will it take the ball to reach the ground? (b) What will its speed be when it strikes the ground?

18. A girl throws a ball vertically downward at 10 m/s from the roof of a building 20 m high. (a) How long will it take the ball to reach the ground? (b) What will its speed be when it strikes the ground?

19. A British parachutist bails out at an altitude of 150 m and accidentally drops his monocle. If he descends at the constant speed of 6 m/s, how much time separates the arrival of the monocle on the ground from the arrival of the parachutist himself?

20. A Russian balloonist floating at an altitude of 150 m accidentally drops his samovar and starts to ascend at the constant speed of 1.2 m/s. How high will the balloon be when the samovar reaches the ground?

21. A rocket is launched upward with an acceleration of 100 m/s^2. Eight seconds later the acceleration stops when the fuel is exhausted. Find (a) the highest speed the rocket attains; (b) the maximum altitude; (c) the total time of flight; and (d) the speed with which the rocket strikes the ground.

22. A person in an elevator drops an apple from a height 2 m above the elevator's floor and, with a stopwatch, times the fall of the apple to the floor. What is found (a) when the

elevator is ascending with an acceleration of 1 m/s^2; (b) when it is descending with an acceleration of 1 m/s^2; (c) when it is ascending at the constant speed of 3 m/s; (d) when it is descending at the constant speed of 3 m/s; and (e) when the cable has broken and it is descending in free fall?

23. A rifle is aimed directly at the bull's-eye of a target 50 m away. If the bullet's speed is 350 m/s, how far below the bull's-eye does the bullet strike the target?

24. A rescue line is to be thrown horizontally from the bridge of a ship 30 m above sea level to a lifeboat 30 m away. What speed should the line have?

25. A ball is thrown horizontally from the roof of a building 20 m high at 30 m/s. What will be the magnitude and direction of the ball's velocity when it strikes the ground?

26. A ball is rolled off the edge of a table with a horizontal velocity of 1 m/s. What will be the magnitude and direction of the ball's velocity 0.1 s later?

27. A ball is thrown horizontally toward the north from a rooftop at 8 m/s. A 10-m/s wind is blowing from the east. (a) What is the speed of the ball relative to the ground after 2 s? (b) What angles does its velocity make relative to the vertical at this time? (c) What angle does its velocity make relative to due north at this time?

28. What percentage increase in initial speed is required to increase the range of a javelin by 20%?

29. A blunderbuss can fire a slug 100 m vertically upward. (a) What is its maximum horizontal range? (b) With what speeds will the slug strike the ground when fired upward and when fired so as to have maximum range?

30. A golf ball leaves a tee at 60 m/s and strikes the ground 200 m away. At what two angles with the horizontal could it have begun its flight? Find the time of flight and maximum altitude in each case.

31. A shell is fired at a velocity of 300 m/s at an angle of 30° above the horizontal. (a) How far does it go? What are its time of flight and maximum altitude? (b) At what other angle could the shell have been fired to have the same range? What would its time of flight and maximum altitude have been in this case?

ANSWERS TO MULTIPLE CHOICE

1. b	**6.** c	**11.** a	**16.** a	**21.** a
2. a	**7.** b	**12.** a	**17.** c	**22.** b
3. b	**8.** c	**13.** c	**18.** a	**23.** b
4. b	**9.** b	**14.** d	**19.** b	**24.** b
5. b	**10.** c	**15.** c	**20.** b	**25.** b

FORCE AND MOTION

3

Thus far we have only discussed how motion is described mathematically. But why does anything move in the first place? Why do some things move faster than others? Why are some accelerated and others not? How can a body traveling in one direction be accelerated in the opposite direction? All these are reasonable questions and ones we must be able to answer if we are to understand the factors at work in the world around us that produce the physical phenomena we observe and if we are to harness these factors to meet our needs. Almost three centuries ago Isaac Newton (1642–1727) formulated three principles based upon observations he and others had made which summarize so much of the behavior of moving bodies that they have become known as the laws of motion. These laws form the subject of this chapter.

3–1 FIRST LAW OF MOTION

In everyday life it is a familiar observation that stationary objects tend to remain stationary and moving objects tend to continue moving. A certain amount of effort is needed to start a cart moving on a level road, and once in motion a certain amount of effort is also required to bring it to a stop. Of course, in the case of a cart friction is a factor in its reluctance to begin to move, but friction is only part of the story. Even without friction, a cart at rest on a level road will remain at rest unless something pushes it. And without friction, a cart moving along a level road at a certain velocity will continue to move at that velocity indefinitely.

Newton's first law of motion is a statement of the above behavior:

An object at rest will remain at rest and an object in motion will continue in motion in a straight line at constant velocity in the absence of any interaction with the rest of the universe.

First law of motion

We might object that, although everyday experience does indicate that things in motion *tend* to remain in motion along a straight line at constant velocity, sooner or later they invariably come to a stop and often deviate from a straight path as well. Even celestial bodies such as the sun, moon, and planets are neither at rest nor pursue straight paths. But these observations do not invalidate the first law of motion; they merely emphasize how difficult it is to avoid interactions between something and its environment. A cart rolling along a smooth, perfectly level road will not continue forever owing to friction and air resistance, but we are at liberty to imagine what would happen if the air were to disappear and the friction were to vanish.

The reluctance of an object to change its state of rest or uniform motion is a property of matter known as *inertia*. When a bus suddenly starts to move, its passengers seem to be pushed backward (Fig. 3–1). What is actually happening is that inertia tends to keep their bodies in place relative to the earth while the bus carries their feet forward. When a bus suddenly stops, on the other hand, the passengers seem to be pushed forward. What is actually happening is that inertia tends to keep their bodies moving while the bus has come to a halt.

All matter exhibits inertia

3–2   MASS

A quantitative meaure of the inertia of an object at rest is its *mass*. The greater the resistance something offers to being set in motion, the greater its mass. The inertia of a lead ball exceeds that of an aluminum ball of the same size, as we can tell by kicking them in turn, so the mass of the lead ball exceeds that of the aluminum one (Fig. 3–2).

We can arrive at a precise definition of mass by using a simple experiment to compare the inertias of two objects. What we do is put a small spring between them, push them together so the spring is compressed, and tie a string between them to hold the assembly in place (Fig. 3–3). Now we cut the string. The compressed spring pushes the objects apart, and A flies off to the left at the speed v_A while B flies off to the right at the speed v_B. Object A has a lower speed than object B, and we interpret this difference to mean that A exhibits more inertia than B.

We repeat the experiment a number of times using springs of different stiffness, so that the recoil speeds are different in each case. What we find each time is that A

Mass is a measure of inertia

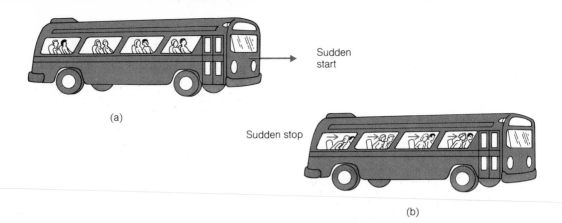

(a)

Sudden start

Sudden stop

(b)

FIG. 3–1 (a) When a bus suddenly starts to move, the inertia of the passengers tends to keep them at rest relative to the earth. (b) When a bus comes to a sudden stop, their inertia tends to keep them moving. Both effects illustrate the first law of motion.

moves more slowly than B, and that, regardless of the exact values of v_A and v_B, their *ratio*

$$\frac{v_B}{v_A}$$

is always the same.

The fact that the speed ratio v_B/v_A is constant gives us a way to specify what we mean by mass unambiguously. If we denote the masses of A and B by the symbols m_A and m_B respectively, we define the ratio of these masses to be

$$\frac{m_A}{m_B} = \frac{v_B}{v_A} \tag{3–1}$$

Operational definition of mass

The object with the greater mass has the lower speed, and vice versa. The above procedure provides a definite, experimental method for finding the ratio between the masses of any two objects—it is an *operational definition*.

The next step is to select a certain object to serve as a standard unit of mass. By international agreement this object is a platinum cylinder at Sèvres, France, called the *standard kilogram*. The mass of any other object in the world can be determined by a recoil experiment with the standard kilogram. (There are easier ways to measure mass, needless to say, but we are interested in basic principles for the present.)

Standard kilogram

FIG. 3–2 The mass of an object determines its inertia

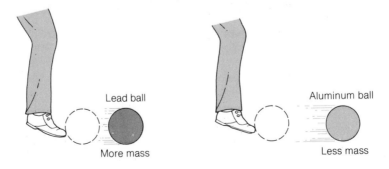

Lead ball

More mass

Aluminum ball

Less mass

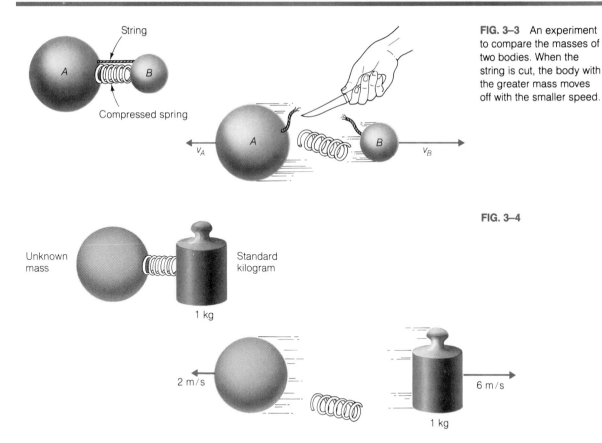

FIG. 3–3 An experiment to compare the masses of two bodies. When the string is cut, the body with the greater mass moves off with the smaller speed.

FIG. 3–4

Example In a recoil experiment with the standard kilogram, an object of unknown mass moves off at 2 m/s and the standard kilogram moves off at 6 m/s (Fig. 3–4). What is the mass of the object?

Solution If we call the unknown mass A and the standard kilogram B, then $m_B = 1$ kg, $v_A = 2$ m/s, and $v_B = 6$ m/s. We find that

$$m_A = (1\,\text{kg}) \left(\frac{6\,\text{m/s}}{2\,\text{m/s}} \right) = 3\,\text{kg} \qquad \blacksquare$$

Why is it necessary to define mass in this seemingly roundabout way? After all, everybody knows that the mass of an object refers to the amount of matter it contains. The trouble is that "amount of matter" is a nebulous concept: it could refer to an object's volume, to the number of atoms it contains, or to yet other properties. It has proved most fruitful to choose the inertia of an object as a measure of the quantity of matter it contains and to define the object's mass in terms of this inertia as manifested in an appropriate experiment.

In the British system of units the *slug* rather than the kilogram is the standard unit of mass. The slug and its relation to the pound (which is a unit of force, not of mass) will be discussed later in this chapter.

3–3 FORCE

**Interactions produce
changes in velocity**

According to the first law of motion, the velocity of an object (which may be 0) remains
constant as long as it is isolated from the rest of the universe. When the object interacts
with something else, its velocity may change. We can interact with a football by kicking
it, and the result is a change in the football's velocity from $\mathbf{v}_1 = 0$ to some value $\mathbf{v}_2$
(Fig. 3–5). Or the interaction can take the form of catching a moving football, in which
case again the football's velocity changes. An interaction can lead to a change in the
direction of $\mathbf{v}$ as well as to a change in its magnitude, as for instance when a football
bounces off a tree.

Force

  Some interactions are more effective than others in causing velocity changes. A
swift kick affects the velocity of a football more than a gentle tap does. The concept
of *force* can be used to put the matter on a precise basis. In general,

A force is any influence that can produce a change in the velocity of an object.

FIG. 3–5 Three
examples of how an
interaction may give rise
to a change in the velocity
of an object.

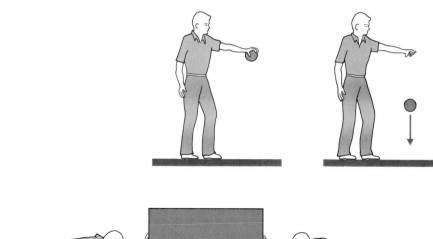

FIG. 3–6 A downward force exerted by the earth causes dropped objects to fall.

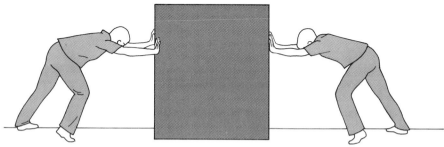

FIG. 3–7 When several forces act on an object, they may cancel one another out to leave no net force.

This definition is in accord with the notion of a force as a "push" or a "pull," but it goes further since no direct contact is necessarily implied. No hand reaches up from the earth to pull a dropped stone downward, yet the increasing downward velocity of the falling stone testifies to the action of a force upon it (Fig. 3–6).

It is entirely possible for two or more forces to act upon an object without affecting its state of motion; the forces may be such as to cancel one another out. What is required for a velocity change is a *net force,* often called an *unbalanced force.* When an object is acted upon by several forces whose vector sum is zero, the forces are said to be *balanced forces* and the object is then in *equilibrium* (Fig. 3–7). But each of the forces acting by itself is capable of accelerating the object.

We can therefore restate the first law of motion in terms of net force:

In the absence of a net force acting on it, an object at rest will remain at rest and an object in motion will continue in motion at constant velocity.

First law of motion

Every force, without exception, arises from one or another of the four fundamental interactions that are possible between the elementary particles of which all matter consists. Two of these interactions are effective only when the particles involved are extremely close together. These are called the "strong" and "weak" *nuclear interactions* because they are responsible for the ability of protons and neutrons to stick together to form stable atomic nuclei, and one is more powerful than the other. The others—the *gravitational* and the *electromagnetic interactions*—are unlimited in range,

The four fundamental interactions

Muscular forces

but their strength decreases with distance. In later chapters we shall explore the properties of these interactions, and in Chapter 32 we shall see how they are related.

The forces an animal exerts are produced by contractions of its skeletal muscles. A muscle is a bundle of parallel fibers that tapers at each end into a tendon, which provides the connection to a bone. In some cases a muscle end forks into two or even three tendons. The bones linked by a muscle are hinged together at a joint, and the motion of the bones relative to the joint is controlled by the muscle, usually in conjunction with another muscle on the opposite side.

A muscle fiber contracts when it is given an electrical stimulus by a nerve ending. The force of the contraction is constant for each fiber; the greater the required total force, the greater the number of fibers that are stimulated. The maximum force a muscle can exert thus depends on the number of fibers it contains, which is proportional to its cross-sectional area. Maximum forces of up to 70 N/cm^2 (100 lb/in^2) have been reported. An athlete might have a biceps muscle in his arm 8 cm (3 in) in diameter, which means it would be capable of producing forces up to 3500 N (790 lb). As will be seen in the next chapter, the geometries of animal skeletons and muscles favor range of movement over force, so the actual forces a person's hands and feet can exert are considerably smaller than those produced by the muscles themseves.

Muscular strength and animal size

An animal of a certain type whose length (or other representative linear dimension) is L has, in general, muscles whose cross-sectional areas and hence strengths are roughly proportional to L^2. Hence another animal of the same type whose length is, say, $2L$, has muscles which are $(2L)^2/L^2 = 4$ times stronger than the corresponding ones in the first animal. To be sure, the mass of an animal depends upon its volume and so upon L^3, which means that the larger it is, the stronger its muscles have to be to carry out the same tasks. Because mass varies as L^3 whereas strength varies as L^2, large animals are weaker in relation to their masses than smaller ones. This is obvious in nature, where many insects, for instance, can carry objects several times their own weights, whereas animals the size of humans are limited to loads comparable to their own weights. Whether the muscles of a certain kind of animal are intrinsically stronger or weaker than those of an animal of a different kind is another matter; human muscles are considerably stronger than those of insects, figured on the basis of force exerted per unit cross-sectional area. Apart from the structural problems a man-sized insect would have, it would be a rather feeble creature.

3–4   SECOND LAW OF MOTION

In the *second law of motion* we have a quantitative definition of force:

Second law of motion

The net force acting upon an object is equal to the product of the mass and the acceleration of the object; the direction of the force is the same as that of the object's acceleration.

In equation form,

$$\mathbf{F} = m\mathbf{a}$$ *Second law of motion*   (3–2)

Force = mass × acceleration

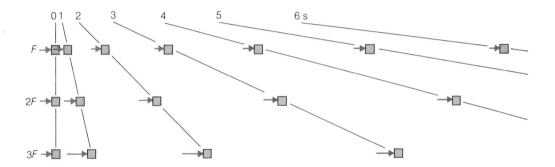

The second law of motion provides us with a way to analyze and compare forces in terms of the accelerations they give rise to. Thus a force which causes something to have twice the acceleration another force produces must be twice as great as the other one (Fig. 3–8). An object moving to the right but going slower and slower is accelerated to the left. Hence there is a force toward the left acting on it (Fig. 3–9). The first law of motion is clearly a special case of the second: when the net force on an object is zero, its acceleration is also zero.

The second law is in accord with the definition of mass given earlier since the smaller the mass of an object acted upon by a given force, the greater its acceleration and hence final speed if it starts from rest (Fig. 3–10).

It is convenient to have a special unit for force. In the SI system of units the appropriate unit is the *newton* (abbreviated N):

A newton is that force which, when applied to a 1-kg mass, gives it an acceleration of 1 m/s^2.

The newton is not a fundamental unit like the meter, second, and kilogram, and in some calculations it may have to be replaced by its equivalent in terms of the latter. This equivalent may be found as follows:

$$F = ma$$
$$1\,\text{N} = (1\,\text{kg})\left(1\,\frac{\text{m}}{\text{s}^2}\right) = 1\,\text{kg·m/s}^2$$

A newton is equivalent to 0.225 lb, a little less than 1/4 lb.

FIG. 3–8 The acceleration of an object is proportional to the net force applied to it. Successive positions of a block are shown at 1-s intervals while forces of F, 2F, and 3F are applied.

The newton is the SI unit of force

FIG. 3–9 A force and the acceleration it produces are always in the same direction.

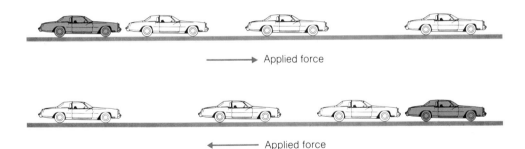

Applied force

Applied force

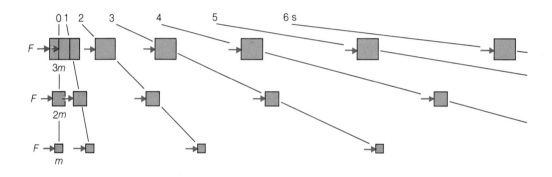

FIG. 3–10 When the same force is applied to objects of different masses, the resulting accelerations are inversely proportional to the masses. Successive positions of blocks of mass m, $2m$, and $3m$ are shown at 1-s intervals while identical forces of F are applied.

Let us examine a few probems in order to become familiar with the application of the second law of motion. Some of these problems are, of course, artificial and over-simplified, but they do illustrate the power of the second law in situations involving force and motion.

Example A 60-g tennis ball approaches a racket at 30 m/s, is in contact with the racket's strings for 5 ms (1 ms = 1 millisecond = 10^{-3} s), and then rebounds at 30 m/s (Fig. 3–11). What was the average force the racket exerted on the ball?

Solution The tennis ball experienced a velocity change of

$$\Delta v = v_f - v_0 = (-30 \text{ m/s}) - (30 \text{ m/s}) = -60 \text{ m/s}$$

so its acceleration was

$$a = \frac{\Delta v}{\Delta t} = \frac{-60 \text{ m/s}}{5 \times 10^{-3} \text{ s}} = -1.2 \times 10^4 \text{ m/s}^2$$

The corresponding force is, since 60 g = 0.060 kg,

$$F = ma = (-0.060 \text{ kg})(1.2 \times 10^4 \text{ m/s}^2) = -720 \text{ N}$$

The minus sign means that the force was in the opposite direction to that of the ball when it approached the racket. The British equivalent of 720 N is 162 lb. ■

Example A force of 10 N is applied to a 4.0-kg block that is at rest on a perfectly smooth, level surface. Find the speed of the block and how far it has gone after 6.0 s.

Solution We start from the second law of motion in scalar form, since only one direction is involved here:

$$F = ma$$

We know what F and m are, so we find the acceleration a of the block as follows:

$$a = \frac{F}{m} = \frac{10 \text{ N}}{4.0 \text{ kg}} = \frac{10 \text{ kg·m/s}^2}{4.0 \text{ kg}} = 2.5 \text{ m/s}^2$$

The direction of the acceleration is the same as that of the force.

FIG. 3–11

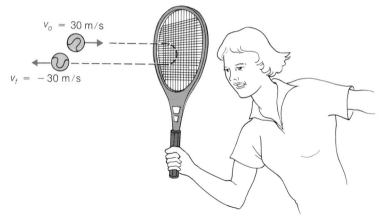

To find the speed of the block after $t = 6.0$ s, we use the formula $v = at$ and obtain

$$v = at = (2.5 \text{ m/s}^2)(6.0 \text{ s}) = 15 \text{ m/s}$$

For the distance the block travels in $t = 6.0$ s at an acceleration of $a = 2.5 \text{ m/s}^2$ we require Eq. (2–8),

$$s = v_0 t + \tfrac{1}{2} at^2$$

The block started from rest, so $v_0 = 0$ and we have

$$s = \tfrac{1}{2} at^2 = \tfrac{1}{2} \times (2.5 \text{ m/s}^2)(6.0\text{s})^2 = 45 \text{ m}$$

After 6.0 s a 4.0-kg mass acted upon by a 10-N force will have gone 45 m and have a speed of 15 m/s (Fig. 3–12). ∎

Example During performances of the Bouglione Circus in 1976, John Tailor was fired from a compressed-air cannon whose barrel was 20 m long. Mr. Tailor emerged from the cannon (twice daily, three times on Saturdays and Sundays) at 40 m/s. If Mr.

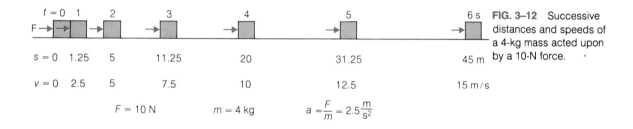

FIG. 3–12 Successive distances and speeds of a 4-kg mass acted upon by a 10-N force.

FIG. 3–13

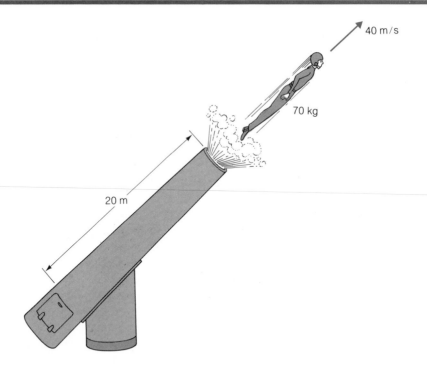

40 m/s

70 kg

20 m

Tailor's mass was 70 kg, find the average force on him during the firing of the cannon (Fig. 3–13).

Solution   We start by finding Mr. Tailor's acceleration with the help of Eq. (2–11),

$$v_f^2 = v_0^2 + 2as$$

Here $v_0 = 0$, $v_f = 40$ m/s, and $s = 20$ m, so

$$v_f^2 = 0 + 2as$$

$$a = \frac{v_f^2}{2s} = \frac{(40\,\text{m/s})^2}{2 \times 20\,\text{m}} = 40\,\text{m/s}^2$$

The corresponding average force is

$$F = ma = 70\,\text{kg} \times 40\,\text{m/s}^2 = 2800\,\text{N}$$

which is about 630 lb.   ■

3–5   WEIGHT

Weight is a force

The force with which an object is attracted to the earth is called its *weight*. Weight is different from mass, which is a measure of the inertia an object exhibits. Although they are different physical quantities, mass and weight are closely related.

The weight of a stone is the force that causes it to be accelerated when it is dropped. All objects in free fall near the earth's surface have a downward acceleration of $g = 9.8$ m/s^2, the acceleration of gravity. (We continue to ignore the small variation in g with geographic position.) If the stone's mass is m, then the downward force on it, which is its weight w, can be found from the second law of motion, $F = ma$, by letting $F = w$ and $a = g$. Evidently

$$w = mg \qquad\qquad Weight \quad (3\text{–}3)$$

Weight $=$ mass $\times$ acceleration of gravity

The weight of any object is equal to its mass multiplied by the acceleration of gravity. Since g is a constant near the earth's surface, the weight w of an object is always directly proportional to its mass m: a large mass is heavier than a small one. **Weight is proportional to mass**

The mass of an object is a more fundamental property than its weight, because its mass when at rest is the same everywhere in the universe whereas the gravitational force on it depends upon its position relative to the earth or to some other astronomical body. A 100-kg person weighs 980 N on the earth, but he or she would weigh 2587 N on Jupiter, 372 N on Mars, 162 N on the moon, and 0 in space far from the sun and other stars. (The mass of an object varies with its speed with respect to an observer. This effect is significant only at speeds approaching that of light, 3×10^8 m/s, and is discussed in Chapter 26.) **The mass of an object is the same everywhere, unlike its weight**

Example A loaded elevator whose total mass is 800 kg is suspended by a cable whose maximum permissible tension is 20,000 N. What is the greatest upward acceleration possible for the elevator under these circumstances? What is the maximum possible downward acceleration?

Solution When the elevator is at rest, or moving at constant speed, the tension in the cable is just its weight of

$$w = mg = (800 \text{ kg}) (9.8 \text{ m/s}^2) = 7840 \text{ N}$$

To accelerate the elevator upward, an additional tension F is required in order to provide a net upward force (Fig. 3–14). Since the total tension cannot exceed 20,000 N, the greatest accelerating force available is

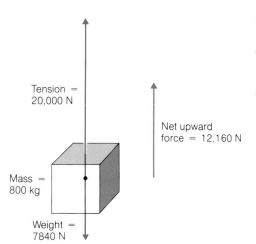

FIG. 3–14 The net upward force on an elevator of mass 800 kg is 12,160 N when the tension in its supporting cable is 20,000 N.

Tension = 20,000 N

Net upward force = 12,160 N

Mass = 800 kg

Weight = 7840 N

$$F = F_{max} - w = 20{,}000\,\text{N} - 7840\,\text{N} = 12{,}160\,\text{N}$$

The elevator's acceleration when this net force is applied is

$$a = \frac{F}{m} = \frac{12{,}160\,\text{N}}{800\,\text{kg}} = 15.2\,\text{m/s}^2$$

For the elevator to exceed the downward acceleration of gravity $g = 9.8\,\text{m/s}^2$, a downward force besides the weight of the elevator is needed. Since this cannot be provided by a supporting cable, the maximum downward acceleration is $9.8\,\text{m/s}^2$. ∎

Example   Figure 3–15 shows a 12-kg block, A, which hangs from a string that passes over a pulley and is connected at its other end to a 30-kg block, B, which rests on a frictionless table. Find the accelerations of the two blocks under the assumption that the string is massless and the pulley is massless and frictionless. What is the tension in the string?

Solution   Because the blocks are joined by the string, their accelerations have the same magnitude a even though different in direction. The net force on B is equal to the tension T in the string; so, from the second law of motion, taking the left as the $+$ direction so that a will come out positive,

$$F_B = T = m_B a$$

In the case of A, the net force is the difference between its weight $m_A g$, which acts downward, and the tension T in the string, which acts upward. Considering the downward direction as positive, so that the two accelerations will have the same sign,

$$F_A = m_A g - T = m_A a$$

Adding these two equations together eliminates T, whose value we do not know at this point, and permits us to find a:

$$F_A + F_B = m_A g - T + T = m_A a + m_B a$$
$$m_A g = (m_A + m_B)a$$
$$a = \frac{m_A g}{m_A + m_B} = \frac{(12\,\text{kg})\,(9.8\,\text{m/s}^2)}{12\,\text{kg} + 30\,\text{kg}} = 2.8\,\text{m/s}^2$$

FIG. 3–15 Both blocks have the same acceleration.

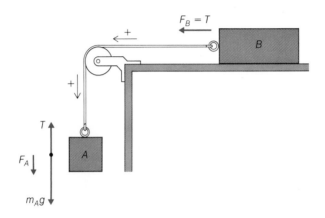

The tension in the string is

$$T = m_B a = (30 \text{ kg})(2.8 \text{ m/s}^2) = 84 \text{ N}$$

If B were fixed in place, the tension would equal the weight of A, or $m_A g = 117.6 \text{ N}$. Here the tension is less because B moves in response to the pull of A's weight, but it is not zero because of B's inertia. ∎

Example Figure 3–16 shows the same two blocks, A and B, suspended by a string on either side of a massless, frictionless pulley. Find the accelerations of the two blocks and the tension in the string.

Solution Here A moves upward and B moves downward, both with accelerations having the same magnitude a. Applying the second law of motion to the two blocks gives

$$F_A = T - m_A g = m_A a$$
$$F_B = m_B g - T = m_B a$$

where we have considered up as $+$ for A and down as $+$ for B. To eliminate T we add these equations, and then solve for a:

$$F_A + F_B = T - m_A g + m_B g - T = m_A a + m_B a$$
$$(m_B - m_A) = (m_A + m_B)a$$
$$a = \frac{(m_B - m_A)g}{m_A + m_B} = \frac{(30 \text{ kg} - 12 \text{ kg})(9.8 \text{ m/s}^2)}{12 \text{ kg} + 30 \text{ kg}} = 4.2 \text{ m/s}^2$$

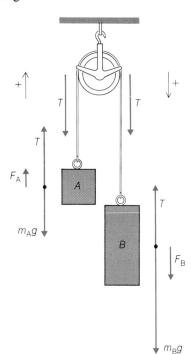

FIG. 3–16 The force on each block is the difference between its weight and the tension in the string.

FIG. 3–17

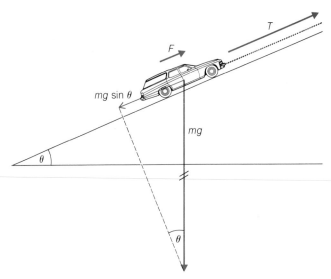

The tension in the string may be found from the first equation above:

$$T = m_A a + m_A g = m_A(a + g) = (12 \text{ kg})(4.2 + 9.8)\text{m/s}^2 = 168 \text{ N}$$

In Section 8-7 we will learn how to take into account the effect of the mass of an actual pulley.   ■

Example   A rope whose working strength is 2000 N is used to tow a 1000-kg car up a 10° incline, as in Fig. 3–17. Find the maximum acceleration that can be given to the car.

Solution   The component of the weight mg of the car that is parallel to the incline is $mg \sin \theta$. If T is the maximum tension in the rope, the maximum net force along the incline that can be applied to the car is

$$F = T - mg \sin \theta$$

From the second law of motion,

$$T - mg \sin \theta = ma$$

$$a = \frac{T}{m} - g \sin \theta = \frac{2000 \text{ N}}{1000 \text{ kg}} - (9.8 \text{ m/s}^2)(\sin 10°)$$

$$= 2.0 \text{ m/s}^2 - 1.7 \text{ m/s}^2 = 0.3 \text{ m/s}^2$$

On a level road, $\theta = 0$ and $\sin \theta = 0$, and the maximum acceleration would be 2.0 m/s^2.   ■

3–6   BRITISH UNITS OF MASS AND FORCE*

The slug and pound are the British units of mass and force

In the British system the unit of mass is the *slug* and the unit of force is the *pound* (lb). A body whose mass is 1 slug experiences an acceleration of 1 ft/s^2 when a net force

*This section may be omitted without loss of continuity.

of 1 lb acts on it. Thus

$$1\,\text{lb} = 1\,\frac{\text{slug·ft}}{\text{s}^2}$$

The slug is an unfamiliar unit because in everyday life weights rather than masses are specified in the British system: we go shopping for 10 lb of apples, not 1/3 slug of apples. In the metric system, on the other hand, masses are normally specified: European grocery scales are calibrated in kilograms, not in newtons.

In order to convert a weight in pounds to a mass in slugs we make use of Eq. (3–3) to obtain

Relation between slugs and pounds

$$m\,(\text{slugs}) = \frac{w\,(\text{lb})}{g\,(\text{ft/s}^2)}$$

Since g, the acceleration of gravity at the earth's surface, has the value 32 ft/s^2,

$$m\,(\text{slugs}) = \frac{w\,(\text{lb})}{32\,\text{ft/s}^2}$$

The *weight* of a 1-slug mass is 32 lb, and the *mass* of a 1-lb weight is 1/32 slug (Table 3–1).

1 kilogram corresponds to 2.21 pounds in the sense that the weight of 1 kilogram is 2.21 pounds.

Relations between kilograms and pounds

1 pound corresponds to 0.454 kilogram in the sense that the mass of 1 pound is 0.454 kilogram.

Example A 3000-lb car has an initial speed of 10 mi/h. How much force is required to accelerate the car to a speed of 50 mi/h in 9.0 s?

Solution The mass of the car is

$$m = \frac{w}{g} = \frac{3000\,\text{lb}}{32\,\text{ft/s}^2} = 94\,\text{slugs}$$

Before we can compute the car's acceleration we must convert the speeds from mi/h to ft/s. From the table inside the back cover of this book, 1 mi/h = 1.47 ft/s, and thus, to two significant figures,

TABLE 3–1
Units of mass and weight

System of units	Unit of mass	Unit of weight	Acceleration of gravity g	To find mass m given weight w	To find weight w given mass m
Metric	Kilogram (kg)	Newton (N)	9.8 m/s^2	$m\,(\text{kg}) = \dfrac{w\,(\text{N})}{9.8\,\text{m/s}^2}$	$w\,(\text{N}) = m\,(\text{kg}) \times 9.8\,\text{m/s}^2$
British	Slug	Pound (lb)	32 ft/s^2	$m\,(\text{slugs}) = \dfrac{w\,(\text{lb})}{32\,\text{ft/s}^2}$	$w\,(\text{lb}) = m\,(\text{slugs}) \times 32\,\text{ft/s}^2$

Conversion of units: 1 slug = 14.6 kg   1 newton = 0.225 lb
1 kg = 0.0685 slug   1 lb = 4.45 newtons

$$v_0 = 10 \, \text{mi/h} \times 1.47 \frac{\text{ft/s}}{\text{mi/h}} = 15 \, \text{ft/s}$$

$$v_f = 50 \, \text{mi/h} \times 1.47 \frac{\text{ft/s}}{\text{mi/h}} = 73 \, \text{ft/s}$$

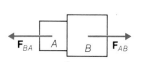

FIG. 3–18 According to the third law of motion, $\mathbf{F}_{AB} = -\mathbf{F}_{BA}$.

The car's acceleration is

$$a = \frac{v_f - v_0}{t} = \frac{(73 - 15)\text{ft/s}}{9.0 \, \text{s}} = 6.4 \, \text{ft/s}^2$$

and the force that must act on it is

$$F = ma = 94 \, \text{slugs} \times 6.4 \, \text{ft/s}^2 = 600 \, \text{lb}$$ ■

3–7   THIRD LAW OF MOTION

The third of Newton's laws of motion states that

Third law of motion

> **When an object exerts a force on another object, the second object exerts on the first a force of the same magnitude but in the opposite direction.**

If we call one of the interacting objects *A* and the other *B,* then according to the third law of motion

$$\mathbf{F}_{AB} = -\mathbf{F}_{BA} \qquad\qquad \textit{Third law of motion} \quad (3\text{–}4)$$

Here $\mathbf{F}_{AB}$ is the force *A* exerts on *B* and $\mathbf{F}_{BA}$ is the force *B* exerts on *A* (Fig. 3–18).

Action and reaction forces

The third law of motion always applies to two different forces on two different bodies—the *action force* that one body exerts on another, and the equal but opposite *reaction force* that the second exerts on the first.

There is no such thing as a single force in the universe. Four examples of action-reaction pairs of forces are shown in Fig. 3–19. (a) We push against a wall; the wall pushes back on us. (b) We throw a ball; as we are pushing it into the air, it is pushing back on our hand. (c) We fire a rifle, and the expanding gases in its barrel push the

FIG. 3–19 Every action force in the universe is accompanied by an equal and opposite reaction force. The two forces act on different bodies.

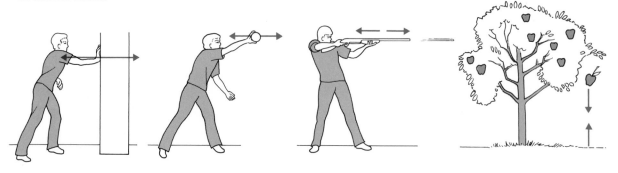

(a)   (b)   (c)   (d)

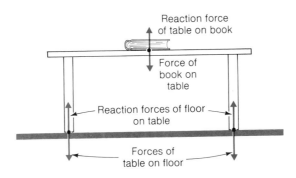

FIG. 3–20 The action-reaction forces between a book and a table and between the table and the floor.

Reaction force of table on book

Force of book on table

Reaction forces of floor on table

Forces of table on floor

bullet out; the gases push back on the rifle, which gives rise to the recoil force we feel with our shoulder. (d) An apple falls because of the downward gravitational pull of the earth; there is an equal upward pull by the apple on the earth which we cannot detect because the earth is so much more massive than the apple, but it nevertheless exists.

Let us apply the third law to a few more situations in order to appreciate its significance. A 1-kg book lies stationary on a table, pressing down on the table with a force of 9.8 N (Fig. 3–20). The table pushes upward on the book with the reaction force of 9.8 N. Why doesn't the book fly upward into the air? The answer is that the upward force of 9.8 N on the book merely balances its weight of 9.8 N, which acts downward. If the table were not there to cancel out the latter 9.8-N force, the book would, of course, be accelerated downward.

Another illustration of the third law of motion is the operation of walking. We push backward with one foot, and the earth pushes forward on us. The forward reaction force exerted by the earth causes us to move forward, and at the same time, the backward force of our foot causes the earth to move backward (Fig. 3–21). Owing to the earth's enormously larger mass, its motion cannot be detected practically, but it is there. Why is it that there is no reaction force on us, responding to the earth's push on our foot, to keep us from moving? The explanation is that every action-reaction pair of forces acts on *different* bodies. We push on the earth, the earth pushes back on us. If there are no *additional* forces present to impede our motion (for instance, pressure by a wall directly in front of us), we proceed to undergo a forward acceleration.

We might conceivably find ourselves on a frozen lake with a perfectly smooth surface. Now we cannot walk because the absence of friction prevents us from exerting a backward force on the ice that would produce a forward force on us. But what we can do is exert a force on some object we may have with us, say a rock. We throw the rock forward by applying a force to it; at the same time the rock is pressing back on us with the identical force but in the opposite direction, and in consequence we find ourselves moving backward (Fig. 3–22).

We shall learn in Chapter 7 how these notions are expressed in the *principle of conservation of momentum,* one of the most useful formulations of the laws of motion.

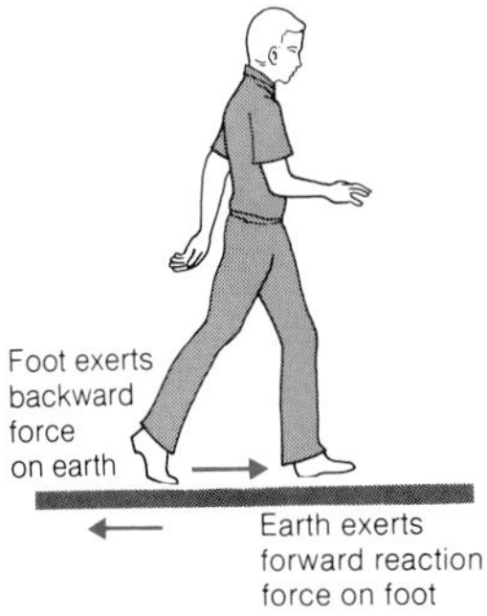

Every action force is accompanied by a reaction force

Walking and reaction forces

Foot exerts backward force on earth

Earth exerts forward reaction force on foot

FIG. 3–21 When we push backward on the earth with one foot, the opposite reaction force of the earth pushes forward on us. The latter force causes us to move forward.

3–8 FRICTION

Frictional forces are of many kinds, but they all act to impede motion. The effects of friction must be distinguished from those of inertia. The term *inertia* refers to the fact

FIG. 3–22 If we throw a rock while standing on a frozen lake, the reaction force pushes us backward.

Frictional forces impede motion

Friction can be useful

How frictional forces behave

Lubricants

that bodies maintain their original states of rest or of motion in the absence of net forces on them; but even the smallest force is sufficient to accelerate a body despite its inertia. The term *friction,* on the other hand, refers to actual forces that come into being when two surfaces are in contact that act to oppose motion between them.

Often friction is desirable: the fastening action of nails, screws, and bolts and the resistive action of brakes depend upon it. Walking would be impossible without friction. In many situations, however, friction merely reduces efficiency, and great efforts are made in industry to minimize it through the use of lubricants—notably grease and oil— and special devices—notably the wheel. About a quarter of the power output of a typical automobile engine is wasted in overcoming friction, most of it due to the sliding of the pistons in the cylinders.

To clearly understand the characteristic properties of frictional forces, let us consider what happens when we attempt to move a box across a level floor (Fig. 3–23). At first the box is stationary; no horizontal forces whatever act on it. As we begin to push, the box remains in place because the floor exerts a force on the bottom of the box which opposes the force we apply. This opposition force is friction, and it arises from the nature of the contact between the floor and the box. As we push harder, the frictional force also increases to match our efforts, until finally we are able to exceed the frictional force and begin to move the box.

Evidently the opposing frictional force has a maximum value that it cannot exceed; and when we apply a force greater than this maximum, the box will experience a net force. As the box moves under the influence of the net force, the frictional force usually drops to slightly less than its maximum value when the box is at rest. Because the net force on the box is the force of our push *minus* the force of friction, it is always less than (or equal to) the force we apply; it may even be zero, as we have seen.

Lubricants reduce friction by separating two contacting surfaces with an intermediate layer of a softer material. Instead of rubbing against each other, the surfaces rub against the lubricant (Fig. 3–24). Depending upon the specific application, the most suitable lubricant may be a gas, a liquid, or a solid. Most lubricants are oils derived from petroleum: grease consists of oil to which a thickening agent has been added to prevent the oil from running out from between the surfaces involved. The joints of the limbs of the human body are lubricated by a substance called *synovial fluid,* which resembles blood plasma.

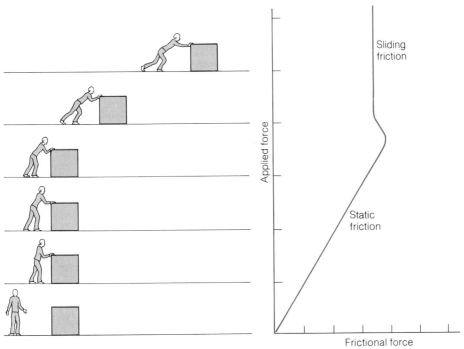

FIG. 3–23 As force is applied to a box on a level floor, the frictional resistive force increases to a certain maximum, decreases somewhat as the box begins to move, and then remains constant.

There is no relative motion between the rim of a wheel and a smooth surface over which it rolls if wheel and surface are both rigid, and there is accordingly no frictional resistance to overcome. By reducing friction so drastically, the wheel makes it possible to transport loads from one place to another without the enormous forces that dragging them would require. Most aspects of our present technological civilization rely upon the wheel in one way or another.

Neither wheels nor the surfaces on which they travel can ever be perfectly rigid. Figure 3–25 shows the flattening of the wheel and the indentation of the surface which both contribute to *rolling friction*. Because the wheel and surface must be constantly deformed as the wheel rolls, a force is needed to keep the wheel rolling. However, this force is usually many times smaller than that needed to overcome sliding friction. Balls and rollers are widely used to reduce the friction between a rotating shaft and the bearings that hold it in place by replacing sliding friction with rolling friction. Tire deformation accounts for most of the frictional force on a car at speeds of up to 50 to 80 km/h, when air resistance starts to dominate.

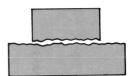

Lubricant

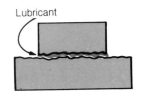

FIG. 3–24 A lubricant reduces friction by providing a soft material able to flow readily to separate surfaces in contact.

3–9 COEFFICIENT OF FRICTION

It is a matter of experience that the frictional force exerted by one surface upon another depends upon two factors: (1) the perpendicular force with which one surface is pressed against the other, and (2) the nature of the surfaces in contact. The perpendicular force is usually called the *normal force*, symbol N.

Normal force

The greater the normal force, the greater the friction, regardless of area in contact

The more tightly two objects are pressed together by a normal force, the greater the friction between them. For this reason an empty box is easier to push across a floor than a similar box loaded with something heavy (Fig. 3–26). Equally familiar is the effect of the nature of the contacting surfaces. For instances, it takes more than three times as much force to push a wooden box across a wooden floor as it does to push a steel box of the same weight across a steel floor. Interestingly enough, the area in contact between the two surfaces is not important: it is just as hard to push a small 50-kg box over a given floor as it is to push a large 50-kg box of the same material over the same floor.

To a good degree of approximation the following formula relates the normal force N pressing one surface against another with the frictional force F_f that results:

$$F_f = \mu N \qquad\qquad \textit{Maximum frictional force} \quad (3\text{–}5)$$

Frictional force = coefficient of friction × normal force

The quantity μ (Greek letter "mu") is called the *coefficient of friction* and is a constant for a given pair of surfaces. The value of F_f given by the above formula represents a maximum. When the applied force is less than F_f, the frictional force always equals the applied force. Otherwise, since F_f acts in the opposite direction to an applied force, things would move *backward* when pushed weakly—which, needless to say, does not happen. Table 3–2 is a list of coefficients of friction for several surfaces. Static friction is discussed later in this section.

When an object is being pushed or pulled horizontally, the normal force N holding it against the surface it is on is simply its weight mg. In such cases,

$$F_f = \mu N = \mu mg$$

We must apply a force greater than μmg when moving a body of mass m across a level surface where the coefficient of friction is μ.

Example   A 100-kg wooden crate is being pushed across a wooden floor with a horizontal force of 350 N (Fig. 3–27). What is its acceleration?

Solution   The frictional force that opposes the applied force F_A here is

$$F_f = \mu mg = (0.3)(100\,\text{kg})\left(9.8\frac{\text{m}}{\text{s}^2}\right) = 294\,\text{N}$$

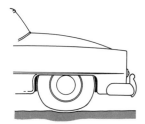

FIG. 3–25   The flattening of a wheel and the indentation of the surface it presses on both contribute to rolling friction.

FIG. 3–26   The greater the normal force **N** with which one surface is pressed against the other, the greater the force of friction between them.

Materials in Contact	Coefficient of Static Friction, μ_s	Coefficient of Sliding Friction, μ
Wood on wood	0.5	0.3
Wood on stone	0.5	0.4
Steel on steel (smooth)	0.15	0.09
Metal on metal (lubricated)	0.03	0.03
Leather on wood	0.5	0.4
Rubber tire on dry concrete	1.0	0.7
Rubber tire on wet concrete	0.7	0.5
Glass on glass	0.94	0.40
Steel on Teflon	0.04	0.04
Bone on bone (dry)		0.3
(lubricated with synovial fluid)		0.003

TABLE 3–2
Approximate coefficients of static and sliding friction for various materials in contact

since the coefficient of friction here is 0.3. Hence the net force acting on the crate is

$$F = F_A - F_f = 350 \text{ N} - 294 \text{ N} = 56 \text{ N}$$

and the crate's acceleration is, from the second law of motion,

$$a = \frac{F}{m} = \frac{56 \text{ N}}{100 \text{ kg}} = 0.56 \text{ m/s}^2$$ ∎

Example The coefficient of friction between a rubber tire and a dry concrete road is 0.7. What is the distance in which a car will skid to a stop on such a road if its brakes are locked when it is moving at 80 km/h (50 mi/h)?

Solution The first step is to calculate the acceleration of the car under these circumstances. The normal force N is the car's weight of mg, and the frictional force is

$$F = \mu N = \mu mg$$

From the second law of motion, $F = ma$, and

$$\mu mg = ma$$
$$a = \mu g = (0.7)(9.8 \text{ m/s}^2) = 6.9 \text{ m/s}^2$$

We can find the distance s from $v_f^2 = v_0^2 + 2as$. Here $v_f = 0$ and

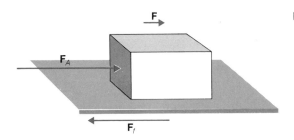

FIG. 3–27

FIG. 3–28 The vertical component of the force applied to the sled reduces the normal force of the sled on the snow, and the horizontal component overcomes the frictional resistance of the snow to the sled's motion.

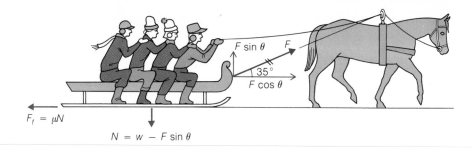

$$v_0 = (80 \text{ km/h})\left(0.278 \frac{\text{m/s}}{\text{km/h}}\right) = 22 \text{ m/s}$$

Therefore, since $a = -6.9 \text{ m/s}^2$ (the minus sign is needed because the acceleration is negative),

$$s = -\frac{v_0^2}{2a} = -\frac{(22 \text{ m/s})^2}{2(-6.9 \text{ m/s}^2)} = 35 \text{ m}$$

which is 115 ft. ■

Example A 300-kg sled is pulled at constant speed over level snow by a rope that makes an angle of 35° with the horizontal. If the coefficient of friction is 0.10, find the force required.

Solution Here the frictional force μN must be overcome by the horizontal component $F \cos \theta$ of the applied force **F** (Fig. 3–28). Since the normal force is the sled's weight mg minus the upward vertical component $F \sin \theta$ of the force **F**, we have

$$\text{Horizontal component of } \mathbf{F} = \text{frictional force,}$$
$$F \cos \theta = \mu(mg - F \sin \theta)$$

or

$$F = \frac{\mu mg}{\mu \sin \theta + \cos \theta} = \frac{(0.10)(300 \text{ kg})(9.8 \text{ m/s}^2)}{(0.10)(\sin 35°) + \cos 35°} = 335 \text{ N}$$

■

3–10 STATIC FRICTION

Coefficient of static friction is usually greater than that of sliding friction

When a body in contact with a surface is pushed, the frictional force resisting motion increases with the applied force until a limiting value is reached. If the applied force exceeds the limiting value of the frictional force, the body begins to move. The value of the coefficient of friction corresponding to the maximum frictional force between two surfaces at rest is called the *coefficient of static friction* and is denoted by the symbol μ_s.

When no lubricant is present, the coefficient of static friction μ_s is greater than that of sliding friction μ: the force needed to set a body in motion against friction is more than that needed to maintain it in motion at constant speed. (Fig. 3–23 shows the

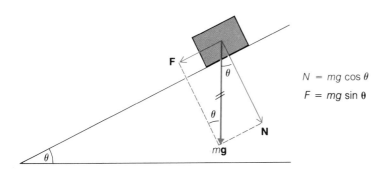

FIG. 3–29 The weight *m***g** of a block on an inclined plane can be resolved into forces parallel and perpendicular to the plane. At an angle θ such that tan θ = μ_s , the block slides down the plane at constant speed.

drop in frictional force when the box has begun to move.) When the surfaces in contact are smooth and well lubricated, μ_s and μ are virtually the same; Table 3–2 contains some typical values of coefficients of static friction.

Example   A wooden chute is being built along which wooden crates of merchandise are to be slid down into the basement of a store. (a) What angle with the horizontal should the chute make if the crates are to slide down at constant speed? (b) With what force must a 100-kg crate be pushed in order to start it sliding down the chute if the angle of the chute is that found in (a)?

Solution   The procedure here is first to resolve the weight of the crate, which is a force of magnitude $w = mg$ that acts downward, into a component **F** parallel to the plane and a component **N** perpendicular to the plane. With the help of Fig. 3–29 we find that

$$F = mg \sin \theta \qquad N = mg \cos \theta$$

When the crate slides down at constant speed, there is no net force acting on it, according to Newton's first law of motion. Hence the downward force along the chute must exactly balance the force of sliding friction, which means that

$$F = \mu N$$
$$mg \sin \theta = \mu mg \cos \theta$$
$$\mu = \frac{\sin \theta}{\cos \theta} = \tan \theta$$

From Table 3–2 the value of μ for wood on wood is 0.3, and so θ = 17°.
   (b) We note that the coefficient of static friction here is $\mu_s = 0.5$. Hence the force of static friction to be overcome is

$$F_f = \mu_s N = \mu_s mg \cos \theta$$

This is greater than the force of sliding friction, and so a crate will not begin to move without a push. The force component along the plane due to the crate's own weight is $F = mg \sin \theta$. Here F is less than F_f, and so, if we call F' the outside force parallel to the plane required to move the crate,

Outside + forward component of weight = backward frictional force

$$F' + mg \sin \theta = \mu_s mg \cos \theta$$

FIG. 3–30 If slipping is not to occur, the angle θ must be smaller than a certain critical angle that depends on the coefficient of static friction between the heel and the floor.

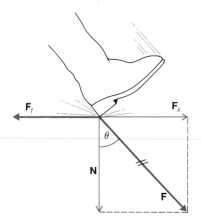

$$F' = mg\,(\mu_s \cos\theta - \sin\theta)$$
$$= (100\,\text{kg})\,(9.8\,\text{m/s}^2)\,(0.5\cos 17° - \sin 17°) = 182\,\text{N} \qquad \blacksquare$$

Example A person whose shoes have leather heels is walking on a wooden floor. (a) Find the maximum angle the forward-swinging leg may make with the vertical in order that the heel not slip on the floor. (b) How is this angle affected if the floor is wet, which reduces the coefficient of static friction?

Solution (a) The geometry of the situation is shown in Fig. 3–30. The force **F** that the leg exerts on the floor can be resolved into a normal component **N** and a component **F**$_x$ parallel to the floor, where

$$N = F\cos\theta \qquad F_x = F\sin\theta$$

The condition for the heel not to slip is that the frictional force **F**$_f$ have the same magnitude as **F**$_x$. Since

$$F_f = \mu N = \mu F\cos\theta$$

we have for the limiting angle

$$F_f = F_x$$
$$\mu F\cos\theta = F\sin\theta$$
$$\tan\theta = \mu$$

From Table 3–2 the coefficient of static friction for leather on wood is 0.5, so

$$\tan\theta = 0.5 \qquad \text{and} \qquad \theta = 27°$$

If θ is equal to or less than 27°, the heel will not slip.

  (b) Reducing μ reduces θ, so smaller steps have to be taken on a slippery surface. Slipping is a sudden process because, as it starts, the smaller coefficient of sliding friction applies, and the frictional force drops sharply. This is the reason it is difficult to keep from falling once slipping begins. $\qquad \blacksquare$

IMPORTANT TERMS

The **inertia** of an object refers to the apparent resistance it offers to changes in its state of motion. The property of matter that manifests itself as inertia is called **mass.** The unit of mass in the SI system is the **kilogram,** in the British system the **slug.**

A **force** is any influence that can cause an object to be accelerated. The unit of force in the SI system is the **newton,** in the British system the **pound.**

The **weight** of an object is the gravitational force exerted on it by the earth. The weight of an object is proportional to its mass.

Newton's **first law of motion** states that, in the absence of a net force acting on it, an object at rest will remain at rest and an object in motion will continue in motion at constant velocity. The **second law of motion** states that a net force acting on an object causes it to have an acceleration proportional to the magnitude of the force and inversely proportional to the object's mass; the acceleration is in the same direction as the force. The **third law of motion** states that when an object exerts a force on another object, the second object exerts on the first a force of the same magnitude but in the opposite direction.

The term **friction** refers to the resistive forces that arise to oppose the motion of an object past another with which it is in contact. **Sliding friction** is the frictional resistance an object in motion experiences, whereas **static friction** is the frictional resistance a stationary object must overcome in order to be set in motion.

The **coefficient of friction** is the constant of proportionality for a given pair of contacting surfaces that relates the frictional force between them to the normal force with which one presses against the other; usually the coefficient of static friction is greater than that of sliding friction.

IMPORTANT FORMULAS

Second law of motion: $\mathbf{F} = m\mathbf{a}$

Weight: $w = mg$

Third law of motion: $\mathbf{F}_{AB} = -\mathbf{F}_{BA}$

Frictional force: $F_f = \mu N$

MULTIPLE CHOICE

1. When an object undergoes an acceleration,
  (a) its mass always increases.
  (b) its speed always increases.
  (c) it always falls toward the earth.
  (d) a force always acts upon it.

2. A force acts on an object that is free to move. If we know the magnitude and direction of the force and the mass of the object, Newton's second law of motion enables us to determine the object's
  (a) weight.
  (b) position.
  (c) speed.
  (d) acceleration.

3. Which of the following is not a unit of mass?
  (a) the gram
  (b) the kilogram
  (c) the pound
  (d) the slug

4. A sheet of paper can be withdrawn from under a bottle of milk without toppling it if the paper is jerked out quickly. This is an example of
  (a) inertia.
  (b) weight.
  (c) acceleration.
  (d) the third law of motion.

5. The weight of an object
  (a) is the quantity of matter it contains.
  (b) refers to its inertia.
  (c) is basically the same quantity as its mass but expressed in different units.
  (d) is the force with which it is attracted to the earth.

6. An automobile that is towing a trailer is accelerating on a level road. The force that the automobile exerts on the trailer is
  (a) equal to the force the trailer exerts on the autombile.
  (b) greater than the force the trailer exerts on the automobile.
  (c) equal to the force the trailer exerts on the road.
  (d) equal to the force the road exerts on the trailer.

7. When a horse pulls a wagon, the force that causes the horse to move forward is the force
  (a) he exerts on the wagon.
  (b) the wagon exerts on him.
  (c) he exerts on the ground.
  (d) the ground exerts on him.

8. The action and reaction forces referred to in Newton's third law of motion
  (a) act upon the same object.
  (b) act upon different objects.
  (c) need not be equal in magnitude but must have the same line of action.
  (d) must be equal in magnitude but need not have the same line of action.

9. To set an object in motion on a surface usually requires
   (a) less force than to keep it in motion.
   (b) the same force as that needed to keep it in motion.
   (c) more force than to keep it in motion.
   (d) only as much force as is needed to overcome inertia.

10. The frictional force between two surfaces in contact does *not* depend on
    (a) the normal force pressing one against the other.
    (b) the areas of the surfaces.
    (c) whether the surfaces are stationary or in relative motion.
    (d) whether a lubricant is used or not.

11. When a 1-N force acts on a 1-kg object that is able to move freely, the object receives
    (a) a speed of 1 m/s.
    (b) an acceleration of 0.102 m/s^2.
    (c) an acceleration of 1 m/s^2.
    (d) an acceleration of 9.8 m/s^2.

12. When a 1-N force acts on a 1-N object that is able to move freely, the object receives
    (a) a speed of 1 m/s.
    (b) an acceleration of 0.102 m/s^2.
    (c) an acceleration of 1 m/s^2.
    (d) an acceleration of 9.8 m/s^2.

13. The weight of 600 g of salami is
    (a) 0.061 N.   (b) 5.9 N.
    (c) 61 N.   (d) 5880 N.

14. A certain force gives a 5-kg object an acceleration of 2.0 m/s^2. The same force would give a 20-kg object an acceleration of
    (a) 0.5 m/s^2.   (b) 2.0 m/s^2.
    (c) 4.9 m/s^2.   (d) 8.0 m/s^2.

15. A force of 10 N gives an object an acceleration of 5 m/s^2. What force would be needed to give it an acceleration of 1 m/s^2?
    (a) 1 N   (b) 2 N
    (c) 5 N   (d) 50 N

16. A 2400-kg car accelerates from 10 m/s to 30 m/s in 6 s. The force on the car is
    (a) 816 N.   (b) 8000 N.
    (c) 12,000 N.   (d) 78,400 N.

17. The coefficient of static friction between two wooden surfaces is
    (a) 0.5.
    (b) 0.5 N.
    (c) 0.5 kg/N.
    (d) 0.5 N/kg.

18. A force of 40 N is needed to set a 10-kg steel box moving across a wooden floor. The coefficient of static friction is
    (a) 0.08.   (b) 0.25.
    (c) 0.4.   (d) 2.5.

19. The coefficient of static friction for steel on ice is 0.1. The force needed to set a 70-kg skater in motion is approximately
    (a) 0.1 N.   (b) 0.7 N.
    (c) 7 N.   (d) 70 N.

20. A horizontal force of 150 N is applied to a 51-kg carton on a level floor. The coefficient of static friction is 0.5 and that of sliding friction is 0.4. The frictional force acting on the carton is
    (a) 150 N.   (b) 200 N.
    (c) 250 N.   (d) 500 N.

21. The coefficients of static and sliding friction for wood on wood are respectively 0.5 and 0.3. If a 100-N wooden box is pushed across a horizontal wooden floor with just enough force to overcome the force of static friction, its acceleration is
    (a) 0.2 m/s^2.   (b) 0.5 m/s^2.
    (c) 2.0 m/s^2.   (d) 5.0 m/s^2.

22. A toboggan reaches the foot of a hill at a speed of 4 m/s and coasts on level snow for 15 m before coming to a stop. The coefficient of sliding friction is
    (a) 0.004.   (b) 0.05.
    (c) 0.16.   (d) 0.27.

EXERCISES

1. When a body is accelerated, a force is invariably acting upon it. Does this mean that, when a force is applied to a body, it is invariably accelerated?

2. The moon revolves around the earth in an approximately circular orbit. Is the moon accelerated in its motion? Does a force act on the moon? If so, in which direction?

3. It is less dangerous to jump from a high wall onto loose earth than onto a concrete pavement. Why?

4. Compare the tension in the coupling between the first two cars in a train with the tension in the coupling between the last two cars (a) when the train's speed is constant; (b) when the train's speed is increasing.

5. Measurements are made of distance versus time for three moving objects. The distances are found to be directly proportional to t, t^2, and t^3, respectively. What can you say about the net force acting on each of the objects?

6. A person in an elevator suspends a 1-kg mass from a spring balance. What is the nature of the elevator's motion when the balance reads 9.0 N? 9.8 N? 10.0 N?

7. When a force equal to its weight is applied to a body free to move, what is its acceleration?

8. Since the opposite forces of the third law of motion are equal in magnitude, how can anything ever be accelerated?

9. Can we conclude from the third law of motion that a single force cannot act upon a body?

10. Two boys wish to break a string. Are they more likely to do this if each takes one end of the string and they pull against each other, or if they tie one end of the string to a tree and both pull on the free end? Why?

11. An engineer designs a propeller-driven spacecraft. Because there is no air in space, he incorporates a supply of oxygen as well as a supply of fuel for the motor. What do you think of the idea?

12. Ships are often built on ways that slope down to a nearby body of water. Normally a ship is launched before most of its interior and superstructure have been installed, and is completed when afloat. Is this done because the additional weight would cause the ship to slide down the ways prematurely?

13. A force of 20 N acts upon a body whose weight is 8 N. (a) What is the mass of the body? (b) What is its acceleration?

14. A force of 20 N acts upon a body whose mass is 4 kg. (a) What is the weight of the body? (b) What is its acceleration?

15. An empty truck whose mass is 2000 kg has a maximum acceleration of 1 m/s^2. What is its maximum acceleration when it is carrying a 1000-kg load?

16. A force of 20 N gives an object an acceleration of 5 m/s^2. (a) What force would be needed to give the same object an acceleration of 1 m/s^2? (b) What force would be needed to give an acceleration of 10 m/s^2?

17. A net horizontal force of 4000 N is applied to a 1400-kg car. What will the car's speed be after 10 s if it started from rest?

18. A 430-g soccer ball lying on the ground is kicked and flies off at 25 m/s. If the duration of the impact was 0.01 s, what was the average force on the ball?

19. A 12,000-kg airplane launched by a catapult from an aircraft carrier is accelerated from 0 to 200 km/h in 3 s. (a) How many times the acceleration of gravity is the airplane's acceleration? (b) What is the average force the catapult exerts on the airplane?

20. A 2000-kg truck is braked to a stop in 15 m from an initial speed of 12 m/s. How much force was required?

21. A mass of 8 kg and another of 12 kg are suspended by a string on either side of a frictionless pulley. Find the acceleration of each mass.

22. A 100-kg man slides down a rope at constant speed. (a) What is the minimum breaking strength the rope must have? (b) If the rope has precisely this strength, will it support the man if he tries to climb back up?

23. A 100-kg wooden crate rests on a level wooden floor. What is the minimum force required to move it at constant speed across the floor?

24. A woman prevents a 2-kg brick from falling by pressing it against a vertical wall. The coefficient of static friction is 0.6. What force must she use? Is this more or less than the weight of the brick?

25. An eraser is pressed against a vertical blackboard with a horizontal force of 10 N. The coefficient of friction between eraser and blackboard is approximately 0.2. Find the force parallel to the blackboard required to move the eraser.

26. A tennis ball rolling along a floor is found to be slowing down with an acceleration of -1.6 ft/s^2. Find the coefficient of rolling friction.

27. A tennis ball whose initial speed is 2 m/s rolls along a floor for 5 m before coming to a stop. Find the coefficient of rolling friction.

28. A 1250-kg car reaches a speed of 2 m/s after being pushed by two people for 12 m starting from rest. If the coefficient of rolling friction is 0.007, find the force the people exerted on the car.

29. A 2-kg wooden block whose initial speed is 3 m/s slides on a smooth floor for 2 m before it comes to a stop. (a) Find the coefficient of friction. (b) How much force would be needed to keep the block moving at constant speed across the floor?

30. The longest recorded skid marks were found on a road in England after an accident in 1960. The marks were 290 m long. If the coefficient of friction between tires and road was 0.7 and the car's brakes were locked, find the minimum initial speed of the car in km/h. Why is it likely that the initial speed was actually higher than this?

PROBLEMS

1. How much force is needed to give a 5-kg box an upward acceleration of 2 m/s^2?

2. How much force must you supply to give a 1-kg object an upward acceleration of $2g$? A downward acceleration of $2g$?

3. A 1000-kg elevator has a downward acceleration of 1 m/s^2. What is the tension in its supporting cable?

4. A 1000-kg elevator has an upward acceleration of 1 m/s^2. What is the tension in its supporting cable?

5. A 1200-kg elevator is supported by a cable that can safely withstand a tension of no more than 15 kN. (a) What is the maximum upward acceleration the elevator can have? (b) The maximum downward acceleration?

6. A 60-kg person stands on a scale in an elevator. How many newtons does the scale read (a) when the elevator is ascending with an acceleration of 1 m/s^2; (b) when it is descending with an acceleration of 1 m/s^2; (c) when it is ascending at the constant speed of 3 m/s; (d) when it is descending at the constant speed of 3 m/s; (e) when the cable has broken and the elevator is descending in free fall?

7. A parachutist whose total mass is 100 kg is falling at 50 m/s when her parachute opens. Her speed drops to 7 m/s in a vertical distance of 40 m. What total force did her harness have to withstand? How many times her weight is this force?

8. A car moving at 10 m/s (22.4 mi/h) strikes a stone wall. (a) The car is very rigid and the 80-kg driver comes to a stop in a distance of 0.2 m. What is his average acceleration and how does it compare with the acceleration of gravity g? How much force acted upon him? Express this force in both newtons and pounds. (b) The car is so constructed that its front end gradually collapses upon impact, and the driver comes to a stop in a distance of 1 m. Answer the same questions for this situation.

9. A 40-kg kangaroo exerts a constant force on the ground in the first 60 cm of her jump, and rises 2.0 m higher. When she carries a baby kangaroo in her pouch, she can rise only 1.8 m higher. What is the mass of the baby kangaroo?

10. A railway boxcar is set in motion along a track at the same initial velocity as a truck on a parallel road. If the only horizontal forces acting on both vehicles are due to rolling friction with the respective coefficients of 0.0045 and 0.04, which vehicle will come to a stop first? How many times farther will the other vehicle travel?

11. A 5-kg block resting on a horizontal surface is attached to a 5-kg block that hangs freely by a string passing over a pulley, an arrangement similar to that shown in Fig. 3–15. (a) If there is no friction between the first block and the surface, what is the block's acceleration? (b) If the coeffi-

cient of friction between the first block and the surface is 0.2, what is the block's acceleration?

12. A 10-kg block resting on a horizontal surface is attached to a 5-kg block that hangs freely by a string passing over a pulley, an arrangement similar to that shown in Fig. 3–15. (a) If there is no friction between the first block and the surface, what is the block's acceleration? What is the block's acceleration if the coefficient of friction between the first block and the surface is (b) 0.4? (c) 0.5? (d) 0.6?

13. A very strong 100-kg man is having a tug-of-war with a 1500-kg elephant. The coefficient of static friction for both on the ground is 1.0. (a) Find the force needed to move the elephant. (b) What would happen if the man could actually exert such a force?

14. A truck moving at 15 m/s is carrying a 2000-kg steel girder that rests on the bed of the truck without any fastenings. The coefficient of static friction between the girder and the truck bed is 0.5. (a) Find the minimum distance in which the truck can come to a stop without having the girder move forward. (b) If the girder had a mass of 3000 kg, would this distance be different?

15. A sprinter presses on the ground with a force equal to three times his own weight at a 50° angle with the horizontal at the start of a race. What is his forward acceleration?

16. A string with a weight at its lower end is suspended inside a car. When the car starts to move, the string makes an angle of 10° with the vertical. What is the car's acceleration? What happens when the car later travels at constant velocity?

17. A 10-kg crate and a 100-kg crate are both sliding without friction down a plane inclined at 20° with the horizontal. What is the acceleration of each crate?

18. A cyclist finds that she is able to coast at constant speed along a road that slopes downward at an angle of 1° with the horizontal. If she and her bicycle together have a mass of 70 kg, find the force required to propel them at constant speed along a level road.

19. A sled slides down a snow-covered hill at constant speed. If the hillside is 10° above the horizontal, what is the coefficient of sliding friction between the runners of the sled and the snow?

20. A 200-kg crate is being slid down a ramp that makes an angle of 20° with the horizontal. The coefficient of friction is 0.3. How much force parallel to the plane must be applied to the crate if it is to slide down at constant speed? In which direction must the force be applied?

21. A block slides down an inclined plane 9 m long that makes an angle of 38° with the horizontal. The coefficient of sliding friction is 0.25. If the block starts from rest, find the time required for it to reach the foot of the plane.

22. If the block of Problem 21 has a mass of 50 kg, find the minimum force required to move it upward along the plane. What should the direction of this force be?

23. A skier starts from rest and slides 50 m down a slope that makes an angle of 40° with the horizontal; he then continues sliding on level snow. (a) If the coefficient of friction between skis and snow is 0.10 and air resistance is neglected, what is the speed of the skier at the foot of the slope? (b) How far away from the foot of the slope does he come to a stop?

24. A block takes twice as long to slide down an inclined plane that makes an angle of 35° with the horizontal as it does to fall freely through the same vertical distance. What is the coefficient of friction?

25. A stick is used to push a block of wood toward the blade of a circular saw. (a) If the weight of the block of wood is w, the angle between the stick and the table is θ, and the coefficient of friction between block and table is μ, verify that the force that must be applied to the stick to move the block at constant speed is $\mu w/(\cos \theta - \mu \sin \theta)$. (b) Show that if the stick is held at too steep an angle, the block cannot be moved, no matter how much force is applied. (c) Find the value of the critical angle for $\mu = 0.25$.

ANSWERS TO MULTIPLE CHOICE

1. d	**6.** a	**11.** c	**15.** b	**19.** d
2. d	**7.** d	**12.** d	**16.** b	**20.** a
3. c	**8.** b	**13.** b	**17.** a	**21.** c
4. b	**9.** c	**14.** a	**18.** c	**22.** b
5. d	**10.** b			

EQUILIBRIUM

4

The next phase of our analysis of force and motion is an inquiry into the conditions under which an object acted upon by two or more forces is nevertheless not accelerated. As we shall find, it is not enough that the vector sum of the forces equal zero, since an object may be set rotating by such forces if their lines of action do not meet at a common point. Although dynamics—the study of moving bodies—is naturally of primary significance to the physicist, an introduction to statics—the study of bodies at rest—is valuable both because statics has important applications in technology and because it affords further practice in the use of vector methods.

4–1 TRANSLATIONAL EQUILIBRIUM

An object that has no net force acting on it is said to be in *translational equilibrium*. The important point is that the object has no linear acceleration.

According to the first law of motion, such an object need not be at rest but may instead be moving along a straight path at constant velocity.

Translational equilibrium

The condition for translational equilibrium may be expressed in the form

$$\Sigma \mathbf{F} = 0 \qquad \qquad \textit{Translational equilibrium} \quad (4–1)$$

where the symbol Σ (Greek capital letter *sigma*) means "sum of" and $\mathbf{F}$ refers to the various forces acting on a specific object. This is simply the mathematical way of stating that, at equilibrium, the forces are such as to cancel one another out.

Force components in each direction must total zero for equilibrium

In many equilibrium situations all the various forces lie in the same plane. When this is the case, we can establish a set of *x-y* coordinate axes wherever convenient in the plane and then resolve each force $\mathbf{F}$ into the components $\mathbf{F}_x$ and $\mathbf{F}_y$. Thus we can replace Eq. (4–1), which is a vector equation, with the two scalar equations

$$\Sigma F_x = 0 \qquad \qquad \qquad \qquad (4–2)$$
$$\qquad \qquad \textit{Translational equilibrium}$$
$$\Sigma F_y = 0 \qquad \qquad \qquad \qquad (4–3)$$

It is usually much easier to calculate the components of each force present and to make use of Eqs. (4–2) and (4–3) than it is to work with the forces themselves in Eq. (4–1).

There are three steps to follow in working out problems concerning the equilibrium of an object:

1. Draw a sketch of the forces that act on the object. (This is called a *free-body diagram*.) Do not show the forces that the object exerts on other objects, since such forces do not affect the equilibrium of the object itself.
2. Choose a convenient set of coordinate axes and resolve the various forces acting on the object into components along these axes.
3. Set the sum of the force components along each axis equal to zero, as specified in Eqs. (4–2) and (4–3). Then solve the resulting equations algebraically for whatever quantities are to be found in the problem.

Several examples will make clear the above procedure. In the first, shown in Fig. 4–1, we have the simple case of a box of weight *w* being supported by a single weightless rope. We note that the only force a rope can exert is a pull along its length in a direction away from the point of attachment at each end. If the weight of the rope is negligible,

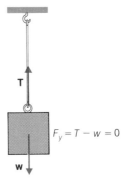

FIG. 4–1 A suspended object is in equilibrium when the tension in the rope is equal in magnitude to the weight of the object.

the two pulls are of equal magnitude. The box will be in equilibrium when all the forces acting on it cancel one another out, which here means that

$$\Sigma F_y = 0$$

since there are no forces in the x-direction. The tension T in the rope acts upward (the $+y$-direction) on the box and the box's weight w acts downward (the $-y$-direction) on it; hence

$$\Sigma F_y = T - w = 0$$

and

$$T = w \qquad\qquad (4\text{--}4)$$

The tension in the rope must equal the weight being supported.

Resolving rope tension into components

In Fig. 4-2 the same box is suspended from two ropes, A and B, which are at the angles θ and ϕ, respectively, with the horizontal. We begin by resolving the tension in each rope into components in the x- and y-directions, so that we have T_{Ay} and T_{By} upward, T_{Ax} to the left, and T_{Bx} to the right. From Fig. 4-2(b) we have

$$T_{Ax} = -T_A \cos \theta \qquad T_{Bx} = T_B \cos \phi$$

$$T_{Ay} = T_A \sin \theta \qquad T_{By} = T_B \sin \phi$$

We may now ignore the actual tensions T_A and T_B, and treat their components as individual forces acting at the same point as the weight w, as in Fig. 4-2(c). For the forces in the vertical and horizontal directions to cancel separately,

$$\Sigma F_x = T_{Bx} + T_{Ax} = 0 \qquad\qquad (4\text{--}5)$$

$$\Sigma F_y = T_{Ay} + T_{By} - w = 0 \qquad\qquad (4\text{--}6)$$

FIG. 4–2 (a) A box of weight w is suspended by two ropes. (b) A free-body diagram of the forces acting on the box. (c) At equilibrium the sum ΣF_x of the horizontal force components and the sum ΣF_y of the vertical force components each equal zero.

Example   In the above situation $w = 100$ N, $\theta = 37°$, and $\phi = 60°$. Find the tension in each rope.

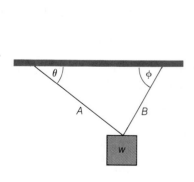

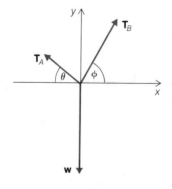

(b)

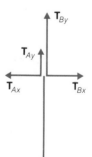

$$T_{Ax} = -T_A \cos \theta$$
$$T_{Ay} = T_A \sin \theta$$
$$T_{Bx} = T_B \cos \phi$$
$$T_{By} = T_B \sin \phi$$

(c)

Solution From Eq. (4–5) we find that

$$T_{Bx} + T_{Ax} = 0$$

$$T_B \cos \phi - T_A \cos \theta = 0$$

$$T_B = T_A \frac{\cos \theta}{\cos \phi} = T_A \frac{\cos 37°}{\cos 60°} = 1.6 T_A$$

From Eq. (4–6) we find that

$$T_{Ay} + T_{By} - w = 0$$

$$T_A \sin \theta + T_B \sin \phi - 100\,\text{N} = 0$$

Substituting $1.6 T_A$ for T_B in the last equation, we obtain

$$T_A \sin \theta + 1.6 T_A \sin \phi - 100\,\text{N} = 0$$

$$T_A(\sin 37° + 1.6 \sin 60°) = 100\,\text{N}$$

$$T_A = 50\,\text{N}$$

The tension in rope A is 50 N. Since $T_B = 1.6 T_A$ here,

$$T_B = 1.6 \times 50\,\text{N} = 80\,\text{N}$$

The tension in rope B is 80 N. The algebraic sum of the tensions in the two ropes is 130 N, which is more than the weight being supported, but the vector sum of $\mathbf{T}_A$ and $\mathbf{T}_B$ is 100 N, acting upward, which cancels the downward force $\mathbf{w} = 100$ N. ■

Example A 100-N box is suspended from the end of a horizontal strut, as in Fig. 4–3. Find the tension in the cable supporting the strut under the assumption that the strut's weight is negligible.

Solution It is easiest to consider the equilibrium of the end of the strut. The three forces that act on the end of the strut are the tension $\mathbf{T}$ in the cable, the outward force $\mathbf{F}_s$ exerted by the strut itself, and the weight $\mathbf{w}$ of the box. The horizontal and vertical components of the tension T are, from the diagram,

$$T_x = -T \cos 30°$$

$$T_y = T \sin 30°$$

FIG. 4–3 (a) A box of weight w is suspended from the end of a horizontal strut held in place by a cable attached to the wall. (b) A free-body diagram of the forces acting on the end of the strut. (c) At equilibrium, $\Sigma F_x = 0$ and $\Sigma F_y = 0$.

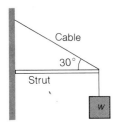

(a)

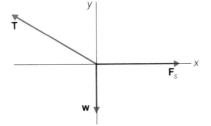

(b)

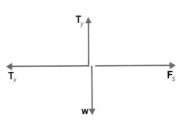

(c)

The end of the strut is in equilibrium when

$$\Sigma F_x = T_x + F_s = 0$$

$$\Sigma F_y = T_y - w = 0$$

All we need is the second of these equations to find T:

$$T_y - w = 0$$

$$T \sin 30° = w$$

$$T = \frac{w}{\sin 30°} = \frac{100\,\text{N}}{0.500} = 200\,\text{N}$$

4–2 TORQUE

Concurrent forces

When the lines of action of the various forces that act on an object intersect at a common point, they do not tend to set the object in rotation. Such forces are said to be *concurrent*.

Nonconcurrent forces

If the lines of action of the various forces do *not* intersect, the forces are *nonconcurrent,* and the body may be set into rotation even though the vector sum of the forces may equal zero. In Fig. 4-4(a) the three applied forces are concurrent and, if their vector sum is zero, the object is in equilibrium. In Fig. 4-4(b) the same forces are nonconcurrent, and their combined effect is to set the object spinning counterclockwise. If we want the term equilibrium to imply the absence of a rotational acceleration as well as the absence of a linear acceleration, we must supplement $\Sigma F = 0$ with another condition that the forces on a body must obey if it is to be in equilibrium.

Condition for balance

A hint as to the nature of this additional condition may be obtained by watching a seesaw in operation at a playground (Fig. 4-5). A small child can balance a large child merely by sitting farther from the pivot. Two children of the same weight will not balance unless they sit the same distance from the pivot, though the exact distance does not matter. Evidently both the magnitudes of the forces (here the weights of the children) and their lines of action determine whether or not the object is in equilibrium. If we were to try various combinations of weights and distances from the pivot, we would find that the seesaw is balanced when the product w_1L_1 of the weight w_1 and distance from the pivot L_1 of one child is equal to the product w_2L_2 of the other child's weight and distance from the pivot.

To make our discussion perfectly general, let us consider a force $\mathbf{F}$ acting upon an object free to rotate about some pivot point O (Fig. 4-6). The perpendicular distance

FIG. 4–4 (a) The lines of action of concurrent forces intersect at a common point. (b) When the lines of action do not intersect at a common point, the forces are nonconcurrent and the object cannot be in rotational equilibrium.

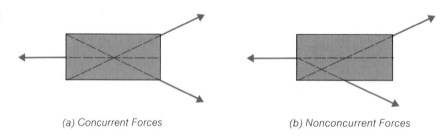

(a) Concurrent Forces   *(b) Nonconcurrent Forces*

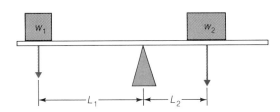

FIG. 4–5 A seesaw is balanced when $w_1L_1 = w_2L_2$.

L from O to the line of action of the force is called the *moment arm* of the force about O. The product of the magnitude F of the force and its moment arm L is known as the *torque* of the force about O. The symbol for torque is τ (Greek letter *tau*), so that

$$\tau = FL \qquad\qquad\qquad Torque \quad (4-7)$$

Torque = force × moment arm

In the SI system torque is expressed in newton·meters (N·m); in the British system, in lb·ft.

The greater the torque applied to an object, the greater the tendency of the object to be set into rotation.

Example The cylinder-head bolts on the diesel engine of a truck are supposed to be tightened to a torque of 70 N·m. If the center of the handle on the wrench used is 35 cm from the axis of the bolt, what force perpendicular to the handle should be applied?

Solution The moment arm here is $L = 35$ cm $= 0.35$ m. Since $\tau = FL = 70$ N·m, the required force is

$$F = \frac{\tau}{L} = \frac{70\,\text{N·m}}{0.35\,\text{m}} = 200\,\text{N} \qquad \blacksquare$$

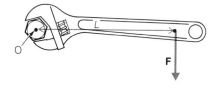

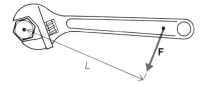

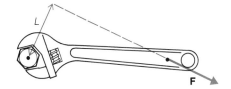

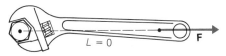

FIG. 4–6 A measure of the turning effect of a force about a pivot point O is its torque τ, which is equal to the product FL of the magnitude F of the force and the moment arm L. If the line of action of the force passes through the pivot point, the moment arm is $L = 0$ and $\tau = 0$ also.

FIG. 4–7 (a) A torque
that tends to produce a
counterclockwise rotation
is considered positive.
(b) A torque that tends to
produce a clockwise
rotation is considered
negative

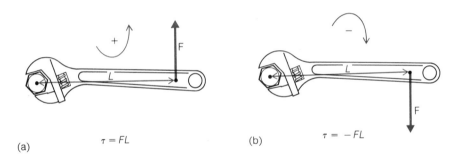

$\tau = FL$ $\tau = -FL$

(a)   (b)

4–3   ROTATIONAL EQUILIBRIUM

**Positive and negative
torques**

By convention a torque that tends to produce a counterclockwise rotation is considered positive and a torque that tends to produce a clockwise rotation is considered negative (Fig. 4-7). Thus the condition for an object to be in rotational equilibrium is that the sum of the torques acting upon it about any point, using the above convention for plus and minus signs, be zero:

$$\Sigma\tau = 0 \qquad\qquad\qquad\qquad\qquad \textit{Rotational equilibrium} \quad (4\text{–}8)$$

Of course, if the various forces that act do not all lie in the same plane, it is necessary that the sum of the torques in each of three mutually perpendicular planes be zero.

**Torques can be
calculated about any
point**

It is possible to prove that if the sum of the torques on an object is zero about any point, it is also zero about all other points. Hence the location of the point about which torques are calculated in an equilibrium problem is completely arbitrary; *any* point will do. (Of course, *all* torques must be calculated about this point.) Let us verify this statement with an example.

Example Figure 4-8 shows a rod 2.0 m long that has weights of 10 N and 30 N at its ends. We assume that the weight of the rod is negligible. At what point should the rod be picked up if it is to have no tendency to rotate? In other words, where is the balance point of the rod?

Solution 1 We first compute torques about the unknown balance point. If x is the distance of the 30-N weight from this point, the 10-N weight is $(2.0 \text{ m} - x)$ from it on the other side. The torques these weights exert are

FIG. 4–8 Torques are
computed about the
unknown balance point of
the rod in solution 1.

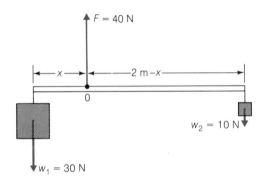

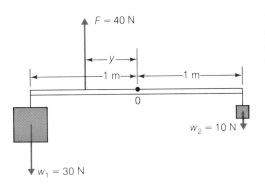

$$\tau_1 = w_1L_1 = +30x\,\text{N}$$

$$\tau_2 = w_2L_2 = -10(2.0\,\text{m} - x)\,\text{N}$$

and equilibrium will result when

$$\Sigma\tau = \tau_1 + \tau_2 = 30x\,\text{N} - 10(2.0\,\text{m} - x)\,\text{N} = 0$$

$$30x = 10(2.0\,\text{m} - x)$$

$$40x = 20\,\text{m}$$

$$x = 0.5\,\text{m}$$

When the rod is picked up 0.5 m from the 30-N weight, the two weights exert opposite torques of the same magnitude (15 N·m) about this point, so the rod is in balance. ∎

Solution 2   Let us solve the same problem by calculating torques about the middle of the rod, as shown in Fig. 4-9. Here y represents the distance between the balance point and the middle of the rod. We have three torques to take into account now:

$$\tau_1 = w_1L_1 = +(30\,\text{N} \times 1.0\,\text{m}) = +30\,\text{N·m}$$

$$\tau_2 = FL_2 = -40y\,\text{N}$$

$$\tau_3 = w_2L_3 = -(10\,\text{N} \times 1.0\,\text{m}) = -10\,\text{N·m}$$

The condition for equilibrium is

$$\Sigma\tau = \tau_1 + \tau_2 + \tau_3 = 30\,\text{N·m} - 40y\,\text{N} - 10\,\text{N·m} = 0$$

from which we obtain

$$y = 0.5\,\text{m}$$

The location of the balance point is the same regardless of the particular point about which torques are calculated. It is usually wise to calculate torques about the point of application of one of the forces that act on an object, since this makes it unnecessary to consider the torque produced by that force and thereby simplifies the arithmetic. ∎

Not all equilibrium situations are necessarily stable. For instance, a cone balanced on its apex is in equilibrium, but it will fall over when disturbed even slightly (Fig. 4-10). This is an example of an *unstable* equilibrium. The same cone on its base will

Types of equilibrium

FIG. 4–10 These cones are all in equilibrium, but only one of them is in a stable position.

Stable
equilibrium

Unstable
equilibrium

Neutral
equilibrium

return to its original position if tipped over a little; hence it is in *stable equilibrium* on its base. There is a third possibility as well, illustrated by a cone lying on its side. If such a cone is displaced, it remains in equilibrium in its new position with no tendency either to move further or to return to where it was before. A cone on its side is said to be in *neutral equilibrium*.

4–4 CENTER OF GRAVITY

Center of gravity

The *center of gravity* (CG) of an object is that point from which it can be suspended in any orientation without tending to rotate (Fig. 4-11). Each of the constituent particles of the object has a certain weight, and therefore exerts a torque about whatever point the object is suspended from. There is only a single point in an object about which all these torques cancel out no matter how it is oriented; this is its center of gravity. For equilibrium purposes we can therefore regard the entire weight of an object as a downward force acting from its center of gravity.

We can now recognize the distinctions between the different kinds of equilibrium shown in Fig. 4-10. The left-hand cone is in stable equilibrium because its CG has to be raised to change its orientation; the center cone is in unstable equilibrium because any change in its orientation lowers its CG; and the right-hand cone is in neutral equilibrium because the height of its CG does not change when it is rolled along on its side.

If the rod of Figs. 4-8 and 4-9 had the weight w instead of being weightless, we could take into account its effect on the location of the balance point by including the torque due to a force of magnitude w acting downward at the center of the rod. The CG of a uniform object of regular shape is located at its geometrical center. The CG of an irregular object need not even be located within the object itself: the CG of a seated person, for example, is a few inches in front of the person's abdomen.

FIG. 4–11 A body suspended from its center of gravity is in equilibrium in any orientation.

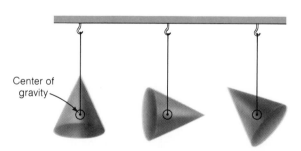

Center of
gravity

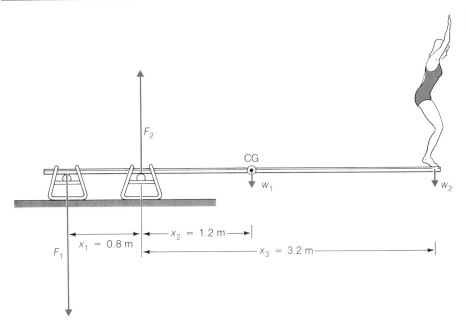

FIG. 4-12

Example   A beam that projects beyond its supports is called a *cantilever;* a diving   **Cantilever**
board is an example. Find the forces exerted by the two supports of the 4-m, 50-kg
uniform diving board shown in Fig. 4-12 when a 60-kg woman stands at its end.

Solution   The weight of the diving board is $w_1 = m_1 g_1 = 490$ N and that of the
woman is $w_2 = m_2 g = 588$ N. It is obvious that the force F_1 exerted by the left-hand
support must be downward. (If we nevertheless were to consider F_1 as upward, the
result would be a negative value for F_1, signifying the opposite direction.) We can
calculate F_1 without knowing F_2 by computing torques about the point of application
of F_2. The CG of the board is at its middle, which is 1.2 m from the pivot point. Hence
the three torques that act about this point are

$$\tau_1 = +F_1 x_1 = +(F_1 \times 0.8\,\text{m}) = 0.8\,F_1\,\text{m}$$

$$\tau_2 = -w_1 x_2 = -(490\,\text{N} \times 1.2\,\text{m}) = -588\,\text{N·m}$$

$$\tau_3 = -w_2 x_3 = -(588\,\text{N} \times 3.2\,\text{m}) = -1882\,\text{N·m}$$

and so

$$\Sigma\tau = \tau_1 + \tau_2 + \tau_3 = 0.8\,F_1\,\text{m} - 588\,\text{N·m} - 1882\,\text{N·m} = 0$$

$$F_1 = 3088\,\text{N}$$

We can find F_2 by considering the translational equilibrium of the loaded board:

$$F_2 = F_1 + w_1 + w_2 = 3088\,\text{N} + 490\,\text{N} + 588\,\text{N} = 4166\,\text{N} \qquad ■$$

Example   A person holds a 10-kg pail of water with his upper arm at his side and his
forearm outstretched, as in Fig. 4-13. The palm of his hand is 35 cm from his elbow,

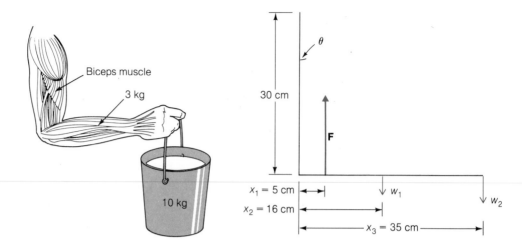

FIG. 4–13 The force exerted by the elbow on the forearm can be disregarded if torques are calculated about the elbow. The horizontal component of the tension in the biceps muscle has no moment arm about the elbow and so is also disregarded.

his upper arm is 30 cm long, and his biceps muscle is attached to his forearm 5 cm from his elbow. The person's forearm (including his hand) has a mass of 3 kg and its CG is 16 cm from the elbow. Find the force the biceps muscle exerts to support the forearm and pail.

Solution   We will calculate torques about the elbow, which simplifies the calculation since we need to consider only the vertical component F_y of the muscular force **F**, the weight $w_1 = m_1 g = 29.4$ N of the forearm, and the weight $w_2 = m_2 g = 98$ N of the pail. Since

$$\tan \theta = \frac{5\,\mathrm{cm}}{30\,\mathrm{cm}} = 0.167 \quad \text{and} \quad \theta = 9.5°$$

we have

$$F_y = F \cos \theta = 0.986\,F$$

The torques about the elbow, τ_1 exerted by the muscle, τ_2 by the forearm's weight, and τ_3 by the pail of water, are respectively

$$\tau_1 = F_y x_1 = 0.986F \times 0.05\ \mathrm{m} = 0.0493\,F\ \mathrm{m}$$
$$\tau_2 = -w_1 x_2 = -29.4\ \mathrm{N} \times 0.16\ \mathrm{m} = -4.7\ \mathrm{N{\cdot}m}$$
$$\tau_3 = -w_2 x_3 = -98\ \mathrm{N} \times 0.35\ \mathrm{m} = -34.3\ \mathrm{N{\cdot}m}$$

The sum of the torques about the elbow must be 0 for equilibrium, hence

$$\Sigma\tau = \tau_1 + \tau_2 + \tau_3 = 0.0493\,F\ \mathrm{m} - 4.7\ \mathrm{N{\cdot}m} - 34.3\ \mathrm{N{\cdot}m} = 0$$

from which we find

$$F = \frac{(4.7 + 34.3)\,\mathrm{N{\cdot}m}}{0.0493\,\mathrm{m}} = 791\ \mathrm{N}$$

This force, which is equivalent to 178 lb, is over six times the combined weights of the forearm and the pail of water. We might regard the body as being inefficiently designed, since such large muscular forces are required for ordinary tasks. However, when we

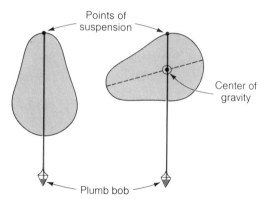

FIG. 4–14 To find the center of gravity of a flat body, suspend it and a plumb bob successively from two different points on its edge. The center of gravity is located at the intersection of the two lines of action of the plumb bob.

reflect upon the large span this arrangement enables the hand to move through and the speed at which it can do so, it is clear that inefficiency in one sense has been traded for efficiency in another. ∎

To obtain the CG of an irregular object, we can use the experimental method shown in Fig. 4-14. An analytical procedure that follows from the definition of CG can be applied to give a more accurate result. What is done is simply to break up the object into two or more separate ones whose CGs are known, to consider each of these component objects as particles joined by weightless rods, and then to calculate the balance point of the resulting assembly. If the object is complex, for example the hull of a ship, a great many separate elements must be considered if the result is to be accurate, but often some degree of regularity is present to facilitate the task.

How to find the CG of an irregular object

To see how such a calculation can be made, let us consider a system of three particles whose masses are m_1, m_2, and m_3 that are located x_1, x_2, and x_3 respectively from one end of a weightless rod, as in Fig. 4–15. The CG of the system is located at some distance X from the end of the rod, which means that the torque exerted about this end by a single particle of mass $M = m_1 + m_2 + m_3$ located at X will be exactly the same as the torque exerted by the actual system of the three particles. Hence

$$\Sigma \tau = m_1 g x_1 + m_2 g x_2 + m_3 g x_3 = MgX = (m_1 + m_2 + m_3)gX$$

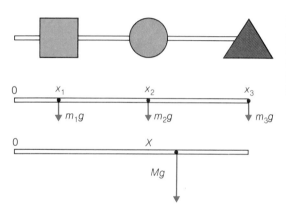

FIG. 4–15 The center of gravity of a complex object can be found by considering it as a system of particles and applying Eq. (4-9).

FIG. 4–16

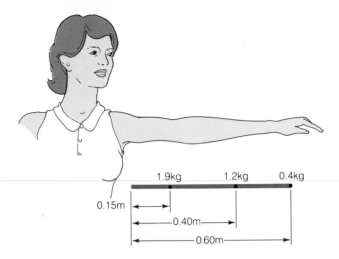

1.9kg   1.2kg   0.4kg

0.15m

0.40m

0.60m

so that

$$X = \frac{m_1 x_1 + m_2 x_2 + m_3 x_3}{m_1 + m_2 + m_3}$$

This formula can be generalized to a system of any number of particles (or objects that can be represented by particles located at their CGs) in a straight line, giving

$$X = \frac{m_1 x_1 + m_2 x_2 + m_3 x_3 + \cdots}{m_1 + m_2 + m_3 + \cdots} \qquad \textit{Center of gravity of composite object} \quad (4–9)$$

Example   The hand, forearm, and upper arm of a certain woman have the respective masses 0.4 kg, 1.2 kg, and 1.9 kg, and their CGs are respectively 0.60 m, 0.40 m, and 0.15 m from her shoulder joint (Fig. 4-16). Find the distance of the CG of her entire unbent arm from the shoulder joint.

Solution   From Eq. (4–9)

$$X = \frac{(0.4\,\text{kg} \times 0.60\,\text{m}) + (1.2\,\text{kg} \times 0.40\,\text{m}) + (1.9\,\text{kg} \times 0.15\,\text{m})}{(0.4 + 1.2 + 1.9)\,\text{kg}}$$

$$= 0.29\,\text{m}$$

The CG of the entire arm is 0.29 m from the shoulder joint.   ◼

If the irregular object whose CG is to be found lies in a plane rather than along a straight line (or is three-dimensional), the same procedure is applied along two (or three) coordinate axes. An example will make this clear.

Example   Find the CG of the L-shaped steel plate shown in Fig. 4-17.

Solution   We imagine the plate to consist of two sections, one a rectangle 4 m long by 1 m wide and the other a square 1 m on a side. The CGs of these sections are at their geometric centers, as shown in Fig. 4-17(b). Now we replace each section by a particle at its CG. If the steel plate has a mass of M per square meter, particle 1 has a

FIG. 4–17

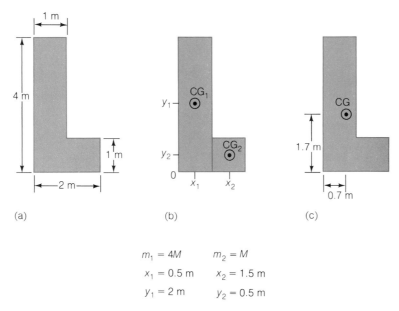

(a)   (b)   (c)

$$m_1 = 4M \qquad m_2 = M$$
$$x_1 = 0.5 \text{ m} \qquad x_2 = 1.5 \text{ m}$$
$$y_1 = 2 \text{ m} \qquad y_2 = 0.5 \text{ m}$$

mass of $4M$ and particle 2 has a mass of M, since their areas are respectively 4 m^2 and 1 m^2. The x- and y-coordinates of the CG of the entire plate are as follows:

$$X = \frac{m_1 x_1 + m_2 x_2}{m_1 + m_2} = \frac{(4M \times 0.5 \text{ m}) + (M \times 1.5 \text{ m})}{4M + M} = 0.7 \text{ m}$$

$$Y = \frac{m_1 y_1 + m_2 y_2}{m_1 + m_2} = \frac{(4M \times 2 \text{ m}) + (M \times 0.5 \text{ m})}{4M + M} = 1.7 \text{ m}$$

The location of this point is shown in Fig. 4-17(c). We did not need to know the value of M in this case since the steel plate is uniform. ■

The response of an object to a net force acting on it depends upon whether the line of action of the force passes through the object's CG or not. In the former case, the force has no effect on the object's rotational motion (Fig. 4-18); if the object is not rotating to begin with, such a force will simply accelerate it in the direction of the force. However, if the line of action of the force does not pass through the CG, the object will start to spin.

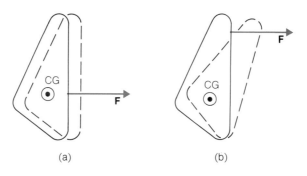

(a)   (b)

FIG. 4–18   (a) A body remains in rotational equilibrium when the line of action of an applied force passes through its center of gravity. (b) When the line of action of an applied force does not pass through its center of gravity, the body is given in a rotational as well as a translational acceleration.

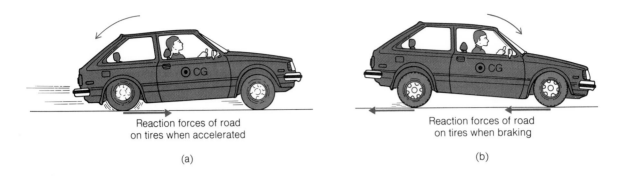

Reaction forces of road
on tires when accelerated

(a)

Reaction forces of road
on tires when braking

(b)

FIG. 4–19 A car tilts backward when accelerated (a) and forward when braked (b) because the lines of action of the applied forces in each case do not pass through the car's center of gravity. In (a) the car is assumed to have rear-wheel drive; if the car has front-wheel drive, the reaction forces act on its front tires, but the situation is otherwise the same.

We have all noticed that a car tilts backward when it is accelerated, and that it tilts forward when the brakes are applied. These effects occur because the forces applied to the car in each case are horizontal ones that act where the tires touch the road, and so are not in line with the car's center of gravity. The car rotates when the resulting torques act on it until the springs of the suspension system restore rotational equilibrium (Fig. 4-19).

4–5 MECHANICAL ADVANTAGE: THE LEVER

A machine is a device that transmits force or torque to accomplish a definite purpose. All machines, however complicated, are actually combinations of only three basic machines: the lever, the inclined plane, and the hydraulic press. Thus the train of gears that transmits power from the engine of a car to its wheels is a development of the lever, the screw jack that can raise one end of the car from the ground is a development of the inclined plane, and the brake system that permits a touch of the foot to stop the car is a development of the hydraulic press.

Mechanical advantage

The *mechanical advantage* (MA) of a machine is the ratio between the output force F_{out} it exerts and the input force F_{in} that is furnished to it:

$$\text{MA} = \frac{F_{out}}{F_{in}} \qquad\qquad \textit{Mechanical advantage} \quad (4\text{–}10)$$

A mechanical advantage greater than 1 signifies that the output force exceeds the input force, while a mechanical advantage less than 1 means that the output force is smaller. Usually the MA is greater than 1, which makes it possible for a relatively small applied force to accomplish a task ordinarily beyond its capacity, but sometimes the reverse is true, as in the case of a pair of scissors where the range of motion is increased at the expense of a reduced force.

Machine efficiency

A distinction must be made between the theoretical mechanical advantage of a machine, which is the value of its MA under ideal circumstances, and its actual mechanical advantage, which takes into account the effects of friction and any other factors present that tend to resist the transformation of F_{in} to F_{out}. The efficiency of a machine is equal to the ratio between its actual and its theoretical mechanical advantages. In

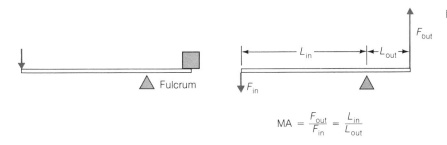

FIG. 4–20 The lever.

$$\text{MA} = \frac{F_{out}}{F_{in}} = \frac{L_{in}}{L_{out}}$$

some machines, such as the simple lever, the efficiency may be close to 100 percent, while in others, such as the screw, it may be less than 10 percent. In the latter case the low efficiency is actually an advantage, since it prevents the screw from backing out by itself. We shall consider only theoretical mechanical advantages here unless otherwise stated.

The *principle of equilibrium* is the basis for calculating mechanical advantage: when it is provided with the input force F_{in}, a machine will exactly balance a load equal to F_{out}. In the case of the lever shown in Fig. 4-20 the condition for equilibrium is that the torque produced by F_{in} about the fulcrum be the same in magnitude as that produced by F_{out}. If we call the lever arms of the respective forces L_{in} and L_{out}, we have

Principle of equilibrium

$$F_{in}L_{in} = F_{out}L_{out}$$
$$\frac{F_{out}}{F_{in}} = \frac{L_{in}}{L_{out}}$$

and

$$\text{MA} = \frac{F_{out}}{F_{in}} = \frac{L_{in}}{L_{out}} \qquad \textit{The lever} \quad (4\text{–}11)$$

The theoretical mechanical advantage of the lever is equal to the inverse ratio of the lever arms. A 4:1 ratio of lever arms, for instance, means an MA of 4 when the input force is applied to the longer arm, an MA of $\frac{1}{4}$ when it is applied to the shorter arm.

Example A steel rod 2.00 m long is to be used to pry up one end of a 500-kg crate. If a workman can exert a downward force of 350 N on one end of the rod, where should he place a block of wood to act as the fulcrum?

Solution The weight of the crate is $w = mg = 4900$ N. A force of half this is needed to lift one end, so the MA must be

$$\text{MA} = \frac{F_{out}}{F_{in}} = \frac{w/2}{F_{in}} = \frac{2450\,\text{N}}{350\,\text{N}} = 7$$

Hence the ratio of the lever arms should be 7:

$$L_{in} = 7L_{out}$$

FIG. 4–21   The three classes of lever.

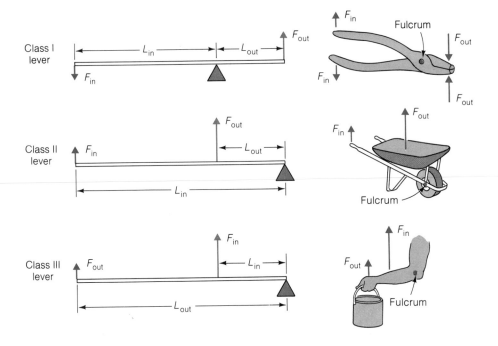

Since the lever is 2.00 m long,

$$L_{in} + L_{out} = 2.00\,m$$

$$7L_{out} + L_{out} = 2.00\,m$$

$$L_{out} = \frac{2.00\,m}{8} = 0.25\,m = 25\,cm$$

The fulcrum should be placed 25 cm from the crate.   ▪

The lever of Fig. 4-20, where the fulcrum is between the load and the applied force, is called a Class I lever. In a Class II lever, of which a wheelbarrow is an example, the load is between the fulcrum and the applied force, as in Fig. 4-21, while in a Class III lever, of which the human forearm is an example, the load and fulcrum are at the ends with the applied force between them.

The various kinds of elementary lever are all handicapped by the limited angle through which they can operate. Certain developments of the lever, however, can readily be used on a continuous basis, and one or another of them is an important element in nearly every motor-driven machine. Perhaps the simplest is the *wheel and axle* (Fig. 4-22). A wheel of radius R is attached to an axle of smaller radius r. The input force acts tangentially on the wheel, and the output force is exerted by the rim of the axle; the center of the axle acts as the fulcrum. By the principle of equilibrium the torque of the applied force F_{in} about the center of the axle must equal that of F_{out}, which means that

$$F_{in}R = F_{out}r$$

Classes of lever

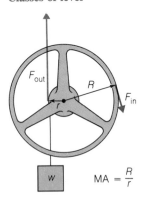

$$MA = \frac{R}{r}$$

FIG. 4–22   The wheel and axle.

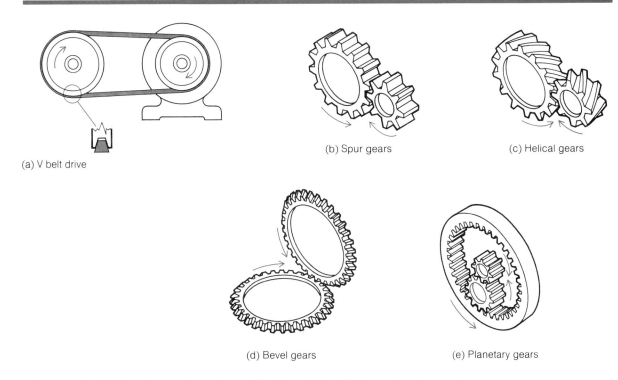

(a) V belt drive

(b) Spur gears

(c) Helical gears

(d) Bevel gears

(e) Planetary gears

FIG. 4–23 Both V belt and gear drive systems utilize the principle of the lever. In each case the mechanical advantage is equal to the ratio of diameters of the driven and driving gears. (a) A V belt has the advantage over a flat belt that it cannot slide sideways on the pulleys; the large area in contact with the sides of the pulleys helps prevent slipping. (b) Spur gears have their teeth cut parallel to the axes of rotation. (c) Helical gears have curved teeth cut in a spiral pattern at an angle to their axes; several teeth of meshing helical gears are in contact at all times, which makes a set of these gears smoother in operation and able to bear greater torque. (d) Bevel gears have straight teeth cut at 45° relative to their axes, and are used to change the direction of rotation by 90° with or without a change of speed. (e) A planetary gear is one whose teeth are cut on the inside of a wheel instead of on the outside.

Hence the theoretical mechanical advantage here is

The wheel and axle is a development of the lever

$$\text{MA} = \frac{F_{\text{out}}}{F_{\text{in}}} = \frac{R}{r} \qquad \textit{Wheel and axle} \quad (4\text{–}12)$$

The larger the wheel is relative to the axle, the greater is the mechanical advantage. For this reason trucks and buses not equipped with power steering have larger steering wheels than those in ordinary cars, which permits the drivers of the larger vehicles to provide the greater steering torque required. Other developments of the lever meant for torque transmission are the belt and gear drive systems (Fig. 4-23).

The ranges of motion of the ends of a lever are in proportion to their lengths: if the left-hand end of the lever of Fig. 4-24 is moved through an arc of s_{in}, the right-hand end will be moved through an arc of s_{out}, where

MA and ranges of motion

$$\frac{s_{\text{in}}}{s_{\text{out}}} = \frac{L_{\text{in}}}{L_{\text{out}}}$$

FIG. 4-24 Each end of a lever moves through a different distance if its arms are not equal in length. The ratio of distances equals the ratio of lever arms.

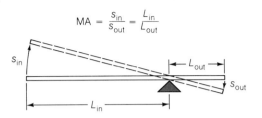

$$MA = \frac{S_{in}}{S_{out}} = \frac{L_{in}}{L_{out}}$$

Hence

$$MA = \frac{F_{out}}{F_{in}} = \frac{S_{in}}{S_{out}}$$ *Mechanical advantage*   (4–13)

In fact, this relationship holds not only for the lever but for all simple machines, and is often easier to apply than the principle of equivalence itself.

Example   Find the mechanical advantage of the block and tackle shown in Fig. 4-25.

Solution   When the free end of the rope is pulled with the force F_{in} through a distance d, the movable block is raised through a height of $\frac{1}{4}d$, since there are four strands that must be shortened. Hence

$$MA = \frac{S_{in}}{S_{out}} = \frac{d}{\frac{1}{4}d} = 4$$

The mechanical advantage of this block and tackle is 4.

FIG. 4–25 The theoretical mechanical advantage of a block and tackle is equal to the number of strands of rope that support the movable block, in this case 4.

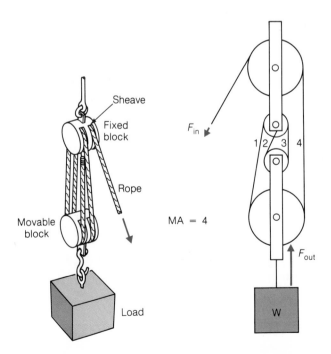

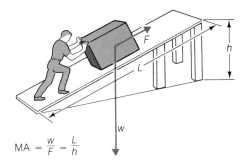

FIG. 4–26 The inclined plane.

$$MA = \frac{w}{F} = \frac{L}{h}$$

In general, the MA of a block and tackle is equal to the number of strands of rope that support the movable block and thereby the load. The strand to which the tension F_{in} is applied in Fig. 4–25 does not contribute to supporting the movable block; the upper pulley merely serves to change the direction of the applied force. ■

Block and tackle

4–6 THE INCLINED PLANE

The inclined plane is the second of the three basic machines; the third, the hydraulic press, will be considered in Chapter 10.

We are so accustomed to using the inclined plane in everyday life that it may be hard to think of it as a "machine." So instinctive an act as choosing a gradual slope of a hill to walk up instead of a steep slope is based upon the principle of the inclined plane. The most familiar adaptation of the inclined plane is a staircase, where the continuous surface of the plane is replaced by a series of steps for convenience in walking.

Figure 4-26 shows an inclined plane along which a crate of weight w is being pushed. The plane is L long and h high. From Eq. (4–13) we have

$$MA = \frac{s_{in}}{s_{out}} = \frac{L}{h} \qquad\qquad \textit{Inclined plane} \quad (4\text{–}14)$$

Example A 150-kg safe on frictionless casters is to be raised 1.2 m off the ground to the bed of a truck. Planks 4.0 m long are available for the safe to be rolled along. How much force is needed to push the safe up to the truck?

Solution Since MA = F_{out}/F_{in}, from Eq. 4–14 we have

$$F_{in} = \frac{F_{out}}{MA} = \frac{mg}{L/h} = \frac{mgh}{L} = \frac{(150\,\text{kg})(9.8\,\text{m/s}^2)(1.2\,\text{m})}{4.0\,\text{m}} = 441\,\text{N} \qquad ■$$

The simplest development of the inclined plane is the *wedge* (Fig. 4-27). The MA of a wedge L long and h thick is L/h just as for an inclined plane, but this mechanical advantage is never even approximately realized owing to friction. Aside from its use in splitting logs, leveling objects, and holding doors open, the wedge provides the operating principle of all cutting tools. A knife or chisel is obviously a wedge, but so are the teeth of a saw and the abrasive chips of a grindstone.

A wedge is an inclined plane

FIG. 4–27 The wedge.

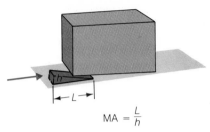

$$\text{MA} = \frac{L}{h}$$

Perhaps the most important application of the inclined plane is the *screw*. In essence a screw is an inclined plane wrapped around a cylinder to form a continuous helix (Fig. 4-28). The corrugations of a screw are called *threads,* and are usually (though not always) triangular in cross section. A *right-hand* screw is one that moves away from the viewer when turned clockwise, as in Fig. 4-29, while a *left-hand* screw moves away when turned counterclockwise. Standard screws are all right-handed, with left-hand ones employed only for special purposes.

The *pitch p* of a screw is the distance between adjacent threads, and the screw travels through this distance in each complete rotation. (In the British system, the number of threads per inch is specified, which is the reciprocal of p. Thus a screw that has 16 threads per inch has a pitch of $\frac{1}{16}$ in. = 0.0625 in.) The theoretical mechanical advantage of a screw depends upon the lever arm L used to turn it. In a screwdriver L is the handle radius, while in a wrench it is the distance from where the wrench is grasped to the center of its jaws. When a screw of pitch p is turned through one rotation, the applied force moves through a circle of circumference $2\pi L$ and the screw advances the distance p. Hence

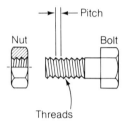

FIG. 4–28 A bolt is an example of the screw.

$$\text{MA} = \frac{s_{\text{in}}}{s_{\text{out}}} = \frac{2\pi L}{p} \qquad\qquad \textit{The screw} \quad (4\text{–}15)$$

Owing to the considerable friction that occurs between threads in contact, a screw is a relatively inefficient machine, but this friction is normally an advantage inasmuch as it prevents the screw from backing out under load.

FIG. 4–29 A right-hand screw moves into the page when turned clockwise, whereas a left-hand screw moves into the page when turned counterclockwise.

Right-hand screw

Left-hand screw

Example   A wrench whose effective lever arm is 20 cm is used to tighten a nut whose pitch is 1.5 mm. The efficiency is 5 percent, and a force of 50 N (about 11 lb) is applied to the wrench. Find the force exerted by the nut as it is screwed down.

Solution   Since 1 cm = 10 mm, $L = 200$ mm here, and

$$F_{out} = (\text{Eff})(\text{MA})(F_{in}) = (\text{Eff})\left(\frac{2\pi L}{p}\right)(F_{in})$$

$$= (0.05)\left(\frac{2\pi \times 200\,\text{mm}}{1.5\,\text{mm}}\right)(50\,\text{N}) = 2094\,\text{N}$$

which is about 740 lb—42 times the force applied to the wrench. Evidently the screw is a very powerful device. ▪

Figure 4-30 shows a worm gear drive. Each turn of the input screw causes the output gear to rotate by one tooth, so that the MA is equal to the number of teeth in the output gear. Worm gears are used when a large MA is required, or when a nonreversible gear drive is required. Often an oil bath is provided to reduce friction, which otherwise is considerable.

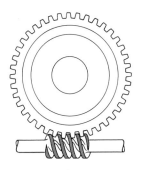

FIG. 4–30   A worm gear drive.

IMPORTANT TERMS

When the net force acting on an object is zero, the object is in **translational equilibrium.** When the net torque acting on an object is zero, the object is in **rotational equilibrium.**

The **torque** of a force about a particular pivot point is the product of the magnitude of the force and the perpendicular distance from the line of action of the force to the pivot point. The latter distance is called the **moment arm** of the force.

The **center of gravity** of an object is that point from which it can be suspended in any orientation without tending to rotate. The weight of an object can be considered as a downward force acting on its center of gravity.

A **machine** is a device that transmits force or torque. The three basic machines are the **lever,** the **inclined plane,** and the **hydraulic press.**

The **mechanical advantage** of a machine is the ratio between the output force (or torque) it exerts and the input force (or torque) that is furnished to it. The **theoretical mechanical advantage** is its value under ideal circumstances, while the **actual mechanical advantage** is its value when friction is taken into account.

The **efficiency** of a machine is the ratio between its actual and theoretical mechanical advantages; it is always less than 100%.

IMPORTANT FORMULAS

Equilibrium of a particle: $\Sigma F_x = 0$
$\Sigma F_y = 0$

Torque: $\tau = FL$

Rotational equilibrium of object:

$\Sigma \tau = 0$ about any point

Mechanical advantage: $\text{MA} = \dfrac{F_{out}}{F_{in}} = \dfrac{S_{in}}{S_{out}}$

MULTIPLE CHOICE

1. Which of the following sets of horizontal forces could leave an object in equilibrium?
   (a) 25, 50, and 100 N
   (b) 5, 10, 20, and 50 N
   (c) 8, 16, and 32 N
   (d) 20, 20, and 20 N

2. Which of the following sets of horizontal forces could not leave an object in equilibrium?
   (a) 6, 8, and 10 N
   (b) 10, 10, and 10 N
   (c) 10, 20, and 30 N
   (d) 20, 40, and 80 N

3. In general, the number of scalar equations that must be satisfied if an object free to move in a plane is to be in equilibrium is

(a) 2. (b) 3.
(c) 4. (d) 6.

4. In general, the number of scalar equations that must be satisfied if an object free to move in three dimensions is to be in equilibrium is

(a) 2. (b) 3.
(c) 4. (d) 6.

5. An object in equilibrium may *not* have

(a) any forces acting upon it.
(b) any torques acting upon it.
(c) velocity.
(d) acceleration.

6. Two ropes are used to support a stationary weight W. The tensions in the ropes must

(a) each be W/2.
(b) each be W.
(c) have a vector sum of magnitude W.
(d) have a vector sum of magnitude greater than W.

7. A weight is suspended from the middle of a rope whose ends are at the same level. In order for the rope to be perfectly horizontal, the forces applied to the ends of the rope

(a) must be equal to the weight.
(b) must be greater than the weight.
(c) might be so great as to break the rope.
(d) must be infinite in magnitude.

8. If the sum of the torques on an object is zero about a certain point,

(a) it is zero about no other point.
(b) it is zero about some other points.
(c) it is zero about all other points.
(d) the body must be in equilibrium.

9. In an equilibrium problem, the axis about which torques are computed

(a) must pass through one end of the object.
(b) must pass through the center of gravity of the object.
(c) must intersect the line of action of at least one force acting on the object.
(d) may be located anywhere.

10. The center of gravity of an object

(a) is always at its geometrical center.
(b) is always in the interior of the object.
(c) may be outside the object.
(d) is sometimes arbitrary.

11. The output force produced by a lever does *not* depend on

(a) the input force.
(b) friction at the fulcrum.
(c) the MA of the arrangement.
(d) the class of the lever.

12. A 100-N box is suspended by a string from an overhead support. If a horizontal force of 58 N is applied to the box, the string will make an angle with the vertical of

(a) 30°. (b) 45°.
(c) 60°. (d) 75°.

13. A 5-N picture is supported by two strings that run from its upper corners to a nail on the wall. If each string makes a 40° angle with the vertical, the tension in each is

(a) 3.3 N. (b) 3.9 N.
(c) 5 N. (d) 10 N.

14. A torque of 20 N·m is required to cut a thread on the end of a pipe. If the handles on the die are 30 cm from its axis on each side, the force each hand must apply is

(a) 0.33 N. (b) 10 N.
(c) 33 N. (d) 66 N.

15. A 60-kg object is attached to one end of a steel tube 2.4 m long whose mass is 40 kg. The distance from the loaded end to the balance point is

(a) 48 cm. (b) 60 cm.
(c) 80 cm. (d) 160 cm.

16. A person pries up one end of a 200-kg crate with a steel pipe 2.0 m long. If the force exerted is 350 N, the distance of the fulcrum from the crate is

(a) 30 cm. (b) 53 cm.
(c) 71 cm. (d) 86 cm.

17. The minimum number of pulleys in a block and tackle needed to achieve an MA of 6 is

(a) 3. (b) 4.
(c) 5. (d) 6.

18. The highest MA that can be obtained with a system of two pulleys is

(a) 1. (b) 2.
(c) 3. (d) 4.

19. A force of 50 N is needed to raise a 240-N load with a pulley system. The load ascends 1 m for every 5 m of rope pulled through the pulleys. The efficiency of the system is

(a) 48%.
(b) 50%.
(c) 96%.
(d) 104%.

EXERCISES

1. A ladder rests against a frictionless wall. (a) In what direction must the force the ladder exerts on the wall be? (b) How is the force the ladder exerts on the ground related to the weight of the ladder? Why?

2. Would you expect any difference in how soon you get tired between standing with your feet together and standing with them some distance apart? Why?

3. What determines the moment arm of a force?

4. In an equilibrium problem, under what circumstances is it necessary to consider the torques exerted by the various forces?

5. About which point should the torques of the various forces that act on a body be calculated when this is necessary?

6. A person on a bicycle presses down on each pedal through the front half of its circle. In which position is the torque a maximum? In which positions is it zero?

7. What must be true of the MA of a machine meant to increase force? Of a machine meant to increase speed?

8. A *couple* consists of two forces whose magnitudes are the same that act in opposite directions along parallel lines of action a certain distance apart. If the magnitude of each force is F and the separation between their lines of action is d, find a formula for the torque exerted by a couple.

9. The torque required to fracture a person's tibia (the large bone in the lower leg) is about 100 N·m. If the distance between the pivoting heel of the safety binding on a ski and the toe release is 35 cm, what is the maximum horizontal force at which the release should be set to open?

10. A 50-kg object is suspended from two ropes which each make an angle of 30° with the vertical. What is the tension in each rope?

11. A horizontal force of 80 N acts upon a 5-kg object suspended by a string. What is the direction and magnitude of the force the string must provide to keep the object at rest?

12. A 500-kg load of bricks is lifted by a crane alongside a building under construction. A horizontal rope is used to pull the load to where it is required, and the supporting cable is then at an angle of 10° from the vertical. What is the tension in the rope? In the supporting cable?

13. A horizontal beam 6 m long projects from the wall of a building. A guy wire that makes a 40° angle with the horizontal is attached to the outer end of the beam. When a weight of 100 N is attached to the end of the beam, what is the tension in the guy wire? (Neglect the weight of the beam.)

14. A 20-kg object is suspended from one end of a 30-kg wooden beam 3 m long. Where should the beam be picked up so that it remains horizontal?

15. Three people are carrying a horizontal ladder 4 m long. One of them holds the front end of the ladder and the other two hold opposite sides of the ladder the same distance from its far end. What is the distance of the latter two people from the far end of the ladder if each person supports one-third of the ladder's weight?

16. A car is stuck in the mud. To get it out, the driver ties one end of a rope to the car and the other to a tree 30 m away. He then pulls sideways on the rope at its midpoint. If he exerts a force of 500 N, how much force is applied to the car when he has pulled the rope 1.5 m to one side?

17. A bird sits on a telephone wire midway between two poles 20 m apart. The wire, assumed weightless, sags by 50 cm. If the tension in the wire is 90 N, find the mass of the bird.

18. The metal plate shown in the figure below has a circular hole 10 cm in radius. Find the location of the center of gravity of the plate.

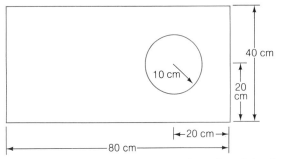

19. Find the location of the center of gravity of the flat object shown in the figure below.

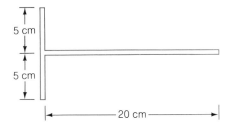

20. A ramp 25 m long slopes down 1.2 m to the edge of a

lake. What force is needed to pull a 300-kg boat on an 80-kg trailer up along the ramp if friction is negligible?

21. The figure below shows a pulley system that has a higher mechanical advantage than a block and tackle using the same number of pulleys. Find its MA. (*Hint:* Consider the tension in each rope segment that supports the mass.)

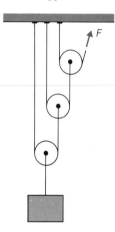

22. A block and tackle is used to pull a car out of the mud. A block with two pulleys is fastened to the car's bumper, and another block with three pulleys is tied to a tree. One end of a rope is made fast to the bumper and threaded through the various pulleys. Neglecting friction, how much force is applied to the car when two men exert a total force of 900 N on the free end of the rope?

23. In a certain winch a spur gear with 10 teeth is used to drive another spur gear with 40 teeth. The efficiency of the winch is 80%. (a) If the input gear is turned at 8 revolutions/min, how fast does the output gear turn? (b) If a torque of 30 N·m is applied to the input gear, what is the output torque?

24. A wood screw with a pitch of 2 mm is being turned by a screwdriver whose handle is 4 cm in diameter. A pilot hole has been drilled for the screw, and grease is used to increase the efficiency to 10%. If a force of 5 N is applied to the handle of the screwdriver, with how much force does the screw advance?

25. A machinist's vise has a well-lubricated screw of 3 mm pitch and a handle 13 cm long. If a force of 15 N applied to the end of the handle produces a force of 3 kN between the jaws of the vise, find its efficiency.

PROBLEMS

1. A 15-kg child and a 25-kg child sit at opposite ends of a 4-m seesaw pivoted at its center. Where should a third child whose mass is 20 kg sit in order to balance the seesaw?

2. A 25-N bag of cement is placed on a 4-m-long plank 1.8 m from one end. The plank itself weighs 8 N. Two men pick up the plank, one at each end. How much weight must each support?

3. The front and rear axles of a 24-kN truck are 4.0 m apart. The center of gravity of the truck is located 2.5 m behind its front axle. Find the weight supported by the front wheels of the truck.

4. The front wheels of a certain car are found to support 600 kg and the rear wheels 400 kg. The car's wheelbase (distance between axles) is 2.2 m. How far from the forward axle is the center of gravity of the car?

5. The center of gravity of a 150-kg polar bear standing on all fours is 100 cm from her feet and 80 cm from her hands. Find the force the ground exerts on each of her hands and feet.

6. A 60-kg woman is sitting with her upper body and lower legs vertical and her thighs horizontal. The mass of her upper body is 40 kg and its CG is 32 cm above her hips; the mass of her thighs is 13 kg and their CG is 16 cm in front of her hips; and the mass of her lower legs and feet is 7 kg and their CG is 23 cm below her knees and 37 cm in front of her hips. Find the location of the CG of the seated woman.

7. When a person stands on one foot with his heel raised, the entire reaction force of the floor, which is equal to his weight w, acts upward on the ball of his foot, as in the figure below. In order to raise his heel, he must apply the upward force F_1 via his Achilles tendon, so the downward force F_2 on his ankle is greater than his weight w. Find the values of F_1 and F_2 for a 75-kg man for whom $L_1 = 5$ cm and $L_2 = 15$ cm. If the distance L_1 were greater, would this mean an increase or a decrease in the muscular force F_1?

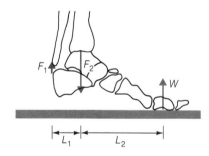

8. A butcher with his upper arm vertically at his side and his forearm horizontal presses down on a scale with his hand, which is 40 cm from his elbow joint. His forearm and hand together weigh 30 N, and their center of gravity is 20 cm from his elbow joint. The triceps muscle on the back of his upper arm is attached 2.5 cm behind the center of the elbow joint and it is exerting an upward force of 1 kN. What is the reading on the scale?

9. The wheels of a certain truck are 1.8 m apart, and the truck falls over when tilted sideways at 30° from the horizontal. How high above the road is its center of gravity?

10. A 60-kg woman stands upright with her feet 20 cm apart and her arms outstretched on both sides. In her right hand she holds a dumbbell 70 cm from the centerline of her body. Find the maximum mass the dumbbell can have if she is not to topple over.

11. A 40 kg boom 3 m long is hinged to a vertical mast and held in position by a rope at its end that is attached to the mast 1 m above the hinge pin. If the boom is horizontal and is uniform, find the tension in the rope.

12. A uniform horizontal beam 4 m long is supported at one end by a rigid post and at the other by a rope that makes an angle of 40° with the beam. A load of 1000 kg is suspended from the outer end of the beam, which itself has a mass of 200 kg. Find the tension in the rope.

13. A fishing rod 3 m long is attached to a pivoting holder at its lower end. The rod is 50° above the horizontal. A fisherman grasps the rod 1 m from its lower end. What horizontal force must the fisherman exert to keep the rod at this angle when reeling in a 10-kg fish from directly below the rod's upper end?

14. The figure below is a simplified diagram of a person bending over with the back horizontal. Typically a person's upper body (head, torso, and arms) comprises 65% of the total weight and has its CG 72% of the height above the feet. In the case of an 80-kg person 1.8 m tall, the upper body weight is 510 N and its CG is 36 cm above the base of the spine. This weight is supported by the back muscles, whose effect is equivalent to that of a single force **F** acting 42 cm above the base of the spine at an angle of 12°. (a) Find the

magnitude of **F**. (b) Find the magnitude of the reaction force **R** that acts on the base of the spine. How many times greater than the weight of the upper body is this force? (Because bending over, particularly when picking something up, puts so much stress on the lower spine, this posture is best avoided. When an object is to be picked up, the back should be kept as vertical as possible and the actual lifting done by the leg muscles.)

15. The arm of a certain person has a mass of 3 kg and its center of gravity is 28 cm from the shoulder joint. When the arm is outstretched so that it is horizontal, the shoulder muscle supporting it has a line of action 15° above the horizontal. (a) If this muscle is attached to the upper arm 13 cm from the shoulder joint, find the force developed by the muscle. (b) Find the force when the outstretched hand 60 cm from the shoulder joint is used to support a 1-kg load.

16. A door 3 m high and 1.2 m wide has hinges on one edge that are 30 cm above the bottom and 30 cm below the top. The entire 500-N weight of the door is supported by the lower hinge. Find the magnitude and direction of (a) the force the door exerts on the lower hinge; (b) the force the upper hinge exerts on the door.

17. A door 2.4 m high and 0.8 m wide has hinges at the top and bottom of one edge. The entire 200-N weight of the door is supported by the upper hinge. Find the magnitude and direction of (a) the force the door exerts on the upper hinge; (b) the force the lower hinge exerts on the door.

18. A uniform 10-kg ladder 2.5 m long is placed against a frictionless wall with its base on the ground 80 cm from the wall. Find the magnitudes of the forces exerted on the wall and on the ground.

19. A uniform 15-kg ladder 3 m long rests against a frictionless wall at a point 2.4 m off the floor. Find the vertical and horizontal components of the force exerted by the ladder on the floor.

20. A uniform 12-kg ladder 2.8 m long rests against a vertical frictionless wall with its lower end 1.1 m from the wall. If the ladder is to stay in place, what must be the minimum coefficient of friction between the bottom of the ladder and the ground?

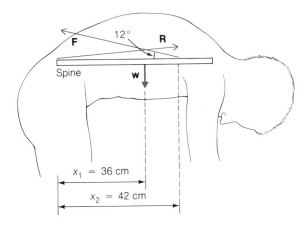

ANSWERS TO MULTIPLE CHOICE

1. d	**6.** c	**11.** d	**16.** b
2. d	**7.** d	**12.** a	**17.** c
3. b	**8.** c	**13.** b	**18.** c
4. d	**9.** d	**14.** c	**19.** c
5. d	**10.** c	**15.** a	

CIRCULAR MOTION

5

Nearly everything in the natural world travels in a curved path. Often these paths are either circles or are very close to being circles. For example, the orbits of the earth and the other planets about the sun are almost circular in shape, as is the orbit of the moon about the earth. On a smaller scale, a handy way to visualize an atom is to imagine it as having a central nucleus with electrons circling around. And, of course, circular motion is no novelty in our own experience; it is hard to think of any important aspect of technology in which circular motion of some kind is not involved. We shall therefore find it both interesting and essential for our later work to consider this kind of motion in some detail.

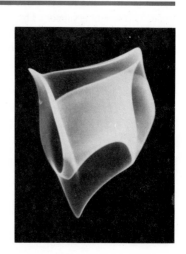

5–1 CENTRIPETAL FORCE

An object traveling in a circle at a constant speed is said to undergo *uniform circular motion*.

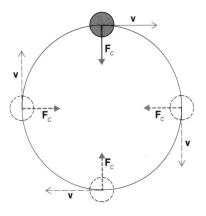

FIG. 5–1 Even though the velocity **v** of a body traveling in a circle at constant speed has the same magnitude along its path, the direction of **v** changes constantly. The inward force that causes this change in direction is called centripetal force, **F**$_c$.

Although the velocity of such an object has the same *magnitude* all along its path, the *direction* of the velocity changes constantly. A changing velocity means an acceleration, which in turn signifies that the object must be acted upon by a force. Since the object's path is a circle, the force on it must be directed toward the center of the circle (Fig. 5–1). This force is called *centripetal force,* literally "force seeking the center." Without it, circular motion cannot occur. In general,

Centripetal force is needed for motion along a curved path

Centripetal force = inward force on an object moving in a curved path

To verify directly the crucial role of centripetal force in circular motion, we can whirl a ball at the end of a string (Fig. 5–2). As the ball swings around, we must continually exert an inward force on it by means of the string. If we let go of the string, the ball flies off tangent to its original circular path. With no centripetal force on it, the ball then proceeds along a straight path at constant velocity as the first law of motion predicts. (Actually, of course, the ball will fall to the ground eventually because of gravity, but this is irrelevant here.)

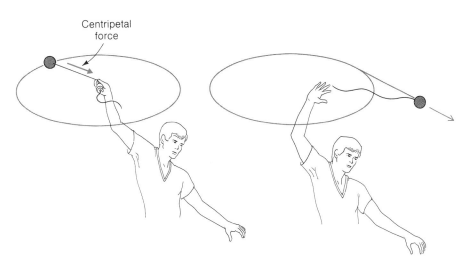

Centripetal force

FIG. 5–2 When a ball is whirled at the end of a string, the tension in the string provides the centripetal force that keeps the ball moving in a circle.

FIG. 5–3 The centripetal force exerted when a car rounds a curve on a level road is provided by friction between its tires and the road.

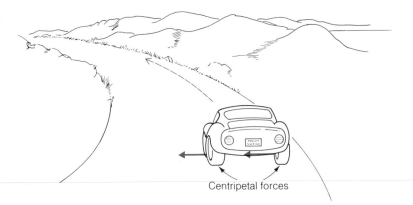

Centripetal forces

A centripetal force is acting whenever rotational motion occurs, since such a force is required to change the direction of motion of a particle from the straight line that it would normally follow to a curved path. Gravitation provides the centripetal forces that keep the planets moving around the sun and the moon around the earth. Friction between its tires and the road provides the centripetal force needed by a car in rounding a curve (Fig. 5–3). If the tires are worn and the road wet or icy, the frictional force is small and may not be enough to permit the car to turn.

5–2 CENTRIPETAL ACCELERATION

How large a centripetal force is needed to keep a given object moving in a circle with a certain speed? To find out, we must first compute the acceleration of such an object. In the following derivation we shall consider the circular motion of a particle; the same arguments and conclusions hold for the circular motion of the center of gravity of an object of definite size.

How formula for centripetal acceleration is derived

In Fig. 5–4(a) a particle is shown traveling along a circular path of radius r at the constant speed v. At $t = 0$ the particle is at the point A, where its velocity is $\mathbf{v}_A$, and at $t = \Delta t$ the particle is at the point B, where its velocity is $\mathbf{v}_B$. The change $\Delta\mathbf{v}$ in the particle's velocity in the time interval Δt is $\Delta\mathbf{v} = \mathbf{v}_B - \mathbf{v}_A$, and its acceleration is

$$\mathbf{a} = \frac{\Delta\mathbf{v}}{\Delta t}$$

The vector triangle whose sides are $-\mathbf{v}_A$, $\mathbf{v}_B$, and $\Delta\mathbf{v}$ is similar to the space triangle whose sides are OA, OB, and s, as we can see from Fig. 5–4(c). Since v is the constant speed of the particle, the magnitudes of $-\mathbf{v}_A$ and $\mathbf{v}_B$ are both v. Also, OA and OB are radii of the circle, so their lengths are both r. Corresponding sides of similar triangles are proportional, hence

$$\frac{\Delta v}{v} = \frac{s}{r} \quad \text{and} \quad \Delta v = \frac{vs}{r}$$

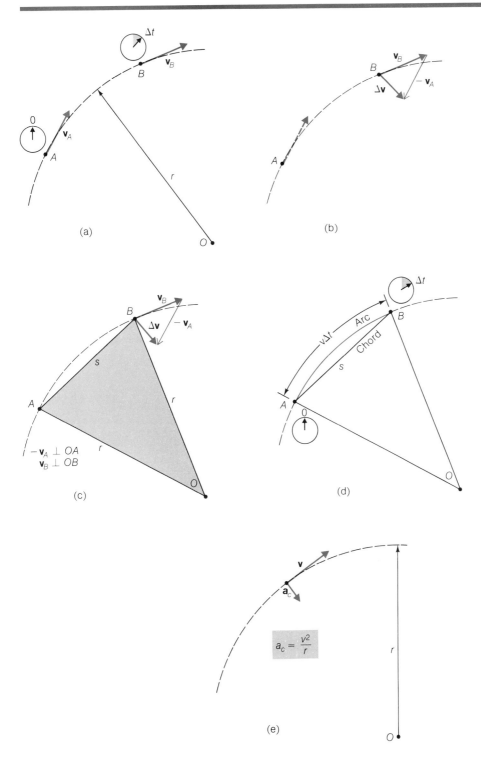

FIG. 5–4 (a) *A* and *B* are two successive positions, Δ*t* apart, of a particle undergoing uniform circular motion at the speed *v* in a circle of radius *r*. (b) The velocity of the particle at *A* is $\mathbf{v}_A$ and at *B* is $\mathbf{v}_B$; the change in its velocity in going from *A* to *B* is $\Delta\mathbf{v} = \mathbf{v}_B - \mathbf{v}_A$. (c) The space and vector triangles are similar because both are isosceles with the long sides of each perpendicular to the corresponding long sides of the other. (d) The chord joining *A* and *B* is *s*, while the actual distance the particle traverses is *v* Δ*t*. In calculating the instantaneous acceleration of the particle we are restricted to having *A* and *B* an infinitesimal distance apart, in which case the chord and arc have the same length. (e) The magnitude of the centripetal acceleration is $a_c = \Delta v/\Delta t = v^2/r$.

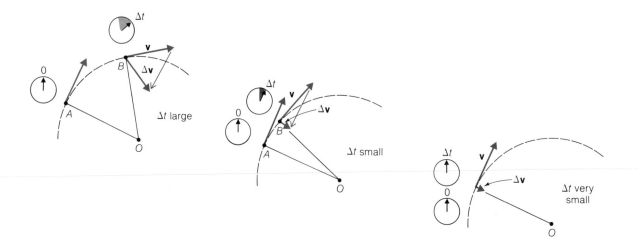

FIG. 5–5 The centripetal acceleration of a particle in uniform circular motion points toward the center of the circle.

The distance the particle actually covers in going from A to B is the arc joining these points, the length of which is $v \, \Delta t$. The distance s, however, is the chord joining A and B, as in Fig. 5–4(d). We are finding the *instantaneous* acceleration of the particle, and we are therefore concerned with the case where A and B are very close together, in which case the chord and arc are equal. Hence

$$s = v \, \Delta t$$

and we have

$$\Delta v = \frac{v^2 \, \Delta t}{r}$$

The magnitude of the particle's acceleration is therefore

$$a_c = \frac{\Delta v}{\Delta t} = \frac{v^2}{r}$$

The direction of centripetal acceleration is toward center of circle

In Fig. 5–4(c) $\Delta \mathbf{v}$ does not quite point toward O, the center of the particle's circular path. When A and B are very close together, however, $\Delta \mathbf{v}$ *does* point toward O (Fig. 5–5). Because the acceleration $\mathbf{a}_c$ is an instantaneous acceleration, we are solely concerned with the case when Δt and hence s are extremely small, and the direction of $\mathbf{a}_c$ is accordingly radially inward.

The inward, or *centripetal*, acceleration of a particle in uniform circular motion is proportional to the square of its speed and inversely proportional to the radius of its path:

$$a_c = \frac{v^2}{r}$$

Centripetal acceleration   (5–1)

Example   A physics teacher swings a pail of water in a vertical circle 1.0 m in radius at constant speed (Fig. 5–6). What is the maximum time per revolution if the water is not to spill?

FIG. 5–6 The water in the pail will not spill if the centripetal acceleration at the top of its circular path equals g, the acceleration of gravity.

Solution The water will not spill if, at the top of its path, its downward acceleration g due to gravity is equal to (or less than) the inward centripetal acceleration v^2/r. Thus the minimum speed is specified by

$$\frac{v^2}{r} = g$$
$$v = \sqrt{rg} = \sqrt{(1.0\,\text{m})(9.8\,\text{m/s}^2)} = 3.1\,\text{m/s}$$

The time needed by an object in uniform circular motion to make a complete revolution is called its *period,* symbol T. The distance the object travels in making a circle of radius r is $2\pi r$, the circle's circumference (Fig. 5–7). The speed of the object is therefore related to T by

Period of circular motion

$$v = \frac{\text{distance}}{\text{time}} = \frac{2\pi r}{T}$$

In this case,

$$T = \frac{2\pi r}{v} = \frac{(2\pi)(1.0\,\text{m})}{(3.1\,\text{m/s})} = 2.0\,\text{s}$$ ■

FIG. 5–7 A body in uniform circular motion whose period is T has a speed of $2\pi r/T$.

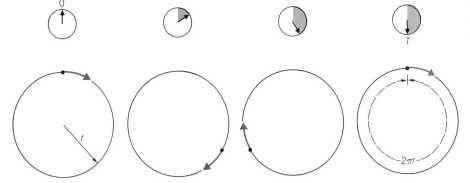

5–3   MAGNITUDE OF CENTRIPETAL FORCE

From the second law of motion $\mathbf{F} = m\mathbf{a}$ we see that the centripetal force $\mathbf{F}_c$ that must be acting on an object of mass m in uniform circular motion is $\mathbf{F}_c = m\mathbf{a}_c$. Since the centripetal acceleration has the magnitude

$$a_c = \frac{v^2}{r}$$

the magnitude of the centripetal force is

$$F_c = \frac{mv^2}{r}$$

Centripetal force   (5–2)

The centripetal force that must be exerted to maintain an object in uniform circular motion increases with increasing mass and with increasing speed, with the force more sensitive to a change in speed since it is the square of the speed that is involved. An increase in the radius of the path, however, reduces the required centripetal force (Fig. 5–8).

Example   Find the centripetal force required by a 1200-kg car that makes a turn of radius 40 m at a speed of 25 km/h (Fig. 5–9).

FIG. 5–8   Centripetal force.

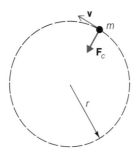

The centripetal force on an object in uniform circular motion is equal in magnitude to mv^2/r.

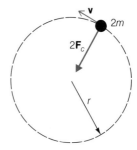

Doubling the mass doubles the required centripetal force.

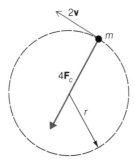

Doubling the speed, however, quadruples the required centripetal force.

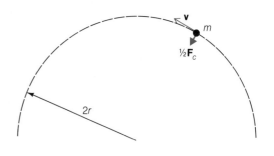

Doubling the radius of the circle halves the required centripetal force.

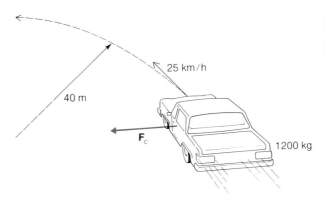

FIG. 5–9 A centripetal force of 1470 N is required by this car to make the turn shown.

25 km/h

40 m

F_c

1200 kg

Solution The car's speed is

$$v = (25\,\text{km/h})\left(0.278\ \frac{\text{m/s}}{\text{km/h}}\right) = 7.0\,\text{m/s}$$

Accordingly the centripetal force is

$$F_c = \frac{mv^2}{r} = \frac{(1200\,\text{kg})(7.0\,\text{m/s})^2}{40\,\text{m}} = 1470\,\text{N}$$

On a level road, the centripetal force of 1470 N must be provided by the pavement acting on the car's tires through the agency of friction. We can easily determine the minimum coefficient of friction μ that must be present if the car is to make the turn on such a road without skidding. The frictional force is

$$F_f = \mu N$$

in general. To find μ, we substitute 1470 N for F_f and the car's weight of $w = mg = 1.18 \times 10^4$ N for the normal force N. Thus

$$\mu = \frac{F_f}{N} = \frac{F_c}{w} = \frac{1470\,\text{N}}{1.18 \times 10^4\,\text{N}} = 0.125$$

which is available under good driving conditions. ■

Example Usually the friction between its tires and the road is enough to provide a car with the centripetal force it needs to make a turn. However, if the car's speed is high or the road surface is slippery, the available frictional force may not be enough and the car will skid. To avoid the likelihood of skids, highway curves are often *banked* so that the roadbed tilts inward. The horizontal component of the reaction force of the road on the car (the action force is the car pressing on the road) then furnishes the required centripetal force. Find the proper banking angle for a car making a turn of radius r at the speed v.

Banked turns

Solution The reaction force **F** of the road on the car is perpendicular to the roadbed, as in Fig. 5–10, since friction is not involved here. This force can be resolved into two

FIG. 5–10 When a car rounds a banked curve, the horizontal component **F**$_x$ of the reaction force **F** of the road on the car provides it with the required centripetal force.

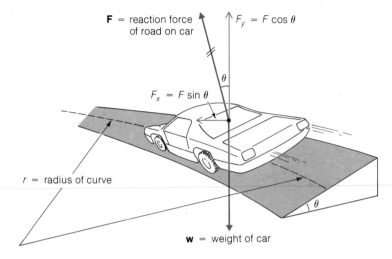

components, **F**$_y$ which supports the weight **w** of the car, and **F**$_x$ which is available to provide centripetal force. From the diagram,

$$F_x = F \sin \theta = \text{horizontal component of reaction force}$$
$$F_y = F \cos \theta = \text{vertical component of reaction force}$$

where θ is the angle between the roadbed and the horizontal. Since F_x furnishes the centripetal force F_c,

$$F_x = F_c$$
$$F \sin \theta = \frac{mv^2}{r}$$

The vertical component of the reaction force equals the car's weight, and so

$$F_y = mg$$
$$F \cos \theta = mg$$

We divide the first of these equations by the second to obtain

$$\frac{F \sin \theta}{F \cos \theta} = \frac{mv^2}{mgr}$$
$$\tan \theta = \frac{v^2}{gr} \qquad\qquad \textit{Banking angle} \quad (5\text{–}3)$$

Banking angle is independent of car's mass

The proper banking angle θ varies directly with the square of the car's speed and inversely with the radius of the curve. The mass of the car does not matter. When a car goes around a curve at precisely the design speed, the reaction force of the road provides the centripetal force. If the car goes more slowly than this, friction tends to keep it from sliding down the inclined roadway; if the car goes faster, friction tends to keep it from skidding outward.

The same considerations apply to an airplane making a turn, in which case Eq. (5–3) specifies the angle its wings should make with the horizontal. ■

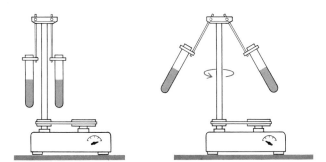

FIG. 5–11 A simple centrifuge.

5–4 THE CENTRIFUGE

A *centrifuge* is a device widely used to separate particles of some kind from a liquid in which they are suspended, for instance blood cells from plasma, or to separate liquids of different density from each other, for instance cream from milk. A simple type of centrifuge is shown in Fig. 5–11. As the centrifuge turns, the tubes swing upward, and the denser material migrates to the outer end of each tube.

A centrifuge speeds up the separation of substances of different density

No force pushes the denser material outward. Rather, each tube is pulled inward as the centrifuge turns, and the denser material responds less readily than the lighter material by virtue of its greater inertia and hence is left behind at the end of the tube. High speeds mean more effectiveness in separating substances having similar densities; some modern centrifuges operate at speeds exceeding 100,000 revolutions per minute.

The faster the centrifuge turns, the greater the angle between its arms and the central shaft. Why? Let us look at Fig. 5–12, which shows a particle of mass m suspended by a massless string and whirled in a horizontal circle. At a given speed, the horizontal component $\mathbf{T}_x$ of the tension $\mathbf{T}$ in the string provides the centripetal force

Why the arms of a centrifuge rise with increasing speed

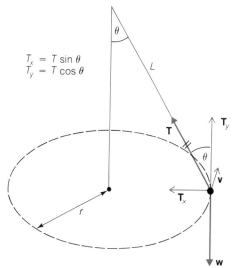

FIG. 5–12 A conical pendulum. When the particle's speed v increases, the angle θ also increases.

$$T_x = T \sin \theta$$
$$T_y = T \cos \theta$$

on the particle and its vertical component T_y is the force that supports the particle's weight w. When the particle's speed is increased, the tension T increases in magnitude so that T_x equals the new, larger centripetal force. But as T increases at a given angle θ, so does T_y. Now there is a net upward force on the particle, and it rises until T_y decreases to again equal w.

Conical pendulum

The arrangement of Fig. 5–12 is called a *conical pendulum* because the string traces out a cone in space as the particle moves in a circle. Reasoning similar to that in the case of a banked turn leads to the formula

$$\tan \theta = \frac{v^2}{gr}$$

for a conical pendulum. Though the string can approach close to the horizontal, it can never quite get there.

5–5   GRAVITATION

The earth and the other planets pursue approximately circular orbits around the sun. We conclude that the planets are being acted upon by centripetal forces that originate in the sun, since the sun is at the center of all the orbits. This much was generally understood by the middle of the seventeenth century, when Newton turned his mind to the question of exactly what the nature of the centripetal forces was.

Gravity holds the planets in orbits around the sun and the moon in an orbit around the earth

Newton proposed that the inward force exerted by the sun that is responsible for the planetary orbits is merely one example of a universal interaction, called *gravitation*, that occurs between all objects in the universe by virtue of their possession of mass. Another example of gravitation, according to Newton, is the attraction of the earth for nearby objects. Thus the centripetal acceleration of the moon and the downward acceleration g of objects dropped near the earth's surface have an identical cause, namely the gravitational pull of the earth.

Newton was able to arrive at the form of the *law of universal gravitation* from an analysis of the motions of the planets about the sun:

Newton's law of gravitation

Every object in the universe attracts every other object with a force directly proportional to each of their masses and inversely proportional to the square of the distance separating them.

The law of gravitation is expressed in equation form as

$$F_{\text{grav}} = G\frac{m_A m_B}{r^2} \qquad \text{\textit{Gravitational force}} \quad (5\text{–}4)$$

where m_A and m_B are the masses of any two objects and r is the distance between them. The quantity G is a universal constant whose value is

$$G = 6.67 \times 10^{-11}\frac{\text{N·m}^2}{\text{kg}^2} \qquad \text{\textit{Gravitational constant}}$$

The direction of the gravitational force is always along a line joining the two objects A and B. The force on A exerted by B is equal in magnitude to that on B exerted

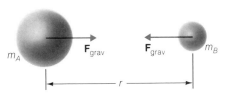

FIG. 5–13 The gravitational forces between two spherical objects.

by A, but is in the opposite direction (Fig. 5–13). A homogeneous spherical object, or one composed of homogeneous spherical shells, behaves gravitationally as if all its mass were concentrated at its center.

Gravity acts along line between centers of mass of two objects

Example   A grocer installs a 100-kg lead block under the pan of his scale. By how much does this increase the reading of the scale when 1 kg of onions are on the pan, if the centers of mass of the lead and of the onions are 0.3 m apart?

Solution   The gravitational force of the lead on the onions is

$$F = G\frac{m_A m_B}{r^2} = \left(6.67 \times 10^{-11}\frac{\text{N·m}^2}{\text{kg}^2}\right)\frac{(100\,\text{kg})(1\,\text{kg})}{(0.3\,\text{m})^2} = 7.4 \times 10^{-8}\,\text{N}$$

The increase in the scale reading is therefore

$$m = \frac{F}{g} = \frac{7.4 \times 10^{-8}\,\text{N}}{9.8\,\text{m/s}^2} = 7.6 \times 10^{-9}\,\text{kg} = 0.0000076\,\text{g}$$

so it is hardly worth the effort. Blowing gently on the onions will increase the reading over a million times more.   ■

It is impossible to determine the gravitational constant G from astronomical data alone, as Newton realized. A direct measurement of the gravitational force between two known masses a known distance apart is required. The difficulty here, as the foregoing example illustrates, is that gravitational forces are minute between objects of laboratory size. The value of G was finally established in 1798, over a century after Newton's work, by Henry Cavendish. He used an instrument called a *torsion balance* (Fig. 5–14), which is the rotational analog of an ordinary spring balance that measures forces in terms of the extension of a calibrated spring. The forces exerted on the small spheres in Cavendish's experiment could be found from the resulting twist in the fine suspending thread.

The Cavendish experiment to find G

What is the justification for assuming that the law of gravitation, obtained from data on the solar system, is also valid for the entire universe, describing the gravitational attraction of objects both larger and smaller than the members of the solar system?

There is no simple answer to this legitimate query; instead we must invoke a broad body of knowledge that bears upon the subject. For example, we observe that all the matter on the earth's surface experiences the same acceleration in free fall, which suggests identical gravitational behavior. Careful analysis of the light reaching us from the stars and galaxies throughout the visible universe indicates that the matter of which these bodies are composed behaves identically with matter found on the earth; and so on. Nowhere do we find reason to suspect there should be any objects in the universe that do not obey Newton's law of gravitation, and it is unreasonable to propose the existence of such objects with no evidence for the need to do so.

Why gravitation is believed to be universal

FIG. 5–14 Torsion balance for measuring gravitational forces.

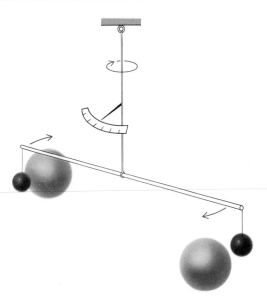

Perfect sphere

Actual shape of earth (exaggerated)

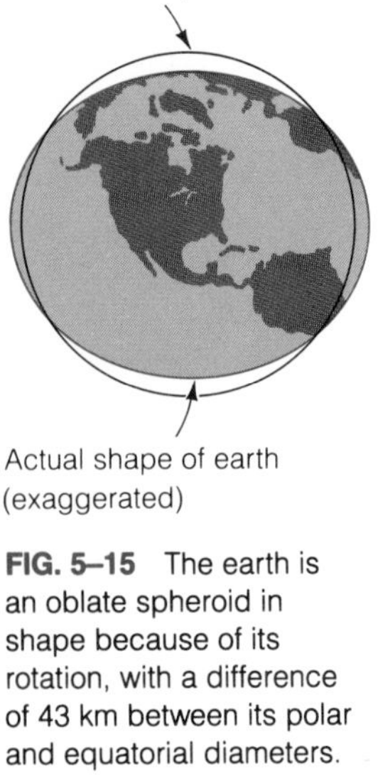

FIG. 5–15 The earth is an oblate spheroid in shape because of its rotation, with a difference of 43 km between its polar and equatorial diameters.

Finding the earth's mass

Strong theoretical reasons exist as well for believing in the unity of gravitational phenomena, but in physics experiment and observation are the final arbiters of the correctness of an idea, so these reasons, though welcome as corroboration, must take second place.

5–6   GRAVITY AND THE EARTH

Why is the earth round? Gravity provides the cause. If a part of the earth were to project by much, the gravitational pull of the remainder would lead to such strong pressures on the underlying material that it would flow out sideways until the protuberance became level, or nearly so, with the rest of the surface. In the same way, pressures around the margin of a deep cavity would cause the underlying material to flow into it. The earth's mountains and ocean basins are actually very small-scale irregularities—the total range from the Pacific depths to the summit of Everest is less than 20 km, not much compared with the earth's radius of 6400 km. The smaller an object is, the more likely its rigidity will be sufficient to withstand the tendency of gravity to impose a spherical form. Thus the satellites Phobos ("fear") and Deimos ("panic") of Mars have been able to remain oblong in shape because their longest dimensions are respectively only 23 and 11 km.

Because the earth is rotating, its equatorial region bulges outward, just as a ball on a string swings outward when whirled around (Fig. 5–15). The earth is about 0.34 percent away from being a perfect sphere (apart from surface irregularities). Venus, whose "day" is 243 of our days, turns so slowly that its distortion is negligible, whereas Saturn, at the other extreme, spins so rapidly that it is out of round by 9.6 percent.

Given the values of G, g, and the earth's radius r_e, we can calculate the earth's mass. Let us consider an object of mass m on the earth's surface, say an apple. The gravitational pull of the earth on the apple is the apple's weight of

FIG. 5–16

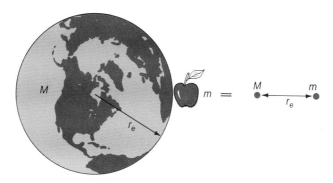

$$w = mg$$

As mentioned earlier, a spherical object behaves gravitationally as though its mass were concentrated at its center. Thus the earth-apple system can be represented by two particles of masses M and m a distance r_e apart, where M is the earth's mass and r_e is its radius (Fig. 5–16). According to Newton's law of gravitation, the force the earth exerts on the apple is

Weight of apple at earth's surface

$$F = G\frac{Mm}{r_e^2}$$

Gravitational force of earth on apple

This force must equal the apple's weight w, and so

$$F = w$$
$$G\frac{Mm}{r_e^2} = mg$$

When we solve this equation for M we see that the apple's mass m drops out. The mass of the earth is

$$M = \frac{gr_e^2}{G} = \frac{(9.8\,\text{m/s}^2)(6.4 \times 10^6\,\text{m})^2}{6.7 \times 10^{-11}\,\text{N}\cdot\text{m}^2/\text{kg}^2} = 6.0 \times 10^{24}\,\text{kg}$$

Figure 5–17 shows the structure of the earth on the basis of indirect, but persuasive, evidence from a number of lines of inquiry. The outer skin is a relatively thin *crust* of rock about 5 km thick under the oceans and an average of 35 km thick under the continents. The *mantle*, about 2900 km thick, consists of dense rock probably similar in composition to certain surface rocks and to stony meteorites. Because the materials of the mantle and crust are relatively light, they provide only 67 percent of the earth's mass although they comprise 80 percent of its volume, and so the *core* must be very dense to make up the rest of the mass. Also, as discussed in Sec. 12-5, earthquake wave studies show that the core must be a liquid, and the existence of the earth's magnetic field means that the liquid must be a metal in order to support the required electric currents (the earth's interior is too hot for permanent magnetism to occur there). Iron, which is abundant in the universe generally, seems to meet all the requirements, and it is accepted by geologists that the earth's core consists chiefly of molten iron.

The earth consists of a thin outer crust, a thick mantle, and a liquid core

FIG. 5–17 Structure of the earth. The mantle is composed of dense rock, the core probably of molten iron.

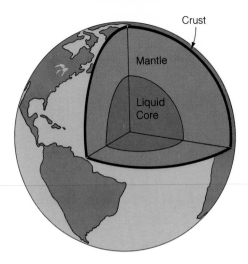

Crust

Mantle

Liquid Core

The earth's gravitational pull on an object varies inversely as the square of its distance from the center of the earth (Fig. 5–18). Let us see whether this holds for the moon.

FIG. 5–18 The gravitational force of the earth on an object varies inversely with the square of the object's distance from the center of the earth. Hence a person's weight at a distance r from the earth's center is $(r_e/r)^2$ of the weight on the earth's surface. Shown here is how the weight of a 100-kg person decreases with distance; the mass of 100 kg, of course, is the same everywhere in the universe.

Example   The moon is 3.84×10^8 m from the earth and circles the earth once every 27.3 days (Fig. 5–19). Compare the moon's centripetal acceleration with the acceleration it would experience on the basis of Newton's law of gravitation.

Solution   The moon's period of revolution is

$$T = 27.3 \text{ d} \times 86{,}400 \text{ s/d} = 2.36 \times 10^6 \text{ s}$$

and so its orbital speed (see Fig. 5–7) is

$$v = \frac{\text{orbit circumference}}{\text{period}} = \frac{2\pi r}{T} = \frac{(2\pi)(3.84 \times 10^8 \text{ m})}{2.36 \times 10^6 \text{ s}} = 1.02 \times 10^3 \text{ m/s}$$

The centripetal acceleration of the moon is accordingly

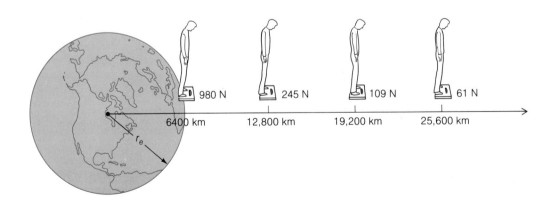

980 N   245 N   109 N   61 N

6400 km   12,800 km   19,200 km   25,600 km

r_e

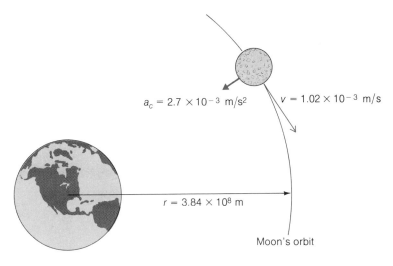

FIG. 5–19 The moon is accelerated toward the center of the earth.

$a_c = 2.7 \times 10^{-3} \text{ m/s}^2$ $v = 1.02 \times 10^{-3} \text{ m/s}$

$r = 3.84 \times 10^8 \text{ m}$

Moon's orbit

$$a_c = \frac{v^2}{r} = \frac{(1.02 \times 10^3 \text{ m/s})^2}{3.84 \times 10^8 \text{ m}} = 2.7 \times 10^{-3} \text{ m/s}^2$$

Centripetal acceleration of moon

and is directed toward the center of the earth.

The ratio between the moon's orbital radius and the earth's radius is

$$\frac{r}{r_e} = \frac{3.84 \times 10^8 \text{ m}}{6.4 \times 10^6 \text{ m}} = 60$$

The gravitational force the earth exerts on the moon is therefore $(1/60)^2 = 1/3600$ as strong as the force the earth would exert on it if the moon were at the earth's surface. The acceleration a of the moon toward the earth in turn ought to be 1/3600 of the acceleration of an object at the earth's surface. Since the latter acceleration is g,

$$a = \frac{g}{3600} = \frac{9.8 \text{ m/s}^2}{3600} = 2.7 \times 10^{-3} \text{ m/s}^2$$

Acceleration of moon due to earth's gravitational pull

which is the same as the moon's observed centripetal acceleration. The correspondence between the two figures was used by Newton as evidence for the universal validity of the law of gravitation. ■

5–7 EARTH SATELLITES

What keeps an artificial earth satellite from falling down? The answer, of course, is that it *is* falling down, but, like the moon, at just such a rate as to circle the earth in a stable orbit. Let us use what we know about gravitation and circular motion to investigate the orbits of earth satellites. In the following discussion the frictional resistance of the atmosphere, which ultimately brings down all artificial satellites, will be neglected.

A stable satellite orbit is possible at any distance from the earth

Near the earth the gravitational force on an object of mass m is its weight

$$w = mg$$

where g is the acceleration of gravity at the location of the object. For uniform circular motion about the earth this force must provide the object with the centripetal force

$$F_c = \frac{mv^2}{r}$$

Hence the condition for a stable orbit is

$$w = F_c$$

$$mg = \frac{mv^2}{r}$$

$$v = \sqrt{rg} \qquad\qquad\qquad (5\text{--}5)$$

Distant satellites move slower than nearby ones

Because the acceleration of gravity g is proportional to the square of the distance r from the center of the earth, the greater the orbital radius of a satellite, the smaller its speed v. If we let g_0 be the value of g at the earth's surface, then at the radius r

$$g = \left(\frac{r_e}{r}\right)^2 g_0 \qquad\qquad\qquad (5\text{--}6)$$

Hence the satellite speed at the radius r is

$$v = \sqrt{rg} = \sqrt{\frac{r_e^2 g_0}{r}} \qquad\qquad \textit{Orbit of earth satellite}\quad (5\text{--}7)$$

For an orbit just above the earth's surface,

$$v_0 = \sqrt{r_e g_0} = \sqrt{(6.4 \times 10^6\,\text{m})(9.8\,\text{m/s}^2)} = 7.9 \times 10^3\,\text{m/s}$$

which is about 28,400 km/h. Anything sent off tangent to the earth's surface at this speed will become a satellite of the earth. If it is sent off at a higher speed, its orbit will be elliptical rather than circular (Fig. 5–20). If the object's speed is great enough, it can escape permanently from the earth. Readers of *Alice in Wonderland* may recall the remark, "Now, here, you see, it takes all the running you can do to stay in the same place. If you want to get somewhere else, you must run at least twice as fast as **Escape speed** that!" The correct ratio is actually $\sqrt{2}$, so that the *escape speed* is $\sqrt{2}v_0 = 11.2 \times 10^3$ m/s in the case of the earth.

Example   Satellites used to relay radio communications are placed in orbits whose period is exactly 1 day so that they remain indefinitely over a particular location on the earth's equator. Find the altitude of such a "synchronous" orbit.

Solution   As we know (see Fig. 5–7), the period of an object in uniform circular motion is given by $T = 2\pi r/v$. Hence

$$v = \frac{2\pi r}{T}$$

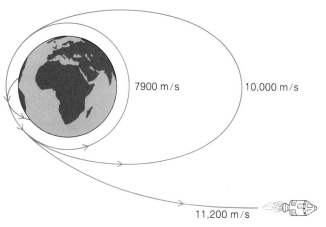

7900 m/s   10,000 m/s

11,200 m/s

FIG. 5–20 Depending upon its speed, an object projected horizontally above the earth's surface may fall back to the earth, revolve around the earth in a circular orbit, revolve around the earth in an elliptical orbit, or escape permanently from the earth into space.

where $T = 1$ day $= 86{,}400$ s. Another formula for the satellite speed is given by Eq. (5–7),

$$v = \sqrt{\frac{r_e^2 g_0}{r}}$$

Setting the two formulas equal enables us to eliminate v:

$$\frac{2\pi r}{T} = \sqrt{\frac{r_e^2 g_0}{r}}$$

$$\frac{4\pi^2 r^2}{T^2} = \frac{r_e^2 g_0}{r}$$

$$r^3 = \frac{r_e^2 g_0 T^2}{4\pi^2}$$

Therefore

$$r = \sqrt[3]{\frac{r_e^2 g_0 T^2}{4\pi^2}} = \sqrt[3]{\frac{(6.4 \times 10^6\,\text{m})^2 (9.8\,\text{m/s}^2)(8.64 \times 10^4\,\text{s})^2}{4\pi^2}}$$

$$= \sqrt[3]{759 \times 10^{20}\,\text{m}} = \sqrt[3]{75.9 \times 10^{21}\,\text{m}} = 4.23 \times 10^7\,\text{m}$$

The corresponding altitude h (Fig. 5–21) above the earth's surface is

$$h = r - r_e = 42.3 \times 10^6 \text{ m} - 6.4 \times 10^6 \text{ m} = 35.9 \times 10^6 \text{ m}$$

Altitude of synchronous orbit

■

FIG. 5–21

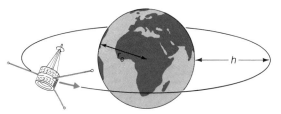

5–8   APPARENT WEIGHT

Weightlessness

Because an earth satellite is always falling toward the earth, an astronaut inside one feels "weightless." In reality, there *is* a gravitational force acting on him; what is missing to his senses is the upward reaction force provided by a stationary platform underneath him—the seat of a chair, the floor of a room, the ground itself. Instead of pushing back, the floor of the satellite falls just as fast as he does toward the earth.

Actual and apparent weight

It is useful to distinguish between the *actual weight* of an object, which is the gravitational force acting on it, and its *apparent weight,* which is the force it exerts on whatever it rests upon. We can think of the apparent weight of a person as the reading on a bathroom spring scale the person is standing on. An astronaut in an earth satellite has no apparent weight because he does not press down on the floor of the satellite.

A person jumping off a diving board is just as "weightless" in his descent as an astronaut, since nothing restricts his acceleration toward the earth either. But a person standing on the ground is acted upon by *both* the downward force of gravity and the upward reaction force of the ground: the latter force is what prevents him from simply dropping all the way down to the center of the earth. The human body (indeed, all living organisms on the earth) evolved in the presence of both these forces, and various body functions, such as blood circulation, do not seem to take place efficiently in a "weightless" state. Future satellites and other spacecraft designed for long journeys may be set in rotation so that inertia will cause astronauts to press against the cabin sides, which will then press back (action-reaction again) and so bring about a situation corresponding to that on the earth's surface.

Example   An airplane pulls out of a dive in a circular arc whose radius is 1000 m. If the speed of the airplane is a constant 200 m/s, find the apparent weight of the 80-kg pilot at the lowest point of the arc.

Solution   The downward force the pilot exerts on his seat has two components. The first is his weight mg, the gravitational pull of the earth on him. The second is the reaction force mv^2/r to the centripetal force that leads to the pilot's curved path through space. If we think of the situation as one of equilibrium within the airplane, we could say that the airplane pushes upward on the pilot, who in turn must push down on the airplane with the same force since he is stationary in his seat (Fig. 5–22).

The total downward force is therefore

$$F = mg + \frac{mv^2}{r} = (80\,\text{kg})(9.8\,\text{m/s}^2) + \frac{(80\,\text{kg})(200\,\text{m/s})^2}{1000\,\text{m}} = 3984\,\text{N}$$

The pilot presses down on his seat with a force of 3984 N. His apparent weight is therefore 3984 N, more than five times his actual weight. Because there has been no compensating increase in his muscular strength, the pilot may be unable to move his arms and legs in order to control the airplane. A further complication is the tendency of the pilot's blood to leave his head because of inertia, leaving him with impaired vision ("blacked out") and perhaps unconscious. Special pressure suits have been devised that prevent disturbances in blood supply and in the positions of internal organs during severe accelerations in flight.

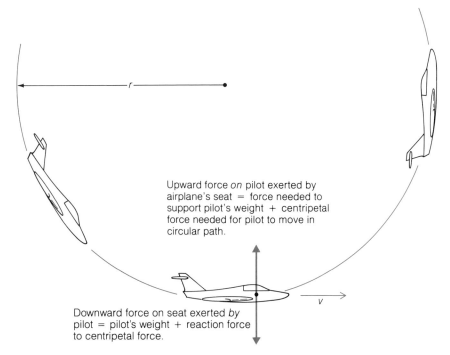

FIG. 5–22 Forces exerted by and on the pilot of an airplane making a vertical circle at constant speed.

Upward force *on* pilot exerted by airplane's seat = force needed to support pilot's weight + centripetal force needed for pilot to move in circular path.

Downward force on seat exerted *by* pilot = pilot's weight + reaction force to centripetal force.

Because the forces exerted by and on a pilot or astronaut vary with his mass, it is often convenient to speak instead of his acceleration, by custom in units of g. The above pilot's acceleration is

Accelerations are often measured in units of g

$$a = \frac{F}{m} = \frac{3984\,\text{N}}{80\,\text{kg}} = 49.8\,\text{m/s}^2 = 5.1\,g \qquad \blacksquare$$

5–9 KEPLER'S LAWS AND GRAVITATION

From astronomical observations which he and others (notably Tycho Brahe) had made over a long period of time, Johannes Kepler (1571–1630) discovered three laws that the planets obey as they move around the sun:

1. Each planet has an elliptical orbit with the sun at one focus.
2. Each planet moves so that a radius vector from the sun to it sweeps out equal areas in equal times.
3. The ratio between the square of a planet's period of revolution and the cube of its average distance from the sun has the same value for all the planets.

These laws were used by Newton to arrive at his law of universal gravitation. Let us see how the formula $F_{\text{grav}} = Gm_A m_B$ follows from Kepler's third law.

For simplicity we will assume that the planets move in circular orbits around a stationary sun. A planet of mass m_A, orbital radius r, and speed v must be acted upon by the centripetal force

$$F_c = \frac{m_A v^2}{r}$$

If the period of the orbit is T, then, as we know, $v = 2\pi r/T$ and

$$F_c = \frac{4\pi^2 m_A r}{T^2}$$

According to Kepler's third law,

Kepler's third law $$\frac{T^2}{r^3} = K$$

where K has the same value for all the planets. Hence $T^2 = Kr^3$ and

$$F_c = \frac{4\pi^2 m_A}{Kr^2}$$

According to Newton's hypothesis, this centripetal force is provided by the gravitational force exerted by the sun, from which we conclude that F_{grav} is directly proportional to the mass of a planet and inversely proportional to the square of its distance from the sun.

To complete the analysis, we refer to Newton's third law of motion, which requires that the force a planet exerts on the sun be equal in magnitude to the force the sun exerts on the same planet. If the above formula is correct, then we should be able to apply it either way for a given planet and get the same value of F_{grav}. Since r is the same in both cases, F_{grav} must be proportional to *both* the planet's mass m_A and the sun's mass m_B. Because the sun's mass is constant, we can express the quantity $4\pi^2/K$ as Gm_B, so that

$$F_{\text{grav}} = G\frac{m_A m_B}{r^2}$$

where G is a universal constant. Extending this formula to *any* two bodies in the universe gives Newton's law of gravitation.

IMPORTANT TERMS

An object traveling in a circle at constant speed is said to be undergoing **uniform circular motion.**

The velocity of an object in uniform circular motion continually changes in direction although its magnitude remains constant. The acceleration that causes the object's velocity to change is called **centripetal acceleration,** and it points toward the center of the object's circular path.

The inward force that provides an object in uniform circular motion with its centripetal acceleration is called **centripetal force.**

Newton's **law of universal gravitation** states that every object in the universe attracts every other object with a force directly proportional to both their masses and inversely proportional to the square of the distance separating them.

IMPORTANT FORMULAS

Period of orbit: $T = \dfrac{2\pi r}{v}$

Centripetal acceleration: $a_c = \dfrac{v^2}{r}$

Centripetal force: $F_c = \dfrac{mv^2}{r}$

Law of gravitation: $\quad F_{\text{grav}} = G\dfrac{m_A m_B}{r^2}$

Satellite orbit: $\quad v = \sqrt{rg} = \sqrt{\dfrac{r^2_{\text{earth}} g_0}{r}}$

MULTIPLE CHOICE

1. In order to cause a moving object to pursue a circular path, it is necessary to apply
  (a) inertial force. (b) gravitational force.
  (c) frictional force. (d) centripetal force.

2. An object traveling in a circle at constant speed
  (a) has a constant velocity.
  (b) is not accelerated.
  (c) has an inward radial acceleration.
  (d) has an outward radial acceleration.

3. The acceleration of an object undergoing uniform circular motion is constant in
  (a) magnitude only.
  (b) direction only.
  (c) both magnitude and direction.
  (d) neither magnitude nor direction.

4. The centripetal force on a car rounding a curve on a level road is provided by
  (a) gravity.
  (b) friction between its tires and the road.
  (c) the torque applied to its steering wheel.
  (d) its brakes.

5. The centripetal force needed to keep the earth in orbit is provided by
  (a) inertia.
  (b) its rotation on its axis.
  (c) the gravitational pull of the sun.
  (d) the gravitational pull of the moon.

6. The radius of the path of an object in uniform circular motion is doubled. The centripetal force needed if its speed remains the same is
  (a) half as great as before.
  (b) the same as before.
  (c) twice as great as before.
  (d) four times as great as before.

7. The gravitational acceleration of an object
  (a) has the same value everywhere in space.
  (b) has the same value everywhere on the earth's surface.
  (c) varies somewhat over the earth's surface.
  (d) is greater on the moon because of its smaller diameter.

8. A hole is drilled to the center of the earth and a stone is dropped into it. When the stone is at the earth's center, compared with the values at the earth's surface
  (a) its mass and weight are both unchanged.
  (b) its mass and weight are both zero.
  (c) its mass is unchanged and its weight is zero.
  (d) its mass is zero and its weight is unchanged.

9. The moon's mass is 1.2 percent of the earth's mass. Relative to the gravitational force the earth exerts on the moon, the gravitational force the moon exerts on the earth
  (a) is smaller.
  (b) is the same.
  (c) is greater.
  (d) depends on the phase of the moon.

10. The shape of the earth is closest to that of
  (a) a perfect sphere. (b) an egg.
  (c) a football. (d) a grapefruit.

11. The reasons why the earth's core is believed to be molten iron do *not* include iron's
  (a) ability to conduct electric current.
  (b) ability to be permanently magnetized.
  (c) high density.
  (d) relative abundance in the universe.

12. A $\frac{1}{2}$-kg ball moves in a circle 0.4 m in radius at a speed of 4 m/s. Its centripetal acceleration is
  (a) 10 m/s^2. (b) 20 m/s^2.
  (c) 40 m/s^2. (d) 80 m/s^2.

13. The centripetal force on the ball of question 12 is
  (a) 10 N. (b) 20 N.
  (c) 40 N. (d) 80 N.

14. A toy cart at the end of a string 0.7 m long moves in a circle on a table. The cart has a mass of 2 kg and the string has a breaking strength of 40 N. The maximum speed of the cart is approximately
  (a) 1.9 m/s. (b) 3.7 m/s.
  (c) 11.7 m/s. (d) 16.7 m/s.

15. On a rainy day the coefficient of friction between a car's tires and a certain level road surface is reduced to half its usual value. The maximum safe velocity for rounding the curve is
  (a) unchanged.
  (b) reduced to 25% of its usual value.
  (c) reduced to 50% of its usual value.
  (d) reduced to 71% of its usual value.

16. A car is traveling at 50 km/h on a road such that the coefficient of friction between its tires and the road is 0.5. The minimum turning radius of the car is
  (a) 9.9 m. (b) 39.4 m.
  (c) 947 m. (d) 5100 m.

17. A 2-kg stone at the end of a string 1 m long is whirled in a vertical circle. The tension in the string is 52 N when the stone is at the bottom of the circle. The stone's speed then is

(a) 4 m/s.

(b) 5 m/s.

(c) 6 m/s.

(d) 7 m/s.

18. A woman has a mass of 60 kg at the earth's surface. At a height of one earth's radius above the surface her mass is

(a) 15 kg.

(b) 30 kg.

(c) 60 kg.

(d) 120 kg.

19. Earth satellite A has an orbit four times greater in radius than satellite B. The orbital speed of A is

(a) $v_B/4$.

(b) $v_B/2$.

(c) $2v_B$.

(d) $4v_B$.

EXERCISES

1. Under what circumstances, if any, can an object move in a circular path without being accelerated?

2. Where should you stand on the earth's surface to experience the most centripetal acceleration? The least?

3. A person swings an iron ball in a vertical circle at the end of a string. At what point in the circle is the string most likely to break? Why?

4. A car makes a clockwise turn on a level road at too high a speed and overturns. Do its left or its right wheels leave the ground first?

5. An airplane makes a vertical circle in which it is upside down at the top of the loop. Will the pilot fall out of his seat if he has no belt to hold him in place?

6. Two satellites are launched from a certain station with the same initial speeds relative to the earth's surface. One is launched toward the west, the other toward the east. Will there be any difference in their orbits? If so, what will the difference be and why?

7. An earth satellite is placed in an orbit whose radius is half that of the moon's orbit. Is its time of revolution longer or shorter than that of the moon?

8. For the moon to have the same orbit it has now, what would its speed have to be if (a) the moon's mass were double its present mass, and (b) the earth's mass were double its present mass?

9. A phonograph record 30 cm in diameter rotates $33\frac{1}{3}$ times per minute. (a) What is the linear speed of a point on its rim? (b) What is the centripetal acceleration of a point on its rim?

10. The minute hand of a large clock is 0.5 m long. (a) What is the linear speed of its tip in meters per second? (b) What is the centripetal acceleration of the tip of the hand?

11. What is the minimum radius at which an airplane flying at 300 m/s can make a U-turn if its centripetal acceleration is not to exceed $4g$?

12. A string 1 m long breaks when its tension is 100 N. What is the greatest speed at which it can be used to whirl a 1-kg stone? (Neglect the gravitational pull of the earth on the stone.)

13. What is the centripetal force needed to keep a 3-kg mass moving in a circle of radius 0.5 m at a speed of 8 m/s?

14. A 2000-kg car is rounding a curve of radius 200 m on a level road. The maximum frictional force the road can exert on the tires of the car is 4000 N. What is the highest speed at which the car can round the curve?

15. A string 0.8 m long is used to whirl a 2-kg stone in a vertical circle. What must be the speed of the stone at the top of the circle if the string is to be just taut? How does this speed compare with that required for a 1-kg stone in the same situation?

16. A string 1 m long is used to whirl a $\frac{1}{2}$-kg stone in a vertical circle. What is the tension in the string when the stone is at the top of the circle moving at 5 m/s?

17. A road has a hump 12 m in radius. What is the minimum speed at which a car will leave the road at the top of the hump?

18. The 200-g head of a golf club moves at 45 m/s in a circular arc of 1 m radius. How much force must the player exert on the handle of the club to prevent it from flying out of his hands at the bottom of the swing? Assume that the shaft of the club has negligible mass.

19. Find the gravitational force between two 1000-kg lead spheres whose centers are 3 m apart.

20. A 2-kg mass is 1 m away from a 5-kg mass. What is the gravitational force (a) that the 5-kg mass exerts upon the 2-kg mass; (b) that the 2-kg mass exerts upon the 5-kg mass? (c) If both masses are free to move, what are their respective accelerations in the absence of other forces?

21. An object dropped near the earth's surface falls 4.9 m in the first second. How far does the moon fall toward the earth in each second? Why doesn't the moon ever reach the earth?

22. What is the acceleration of a meteor when it is one earth's radius above the surface of the earth? Two earth's radii?

23. The radius of the earth is 6.4×10^6 m. What is the acceleration of a meteor when it is 8×10^6 m from the center of the earth?

PROBLEMS

1. A dime is placed 10 cm from the center of a record. The coefficient of friction between coin and record is 0.3. Will the coin remain where it is or will it fly off when the record turns at $33\frac{1}{3}$ rev/min? At 78 rev/min?

2. A box is resting on the flat floor in the rear of a station wagon moving at 15 m/s. What is the minimum radius of a turn the station wagon can make if the box is not to slip? Assume that $\mu_s = 0.4$.

3. An airplane traveling at 500 km/h banks at an angle of 45° as it makes a turn. What is the radius of the turn in kilometers? Assume that the rudder is not used in making the turn.

4. A highway curve has a radius of 300 m. (a) At what angle should it be banked for a traffic speed of 100 km/h? (b) If the curve is not banked, what is the minimum coefficient of friction required between tires and road?

5. A curve in a road 8 m wide has a radius of 60 m. How much higher than its inner edge should the outer edge of the road be if it is to be banked properly for cars traveling at 30 km/h?

6. A car whose speed is 90 km/h rounds a curve 180 m in radius which is properly banked for a speed of 45 km/h. Find the minimum coefficient of friction between tires and road that will permit the car to make the turn.

7. Show that $\cos \theta = g/4\pi^2 f^2 L$ for the conical pendulum of Fig. 5–12, where f is the number of revolutions per second the particle makes.

8. (a) Find the tension T in the wire and the force F the horizontal beam exerts when the structure of Fig. 5–23 is at

rest. What is the direction of **F**? (b) Find T and F when the structure is rotating twice per second. What is the direction of **F** now? Neglect the masses of the beam and of the wire.

9. The moon's mass is 7.3×10^{22} kg and the average radius of its orbit is 3.8×10^8 m. At what point could an object be placed between the earth and the moon where it would experience no resultant force? (Neglect the gravitational attractions of the sun and the other planets.)

10. The mass of the planet Jupiter is 1.9×10^{27} kg and that of the sun is 2.0×10^{30} kg. The average distance between them is 7.8×10^{11} m. (a) What is the gravitational force the sun exerts on Jupiter? (b) Assuming that Jupiter has a circular orbit, what must its speed be for the orbit to be a stable one?

11. The moon's radius is 27% of the earth's radius and its mass is 1.2% of the earth's mass. (a) What is the acceleration of gravity on the surface of the moon? (b) How much would a 60-kg person weigh there?

12. The mass of the planet Jupiter is 1.9×10^{27} kg and its radius is 7.0×10^7 m. What is the acceleration of gravity on the surface of Jupiter?

13. If a planet existed whose mass and radius were both twice those of the earth, what would the acceleration of gravity at its surface be in terms of g?

14. If a planet existed whose mass and radius were both half those of the earth, what would the acceleration of gravity at its surface be in terms of g?

15. Find the speed of an earth satellite whose orbit is 400 km above the earth's surface. What is the period of the orbit?

16. A satellite is to be put into orbit around the moon just above its surface. What should its speed be? Assume that the moon's radius is half that of the earth and that the acceleration of gravity at its surface is $g/6$.

17. The earth's average orbital radius is 1.5×10^{11} m and its average orbital speed is 3.0×10^4 m/s. From these figures, together with the value of G, find the mass of the sun.

18. Most of the stars in the galaxy of which the sun is a member (the "Milky Way") are concentrated in an assembly about 100,000 light-years across whose shape is roughly that of a fried egg. The sun is about 30,000 light-years from the center of the galaxy, and revolves around it with a period of about 2×10^8 years. A reasonable estimate for the mass of the galaxy may be obtained by considering this mass to be concentrated at the galactic center with the sun revolving around it like a planet around the sun. On this basis, calculate the mass of the galaxy. How many stars having the

FIG. 5–23

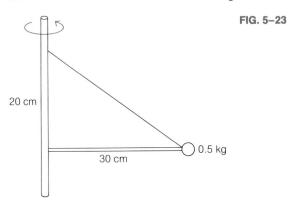

20 cm

30 cm

0.5 kg

mass of the sun is this equivalent to? (1 year $= 3.16 \times 10^7$ s, 1 light year $= 9.46 \times 10^{15}$ m. and $m_{sun} = 1.99 \times 10^{30}$ kg.)

19. The gravitational force the sun exerts on the earth is nearly 180 times greater than the force the moon exerts on the earth, yet the moon is more effective in producing the tides than the sun. To see why, perform the following calculations. First find the difference between the force the sun exerts on 1 kg of water at a point on the equator nearest the sun and the force the sun would exert on 1 kg of water at the earth's center. Then make the same calculation for the forces the moon exerts on 1 kg of water at these locations, and compare the results. The moon's mass is 7.3×10^{22} kg, the sun's mass is 2.0×10^{30} kg, the earth's radius is 6.4×10^6 m, the earth's orbital radius is 1.5×10^{11} m, and the moon's orbital radius is 3.8×10^8 m.

Note: There is an easy way to make these calculations. When $x \ll 1$, $1/(1 - x)^2 \approx 1 + 2x$. Hence if R is the distance from the sun (or moon) to the earth's center and r is the earth's radius, then

$$\frac{1}{(R - r)^2} = \frac{1}{R^2\left(1 - \dfrac{r}{R}\right)^2} \approx \frac{1}{R^2}\left(1 + 2\frac{r}{R}\right)$$

ANSWERS TO MULTIPLE CHOICE

1. d	**6.** a	**11.** b	**16.** b	**18.** c
2. c	**7.** c	**12.** c	**17.** a	**19.** b
3. a	**8.** c	**13.** b		
4. b	**9.** b	**14.** b		
5. c	**10.** d	**15.** d		

ENERGY

6

We all use the word "energy," but how many of us know exactly what it means? We speak of the energy of a lightning bolt or of an ocean wave; we say that an active person is energetic; we hear a candy bar described as being full of energy; we read that most of the world's electricity will come from nuclear energy in the years to come. What do a lightning bolt, an ocean wave, an active person, a candy bar, and an atomic nucleus have in common?

In general terms, energy refers to an ability to accomplish change. All changes in the physical world involve energy, usually with energy being transformed from one sort into another. But "change" is not a very precise concept, and we must clarify our ideas before going further. What we shall do is first define a quantity called work, and then see how it permits us to discuss energy and its relation to change in the orderly manner of science.

6–1   WORK

Force and change

All changes in the physical universe are the result of forces. Forces set things in motion, change their paths, and bring them to a stop; forces pull things together and push them apart. The physical quantity called *work* is a measure of the amount of change (in a general sense) a force gives rise to when it acts upon something.

When we push against a brick wall, nothing happens. We have applied a force, but the wall has not yielded and shows no effects. On the other hand, when we apply exactly the same force to a ball, the ball flies through the air for some distance. In the latter case something has been accomplished because of our push, whereas in the former there has been no result (Fig. 6–1).

What is the essential difference between the two situations? In the first case, where we pushed against a wall, the wall did not move. But in the second case, where we threw the ball, the ball *did* move while the force was being applied and before it left our hand. The displacement of the body while the force acted on it was what made the difference.

If we think carefully along these lines, we will see that whenever a force acts so as to produce motion in an object, the force acts during a displacement of the object. In order to make this notion definite, we may make a preliminary definition of work as follows:

Preliminary definition of work

The work done by a force acting on an object that moves in the same direction as the force is equal to the magnitude of the force multiplied by the distance through which the force acts.

In equation form,

$$W = Fs \qquad (\textbf{F parallel to s}) \tag{6–1}$$
Work = force × distance

No work is done without both a force and a displacement

This definition is a great help in clarifying the effects of forces. Unless a force acts through a distance, no work is done no matter how great the force. And even if an object moves through a distance, no work is done unless a force is acting upon it or it exerts a force on something else (Fig. 6–2).

FIG. 6–1 Work is done by a force when the object it acts upon is displaced while the force is applied.

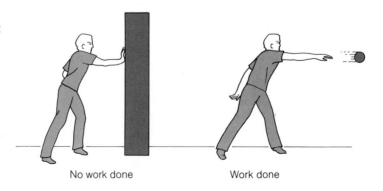

No work done   Work done

Work = force × distance through which force acts (parallel case).

$W = Fs$

FIG. 6–2

When there is no net applied force, no work is done even though the object may move.

$F = 0 \qquad W = 0 \qquad F = 0$

When the object acted upon by a force remains at rest, no work is done

$W = 0$

$s = 0$

Our intuitive concept of work is in accord with the precise definition above: when something happens because a person applies a force of some kind, we say that he has done work. Here we have simply broadened the concept to include inanimate forces. (We still must be careful, though; while we may become tired after pushing against a brick wall for a long time, we still have done no work on the wall if it remains in place.)

The above definition of work has an important qualification: the force **F** must be in the same direction as the vector displacement **s.** If **F** and **s** are not parallel, we must replace F in the formula $W = Fs$ by the magnitude of its component $\mathbf{F}_s$ in the direction of the displacement s (Fig. 6–3). The magnitude of this component is

$$F_S = F \cos \theta$$

where θ is the angle between **F** and **s.** Hence the most general definition of work is

$$W = Fs \cos \theta \qquad\qquad Work \quad (6\text{–}2)$$

Work is a scalar quantity; there is no direction associated with work even though it depends upon two vector quantities, force and displacement.

Equation (6–2) is always correct, since when the force and the displacement are parallel, $\theta = 0$ and $\cos 0 = 1$. When the force and the displacement are perpendicular,

General definition of work

FIG. 6–3 The work done by a force depends upon the angle θ between the force and its displacement. (a) In general, $W = Fs \cos \theta$. (b) When **F** is parallel to **s,** $W = Fs$. (c) When **F** is perpendicular to **s,** $W = 0$ since **F** then has no component in the direction of **s** and therefore does not affect its motion.

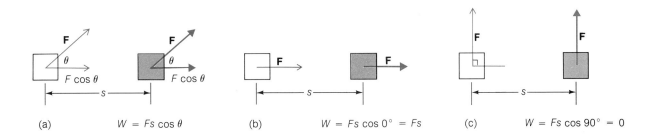

(a) $\qquad W = Fs \cos \theta$ (b) $\qquad W = Fs \cos 0° = Fs$ (c) $\qquad W = Fs \cos 90° = 0$

FIG. 6–4

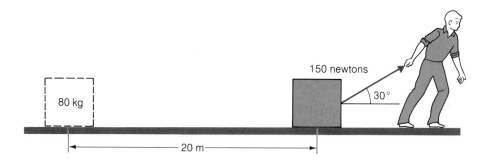

No work is done when F is perpendicular to s

$\theta = 90°$ and $\cos 90° = 0$, so $W = 0$. In order for work to be done, the applied force must have a component in the direction of the displacement.

In the SI system, work is given a special unit, the *joule*, abbreviated J. One joule is equal to the work done by a force of 1 N acting through a distance of 1 m. That is,

Units of work

$$1 \text{ joule} = 1 \text{ J} = 1 \text{ N·m} \qquad\qquad\qquad \textit{The joule}$$

In the British system, the unit of work is the *foot-pound,* abbreviated ft·lb. One ft·lb is equal to the work done by a force of 1 lb acting through a distance of 1 ft.

Example   A person pulls an 80-kg crate for 20 m across a level floor using a rope that is 30° above the horizontal. The person exerts a force of 150 N on the rope (Fig. 6–4). How much work is performed?

Solution   The work done is

$$w = Fs \cos \theta = (150 \text{ N})(20 \text{ m})(\cos 30°)$$
$$= (150)(20)(0.866) \text{ N} = 2.6 \times 10^3 \text{ J}$$

Evidently the 80-kg mass of the crate has no significance here—it is the force exerted on the crate that determines how much work is done.   ∎

6–2   WORK DONE AGAINST GRAVITY

It is easy to compute the work done in lifting an object against gravity. The force of gravity on an object of mass m is the same as its weight $w = mg$. Hence in order to raise the object to a height h above its original position, a force of mg must be exerted on it. Since $F = mg$ and $s = h$ here, the work done is

$$W = Fs = mgh \qquad\qquad\qquad\qquad\qquad (6\text{–}3)$$

Thus to lift an object of mass m to a height h requires the performance of the amount of work mgh (Fig. 6–5).

Work done against gravity is independent of path

It is important to note that only the height h is involved in work done against the force of gravity. The particular route taken by an object being raised is not significant (Fig. 6–6); excluding any frictional effects, exactly as much work must be expended to climb a flight of stairs as to go up in an elevator to the same floor (though not by the person involved!).

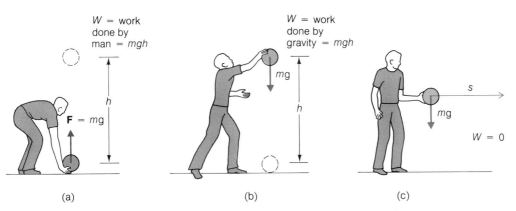

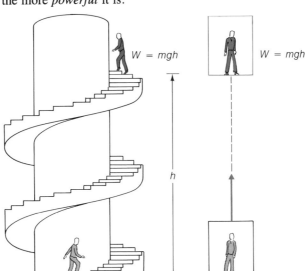

Example Eating a banana enables a person to perform about 4×10^4 J of work, figuring his or her efficiency at 10 percent. To what height does eating a banana enable a 60-kg woman to climb?

Solution From $W = mgh$ we have

$$h = \frac{W}{mg} = \frac{4 \times 10^4 \text{ J}}{(60 \text{ kg})(9.8 \text{ m/s}^2)} = 68 \text{ m}$$

FIG. 6–5 (a) The work mgh must be done to lift an object of mass m to a height h. (b) When an object of mass m falls from a height h, the force of gravity does the work mgh on it. (c) The force of gravity does no work on objects that move parallel to the earth's surface.

6–3 POWER

Often the time needed to perform a task is just as significant as the actual amount of work required. Given enough time, even the feeblest motor can raise the Sphinx. However, if we want to carry out a certain operation quickly, we try to obtain a motor whose output of work is rapid in terms of the total required. The rate at which work is done is therefore an important engineering quantity. This rate is called *power*: the faster some agency can do work, the more *powerful* it is.

Power is rate of doing work

FIG. 6–6 In the absence of friction, the work done in lifting a mass m to a height h is mgh regardless of the exact path taken.

If an amount of work W is performed in a time interval $t,$ the power involved is

$$P = \frac{W}{t}$$ *Power* (6–4)

$$\text{Power} = \frac{\text{work done}}{\text{time interval}}$$

SI units of work

In the SI system, where work is measured in joules and time in seconds, the unit of power is the *watt*:

1 watt $= 1\,\text{W} = 1\,\text{J/s}$ *The watt*

The watt is rather small for most industrial purposes; even an electric clock requires several watts of power. The larger *kilowatt* (kW) is accordingly in common use, where

1 kW $= 1000$ watts

The *kilowatt-hour* (kWh) is often used as a unit of work. Since 1 kW $= 1000$ W $= 1000$ J/s and 1 h $= 3600$ s,

1 kilowatt-hour $= 1\,\text{kWh} = 3.60 \times 10^6\,\text{J}$

British units of work

In the British system, where energy is measured in foot-pounds and time in seconds, the unit of power is the *ft·lb/s*. The ft·lb/s is also inconveniently small and has been replaced by the larger *horsepower* (hp) in engineering practice. The horsepower was introduced two centuries ago by James Watt to compare the output of the steam engine he had perfected with a more familiar source of power. Today the horsepower is defined as

1 hp $= 550$ ft·lb/s

In the United States the watt and kilowatt are employed in connection with electric power, while mechanical power is customarily specified in horsepower. To convert power figures from one system to the other, we note that

1 hp $= 746\,\text{W} = 0.746\,\text{kW}$
1 kW $= 1.34\,\text{hp}$

Example An electric motor with an output of 15 kW provides power for the elevator of a 6-story building. If the total mass of the loaded elevator is 1000 kg, what is the minimum time needed for it to rise the 30 m from the ground floor to the top floor (Fig. 6–7)?

Solution The work done in raising the elevator through a height h is

$W = mgh$

Since $P = W/t,$ the time needed for the motor to raise the elevator by 30 m is

$$t = \frac{W}{P} = \frac{mgh}{P} = \frac{(1000\,\text{kg})(9.8\,\text{m/s}^2)(30\,\text{m})}{15 \times 10^3\,\text{W}} = 20\,\text{s}$$ ∎

Force, speed, and power

When a constant force performs work on an object moving in the same direction as that of the force, the power delivered is equal to the product of the force F and the

FIG. 6–7

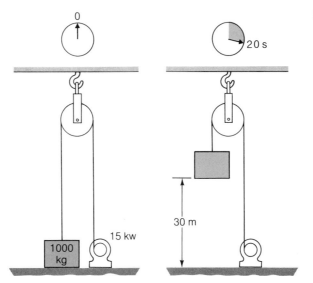

speed v of the object. We can see why when we express the power in terms of the work done by the force in the time t:

$$P = \frac{W}{t} = \frac{Fs}{t}$$

But the speed of an object that travels the distance s in the time t is

$$v = \frac{s}{t}$$

Hence

$$P = Fv$$
Power = force × speed

In the general case where the angle between **F** and **v** is θ, from Eq. (6–2) we have

$$P = Fv \cos \theta \qquad\qquad\qquad Power \quad (6\text{–}5)$$

Example A swimmer develops an average power of 200 W as she covers 100 m in 80 s. What is the resistive force exerted by the water on her?

Solution The swimmer's speed is $v = 100\text{ m}/80\text{ s} = 1.25$ m/s. Since $P = Fv$ here,

$$F = \frac{P}{v} = \frac{200\,\text{W}}{1.25\,\text{m/s}} = 160\,\text{N}$$

which is about 36 lb.

A person in good physical condition is usually capable of a continuous power output of about 75 W, which is 0.1 hp. An athlete such as a runner or swimmer may have a power output several times greater during a distance event. What limits the power

Human power output

output of a trained athlete is not muscular development but the supply of oxygen via the bloodstream from the lungs to the muscles, where it is needed for the metabolic processes that extract work from nutrients. However, for a momentary effort such as that of a weight lifter or a jumper, an athlete's power output may exceed 5 kW.

6–4   ENERGY

From the straightforward notion of work we proceed to the complicated and many-sided concept of *energy:*

Energy and work

Energy is that property whose possession enables something to perform work.

When we say that something has energy, we mean it is capable (directly or indirectly) of exerting a force on something else and doing work on it. On the other hand, when we do work on something, we have added to it an amount of energy equal to the work done. The units of energy are the same as those of work, the joule and the foot-pound.

There are three broad categories of energy:

Categories of energy

1. *Kinetic energy,* which is the energy something possesses by virtue of its motion.
2. *Potential energy,* which is the energy something possesses by virtue of its position.
3. *Rest energy,* which is the energy something possesses by virtue of its mass.

In the above descriptions the word "something" was used instead of "object" because, as we shall see later, such nonmaterial entities as force fields and massless particles may also possess energy.

Heat is a form of energy

All modes of energy possession fit into one or another of these three categories. For instance, it is convenient for many purposes to think of heat as a separate form of energy, but what this term actually refers to is the sum of the kinetic energies of the randomly moving atoms and molecules in a body of matter.

6–5   KINETIC ENERGY

When we perform work on a ball by throwing it, what becomes of this work?

Let us suppose we apply the uniform force **F** to the ball for a distance s before it leaves our hand, as in Fig. 6–8(a). The work done on the ball is therefore, since cos $\theta = 1$,

$$W = Fs \tag{6–6}$$

The mass of the ball is m. As we throw it, its acceleration has the magnitude

$$a = \frac{F}{m} \tag{6–7}$$

according to the second law of motion, $\mathbf{F} = m\mathbf{a}$ (Fig. 6–8(b)).

We know from the formula

$$v_f^2 = v_0^2 + 2as \tag{2–11}$$

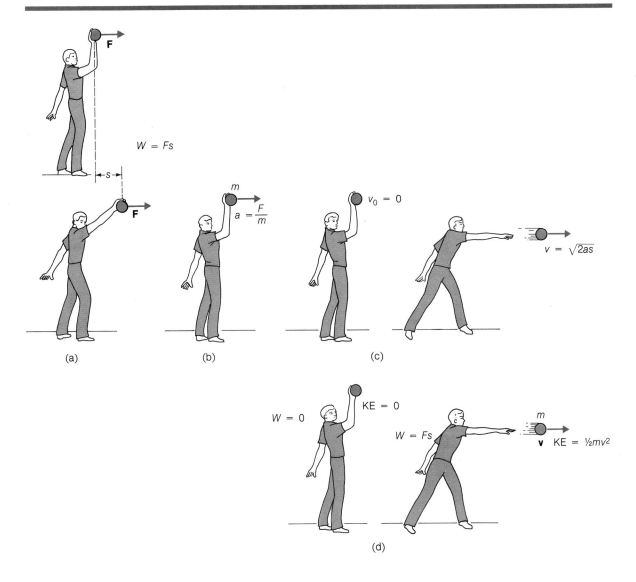

$W = Fs$

(a)   (b)   (c)

(d)

FIG. 6–8 Successive steps in the derivation of the formula KE = ½mv^2 for the kinetic energy of a moving body.

that when an object starting from rest ($v_0 = 0$) undergoes an acceleration of magnitude a through a distance s, its final speed v is related to a and s by

$$v_f^2 = 2as \qquad (6\text{--}8)$$

This relationship is pictured in Fig. 6–8(c).

If we now substitute F/m for a in Eq. (6–8), we find that

$$v_f^2 = 2as = 2\left(\frac{F}{m}\right)s$$

which we can rewrite as

$$Fs = \tfrac{1}{2}mv_f^2 \qquad (6\text{--}9)$$

FIG. 6–9

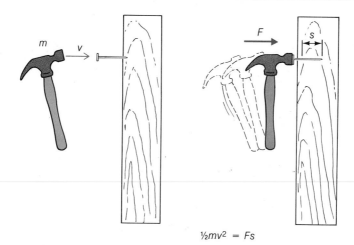

$$\tfrac{1}{2}mv^2 = Fs$$

All moving objects possess kinetic energy

The quantity on the left-hand side, Fs, is the work our hand has done in throwing the ball, as in Eq. (6–6). The quantity on the right-hand side, $\frac{1}{2}mv_f^2$, must therefore be the energy acquired by the ball as a result of the work we did on it. This energy is *kinetic energy*, energy of motion. That is, we interpret the preceding equation as follows (Fig. 6–8(d)):

$$Fs = \tfrac{1}{2}mv_f^2$$

Work done on ball = kinetic energy of ball

The symbol for kinetic energy is KE. The kinetic energy of an object of mass m and velocity v is therefore

$$\mathrm{KE} = \tfrac{1}{2}mv^2 \qquad\qquad\qquad \textit{Kinetic energy} \quad (6\text{–}10)$$

A moving object is able to perform an amount of work equal to $\frac{1}{2}mv^2$ in the course of being stopped.

Example   A 600-g hammer head strikes a nail at a speed of 4 m/s and drives it 5 mm into a wooden board (Fig. 6–9). What is the average force on the nail?

Solution   The initial kinetic energy of the hammer head is $\frac{1}{2}mv^2$ and the work done on the nail is Fs. Hence $Fs = \frac{1}{2}mv^2$ and

$$F = \frac{mv^2}{2s} = \frac{(0.6\,\mathrm{kg})(4\,\mathrm{m/s})^2}{2(0.005\,\mathrm{m})} = 960\,\mathrm{N}$$

which is 216 lb.   ■

Example   Find the kinetic energy of a 1200-kg car when it is moving at 25 km/h (16 mi/h) and when it is moving at 100 km/h (62 mi/h).

Solution   Since

$$v_1 = (25\,\mathrm{km/h})\left(0.278\,\frac{\mathrm{m/s}}{\mathrm{km/h}}\right) = 7.0\,\mathrm{m/s}$$

FIG. 6–10

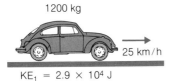

1200 kg

25 km/h

$KE_1 = 2.9 \times 10^4$ J

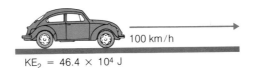

100 km/h

$KE_2 = 46.4 \times 10^4$ J

$$v_2 = (100\,\text{km/h})\left(0.278\,\frac{\text{m/s}}{\text{km/h}}\right) = 27.8\,\text{m/s}$$

we have (Fig. 6–10)

$$KE_1 = \tfrac{1}{2}mv_1^2 = \tfrac{1}{2}(1200\,\text{kg})(7.0\,\text{m/s})^2 = 2.9 \times 10^4\,\text{J}$$

$$KE_2 = \tfrac{1}{2}mv_2^2 = \tfrac{1}{2}(1200\,\text{kg})(27.8\,\text{m/s})^2 = 46.4 \times 10^4\,\text{J}$$

At 100 km/h the car has 16 times as much kinetic energy as it does at 25 km/h. The fact that kinetic energy, and hence ability to do work (that is, damage), is proportional to the *square* of the speed is responsible for the severity of automobile accidents at high speeds. ■

Example   What is the power output of the engine of a 1200-kg car if the car can go from 25 km/h to 100 km/h in 12 s?

Solution   The work needed to accelerate the car is, from the preceding problem,

$$W = KE_2 - KE_1 = 46.4 \times 10^4\,\text{J} - 2.9 \times 10^4\,\text{J} = 43.5 \times 10^4\,\text{J}$$

The power needed to provide this amount of work in 12 s is

$$P = \frac{W}{T} = \frac{43.5 \times 10^4\,\text{J}}{12\,\text{s}} = 3.63 \times 10^4\,\text{W} = 36.3\,\text{kW}$$

which is equivalent to

$$P = 36.3\,\text{kW} \times 1.34\,\frac{\text{hp}}{\text{kW}} = 48.6\,\text{hp}$$

■

We can use the relationship $Fs = \tfrac{1}{2}mv^2$ between work done and the resulting kinetic energy to arrive at an interesting conclusion about animal running speeds. According to this formula,

The running speed of an animal is independent of its size

$$v = \sqrt{\frac{2Fs}{m}}$$

Let us interpret v as an animal's speed, F as the force its leg muscles exert over the distance s, and m as its mass. As we saw in Sec. 3–3, the mass of an animal is approximately proportional to L^3, where L is a representative linear dimension such as its length, and the forces its muscles can exert are approximately proportional to L^2. The distance through which a muscle acts is proportional to L. Hence the quantity Fs/m depends on L as $L^2 \times L/L^3 = 1$, which means that Fs/m, and hence v, should not vary with L at all! In fact, although different animals have different running abilities, there is indeed little correlation with size over a wide span. A hare can run as fast as a horse.

6–6   POTENTIAL ENERGY

A raised object has the potential of doing work

When we drop a stone from a height h, it falls faster and faster and finally strikes the ground. In striking the ground the stone does work; if it is sufficiently heavy and has fallen from a great enough height, the work done by the stone is manifest as a hole (Fig. 6–11). Evidently the stone at its original location h above the ground had a capacity to do work, even though it was stationary at the time. The work the stone can perform in falling to the ground is called its *potential energy,* symbol PE.

We have already calculated that the work we must do to raise the stone to the height h is

$$W = mgh \tag{6–3}$$

where m is the stone's mass. This amount of work can also be done *by* the stone after dropping from the height h. Since the raised stone has the potentiality of doing the amount of work mgh, we define its *potential energy* as

$$PE = mgh \qquad \qquad \textit{Gravitational potential energy} \quad (6–11)$$

the product of its mass, the acceleration of gravity, and its height. In the British system of units, weights rather than masses are usually specified. Since $w = mg$, we may also write Eq. (6–11) as

$$PE = wh \tag{6–12}$$

which is more convenient in treating problems in this system of units.

Potential energy depends upon reference level

The gravitational potential energy of an object depends upon the reference level from which its height h is measured. For example, the potential energy of a 1.0-kg book held 10 cm above a desk is

$$PE = mgh = (1.0\,\text{kg})\left(9.8\,\frac{\text{m}}{\text{s}^2}\right)(0.10\,\text{m}) = 0.98\,\text{J}$$

with respect to the desk (Fig. 6–12). However, if the book is 1.0 m above the floor of the room, its potential energy is

FIG. 6–11

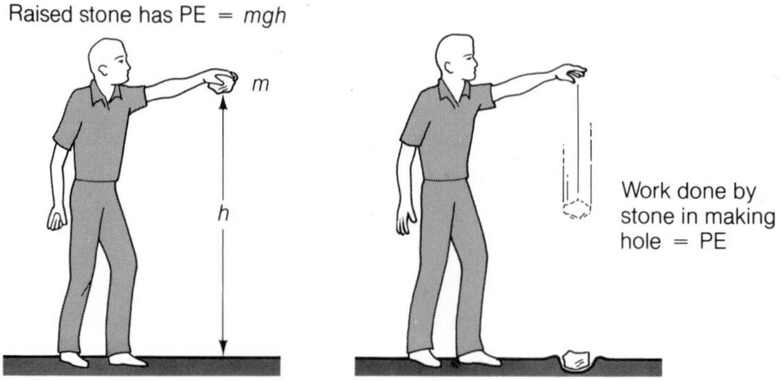

Raised stone has PE = mgh

m

h

Work done by stone in making hole = PE

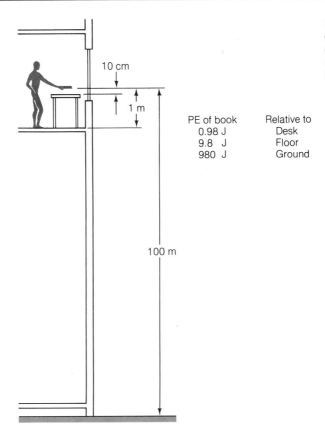

FIG. 6–12 The potential energy of an object depends upon the reference level from which its height h is measured.

10 cm

1 m

PE of book	Relative to
0.98 J	Desk
9.8 J	Floor
980 J	Ground

100 m

$$\text{PE} = mgh = (1.0\,\text{kg})\left(9.8\,\frac{\text{m}}{\text{s}^2}\right)(1.0\,\text{m}) = 9.8\,\text{J}$$

with respect to the floor. And the book may conceivably be 100 m above the ground, so its potential energy is

$$\text{PE} = mgh = (1.0\,\text{kg})\left(9.8\,\frac{\text{m}}{\text{s}^2}\right)(100\,\text{m}) = 980\,\text{J}$$

with respect to the ground. The height h in the formula $\text{PE} = mgh$ means nothing unless the base height $h = 0$ is specified.

PE is a property of system of interacting objects

In general, potential energy is a relative quantity. Just as the KE of a moving object depends upon the frame of reference in which its velocity is measured, so the PE of an object subject to a force depends upon the reference position chosen. Further, the potential energy is not a property of the object by itself but of the *system* of the object and the body that exerts the force on it. The PE of a stone held above the earth's surface is shared by both the stone and the earth. When the stone is released, both it and the earth move toward each other. However, the earth's motion is imperceptibly small because of its immense mass relative to that of the stone. It is therefore appropriate to attribute the entire PE of the system to the stone, but in the case of two objects more

FIG. 6–13

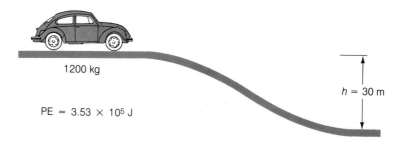

1200 kg

PE = 3.53 × 10⁵ J

$h = 30$ m

nearly comparable in mass, it must be kept in mind that the PE belongs to the system, not just to one of them.

Example   Compare the potential energy of a 1200-kg car at the top of a hill 30 m high (Fig. 6–13) with its kinetic energy when moving at 100 km/h.

Solution   The car's potential energy relative to the bottom of the hill is

$$PE = mgh = (1200\,\text{kg})(9.8\,\text{m/s}^2)(30\,\text{m}) = 3.53 \times 10^5\,\text{J}$$

In an earlier example we found that the kinetic energy of the car at 100 km/h is 4.64 × 10⁵ J, which is greater than its potential energy at the top of the hill. This means that a crash at 100 km/h (62 mi/h) into a stationary obstacle will yield more work— that is, do more damage—than dropping the car 30 m (98 ft).   ■

Potential energy is a general concept

We have spoken of one type of potential energy only—namely, that possessed by an object by virtue of being raised above some reference level in the earth's gravitational field. The concept of potential energy is a much more general one, however, for it refers to the energy something has as a consequence of its position regardless of the nature of the force acting on it. The earth itself, for instance, has potential energy with respect to the sun, since if its orbital motion were to cease it would fall toward the sun (Fig. 6–14). An iron nail has potential energy with respect to a nearby magnet, since it will fly to the magnet if released. An object at the end of a stretched spring has potential energy with respect to its position when the spring has its normal extension, since if let go the object will move as the spring contracts. In each of these cases the object in question has the potentiality of doing work in its original position.

6–7 REST ENERGY

All objects have energy by virtue of their mass alone

Every body of matter possesses a certain inherent amount of energy called *rest energy* even if it is not moving (so that KE = 0) and is not being acted upon by a force (so that PE = 0). A body whose mass when it is at rest is m_0 has a rest energy E_0 of

$$E_0 = m_0 c^2 \qquad\qquad \textit{Rest energy}\quad (6\text{–}13)$$

where c is the speed of light, 3×10^8 m/s. (As we shall learn in Chapter 26, the mass of a moving body increases with its speed, so that m is not always equal to m_0; the

High potential
energy

Low potential
energy

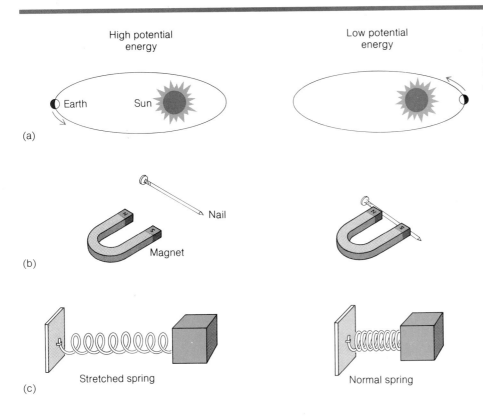

(a)

(b)

Nail

Magnet

(c)

Stretched spring

Normal spring

FIG. 6–14 Three
examples of potential
energy. In each case the
PE is a property of the
entire system.

difference is only significant at speeds near that of light, however.) Equation (6–13) was discovered by Albert Einstein in the early years of this century and has been verified by many experiments since then.

Why are we not aware of rest energy as we are aware of kinetic and potential energies? After all, a 1-kg object—such as this book—contains the rest energy

$$m_0 c^2 = (1 \text{ kg})(3 \times 10^8 \text{ m/s})^2 = 9 \times 10^{16} \text{ J}$$

which is enough energy to send a payload of perhaps a million tons to the moon. How can so much energy be bottled up without revealing itself in some manner?

In fact, all of us *are* familiar with processes in which rest energy is liberated, only we do not usually think of them in these terms. In every chemical reaction in which energy is given off, for instance a fire, a certain amount of matter is being converted into energy in the form of heat, which is molecular kinetic energy. But the amount of matter that vanishes in such reactions is so small that it escapes our notice. When 1 kg of dynamite explodes, 6×10^{-11} kg of matter is transformed into energy. The lost mass is so minute a fraction of the total mass involved as to be impossible to detect directly (hence the "law" of conservation of mass in chemistry), but it results in the evolution of

$$m_0 c^2 = (6 \times 10^{-11} \text{ kg})(3 \times 10^8 \text{ m/s})^2 = 5.4 \times 10^6 \text{ J}$$

of energy, which is very hard to avoid detecting.

Rest energies are very large

Rest energy is liberated in many familiar processes

Example   Solar energy reaches the earth at the rate of about 1.4 kW per m^2 of surface perpendicular to the direction of the sun. By how much does the mass of the sun decrease per second? The mean radius of the earth's orbit is 1.5×10^{11} m.

Solution   The surface area of a sphere of radius r is

$$A = 4\pi r^2$$

so the total power radiated by the sun, which is equal to the power received by a sphere whose radius is that of the earth's orbit, is given by

$$P = \frac{P}{A} \times A = \frac{P}{A} \times 4\pi r^2$$
$$= (1.4 \times 10^3 \text{ W/m}^2)(4\pi)(1.5 \times 10^{11} \text{ m})^2 = 3.96 \times 10^{26} \text{ W}$$

Thus the sun loses $E_0 = 3.96 \times 10^{26}$ J of rest energy per second, which means that its rest mass decreases by

$$m_0 = \frac{E_0}{c^2} = \frac{3.96 \times 10^{26} \text{ J}}{(3 \times 10^8 \text{ m/s})^2} = 4.4 \times 10^9 \text{ kg}$$

per second. Since the sun's mass is 2.0×10^{30} kg, it is in no immediate danger of running out of matter. The mechanism by which rest mass is converted into energy in the sun is discussed in Chapter 31. ∎

6–8   CONSERVATION OF ENERGY

Conservation principles

One of the chief distinctions between the physical sciences and nearly all other scientific disciplines is that in the former certain very general conservation principles have been found valid. A conservation principle states that no matter what changes a system of some kind that is isolated from the rest of the universe undergoes, a certain quantity keeps the same value it had originally. For example, the law of conservation of mass revolutionized chemistry by holding that the total mass of the products of a chemical reaction is the same as the total mass of the original substances. The increase in mass of a piece of iron when it rusts therefore indicates that the iron has combined with some other material, rather than having decomposed, as the early chemists believed. In fact, the gas oxygen was discovered in the course of seeking this other material.

Usefulness of conservation principles

Given one or more conservation principles that apply to a given system, we can immediately determine which classes of events can take place in the system and which cannot. Thus when iron rusts, the gain in mass means that it has combined chemically with something else. In physics it is often possible to draw some conclusions about the behavior of the particles that make up a system without a detailed investigation, basing our analysis simply upon the conservation of some particular quantities. The power of this method of approach is exemplified by the great success of physics in understanding natural phenomena, a success largely due to the variety of conservation principles that have been discovered.

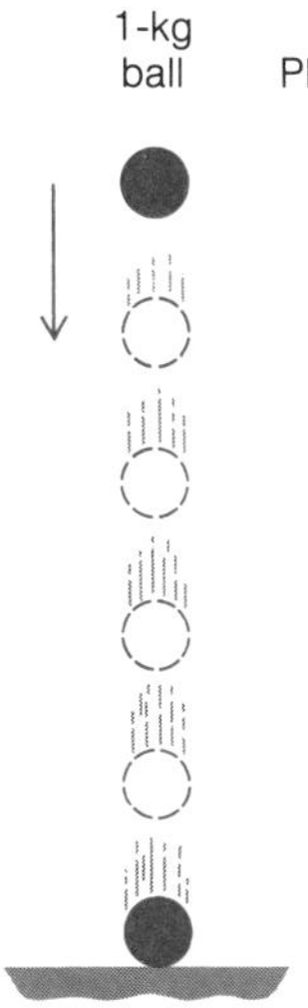

Height	1-kg ball	PE = mgh	KE = $\frac{1}{2}mv^2$	PE + KE
50 m		490 J	0 J	490 J
40		392	98	490
30		294	196	490
20		196	294	490
10		98	392	490
0		0	490	490

FIG. 6–15 The total energy of a falling ball remains constant as its potential energy is transformed into kinetic energy.

The first conservation principle we shall study is that of *conservation of energy:*

The total amount of energy in a system isolated from the rest of the universe always remains constant, although energy transformations from one form to another, including rest energy, may occur within the system.

Conservation of energy

This principle is perhaps the most fundamental generalization in all of science, and no violation of it has ever been found.

In a great many physical processes the rest masses, and hence the rest energies, of the participating objects do not change. In such processes mechanical energy is conserved: the sum of the kinetic and potential energies of the objects involved is constant. An increase in potential energy means a decrease in kinetic energy, and vice versa.

Conservation of mechanical energy

A falling ball provides a simple example of conservation of mechanical energy. As it falls, its initial potential energy is converted into kinetic energy, so that the total energy of the stone remains the same. The potential energy of a 1-kg ball 50 m above the ground is $mgh = 490$ J, and its total mechanical energy is 490 J until it interacts with the ground and transfers energy to it (Fig. 6–15).

Another example is the motion of a planet about the sun. Planetary orbits are elliptical, so that at different points in its orbit the planet is at different distances from the sun. When the planet is close to the sun, it has a low potential energy, just as a stone near the ground has a low potential energy; when the planet is far from the sun, it has a high potential energy (Fig. 6–14(a)). Since the sum of the planet's PE and KE must be constant, we conclude that the kinetic energy of the planet is at a maximum when it is nearest the sun and at a minimum when it is farthest from the sun.

Energy conservation in planetary motion

Newton's laws of motion enable us—in theory—to solve all mechanical problems, that is, problems that involve forces and moving objects. However, these laws are

FIG. 6–16 When a body of mass m falls from a height h, all its initial potential energy mgh has been converted to kinetic energy just as it strikes the ground. Its final velocity is therefore $\sqrt{2gh}$ (in the absence of friction) regardless of the precise path it takes.

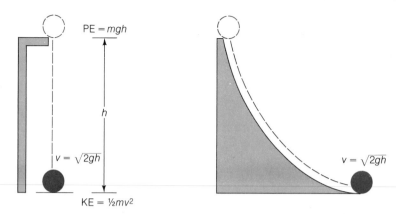

actually useful only in the simplest cases because in order to apply them we must take into detailed account all the various forces acting on each object at every point in its path, which is usually a difficult and complicated procedure. The great advantage of the principle of conservation of mechanical energy is that it permits us to draw definite conclusions about the relationship between the initial and final states of motion of some object or system of objects without having to investigate exactly what happens in between.

Example A ball slides down a smooth, curved track so that it is moving horizontally when it reaches the bottom, as in Fig. 6–16. What is its final speed?

Solution Because the path of the ball is curved, to apply the laws of motion directly means an involved calculation, but conservation of mechanical energy makes the problem ridiculously easy. When it is let go, the ball has a potential energy relative to the bottom of its path of

$$PE_{top} = mgh$$

At the bottom the kinetic energy of the ball is

$$KE_{bottom} = \tfrac{1}{2}mv^2$$

Conservation of mechanical energy requires that

$$KE_{bottom} = PE_{top}$$
$$\tfrac{1}{2}mv^2 = mgh$$
$$v = \sqrt{2gh}$$

This is the same speed the ball would have if it were simply dropped. ∎

Conservative forces

If the work performed in taking a body from a to b does not depend upon the path taken but only on the locations of a and b, the force acting is said to be *conservative*. Work done against a conservative force can be recovered by returning the body from b to a. Gravity is an example of a conservative force, as we have seen: if we do the amount of work $W = mgh$ to lift a body through the height h, the body can do the same amount of work when it is allowed to fall to the ground. A 2-kg brick 5 m above the ground has a PE of 98 J relative to the ground regardless of how it got there, and all this PE can be turned into work equally well by dropping the brick or by allowing it to slide down a frictionless track.

On the other hand, when the work done *does* depend on the exact path taken, the force acting is said to be *nonconservative* (or *dissipative*), and the work cannot be recovered by reversing the path. Friction is an example of a nonconservative force: the longer the path, the more the work needed to overcome friction regardless of where the end points *a* and *b* are. Returning the body from *b* to *a* involves further work, not the recovery of the original work done.

Nonconservative forces

What happens to the energy used to overcome the effects of friction? To find the answer, all we need do is rub one piece of wood against another. After a short time, it is obvious that the contacting surfaces of the pieces of wood are warmer than before. This is a quite general observation: work done against frictional forces produces a rise in the temperature of the objects involved. What is happening is that the energy that has disappeared on a macroscopic level reappears on a microscopic level as additional molecular kinetic energy, which is manifested in a rise in temperature. Temperature and heat are examined in detail in later chapters.

Work done against friction becomes heat

IMPORTANT TERMS

Work is a measure of the change (in a general sense) a force gives rise to when it acts upon something. When an object undergoes a displacement while a force acts on it, the work done by the force is equal to the product of the displacement and the component of the force in the direction of the displacement. In the SI system the unit of work is the **joule** and in the British system it is the **foot-pound.**

The rate at which work is done is called **power.** The unit of power in the metric system is the **watt,** which is equal to 1 J/s, and in the British system it is the **foot-pound per second.** The **horsepower** is a unit of power equal to 550 ft·lb/s, which is 746 watts.

Energy is that which may be converted into work. When something possesses energy, it is capable of performing work or, in a general sense, of accomplishing a change in some aspect of the physical world. The units of energy are those of work.

The three broad categories of energy are **kinetic energy,** which is the energy something possesses by virtue of its motion; **potential energy,** which is the energy something possesses by virtue of its position in a force field; and **rest energy,** which is the energy something possesses by virtue of its mass.

The principle of **conservation of energy** states that the total amount of energy in a system isolated from the rest of the universe always remains constant, although energy transformations from one form to another, including rest energy, may occur within the system.

IMPORTANT FORMULAS

Work: $W = Fs \cos \theta$

Work in lifting object: $W = wh = mgh$

Power: $P = \dfrac{W}{t} = Fv \cos \theta$

Kinetic energy: $KE = \frac{1}{2}mv^2$

Gravitational potential energy: $PE = wh = mgh$

Rest energy: $E_0 = m_0 c^2$

MULTIPLE CHOICE

1. According to the principle of conservation of energy (with energy interpreted as including rest energy), energy can be

    (a) created but not destroyed.
    (b) destroyed but not created.
    (c) both created and destroyed.
    (d) neither created nor destroyed.

2. A golf ball and a ping-pong ball are dropped in a vacuum chamber. When they have fallen halfway down, they have the same

    (a) speed.
    (b) potential energy.
    (c) kinetic energy.
    (d) rest energy.

3. In the formula $E = m_0 c^2$, the symbol c represents
   (a) the speed of the body.
   (b) the speed of sound.
   (c) the speed of light.
   (d) the rest energy of 1 kg of matter.

4. Which of the following is not a unit of power?
   (a) joule-second
   (b) watt
   (c) newton-meter per second
   (d) horsepower

5. To keep a vehicle moving at the speed v requires a force F. The power needed is
   (a) Fv.
   (b) $\frac{1}{2}Fv^2$.
   (c) F/v.
   (d) F/v^2.

6. A 2-kg book is held 1 m above the floor for 50 s. The work done is
   (a) 0.
   (b) 10.2 J.
   (c) 100 J.
   (d) 980 J.

7. A 40-kg boy runs up a staircase to a floor 5 m higher in 7 s. His power output is
   (a) 29 W.
   (b) 280 W.
   (c) 1400 W.
   (d) 1.37×10^4 W.

8. A 1-kg mass has a potential energy of 1 joule relative to the ground when it is at a height of
   (a) 0.012 m.
   (b) 1 m.
   (c) 9.8 m.
   (d) 32 m.

9. A 1-N weight has a potential energy of 1 joule relative to the ground when it is at a height of
   (a) 0.102 m.
   (b) 1 m.
   (c) 9.8 m.
   (d) 32 m.

10. A total of 4900 joules is expended in lifting a 50-kg mass. The mass was raised to a height of
   (a) 10 m.
   (b) 98 m.
   (c) 960 m.
   (d) 245,000 m.

11. A 1-kg mass has a kinetic energy of 1 joule when its speed is
   (a) 0.45 m/s.
   (b) 1 m/s.
   (c) 1.4 m/s.
   (d) 4.4 m/s.

12. A 1-N weight has a kinetic energy of 1 joule when its speed is
   (a) 0.45 m/s.
   (b) 1 m/s.
   (c) 1.4 m/s.
   (d) 4.4 m/s.

13. A 1000-kg car whose speed is 80 km/h has a kinetic energy of
   (a) 2.52×10^4 J.
   (b) 2.47×10^5 J.
   (c) 2.42×10^6 J.
   (d) 3.20×10^6 J.

14. The height above the ground of a child on a swing varies from 0.5 m at the lowest point to 2.0 m at the highest point. The maximum speed of the child is
   (a) about 5.4 m/s.
   (b) about 7.7 m/s.
   (c) about 29.4 m/s.
   (d) dependent on the child's mass.

15. Car A has a mass of 1000 kg and a speed of 60 km/h, and car B has a mass of 2000 kg and a speed of 30 km/h. The kinetic energy of car A is
   (a) half that of car B.
   (b) equal to that of car B.
   (c) twice that of car B.
   (d) four times that of car B.

16. A sedentary person requires about 6 million J of energy per day. This rate of energy consumption is equivalent to about
   (a) 70 W.
   (b) 335 W.
   (c) 600 W.
   (d) 250,000 W.

17. The mass equivalent of 6 million J is
   (a) 6.7×10^{-11} kg.
   (b) 5.4×10^{-9} kg.
   (c) 6.7×10^{-3} kg.
   (d) 2×10^{-2} kg.

EXERCISES

1. Under what circumstances (if any) is no work done on a moving body even though a net force acts upon it?

2. The potential energy of a golf ball in a hole is negative relative to the ground. Under what circumstances (if any) is its kinetic energy negative? Its rest energy?

3. The energy used to lift a 30-kg mass is 4000 J. If the mass is at rest before and after its elevation, how high does it go?

4. The sun exerts a force of 4×10^{28} N on the earth, and the earth travels 9.4×10^{11} m in its annual orbit of the sun. How much work is done by the sun on the earth in the course of a year?

5. (a) A force of 130 N is used to lift a 12-kg mass to a height of 8 m. How much work is done by the force? (b) A force of 130 N is used to push a 12-kg mass on a horizontal, frictionless surface for a distance of 8 m. How much work is done by the force?

6. A 20-kg wooden box is pushed a distance of 15 m on a horizontal stone floor by a force just sufficient to overcome the friction between box and floor. The coefficient of friction is 0.4. (a) What is the required force? (b) How much work does the force do?

7. In 1970 approximately 2×10^{20} J of work were performed throughout the world by inanimate devices of all kinds, perhaps 15 times as much as the muscle power provided in that year. The work was used for heat, light, transport, manufacturing, and so forth. About 98% of the work was ultimately derived from the fossil fuels coal, natural gas, and oil, the rest mainly from water power with a small (0.25% of the total) contribution from nuclear power stations. (a) Express the power consumption in 1970 in watts. (b) Find the average power consumption per person in watts and in horsepower on the assumption that the world's population in 1970 was 3.5×10^{9}.

8. A 75-kg man carrying a 10-kg pack climbs a mountain 2800 m high in 10 h. (a) What is his average power output? (b) The efficiency with which his body utilizes food energy for climbing is 15%. How many joules of food energy were needed for the climb?

9. A weightlifter raises a 150-kg barbell from the floor to a height of 2.2 m in 0.8 s. What is his average power output during the lift?

10. A crane whose motor has a power input of 5 kW raises a 1200-kg beam through a height of 30 m in 90 s. Find the efficiency.

11. The anchor windlass of a boat must be able to raise a total load (anchor plus chain) of 800 kg at a speed of 0.5 m/s. What should the minimum rating of the motor be, in kilowatts?

12. A white horse has a power output of 1 hp. What is the maximum force it can exert at a speed of 3 m/s?

13. Each of the four engines of a DC-8 airplane develops 7500 hp when the cruising speed is 240 m/s. How much thrust does each engine produce under these circumstances?

14. A motorboat requires 160 hp to move at the constant speed of 8 m/s. How much resistive force does the water exert on it at that speed?

15. Find the average kinetic energy of a 70-kg runner who covers 400 m in 45 s.

16. Find the speed of a 2-g insect whose kinetic energy is 0.01 J and whose potential energy is 0.04 J.

17. A 60-kg woman stands on a diving board 2 m above the surface of a lake at a place where it is 4 m deep. What is her potential energy with respect to the water surface and with respect to the lake bottom?

18. Is the work needed to bring a car's speed from 10 to 20 km/h less than, equal to, or more than the work needed to bring its speed from 90 to 100 km/h? If the amounts of work are different, what is the ratio between them?

19. A 2-kg ball is at rest when a horizontal force of 5 N is applied. In the absence of friction, what is the speed of the ball after it has gone 10 m?

20. A 1-kg trout is hooked by a fisherman and swims off at 2.5 m/s. The fisherman stops the trout in 50 cm by braking his reel. How much tension is exerted on the line?

21. At her highest point, a 40-kg girl on a swing is 2 m from the ground while at her lowest point she is 0.8 m from the ground. What is her maximum speed? On another swing a 50-kg boy undergoes exactly the same motion. What is his maximum speed?

22. A 3-kg stone is dropped from a height of 100 m. Find its kinetic and potential energies when it is 50 m from the ground.

23. Find the height of the bar a pole-vaulter can clear on the basis of the following assumptions: his running speed is 8 m/s; his center of gravity is initially 1.1 m above the ground and he pulls himself upward along the pole 0.6 m as the pole swings into a vertical position; and all his initial KE is converted into work done to raise his CG sufficiently to clear the bar.

24. A force of 500 N is used to lift a 20-kg object to a height of 10 m. There is no friction present. (a) How much work is done by the force? (b) What is the change in the potential energy of the object? (c) What is the change in the kinetic energy of the object?

25. A woman skis down a slope 100 m high. Her speed at the foot of the slope is 20 m/s. What percentage of her initial potential energy was dissipated?

26. The source of the sun's energy (and therefore, directly or indirectly, of nearly all energy available on the earth) is the conversion of hydrogen to helium. As described later in the book, the nuclei of four hydrogen atoms, each of mass 1.673×10^{-27} kg, join together in a series of separate reactions to yield a helium nucleus of mass 6.646×10^{-27} kg. How much energy is liberated each time a helium nucleus is formed? How many helium nuclei are formed to produce the 10^{7} J a moderately active person requires per day?

PROBLEMS

1. In the operation of a certain pile driver, a 500-kg hammer is dropped from a height of 5 m above the head of a pile. If the pile is driven 20 cm into the ground with each impact of the hammer, what is the average force on the pile when struck?

2. A horizontal force of 5 N is used to push a box up a ramp 5 m long that is at an angle of 15° above the horizontal. How much work is done?

3. A horse is towing a barge with a rope that makes an angle of 20° with the canal. If the horse exerts a force of 400 N, how much work does it do in moving the barge 1 km?

4. A person pulls a 100-kg crate for 20 m across a level floor using a rope that is 30° above the horizontal. If the coefficient of friction between crate and floor is 0.30 and just enough force is used to move the crate without accelerating it, how much work is performed? (Assume the rope is attached to the center of gravity of the crate.)

5. Two people set out to climb to the summit of a 3000-m mountain starting from sea level. One of them sets out along a slope that averages 30° above the horizontal, the other along a slope that averages 40° above the horizontal. Each person has a mass of 80 kg and carries a 10-kg knapsack. Find the work done by each of them.

6. A certain frog's hind legs produce a force of 2.5 times its weight through a vertical distance of 9 cm, at which point the frog becomes airborne. (a) What is the frog's speed at takeoff? (b) What height above the ground does the frog reach? Assume a vertical jump.

7. In 1932 five members of the Polish Olympic ski team climbed from the 5th to the 102nd floor of the Empire State Building, a distance of approximately 350 m, in 21 min. If one of these men had a mass of 70 kg, how much power did he develop during the ascent?

8. A 15-kW motor is used to hoist an 800-kg bucket of concrete to the twentieth floor of a building under construction, a height of 90 m. If no power is lost, how much time is required?

9. A certain 800-kg car has a motor whose power output is 30 kW. If the car is carrying two passengers whose total mass is 150 kg, how long will it need to accelerate from 70 to 110 km/h?

10. A 70-kg sprinter pushes on the starting blocks for 0.2 s and leaves them with a speed of 5 m/s. He then continues to accelerate for a further 5 s until his speed is 12 m/s. Find the reaction force exerted by the starting blocks on the sprinter and his average power output in each of the accelerations.

11. A person's metabolic processes can usually operate at a power of 6 W/kg of body mass for several hours at a time. If a 60-kg woman carrying a 12-kg pack is walking uphill

with an energy-conversion efficiency of 20%, at what rate, in meters per hour, does she ascend?

12. A man uses a rope and system of pulleys to lift an 80-kg object to a height of 2 m. He exerts a force of 220 N on the rope and pulls a total of 8 m of rope through the pulleys in the course of raising the object, which is at rest afterward. (a) How much work does the man do? (b) What is the change in the object's PE? (c) If the answers to (a) and (b) are different, explain.

13. (a) A force of 8 N is used to push a 0.5-kg ball over a horizontal, frictionless table a distance of 3 m. If the ball starts from rest, what is its final kinetic energy? (b) The same force is used to lift the same ball a height of 3 m. If the ball starts from rest, what is its final kinetic energy?

14. A 60-kg woman is riding a 10-kg bicycle at a constant speed of 20 km/h. Friction and air resistance total 25 N. Find her power output (a) when the road is horizontal and (b) when it goes up a 5° hill.

15. The power required to propel a 1200-kg car at 40 km/h on a level road is 30 kW. (a) How much resistance must the car overcome at this speed? (b) How much power is needed for the car to ascend an 8° hill at the same speed?

16. A ball is dropped from a height of 1 m and loses 10% of its kinetic energy when it bounces on the ground. To what height does it rise?

17. A waterfall is 30 m high and 10^4 kg of water flows over it per second. (a) How much power does this flow represent? (b) If all this power could be converted to electricity, how many 100-W light bulbs could be supplied?

18. A method for storing large amounts of energy involves pumping water from a low reservoir to a high one. The pumping is done by turbines powered by electric motors; when the water is allowed to descend through the same turbines, the motors act as generators to supply electricity during times of high demand. A system of this kind under construction in Wales will have a difference in water level of 440 m and can provide 1320 MW. If the efficiency with which the energy of the falling water is converted into electricity is 85%, how much water must pass through the turbines per minute? Neglect changes in the water levels of the reservoirs.

19. The bilge pump of a boat is able to raise 200 liters of water per minute through a height of 1.2 m. If the pump is 60% efficient, how much power must be supplied to the pump? The mass of 1 liter of water is 1 kg.

20. Steam enters a 10-MW-output turbine at 800 m/s and

emerges at 100 m/s. Assuming 90% mechanical efficiency, what mass of steam passes through the turbine per second?

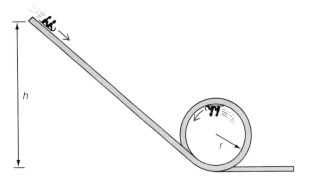

21. A sled slides down the frictionless track shown in the figure above and loops the loop without falling off. What is the minimum value of h in terms of r?

22. A ball is swung in a vertical circle on an 80-cm string just fast enough for the string to be barely taut at the top of the circle. If the total mechanical energy (PE + KE) of the ball is constant, find the speed of the ball (a) at the top of the circle, (b) at the bottom of the circle, and (c) when the string is horizontal.

ANSWERS TO MULTIPLE CHOICE

1. d	**6.** a	**11.** c	**16.** a
2. a	**7.** b	**12.** d	**17.** a
3. c	**8.** a	**13.** b	
4. a	**9.** b	**14.** a	
5. a	**10.** a	**15.** c	

MOMENTUM

7

So complex is the physical universe that many different quantities turn out to be useful in describing its various aspects. We have already been introduced to length, time, mass, force, torque, work, and energy, and more are to come. There is nothing sacred about any of these quantities—it is entirely possible to dispense with any of them, but only at the expense of making physics a good deal more complicated than it already is. The idea behind the definition of each of the various physical quantities is to single out something that unifies a wide range of observations, so that it is then possible to boil down to a brief, clear statement a large number of separate discoveries about nature. In this chapter we shall learn how the concepts of linear momentum and impulse supplement those of work and energy to provide a particularly simple theoretical framework for analyzing the behavior of moving bodies.

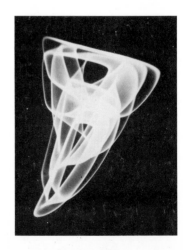

7–1   LINEAR MOMENTUM

We all know that a baseball struck squarely by a bat is harder to stop than the same baseball thrown gently, and that the heavy iron ball used for the shotput is harder to stop than a baseball whose velocity is the same (Fig. 7–1). These observations suggest that a measure of the tendency of a body to continue in motion at constant velocity is the product $m\mathbf{v}$ of its mass m and velocity $\mathbf{v}$.

The quantity $m\mathbf{v}$ is called the *linear momentum* of a moving body:

Linear momentum is a vector quantity, unlike energy

$$\text{Linear momentum} = m\mathbf{v} \qquad (7\text{–}1)$$

The symbol $\mathbf{p}$ is sometimes used to represent linear momentum. Linear momentum is a vector quantity whose direction is the direction of $\mathbf{v}$. The kinetic energy of a moving body, which also depends upon its mass and velocity since $\text{KE} = \frac{1}{2}mv^2$, is a scalar quantity with magnitude only. The different significances of linear momentum and kinetic energy will be discussed later in this chapter.

Because $m\mathbf{v}$ describes the tendency of a moving body to pursue a straight path at constant velocity, it is referred to as *linear* momentum. A different quantity, *angular momentum,* describes the tendency of a spinning body such as a top to continue to spin. When there is no question as to which is meant, linear momentum is usually referred to simply as momentum.

7–2   IMPULSE

To set something in motion from rest, a force must be applied for a period of time. We might expect that the greater the force and the longer the time, the more momentum the object will have. This expectation is correct, and the product $\mathbf{F}\,\Delta t$ of a constant force $\mathbf{F}$ and the time interval Δt during which it acts is accordingly given the status of a physical quantity in its own right. This quantity is called *impulse:*

An impulse produces a change in momentum

$$\text{Impulse} = \mathbf{F}\,\Delta t \qquad (7\text{–}2)$$

Impulse, like momentum, is a vector quantity. Let us see how the momentum of an object is affected when it receives a certain impulse.

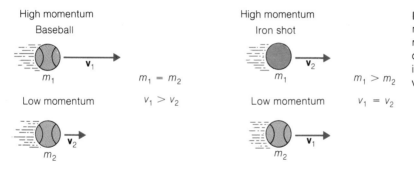

High momentum
Baseball
m_1 $\mathbf{v}_1$ $m_1 = m_2$

Low momentum
m_2 $\mathbf{v}_2$ $v_1 > v_2$

High momentum
Iron shot
m_1 $\mathbf{v}_2$ $m_1 > m_2$

Low momentum
m_2 $\mathbf{v}_1$ $v_1 = v_2$

FIG. 7–1 The linear momentum $m\mathbf{v}$ of a moving body is a measure of its tendency to continue in motion at constant velocity.

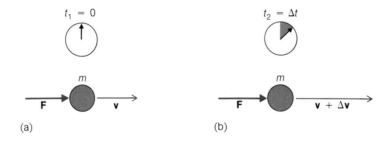

(a)   (b)

The second law of motion states that the force **F** applied to an object of constant mass m that undergoes the acceleration **a** is given by

Second law of motion

$$\mathbf{F} = m\mathbf{a} \tag{7–3}$$

When a force **F** is applied at the time $t_1 = 0$ to an object whose initial velocity is **v**, at the later time $t_2 = \Delta t$ its velocity will have changed to **v** + Δ**v** (Fig. 7–2). The object's acceleration in this time interval is

Definition of acceleration

$$\mathbf{a} = \frac{\text{velocity change}}{\text{time interval}} = \frac{\Delta \mathbf{v}}{\Delta t}$$

and so Eq. (7–3) becomes

$$\mathbf{F} = m\mathbf{a} = m\frac{\Delta \mathbf{v}}{\Delta t}$$

which we can rewrite as

$$\mathbf{F}\,\Delta t = m\,\Delta \mathbf{v}$$

Evidently the impulse provided by the force equals the momentum change of the object:

The impulse given to a body equals its momentum change

$$\mathbf{F}\,\Delta t = \Delta(m\mathbf{v}) \tag{7–4}$$

Impulse = momentum change

Figure 7–3 shows the effect of applying the constant force **F** for the time Δt to several objects with different momenta $m\mathbf{v}_1$. In each case, the final momentum $m\mathbf{v}_2$ is obtained by finding the vector sum $m\mathbf{v}_2 = m\mathbf{v}_1 + \mathbf{F}\,\Delta t$.

Units of impulse and momentum

In the SI, the unit of impulse is the *newton-second* (N·s), and the unit of momentum is the kg·m/s; they are actually the same, of course, but it is often convenient to distinguish between them in this way. (The corresponding British units are the lb·s and (slug·ft)/s.)

Example   The head of a golf club is in contact with a 46-g golf ball for 0.50 ms (1 ms = 1 millisecond = 10^{-3} s), and as a result the ball flies off at 70 m/s. Find the average force that was acting on the ball during the impact.

Solution   The ball starts from rest, hence its momentum change is

$$\Delta m v = (0.046 \text{ kg})(70 \text{ m/s}) = 3.22 \text{ kg·m/s}$$

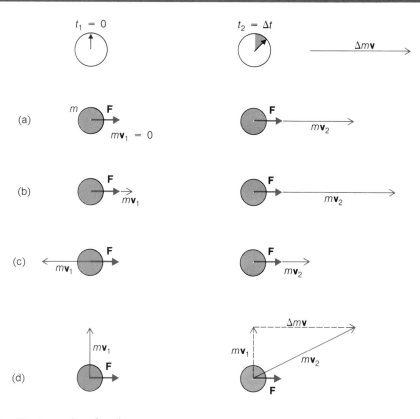

FIG. 7–3 Applying a constant force **F** to a mass m for a time Δt changes its momentum by $\Delta m\mathbf{v} = \mathbf{F} \, \Delta t$. At (a) the mass is initially at rest, at (b) its initial momentum is in the same direction as **F**, at (c) its initial momentum is in the opposite direction to **F**, and at (d) its initial momentum is perpendicular to **F**. Since momentum is a vector quantity, the momentum change $\Delta m\mathbf{v}$ must be added to the initial momentum $m\mathbf{v}$ by the process of vector addition.

From Eq. (7–4) we therefore have

$$F = \frac{\Delta mv}{\Delta t} = \frac{3.22 \, \text{kg·m/s}}{5.0 \times 10^{-4} \, \text{s}} = 6.44 \times 10^3 \, \text{N}$$

whose British equivalent is 1450 lb. A force of the same magnitude but acting in the opposite direction (the *recoil force*) acts on the club's head during the impact, in accordance with the third law of motion. No golf club could withstand such a static load, but the impact is so brief that its only effect on the shaft is to temporarily bend it by a few cm. ■

7–3 CONSERVATION OF MOMENTUM

Energy and work are scalar quantities, having magnitude only. Despite the fundamental and all-inclusive character of the law of conservation of energy, it cannot by itself provide complete solutions to most problems that involve interacting bodies. A simple example is the firing of a rifle: the requirement that energy be conserved means that the kinetic energies of the bullet and the recoiling rifle, plus the heat and sound energy that are liberated, must equal the chemical energy of the detonated explosive, but this does not tell us how the total energy is divided among the rifle, the bullet, and the atmo-

Energy and work are scalar quantities; momentum and impulse are vector quantities

sphere. Indeed, because energy is a scalar quantity, its conservation does not even imply that the bullet and rifle must move in opposite directions. To complete the solution of many problems in dynamics in which a detailed knowledge of the active forces is lacking, an additional principle of a vector nature is required.

Exchanging momenta among the members of a system does not change the total momentum of the system

Let us consider a system of two or more particles instead of a single particle. If no forces from outside the system act upon its component particles, the total linear momentum of the system, which is the sum

$$MV = m_1v_1 + m_2v_2 + m_3v_3 + \cdots \tag{7-5}$$

of the individual momenta of its particles, cannot change; with no force there is no impulse, hence no change in momentum. However, the *distribution* of the total momentum MV among the various particles in the system may change without MV changing in the absence of an external force.

Two particles A and B might collide and thereby exert forces upon each other. At every instant during their interaction these forces, $\mathbf{F}_{AB}$ acting on B and $\mathbf{F}_{BA}$ acting on A, obey Newton's third law of motion,

$$\mathbf{F}_{AB} = -\mathbf{F}_{BA}$$

This is shown in Fig. 7–4. Hence the impulses exchanged must be equal and opposite,

$$\mathbf{F}_{AB}\,\Delta t = -\mathbf{F}_{BA}\,\Delta t$$

and the *total* momentum of the system of A and B together is the same after the collision as it was before.

Since we almost always can include the sources of all forces relevant to a particular process within what we choose to be our "system," in such cases we have the condition that, no matter what interactions take place within the system, its total momentum never changes. Thus we have the principle of *conservation of linear momentum*:

Conservation of linear momentum

When the vector sum of the external forces acting upon a system of particles equals zero, the total linear momentum of the system remains constant.

Analyzing an explosion

Let us consider a specific example readily treated with the help of conservation of momentum. Suppose that we have an isolated particle of mass m, initially at rest, that

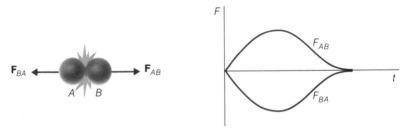

FIG. 7–4   Particles A and B collide and exert the forces F_{AB} and F_{BA} on each other. As shown in the graph, these forces are equal in magnitude and opposite in direction at all times. Their impulses are also equal and opposite, so the total momentum of A and B is left unchanged by the collision although it may be distributed differently between them.

suddenly explodes into two particles of masses m_1 and m_2, which fly apart (Fig. 7–5). The forces acting on the original particle that caused it to break up were internal ones, and no external force was present. Since m has the initial momentum of zero, the final momentum of m_1 and m_2, when added together, must also be zero. Hence

$$m\mathbf{v} = 0 = m_1\mathbf{v}_1 + m_2\mathbf{v}_2$$

and

$$\mathbf{v}_2 = -\frac{m_1}{m_2}\mathbf{v}_1$$

where $\mathbf{v}_1$ and $\mathbf{v}_2$ are the final velocities of the two fragments. We note immediately that these velocities must be in opposite directions along the same line. This problem could *not* be solved starting from $\mathbf{F} = m\mathbf{a}$, since we do not know explicitly what forces were acting during the explosion.

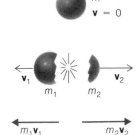

FIG. 7–5 The total momentum of a system of particles remains constant if no external forces act on the system.

Example An astronaut in orbit outside an orbiting space station throws his 0.8-kg camera away in disgust when it jams (Fig. 7–6). If he and his space suit together have a mass of 100 kg and the speed of the camera is 12 m/s, how far away from the space station will he be in 1 h?

Solution Here $m_1 = 0.8$ kg, $m_2 = 100$ kg, and $v_1 = 12$ m/s. From Eq. (7–6),

$$v_2 = -\frac{m_1}{m_2}v_1 = -\left(\frac{0.8\,\text{kg}}{100\,\text{kg}}\right)(12\,\text{m/s}) = -0.096\,\text{m/s}$$

After $t = 1\,\text{h} = 3600$ s the astronaut will be

$$s = v_2 t = (0.096\,\text{m/s})(3600\,\text{s}) = 346\,\text{m}$$

away from the space station. The total momentum of the system of astronaut plus camera remains zero until other forces act on them. ∎

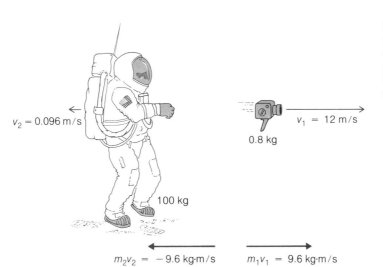

FIG. 7–6 The forward momentum of the thrown camera is equal in magnitude to the backward momentum of the astronaut who threw it.

**Conservation principles
and the laws of nature**

The conservation of linear momentum is a generalization based upon innumerable experiments and observations, and no exception to it has ever been found. With the help of advanced mathematics it is possible to show that if the laws of nature are the same at every point in space, then the principle of conservation of linear momentum must follow as an inevitable consequence. It is also possible to show that if the laws of nature do not change with time, so that they were always the same as they are now and always will remain the same, then energy must be conserved in all interactions. Thus these principles, as well as being useful relationships for solving practical problems, give us a hint of a profound order underlying the physical universe.

7–4   ROCKET PROPULSION

**Rocket propulsion is
based on momentum
conservation**

The principle underlying rocket flight is conservation of momentum. The total momentum of a rocket on its launching pad is zero. When it is fired, the exhaust gases shoot downward at high speed, and the rocket moves upward to balance the momentum of the gases (Fig. 7–7). Rockets do not operate by "pushing" against their launching pads, the air, or anything else; in fact, they perform best in space, where there is no atmosphere to impede their motion. The energy of the rocket and its exhaust comes from chemical energy stored in the fuel. The total momentum of the system of rocket plus exhaust, which is initially zero, does not remain constant after a launch from the earth because of the impulses provided by air resistance and the earth's gravitational pull.

FIG. 7–7 Conservation of momentum in rocket flight. The downward momentum of the exhaust gases is exactly balanced by the upward momentum of the rocket itself.

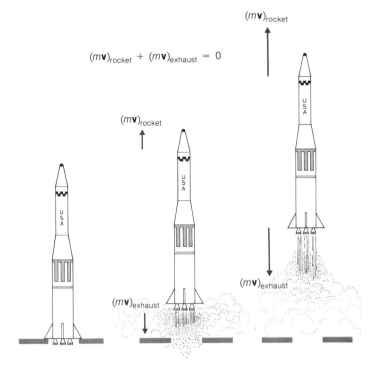

$$(m\mathbf{v})_{rocket} + (m\mathbf{v})_{exhaust} = 0$$

Rocket propulsion is a gradual rather than an instantaneous process, with the fuel burned and ejected as exhaust gases at a certain rate instead of in one lump. As a result, part of the momentum of the exhaust is wasted in pushing forward unburned fuel. When this factor is taken into account, the ultimate speed of a rocket (neglecting air resistance and gravity) turns out to be directly proportional to the speed of the exhaust gases and to the logarithm of the ratio of the rocket's initial mass to its final mass after the fuel has been consumed. A typical modern rocket might have an exhaust speed of 3000 m/s with 75 percent of its initial mass consisting of fuel, which gives a final speed of 4155 m/s.

Factors that determine the final speed of a rocket

Example A spacecraft's motor consumes 100 kg of fuel per second, which it exhausts at 3000 m/s. Find the thrust (upward force) exerted on the spacecraft.

Solution If a mass Δm of exhaust gas is ejected at the speed v in the time interval Δt, its momentum change is

$$\Delta(mv) = v\,\Delta m$$

From Eq. (7–4) $F\,\Delta t = \Delta(mv)$, so the associated force is

$$F = \frac{\Delta(mv)}{\Delta t} = v\,\frac{\Delta m}{\Delta t}$$

The thrust of a rocket is the product of the exhaust speed and the rate at which fuel is consumed. Here

$$F = v\frac{\Delta m}{\Delta t} = (3000\,\text{m/s})(100\,\text{kg/s}) = 3 \times 10^5\,\text{N}$$

When the spacecraft is launched from the earth's surface, the thrust of its motor must exceed its initial weight $m_0\,g$ for it to rise from the ground. If the thrust remains constant, the acceleration of the spacecraft increases as its mass diminishes owing to the consumption of fuel. A final acceleration of several times g in magnitude is usual. ∎

To attain higher speeds than a single rocket is capable of, two or more rocket stages can be used. The first stage is a large rocket whose payload is another, smaller rocket. When the fuel of the first stage has been consumed, its fuel tanks and engine are cast loose. Then the second stage is fired starting from a high initial speed instead of from rest and without the burden of the fuel tanks and engine of the first stage. This process can be repeated a number of times, depending upon the final speed required. The Saturn V launch vehicle that propelled the Apollo 11 spacecraft to the moon in July 1969 employed three stages, as shown in Fig. 7–8. At original ignition the entire assembly was 111 m long and had a mass of 2.9×10^6 kg (3240 tons).

Multistage rockets

7–5 COLLISIONS

The law of conservation of momentum is indispensable in dealing with collisions between two or more objects. In such cases no external forces act on the participants, and therefore their total momentum before they collide equals their total momentum

Momentum is redistributed in a collision

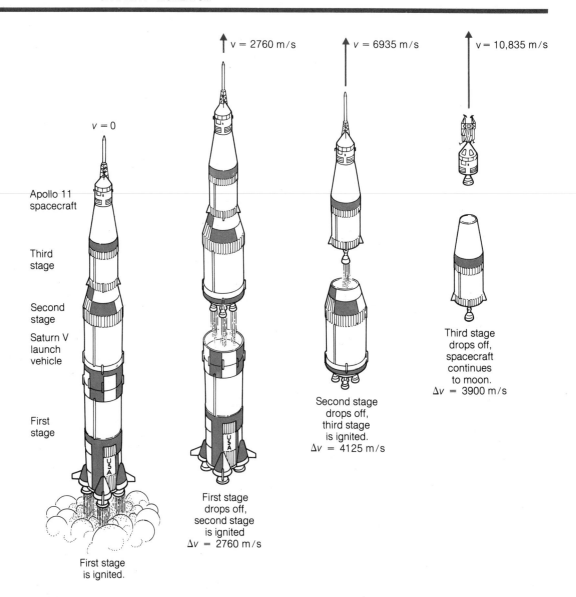

$v = 0$

$v = 2760$ m/s

$v = 6935$ m/s

$v = 10,835$ m/s

Apollo 11
spacecraft

Third
stage

Second
stage

Saturn V
launch
vehicle

First
stage

First stage
is ignited.

First stage
drops off,
second stage
is ignited
$\Delta v = 2760$ m/s

Second stage
drops off,
third stage
is ignited.
$\Delta v = 4125$ m/s

Third stage
drops off,
spacecraft
continues
to moon.
$\Delta v = 3900$ m/s

FIG. 7–8 The Saturn V
launch vehicle that
propelled the Apollo 11
spacecraft to the moon for
the first manned landing
used three rocket stages.

**KE is conserved in an
elastic collision**

afterward. *The essential effect of the collision is to redistribute the total momentum of
the objects*.

If we only know the masses and initial velocities of the objects involved in a
collision, however, momentum conservation by itself does not yield a unique result for
their subsequent motion. An unknown amount of kinetic energy may be lost to heat,
sound, or other forms of energy when the objects interact, and, though the requirement
that momentum be conserved does set limits to the result of the collision, it can go no
further without additional information. These limiting cases, however, are worth examining.

At one extreme are completely *elastic collisions* in which kinetic energy is conserved. During the actual collision, to be sure, some kinetic energy becomes elastic

potential energy as the objects are deformed by the impact, but all of this energy is returned as the objects move apart.

At the other extreme are *completely inelastic collisions* in which the objects stick together permanently upon impact. The kinetic energy loss in a completely inelastic collision is the maximum possible consistent with momentum conservation. Such a collision can be analyzed on the basis of momentum conservation only.

KE is not conserved in an inelastic collision; the maximum KE loss occurs when the bodies stick together

Example   A 5-kg lump of clay that is moving at 10 m/s to the left strikes a 6-kg lump of clay moving at 12 m/s to the right. The two lumps stick together after they collide. Find the final speed of the composite object and the kinetic energy dissipated in the collision.

Solution   This is an example of a completely inelastic collision. If we call the mass of the final object M and its velocity V, conservation of linear momentum requires that

Momentum afterward = momentum before
$$MV = m_1 v_1 + m_2 v_2$$

Adopting the convention that motion to the right is $+$ and to the left is $-$, we have

$$m_1 = 5 \, \text{kg} \qquad m_2 = 6 \, \text{kg} \qquad M = m_1 + m_2 = 11 \, \text{kg}$$
$$v_1 = -10 \, \text{m/s} \qquad v_2 = +12 \, \text{m/s} \qquad V = ?$$

Solving for V yields
$$V = \frac{m_1 v_1 + m_2 v_2}{M} = \frac{(5 \, \text{kg})(-10 \, \text{m/s}) + (6 \, \text{kg})(12 \, \text{m/s})}{11 \, \text{kg}} = 2 \, \text{m/s}$$

Since V is positive, the composite body moves off to the right (Fig. 7–9).

Energy and momentum are independent concepts. The lumps of clay before the collision have the kinetic energies

$$KE_1 = \tfrac{1}{2} m_1 v_1^2 = \tfrac{1}{2}(5 \, \text{kg})(-10 \, \text{m/s})^2 = 250 \, \text{J}$$

$$KE_2 = \tfrac{1}{2} m_2 v_2^2 = \tfrac{1}{2}(6 \, \text{kg})(12 \, \text{m/s})^2 = 432 \, \text{J}$$

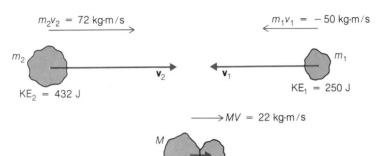

$m_2 v_2 = 72$ kg·m/s

m_2

v_2

$KE_2 = 432$ J

$m_1 v_1 = -50$ kg·m/s

m_1

v_1

$KE_1 = 250$ J

$MV = 22$ kg·m/s

M

V

$KE_3 = 22$ J

FIG. 7–9 In a completely inelastic collision, the colliding bodies stick together. Kinetic energy is not conserved in such an event. Linear momentum is conserved in all collisions.

After the collision the new lump of clay has the kinetic energy

$$KE_3 = \tfrac{1}{2}MV^2 = \tfrac{1}{2}(11\,\text{kg})(2\,\text{m/s})^2 = 22\,\text{J}$$

The total kinetic energy prior to the collision was 432 + 250 or 682 J, while afterward it is only 22 J. The difference of 660 J was dissipated largely into heat energy in the collision, with some probably being lost to sound energy as well. ∎

Momentum is a vector quantity and must be conserved in every direction

It is essential to keep in mind the directional character of linear momentum. Sometimes the problem under consideration involves bodies that move along the same straight line, as in the preceding example, but in general the bodies may move in two or three dimensions and we must be sure to take this into account by a vector calculation.

Example   A 60-kg man is sliding east on the frictionless surface of a frozen pond at a velocity of 0.50 m/s. He is struck by a 1.0-kg snowball whose velocity is 20 m/s toward the north. If the snowball sticks to the man, what is his final velocity?

Solution   Here linear momentum must be conserved separately in both the east-west and north-south directions, which we shall call the x and y axes respectively. Since

$$m_1 = 60\,\text{kg} \qquad m_2 = 1.0\,\text{kg} \qquad M = m_1 + m_2 = 61\,\text{kg}$$

$$v_{1x} = 0.50\,\text{m/s} \qquad v_{2x} = 0 \qquad V_x = ?$$

$$v_{1y} = 0 \qquad v_{2y} = 20\,\text{m/s} \qquad V_y = ?$$

we have (see Fig. 7–10)

$$\text{Momentum afterward} = \text{momentum before}$$

x direction: $MV_x = m_1 v_{1x} + m_2 v_{2x} = (60\,\text{kg})(0.5\,\text{m/s}) = 30\,\text{kg·m/s}$

y direction: $MV_y = m_1 v_{1y} + m_2 v_{2y} = (1.0\,\text{kg})(20\,\text{m/s}) = 20\,\text{kg·m/s}$

Hence the magnitude MV of the momentum of man + snowball after the collision is

$$MV = \sqrt{(MV_x)^2 + (MV_y)^2} = \sqrt{(30\,\text{kg·m/s})^2 + (20\,\text{kg·m/s})^2} = 36\,\text{kg·m/s}$$

FIG. 7–10

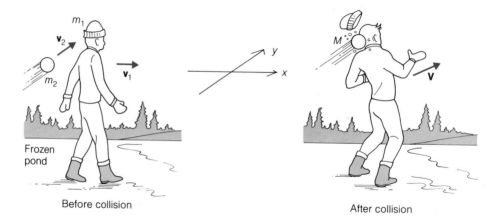

Frozen pond

Before collision

After collision

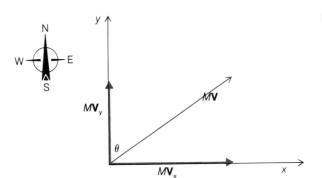

FIG. 7–11

and the corresponding final speed is

$$V = \frac{MV}{M} = \frac{36\,\text{kg·m/s}}{61\,\text{kg}} = 0.59\,\text{m/s}$$

We can specify the direction in which the man + snowball combination moves after the collision in terms of the angle θ between the $+y$ direction (which is north) and *MV*:

$$\tan\theta = \frac{MV_x}{MV_y} = \frac{30\,\text{kg·m/s}}{20\,\text{kg·m/s}} = 1.5$$

$$\theta = 56°$$

Thus man + snowball move in a direction 56° to the east of north (Fig. 7–11). ■

In a completely elastic collision, no kinetic energy is lost. An example is a collision between a moving billiard ball and a stationary one on a level table. The potential energies of the balls remain the same and the energy lost to heat and sound is negligible, so the sum of their kinetic energies before the collision must equal the sum of their kinetic energies afterward.

Billiard-ball collisions are very nearly elastic

Example   A ball rolling on a level table strikes head-on another identical ball that is stationary. What is the result of the collision?

Solution   Let us call the initial and final velocities of the balls v_1, v_2 and $v_1{}'$, $v_2{}'$. Conservation of momentum and of energy require that

	Before		After
Momentum:	$m_1 v_1 + m_2 v_2$	$=$	$m_1 v_1{}' + m_2 v_2{}'$
Kinetic energy:	$\frac{1}{2}m_1 v_1^2 + \frac{1}{2}m_2 v_2^2$	$=$	$\frac{1}{2}m_1 v_1'^2 + \frac{1}{2}m_2 v_2'^2$

Since we have said that the balls are identical, $m_1 = m_2$, and since the second ball was originally at rest, $v_2 = 0$; hence

$$v_1 = v_1{}' + v_2{}'$$

$$v_1^2 = v_1'^2 + v_2'^2$$

The *only* way to solve these equations is to have either $v_1{}'$ or $v_2{}'$ equal zero. If $v_2{}'$ were zero, it would mean that the first ball traveled completely *through* the second ball.

FIG. 7–12 A rolling ball makes a head-on collision with an identical stationary ball; the first ball stops and the second begins moving with the first's initial velocity.

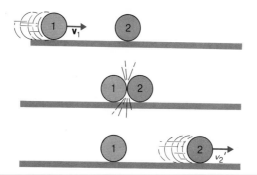

Because this is impossible, we must have as the solution

$$\mathbf{v}_1' = 0 \qquad \mathbf{v}_2' = \mathbf{v}_1$$

the first ball stops, and the second begins to move with the original speed of the first ball (Fig. 7–12). ■

The results of the preceding problem help us to understand the operation of the toy shown in Fig. 7–13, which consists of a number of identical steel balls suspended by strings. When one ball at the left is pulled out and released, it swings to strike the row of stationary balls, and one ball at the right swings out in response. Why not two balls, or indeed all the others? The answer is that both momentum and KE can be conserved *only* if a single ball swings out on the right. If two balls were to swing out, then their joint mass is twice that of the ball on the left, and to conserve momentum their initial speeds must be half that of the ball on the left. But the combined kinetic energies of the two balls would then be half that of the ball on the left:

	One ball on left		Two balls on right
Momentum:	mv	$=$	$m\left(\dfrac{v}{2}\right) + m\left(\dfrac{v}{2}\right) = mv$
Kinetic energy:	$\tfrac{1}{2}mv^2$	$\neq$	$\tfrac{1}{2}m\left(\dfrac{v}{2}\right)^2 + \tfrac{1}{2}m\left(\dfrac{v}{2}\right)^2 = \tfrac{1}{4}mv^2$

Only if half the initial KE is lost can two balls swing out. The same reasoning shows why, if two balls are pulled out at the left and released, as in Fig. 7–13(b), two balls at the right will swing out after the collision, and so on.

What about the case when more than half the balls are pulled out to the left? In that event, the balls at the left cannot give up all their momentum and kinetic energy to the balls at the right, and so several of the pulled-out balls continue to move to the right after the collision. Situations of this kind are discussed in the next section.

7–6   MORE ABOUT COLLISIONS

A moving object strikes a stationary one, and as a result the second object is set in motion. What is the mass ratio between the two that will cause the struck object to

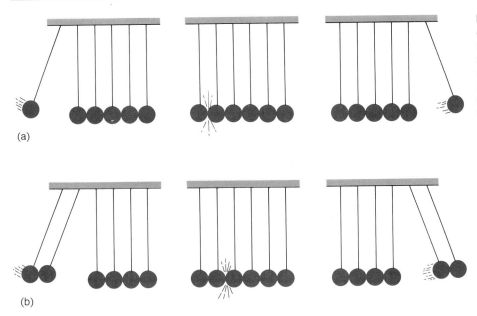

FIG. 7–13 The behavior of this arrangement of suspended steel balls is determined both by conservation of momentum and by conservation of kinetic energy.

(a)

(b)

have the highest possible speed after the impact? What mass ratio will lead to the greatest transfer of energy to the struck object? These are not questions of abstract interest only but are closely connected with a variety of actual problems that range from the design of golf clubs to the design of nuclear reactors.

For simplicity we will confine ourselves to head-on collisions in which both objects move along the same straight line. Let us consider an object of mass m_1 and initial speed v_1 that strikes a stationary object of mass m_2, after which their respective speeds are v_1' and v_2'. From conservation of momentum,

$$m_1 v_1 = m_1 v_1' = m_2 v_2'$$
$$m_1(v_1 - v_1') = m_2 v_2' \tag{7–7}$$

If the collision is completely elastic, kinetic energy is conserved, and

$$\tfrac{1}{2}m_1 v_1{}^2 = \tfrac{1}{2}m_1 v_1'{}^2 + \tfrac{1}{2}m_2 v_2'{}^2$$
$$m_1(v_1{}^2 - v_1'{}^2) = m_2 v_2'{}^2$$
$$m_1(v_1 + v_1')(v_1 - v_1') = m_2 v_2'{}^2 \tag{7–8}$$

Now we divide Eq. (7–8) by Eq. (7–7) to obtain

$$v_1 + v_1' = v_2'$$
$$v_1 = v_2' - v_1' \tag{7–9}$$

Since $v_2' - v_1'$ is the velocity of m_2 relative to m_1 after the collision and $-v_1$ is the same relative velocity before it, this result means that the effect of the collision is to reverse the direction of the relative velocity without changing its magnitude. Thus the relative velocity of approach is equal to the relative velocity of recession, a conclusion

Relative velocity of approach equals relative velocity of recession

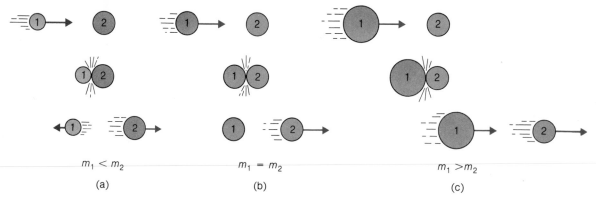

$m_1 < m_2$

(a)

$m_1 = m_2$

(b)

$m_1 > m_2$

(c)

FIG. 7–14 The result of an elastic head-on collision between a moving object of mass m_1 and a stationary object of mass m_2 depends upon the ratio of their masses.

that holds even if m_2 had been moving before the collision (see Prob. 13). In the latter event, $-(v_2 - v_1) = v_2' - v_1'$.

Combining Eqs. (7–7) and (7–9) yields for the final velocity v_1' of m_1

$$v_1' = \frac{m_1 - m_2}{m_1 + m_2} v_1 \qquad (7\text{–}10)$$

and for the final velocity v_2' of m_2

$$v_2' = \frac{2m_1}{m_1 + m_2} v_1 \qquad (7\text{–}11)$$

How the effects of a collision depend on the relative masses of the colliding objects

Formulas (7–10) and (7–11) permit us to draw some general conclusions. If m_1 is less than m_2, v_1' is in the opposite direction to v_1: the lighter object rebounds from the heavier one (Fig. 7–14a). A ball striking a wall is an extreme example, where, since m_2 is virtually infinite compared with m_1, $v_1' = -v_1$. In the event that $m_1 = m_2$, $v_1' = 0$ and $v_2 = v_1$: the colliding object stops, while the struck one moves off with the same velocity it had (Fig. 7–14b). When m_1 is greater than m_2, as in the case of a table-tennis serve, the colliding object continues on in the same direction after the impact but with reduced speed while the struck object moves ahead of it at a faster pace (Fig. 7–14c). When m_1 is much greater than m_2, the colliding object loses little speed while the struck one is given a speed nearly twice v_1.

The ratio between the kinetic energy KE_2' transferred to the initially stationary object and the kinetic energy KE_1 of the colliding object can be found with the help of Eq. (7–11):

Energy transfer is a maximum when the objects have the same masses.

$$\frac{KE_2'}{KE_1} = \frac{\frac{1}{2}m_2v_2'^2}{\frac{1}{2}m_1v_1^2} = \frac{4m_1m_2}{(m_1 + m_2)^2} = \frac{4(m_2/m_1)}{(1 + m_2/m_1)^2} \qquad (7\text{–}12)$$

This formula is plotted in Fig. 7–15. Evidently the transfer of energy is a maximum for $m_1 = m_2$, when *all* the energy of m_1 is given to m_2. This is the situation illustrated in Fig. 7–14b.

From Eq. (7–12) it would seem that the best mass for the head of a golf club would be the same as that of a golf ball, in order that all the energy of the club be given to the ball (assuming an elastic collision). In this case $v_2' = v_1$. However, according to Eq. (7–11), the greater the mass m_1 of the clubhead, the greater the ball's velocity

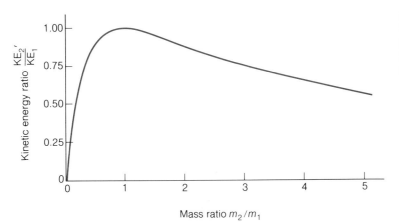

v_2' will exceed v_1, up to a limit of $2v_1$. The trouble with a heavy clubhead is twofold: it is hard to swing a heavy golf club as fast as a light one, and the more m_1 exceeds m_2, the smaller is the proportion of KE that is transferred to the ball—the extra effort does not provide a commensurate return. Experience has led golfers to use clubheads whose masses are typically about four times the 46-g mass of a golf ball where maximum distance is required, although this ratio does not seem to be very critical.

If a collision is not perfectly elastic, the relative velocity of recession will be reduced by some fraction e, which is called the *coefficient of restitution*. In general, when both objects are in motion before and after the collision,

Coefficient of restitution

$$\text{Coefficient of restitution} = e = \frac{v_2' - v_1'}{v_1 - v_2} \qquad (7\text{–}13)$$

In a perfectly elastic collision, $e = 1$; and in a perfectly inelastic collision, when the objects stick together and $v_2' = v_1'$, $e = 0$. When a ball is dropped on a horizontal surface from a height h, the height of rebound is $h' = e^2 h$. (To verify this, note that $v_2 = v_2' = 0, v_1 = -\sqrt{2\,gh}$, and $v_1' = \sqrt{2\,gh'}$, where upward is considered the positive direction.) Thus if a ball dropped from a height of 100 cm to the floor rebounds to a height of 60 cm, the coefficient of restitution must be $\sqrt{h'/h} = \sqrt{0.6} = 0.77$.

When the coefficient of restitution is taken into account in a collision between a moving object and a stationary one,

$$v_2 = \frac{(1 + e)m_1}{m_1 + m_2}\, v_1 \qquad (7\text{–}14)$$

$$\frac{\text{KE}_2'}{\text{KE}_1} = \frac{(1 + e)^2 m_1 m_2}{(m_1 + m_2)_2} = \frac{(1 + e)^2 (m_2/m_1)}{(1 + m_2/m_1)^2} \qquad (7\text{–}15)$$

In the case of a golf club striking a golf ball, typical values might be $m_1 = 200$ g, $m_2 = 46$ g, and $e = 0.7$, from which we find that the ball moves off with a velocity 38 percent greater than the velocity of the clubhead and carries with it 44 percent of the clubhead's original energy. If the collision were perfectly elastic, $e = 1$ and these figures would be 63 percent and 61 percent respectively.

In Chapter 31 we will see how the theory of collisions is applied to the design of nuclear reactors.

IMPORTANT TERMS

The **linear momentum** of an object is the product of its mass and velocity. Linear momentum is a vector quantity having the direction of the object's velocity.

The **impulse** of a force is the product of the force and the time during which it acts. Impulse is a vector quantity having the direction of the force. When a force acts on an object that is free to move, its change in momentum equals the impulse given it by the force.

The law of **conservation of momentum** states that when the vector sum of the external forces acting upon a system of particles equals zero, the total linear momentum of the system remains constant.

The **thrust** of a rocket is the force that results from the expulsion of exhaust gases.

A **completely elastic collision** is one in which kinetic energy is conserved. A **completely inelastic collision** is one in which the objects stick together upon impact, which results in the maximum possible kinetic energy loss. Linear momentum is conserved in all collisions.

IMPORTANT FORMULAS

Linear momentum: $\mathbf{p} = m\mathbf{v}$

Impulse and momentum change: $\mathbf{F}\,\Delta t = \Delta(m\mathbf{v})$

MULTIPLE CHOICE

1. An object at rest may possess
   (a) velocity.   (b) momentum.
   (c) kinetic energy.   (d) potential energy.

2. An object in motion need not possess
   (a) velocity.   (b) momentum.
   (c) kinetic energy.   (d) potential energy.

3. An object which has momentum must also have
   (a) acceleration.   (b) impulse.
   (c) kinetic energy.   (d) potential energy.

4. Momentum is most closely related to
   (a) kinetic energy.   (b) potential energy.
   (c) impulse.   (d) power.

5. The impulse given to an object is equal to the consequent change in its
   (a) velocity.   (b) momentum.
   (c) kinetic energy.   (d) potential energy.

6. When the velocity of a moving object is doubled,
   (a) its acceleration is doubled.
   (b) its momentum is doubled.
   (c) its kinetic energy is doubled.
   (d) its potential energy is doubled.

7. If a shell fired from a cannon explodes in midair,
   (a) its total momentum increases.
   (b) its total momentum decreases.
   (c) its total kinetic energy increases.
   (d) its total kinetic energy decreases.

8. When two or more objects collide, it is always true that
   (a) the momentum of each one remains unchanged.
   (b) the kinetic energy of each one remains unchanged.
   (c) the total momentum of all the objects remains unchanged.
   (d) the total kinetic energy of all the objects remains unchanged.

9. An elastic collision conserves
   (a) kinetic energy but not momentum.
   (b) momentum but not kinetic energy.
   (c) neither momentum nor kinetic energy.
   (d) both momentum and kinetic energy.

10. A ball whose momentum is $\mathbf{p}$ strikes a wall and bounces off. The change in the ball's momentum is
    (a) 0.   (b) $\mathbf{p}/2$.
    (c) $\mathbf{p}$.   (d) $2\mathbf{p}$.

11. An iron sphere of mass 30 kg has the same diameter as an aluminum sphere of mass 10.5 kg. The spheres are simultaneously dropped from a cliff. When they are 10 m from the ground, they have identical
    (a) accelerations.   (b) momenta.
    (c) potential energies.   (d) kinetic energies.

12. A 30-kg girl and a 25-kg boy face each other on frictionless roller skates. The girl pushes the boy, who moves away at a speed of 1.0 m/s. The girl's speed is
    (a) 0.45 m/s.   (b) 0.55 m/s.
    (c) 0.83 m/s.   (d) 1.2 m/s.

13. An astronaut whose total mass is 100 kg ejects 1 gm of gas from his propulsion pistol at a speed of 50 m/s. His recoil speed is
    (a) 0.5 mm/s.   (b) 5 mm/s.
    (c) 5 cm/s.   (d) 50 cm/s.

14. If a rocket of initial mass m is to rise from its launching pad, its initial thrust must exceed
    (a) $\frac{1}{2}\,mg$.   (b) mg.
    (c) $2\,mg$.   (d) $\frac{1}{2}\,mg^2$.

EXERCISES

1. Is it possible for an object to have more kinetic energy but less momentum than another object? Less kinetic energy but more momentum?

2. When the momentum of an object is doubled in magnitude, what happens to its kinetic energy?

3. When the kinetic energy of an object is doubled in magnitude, what happens to its momentum?

4. When a rocket explodes in midair, how are its total momentum and total kinetic energy affected?

5. How is the principle of conservation of linear momentum related to the definition of mass given in Chapter 3 and to Newton's first law of motion? In what way does this principle go beyond the definition of mass and the first law of motion?

6. (a) When an object at rest breaks up into two parts which fly off, must they move in exactly opposite directions? (b) When a moving object strikes a stationary one and the two do not stick together, must they move off in exactly opposite directions?

7. A railway car is at rest on a frictionless track. A man at one end of the car walks to the other end. (a) Does the car move while he is walking? (b) If so, in which direction? (c) What happens when the man comes to a stop?

8. An empty coal car coasts at a certain speed along a level railroad track without friction. (a) It begins to rain. What happens to the speed of the car? (b) The rain stops, and the collected water gradually leaks out. What happens to the speed of the car now?

9. Find the average momentum of a 70-kg runner who covers 400 m in 45 s.

10. A 1000-kg car strikes a tree at 30 km/h and comes to a stop in 0.15 s. Find its initial momentum and the average force on the car while it is being stopped.

11. A 160,000-kg DC-8 airplane is flying at 870 km/h. (a) Find its momentum. (b) If the thrust its engines develop is 340,000 N, how much time is needed for the airplane to reach this speed starting from rest? Neglect air resistance, changes in altitude, and the fuel consumed by the engines.

12. A 170-g softball moving at 30 m/s is caught by a player. If the average force on the player's hands during the catch is 500 N, find the time in which the ball came to a stop.

13. Water emerging from a hose at a rate of 2 liters/s and a speed of 8 m/s strikes a person. If the water loses all its momentum on impact, find the force on the person. (The mass of 1 liter of water is 1 kg.)

14. The motors of a spacecraft provide a total thrust of 1.8 MN at takeoff. If the exhaust speed is 2.5 km/s, find the rate at which fuel is being consumed.

15. A 30-kg girl who is running at 3 m/s jumps on a stationary 10-kg sled on a frozen lake. How fast does the sled then move?

16. The 176-g head of a golf club is moving at 45 m/s when it strikes a 46-g golf ball and sends it off at 65 m/s. Find the final speed of the clubhead after the impact, assuming that the mass of the club's shaft can be neglected.

17. A hunter has a rifle that can fire 60-g bullets with a speed of 900 m/s. A 40-kg leopard springs at him at 10 m/s. How many bullets must the hunter fire into the leopard in order to stop him in his tracks?

18. A hunter in a rowboat loses the oars and decides to set the boat in motion by firing his rifle astern five times in succession. The total mass of the boat and its contents is 150 kg, the mass of each bullet is 15 g, and the speed of the bullets is 600 m/s. What would the speed of the boat be if there were no water resistance? How far would it go in 1 h?

19. A steel ball bearing is dropped on a steel plate from a height of 2 m. If the coefficient of restitution is 0.97, find the height to which the ball rebounds.

20. A rubber ball is dropped on the ground from a height of 150 cm and on its second rebound reaches a height of 50 cm. Find the coefficient of restitution.

PROBLEMS

1. A certain cannon has a range of 2 km. One day a shell which it fires explodes at the top of its path into two equal fragments, one of which falls vertically downward. If there is no air resistance, how far away from the cannon does the other fragment land?

2. The cannon of the previous problem fires another shell that also explodes into two equal fragments at the top of its path. One of these fragments lands next to the cannon. How far away from the cannon does the other fragment land?

3. A rocket fired from the earth's surface ejects 2% of its mass at a speed of 2 km/s in the first second of its flight. Find the initial acceleration of the rocket.

4. A 0.5-kg stone moving at 4 m/s overtakes a 4-kg lump of clay moving at 1 m/s. The stone becomes embedded in the clay. (a) What is the speed of the composite body after the collision? (b) How much KE is lost in the collision?

5. A 1000-kg car moving east at 80 km/h overtakes a 1500-kg car moving east at 40 km/h and collides with it. The two cars stick together. (a) What is the initial speed of the wreckage? (b) How much KE is lost in the collision?

6. A 1000-kg car moving east at 80 km/h collides head-on with a 1500-kg car moving west at 40 km/h, and the two cars stick together. (a) Which way does the wreckage move and with what initial speed? (b) How much KE is lost in the collision?

7. A neutron of mass 1.67×10^{-27} kg and speed 10^5 km/s collides with a stationary deuteron of mass 3.34×10^{-27} kg. The two particles stick together. What is the speed of the composite particle (called a *triton*)?

8. A neutron of mass 1.67×10^{-27} kg and speed 10^5 m/s collides head-on with a stationary deuteron of mass 3.34×10^{-27} kg. The particles do not stick together, and the deuteron moves off at 6.67×10^4 m/s. What is the speed of the neutron? Is the collision elastic?

9. A 0.5-kg stone moving north at 4 m/s collides with a 4-kg lump of clay moving west at 1 m/s. The stone becomes embedded in the clay. What is the velocity (magnitude and direction) of the composite body after the collision?

10. A 1200-kg car traveling east at 30 km/h collides with an 1800-kg car traveling north at 20 km/h. The cars stick together after the collision. What is the velocity (magnitude and direction) of the wreckage?

11. A billiard ball at rest is struck by another ball of the same mass whose speed is 5 m/s. After an elastic collision the incident ball goes off at an angle of 40° with respect to its original direction of motion and the struck ball goes off at an angle of 50° with respect to this direction. Find the final speeds of both balls.

12. A billiard ball at rest is struck by another billiard ball of the same mass whose speed is 4 m/s. After an elastic

collision the incident ball goes off at an angle of 25° with respect to its original direction of motion. Find the angle the struck ball makes with this direction and the final speeds of both balls.

13. Verify that, in an elastic collision between two bodies moving along the same straight line, the relative velocity with which they move apart after the collision is equal to the relative velocity of approach before it. That is, if v_1 and v_2 are the initial velocities of the bodies and v_1' and v_2' their final velocities, show that $(v_2' - v_1') = -(v_2 - v_1)$ is in agreement with the conservation of both momentum and kinetic energy.

14. A 1-kg ball moving at 2 m/s in the $+x$-direction collides head-on with a 2-kg ball moving at 3 m/s in the $-x$-direction. The coefficient of restitution is 0.8. Find the velocities of the balls after the collision.

15. A ballistic pendulum consists of a wooden block of mass M suspended by long cords from the ceiling. A bullet of mass m and velocity v is fired horizontally into the block, which swings away until its height is the amount h above its original height. Find a formula that gives v in terms of g and the readily measurable quantities m, M, and h.

ANSWERS TO MULTIPLE CHOICE

1. d	**4.** c	**7.** c	**10.** d	**13.** a
2. d	**5.** b	**8.** c	**11.** a	**14.** b
3. c	**6.** b	**9.** d	**12.** c	

ROTATIONAL MOTION

8

Until now we have been discussing only translational motion, motion in which the position of something changes from one moment to the next. But rotational motion is just as common as translational motion. Wheels, pulleys, propellers, drills, and phonograph records all rotate while carrying out their functions. In the atomic world protons, neutrons, and electrons all rotate; and their rotations in part govern how they interact to form nuclei and atoms and how atoms interact to form molecules, liquids, and solids. In this chapter our chief concern will be the rotational motion of a rigid body about a fixed axis. As we shall find, all of the formulas that describe such motion are exact analogs of the formulas we have already used to describe translational motion.

8–1 ANGULAR MEASURE

We are accustomed to measuring angles in degrees, where 1° is defined as 1/360 of a full rotation; that is, a complete turn represents 360°. A more

FIG. 8–1 The ratio of arc to radius gives the magnitude of an angle in radians.

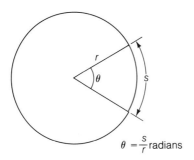

$$\theta = \frac{s}{r}\,\text{radians}$$

The radian is a unit of angular measure

suitable unit for our present purposes is the *radian* (rad). The radian is defined with the help of a circle drawn with its center at the vertex of the angle in question. If the circle's radius is r and the arc cut by the angle is s as in Fig. 8–1, then the angle in radians is given by

$$\theta = \frac{s}{r} = \frac{\text{arc length}}{\text{radius}} \qquad\qquad \textit{Radian measure} \quad (8\text{–}1)$$

That is, the angle θ between two radii of a circle, in radian measure, is the ratio of the arc s to the radius r. Evidently an angle of 1 rad has an arc length that is the same as the radius.

Conversion between degrees and radians

It is easy to find the conversion factor between degrees and radians and vice versa. We observe that there are 360° in a complete circle, while the number of radians in a complete circle is

$$\theta = \frac{s}{r} = \frac{2\pi r}{r} = 2\pi$$

because the circumference of a circle of radius r is $2\pi r$ (Fig. 8–2). Hence

$$360° = 2\pi \text{ rad}$$

from which we find that

$$1° = 0.01745 \text{ rad} \qquad \text{and} \qquad 1 \text{ rad} = 57.30°$$

The radian has no dimensions

The radian is an odd kind of unit because it has no dimensions—an angle expressed in radians is specified by a ratio of lengths, so it is really a pure number. In calculations the unit "rad" is always dropped at the end unless the result is an angular quantity.

FIG. 8–2 $360° = 2\pi$ rad, so 1 rad = 57.30°.

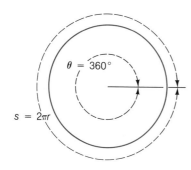

$$\theta = 360°$$

$$s = 2\pi r$$

$$
\begin{array}{rcl}
0° & = & 0 \text{ rad} \\
30° & = & \pi/6 \text{ rad} = 0.524 \text{ rad} \\
45° & = & \pi/4 \text{ rad} = 0.785 \text{ rad} \\
60° & = & \pi/3 \text{ rad} = 1.047 \text{ rad} \\
90° & = & \pi/2 \text{ rad} = 1.571 \text{ rad} \\
180° & = & \pi \text{ rad} = 3.142 \text{ rad} \\
270° & = & 3\pi/2 \text{ rad} = 4.712 \text{ rad} \\
360° & = & 2\pi \text{ rad} = 6.283 \text{ rad} \\
720° & = & 4\pi \text{ rad} = 12.566 \text{ rad}
\end{array}
$$

TABLE 8–1
Degrees to radians

Example A phonograph record 30 cm in diameter turns through an angle of 120°. How far does a point on its rim travel?

Solution First the angle is converted from degrees to radians:

$$
\theta = (120°)\left(0.01745 \frac{\text{rad}}{\text{degree}}\right) = 2.09 \text{ rad}
$$

The radius of the record is 15 cm, and so, from Eq. (8–1),

$$
s = r\theta = (15 \text{ cm})(2.09 \text{ rad}) = 31.4 \text{ cm}
$$
∎

Example A television image with 525 horizontal lines is being displayed on a picture tube whose screen is 50 cm high. If a viewer's eyes can resolve detail to 0.0003 rad—about 1′ (one minute of arc), where $60′ = 1°$—how far away should he be from the screen in order to just be able to see the separate lines?

Solution The distance between adjacent lines is $s = 0.5$ m/525, hence

$$
r = \frac{s}{\theta} = \frac{0.5 \text{ m}}{525 \times 0.0003} = 3.17 \text{ m}
$$
∎

Sometimes it is useful to express angles in radian measure in terms of π itself. For example, an angle of 90° is $\frac{1}{4}$ of a complete circle, and so

Expressing angles in terms of π

$$
90° = \tfrac{1}{4} \text{ circle} \times 2\pi \frac{\text{rad}}{\text{circle}} = \frac{\pi}{2} \text{ rad}
$$

Of course, this has the same numerical value as

$$
90° \times 0.01745 \frac{\text{rad}}{\text{degree}} = 1.571 \text{ rad}
$$

since $\pi/2 = 1.571$. Table 8–1 gives some other examples of radian measure in terms of π and as decimals.

8–2 ANGULAR VELOCITY

If a rotating body turns through the angle θ in the time t, its average *angular velocity* ω (Greek letter *omega*) is

Angular velocity

FIG. 8–3 The angular velocity of a particle in uniform circular motion is $\omega = v/r$.

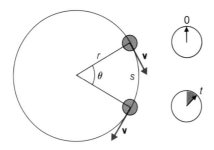

Angular velocity (8–2)

If θ is in radians and t in seconds, which are the usual units for these quantities, the unit of ω is the rad/s.

Two other common units of angular velocity are the revolution per second (rps) and the revolution per minute (rpm), where

Revolutions per second and per minute

$$1 \frac{\text{rev}}{\text{s}} = \left(1 \frac{\text{rev}}{\text{s}}\right)\left(2\pi \frac{\text{rad}}{\text{rev}}\right) = 2\pi \, \text{rad/s} = 6.28 \, \text{rad/s}$$

$$1 \frac{\text{rev}}{\text{min}} = \left(1 \frac{\text{rev}}{\text{min}}\right)\left(2\pi \frac{\text{rad}}{\text{rev}}\right)\left(\frac{1}{60 \, \text{s/min}}\right) = \frac{\pi}{30} \, \text{rad/s} = 0.105 \, \text{rad/s}$$

Relation between linear speed and angular velocity

Let us consider a particle moving with the uniform speed v in a circle of radius r, as in Fig. 8–3. This particle travels the distance $s = vt$ in the time t. The angle through which it moves in that time is

$$\theta = \frac{s}{t} = \frac{vt}{r}$$

so that its angular velocity is

$$\omega = \frac{\theta}{t} = \frac{vt}{rt}$$

or

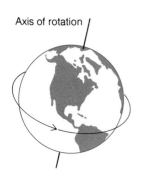

Axis of rotation

$$\omega = \frac{v}{r} \qquad\qquad\qquad \textit{Angular velocity} \quad (8\text{–}3)$$

$$\text{Angular velocity} = \frac{\text{linear speed}}{\text{path radius}}$$

The above relationship can be written in another way:

$$v = \omega r \qquad\qquad\qquad (8\text{–}4)$$

Linear speed = angular velocity × path radius

The formulas of this section are valid only when ω is expressed in radian measure.

FIG. 8–4 The axis of rotation is that line of particles which does not move in a body rotating in space. (It may be a line in space.)

The *axis of rotation* of a rigid body turning in place is that line of particles which does not move (Fig. 8–4). Sometimes the axis of rotation is a line in space. All other

Axis of rotation

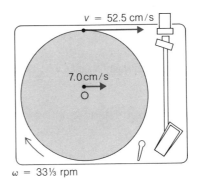

FIG. 8–5

particles of the body move in circles about the axis. Since $v = \omega r$, the farther a particle is from the axis, the greater its linear velocity, although all the particles of the body (except those on the axis) have the same angular velocity.

Example Find the linear speeds of points 2 cm and 15 cm from the axis of a phonograph record rotating at $33\frac{1}{3}$ rpm (Fig. 8–5).

Solution The angular velocity of the record is

$$\omega = (33\tfrac{1}{3}\,\text{rpm})\left(0.105\,\frac{\text{rad/s}}{\text{rpm}}\right) = 3.50\,\text{rad/s}$$

Hence a point 2 cm from the axis has a linear velocity of

$$v = \omega r = \left(3.5\,\frac{\text{rad}}{\text{s}}\right)(2\,\text{cm}) = 7.0\,\text{cm/s}$$

while a point on the record's rim, where $r = 15$ cm, has a velocity of

$$v = \left(3.5\,\frac{\text{rad}}{\text{s}}\right)(15\,\text{cm}) = 52.5\,\text{cm/s}$$ ■

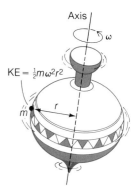

FIG. 8–6 The kinetic energy of each particle of a rotating body depends upon the square of its distance r from the axis of rotation.

8–3 ROTATIONAL KINETIC ENERGY

A rotating body possesses kinetic energy because its constituent particles are in motion, even though the body as a whole remains in place. The speed of a particle that is the distance r from the axis of a rigid body rotating with the angular velocity ω is, as we know, $v = \omega r$ (Fig. 8–6). If the particle's mass is m, its kinetic energy is therefore

$$\text{KE} = \tfrac{1}{2}mv^2 = \tfrac{1}{2}m\omega^2 r^2 \qquad (8\text{–}5)$$

KE of a particle in a rotating body

The body consists of numerous particles which need not have the same mass or be the same distance from the axis. However, all the particles have the common angular velocity ω, which enables us to formulate the mechanics of rotating bodies in a very convenient manner. We start by noting that the total kinetic energy of all the particles may be written

$$\text{KE} = \Sigma \tfrac{1}{2}mv^2 = \tfrac{1}{2}(\Sigma mr^2)\omega^2 \qquad (8\text{–}6)$$

KE of the entire rotating body

where the symbol Σ means, as mentioned before, "sum of." Equation (8–6) states that the kinetic energy of a rotating rigid body is equal to one-half the sum of the mr^2 values of its constituent particles multiplied by the square of its angular velocity ω.

The quantity

$$I = \Sigma mr^2 \qquad\qquad \textit{Moment of inertia} \quad (8\text{–}7)$$

Moment of inertia is rotational analog of mass

is known as the *moment of inertia* of the body. It has the same value regardless of the body's state of motion. The farther a given particle is from the axis of rotation, the faster it moves and the greater is its contribution to the kinetic energy of the body. The moment of inertia of a body depends upon the way in which its mass is distributed relative to its axis of rotation; it is perfectly possible for one body to have a greater moment of inertia than another even though its mass may be much the smaller of the two.

The kinetic energy of a body of moment of inertia I rotating with the angular velocity ω is therefore

$$KE = \tfrac{1}{2}I\omega^2 \qquad\qquad \textit{Rotational kinetic energy} \quad (8\text{–}8)$$

KE of rotating body in terms of moment of inertia

Evidently the rotational analog of mass is moment of inertia, just as the rotational analog of velocity is angular velocity. We shall find further support for the correspondence of mass and moment of inertia later in this chapter.

8–4 MOMENT OF INERTIA

How to calculate moments of inertia

A rigid body may be considered to be made up of a large number of separate particles whose masses are m_1, m_2, m_3, and so on (Fig. 8–7). To compute the moment of inertia of such a body about a specified axis, we multiply the mass of each of these particles by the square of its distance from the axis (r_1^2, r_2^2, r_3^2, and so on) and add all the mr^2 values. That is,

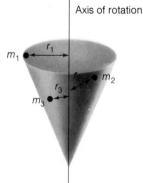

Axis of rotation

$$I = \Sigma mr^2 = m_1 r_1^2 + m_2 r_2^2 + m_3 r_3^2 + \cdots \qquad (8\text{–}9)$$

The unit of I is the $kg \cdot m^2$ in the metric system and the $slug \cdot ft^2$ in the British system.

Example Determine the moment of inertia of a thin ring of mass M and average radius R about an axis passing through its center and perpendicular to the plane in which it lies.

Solution As in Fig. 8–8, we proceed by subdividing the ring into n segments, each of which is at a distance R from the axis. Hence

$$I = m_1 R^2 + m_2 R^2 + m_3 R^2 + \cdots + m_n R^2$$

$$= (m_1 + m_2 + m_3 + \cdots + m_n)R^2$$

FIG. 8–7 A rigid body consists of a large number of particles each of which has a certain mass m and distance r from the axis.

But the sum of the masses of the segments is the same as the total mass M of the ring, and so

$$I = MR^2 \qquad\qquad \blacksquare$$

When a body consists of a continuous distribution of matter, the more particles we imagine it to contain, the more accurate will be our value of its moment of inertia. While I can be calculated for a few simple bodies without difficulty by Eq. (8–9), in general either considerable labor or the use of integral calculus is required. Figure 8–9 gives the moments of inertia of several regularly shaped bodies in terms of the total mass M and dimensions of each. In Sec. 11–6 we shall learn of a simple experimental way to determine the moment of inertia of an object whose shape is too complex to permit I to be calculated readily.

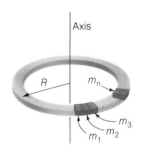

FIG. 8–8

Example The earth's mass is 6×10^{24} kg, and its radius is 6.4×10^{6} m. Find its rotational kinetic energy.

Solution Considering the earth to be a uniform sphere, its moment of inertia is

$$I = \tfrac{2}{5}MR^2 = \tfrac{2}{5}(6 \times 10^{24}\,\text{kg})(6.4 \times 10^{6}\,\text{m})^2 = 9.8 \times 10^{37}\,\text{kg}\cdot\text{m}^2$$

The angular velocity of the earth is 7.3×10^{-5} rad/s, corresponding to one rotation per day. Hence its rotational kinetic energy is

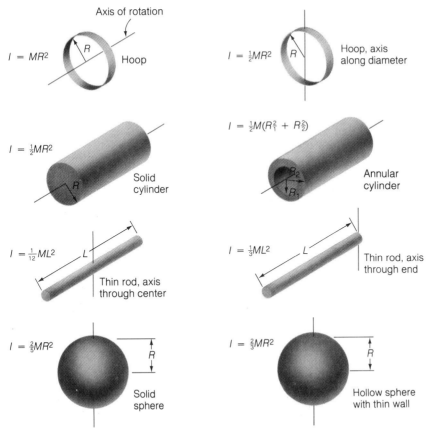

FIG. 8–9 Moments of inertia of various bodies each of mass M, about indicated axes.

$$KE = \tfrac{1}{2}I\omega^2 = \tfrac{1}{2}(9.8 \times 10^{37}\,\text{kg}\cdot\text{m}^2)\left(7.3 \times 10^{-5}\,\frac{\text{rad}}{\text{s}}\right)^2 = 2.6 \times 10^{29}\,\text{J}$$

The kinetic energy of the earth's orbital motion is about 10,000 times greater than that of its rotational motion. ■

8–5   COMBINED TRANSLATION AND ROTATION

Total KE is sum of translational KE and rotational KE

When a rigid body is both moving through space and undergoing rotation, its total kinetic energy is the sum of its translational and rotational kinetic energies. The translational KE is calculated on the basis that the body is a particle whose linear speed is the same as that of the body's center of gravity; the rotational KE is calculated on the basis that the body is rotating about an axis that passes through the center of gravity. Thus

$$KE = KE_{\text{translation}} + KE_{\text{rotation}} = \tfrac{1}{2}mv^2 + \tfrac{1}{2}I\omega^2 \qquad (8\text{–}10)$$

where m is the body's mass, v is the velocity of its center of gravity, I is its moment of inertia about an axis through the center of gravity, and ω is its angular velocity about that axis.

Example   Consider a cylinder of radius R and mass m that is poised at the top of an inclined plane (Fig. 8–10). Will it have a greater speed at the bottom if it slides down without friction or if it rolls down?

Solution   In the first case, we set the cylinder's initial potential energy of mgh equal to its final kinetic energy of $\tfrac{1}{2}mv^2$, and find that

$$PE = KE$$
$$mgh = \tfrac{1}{2}mv^2$$
$$v = \sqrt{2gh}$$

In the second case the cylinder has both translational and rotational kinetic energy at the bottom, so that

FIG. 8–10

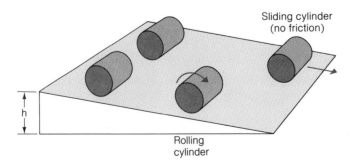

Sliding cylinder
(no friction)

h

Rolling
cylinder

PE = KE

$$mgh = \tfrac{1}{2}mv^2 + \tfrac{1}{2}I\omega^2$$

The moment of inertia of a cylinder is $I = \tfrac{1}{2}mR^2$; if it rolls without slipping, its linear and angular velocities are related by the formula $\omega = v/R$. Hence

$$mgh = \tfrac{1}{2}mv^2 + \tfrac{1}{2}(\tfrac{1}{2}mR^2)\left(\frac{v^2}{R^2}\right) = \tfrac{1}{2}mv^2 + \tfrac{1}{4}mv^2 = \tfrac{3}{4}mv^2$$

$$v = \sqrt{\tfrac{4}{3}gh}$$

The cylinder moves more slowly when it rolls down the plane than when it slides without friction because some of the available energy is absorbed by its rotation. ∎

8–6 ANGULAR ACCELERATION

A rotating body need not have a uniform angular velocity ω, just as a moving particle need not have a uniform linear velocity v. If the angular velocity of a body changes by an amount $\Delta\omega$ in the time interval Δt, its average *angular acceleration* α (Greek letter "alpha") is

$$\alpha = \frac{\Delta\omega}{\Delta t} \qquad\qquad \textit{Angular acceleration} \quad (8\text{–}11)$$

Angular acceleration is analog of linear acceleration

The unit of angular acceleration is the rad/s^2.

A particle moving in a circle of radius r that experiences an angular acceleration α also experiences a linear acceleration a_T tangential to its path, since its orbital speed is changing (Fig. 8–11). From the definition $\omega = v/r$, if r is constant

Tangential acceleration

$$\Delta\omega = \frac{\Delta v}{r} \qquad \text{and} \qquad \alpha = \frac{\Delta\omega}{\Delta t} = \frac{\Delta v}{r\Delta t}$$

Because the magnitude of the particle's linear acceleration along the direction of its velocity $\mathbf{v}$ is $a_T = \Delta v/\Delta t$, we see that

$$\alpha = \frac{a_T}{r} \tag{8–12}$$

$$\text{Angular acceleration} = \frac{\text{tangential acceleration}}{\text{path radius}}$$

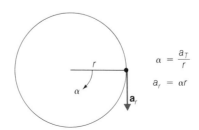

FIG. 8–11 The angular and tangential accelerations of a particle moving in a circle are proportional to each other.

$$\alpha = \frac{a_T}{r}$$

$$a_r = \alpha r$$

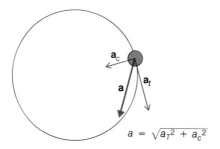

FIG. 8–12 The tangential and centripetal accelerations of a particle in circular motion are always perpendicular.

$$a = \sqrt{a_T{}^2 + a_c{}^2}$$

Conversely,

$$a_T = \alpha r \tag{8-13}$$

Tangential acceleration = angular acceleration × path radius

We must be careful to distinguish between the *tangential acceleration* a_T of a particle, which represents a change in its speed, and its *centripetal acceleration* a_c, which represents a change in its direction of motion. A particle in circular motion with the speed v has, as we know, the centripetal acceleration

$$a_C = \frac{v^2}{r}$$

Tangential and centripetal accelerations are always perpendicular to each other

directed toward the center of its circular path. The accelerations $\mathbf{a}_T$ and $\mathbf{a}_C$ are therefore always perpendicular (Fig. 8–12). *All* particles in circular motion have centripetal accelerations. Only those particles whose speed changes in magnitude, however, have tangential accelerations. The centripetal acceleration of a particle moving in a circle of radius r can be expressed in terms of its angular velocity ω as

$$a_C = \omega^2 r \qquad\qquad \textit{Centripetal acceleration} \quad (8-14)$$

because $v = \omega r$.

Example   Find the total linear acceleration of a particle moving in a circle of radius 0.4 m, at the instant when the angular velocity is 2 rad/s and the angular acceleration is 5 rad/s^2.

Solution   From Eqs. (8–13) and (8–14), for the particle's tangential and centripetal accelerations, we have

$$a_T = \alpha r = \left(5\ \frac{\text{rad}}{\text{s}^2}\right)(0.4\,\text{m}) = 2\,\text{m/s}^2$$

$$a_C = \omega^2 r = \left(2\ \frac{\text{rad}}{\text{s}}\right)^2 (0.4\,\text{m}) = 1.6\,\text{m/s}^2$$

We must add a_T and a_C vectorially, since they are vector quantities and are in different directions. As shown in Fig. 8–12, the magnitude a of the vector sum $\mathbf{a}$ of $\mathbf{a}_T$ and $\mathbf{a}_C$ is

$$a = \sqrt{a_T^2 + a_c^2} = \sqrt{2^2 + 1.6^2}\,\text{m/s}^2 = 2.6\,\text{m/s}^2 \qquad \blacksquare$$

Let us consider a rigid body whose angular acceleration has the constant value α. The body's angular velocity is ω_0 at $t = 0$ and at the later time $t = t$ it has changed to

$$\omega_f = \omega_0 + \alpha t \qquad \begin{array}{c}\textit{Angular velocity under}\\ \textit{constant acceleration}\end{array} \quad (8\text{--}15)$$

Angular displacement and velocity under constant acceleration

In Chapter 2 we obtained the formula

$$s = v_0 t + \tfrac{1}{2}\alpha t^2$$

for the displacement of an accelerated body. By means of the same reasoning, the angular displacement θ of a rigid body undergoing an angular acceleration turns out to be

$$\theta = \omega_0 t + \tfrac{1}{2}\alpha t^2 \qquad\qquad \textit{Angular displacement} \quad (8\text{--}16)$$

The angular analog of the linear formula

$$v_f^2 = v_0^2 + 2as$$

is the equally useful

$$\omega_f^2 = \omega_0^2 + 2\alpha\theta \qquad \begin{array}{c}\textit{Angular velocity under}\\ \textit{constant acceleration}\end{array} \quad (8\text{--}17)$$

Example   The speed of a motor increases from 120 rad/s to 180 rad/s in 20 s. How many revolutions does it make in this period of time?

Solution   The angular acceleration of the motor is

$$\alpha = \frac{\Delta\omega}{\Delta t} = \frac{\omega_f - \omega_0}{\Delta t} = \frac{(180 - 120)\ \text{rad/s}}{20\,\text{s}} = 3.0\,\text{rad/s}^2$$

We can use either Eq. (8–16) or (8–17) to find θ, since we know ω_0, ω_f, α, and t. From Eq. (8–16) we have

$$\theta = \omega_0 t + \tfrac{1}{2}\alpha t^2 = \left(120\ \frac{\text{rad}}{\text{s}}\right)(20\,\text{s}) + \tfrac{1}{2}\left(3.0\ \frac{\text{rad}}{\text{s}^2}\right)(20\,\text{s})^2 = 3000\,\text{rad}$$

If we were to use Eq. (8–17), we would rewrite it as

$$\theta = \frac{\omega_f^2 - \omega_0^2}{2\alpha}$$

and get the same result. Since there are 2π rad in a revolution,

$$\theta = \frac{3000\,\text{rad}}{2\pi\ \text{rad/rev}} = 477\,\text{revolutions} \qquad \blacksquare$$

8–7   TORQUE AND ANGULAR ACCELERATION

According to Newton's second law of motion, a net force applied to a body causes it to be accelerated. What can cause a body capable of rotation to experience an angular acceleration?

To fix our ideas, let us look once more at a single particle of mass m restricted to motion in a circle of radius r (Fig. 8–13). A force $\mathbf{F}$ that acts upon the particle tangent to the particle's path gives it the acceleration a_T according to the formula

$$F = ma_T \tag{8–18}$$

The tangential acceleration a_T here is equal to αr, and so

$$F = mr\alpha$$

Multiplying both sides of this equation by r gives

$$Fr = mr^2\alpha \tag{8–19}$$

Torque produces angular acceleration

We recognize Fr as the torque τ of the force F about the axis of the particle's rotation (see Sec. 4–2) and mr^2 as the particle's moment of inertia I. Equation (8–19) therefore states that

$$\tau = I\alpha \tag{8–20}$$

Torque = moment of inertia × angular acceleration

Although we derived Eq. (8–20) for the case of a single particle, it is also valid for any rotating body, provided that the torque and moment of inertia are both calculated about the same axis.

Torque is rotational analog of force

The formula $\tau = I\alpha$ is the fundamental law of motion for rotating bodies in the same sense that $F = ma$ is the fundamental law of motion for bodies moving through space. In rotational motion, torque plays the same role that force does in translational motion. The angular acceleration experienced by a body when the torque τ acts upon it is

$$\alpha = \frac{\tau}{I}$$

proportional to the torque and inversely proportional to the body's

FIG. 8–13

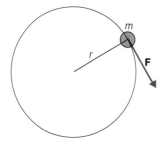

Example A 2-kg grindstone 10 cm in radius is turning at 120 rad/s. The motor is switched off and a chisel is pressed against the grindstone with a force whose tangential component is 2 N. How long will it take the grindstone to come to a stop?

Solution The grindstone is a solid cylinder and so its moment of inertia is

$$I = \tfrac{1}{2}MR^2 = \tfrac{1}{2}(2\,\text{kg})(0.1\,\text{m})^2 = 0.01\,\text{kg}\cdot\text{m}^2$$

The torque the chisel exerts on the grindstone is

$$\tau = -Fr = -(2\,\text{N})(0.1\,\text{m}) = -0.2\,\text{N}\cdot\text{m}$$

where the minus sign is required because the torque acts to slow down the grindstone. The angular acceleration of the grindstone is

$$\alpha = \frac{\tau}{I} = \frac{-0.2\,\text{N}\cdot\text{m}}{0.01\,\text{kg}\cdot\text{m}^2} = -20\,\text{rad/s}^2$$

We now call upon Eq. (8–15), $\omega_f = \omega_0 + \alpha t$, and set $\omega_f = 0$ to correspond to the grindstone coming to a stop. The result is

$$\alpha t = -\omega_0$$
$$t = \frac{-\omega_0}{\alpha} = \frac{-120\,\text{rad/s}}{-20\,\text{rad/s}^2} = 6\,\text{s}$$ ∎

Example Figure 8–14 shows a 12-kg block, A, and a 30-kg block, B, connected by a string that passes over a pulley, just as in the illustrative problem at the end of Sec.

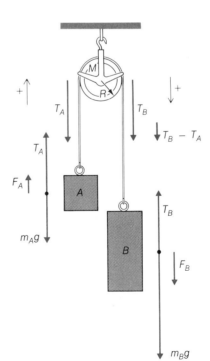

FIG. 8–14 The inertia of the pulley reduces the accelerations of the suspended blocks.

3–5 (Fig. 3–16). Here, however, the pulley is not massless but is a uniform disk whose mass is 10 kg. Find the accelerations of the blocks.

Solution In the earlier problem, where the pulley was massless, the string had the same tension on both sides of the pulley. Here, owing to the inertia of the pulley, the tensions are different. The net force F_A on block A is the difference between the tension T_A in the string supporting it and its weight of $m_A g$, so if its acceleration is a (considered positive as in the figure),

$$F_A = T_A - m_A g = m_A a$$

$$T_A = m_A(g + a)$$

The net force $F_B = m_B g - T_B$ on block B is directed downward (considered positive), so its acceleration is also a and

$$F_B = m_B g - T_B = m_B a$$

$$T_B = m_B(g - a)$$

The difference $T_B - T_A$ between the tensions equals the net force on the pulley's rim, which means that the torque on the pulley is

$$\tau = (T_B - T_A)R = m_B(g - a)R - m_A(g + a)R$$

$$= (m_B - m_A)gR - (m_A + m_B)aR$$

From Fig. 8–9 the moment of inertia of the pulley is $I = \frac{1}{2}MR^2$. Since $\tau = I\alpha$ and $\alpha = a/R$, we have

$$(m_B - m_A)gR - (m_A + m_B)aR = I\alpha = \frac{1}{2}MR^2\left(\frac{a}{R}\right) = \frac{1}{2}MRa$$

from which we obtain

$$a = \frac{(m_B - m_A)g}{M/2 + (m_A + m_B)} = \frac{18\,\text{kg} \times 9.8\,\text{m/s}^2}{(10\,\text{kg})/2 + 42\,\text{kg}} = 3.75\,\text{m/s}^2$$

This is less than the 4.2 m/s² acceleration found earlier when the pulley was considered massless. It is not necessary here to know the pulley's radius. ■

Center of gravity and line of action of applied force

When a net force acts on a body that is able to move freely, the body will experience both linear and angular accelerations unless the line of action of the force passes through the body's center of gravity (Fig. 8–15). In the latter case the body will be in rotational equilibrium, and if it was not rotating initially, it will keep its original orientation during its motion.

8–8 POWER

Mechanical energy is usually transmitted by rotary motion. The power output of almost every modern engine emerges via a rotating shaft, and more often than not this power is expended in some form of rotation as well: the tires of a car, the propeller of a ship,

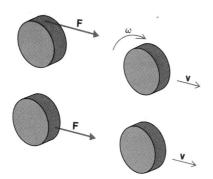

FIG. 8–15 A force whose line of action passes through the center of gravity of a body cannot cause it to rotate.

the bit of a drill, the vanes of a centrifugal pump, and the rotor of a dynamo all function by turning.

The relationships among power, torque, and angular velocity can be derived directly from the definitions of work and power. Let us consider a shaft of radius r on which a tangential force F acts, as in Fig. 8–16. After a time t the shaft has turned through the angle θ, the point at which the force is applied has moved through the distance s, and the force has done the amount of work **Work and power in angular motion**

$$W = Fs = Fr\theta$$

Power is the rate at which work is done, and thus

$$P = \frac{W}{t} = \frac{Fr\theta}{t} = Fr\,\frac{\theta}{t}$$

We recognize that

$$Fr = \tau = \text{torque applied to shaft}$$
$$\frac{\theta}{t} = \omega = \text{angular velocity of shaft}$$

and therefore

$$W = \tau\theta \qquad\qquad \textit{Rotational work} \quad (8\text{–}21)$$

Work $=$ torque $\times$ angular displacement

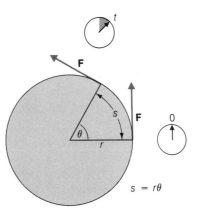

FIG. 8–16 The power transmitted by a rotating shaft is the product of the torque τ on the shaft and its angular velocity ω, hence $P = \tau\omega$.

$s = r\theta$

$$P = \tau\omega \qquad\qquad\qquad \textit{Rotational power} \quad (8\text{--}22)$$

Power = torque × angular velocity

When energy is to be transmitted at a certain rate P by means of a rotating shaft, the higher the angular velocity the lower the torque needed, and vice versa. Equation (8–22) is the rotational analog of the formula

$$P = Fv$$

Power = force × speed

Example   A truck engine develops 230 hp at 5000 rpm. How much torque is exerted on the crankshaft?

Solution   We begin by converting 230 hp and 5000 rpm to their equivalents in watts and radians per second:

$$P = (230\,\text{hp})(746\,\text{W/hp}) = 1.72 \times 10^5\,\text{W}$$

$$\omega = (5000\,\text{rpm})\left(0.105\,\frac{\text{rad/s}}{\text{rpm}}\right) = 525\,\text{rad/s}$$

From Eq. (8–22) we obtain

$$\tau = \frac{P}{\omega} = \frac{1.72 \times 10^5\,\text{W}}{525\ \text{rad/s}} = 328\,\text{N} \cdot \text{m} \qquad\qquad \blacksquare$$

8–9   ANGULAR MOMENTUM

The rotational analog of linear momentum is *angular momentum*. The angular momentum L of a body depends upon its moment of inertia I and angular velocity ω in the same way that its linear momentum depends upon its mass m and linear velocity v:

$$L = I\omega \qquad\qquad\qquad \textit{Angular momentum} \quad (8\text{--}23)$$

Angular momentum = moment of inertia × angular velocity

When there is no net torque on a rigid body, both its angular velocity ω and its angular momentum are constant. A deeper analysis shows that the angular momentum of a body does not change in the absence of a net torque on it even if it is *not* a rigid body, but is so altered during its motion that its moment of inertia changes. In a situation of this kind the angular velocity of the body also changes so that L stays the same. Thus we have the useful theorem of *conservation of angular momentum*:

Conservation of angular momentum

When the sum of the external torques acting upon a system of particles equals zero, the total angular momentum of the system remains constant.

A skater or ballet dancer doing a spin capitalizes upon conservation of angular momentum. In Fig. 8–17 a skater is shown starting her spin with her arms and one leg outstretched. By bringing her arms and extended leg inward, she reduces her moment of inertia considerably and consequently spins faster.

I large, ω small   *I* small, ω large

Example   A skater has the moment of inertia 150 kg·m² when her arms are out-stretched and 50 kg·m² when her arms are brought to her sides. She starts to spin at the rate of 1 revolution per second when her arms are outstretched, and then pulls her arms to her sides. (a) What is her final angular velocity? (b) What are her initial and final kinetic energies?

Solution   (a) Since the conversion factors would only cancel, it is simplest to use the rev/s as the unit of angular velocity. Since $I_1 = 150$ kg·m², $I_2 = 50$ kg·m², and $\omega_1 = 1$ rev/s, from conservation of angular momentum

$$I_1\omega_1 = I_2\omega_2$$

$$\omega_2 = \frac{I_1}{I_2}\omega_1 = \left(\frac{150\,\text{kg}\cdot\text{m}^2}{50\,\text{kg}\cdot\text{m}^2}\right)(1\,\text{rev/s}) = 3\,\text{rev/s}$$

The skater spins three times faster than she did with her arms outstretched.

(b) For calculating KE we must use the rad/s as the unit of angular velocity. Since 1 rev = 2π rad, $\omega_1 = 2\pi$ rad/s and $\omega_2 = 6\pi$ rad/s. The skater's initial and final kinetic energies are therefore

$$\text{KE}_1 = \tfrac{1}{2}I_1\omega_1{}^2 = \tfrac{1}{2}(150\,\text{kg}\cdot\text{m}^2)(2\pi\,\text{rad/s})^2 = 2961\,\text{J}$$

$$\text{KE}_2 = \tfrac{1}{2}I_2\omega_2{}^2 = \tfrac{1}{2}(50\,\text{kg}\cdot\text{m}^2)(6\pi\,\text{rad/s})^2 = 8883\,\text{J}$$

The increase in KE is due to the work the skater had to do to pull in her arms.   ■

Because angular momentum is a vector quantity (Fig. 8–18), a torque must be applied to change the orientation of the axis of rotation of a spinning body as well as   **Spin stabilization**

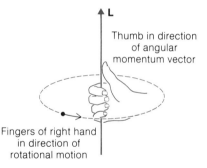

L

Thumb in direction of angular momentum vector

Fingers of right hand in direction of rotational motion

FIG. 8–18 Angular momentum is a vector quantity whose direction is given by the right-hand rule shown here.

FIG. 8-19 The faster a top spins, the more stable it is. When its angular momentum has been lost through friction, the top falls over.

Angular momentum conservation and the laws of nature

to change the magnitude of its angular velocity. The greater the magnitude of **L**, the more torque is needed for it to deviate from its original direction. This is the principle behind the spin stabilization of projectiles such as footballs and bullets. Such projectiles are set spinning about axes in their directions of motion so that they do not tumble and thereby offer excessive air resistance. A top is another illustration of the vector nature of angular momentum. A stationary top set on its tip falls over at once, but a rotating top stays upright until its angular momentum is dissipated by friction between its tip and the ground (Fig. 8-19).

A *gyroscope* is a disk with a high moment of inertia whose axis is mounted on gimbals, as in Fig. 8-20. When it is set in motion, conservation of angular momentum leads to the disk's maintaining a fixed orientation in space regardless of how its support is moved. Thus a gyroscope can be used on an airplane as an artificial horizon when the true horizon is not visible, and as the basis of a compass that operates independently of the earth's magnetic field.

Like the conservation principles of energy and of linear momentum, the conservation of angular momentum turns out to be a consequence of a symmetry property of the universe. If the laws of nature are independent of direction in space—that is, if the laws of nature are the same regardless of how an observer is oriented with respect to an event of some kind—then angular momentum must be conserved in all interactions, as observed.

8-10   COMPARISON WITH LINEAR MOTION

As we have seen, there are many points of correspondence between angular and linear motion. Table 8-2 lists the principal ones. Although the symbols are different, the formulas are the same because the basic concepts involved are so closely related.

FIG. 8-20 A gyroscope.

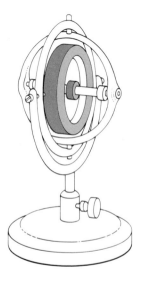

	Linear quantity		Angular quantity	
Distance	$s = v_0 + \frac{1}{2}at^2$	Angle	$\theta = \omega_0 t + \frac{1}{2}\alpha t^2$	
Speed	$v = v_0 + at$	Angular velocity	$\omega = \omega_0 + \alpha t$	
	$v^2 = v_0^2 + 2as$		$\omega^2 = \omega_0^2 + 2\alpha\theta$	
Acceleration	a	Angular acceleration	α	
Mass	m	Moment of inertia	I	
Force	$F = ma$	Torque	$\tau = I\alpha$	
Momentum	$p = mv$	Angular momentum	$L = I\omega$	
Work	$W = Fs$	Work	$W = \tau\theta$	
Power	$P = Fv$	Power	$P = \tau\omega$	
Kinetic energy	$KE = \frac{1}{2}mv^2$	Kinetic energy	$KE = \frac{1}{2}I\omega^2$	

TABLE 8–2
Comparison of linear and angular quantities

8–11 KEPLER'S SECOND LAW

Kepler's laws of planetary motion were mentioned in Sec. 5–9. The second of these laws states that the speed of a planet in its elliptical orbit around the sun varies so that a radius vector from the sun to it sweeps out equal areas in equal times (Fig. 8–21). Let us see how this law follows from the conservation of angular momentum, which must be obeyed by a planet since the gravitational force the sun exerts on it acts along the line joining the centers of the two bodies and so results in no torque on the planet.

Kepler's equal-area law is a consequence of angular momentum conservation

Figure 8–22 shows the position of a planet at two different times Δt apart. In this interval the planet's radius vector moves through the angle $\Delta\theta = \omega \, \Delta t$, where ω is the planet's angular velocity in that part of its orbit. In the same time interval, the radius vector increases from r to $r + \Delta r$, so that it sweeps out a triangle whose base is $r + \Delta r$ and whose altitude is $r \, \Delta\theta$. The area ΔA of this triangle is

$$\text{Area} = \tfrac{1}{2}(\text{base} \times \text{altitude})$$

$$\Delta A = \tfrac{1}{2}(r + \Delta r)(r \, \Delta\theta) = \tfrac{1}{2}r^2 \, \Delta\theta + \tfrac{1}{2}r \, \Delta r \, \Delta\theta$$

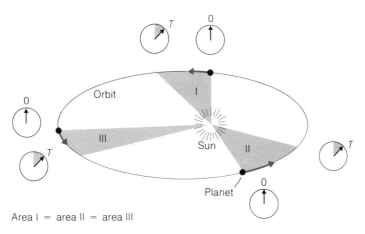

Area I = area II = area III

FIG. 8–21 Kepler's second law.

FIG. 8–22 Kepler's second law is a consequence of the conservation of angular momentum.

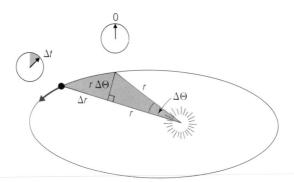

Because both Δr and $\Delta\theta$ are small, the second term of ΔA, which contains their product $\Delta r\,\Delta\theta$, is much smaller than the first term and can be ignored. Hence

$$\Delta A = \tfrac{1}{2}r^2\Delta\theta$$

The rate at which the radius vector sweeps out area is therefore

$$\frac{\Delta A}{\Delta t} = \tfrac{1}{2}r^2\frac{\Delta\theta}{\Delta t} = \tfrac{1}{2}r^2\omega$$

The moment of inertia of the planet about an axis through the sun is $I = mr^2$, so its angular momentum is

$$L = I\omega = mr^2\omega$$

and the law of areas becomes just

$$\frac{\Delta A}{\Delta t} = \frac{L}{2m}$$

The orbital speed of a planet is greatest when it is nearest the sun

Because L must be constant, $\Delta A/\Delta t$ must also be constant, and planets move faster when they are near the sun than they do when they are far from it. In the case of the earth, the difference in speed is about 1 km/s, somewhat over 3 percent.

IMPORTANT TERMS

The **radian** is a unit of angular measure equal to 57.30°. If a circle is drawn whose center is at the vertex of an angle, the angle in radian measure is equal to the ratio between the arc of the circle cut by the angle and the radius of the circle. A full circle contains 2π radians.

The **angular velocity** ω of a rotating body is the angle through which it turns per unit time. The **angular acceleration** α of a rotating body is the rate of change of its angular velocity with respect to time.

The **axis of rotation** of a rigid body turning in place is that line of particles which does not move.

All particles in circular motion experience centripetal accelerations, but only those particles whose angular velocity changes have **tangential accelerations.**

The **moment of inertia** I of a body about a given axis is the rotational analog of mass in linear motion. Its value depends upon the way in which the mass of the body is distributed about the axis.

The **angular momentum** L of a rotating body is the product $I\omega$ of its moment of inertia and angular velocity. The principle of **conservation of angular momentum** states that the total angular momentum of a system of particles remains constant when no net external torque acts upon it.

IMPORTANT FORMULAS

Radian measure: $\theta = \dfrac{s}{r}$

Angular velocity: $\omega = \dfrac{\theta}{t}$

Linear speed: $v = \omega r$

Moment of inertia: $I = \Sigma mr^2$

Rotational kinetic energy: $KE = \frac{1}{2}I\omega^2$

Motion under constant angular acceleration:

$$\theta = \omega_0 t + \frac{1}{2}\alpha t^2$$
$$\omega_f = \omega_0 + \alpha t$$
$$\omega_f^2 = \omega_0^2 + 2\alpha\theta$$

Angular acceleration: $\alpha = \dfrac{\Delta\omega}{\Delta t} = \dfrac{a_T}{r}$

Torque: $\tau = I\alpha$

Work and power:

$$W = \tau\theta$$
$$P = \tau\omega$$

Angular momentum: $L = I\omega$

MULTIPLE CHOICE

1. In a rigid object undergoing uniform circular motion, a particle that is a distance R from the axis of rotation
(a) has an angular velocity proportional to R.
(b) has an angular velocity inversely proportional to R.
(c) has a linear speed proportional to R.
(d) has a linear speed inversely proportional to R.

2. The centripetal acceleration of a particle in circular motion
(a) is less than its tangential acceleration.
(b) is equal to its tangential acceleration.
(c) is more than its tangential acceleration.
(d) may be more or less than its tangential acceleration.

3. The rotational analog of force in linear motion is
(a) moment of inertia.
(b) angular momentum.
(c) torque.
(d) weight.

4. The rotational analog of mass in linear motion is
(a) moment of inertia.
(b) angular momentum.
(c) torque.
(d) angular velocity.

5. A quantity not directly involved in the rotational motion of an object is
(a) mass.   (b) moment of inertia.
(c) torque.   (d) angular velocity.

6. All rotating objects at sea level that have the same mass and angular velocity also have the same
(a) angular momentum.
(b) moment of inertia.
(c) kinetic energy.
(d) potential energy.

7. The moment of inertia of an object does not depend upon
(a) its mass.
(b) its size and shape.
(c) its angular velocity.
(d) the location of the axis of rotation.

8. Of the following properties of a yo-yo moving in a circle, the one that does not depend upon the radius of the circle is the yo-yo's
(a) angular velocity.
(b) angular momentum.
(c) linear velocity.
(d) centripetal acceleration.

9. A yo-yo being swung in a circle need not possess
(a) angular velocity.
(b) angular momentum.
(c) angular acceleration.
(d) centripetal acceleration.

10. The total angular momentum of a system of particles
(a) remains constant under all circumstances.
(b) changes when a net external force acts upon the system.
(c) changes when a net external torque acts upon the system.
(d) may or may not change under the influence of a net external torque, depending on the direction of the torque.

11. A full circle contains
(a) $\pi/4$ radians.
(b) $\pi/2$ radians.
(c) π radians.
(d) 2π radians.

12. A radian is approximately equal to
   (a) 1°.
                    (b) 5°.
   (c) 12°.
                  (d) 60°.

13. A wheel is 1 m in diameter. When it makes 30 rev/min, the linear speed of a point on its circumference is
   (a) $\pi/2$ m/s.
   (b) π m/s.
   (c) 30π m/s.
   (d) 60π m/s.

14. An object undergoes a uniform angular acceleration. In the time t since the object started rotating from rest, the number of turns it makes is proportional to
   (a) $\sqrt{t}$.
                 (b) t.
   (c) t^2.
                 (d) t^3.

15. A wheel that starts from rest has an angular velocity of 20 rad/s after being uniformly accelerated for 10 s. The total angle through which it has turned in these 10 s is
   (a) 2π radians.
          (b) 40π radians.
   (c) 100 radians.
          (d) 200 radians.

16. A flywheel rotating at 10 rev/s is brought to rest by a constant torque in 15 s. In coming to a stop the flywheel makes
   (a) 75 rev.
           (b) 150 rev.
   (c) 472 rev.
          (d) 600 rev.

17. A hoop rolls down an inclined plane. The fraction of its total kinetic energy that is associated with its rotation is
   (a) $\frac{1}{4}$.
            (b) $\frac{1}{3}$.
   (c) $\frac{1}{2}$.
           (d) $\frac{2}{3}$.

18. A solid lead cylinder of radius R, a solid aluminum cylinder of radius R, a hollow lead cylinder of radius $R/2$, and a solid lead sphere of radius R all start rolling down an inclined plane at the same time. The one that reaches the bottom first is the
   (a) solid lead cylinder.
   (b) solid aluminum cylinder.
   (c) hollow lead cylinder.
   (d) solid lead sphere.

19. A solid iron sphere A rolls down an inclined plane, while an identical sphere B slides down the plane in a frictionless manner.
   (a) Sphere A reaches the bottom first.
   (b) Sphere B reaches the bottom first.
   (c) They reach the bottom together.
   (d) Which one reaches the bottom first depends on the angle of the plane.

20. A solid iron sphere A rolls down an inclined plane, while an identical sphere B slides down the plane in a fric-

tionless manner. At the bottom the kinetic energy of sphere A is
   (a) less than that of sphere B.
   (b) equal to that of sphere B.
   (c) more than that of sphere B.
   (d) more or less than that of sphere B, depending on the angle of the plane.

EXERCISES

1. A hollow cylinder and a solid cylinder having the same mass and diameter are released from rest simultaneously at the top of an inclined plane. Which reaches the bottom first?

2. Many flywheels have most of their mass concentrated around their rims. What is the advantage of this?

3. Will a car coast down a hill faster if it has heavy tires or light tires?

4. All helicopters have two propellers. Some have both propellers on vertical axes but rotating in opposite directions, and others have one on a vertical axis and one on a horizontal axis perpendicular to the helicopter body at the tail. Why is a single propeller never used?

5. If the polar ice caps melt, how will the length of the day be affected?

6. The density of a body of matter is its mass divided by its volume. Aluminum has a density of 2.7×10^3 kg/m^3 and iron has a density of 7.8×10^3 kg/m^3. If an aluminum cylinder and an iron cylinder have the same length and the same mass, which has the greater moment of inertia?

7. The dimensions of all mechanical quantities can be expressed in terms of the fundamental dimensions of length L, time T, and mass M. For example, the dimensions of acceleration are LT^{-2}. Find the dimensions in terms of L, T, and M of force, torque, energy, power, linear momentum, angular momentum, and impulse.

8. The lower the center of gravity of an object, the more stable it is. However, it is easier to balance a billiard cue vertically when its tip is on one's finger with its heavy handle up in the air rather than with its handle on the finger. Why?

9. An apple pie is cut into nine equal pieces. What angle (in radians) is included between the sides of each piece?

10. How many radians does the second hand of a clock turn through in 30 s? In 90 s? In 105 s? Express the answers in terms of π.

11. The resolution of the human eye is about $1'$, where $60'$ $= 1°$. (a) Express an angle of $1'$ in radians. (b) What is the length of the smallest detail that can be discerned when an object 25 cm away is being examined? (This is the distance of most distinct vision.)

12. In a good light, a golden eagle can detect a 50-cm hare at a distance of 2 km. What angle does this represent, in radians?

13. What is the angular velocity in radians per second of the hour, minute, and second hands of a clock?

14. A car makes a U-turn in 5 s. What is its average angular velocity?

15. In 1976, Kazuya Shiozaki made 49,299 turns in 5 h 37 min of skipping rope. Find the average angular velocity of the rope.

16. The shaft of a motor rotates at the constant angular velocity of 3000 rev/min. How many radians will it have turned through in 10 s?

17. The blades of a rotary lawnmower are 30 cm long and rotate at 315 rad/s. Find the linear speed of the blade tips and their angular velocity in rpm.

18. In 1977, Dr. Allen Bussey made 20,302 loops in 3 h with a yo-yo whose string was 87.6 cm long. Find the average linear speed of the yo-yo spool.

19. In 1941 a wind-driven electric power station was set up at Grandpa's Knob, Vermont. The propeller was 40 m in diameter. When the propeller was turning at 30 rev/min, what was the linear speed of one of the blade tips in kilometers per hour?

20. A steel cylinder 4 cm in radius is to be machined in a lathe. At how many revolutions per second should it rotate in order that the linear speed of the cylinder's surface be 0.7 m/s?

21. A truck undergoes an acceleration of 0.25 m/s^2. If its wheels are 1 m in diameter, what is their angular acceleration?

22. A rotating platform is to be used to test aircraft equipment under accelerations of 6g. If the equipment is 0.5 m from the axis, what angular velocity is required?

23. A wheel starts from rest under the influence of a constant torque and turns through 500 radians in 10 s. (a) What is its angular acceleration? (b) What is its angular velocity at the end of these 10 s?

24. A phonograph turntable slows down to a stop from an initial angular velocity of 3.5 rad/s in 20 s. (a) What is its acceleration? (b) How many turns does it make while slowing down?

25. An engine idling at 10 rev/s is accelerated at 2.5 rev/s^2 to 20 rev/s. How many revolutions does it make during this acceleration?

26. A circular saw blade rotating at 15 rev/s is brought to a stop in 125 revolutions. How much time did this take?

27. The baton of a drum majorette is a 300-g uniform rod 80 cm long. What is its kinetic energy when it is twirled at an angular velocity of 10 rad/s?

28. A 7-kg bowling ball 30 cm in diameter rolls at a speed of 5 m/s. What is its total kinetic energy?

29. A 1-kW motor rotates at 125 rad/s. How much torque can it exert?

30. A diesel engine can exert 140 N·m of torque when it is turning at 250 rad/s. Find the corresponding power output.

31. A torque of 500 N·m is applied to a turbine rotating at 200 rad/s and after 40 s its speed has doubled. What is the turbine's moment of inertia?

32. What is the angular momentum of a particle of mass m that moves in a circle of radius r at the speed v?

PROBLEMS

1. A barrel 80 cm in diameter is rolling with an angular velocity of 5 rad/s. What is the instantaneous velocity of (a) its top, (b) its center, and (c) its bottom, all with respect to the ground?

2. The propeller of a boat rotates at 100 rad/s when the speed of the boat is 6 m/s. The diameter of the propeller is 40 cm. What is the speed of the tip of the propeller?

3. Rotating flywheels have been proposed for energy storage in electric power plants. The flywheels would be set in motion during off-peak periods by electric motors, which would act as generators to return the stored energy during periods of peak loads. Find the kinetic energy of a 10^5-kg flywheel in the form of a uniform cylinder of radius 2 m that rotates at 400 rad/s. For how many hours could the flywheel supply energy at the rate of 1 MW?

4. A solid sphere 10 cm in radius starts from rest at the top of an inclined plane 10 m long and reaches the bottom in 7 s. What angle does the plane make with the horizontal?

5. A 3-kg hoop 1 m in diameter rolls down an inclined plane 10 m long that is at an angle of 20° with the horizontal. (a) What is the angular velocity of the hoop at the bottom of the plane? (b) What is its linear speed? (c) What is its rotational kinetic energy? (d) What is its total kinetic energy?

6. The pulley in the arrangement described in the second example of Sec. 3–5 (Fig. 3–15) is a uniform disk whose mass is 10 kg. Find the acceleration of the blocks.

7. A rope 1 m long is wound around the rim of a drum of radius 12 cm and moment of inertia 0.02 kg·m^2. The rope is pulled with a force of 2.5 N. (a) Assuming that the drum is free to rotate without friction, what is its final angular velocity? (b) What is its final kinetic energy? (c) How much work is done by the force?

8. A uniform knitting needle 20 cm long is balanced on its point on a table. After a moment it falls over. Find the speed with which the end of the needle strikes the table.

9. A uniform thin rod of length L is pivoted about a horizontal axis at one end. If it is released from a horizontal position, what will its maximum angular velocity be?

10. A cylinder whose axis is fixed has a string wrapped around it which is pulled with a force equal to the cylinder's weight. Show that the acceleration of the string is equal to $2g$.

11. A string is wrapped around a thin-walled hollow cylindrical spool of mass M and radius R. The outer end of the string is held in place and the spool is allowed to fall with the string unwinding as it descends. (a) Show that the linear acceleration a of the spool is $g/2$ by starting from the fact that the tension in the string is $M(g - a)$. (b) Verify this result by using conservation of energy to find the linear speed v of the spool after it has fallen a distance h and then obtaining a from v and h.

12. Find the linear acceleration of the spool of Problem 11 if it is a solid cylinder.

13. The *radius of gyration* of an object is the distance from a specified axis of rotation to a point at which the object's entire mass may be considered to be concentrated from the point of view of rotational motion about that axis. Thus the moment of inertia of an object of mass M and radius of gyration k is $I = Mk^2$. (a) The radius of gyration about its center of a hollow sphere of radius R and mass M is $k = \frac{2}{3}R$. Find its moment of inertia. (b) Find the radius of gyration of a solid sphere about its center.

14. The radius of gyration of an 80-kg flywheel 35 cm in radius is 30 cm. (a) Find its moment of inertia. (b) Find the radius of a solid disk with the same mass and moment of inertia. What does this suggest about the cross-sectional form of the flywheel?

15. A 200-kg cylindrical flywheel 0.3 m in radius is acted upon by a torque of 20 N·m. (a) If it starts from rest, how much time is required to accelerate it to an angular velocity of 10 rad/s? (b) What is its kinetic energy at this angular velocity?

16. A V-8 engine that develops 200 kW at 90 rev/s is coupled through a frictionless 5:1 reduction gear to a windlass

drum 80 cm in diameter. (a) What is the heaviest mass the windlass can raise? (b) At what linear speed will it raise such a load?

17. A high-speed elevator is being planned to lift a total load of 4000 kg at 6 m/s. The winding drum is to be 1.8 m in diameter. Neglecting losses, how much power is required from the driving motor? At what angular velocity should the power be developed?

18. An electric motor develops 5 kW at 2000 rev/min. The motor delivers its power through a spur gear 20 cm in diameter. If two teeth of the gear transmit torque to another gear at the same instant, find the force exerted by each gear tooth.

19. A 30-rev/s motor operates a pump through a V-belt drive. The motor pulley is 20 cm in diameter and the tension in the belt is 180 N on one side and 70 N on the other. What is the power output of the motor?

20. The alternator of a car engine produces 250 W of electric power when it turns at 4000 rev/min. The alternator's pulley has a radius of 4 cm. If the alternator is 95 percent efficient, what is the difference between the tensions in the taut and slack parts of the V-belt connecting it to the engine?

21. Two 0.4-kg balls are joined by a 1-m string and set whirling through the air at 5 rev/s about a vertical axis through the center of the string. After a while the string stretches to 1.2 m. (a) What is the new angular velocity? (b) What are the initial and final kinetic energies of each ball? If these are different, account for the difference.

22. A disk of moment of inertia 1 kg·m^2 that is rotating at 100 rad/s is pressed against a similar disk that is initially at rest but is able to rotate freely. The two disks stick together and rotate as a unit. (a) What is the final angular velocity of the combination? (b) If any kinetic energy was lost, where did it go? (c) The two disks are now separated. What are their new angular velocities? (d) If any kinetic energy was lost in the separation, where did it go?

ANSWERS TO MULTIPLE CHOICE

1. c	6. d	11. d	16. a
2. d	7. c	12. d	17. c
3. c	8. a	13. a	18. d
4. a	9. c	14. c	19. b
5. a	10. c	15. c	20. b

MECHANICAL PROPERTIES OF MATTER

9

The usual procedure of the physicist in approaching a complex situation is to first construct an abstract model that represents the essential features of the situation. Then, if this model proves successful in the sense that predictions obtained from it agree reasonably well with the results of observation and experiment, it is further refined until the agreement is even better. Our work in physics thus far illustrates this procedure. We began by treating objects as though they are simply particles, and went on to extend our analysis by considering them as rigid bodies. Actually, there is no such thing as a rigid body; the strongest block of steel can be stretched, compressed, or twisted by applying suitable forces. In this chapter we shall pursue reality further by examining some mechanical properties of matter that are of importance in technology.

9–1 DENSITY

A characteristic property of every substance is its *density,* which is its mass per unit volume. When we speak of lead as a "heavy" metal and of aluminum

Density is mass per unit volume

as a "light" one, what we really mean is that lead has a higher density than aluminum: a cubic meter of lead has a mass of 11,300 kg, whereas a cubic meter of aluminum has a mass of only 2700 kg.

Although in the SI system the proper unit of density is the kg/m^3, densities are frequently given in g/cm^3. Since there are 10^3 g in a kilogram and 100 cm in a meter,

$$1 \text{ g/cm}^3 = 10^3 \text{ kg/m}^3$$

The densities of various common substances are given in Table 9–1. The symbol for density is d, so that if a volume V of a certain substance has the mass m,

$$d = \frac{m}{V} \hspace{4cm} \textit{Density} \quad (9\text{–}1)$$

Example   A 50-g bracelet is suspected of being gold-plated lead instead of pure gold. When it is dropped in a full glass of water, 4.0 cm^3 of water overflows. Is the bracelet pure gold? If not, what proportion of its mass is gold?

Solution   The density of the bracelet is

$$d = \frac{m}{V} = \frac{50 \text{ g}}{4.0 \text{ cm}^3} = 12.5 \text{ g/cm}^3$$

This is less than the density of gold, which is 19 g/cm^3, so the bracelet is not pure gold. To find the proportion of gold it contains, we note that

$$m_{gold} + m_{lead} = 50 \text{ g}$$

$$V_{gold} + V_{lead} = \frac{m_{gold}}{d_{gold}} + \frac{m_{lead}}{d_{lead}} = 4.0 \text{ cm}^3$$

From the first equation, $m_{lead} = 50 \text{ g} - m_{gold}$. Substituting this expression for m_{lead} and the values $d_{gold} = 19 \text{ g/cm}^3$ and $d_{lead} = 11 \text{ g/cm}^3$ in the second equation yields

$$\frac{m_{gold}}{19 \text{ g/cm}^3} + \frac{50 \text{ g}}{11 \text{ g/cm}^3} - \frac{m_{gold}}{11 \text{ g/cm}^3} = 4.0 \text{ cm}^3$$

$$m_{gold} = 14 \text{ g}$$

Hence the proportion of gold in the bracelet is 14 g/50 g $= 0.28 = 28\%$.   ■

Example   We want to pump water from a well 20 m deep at the rate of 500 liters/min. Assuming 60% efficiency, how powerful should the pump motor be?

Solution   We begin with the definition of efficiency,

$$\text{Efficiency} = \frac{\text{power output}}{\text{power input}}$$

which means that

$$\text{Power input} = \frac{\text{power output}}{\text{efficiency}}$$

| Substance | Mass Density | | | Weight Density |
	kg/m^3	g/cm^3	$slugs/ft^3$	lb/ft^3
Air	1.3	1.3×10^{-3}	2.5×10^{-3}	8×10^{-2}
Alcohol (ethyl)	7.9×10^2	0.79	1.5	48
Aluminum	2.7×10^3	2.7	5.3	1.7×10^2
Balsa wood	1.3×10^2	0.13	0.25	8
Blood (37°C)	1.06×10^3	1.06	2.05	66
Bone	1.6×10^3	1.6	3.1	9.9×10^2
Carbon dioxide	2.0	2.0×10^{-3}	3.8×10^{-3}	0.12
Concrete	2.3×10^3	2.3	4.5	1.4×10^2
Gasoline	6.8×10^2	0.68	1.3	42
Gold	1.9×10^4	19	38	1.2×10^3
Helium	0.18	1.8×10^{-4}	3.5×10^{-4}	1.1×10^{-2}
Hydrogen	0.09	9×10^{-5}	1.8×10^{-4}	5.4×10^{-2}
Ice	9.2×10^2	0.92	1.8	58
Iron	7.8×10^3	7.8	15	4.8×10^2
Lead	1.1×10^4	11	22	7×10^2
Mercury	1.4×10^4	14	26	8.3×10^2
Nickel	8.9×10^3	8.9	17	5.5×10^2
Nitrogen	1.3	1.3×10^{-3}	2.4×10^{-3}	7.7×10^{-2}
Oak	7.2×10^2	0.72	1.4	45
Oxygen	1.4	1.4×10^{-3}	2.8×10^{-3}	9×10^{-2}
Water, pure	1.00×10^3	1.00	1.94	62
Water, sea	1.03×10^3	1.03	2.00	64

TABLE 9–1
Densities of various substances at atmospheric pressure and room temperature

Here

$$\text{Power output} = \frac{\text{work}}{\text{time}} = \left(\frac{\text{work}}{\text{liter}}\right)\left(\frac{\text{liters}}{\text{s}}\right)$$

Since 1 liter $= 10^{-3}$ m^3, the work needed to raise a liter of water through 20 m is

$$\frac{\text{Work}}{\text{liter}} = mgh = dVgh = \left(1.00 \times 10^3 \frac{\text{kg}}{\text{m}^3}\right)\left(10^{-3}\frac{\text{m}^3}{\text{liter}}\right)\left(9.8\frac{\text{m}}{\text{s}^2}\right)(20\,\text{m})$$

$$= 196\,\text{J/liter}$$

The rate of flow is 500 liters/min, and so

$$\text{Power output} = \left(196\,\frac{\text{J}}{\text{liter}}\right)\left(500\,\frac{\text{liters}}{\text{min}}\right)\left(\frac{1}{60\,\text{s/min}}\right) = 1633\,\text{W}$$

Since the efficiency of the system is 60% = 0.60, the required power input is

$$\text{Power input} = \frac{\text{power output}}{\text{efficiency}} = \frac{1633\,\text{W}}{0.60} = 2722\,\text{W}$$

which is 3.6 hp.

Density in the British system

Densities in the British system are properly expressed in slugs/ft^3. In practice, *weight density, dg,* the weight of a substance per unit volume, is more commonly used than its density *d*, which is its mass per unit volume. The difference is a factor of *g* = 32 ft/s^2. Thus the density of pure water is 1.94 slug/ft^3 and its weight density is 62 lb/ft^3.

Specific gravity is density relative to water

The *specific gravity* of a substance is its density relative to that of water and so is a pure number. Since the density of water is almost exactly 1 g/cm^3, the specific gravity of a substance is very nearly equal to the numerical value of its density when expressed in g/cm^3. Thus, since the density of aluminum is 2.7 g/cm^3, its specific gravity is 2.7.

9–2 ELASTICITY

Categories of stress

While solid bodies generally seem perfectly rigid and unyielding, it is nevertheless possible to deform them either temporarily or permanently by applying stresses. Stress forces fall into three categories: *tensions, compressions,* and *shears*. These are illustrated in Fig. 9–1.

A *tensile* stress is applied to an object when equal and opposite forces that act away from each other are exerted on its ends along the same line of action, thereby tending to elongate the object. A *compressive* stress is applied to an object when equal and opposite forces that act toward each other are exerted on its ends along the same line of action, thereby tending to decrease its length. A *shearing* stress is applied to an object when equal and opposite forces are exerted on its ends along different lines of action, thereby tending to change the shape of the object without changing its volume.

In simpler terms, tensions stretch objects on which they act, compressions shrink objects on which they act, and shears twist objects on which they act.

Elastic deformation is proportional to applied stress

The response of an object to any of the above stresses depends upon its composition, shape, temperature, and so on, but as a general rule the amount of deformation of a crystalline solid is directly proportional to the applied stress provided that the force does not exceed a certain limit. In the case of tension, for example, we might find that supporting a 20-kg mass with a certain thin wire causes the wire to stretch 1 mm. (Fig. 9–2). Doubling the mass to 40 kg therefore will produce a total elongation of 2 mm, tripling the mass to 60 kg will produce a total elongation of 3 mm, and so on. When the force is removed, the wire returns to its original length.

Hooke's law

The above proportionality is called *Hooke's law,* and may be written

$$F = ks \qquad\qquad\qquad \textit{Hooke's law} \quad (9\text{–}2)$$

FIG. 9–1 The three types of stress.

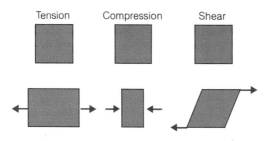

Tension　　Compression　　Shear

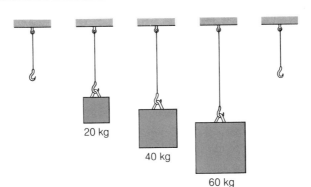

FIG. 9–2 The elongation of a wire is proportional to the stress applied to it, provided that the elastic limit is not exceeded. When the stress is removed, the wire returns to its original length.

20 kg

40 kg

60 kg

where F is the applied tension force, s the resulting elongation, and k a constant whose value depends upon the nature and dimensions of the object under stress. The force constant k is higher for materials such as steel than it is for materials such as lead; it is directly proportional to the cross-sectional area of the object, so that a thick wire of a given material has a higher value of k than a thin wire of the same material. Relationships similar to Eq. (9–2) are found to apply to the behavior of solids under shear stresses and to all states of matter under compressive stresses.

The term *elastic limit* refers to the maximum stress that can be applied to an object without it being permanently deformed as a result. When its elastic limit is exceeded, the object may or may not be far from rupture. Brittle substances like glass or cast iron break at or near their elastic limits. Bone is another material that remains elastic until its breaking point is reached, and so it cannot be permanently altered in size or shape by applying a force.

Elastic limit

Most metals (cast iron is an exception) can be deformed considerably beyond their elastic limits, a property known as *ductility*. Copper is a very ductile metal, and while a copper wire whose cross-sectional area is 1 mm^2 reaches its elastic limit when a force of about 150 N is applied, it will not rupture until the force has more than doubled.

Ductility

A graph of the elongation of an iron rod as a function of the tension applied to it is shown in Fig. 9–3. At first the graph is a straight line, which corresponds to Hooke's

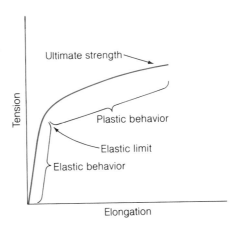

Ultimate strength

Plastic behavior

Elastic limit

Tension

Elastic behavior

Elongation

FIG. 9–3 Graph of the elongation of an iron rod as more and more tension is applied to it. When a tension below the elastic limit is applied and then removed, the rod returns to its original length, which is elastic behavior. When a tension exceeding the elastic limit is applied and removed, the rod does not contract fully but remains permanently longer, which is plastic behavior.

Plastic deformation and ultimate strength

law. Past the elastic limit the graph flattens out, which means that each increase in tension by a given amount produces a proportionately greater increase in length than it did below the elastic limit: the rod stretches more readily. If the tension is removed after having exceeded the elastic limit, the rod will be permanently longer than it was originally; it has undergone *plastic deformation*. The *ultimate strength* of the rod is the greatest tension it can withstand, and it corresponds to the highest point on the curve.

Fatigue

An object may fail through *fatigue* after repeated applications of stresses that are well under its original breaking strength. Minute defects in the internal structure of the material grow a little each time a stress acts, and eventually cracks appear that lead to rupture. Three factors are involved: the level of stress, the total number of stress cycles that occur, and corrosion. The number of cycles needed for failure decreases rapidly with increasing stress. A metal bar that can tolerate a million applications of a force equal to 20% of its ultimate strength may break after only 10,000 or 20,000 applications of a force twice as great. The problem is clearly most severe in machinery—a gasoline or diesel engine operating at 3000 rpm undergoes 180,000 stress cycles per hour. Stress concentrations occur wherever there is a sharp corner or notch, even a scratch, so parts designed for high-speed machines are always highly polished. Inhomogeneities in a weld produce stress concentrations that promote fatigue failure, and welded parts therefore need generous margins of safety. Corrosion can accelerate fatigue by producing pits in a surface from which cracks can develop, and also by causing the cracks to grow faster.

9–3   YOUNG'S MODULUS

Stress and strain

Hooke's law for each kind of stress can be expressed in such a way that only a single constant need be known for a particular material in order to relate the force applied to *any* object of this material to the resulting deformation, regardless of its size and shape. Experimentally it is found that the relative change in size of an object is proportional to the ratio between the applied force and its cross-sectional area. This ratio is called the *stress* on the object: stress is applied force per unit area. The resulting relative change in size is called *strain*: strain is change in length per unit length, change in volume per unit volume, and so on.

Strain is proportional to stress below the elastic limit

Thus we can summarize Hooke's law by saying that, below the elastic limit, *strain is proportional to stress*. Under a given force a thin rod will stretch more than a thick one, and a long rod will stretch more than a short one. The *modulus of elasticity* of a material subjected to a certain kind of stress is defined as the ratio between the stress and the strain that occurs because of it:

$$\text{Modulus of elasticity} = \frac{\text{stress}}{\text{strain}}$$

Young's modulus applies to tensile and compressive stresses

For tension or compression stresses, the modulus of elasticity is called *Young's modulus*. In the case of a rod of initial length L_0 and cross-sectional area A in which a tension or compression force F produces a change ΔL in its length (Fig. 9–4),

$$\text{Stress} = \frac{\text{force}}{\text{area}} = \frac{F}{A}$$

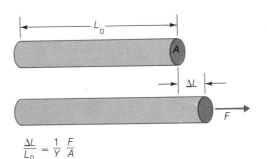

$$\frac{\Delta L}{L_0} = \frac{1}{Y}\frac{F}{A}$$

FIG. 9–4 An object under tension stretches by an amount that depends upon the value of Y for the material of which it is composed. Y is called *Young's modulus.*

$$\text{Strain} = \frac{\text{change in length}}{\text{original length}} = \frac{\Delta L}{L_0}$$

and so

$$\text{Young's modulus} = \frac{\text{stress}}{\text{strain}}$$

$$Y = \frac{F/A}{\Delta L/L_0} \qquad\qquad (9\text{–}3)$$

Hence we have

$$\frac{\Delta L}{L_0} = \frac{1}{Y}\frac{F}{A} \qquad\qquad \textit{Tension or compression} \quad (9\text{–}4)$$

The rod increases in length by ΔL if it is in tension and decreases by that amount if it is in compression. The value of Y depends upon the composition of the rod. Young's moduli for a number of common substances are given in Table 9–2.

According to Hooke's law, $F = k\,\Delta L$. We can therefore express the constant k for an object under tension in terms of Y and the dimensions of the object as $k = YA/L_0$.

Hooke's law and Young's modulus

The proportionality between strain and stress is valid only when the elastic limit is not exceeded. Table 9–3 is a list of the elastic limits and ultimate strengths under

Material	Young's Modulus, Y ($\times 10^{10}$ N/m^2)	Shear Modulus, S ($\times 10^{10}$ N/m^2)	Bulk Modulus, B ($\times 10^{10}$ N/m^2)
Aluminum	7.0	3.0	7.0
Bone	1.5	8.0	
Concrete	2.0		
Copper	11	4.2	14
Glass	5.5	2.3	3.7
Granite	4.5		4.5
Iron	19	7.0	10
Lead	1.6	0.56	0.77
Pine wood (parallel to grain)	1.0		
Steel	20	8.4	16

TABLE 9–2
Typical elastic moduli

TABLE 9–3
Typical elastic limits and
ultimate strengths

Material	Elastic limit ($\times$ 10^8 N/m^2)	Ultimate Strength ($\times$ 10^8 N/m^2)	
		Tension	*Compression*
Aluminum	1.8	2.0	2.0
Bone		1.3	1.7
Concrete		0.02	0.2
Copper	1.5	3.4	3.4
Granite			1.7
Iron (cast)	1.6	1.7	5.5
Pine wood (parallel to grain)		0.4	0.35
Steel	2.5	5.0	5.0

Compressive strength usually exceeds tensile strength

tension of several materials; the elastic limit is the same in both tension and compression, but the ultimate strength may be much greater in compression. The compressive strength of cast iron, for instance, is more than three times its tensile strength, and the difference is a factor of ten in the case of concrete. A material otherwise ideal for a certain purpose may have to be reinforced with something else to provide the missing tensile strength. Thus steel rods are commonly embedded in the concrete used for buildings and bridges, and glass fibers are embedded in the polyester resin used to make boat hulls.

In general, elastic limit and ultimate strength depend upon the previous history of an object as well as upon its composition: hot-rolled and cold-rolled steel have different properties, as do annealed and tempered steel. (An insight into the reasons for these differences is given in Chapter 30.) Further, repeated cycles of stress tend to weaken an object through fatigue, as mentioned earlier. For these reasons the values in Table 9–3 are only approximate, although they are typical of each material.

Example   A copper wire 1 mm in diameter and 2 m long is used to support a mass of 5 kg. By how much does the wire stretch under this load? What is the minimum diameter the wire can have if its elastic limit is not to be exceeded?

Solution   From Eq. (9–4),

$$\Delta L = \frac{L_0}{Y}\frac{F}{A}$$

Here, since the radius of a wire 1 mm in diameter is 5×10^{-4} m, we have

$$L_0 = 2\,\text{m}$$
$$F = mg = (5\,\text{kg})(9.8\,\text{m/s}^2) = 49\,\text{N}$$
$$Y = 1.1 \times 10^{11}\,\text{N/m}^2$$
$$A = \pi r^2 = \pi\,(5 \times 10^{-4}\,\text{m})^2 = 7.85 \times 10^{-7}\,\text{m}^2$$

and so

$$\Delta L = \frac{(2\text{m})(49\,\text{N})}{(1.1 \times 10^{11}\,\text{N/m}^2)(7.85 \times 10^{-7}\,\text{m}^2)} = 1.1 \times 10^{-3}\,\text{m} = 1.1\,\text{mm}$$

To answer the second question, we note from Table 9–3 that the elastic limit of copper is $1.5 \times 10^8\,\text{N/m}^2$. Hence

$$\frac{F}{A} = 1.5 \times 10^8\,\text{N/m}^2$$

$$A = \frac{49\,\text{N}}{1.5 \times 10^8\,\text{N/m}^2} = 3.27 \times 10^{-7}\,\text{m}^2$$

Since the cross-sectional area of a wire of radius r is $A = \pi r^2$, we have

$$r = \sqrt{\frac{A}{\pi}} = \sqrt{\frac{3.27 \times 10^{-7}\,\text{m}^2}{\pi}}$$

$$= \sqrt{1.04 \times 10^{-7}\,\text{m}^2} = \sqrt{10.4 \times 10^{-8}\,\text{m}^2} = 3.2 \times 10^{-4}\,\text{m}$$

and the corresponding diameter is 6.4×10^{-4} m, which is 0.64 mm. ∎

Bone

It is clear from Table 9–3 that bone is an excellent structural material. Bone is a heterogeneous substance in which fibers of a protein called collagen provide most of the tensile strength and inorganic salt crystals provide most of the compressive strength. The different properties of these components lead to different values of Young's modulus for bone in tension and in compression as well as to different ultimate strengths.

Bone size

The compressive load on the leg bones of an animal depends upon its weight, which in turn varies as the cube L^3 of a representative linear dimension L such as its length or height. An animal three times as long as another of the same form will weigh about nine times as much. However, the strength of a bone depends upon its cross-sectional area, which for similar animals varies as L^2. Thus animals widely different in size cannot resemble one another; for example, a large animal must have relatively thicker leg bones than a small one since L^3 increases faster than L^2. It is no accident that a hippopotamus has thicker legs for its size than a mouse does, nor that the largest animals of all, the whales, live in the oceans, where their immense body weights are supported by buoyancy rather than by legs.

9–4 SHEAR

Shear stresses change the shape of an object upon which they act. The volume of the object is not affected. The situation is much like that of a book whose covers are pushed out of alignment, as in Fig. 9–5: the layers of atoms, which are analogous to the pages of the book, are displaced sideways, but the spacing of the layers, which corresponds to the thickness of the pages, remains the same.

Let us consider a block of thickness d whose lower face is fixed in place and upon whose upper face the force F acts (Fig. 9–6). A measure of the relative distortion of the block caused by the shear stress is the angle ϕ, called the *angle of shear*. Because

FIG. 9–5 In shear there is a change in shape without a change in volume. The angle ϕ is the angle of shear.

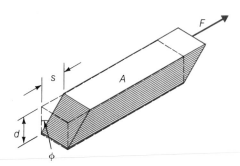

FIG. 9–6 The angle of shear ϕ (in radians) is equal to s/d. The greater the shear modulus S, the more rigid the material.

Angle of shear is a measure of shear strain

this angle is always small, its value in radians is equal to the ratio s/d between the displacement s of the block's faces and the distance d between them. The greater the area A of these faces, the less they will be displaced by the shear force F. Therefore

$$\text{Shear stress} = \frac{\text{force}}{\text{area}} = \frac{F}{A}$$

$$\text{Shear strain} = \text{angle of shear} = \phi = \frac{s}{d}$$

and the stress-strain equation for shear is

$$\text{Shear modulus} = \frac{\text{shear stress}}{\text{shear strain}}$$

$$S = \frac{F/A}{\phi} = \frac{F/A}{s/d}$$

Hence we have

$$\phi = \frac{s}{d} = \frac{1}{S}\frac{F}{A} \qquad\qquad\qquad \textit{Shear} \quad (9\text{–}5)$$

Here the applied forces are *parallel* to the faces upon which they act and not perpendicular as in the case of tension and compression.

Shear modulus

The quantity S is called the *shear modulus,* and the values of S for various substances are given in Table 9–2. The higher the value of S, the more rigid the material; S is sometimes referred to as the *modulus of rigidity* for this reason. It is interesting to note that the shear modulus of any material is usually a good deal less than its Young's modulus. This means that it is easier to slide the atoms of a solid past one another than it is to pull them apart or squeeze them together.

Shear strength

The shear strength of a material is the maximum shear stress an object of that material can withstand before breaking. The greater the shear strength of a material, the more force must be applied to cut a sheet of it with a pair of scissors (or their industrial equivalent) or to punch a hole in the sheet.

Example   Ordinary mild steel ruptures when a shear stress of about 3.5×10^8 N/m^2 is applied. Find the force needed to punch a 1 cm diameter hole in a steel sheet 3 mm thick.

Solution The area across which the shear stress is exerted here is the cylindrical inner surface of the hole (Fig. 9–7), so that

$$A = \pi dh = \pi(10^{-2} \text{ m})(3 \times 10^{-3} \text{ m}) = 9.4 \times 10^{-5} \text{ m}^2$$

Since we are given that $(F/A)_{max} = 3.5 \times 10^8 \text{ N/m}^2$, we have

$$F_{max} = (3.5 \times 10^8 \text{ N/m}^2)(9.4 \times 10^{-5} \text{ m}^2) = 3.3 \times 10^4 \text{ N}$$

which is nearly 4 tons.

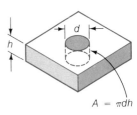

■ **FIG. 9–7**

9–5 BULK MODULUS

When inward forces act over the entire surface of an object, its volume decreases by some amount ΔV from its original volume of V_0. Only those force components that are perpendicular to the object's surface where they act are effective in compressing it, since the parallel components lead only to shear stresses. If the compression force per unit area F/A is the same over the entire surface of the object, as in Fig. 9–8, then

Uniform compression

$$\text{Volume stress} = \frac{\text{force}}{\text{area}} = \frac{F}{A}$$

$$\text{Volume strain} = \frac{\text{change in volume}}{\text{original volume}} = \frac{\Delta V}{V_0}$$

and

$$\text{Bulk modulus} = -\frac{\text{volume stress}}{\text{volume strain}}$$

$$B = -\frac{F/A}{\Delta V/V_0}$$

Thus we have for the relative change in volume

$$\frac{\Delta V}{V_0} = -\frac{1}{B}\frac{F}{A} \qquad \textit{Uniform compression} \quad (9–6)$$

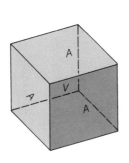

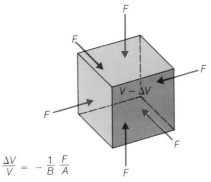

$$\frac{\Delta V}{V} = -\frac{1}{B}\frac{F}{A}$$

FIG. 9–8 Bulk compression.

TABLE 9–4
Bulk moduli of liquids at room temperature

Liquid	Bulk Modulus, B ($\times 10^9$ N/m^2)
Alcohol, ethyl	0.90
Benzene	1.05
Kerosene	1.3
Mercury	26
Oil, lubricating	1.7
Water	2.3

The minus sign corresponds to the fact that an increase in force leads to a decrease in volume. The quantity *B* is called the *bulk modulus*: typical values of *B* are given in Table 9–2.

Pressure

The perpendicular stress F/A is usually called *pressure*, symbol *p*, as discussed in Chapter 10. Hence we can also write

$$\frac{\Delta V}{V_0} = -\frac{p}{B} \tag{9–7}$$

Liquids can support neither tensions nor shears, but they do tend to resist compression. Bulk moduli for several liquids are given in Table 9–4. Interatomic forces within a liquid are smaller than within a solid, which is reflected in the considerably smaller bulk moduli of liquids. Substantial forces are nevertheless needed to compress a liquid by more than a slight amount; to compress a volume of water by 1% requires an inward force per unit area of 2.3×10^7 N/m^2, which is equivalent to 3300 lb/in.^{2}.

Example   Verify the above statement.

Solution   The bulk modulus of water is 2.3×10^9 N/m^2 and, by hypothesis, $\Delta V/V_0 = -0.01$. Hence

$$\frac{F}{A} = -B\frac{\Delta V}{V_0} = -(2.3 \times 10^9 \text{N/m}^2)(-0.01) = 2.3 \times 10^7 \text{N/m}^2 \qquad \blacksquare$$

9–6   BUILDING OVER SPACE

Three successive inventions for building over space have shaped the course of architecture. The earliest, and still the most widely used, is the *post-and-beam* arrangement of Fig. 9–9(a), in which two vertical posts hold up a horizontal beam. Before steel came into general use in the last century, the width that could be spanned by a beam was severely limited by the nature of the materials available, which were mainly wood and stone. Wooden beams cannot support really large loads unless the span is narrow, and obtaining timber of adequate size and quality has always been a problem in much of the world. Stone can be quite strong in compression but is very much less so in tension, and the lower part of a beam is under tension. Narrow doorways and numerous

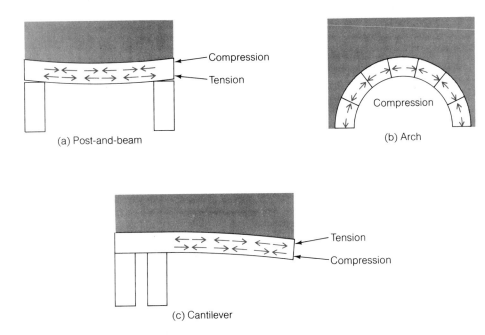

(a) Post-and-beam

(b) Arch

(c) Cantilever

FIG. 9–9 Three ways to build over space. (a) A horizontal beam experiences tensile as well as compressive stresses. (b) The principal stresses in an arch are compressive, so an arch made of stones or bricks, which are weak in tension but strong in compression, can support a considerable load. (c) Modern structural materials such as steel and reinforced concrete are strong in tension as well as compression and so can be used as cantilevers supported only at one end.

interior supports were accordingly necessary until the nineteenth century in buildings of even modest size that used post-and-beam construction.

The great advantage of the *arch* is that the principal stresses imposed on it by a load are compressive (Fig. 9–9(b)). Stone and brick are sufficiently strong in compression for quite large arches, and also for domes, which are the three-dimensional equivalent of arches. The arch first came into wide use in Roman times, although it had been known thousands of years earlier. Innumerable ancient structures based on the arch and the dome still stand, testimony to the sound engineering principle behind these innovations.

The third, and most recent, development in building over space is the *cantilever,* a beam held in place at one end only (Fig. 9–9(c)). Reinforced concrete as well as steel can be used in cantilever construction. The tall buildings of today do not have the load-bearing walls of the past but have internal skeletons instead, with the outer parts of their floors and their thin "curtain" walls, often almost entirely glass, supported by cantilever beams that extend outward from central frames.

IMPORTANT TERMS

The **density** of a substance is its mass per unit volume.

The three categories of stress forces are **tension,** in which equal and opposite forces that act away from each other are applied to a body; **compression,** in which equal and opposite forces that act toward each other are applied to a body; and **shear,** in which equal and opposite forces that do not act along the same line of action are applied to a body. A tensile stress tends to elongate a body, a compressive stress to shorten it, and a shearing stress to change its shape without changing its volume.

Hooke's law states that the amount of deformation experienced by a body under stress is proportional to the magnitude of the stress. Thus the elongation of a wire is proportional to the tension applied to it.

The **elastic limit** is the maximum stress a solid can be subjected to without being permanently altered. Hooke's law is valid only when the elastic limit is not exceeded.

The **stress** on an object is the applied force per unit area; the **strain** is the resulting change in a dimension of the object relative to its original value. A **modulus of elasticity** of a material is the ratio between a particular kind of applied stress and the resulting strain, provided that the elastic limit is not exceeded.

IMPORTANT FORMULAS

Hooke's law:　$F = ks$

Tension or linear compression:　$\dfrac{\Delta L}{L_0} = \dfrac{1}{Y}\dfrac{F}{A}$

Shear:　$\phi = \dfrac{s}{d} = \dfrac{1}{S}\dfrac{F}{A}$

Uniform compression:　$\dfrac{\Delta V}{V_0} = -\dfrac{1}{B}\dfrac{F}{A} = -\dfrac{p}{B}$

MULTIPLE CHOICE

1. The properties of several different materials are being compared. If the samples all have the same size and shape, the one with the greatest mass also has the greatest
(a) density.
(b) elastic limit.
(c) Young's modulus.
(d) bulk modulus.

2. When equal and opposite forces are exerted on an object along different lines of action, the object is said to be under
(a) tension.　　　　　(b) compression.
(c) shear.　　　　　(d) elasticity.

3. Ductility refers to the ability of a metal to
(a) be deformed temporarily.
(b) be deformed permanently.
(c) shrink under compression.
(d) break under tension.

4. The stress on an object when a force acts on it is equal to
(a) the relative change in its dimensions.
(b) the applied force per unit area.
(c) Young's modulus.
(d) the elastic limit.

5. A shearing stress that acts on an object affects its
(a) length.　　　　　(b) width.
(c) volume.　　　　　(d) shape.

6. Another name for the shear modulus of a material is
(a) Young's modulus.
(b) modulus of rigidity.
(c) bulk modulus.
(d) ductility.

7. The only elastic modulus that applies to liquids is
(a) Young's modulus.
(b) shear modulus.
(c) modulus of rigidity.
(d) bulk modulus.

8. According to Hooke's law, the force needed to elongate an elastic object by an amount s is proportional to
(a) s.　　　　　(b) $1/s$.
(c) s^2.　　　　　(d) $1/s^2$.

9. Two wires are made of the same material, but wire A is half as long as and has twice the diameter of wire B. If they are to be stretched by the same amount, the required force on wire A must be
(a) one-eighth that on B.
(b) twice that on B.
(c) four times that on B.
(d) eight times that on B.

10. An iron wire 1 m long with a square cross section 2 mm on a side is used to support a 100 kg load. Its elongation is
(a) 0.0027 mm.　　　　　(b) 0.27 mm.
(c) 2.7 mm.　　　　　(d) 3.7 mm.

11. A wire 10 m long with a cross-sectional area of 0.1 cm^2 stretches by 13 mm when a load of 100 kg is suspended from it. The Young's modulus for this wire is
(a) 0.77×10^{10} N/m^2.
(b) 7.5×10^{10} N/m^2.
(c) 7.7×10^{10} N/m^2.
(d) 9.3×10^{10} N/m^2.

12. The ultimate strength in compression of aluminum is 2×10^8 N/m^2. The maximum mass an aluminum cube 1 cm on each edge can support is approximately
(a) 10^2 kg.　　　　　(b) 10^3 kg.
(c) 10^4 kg.　　　　　(d) 10^5 kg.

EXERCISES

1. A room is 5 m long, 4 m wide, and 3 m high. What is the mass of the air it contains?

2. If gold costs $500/oz, how many mm on a side does a $10,000 cube of gold measure?

3. Mammals have approximately the same density as fresh water. Find the volume in liters of a 55-kg woman and the volume in cubic meters of a 140,000-kg blue whale.

4. A 200-g bottle has a mass of 340 g when filled with water and 344 g when filled with blood plasma. What is the density of the plasma?

5. The radius of the earth is 6.37×10^6 m and its mass is 5.98×10^{24} kg. (a) Find the average density of the earth. (b) The average density of rocks at the earth's surface is 2.7×10^3 kg/m^3. What must be true of the matter of which the earth's interior is composed? Is it likely that the earth is hollow and peopled by another species, as the ancients believed?

6. A three-legged stool has one leg of aluminum, one of brass, and one of steel. The legs have the same dimensions. If the load on the stool is on its exact center, which leg is under the greatest stress and which under the least stress? Which leg experiences the greatest strain and which the least strain?

7. Two wires are made of the same material, but wire A is twice as long and has twice the diameter of wire B. Find the elongation of wire B relative to that of wire A when both are subjected to the same load.

8. The elastic moduli of a particular material are related to one another, and each of them can be expressed in terms of the other two. For instance, the theory of elasticity shows that Young's modulus Y can be expressed in terms of the shear modulus S and the bulk modulus B by the formula

$$Y = \frac{9SB}{(3B + S)}$$

Check this formula for three of the materials listed in Table 9–2. Give several reasons why you would not expect perfect agreement.

9. A nylon rope 10 mm in diameter breaks when a load of 25 kN is applied to it. What would you estimate for the breaking strength of a nylon rope 6 mm in diameter? 14 mm in diameter?

10. How safe is it for a 75-kg circus acrobat to balance on the index finger of his right hand, whose bones have a minimum cross-sectional area of 0.5 cm^2?

11. A human hair 60 μm in diameter can support a mass of 30 g (equivalent to about an ounce of weight). How does the stress in the hair compare with the ultimate strength in tension of aluminum?

12. In a test firing, the three main engines of the space shuttle *Columbia* developed a total thrust of 4.5×10^6 N. Eight steel bolts 90 mm in diameter were used to hold *Columbia* to its launch pad during the firing. What was the ratio between the ultimate strength of the bolts and the applied stress?

13. A coil spring has a force constant of 1000 N/m. How much will it stretch when it is used to support an object whose mass is 8 kg?

14. When a coil spring is used to support a 12-kg object, the spring stretches by 4 cm. What is the force constant of the spring?

15. A steel wire 1 m long and 1 mm square in cross section supports a mass of 6 kg. By how much does it stretch?

16. A steel post 4 m long and 3 cm in radius supports a load of 3000 kg. By how much is it shortened?

17. An 80-kg man has femurs (the femur is the bone of the thigh) 42 cm long and 11 cm^2 in average cross-sectional area. Find the change in length of each femur as the man walks.

18. A brass wire 2 m long whose cross-sectional area is 5 mm^2 stretches by 2.6 mm when a force of 600 N is applied. Find the value of Young's modulus for the wire.

19. A 3-cm cube of raspberry gelatin on a table is subjected to a shearing force of 0.5 N. The upper surface is displaced by 5 mm. What is the shear modulus of the gelatin?

20. If an aluminum object is placed in a vacuum, by what percentage will its volume increase? Atmospheric pressure is 1.013×10^5 N/m^2.

21. The pressure at a depth of 1 km in the ocean exceeds sea-level atmospheric pressure by about 10^7 N/m^2. If an iron anchor whose volume at the surface is 400 cm^3 is lowered to a depth of 1 km, by how much does its volume decrease?

PROBLEMS

1. Suppose you have cylinders of the same length and diameter of bone, aluminum, and steel. For each cylinder find the ratio between the maximum tensile and compressive forces it can withstand and its mass. How does bone compare with aluminum and steel as a structural material?

2. A certain material has a density of d and a tensile strength of U. How long a rod of this material can be suspended from one end without breaking under its own weight? What is this length in the case of aluminum? Steel? (Assume the density of steel to be the same as that of iron.)

3. One gram of gold can be beaten out into a foil 1 m^2 in area. (Thus an ounce of gold can yield 300 ft^2 of foil.) How many atoms thick is such a foil? The mass of a gold atom is 3.27×10^{-25} kg.

4. A copper wire and a steel wire are being used side by side to support a load. (a) If they have the same diameter, what proportion of the load does each one support? (b) What should the ratio of their diameters be if they are each to support half the load?

5. The parachute of a 60-kg woman fails to open and she falls into a snowbank at a terminal speed of 60 m/s, coming to a stop 1.5 m below the surface of the snow. (a) Find the average force on her during the impact. (b) If the area of the woman's body that strikes the snow is 0.2 m^2 and the stress required for serious injury to body tissues is 5×10^5 N/m^2, is she likely to survive the impact?

6. A steel cable whose cross-sectional area is 2.5 cm^2 supports a 1000-kg elevator. The elastic limit of the cable is 3×10^8 N/m^2. What is the maximum upward acceleration that can be given the elevator if the tension in the cable is to be no more than 20% of the elastic limit?

7. An iron pipe 3 m long is used to support a sagging floor. The inside diameter of the pipe is 10 cm and its outside diameter is 11 cm. When the force on it is 15 kN, by how much is it compressed?

8. A wall of lead bricks 1 m high is used to shield a sample of radium. Each brick was originally a cube 10 cm on an edge. What is the height of the lowest brick when the wall has been erected?

9. A vise is used to hold a 5 cm cube of pine wood while it is being worked on. The cube is in contact with the jaws of the vise, whose screw has a pitch of 4 mm and whose handle is grasped 10 cm from the screw axis. If there is no friction and a force of 60 N is applied to the handle, find the amount by which the cube is compressed.

10. By how much can a copper wire 2 m long be stretched before its elastic limit is exceeded?

11. A punch press that exerts a force of 20 kN is employed to punch 1-cm-square holes in sheet aluminum. If the shear strength of aluminum is 70 MN/m^2, find the maximum thickness of aluminum sheet that can be used.

12. Two steel plates are riveted together with ten rivets each 5 mm in diameter. If the maximum shear stress the rivets can withstand is 3.5×10^8 N/m^2, how much force applied parallel to the plates is needed to shear off the rivets?

13. When a torque τ is applied to a cylinder, the angle θ (in radians) through which it twists depends upon the cylinder's length L, radius r, and shear modulus S according to the formula

$$\theta = \frac{2\tau L}{\pi S r^4}$$

The geometry of the situation is shown in the figure below. The above formula makes it possible to determine S experimentally by applying a torque to a wire of a given material and measuring the twist that results, a much simpler procedure than one based directly on Eq. (8–9). Consider a rod 30 cm long that is suspended horizontally at its center by a wire 70 cm long and 1 mm in diameter. When a horizontal force of 0.02 N is applied to one end of the rod, it turns through 25°. Find the shear modulus of the wire.

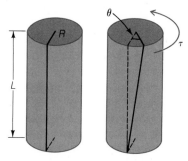

14. A solid steel drive shaft is 5 cm in diameter and 2.5 m long. If the angle of twist is not to exceed 2°, find (a) the maximum torque that can be applied to the shaft, and (b) the maximum power the shaft can transmit at 900 rpm.

ANSWERS TO MULTIPLE CHOICE

FLUIDS

10

As the name implies, a fluid is a substance that flows readily. Gases and liquids are fluids, although the dividing line between solids and liquids is not always a sharp one. Because of its ability to flow, a fluid can exert a buoyant force on an immersed body, multiply an applied force, and provide "lift" to a properly shaped object moving through it—properties that have made possible such diverse applications as the ship, the hydraulic press, and the airplane. In what follows we shall learn that there is nothing mysterious about these properties, which are no more than natural consequences of the laws of physics.

10–1 PRESSURE

When a force **F** acts perpendicular to a surface whose area is A, the pressure p exerted on the surface is defined as the ratio between the magnitude F of the

FIG. 10–1 Three types of pressure gauge. (a) An aneroid measures pressure in terms of the amount by which the thin, flexible ends of an evacuated metal chamber are pushed in or out by the external pressure. (b) A manometer measures pressure in terms of the difference in height *h* of two mercury columns, one open to the atmosphere and the other connected to the source of the unknown pressure. (c) A Bourdon tube straightens out when the internal pressure exceeds the external pressure.

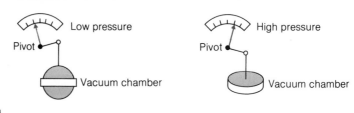

(a) Aneroid

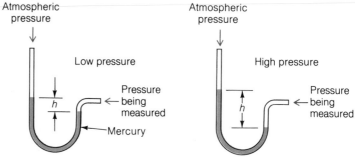

(b) Manometer

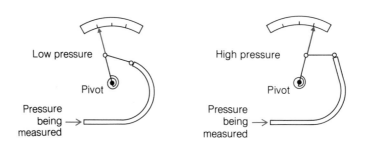

(c) Bourdon tube

Pressure is normal force per unit area

force and the area:

$$p = \frac{F}{A}$$

$$\text{Pressure} = \frac{\text{force}}{\text{area}}$$

Pressure   (10–1)

Pressure is a scalar quantity. Often a perpendicular force is described as *normal*, which allows us to say that *pressure is the magnitude of normal force per unit area*.

Pressures may be measured in a number of ways, three of which are illustrated in Fig. 10–1. Usually what is directly determined is the difference between the unknown pressure and atmospheric pressure. This difference is *gauge pressure,* whereas the true pressure is called the *absolute pressure*. That is,

$$p = p_{\text{gauge}} + p_{\text{atm}} \tag{10-2}$$

Absolute pressure = gauge pressure + atmospheric pressure

Thus a tire inflated to a gauge pressure of 28 lb/in.^{2} contains air at an absolute pressure of 43 lb/in.^{2}, since sea-level atmospheric pressure is 14.7 lb/in.^{2}

The SI unit of pressure is the N/m^2, formally called the *pascal* (Pa). Unfortunately, a number of other pressure units are in common use. The chief ones are listed below.

The *atmosphere* (atm), which represents the average pressure exerted by the earth's atmosphere at sea level; it is equal to 1.013×10^5 N/m^2 (or 14.7 lb/in.^{2} in British units).

The *bar,* equal to 10^5 N/m^2. The *millibar* (mb), which is widely used in meteorology, is equal to 10^{-3} bar or 100 N/m^2. Atmospheric pressure at sea level averages 1013 mb.

The *lb/in.*^{2} (or *psi,* for "pound per square inch") is equal to 6.89×10^3 N/m^2.

The *torr,* which represents the pressure exerted by a column of mercury 1 mm high and is equal to 133 N/m^2. The torr was formerly referred to as the "millimeter of mercury," abbreviated mm Hg.

Example The weight of a 1000-kg car is supported equally by its four tires. The gauge pressure of the air in the tires is 1.8 bars. Find the area of each tire that is in contact with the ground.

Solution The load on each tire consists of the portion of the car's weight it supports plus the weight of a column of the atmosphere whose cross-sectional area equals the area of the tire in contact with the ground, since this part of the tire has no air at atmospheric pressure under it to provide an equal upward force. Hence only the gauge pressure of the air in the tires, which is the excess over atmospheric pressure, is effective in supporting the car's weight. Since the car's weight is

$$w = mg = (1000 \text{ kg})(9.8 \text{ m/s}^2) = 9800 \text{ N}$$

each tire must support 2450 N. The area of each tire in contact with the ground is therefore, since 1 bar = 10^5 N/m^2,

$$A = \frac{F}{p} = \frac{2450 \text{ N}}{1.8 \times 10^5 \text{ N/m}^2} = 0.0136 \text{ m}^2 = 136 \text{ cm}^2 \qquad \blacksquare$$

Pressure is a useful quantity because fluids flow under stress instead of being deformed elastically as solids are. The characteristic lack of rigidity exhibited by fluids has three significant consequences:

(1) The forces a fluid at rest exerts on the walls of its container, and vice versa, always act perpendicular to the walls.

If this were not so, any sideways force by a fluid on a wall would be accompanied, according to the third law of motion, by a sideways force back on the fluid, which would cause the fluid to move parallel to the wall. But the fluid is at rest, so the force must be perpendicular to the container walls. A *moving* fluid is another matter; as we shall learn in Sec. 10–8, frictional forces act between a moving fluid and, for instance, the walls of a pipe or the bank of a river. The difference between a body of liquid in contact with a solid and two solids in contact is that in the former case there is no static friction between them.

(2) An external pressure exerted on a fluid is transmitted uniformly throughout the volume of the fluid.

If this were not so, the fluid would flow from a region of high pressure to one of low pressure, thereby equalizing the pressure. We must keep in mind, however, that the above statement refers to a pressure imposed from outside the fluid. The fluid at the bottom of a container is always under greater pressure than that at the top owing to the weight of the overlying fluid. A notable example is the earth's atmosphere, although such pressure differences are ordinarily significant only for liquids.

(3) The pressure on a small surface in a fluid is the same regardless of the orientation of the surface.

If this were not so, again, the fluid would flow in such a way as to equalize the pressure (Fig. 10–2).

Pumps

A *pump* is a device for creating a pressure difference. The operation of the familiar piston pump is illustrated in Fig. 10–3. A pump of this kind can be used to provide high pressure (to inflate a tire or a football, for instance) or low pressure (to pump water from a well or the bilge of a boat, for instance). The heart circulates blood through the lungs and through the body by means of pumps of this kind, with the contraction and relaxation of the muscular walls of the heart chambers taking the place of piston strokes. Rotary pumps are more suited to motor drive; two examples are shown in Fig. 10–4.

FIG. 10–2 The force exerted by the pressure in a fluid is the same in all directions at any depth.

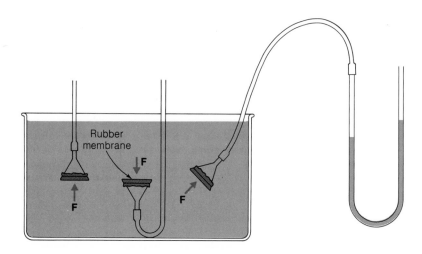

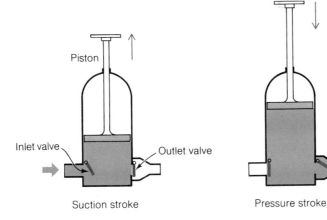

FIG. 10–3 A piston pump. During the suction stroke, the inlet valve opens and the fluid is drawn into the cylinder. When the piston is pushed down, the inlet valve closes and the outlet valve opens, and the fluid is forced out.

Piston

Inlet valve

Outlet valve

Suction stroke

Pressure stroke

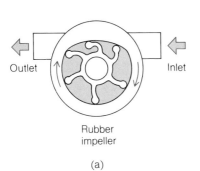

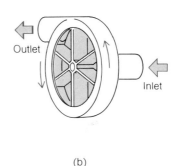

FIG. 10–4 Two types of rotary pump. (a) Impeller pump. The fluid is compressed when the rubber fins of the impeller are bent over at the flattened top of the chamber. (b) Centrifugal pump. The rotating blades set the fluid in motion.

Outlet

Inlet

Rubber impeller

(a)

Outlet

Inlet

(b)

10–2 HYDRAULIC PRESS

We have already considered two of the three basic machines, the lever and the inclined plane. The third, the *hydraulic press,* is based upon property (2) of the previous section: An external pressure exerted on a fluid is transmitted uniformly throughout the volume of the fluid. This statement is known as *Pascal's principle*.

Pascal's principle

Figure 10–5 is a schematic diagram of a hydraulic press. A force F_{in} acts upon a piston of area A_{in} to produce the pressure

The pressure is the same on both pistons, but the force is not

$$p = \frac{F_{in}}{A_{in}}$$

on the confined fluid. This pressure is transmitted by the fluid to the output piston upon which it acts upward, where

$$p = \frac{F_{out}}{A_{out}}$$

FIG. 10–5 The hydraulic press makes use of the fact that pressure exerted on a fluid is transmitted equally throughout the fluid. Valve 1 is closed and valve 2 open on the downstroke of the input piston; valve 1 is open and valve 2 closed on the upstroke.

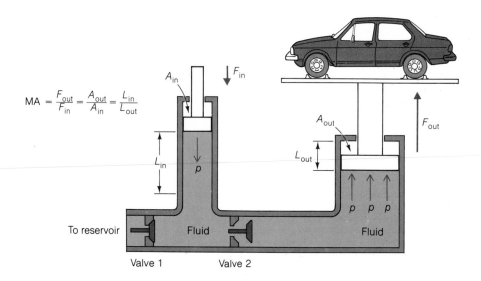

Since the pressure is the same at both pistons,

$$\frac{F_{out}}{A_{out}} = \frac{F_{in}}{A_{in}} \tag{10-3}$$

From the principle of equilibrium the mechanical advantage of the hydraulic press is

$$\mathrm{MA} = \frac{F_{out}}{F_{in}} = \frac{A_{out}}{A_{in}} \qquad \textit{Hydraulic press} \quad (10\text{-}4)$$

A small input force can be considerably increased merely by having the output piston much larger in area than the input piston.

Since the fluid is assumed incompressible, the volume of it transferred from one cylinder to the other as the pistons move must be the same. Hence the piston displacements are inversely proportional to the forces on them: the input piston must move through a large range in order to move the output piston through a small one. If the piston areas are in the ratio 100:1, a force of 1 N can lift a weight of 100 N, but for every cm the weight is raised the input piston must move down 100 cm.

Increased force is obtained at the cost of increased motion

The purpose of the valves indicated in Fig. 10–5 is to permit the output piston to be raised by a series of short strokes of the input piston. When the latter moves downward, valve 1 closes and valve 2 opens, which allows fluid under pressure to move into the large cylinder. When the input piston is pulled upward, valve 1 opens and valve 2 closes, and additional fluid is drawn into the input cylinder to enable it to make another stroke.

The principle of the hydraulic press can be applied in a variety of ways. Common industrial uses include presses of various kinds, garage lifts, control systems in airplanes, and vehicle brakes.

Example The input and output pistons of a hydraulic jack are respectively 1 cm and 4 cm in diameter. A lever with a mechanical advantage of 6 is used to apply force to

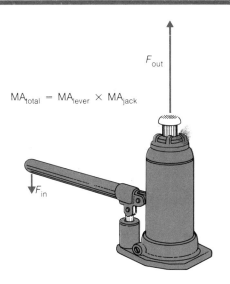

FIG. 10–6 Hydraulic jack.

F_{out}

$MA_{total} = MA_{lever} \times MA_{jack}$

F_{in}

the input piston (Fig. 10–6). How much mass can the jack lift if a force of 180 N (about 40 lb) is applied to the lever and friction is negligible in the system?

Solution The mechanical advantage of the jack is

$$MA = \frac{A_{out}}{A_{in}} = \frac{(\pi r_{out})^2}{(\pi r_{in})^2} = \left(\frac{d_{out}}{d_{in}}\right)^2 = \left(\frac{4\,cm}{1\,cm}\right)^2 = 16$$

The total mechanical advantage of the lever + jack system is

$$MA_{total} = MA_{lever} \times MA_{jack} = 6 \times 16 = 96$$

Hence the output force of the system when the input force is 180 N is

$$F_{out} = MA \times F_{in} = 96 \times 180\,N = 1.73 \times 10^4\,N$$

If the force equals a weight that the jack is lifting, the corresponding mass is

$$M = \frac{w}{g} = \frac{1.73 \times 10^4\,N}{9.8\,m/s^2} = 1.76 \times 10^3\,kg$$

10–3 PRESSURE AND DEPTH

The pressure inside a volume of fluid depends upon the depth below the surface, since the deeper we descend, the greater the weight of the overlying fluid.

Pressure in a fluid increases with depth

Suppose we have a tank of height h and cross-sectional area A which is filled with a fluid of density d (Fig. 10–7). The volume of the tank is

$$V = Ah$$

and the mass of fluid it contains is

$$m = dV = dAh$$

FIG. 10–7 (a) The gauge pressure a fluid in a tank exerts on the bottom is equal to *dgh*. The same formula holds for the fluid pressure at any depth *h* below the surface.

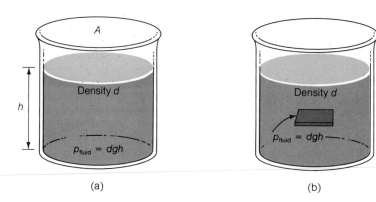

(a)   (b)

The weight of the fluid in the tank is therefore

$$w = mg = dgAh$$

The pressure p_{fluid} the fluid exerts on the bottom of the tank is its weight divided by the area of the bottom, with the result that

$$p_{\text{fluid}} = \frac{F}{A} = \frac{w}{A} = dgh \qquad (10\text{–}5)$$

The pressure difference between the top and the bottom of the tank is directly proportional to the height of the fluid column and to the fluid density.

The above result also applies to *any* depth *h* in a fluid, whether at the bottom or not, since the fluid beneath that depth does not contribute to the weight pressing down there.

Total pressure includes external pressure

The *total* pressure within a fluid, of course, also depends upon the pressure p_{external} exerted on its surface by the atmosphere or, perhaps, by a piston (Fig. 10–8). Thus, in general, the total pressure at a depth *h* in a fluid of density *d* is $p = p_{\text{external}} + p_{\text{fluid}}$:

$$p = p_{\text{external}} + dgh \qquad \textit{Pressure at depth h in a fluid} \quad (10\text{–}6)$$

FIG. 10–8 The total pressure at a depth *h* in a fluid is the sum of the fluid pressure and the external pressure exerted on the fluid surface.

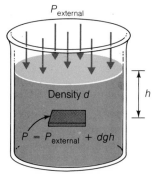

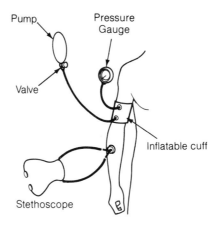

FIG. 10–9 Arterial blood pressures are measured with the help of an inflatable cuff that is wrapped around the upper arm. A pump is used to inflate the cuff until the flow of blood stops, and air is then let out by means of the valve until the flow begins again. The stethoscope is used to monitor the blood flow.

Example Find the pressure on a scuba diver when she is 15 m below the surface of the sea.

Solution The density of seawater is 1.03×10^3 kg/m³. Hence the pressure on the diver due to the sea alone is

$$dgh = \left(1.03 \times 10^3 \frac{\text{kg}}{\text{m}^3}\right)\left(9.8 \frac{\text{m}}{\text{s}^2}\right)(15\,\text{m}) = 1.51 \times 10^5 \frac{\text{N}}{\text{m}^2}$$

Since atmospheric pressure is 1.01×10^5 N/m², the total pressure on the diver is

$$p = p_{\text{external}} + dgh = (1.01 \times 10^5 + 1.51 \times 10^5)\text{N/m}^2$$
$$= 2.52 \times 10^5\,\text{N/m}^2$$

which is about $2\frac{1}{2}$ times atmospheric pressure. She is not crushed because the pressure within her body increases as she descends to match the pressure exerted on it. ∎

Blood pressure

A person's arterial blood pressures are usually measured with the help of an inflatable cuff wrapped around the upper arm at the level of the heart (Fig. 10–9). A stethoscope is used to monitor the sound of the blood flowing through an artery below the cuff. The cuff is first inflated until the flow of blood stops. Then the pressure of the cuff is gradually reduced until the blood just begins to flow, which is recognized by a gurgling sound in the stethoscope. This pressure, called *systolic,* represents the maximum pressure the heart produces in the artery. The pressure in the cuff is then further reduced until the gurgling stops, which corresponds to the restoration of normal blood flow. The pressure at this time, called *diastolic,* represents the pressure in the artery between the strokes of the heart. In a healthy person the systolic and diastolic pressures are respectively about 120 and 80 torr.

Example The average pressure of the blood in a person's arteries is 100 torr at the same elevation as the heart. Find the average pressure in an artery in the head of a standing person, say 40 cm above the heart, and in an artery in the foot, say 120 cm below the heart. The density of blood is 1.06×10^3 kg/m³.

Solution   The difference in pressure in each case is $\Delta p = dgh$. Here

$$\Delta p_1 = dgh_1 = (1.06 \times 10^3 \,\text{kg/m}^3)(9.8 \,\text{m/s}^2)(0.4 \,\text{m}) = 4.16 \times 10^3 \,\text{N/m}^2$$

$$\Delta p_2 = dgh_2 = (1.06 \times 10^3 \,\text{kg/m}^3)(9.8 \,\text{m/s}^2)(1.2 \,\text{m}) = 12.5 \times 10^3 \,\text{N/m}^2$$

Since 1 torr = 133 N/m², Δp_1 = 31 torr and Δp_2 = 94 torr. Hence the pressure in the artery in the head is (100 − 31) torr = 69 torr and that in the artery in the foot is (100 + 94) torr = 194 torr. The arteries that lead to the head expand and contract as needed to keep the flow of blood to the brain constant despite changes in the elevation of the head relative to the heart. Such expansions and contractions require a few seconds to be completed, which explains why sitting up suddenly from a horizontal position may lead to a momentary dizzy sensation.   ■

10–4   ARCHIMEDES' PRINCIPLE

Buoyancy

An object immersed in a fluid seems to weigh less than it does outside. This effect, known as *buoyancy,* permits people to swim, ships to float, and hot air or helium-filled balloons to rise through the atmosphere.

A very simple argument permits us to determine the buoyant force on an object. Suppose we have a solid object of volume V submerged in a fluid of density d. We begin by considering instead a body of fluid of the same size and shape as the object and located at the same depth, as in Fig. 10–10(a). This body of fluid is in equilibrium, which means that its weight of

$$w = Vdg$$

is supported by a buoyant force of this magnitude exerted by the rest of the fluid. The buoyant force is the vector sum of all the forces the rest of the fluid exerts on the body, and this force must be upward because the pressure (and hence the upward force) on the bottom of the body is greater than the pressure (and hence the downward force) on

FIG. 10–10   The buoyant force on a submerged object is equal to the weight of the fluid it displaces.

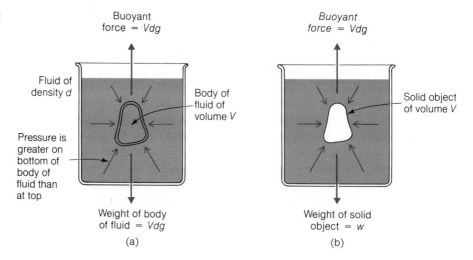

its top. (The forces on the sides of the body as a rule cancel out.) Therefore the buoyant force is Vdg, the weight of the body of fluid.

Now we replace the body of fluid by the solid object, as in Fig. 10–10(b). The various pressures remain the same, so the buoyant force of Vdg also remains the same. We conclude that

Origin of buoyant force

$$F_{\text{buoyant}} = Vdg \qquad \qquad \textit{Archimedes' principle} \quad (10\text{–}7)$$

Buoyant force = weight of displaced fluid

This result is known as Archimedes' principle:

The buoyant force on a submerged object is equal to the weight of fluid displaced by the object.

Archimedes' principle enables us to determine only the buoyant force on a submerged object, not the net force on it. If the weight of the object is greater than the buoyant force on it, it will sink; if the weight is less than the buoyant force, it will rise; if the weight is equal to the buoyant force, it will float in equilibrium.

Net force on object is difference between buoyant force and its weight

Example An iceberg is a chunk of freshwater ice that has broken off from an ice cap, such as those that cover Greenland and Antarctica, or from a glacier at the edge of the sea. (More than three-quarters of the world's fresh water is in the form of ice.) Find the proportion of the volume of an iceberg that is submerged.

Solution If V_{ice} is the iceberg's total volume and V_{sub} is its submerged volume, then

Weight of iceberg = weight of displaced water,

$$V_{\text{ice}}d_{\text{ice}}g = V_{\text{sub}}d_{\text{seawater}}g$$

$$\frac{V_{\text{sub}}}{V_{\text{ice}}} = \frac{d_{\text{ice}}}{d_{\text{seawater}}} = \frac{9.2 \times 10^2\,\text{kg/m}^3}{1.03 \times 10^3\,\text{kg/m}^3} = 0.89$$

Thus 89% of the volume of an iceberg is below sea level (Fig. 10–11). ■

Example A bracelet that appears to be gold ($d_g = 19\ \text{g/cm}^3$) is suspended from a spring scale that reads 50 g in air and 46 g when the bracelet is immersed in a glass of water (Fig. 10–12). Is the bracelet pure gold?

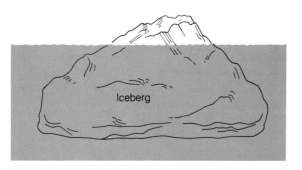

FIG. 10–11 Only 11% of the volume of an iceberg is above sea level.

Iceberg

FIG. 10–12 The volume and density of the bracelet can be determined from these measurements and a knowledge of the density of water.

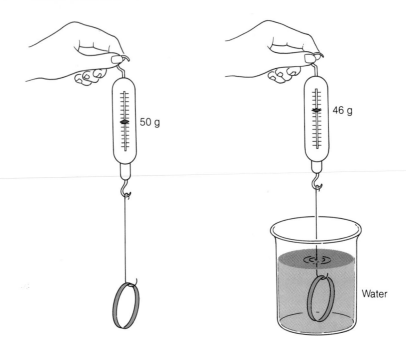

Water

Solution   The first step is to find the volume V of the bracelet. The buoyant force on it in the water is

$$F_b = [m \text{ (in air)} - m \text{ (in water)}]g$$

This force is equal to the weight $Vd_w g$ of the water displaced by the bracelet, so

$$F_b = Vd_w g$$

$$V = \frac{F_b}{d_w g} = \frac{[m \text{ (in air)} - m \text{ (in water)}]g}{d_w g}$$

$$= \frac{m \text{ (in air)} - m \text{ (in water)}}{d_w} = \frac{50 \text{ g} - 46 \text{ g}}{1 \text{ g/cm}^3} = 4 \text{ cm}^3$$

But V is also the bracelet's volume, which means that its density is

$$d_b = \frac{m \text{ (in air)}}{V} = \frac{50 \text{ g}}{4 \text{ cm}^3} = 12.5 \text{ g/cm}^3$$

Evidently the bracelet is not pure gold.   ■

10–5   FLUID FLOW

The study of fluids in motion is one of the more difficult branches of mechanics because of the diversity of phenomena that may occur. However, the fundamental aspects of fluid flow can be understood on the basis of a simple model that is reasonably realistic in many cases. This model concerns liquids that are incompressible and exhibit no

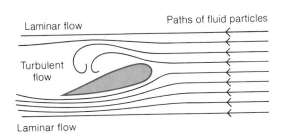

FIG. 10–13 Laminar and turbulent flows around an obstacle.

Laminar flow

Paths of fluid particles

Turbulent flow

Laminar flow

viscosity. (Viscosity is the term used to describe internal friction in a fluid.) In the absence of viscosity, layers of fluid slide freely past one another and past other surfaces, so that it is reasonable to apply the model to such liquids as water but not to such liquids as molasses.

Another approximation we shall make is that the fluid undergoes *laminar* (or *streamline*) *flow* exclusively. In streamline flow, which is illustrated in Fig. 10–13, every particle of liquid passing a particular point follows the same path (called a *streamline*) as the particles that passed that point previously. Furthermore, the direction in which the individual fluid particles move is always the same as the direction in which the fluid as a whole moves.

Laminar flow

At the other extreme is *turbulent flow,* which is characterized by the presence of whirls and eddies, such as those in a cloud of cigarette smoke or at the foot of a waterfall. Turbulence generally occurs at high speeds and when there are obstructions or sharp bends in the path of the fluid.

Turbulent flow

The volume of liquid that flows through a pipe per unit time is easy to compute. If the average speed of the liquid in the pipe of Fig. 10–14 is v, each part of the stream travels the distance vt in the time interval t. The volume of liquid transported the distance vt in the time t is vt multiplied by the pipe's cross-sectional area A, or vtA. Therefore the rate of flow R of liquid through the pipe is

Rate of flow of liquid in a pipe

$$R = \frac{vtA}{t} = vA \qquad \qquad \textit{Rate of flow} \quad (10–8)$$

the product of the liquid speed and the cross-sectional area of the pipe. R is often expressed in units such as liters/min and gal/min instead of ft^3/s or m^3/s.

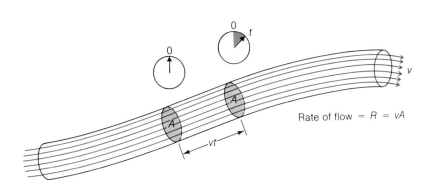

Rate of flow $= R = vA$

FIG. 10–14 The rate of flow of liquid through a pipe is equal to the product of the cross-sectional area of the pipe and the speed of the liquid.

FIG. 10–15 In laminar flow, liquid speed is inversely proportional to the cross-sectional area of the pipe.

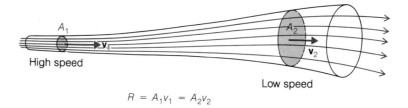

$$R = A_1v_1 = A_2v_2$$

If the pipe size varies, the speed of the liquid also varies so as to keep R constant, so that

$$v_1A_1 = v_2A_2 \qquad\qquad \textit{Equation of continuity} \quad (10\text{–}9)$$

Hence a liquid flows faster through a constriction in a pipe and slower through a dilation. As in Fig. 10–15, streamlines drawn close together signify rapid motion while streamlines far apart signify slow motion.

Example Oxygenated blood from the lungs is pumped by the left ventricle of the heart into a large artery called the aorta, whose diameter is typically 2 cm. When a person is resting, the rate of flow of blood might be 6 liters/min. (a) Find the average speed of blood in the aorta under these circumstances. (b) Find the power output of the left ventricle of a person at rest assuming an average blood pressure of 100 torr at the aorta. (The right ventricle, which pumps blood at the same rate to the lungs where it gives up carbon dioxide and absorbs oxygen, has a smaller power output since there is less resistance to the flow of blood through the lungs than through the rest of the body.)

Solution (a) Since 1 liter $= 10^{-3}$ m^3 and 1 min $= 60$ s, 1 liter/min $= 1.667 \times 10^{-5}$ m^3/s, and $R = 1.00 \times 10^{-4}$ m^3/s here. The speed is therefore

$$v = \frac{R}{A} = \frac{1.00 \times 10^{-4}\,\text{m}^3/\text{s}}{\pi \times 10^{-4}\,\text{m}^2} = 0.318\,\text{m/s}$$

(b) The rate at which the left ventricle does work on the blood passing through it is $P = Fv$, where the force applied is pA. Hence

$$P = pAv = (100\,\text{torr})\left(133\frac{\text{N/m}^2}{\text{torr}}\right)(\pi \times 10^{-4}\,\text{m}^2)(0.318\,\text{m/s}) = 1.33\,\text{W}$$

Both the pressure and the rate of flow increase during physical activity, with a corresponding increase in the power output. ■

10–6 BERNOULLI'S EQUATION

Changes in liquid speed arise from pressure differences

When a liquid flowing through a pipe enters a region where the pipe diameter is reduced, its speed increases. A change in speed involves an acceleration, which means that a net force must be acting upon the liquid. This force can only arise from a difference in pressure between the different parts of the pipe. Evidently the pressure in the part of

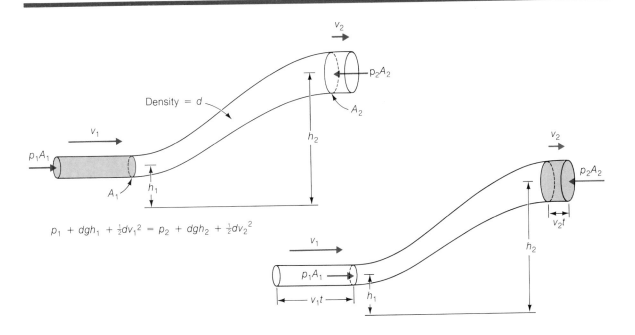

$$p_1 + dgh_1 + \tfrac{1}{2}dv_1{}^2 = p_2 + dgh_2 + \tfrac{1}{2}dv_2{}^2$$

the pipe having a large diameter is the greater, since the liquid increases in speed on its way to the constriction. Thus we expect a relationship between the pressure in a moving liquid and its speed, which turns out to be

FIG. 10–16 Bernoulli's equation.

$$p_1 + dgh_1 + \tfrac{1}{2}dv_1{}^2 = p_2 + dgh_2 + \tfrac{1}{2}dv_2{}^2 \qquad \textit{Bernoulli's equation} \quad (10\text{–}10)$$

Here p_1, h_1, and v_1 are respectively the pressure, height above some reference level, and speed of a liquid of density d at the point 1 in a body of the liquid, and p_2, h_2, and v_2 are the values of these quantities at another point 2.

Equation (10–10) is known as *Bernoulli's equation,* after Daniel Bernoulli (1700–1782), who first derived it. According to Bernoulli's equation the quantity $p + dgh + \tfrac{1}{2}dv^2$ has the same value at all points in an incompressible liquid with negligible viscosity that undergoes laminar flow. The effect of viscosity is to dissipate mechanical energy into heat. If the viscosity of the liquid is not negligible, the quantity $p + dgh + \tfrac{1}{2}dv^2$ decreases in the direction of flow.

Bernoulli's equation expresses conservation of energy in liquid flow

To derive Bernoulli's equation we consider a curved pipe of non-uniform cross section through which a liquid flows, as in Fig. 10–16. Let us apply the principle of conservation of energy to a parcel of the liquid of volume v_1tA_1 as it enters at the left in the time t and to the same parcel as it leaves at the right. The mass of the parcel is

$$m = dV = dv_1tA_1 = dv_2tA_2$$

Mass of liquid parcel

since vA has the same value at 1 and 2. The net amount of work ΔW done on the liquid parcel as it passes from 1 to 2 must be equal to the net change in its potential energy ΔPE as its height goes from h_1 to h_2, plus the net change in its kinetic energy ΔKE as its speed goes from v_1 to v_2. That is,

$$\Delta W = \Delta PE + \Delta KE$$

The work done *on* the parcel at 1 is the force p_1A_1 on it multiplied by the distance v_1t through which the force acts. The work done *by* the parcel at 2 is the force p_2A_2 multiplied by the distance v_2t through which the force acts. The *net* work done on the parcel is therefore

Work done on parcel

$$\Delta W = p_1A_1v_1t - p_2A_2v_2t = \frac{p_1m}{d} - \frac{p_2m}{d}$$

The change in the potential energy of the parcel in going from 1 to 2 is

Change in PE of parcel

$$\Delta PE = mgh_2 - mgh_1$$

and the change in its kinetic energy is

Change in KE of parcel

$$\Delta KE = \tfrac{1}{2}mv_2^2 - \tfrac{1}{2}mv_1^2$$

Therefore

Conservation of energy

$$\Delta W = \Delta PE + \Delta KE$$

$$\frac{p_1m}{d} - \frac{p_2m}{d} = mgh_2 - mgh_1 + \tfrac{1}{2}mv_2^2 - \tfrac{1}{2}mv_1^2$$

When we divide through by the common factor m, multiply by d, and rearrange terms, the result is Bernoulli's equation,

$$p_1 + dgh_1 + \tfrac{1}{2}dv_1^2 = p_2 + dgh_2 + \tfrac{1}{2}dv_2^2$$

10–7 APPLICATIONS OF BERNOULLI'S EQUATION

In many situations the speed, pressure, or height of a liquid is constant, and simplified forms of Bernoulli's equation hold. Thus when a liquid column is stationary, we see that the pressure difference between two depths in it is

Liquid at rest

$$p_2 - p_1 = dg(h_1 - h_2) \tag{10–11}$$

which is just what Eq. (10–5) states. Evidently the latter formula is included in Bernoulli's equation.

Another straightforward result occurs when $p_1 = p_2$. As an example, Fig. 10–17 illustrates a liquid emerging from an orifice in a tank. The liquid pressure equals atmospheric pressure both at the top of the tank and at the orifice. If the orifice is small

Torricelli's theorem

FIG. 10-17 Torricelli's theorem.

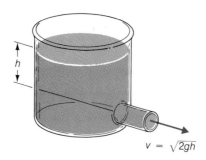

$$v = \sqrt{2gh}$$

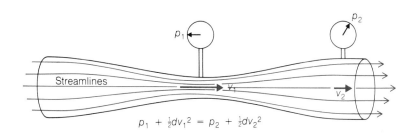

compared with the cross section of the tank, the liquid level in the tank will fall slowly enough for the liquid speed at the top of the tank to be assumed zero. If the speed of the liquid as it leaves the orifice is v and the difference in height between the top of the liquid and the orifice is h, Bernoulli's equation reduces to

$$\tfrac{1}{2}dv^2 = dgh$$
$$v = \sqrt{2gh} \qquad\qquad \textit{Torricelli's theorem} \quad (10\text{–}12)$$

The speed with which the liquid is discharged is the same as the speed of a body falling from rest from the height h. This result is called *Torricelli's theorem,* and, like the relationship between pressure and depth, it is a special case of Bernoulli's equation. The rate at which liquid flows through the orifice may be found from Eq. (10–8) if the orifice area A is known. The volume of liquid being discharged per unit time is **Constant external pressure**

$$R = vA = A\sqrt{2gh} \qquad\qquad \textit{Flow from orifice} \quad (10\text{–}13)$$

Example How fast will water leak through a hole 1 cm^2 in area at the bottom of a tank in which the water level is 3 m high?

Solution From the above formula

$$R = (1 \text{ cm}^2)\left(\frac{1}{10^4 \text{ m}^2/\text{cm}^2}\right)\sqrt{2 \times 9.8 \text{ m/s}^2 \times 3 \text{ m}} = 7.7 \times 10^{-4} \text{ m}^3/\text{s}$$

which is almost a liter per second, an appreciable amount. ∎

The most interesting special case of Bernoulli's equation occurs when there is no change in height during the motion of the liquid (Fig. 10–18). Here

$$p_1 + \tfrac{1}{2}dv_1{}^2 = p_2 + \tfrac{1}{2}dv_2{}^2 \qquad\qquad (10\text{–}14) \quad \textbf{Constant height}$$

which means that the pressure in the liquid is least where the speed is greatest, and vice versa.

A familiar application of Eq. (10–14) is the lifting force produced by the flow of air past the wing of an airplane, as in Fig. 10–19. (Air is, of course, a compressible fluid and so does not fit our model exactly, but the behavior predicted by Bernoulli's equation is not a bad approximation for gases at moderate speeds.) Air moving past the upper surface of the wing must travel faster than air moving past the lower surface; this is indicated by the closeness of the streamlines near the former. The difference in speed leads to a decreased pressure over the top of the wing, a pressure which is equivalent **How an airplane wing develops lift**

FIG. 10–19 Air flow past a wing. At (a) the flow is the same on both surfaces, so that no lift results. The lift at (c) is greater than that at (b) because of the greater pressure differences between upper and lower surfaces (the pressure is least where the streamlines are closest together). At (d) turbulence reduces the available lift.

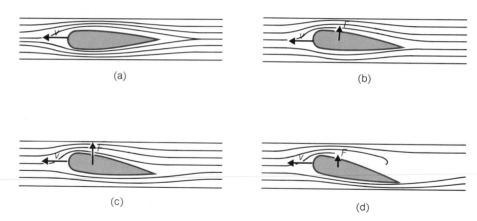

(a)   (b)

(c)   (d)

to a suction force that lifts the wing. The greater the difference in air speeds around the upper and lower surfaces, the greater the lift that is produced, provided the wing shape is not so extreme that turbulence results (Fig. 10–19(d)).

The lift force F_{lift} on a wing of area A is, from Eq. (10–14),

$$F_{\text{lift}} = (p_2 - p_1)A = \tfrac{1}{2}dA(v_2^2 - v_1^2)$$

Here v_1 is the air speed below the wing and v_2 is its speed above the wing. The relationships between these speeds and the speed v of the airplane depend on the shape of the wing and on the angle the wing makes with the direction of the airflow ahead of it, but in general $(v_2^2 - v_1^2)$ is proportional to v^2, so F_{lift} is proportional to Av^2. Thus the faster an airplane moves, the smaller the wing area needed to provide it with lift. Because a small wing also exhibits less resistance to motion through the air ("drag") than a large one, it is clear why high-speed airplanes have relatively small wings. The penalty such airplanes must pay is that their takeoff and landing speeds need to be correspondingly high.

Similar considerations apply to the wings of birds and other flying animals such as bats, even though here the wings provide propulsion as well as lift. A bird's weight w is proportional to its volume and hence to L^3, where L is a representative linear dimension such as its length. The area of the bird's wings is proportional to L^2, so in order that F_{lift} be equal to w—the condition for level flight—v^2L^2 must be proportional to L^3. Hence v^2 must be proportional to L, and v proportional to $\sqrt{L}$; evidently the larger the bird, the greater its minimum flying speed. Of course, different birds have different aerodynamic properties, so we would not expect strict accordance to this relationship, but it does seem to be followed at least in an approximate way. Anyone who has watched birds has observed that small ones can take off quickly with only a few flaps of their wings, whereas large ones, such as ducks, geese, and swans, which require more initial speed, have more trouble in becoming airborne. The huge flying reptiles that lived in the time of the dinosaurs probably had to launch themselves from cliffs and, like present-day sailplanes, could stay aloft for long periods only with the help of rising air currents.

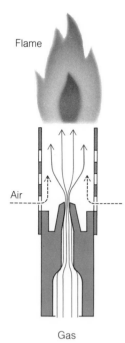

Flame

Air

Gas

FIG. 10–20 Blowtorch nozzle.

The nozzle of a blowtorch (Fig. 10–20) provides another application of the reduced pressure that accompanies an increase in fluid speed. The gas leaves the supply tube

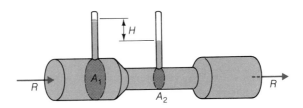

FIG. 10–21 The Venturi meter.

through a small orifice, which means that its speed is high as it escapes. The resulting low pressure sucks in air through perforations in a sleeve around the end of the tube to mix the oxygen needed for combustion with the gas. In this arrangement the flame is hotter than if the gas simply burns at the orifice without being mixed with air first. The laboratory Bunsen burner has a similar construction.

Blowtorch nozzle

Example The device shown in Fig. 10–21, called a *Venturi meter,* provides a convenient method for determining the rate of flow R of a liquid through a pipe. Derive an equation that gives R in terms of the difference in height H between the manometer levels and the cross-sectional areas A_1 and A_2.

Venturi meter

Solution The first step is to apply Bernoulli's equation to the liquid flowing through the pipe. Since the pipe is horizontal, $h_1 = h_2$, and

$$p_1 + \tfrac{1}{2}dv_1^2 = p_2 + \tfrac{1}{2}dv_2^2$$
$$p_1 - p_2 = \tfrac{1}{2}d(v_2^2 - v_1^2)$$

The difference between the heights of the liquid in the vertical manometer tubes reflects the greater pressure at A_1 than at A_2. From Eq. (10–11) we have

$$p_1 - p_2 = dgH$$

Setting equal the above two expressions for $p_1 - p_2$ gives

$$dgH = \tfrac{1}{2}d(v_2^2 - v_1^2)$$
$$2gH = v_2^2 - v_1^2$$

The speeds v_1 and v_2 are related to the cross-sectional areas A_1 and A_2 by $v_1A_1 = v_2A_2$, and so

$$v_2 = v_1\frac{A_1}{A_2} \qquad v_2^2 = v_1^2\left(\frac{A_1}{A_2}\right)^2$$

Therefore

$$2gH = v_2^2 - v_1^2 = v_1^2\left[\left(\frac{A_1}{A_2}\right)^2 - 1\right]$$

Solving for v_1 yields

$$v_1 = \sqrt{\frac{2gH}{(A_1/A_2)^2 - 1}}$$

The rate of flow R is given by $R = v_1A_1$ and so

$$R = A_1 \sqrt{\frac{2gH}{(A_1/A_2)^2 - 1}}$$

which is what is required. Since A_1 and A_2 are fixed quantities, a measurement of the difference H between the manometer levels is all that is needed to find R. ∎

10–8   VISCOSITY

Viscosity refers to internal friction in a fluid

The viscosity of a fluid is a kind of internal friction that prevents neighboring layers of the fluid from sliding freely past one another. Figure 10–22 shows how the speed of a fluid in a pipe varies with distance from the axis. The fluid in contact with the pipe's wall is stationary and, as we would expect, the speed increases to a maximum along the pipe's axis. The smaller the viscosity of the fluid, the greater the various speeds will be, but the characteristic parabolic shape of the speed profile will be maintained as long as the flow remains laminar.

The viscosity of a fluid is given a quantitative definition in terms of the experiment shown in Fig. 10–23, in which a plate of area A is being pulled across a layer of fluid s thick. (A hollow cylinder having a smaller rotating cylinder inside it, with a fluid layer between them, is a better arrangement for actual measurements.) For most fluids it is found that the force F required to pull the plate at the constant speed v is proportional to A and v and inversely proportional to s: the faster the motion and the thinner the layer of fluid, the more force is needed for a given plate area. Different fluids offer different degrees of resistance to the motion, but the force needed varies as Av/s for most of them provided that the speed is not so great that turbulence occurs. Thus we can write

$$F = \frac{\eta Av}{s} \tag{10–15}$$

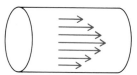

FIG. 10–22 Because of viscosity, the velocity of fluid in a pipe varies from 0 at the pipe wall to a maximum along the axis.

How viscosity is defined

where η, the constant of proportionality, is called the *viscosity* of the fluid (η is the Greek letter *eta*). The SI unit of viscosity is evidently the $N \cdot s/m^2$, which is known as the *poiseuille* (Pl). An older unit, the *poise*, remains in common use, where 10 poise $= 1 \ N \cdot s/m^2$; the *centipoise*, 0.01 poise, is equal to $10^{-3} \ N \cdot s/m^2$.

FIG. 10–23 The viscosity of a fluid is defined in terms of the force needed to pull a flat plate at constant speed across a layer of the fluid.

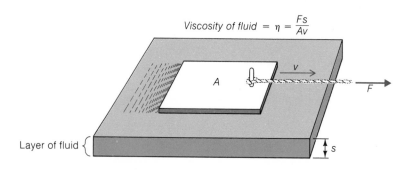

Viscosity of fluid $= \eta = \dfrac{Fs}{Av}$

Layer of fluid

Substance	Temperature (°C)	Viscosity (Pl)	
Gases			**TABLE 10–1**
Air	0	1.7×10^{-5}	Viscosities of various
	100	2.2×10^{-5}	fluids at atmospheric
Water vapor	100	1.3×10^{-5}	pressure and the indicated temperature
Liquids			
Alcohol (ethyl)	20	1.2×10^{-3}	
Blood plasma	37	1.3×10^{-3}	
Blood, whole (varies with speed)	37	$\sim 2 \times 10^{-3}$	
Glycerin	20	0.83	
Water	0	1.8×10^{-3}	
	20	1.0×10^{-3}	
	40	0.66×10^{-3}	
	60	0.47×10^{-3}	
	80	0.36×10^{-3}	
	100	0.28×10^{-3}	

The viscosities of some common fluids are listed in Table 10–1. The viscosity of a liquid decreases with temperature as its molecules become less and less tightly bound to one another (see Chapter 13), a phenomenon familiar to anyone who has heated honey or molasses. One of the reasons a drop in body temperature is so dangerous is that the viscosity of the blood is thereby increased, impeding its flow. The viscosity of a gas, in contrast to that of a liquid, increases with temperature because the higher the temperature, the faster the gas molecules move and the more often they collide with one another (again, see Chapter 13).

Viscosity varies with temperature

Let us return to a fluid moving through a pipe. If the pipe is cylindrical with the length L and inside radius r, and a fluid of viscosity η is flowing through it under the influence of a pressure difference $\Delta p = p_1 - p_2$ as in Fig. 10–24, the rate of flow is

Poiseuille's law gives rate of flow in a pipe

$$R = \frac{\pi r^4 \Delta p}{8\eta L} \qquad \textit{Poiseuille's law} \quad (10\text{--}16)$$

Equation (10–16) is known as *Poiseuille's law*. The dependence of the rate of flow on the viscosity η and on the pressure gradient $\Delta p/L$ are both about what we might expect, but the variation with r^4 is remarkable: halving the radius of a pipe reduces R by a factor of 16 if Δp stays the same. The radius of the pipe plays a far more important role than its length does with respect to viscous resistance.

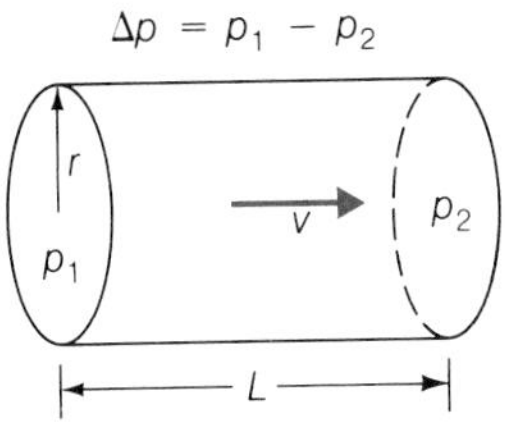

Example Atherosclerosis is the medical term for a common condition in which arteries are narrowed by deposits of tissue called plaque. If the flow of blood is to continue at its usual rate, a higher pressure is required. A consequence of atherosclerosis is therefore an elevated blood pressure, which has many undesirable effects on the body, one of them being that the heart must work harder to circulate the blood. (a) Find the increase in pressure needed to maintain R constant when the radius of an artery is

$$\Delta p = p_1 - p_2$$

FIG. 10–24

decreased by 10%. (b) What is the corresponding increase in the power needed to maintain the flow of blood through that artery?

Solution   (a) For a constant rate of flow R, the product $r^4 \Delta p$ must be constant. Here $r_2 = 0.9r_1$, so

$$r_1^4 \Delta p_1 = r_2^4 \Delta p_2$$

$$\frac{\Delta p_2}{\Delta p_1} = \left(\frac{r_1}{r_2}\right)^4 = \left(\frac{r_1}{0.9r_1}\right)^4 = 1.52$$

The blood pressure must increase by 52% if R is to be unchanged.

(b) We begin by noting that power = force × speed. Since $F = A\Delta p$ and $v = R/A$ here, where A is the cross-sectional area of the artery,

$$P = Fv = (A\Delta p)\left(\frac{R}{A}\right) = R\Delta p$$

and

$$\frac{P_2}{P_1} = \frac{\Delta p_2}{\Delta p_1} = 1.52$$

The power output of the heart must increase by 52% also. Narrowing of the arteries usually means a reduced flow of blood as well as an increase in blood pressure and in the work done by the heart.   ■

10–9   SURFACE TENSION

The surface of a liquid behaves remarkably like a membrane under tension. Thus a steel sewing needle placed horizontally on the surface of some water in a dish does not sink even though its density is nearly eight times that of water (Fig. 10–25). The needle rests in a depression in the water surface just as if that surface were a sheet of rubber stretched across the dish. The term *surface tension* for this effect is quite appropriate. Another manifestation of surface tension is the tendency of a liquid drop to assume a spherical shape, just as an inflated balloon does.

The origin of surface tension lies in the fact that molecules on the surface of a body of liquid are acted upon by a net inward force whereas those in the interior are acted upon by forces in all directions (Fig. 10–26). Since the surface molecules are all being pulled inward, the surface tends to contract to the minimum possible area. A sphere has the least area relative to its volume of any object, and so a liquid sample not acted upon by any external forces (such as gravitation or air resistance if it is moving) will take on a spherical form. A liquid surface can support a small object such as a needle because the weight of the object is insufficient to rupture the surface, which would involve first stretching it to a greater degree than that shown in Fig. 10–25.

A liquid surface has a certain potential energy due to surface tension, just as anything else under tension does. This energy is proportional to the surface area A, so that for a particular liquid

$$PE = \gamma A$$

FIG. 10–25 The surface tension of a water surface supplements buoyancy in supporting a steel needle. The elastic character of a liquid surface resembles that of a membrane under tension.

The constant of proportionality γ (Greek letter *gamma*) is defined as the surface tension of the liquid.

In order to expand a liquid surface by an area ΔA, we might imagine applying a force **F** parallel to the surface along a line of length L perpendicular to **F**. When we have moved the line through the distance Δs, $\Delta A = L \Delta s$ and the work done $F \Delta s$ will have increased the potential energy of the surface by ΔPE. Since $\Delta PE = \gamma \Delta A$,

$$F \Delta s = \gamma L \Delta s$$

$$\gamma = \frac{F}{L}$$

FIG. 10–26 Because a molecule on the surface of a body of liquid has no molecules above it to interact with, the net force on it is inwards.

Thus we can interpret γ as the force of contraction per unit length. The unit of γ can evidently be either the N/m or the J/m^2.

In the case of water, $\gamma = 0.073$ N/m at 20° C. Surface tensions generally decrease with temperature, so that $\gamma = 0.059$ N/m for water at 100°C. Among common liquids, only mercury has a higher surface tension than water, with $\gamma = 0.44$ N/m. At 20°C the surface tension of ethyl alcohol is a third that of water. The high surface tension of water is useful for many small creatures, such as the insects that walk on the surface of a pond and the larvae that are suspended from the surface while they develop. In other places the high γ of water is a handicap. An example is the alveoli of the lungs, which are tiny sacs at the end of the bronchial tubes in whose walls oxygen from the air enters the bloodstream while carbon dioxide leaves. The alveoli are lined with mucus, and if this mucus had the normal surface tension of other tissue fluids, normal breathing would not provide sufficient pressure to expand the alveoli and fill them with air. However, the alveoli secrete a substance that lowers the surface tension of the mucus enough to enable the expansion to occur readily. This substance is an example of a *surfactant*, whose addition to a liquid reduces its surface tension. Detergents are efficient surfactants, which enables even an extremely dilute solution of detergent in water to penetrate small crevices and to be absorbed by tightly-woven textiles. Surfactants are also called "wetting agents" because they allow a water solution to spread uniformly over a surface instead of forming into separate droplets.

Surfactants reduce surface tension

A familiar phenomenon is the rise of most liquids in a capillary tube. Capillarity is responsible for many familiar effects, such as the ability of paper and cloth fibers to absorb water. Two factors are involved: the *cohesion* of the liquid, which refers to the attractive forces its molecules exert on one another, and the *adhesion* of the liquid to the surface of a solid, which refers to the attractive forces the solid exerts on the liquid molecules. If the adhesive forces exceed the cohesive ones, as they do in the case of water and glass, then the liquid tends to stick to the solid and will rise in a capillary tube of that material since the attraction of a liquid molecule to the wall of the tube is greater than the attraction of this molecule to its brethren. In such an event the liquid surface is concave, as in Fig. 10–27(a), with an angle of contact θ that is characteristic of the liquid-solid combination. For water and clean glass, $\theta \approx 0$, as it is for any liquid that "wets" a particular solid, because adhesion must be much greater than cohesion for this to occur. For kerosene and glass, $\theta = 26°$, which reflects the smaller amount by which adhesion exceeds cohesion for these substances. If cohesion is greater than adhesion, as it is for mercury and glass, the liquid level in a capillary tube is lower than the level of the surrounding liquid, and the liquid surface in the tube is convex, with an angle of contact greater than 90°; $\theta = 140°$ for mercury and glass.

Capillarity

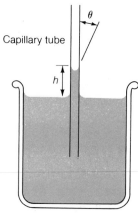

Capillary tube

(a)   Adhesion > cohesion

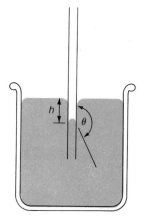

(b) Cohesion < adhesion

It is straightforward to calculate the height h to which a liquid rises (or falls) in a capillary tube. Let us consider the situation shown in Fig. 10–27(a). The upward force on the liquid column is exerted by the vertical component $F \cos \theta$ of the surface tension of the ring of liquid that is adhering to the tube at the top of the column. If the tube has an inner radius of r, the ring is $2\pi r$ in length, and the surface tension force is $F = \gamma L = 2\pi\gamma r$. The downward force is the weight of the elevated liquid, which is $w = dgV = \pi dgr^2 h$ where d is the density of the liquid and $\pi r^2 h$ its volume. Hence

$$F \cos \theta = w$$
$$2\pi\gamma r \cos \theta = \pi dgr^2 h$$
$$h = \frac{2\gamma \cos \theta}{rdg} \qquad\qquad \textit{Capillarity} \quad (10\text{–}17)$$

This formula agrees with our intuitive expectations: the greater the surface tension γ of the liquid, the greater its adhesion to the tube (which means a small θ and a large $\cos \theta$), the narrower the tube, and the less the liquid density, the higher the liquid column. If cohesion exceeds adhesion, as in Fig. 10–27(b), the same formula applies. Now $\theta > 90°$ so $\cos \theta$ is negative, and h is negative also.

Example   The capillaries in a tree trunk are typically 0.02 mm in radius. If the contact angle is 0, find the maximum height to which water can rise in a tree under the influence of surface tension alone.

Solution   At 20°C, $\gamma = 0.073$ N/m. Since $\theta = 0$ here, $\cos \theta = 1$ and, from Eq. (10–17),

$$h = \frac{2\gamma \cos \theta}{rdg} = \frac{(2)(0.073 \text{ N/m})}{(2 \times 10^{-5} \text{ m})(10^3 \text{ kg/m}^3)(9.8 \text{ m/s}^2)} = 0.74 \text{ m}$$

Since trees grow to much greater heights than 74 cm, some mechanism other than capillarity must be involved. It is believed that cohesive forces between water molecules are responsible for the rise of water in trees. As water molecules evaporate from the leaves, other water molecules move to the surface, and these molecules pull the water columns in the capillaries upward behind them. The water columns can thus be regarded as being in tension, so that their internal pressures are negative.   ■

IMPORTANT TERMS

The **pressure** on a surface is the perpendicular force per unit area that acts upon it. **Gauge pressure** is the difference between true pressure and atmospheric pressure.

Pascal's principle states that an external pressure exerted on a fluid is transmitted uniformly throughout its volume.

The **hydraulic press** is a machine consisting of two fluid-filled cylinders of different diameters connected by a tube. The input force is applied to a piston in one of the cylinders, and the output force is exerted by a piston in the other cylinder. The MA of a hydraulic press is equal to the inverse ratio of the cylinder diameters.

Archimedes' principle states that the buoyant force on a submerged object is equal to the weight of fluid it displaces.

In **laminar** (or **streamline**) **flow** every particle of fluid passing a particular point follows the same path, whereas in **turbulent flow** irregular whirls and eddies occur. The greater the speed of a fluid in streamline flow, the lower its pressure.

The **viscosity** of a fluid is a measure of its internal friction.

The **surface tension** of a liquid refers to the tendency of its surface to contract to the minimum possible area in any situation.

IMPORTANT FORMULAS

Pressure: $p = \dfrac{F}{A} = p_{\text{gauge}} + p_{\text{atm}}$

Pressure at depth h in a fluid: $p = p_{\text{external}} + dgh$

Archimedes' principle: $F_{\text{buoyant}} = Vdg$

Equation of continuity: $R = v_1 A_1 = v_2 A_2$

Bernoulli's equation: $p + dgh + \frac{1}{2}dv^2 = \text{constant}$

Poiseuille's law: $R = \dfrac{\pi r^4 \, \Delta p}{8 \eta L}$

Capillarity: $h = \dfrac{2\gamma \cos \theta}{rdg}$

MULTIPLE CHOICE

1. The fluid at the bottom of a container is
  (a) under less pressure than the fluid at the top.
  (b) under the same pressure as the fluid at the top.
  (c) under more pressure than the fluid at the top.
  (d) any of the above, depending upon the circumstances.

2. The pressure at the bottom of a vessel filled with liquid does *not* depend on the
  (a) acceleration of gravity.
  (b) liquid density.
  (c) height of the liquid.
  (d) area of the liquid surface.

3. A man stands on a very sensitive scale and inhales deeply. The reading on the scale
  (a) does not change.
  (b) increases.
  (c) decreases.
  (d) depends on the expansion of his chest relative to the volume of air inhaled.

4. Bernoulli's equation is based upon
  (a) the second law of motion.
  (b) the third law of motion.
  (c) conservation of momentum.
  (d) conservation of energy.

5. An express train goes past a station platform at high speed. A person standing at the edge of the platform tends to be
  (a) attracted to the train.
  (b) repelled from the train.
  (c) attracted or repelled, depending on the ratio between the speed of the train and the speed of sound.
  (d) unaffected by the train's passage.

6. The volume of liquid flowing per second out of an orifice at the bottom of a tank does *not* depend on
  (a) the area of the orifice.
  (b) the height of liquid above the orifice.
  (c) the density of the liquid.
  (d) the value of the acceleration of gravity.

7. The hydraulic press is able to produce a mechanical advantage because
  (a) the force a fluid exerts on a piston is always parallel to its surface.
  (b) an external pressure exerted on a fluid is transmitted uniformly throughout its volume.
  (c) at any depth in a fluid the pressure is the same in all directions.
  (d) the pressure in a fluid varies with its speed.

8. In the operation of a hydraulic press, it is impossible for the output piston to exceed the input piston's
  (a) displacement.   (b) speed.
  (c) force.   (d) work.

9. The input piston of a hydraulic press is 2 cm in diameter and the output piston is 1 cm in diameter. An input force of 1 N will produce an output force of
  (a) 0.25 N.   (b) 0.50 N.
  (c) 2 N.   (d) 4 N.

10. A manometer whose upper end is evacuated and sealed can be used to measure atmospheric pressure. Such an instrument is called a mercury barometer, and the average height of the mercury column in it is 760 mm. If water were used in a barometer instead of mercury, the height of the column of water would be about
  (a) 56 mm.   (b) 760 mm.
  (c) 10 m.   (d) 20 m.

11. Atmospheric pressure does not correspond to approximately
  (a) 14.7 lb/in.^{2} (b) 98 N/m^2.
  (c) 1013 mb.   (d) 1.013×10^5 Pa.

12. A viewing window 30 cm in diameter is installed 3 m below the surface of an aquarium tank filled with sea water. The force the window must withstand is approximately
  (a) 22 N.   (b) 218 N.
  (c) 2140 N.

13. A force of 1000 N is required to raise a concrete block to the surface of a freshwater lake. The force required to lift it out of the water is approximately

(a) 700 N.   (b) 1062 N.

(c) 1140 N.   (d) 1800 N.

14. The depth in fresh water at which the water density is 1% greater than its value at the surface is approximately

(a) 2.3×10^2 m.   (b) 2.3×10^3 m.

(c) 2.3×10^4 m.   (d) 2.3×10^5 m.

15. The total cross-sectional area of all the capillaries of a certain person's circulatory system is 0.25 m^2. If blood flows through the system at the rate of 100 cm^3/s, the average speed of blood in the capillaries is

(a) 0.4 mm/s.   (b) 4 mm/s.

(c) 25 mm/s.   (d) 400 mm/s.

16. Water leaves the safety valve of a boiler at a speed of 30 m/s. The gauge pressure inside the boiler is

(a) 4.5 millibars.   (b) 1.5 bars.

(c) 4.5 bars.   (d) 450 bars.

17. Of the following liquids, the one with the highest surface tension is

(a) cold water.   (b) hot water.

(c) soapy water.   (d) alcohol.

18. Water neither rises nor falls in a silver capillary. This suggests that the contact angle between water and silver is

(a) 0   (b) 45°.

(c) 90°.   (d) 180°.

EXERCISES

(Assume laminar flow and negligible viscosity unless otherwise noted.)

1. A little water is boiled for a few minutes in a tin can, and the can is sealed while it is still hot. Why does the can collapse as it cools?

2. A helium-filled balloon rises to a certain altitude in the atmosphere and floats there instead of rising indefinitely. Why?

3. The height of water at two identical dams is the same, but dam A holds back a lake containing 2 km^3 of water, while dam B holds back a lake containing 1 km^3 of water. What is the ratio of the total force exerted on dam A to that exerted on dam B?

4. An aluminum canoe is floating in a swimming pool. After a while it begins to leak and sinks to the bottom of the pool. What happens to the water level in the pool?

5. An ice cube floats in a glass of water filled to the brim. What will happen when the ice melts?

6. A wooden block is in such perfect contact with the bottom of a water tank that there is no water beneath it. Is there a buoyant force on the block?

7. Will there be any difference between the rates of flow from pipes A and B in the figure below? If so, from which pipe will the liquid flow faster?

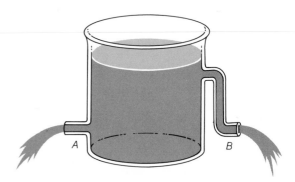

8. A bird is in a closed box that rests on a scale. Does the reading on the scale depend on whether the bird is standing on the bottom of the box or is flying around inside it? What would happen if the bird were in a cage instead?

9. Why is it more dangerous when ice forms on an airplane's wings than when it forms on its fuselage?

10. Does a fluid whose density is greater than that of another fluid necessarily have a greater viscosity as well?

11. Two spheres of the same diameter but of different mass are dropped from a tower. If air resistance is the same for both, which will reach the ground first? Why?

12. A 2-kg brick has the dimensions 7.5 cm × 15 cm × 30 cm. Find the pressures exerted by the brick on a table when it is resting on its various faces.

13. The force on a phonograph needle whose point is 0.1 mm in radius is 0.2 N. What is the pressure in atmospheres that it exerts on the record?

14. A piston weighing 12 N rests on a sample of gas in a cylinder 5 cm in diameter. (a) What is the gauge pressure in the gas? (b) What is the absolute pressure in the gas?

15. A cork 2 cm in radius is used to close one end of a tube whose other end is connected to a vacuum pump. The pump removes virtually all the air from the tube. How much force would be needed to pull the cork out?

16. A Super Constellation airplane whose mass is 50,000 kg is in level flight. The area of its wings is 153 m^2. What is the average difference in pressure between the upper and lower surfaces of its wings?

17. The input and output pistons of a hydraulic jack are respectively 5 mm and 30 mm in diameter. A lever with a mechanical advantage of 8 is used to apply force to the input piston. How much mass can the jack lift if a force of 150 N is applied to the lever and friction is negligible in the system?

18. A lever with a mechanical advantage of 5 is attached to the pump piston of a hydraulic press. The area of the pump piston is 10 cm^2 and that of the output piston is 125 cm^2. (a) If the press is perfectly efficient, find the force the output piston exerts when a force of 120 N is applied to the pump lever. (b) If each stroke of the lever moves the pump piston 3 cm, how many strokes are needed to move the output piston 30 cm?

19. In 1960 the U.S. Navy bathyscaphe *Trieste* descended to a depth of 10,920 m in the Pacific Ocean near Guam. Neglecting the increase in water density with depth, find the pressure on the *Trieste* at the bottom of its dive.

20. If the density of the atmosphere were uniform instead of decreasing with altitude, how thick a layer would it form?

21. A person sucking hard on a thin tube can reduce the pressure in it to 90% of atmospheric pressure. How high can the person suck water up the tube?

22. A standing woman whose head is 35 cm above her heart bends over so that her head is 35 cm below her heart. What is the change (in torr) in the blood pressure in her head?

23. The densities of people are slightly less than that of water. Assuming that these densities are the same, find the buoyant force of the atmosphere on a 70-kg man.

24. What is the minimum area of an ice floe 7 cm thick that can support a 50-kg woman without getting her feet wet? The floe is in a freshwater lake.

25. An 800-kg balloon is drifting horizontally when 20 kg of ballast is dropped in order to start it climbing. What is the balloon's acceleration?

26. Water flows through a hose whose internal diameter is 1 cm at a velocity of 1 m/s. What should the diameter of the nozzle be if the water is to emerge at 5 m/s?

27. During the pumping phase of the heart's action the pressure of blood in the major arteries of a normal person at the level of the heart is about 120 torr. If one of these arteries is cut and blood spurts out vertically, how high will it go?

PROBLEMS

1. A submarine is at a depth of 30 m in sea water. The interior of the submarine is maintained at normal atmospheric pressure. Find the force that must be withstood by a square hatch 80 cm on a side.

2. Calculate the density of sea water at a depth of 5 km. Use the bulk modulus for water given in Table 9–4.

3. Blood plasma has a density of 1026 kg/m^3. What is the minimum height above a vein in which the blood pressure is 16 torr that a container of blood plasma should be placed in order that the plasma enter the vein?

4. A person's head is 40 cm above his heart. The average arterial blood pressure at the level of the heart is 100 torr. Find the average arterial pressure in the person's head when standing and when lying down in an elevator accelerated upward at 3 m/s^2.

5. A steel tank whose capacity is 200 liters has a mass of 36 kg. (a) Will it float in seawater when empty? (b) When filled with fresh water? (c) When filled with gasoline?

6. A sailboat has 8000 kg of lead ballast attached to the bottom of its keel, the center of gravity of which is 1.5 m below the water surface when the boat is level. What is the torque exerted by this ballast when the boat is heeled by 20° from the vertical? (*Hint:* Why is this problem here instead of in Chapter 4?)

7. A 30-kg balloon is filled with 100 m^3 of hydrogen. How much force is needed to hold it down?

8. The largest rigid airship ever built was the German *Hindenburg*, which was 245 m long and had a capacity of 2 × 10^5 m^3 of hydrogen. It was destroyed in a fire at Lakehurst, N.J., in 1937. What was the total payload of the *Hindenburg* including its structure but not including the hydrogen it contained?

9. When a 500-g statue of a falcon suspended from a spring scale is immersed in water, the scale reads 400 g, and when it is suspended in benzene, the scale reads 412 g. What is the density of benzene?

10. A lead fishing sinker suspended from a spring scale is immersed in oil whose density is 820 kg/m^3. The reading on the scale is 0.2 N. Find the mass of the sinker.

11. A pipe whose inside diameter is 30 mm is connected to three smaller pipes whose inside diameters are 15 mm each. If the liquid speed in the larger pipe is 1 m/s, find its speed in the smaller pipes.

12. A tank of height H is filled with water; it is open at the top. A hole is made a distance y from the top. How far from the tank does the water strike the ground?

13. A horizontal stream of water leaves an orifice 1 m above the ground and strikes the ground 2 m away. (a) What is the speed of the water when it leaves the orifice? (b) What is the gauge pressure behind it?

14. The figure on page 238 shows a hose whose internal cross-sectional area is 2 cm^2 being used to siphon water out of a tank. Find the speed with which the water emerges and the rate of flow in liters per second.

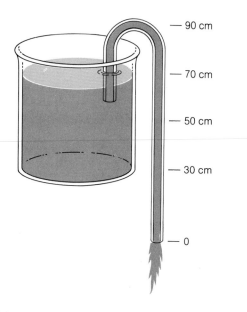

— 90 cm

— 70 cm

— 50 cm

— 30 cm

— 0

15. A boat strikes an underwater rock and punctures a hole 20 cm^2 in area in its hull 65 cm below the waterline. How many liters of water enter the hull per second?

16. A barrel filled with ethyl alcohol is 1.2 m high. A crack 15 mm long and 2 mm wide appears at its base. How many kg of alcohol flow out per minute?

17. The left ventricle of a certain running man pumps 20 liters of blood per minute into his aorta at an average pressure of 140 torr. Find the total power output of his heart under the assumption that his right ventricle has an output 20% as great as that of his left ventricle.

18. Water emerges from a fire hose at a speed of 20 m/s and a rate of flow of 50 liters/s. (a) Find the force with which the nozzle must be held. (b) Find the required power of the pump motor assuming 50% overall efficiency.

19. The bilge pump on a boat is able to raise 100 liters of seawater per minute through a height of 1.5 m. If the overall efficiency of the system is 30%, what is the power output of the pump motor?

20. A 1360-hp pump throws a jet of water 130 m into the air in Geneva, Switzerland. (a) With what speed does the water leave the mouth of the fountain? (b) If the overall efficiency is 60%, how many kilograms of water per minute are thrown into the air?

21. A horizontal pipe 2 cm in radius at one end gradually increases in size so that it is 5 cm in radius at the other end.

The pipe is 4 m long. Water is pumped into the small end of the pipe at a speed of 8 m/s and a pressure of 2 bars. Find the speed and pressure of the water at the pipe's large end.

22. The pipe of Problem 21 is turned so as to be vertical with the small end underneath, so the water flows upward. Find the speed and pressure of the water at the large end of the pipe now.

23. (a) Milk flows through a Venturi meter whose cross-sectional areas are 6 cm^2 and 2 cm^2. Find the rate of flow in liters per second when the difference between the manometer heights is 10 cm. (b) Find the rate of flow when the difference between the manometer heights is the same but the liquid is water.

24. A grease nipple of a car has a hole 0.5 mm in diameter and 5 mm long. If the viscosity of the grease used is 80 N·s/m^2, find the pressure needed to force 1 cm^3 of grease into the nipple in 10 s. How many times atmospheric pressure is this?

25. Find the pressure gradient needed to pump 16 liters of water at 20°C per minute through a pipe 6 mm in radius. Take the viscosity of the water into account.

26. A typical capillary is 1 mm long and has radius of 2 μm. (a) If the pressure difference between its ends is 20 torr, find the average speed of the blood that flows through such a capillary. (b) If the heart pumps 80 cm^3 of blood through a person's body per second, how many capillaries does the body contain?

27. Find the radius of a capillary in which water at 20°C rises 20 mm.

28. Ethyl alcohol rises 14.2 mm in a capillary of radius 0.4 mm. The contact angle is 0. Find the surface tension of ethyl alcohol.

29. A glass capillary whose radius is 0.5 mm is dipped in a dish of mercury, whose surface tension is 0.44 N/m. The contact angle is 140°. Find the height of the mercury in the capillary relative to the mercury surface in the dish.

ANSWERS TO MULTIPLE CHOICE

1. c	**4.** d	**7.** b	**10.** c	**13.** d	**16.** c
2. d	**5.** a	**8.** d	**11.** b	**14.** b	**17.** a
3. d	**6.** c	**9.** a	**12.** c	**15.** a	**18.** c

HARMONIC MOTION

11

Many events in both nature and technology are periodic, with a certain motion repeating itself over and over again. The term simple harmonic motion describes the most fundamental kind of oscillatory behavior, and all periodic events are either examples of simple harmonic motion or else the result of several such motions superimposed upon one another. In harmonic motion, the energy of the vibrations is continually transformed from kinetic to potential and back again. The potential energy may be elastic rather than gravitational in character; we shall find later that elasticity and gravitation are not the only phenomena in which potential energy is a useful concept. The electrical equivalent of potential energy, in particular, makes possible electrical oscillations that closely resemble harmonic motion.

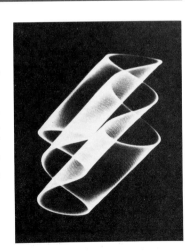

11–1   ELASTIC POTENTIAL ENERGY

When we stretch a spring, it resists being lengthened, and if we then let it go, the spring returns to its original length. As we know, this is an example of

FIG. 11–1 When a spring (or other elastic body) is stretched or compressed, a restoring force comes into being that tries to return the spring to its normal length.

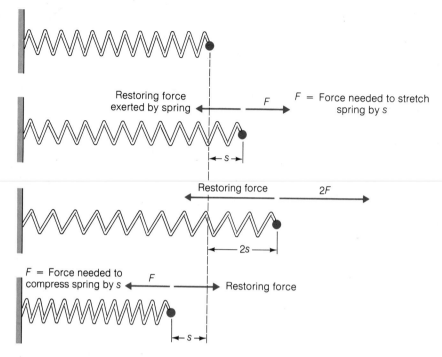

Elastic and plastic behavior

elastic behavior. On the other hand, when we stretch a piece of taffy, it also resists being lengthened, but if we then let it go, nothing happens: the deformation is permanent. This is an example of *plastic* behavior.

In the case of the stretched spring, a *restoring force* comes into being that tries to return the spring to its normal length. The farther we stretch the spring, the greater the restoring force we must overcome (Fig. 11–1). Exactly the same phenomenon occurs

A restoring force occurs when an elastic object is deformed

when we compress the spring: it resists being shortened, and if we let it go, the spring returns to its normal length. Again a restoring force arises, and again the more the compression, the stronger the restoring force to be overcome.

The amount *s* by which an elastic solid is stretched or compressed by a force is directly proportional to the magnitude *F* of the force, provided the elastic limit is not exceeded. This proportionality is called Hooke's law, as mentioned in Chapter 9. Thus we can write

$$F = ks \qquad\qquad\qquad\qquad Hooke's\ law \quad (11\text{–}1)$$

where *k* is a constant whose value depends upon the nature and dimensions of the object. A stiff spring has a higher value of *k* than a weak one.

The work done in stretching (or compressing) an object that obeys Hooke's law is easy to calculate. The work done by a force is the product of the magnitude of the force and the distance through which it acts. Here the force used in stretching the object

Work must be done to overcome the restoring force

is not constant but is proportional to the elongation *s* at each point in the stretching process. Because *F* is proportional to *s*, the *average* force $\overline{F}$ applied while the body is stretched from its normal length by an amount *s* to its final length is

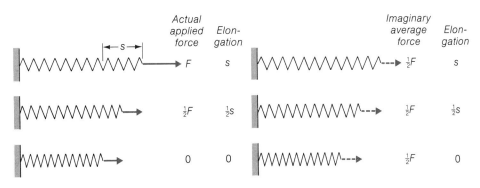

FIG. 11–2 To compute the work done in stretching a body that obeys Hooke's law, the varying force that actually acts during the expansion may be replaced by the average force.

$$\overline{F} = \frac{F_{\text{initial}} + F_{\text{final}}}{2} = \frac{0 + ks}{2} = \frac{1}{2}ks$$

since the initial force is 0 and final force is ks (Fig. 11–2). The work done in stretching the spring is the product of the average force $\overline{F} = \frac{1}{2}ks$ and the total elongation s, so that

$$W = \text{PE} = \frac{1}{2}ks^2 \qquad\qquad \textit{Elastic potential energy} \quad (11\text{--}2)$$

This formula is most often used in connection with springs: to stretch (or compress) a spring whose force constant is k by an amount s from its normal length requires $\frac{1}{2}ks^2$ of work to be done. This work goes into *elastic potential energy*. When the spring is released, its potential energy of $\frac{1}{2}ks^2$ is transformed into kinetic energy or into work done on something else (Fig. 11–3); work done against frictional forces within the spring itself always absorbs some fraction of the available potential energy.

Work done to deform an elastic object equals PE of deformed object

Example   The horizontal spring shown in Fig. 11–4 has a force constant k of 90 N/m. Attached to the free end of the spring is a 1.4 kg block. If the spring is pulled

FIG. 11–3 Some devices that make use of elastic potential energy in their operation.

FIG. 11–4 A 1.4-kg block attached to a spring whose force constant is 90 N/m is pulled 50 cm from its equilibrium position. When the spring is released, its elastic potential energy of $\frac{1}{2}ks^2$ is converted into kinetic energy $\frac{1}{2}mv^2$, and as the block's momentum compresses the spring on the other side of the equilibrium position, the kinetic energy is converted back into elastic potential energy.

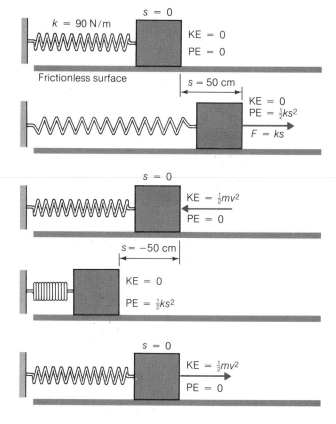

out 50 cm from its equilibrium position and then released, what will the block's speed be when it returns to the equilibrium position?

Solution   When the spring is released, its elastic potential energy starts to be converted into kinetic energy of the block. We shall assume that the spring's mass is small compared with that of the block and that its internal friction may be neglected. At the equilibrium position of the spring, $s = 0$, and all the initial potential energy of $\frac{1}{2}ks^2$ is now kinetic energy $\frac{1}{2}mv^2$. Hence

$$\tfrac{1}{2}mv^2 = \tfrac{1}{2}ks^2$$

$$v = \sqrt{\frac{k}{m}}\,s = \sqrt{\frac{90\,\text{N/m}}{1.4\,\text{kg}}} \times 0.50\,\text{m} = 4.0\,\text{m/s} \qquad ■$$

11–2   SIMPLE HARMONIC MOTION

The energy of an oscillator shifts back and forth between PE and KE

When a spring with an object attached to it is stretched and then released, it does not simply return to its equilibrium position and come to a stop there. What happens is that the elastic potential energy $\frac{1}{2}ks^2$ of the spring is converted into kinetic energy $\frac{1}{2}mv^2$ of the moving object, and as the latter's momentum compresses the spring on the other side of the equilibrium position, this kinetic energy is converted back into elastic potential energy (Fig. 11–4). The amount of compression $-s$ will have the same magnitude as the original extension s, since

$$\tfrac{1}{2}ks^2 = \tfrac{1}{2}k(-s)^2$$

Left to itself in the absence of friction, the spring-object combination will continue oscillating back and forth indefinitely. The behavior of a system oscillating in this way is called *simple harmonic motion*.

Condition for simple harmonic motion

Simple harmonic motion occurs whenever a force acts on a body in the opposite direction to its displacement from its normal position, with the magnitude of the force proportional to the magnitude of the displacement. The elastic restoring force of a stretched or compressed spring always tends to return the spring to its normal length, but the momentum associated with the moving mass compels it to overshoot and thus to oscillate.

Period of harmonic motion

The *period* of a body undergoing simple harmonic motion is the time required for it to make one complete oscillation. (A complete oscillation is often called a *cycle*.) In the case of a spring, the period is the time the spring spends in going from its maximum extension, say, through its maximum compression and back to its maximum extension once more (Fig. 11–5). For all types of simple harmonic motion, the period T is

$$T = 2\pi\sqrt{-\frac{s}{a}} \qquad \qquad \textit{Simple harmonic motion} \quad (11\text{–}3)$$

$$\text{Period} = 2\pi\sqrt{-\frac{\text{displacement}}{\text{acceleration}}}$$

where the acceleration is that experienced by the body when it is at the specified displacement from its equilibrium position. This formula is derived in Sec. 11–3.

To calculate the acceleration of a stretched spring we start with the second law of motion,

$$F = ma$$

and substitute the restoring force $F_r = -ks$ since it is the restoring force that causes the body to be accelerated. This procedure yields

$$F_r = ma$$

$$a = \frac{F_r}{m} = \frac{-ks}{m} \qquad \qquad (11\text{–}4)$$

With this result we find that the period of a body of mass m attached to a spring of force constant k is

Period of system of spring + mass

$$T = 2\pi\sqrt{-\frac{s}{a}} = 2\pi\sqrt{-\frac{s}{-ks/m}}$$

$$T = 2\pi\sqrt{\frac{m}{k}} \qquad \qquad \textit{Oscillating spring} \quad (11\text{–}5)$$

It is worth noting that the period T does not depend upon the maximum displacement s; no matter how much or how little the spring is initially pulled out, precisely

FIG. 11–5 (a) The period *T* of a body undergoing simple harmonic motion is the time required for it to make one complete oscillation. (b) A computer-generated depiction of the motion of a pendulum in a vacuum. Each curve shows displacement angle as a function of time.

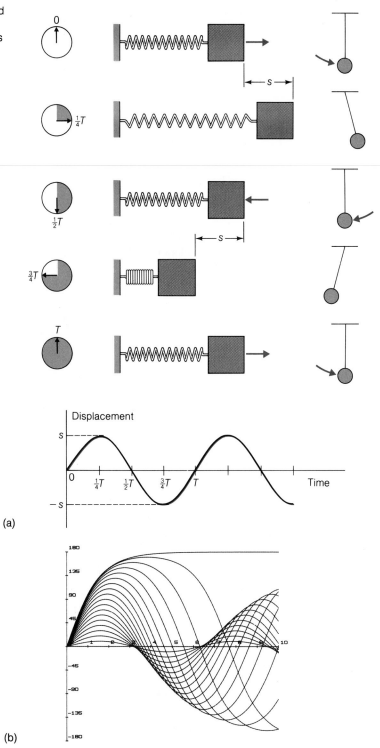

FIG. 11–6

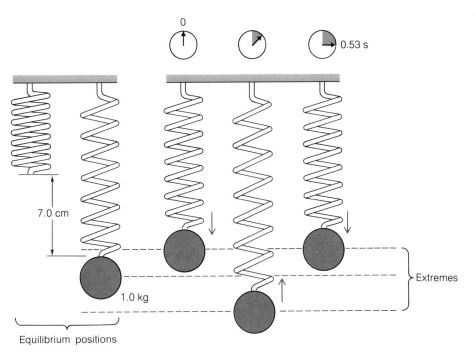

the same amount of time is required for each cycle. If s is small, the maximum acceleration is also small and the body moves back and forth very slowly through its range, while if s is large, the acceleration is also large and the body moves correspondingly rapidly through the larger range. (The maximum displacement A of a body undergoing harmonic motion on either side of its equilibrium position is called the *amplitude* of the motion.) This peculiarity of simple harmonic motion is capitalized upon in the design of mechanical clocks and watches, which use the rotational oscillations of a coil spring or the swings of a pendulum—both essentially simple harmonic motions—to maintain a constant rate independent of any fluctuations in amplitude.

> **Period is independent of amplitude of motion**

A quantity often used in describing harmonic motion is *frequency*. The frequency is the number of cycles executed per unit time. Hence frequency, whose symbol is f, is the reciprocal of period T,

> **The unit of frequency is the hertz**

$$f = \frac{1}{T} \qquad\qquad \textit{Frequency} \quad (11\text{–}6)$$

The unit of frequency is the *hertz* (Hz), where 1 Hz = 1 cycle/s.

The position of a particle undergoing simple harmonic motion varies with time as shown on the graph in Fig. 11–5. The curve has exactly the same shape as a curve of $\sin \theta$ plotted versus θ.

Example   When a 1.0-kg ball is suspended from a spring, the spring stretches by 7.0 cm (Fig. 11–6). If the ball oscillates up and down, what is its period? What is its frequency?

Solution The force exerted on the spring is the ball's weight of

$$mg = (1.0 \text{ kg})(9.8 \text{ m/s}^2) = 9.8 \text{ N}$$

Since $F = ks$, the force constant of the spring is

$$k = \frac{F}{s} = \frac{mg}{s} = \frac{9.8 \text{ N}}{0.070 \text{ m}} = 140 \text{ N/m}$$

The period of the oscillations is therefore

$$T = 2\pi \sqrt{\frac{m}{k}} = 2\pi \sqrt{\frac{1.0 \text{ kg}}{140 \text{ N/m}}} = 0.53 \text{ s}$$

and their frequency is

$$f = \frac{1}{T} = \frac{1}{0.53 \text{ s}} = 1.9 \text{ cycles/s} = 1.9 \text{ Hz} \qquad ■$$

11–3 A MODEL OF SIMPLE HARMONIC MOTION

This model makes it easy to find the period of simple harmonic motion

Figure 11–7 shows a particle moving in a vertical circle at constant speed. The particle is illuminated from above, and it casts a shadow on a horizontal screen below its orbit. As the particle travels around the circle, its shadow oscillates back and forth. The shadow moves fastest at the center, slows down as it approaches each end of the path, comes to a stop, and then reverses its direction. We might suspect that the shadow is executing simple harmonic motion. To verify this suspicion, we must show that the acceleration a of the shadow at any time is proportional to its displacement s from the center of its path and opposite in direction.

Proof that the shadow executes simple harmonic motion

The acceleration of the shadow at any point is simply the horizontal component of the particle's acceleration a_c. As we know, a particle in uniform circular motion at the speed V in a circle of radius R experiences the centripetal acceleration

$$a_c = -\frac{V^2}{R}$$

where the minus sign indicates that the acceleration is inward toward the center of the circle. The horizontal component of this acceleration is the acceleration a of the shadow. Since corresponding sides of similar triangles are proportional, we see from Fig. 11–8 that

$$\frac{a}{a_c} = \frac{s}{R} \qquad \text{and so} \qquad a = \frac{s}{R} a_c$$

Because $a_c = -V^2/R$, the acceleration of the shadow is

$$a = -\frac{s}{R}\frac{V^2}{R} = -\frac{V^2}{R^2} s \qquad (11\text{–}7)$$

Thus the shadow's acceleration is proportional to its displacement s and in the opposite direction, which means that the shadow is indeed executing simple harmonic motion.

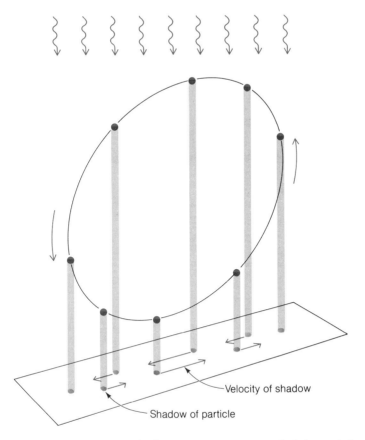

FIG. 11–7 The shadow of a particle undergoing uniform circular motion executes simple harmonic motion.

Velocity of shadow

Shadow of particle

The value of the above analysis is that it gives us an easy way to find the period of an object in simple harmonic motion. The circumference of a circle of radius R is $2\pi R$, and a particle moving around the circle with the constant speed V covers this distance in the time

The particle and its shadow have the same period

$$T = \frac{2\pi R}{V}$$

$$\text{Period} = \frac{\text{distance}}{\text{speed}}$$

From Eq. (11–7) we find that

$$\frac{R}{V} = \sqrt{-\frac{s}{a}}$$

and so

$$T = 2\pi \sqrt{-\frac{s}{a}} \qquad\qquad (11\text{–}3)$$

This is the general formula for the period of simple harmonic motion given in the preceding section.

FIG. 11–8 The acceleration of the shadow is the horizontal component of the particle's centripetal acceleration. The shadow's acceleration is proportional to its displacement s and is in the opposite direction.

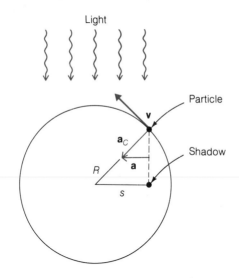

Light

Particle

Shadow

11–4   POSITION, SPEED, AND ACCELERATION

How to find speed of oscillating body

The principle of conservation of energy permits us to express the speed of an object in simple harmonic motion in terms of its frequency f, amplitude A, and displacement s. The total energy of the oscillator (object plus spring) is the sum of its kinetic and potential energies at any time, which are respectively $\frac{1}{2}mv^2$ and $\frac{1}{2}ks^2$. At either extreme

Total energy of harmonic oscillator

of the motion, when $s = +A$ or $s = -A$, the object is stationary and has only the potential energy $\frac{1}{2}kA^2$. Hence

$$\text{Total energy} = \text{KE} + \text{PE}$$
$$\tfrac{1}{2}kA^2 = \tfrac{1}{2}mv^2 + \tfrac{1}{2}ks^2$$
$$mv^2 = k(A^2 - s^2)$$
$$v = \sqrt{k/m}\,\sqrt{A^2 - s^2}$$

From Eqs. (11–5) and (11–6) we know that

$$f = \frac{1}{T} = \frac{1}{2\pi}\sqrt{\frac{k}{m}}$$

which can be rewritten as

$$\sqrt{k/m} = 2\pi f \qquad\qquad\qquad (11\text{–}8)$$

The speed of the object when it has the displacement s is accordingly

$$v = 2\pi f \sqrt{A^2 - s^2} \qquad\qquad \textit{Speed at given displacement}\quad (11\text{–}9)$$

This formula gives only the absolute value of v; whether the sign of v is $+$ or $-$ depends upon whether the body is at $+s$ or $-s$ and upon whether it is on its way toward or away from the equilibrium position from there.

Maximum speed occurs at $s = 0$

From Eq. (11–9) we see that the maximum speed v_{max} of the object, which occurs at the equilibrium position when $s = 0$, is

$$v_{max} = 2\pi f A \hspace{4cm} \textit{Maximum speed} \hspace{0.5cm} (11\text{--}10)$$

The maximum speed is proportional to both the frequency and the amplitude of the motion.

The energy of an oscillating object shifts back and forth between kinetic and potential forms. To find the total energy, we can calculate either KE_{max} or PE_{max}, with the help respectively of Eq. (11–10) or (11–8):

$$KE_{max} = \tfrac{1}{2}mv_{max}^2 = 2\pi^2 mf^2 A^2 \hspace{2cm} \textit{Total energy} \hspace{0.5cm} (11\text{--}11)$$

$$PE_{max} = \tfrac{1}{2}kA^2 = 2\pi^2 mf^2 A^2 \hspace{4cm} (11\text{--}12)$$

The total energy depends upon the square of the frequency and the square of the amplitude. Note that the total energy is not the sum of Eqs. (11–11) and (11–12) because when KE or PE reaches its maximum, the other is zero.

To determine the acceleration of an object in simple harmonic motion, we refer back to Eq. (11–4), which states that

$$a = \frac{-ks}{m}$$

On the basis of Eq. (11–8) this formula becomes

$$a = -4\pi^2 f^2 s \hspace{2cm} \textit{Acceleration at given displacement} \hspace{0.3cm} (11\text{--}13)$$

The acceleration is always opposite in direction to the displacement, which, of course, is one of the conditions for simple harmonic motion to occur. The maximum acceleration occurs at either extreme, when $s = \pm A$, and has the magnitude

Maximum acceleration occurs at $s = \pm A$

$$a_{max} = 4\pi^2 f^2 A \hspace{3cm} \textit{Maximum acceleration} \hspace{0.5cm} (11\text{--}14)$$

The maximum acceleration is proportional to the square of the frequency and to the amplitude.

Although the above formulas were derived for the case of a vibrating spring, their validity is perfectly general and they apply to any type of harmonic oscillator, from the bob of a pendulum to an atom in a molecule.

Example In an automobile engine, each piston moves up and down in an approximation of simple harmonic motion. If a certain piston has a mass of 0.5 kg, has a total travel ("stroke") of 12 cm, and has a frequency of oscillation of 60 Hz (corresponding to 3600 rpm), find the maximum force it experiences assuming simple harmonic motion.

Solution The maximum force is given by $F_{max} = ma_{max}$. Here, since the amplitude is half the total travel, $A = 6$ cm $= 0.06$ m, and

$$F_{max} = ma_{max} = 4\pi^2 mf^2 A = 4\pi^2(0.5\,\text{kg})(60\,\text{Hz})^2(0.06\,\text{m}) = 4264\,\text{N}$$

which is nearly 1000 lb. ■

Example A piston undergoes simple harmonic motion in a vertical direction with an amplitude of 6 cm. A coin is placed on top of the piston (Fig. 11–9). What is the lowest frequency at which the coin will be left behind by the piston on its downstroke?

FIG. 11–9 When the downward acceleration of the piston exceeds g, the coin will be left behind by the piston on its downstroke.

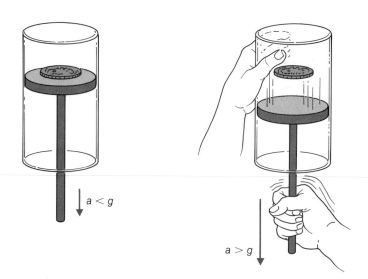

Solution The coin will leave the piston when the downward acceleration of the latter exceeds the acceleration of gravity g. The maximum downward acceleration of piston occurs at the highest point of its motion, when $a_{max} = 4\pi^2 f^2 A$ according to Eq. (11–14). Hence we set a_{max} equal to g and solve for the frequency f:

$$a_{max} = g = 4\pi^2 f^2 A$$

$$f = \frac{1}{2\pi}\sqrt{\frac{g}{A}} = \frac{1}{2\pi}\sqrt{\frac{9.8\,\text{m/s}^2}{0.06\,\text{m}}} = 2.0\,\text{Hz} \qquad \blacksquare$$

How *s*, *v*, and *a* vary with time in harmonic motion

The variations of the displacement, velocity, and acceleration of a particle undergoing simple harmonic motion are plotted versus time in Fig. 11–10. The graphs are plotted on the assumption that the particle is at $s = +A$ when $t = 0$. This corresponds to pulling out the particle and letting it go at $t = 0$. At this instant the particle's acceleration is a maximum and is opposite in direction to s, while the velocity is zero since the particle has not yet started to move.

When the particle is at the origin, $s = 0$ and the spring is at its normal length. Because the spring exerts no force on the particle at this time, its acceleration is zero. The speed of the particle is now a maximum, which follows from the absence of the potential energy associated with a deformed spring.

When $s = -A$, the particle is at the other extreme of its range, and its acceleration, again a maximum, is positive, which means that it is once more in the direction of the origin, though now from the other side. All the energy of oscillation is potential, and the body is accordingly stationary at this instant.

11–5 THE SIMPLE PENDULUM

A pendulum executes simple harmonic motion as it swings back and forth, provided that the arc through which the pendulum bob moves is a fairly small one. We shall see why this limitation arises if we use Eq. (11–3) to calculate the period of a pendulum.

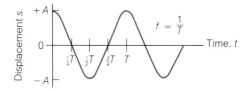

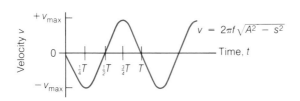

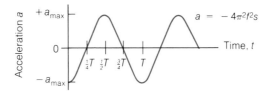

FIG. 11–10 Displacement, velocity, and acceleration in simple harmonic motion.

Figure 11–11 shows a pendulum of length L whose bob has a mass m, together with a diagram of the forces acting on the bob. (It is assumed that the entire mass of the pendulum is concentrated in the bob.) The weight of the bob, $m\mathbf{g}$, which acts vertically downward, may be resolved into two forces, $\mathbf{T}$ and $\mathbf{F}$, which act respectively parallel to and perpendicular to the supporting string L. That is,

A pendulum undergoes simple harmonic motion when it swings through a small arc

$$\mathbf{T} + \mathbf{F} = m\mathbf{g}$$

The force $\mathbf{F}$ is the restoring force that acts to return the bob to the midpoint of its motion. The space triangle hLx and the vector triangle $\mathbf{T}m\mathbf{g}\mathbf{F}$ are similar, since each contains a right angle and two sides of one are parallel to the two corresponding sides of the other, and so

The restoring force is F

$$\frac{F}{x} = \frac{mg}{L}$$

The restoring force acting on the bob is therefore

$$F = -\frac{mgx}{L}$$

where the minus sign indicates that $\mathbf{F}$ points in the direction of decreasing x.

If the bob is not far from the midpoint of its motion, the horizontal distance x is almost exactly equal to the actual path length s, and F then is given by

The condition for simple harmonic motion is that $x = s$

$$F = -\frac{mgs}{L}$$

FIG. 11–11 A pendulum executes simple harmonic motion when its oscillations are so small in amplitude that the chord x is very nearly equal in length to the arc s.

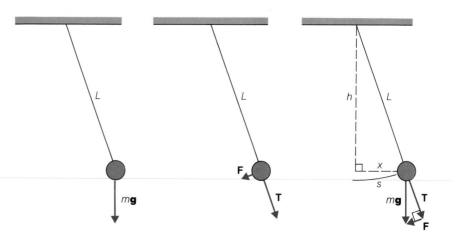

Since F is proportional to $-s$, the motion is simple harmonic. The acceleration of the bob that results from this force is

$$a = \frac{F}{m} = -\frac{gs}{L}$$

Substituting in Eq. (11–3) we find that

$$T = 2\pi \sqrt{-\frac{s}{a}}$$

$$T = 2\pi \sqrt{\frac{L}{g}} \qquad\qquad\qquad\qquad\qquad\qquad \textit{Simple pendulum} \quad (11\text{–}15)$$

The period of a simple pendulum is independent of its mass

Provided that s is small enough so that it is close to x, the motion of a pendulum is simple harmonic in character with a period proportional to the square root of the pendulum's length and independent of the mass of the bob. If the arc through which the pendulum swings on either side of the vertical is 5°, a detailed calculation shows that the actual period will exceed that predicted by Eq. (11–15) by only 0.05%; if the arc is 10° the discrepancy will be 0.2%; and even if the arc is as much as 20° the discrepancy is only 0.8%. Only when the arc on either side of the vertical is about 50° does the discrepancy reach 5%.

Example   How long should a pendulum be for it to have a period of exactly 1 s?

Solution   We first solve Eq. (11–15) for L:

$$T = 2\pi \sqrt{\frac{L}{g}} \qquad T^2 = \frac{4\pi^2 L}{g} \qquad L = \frac{gT^2}{4\pi^2}$$

Inserting the values $g = 9.8 \text{ m/s}^2$ and $T = 1 \text{ s}$, we find that

$$L = \frac{9.8 \text{ m/s}^2 \times 1 \text{ s}^2}{4\pi^2} = 0.25 \text{ m} \qquad\qquad \blacksquare$$

11–6 THE TORSION PENDULUM

A *torsion pendulum* consists of an object suspended by a wire or thin rod, as in Fig. 11–12. When the object is turned through an angle and released, it will oscillate back and forth. The analogies between linear and angular quantities make it easy to find a formula for the period of these oscillations.

In a torsion pendulum, the motion is angular instead of linear

From Hooke's law the restoring torque τ that comes into being when the wire is twisted through an angle θ is

Hooke's law for angular motion

$$\tau = -K\theta$$

where the value of the torsion constant K depends on the material and dimensions of the wire. If the moment of inertia of the object about its axis of suspension is I (the moment of inertia of the suspending wire is usually negligible), then the angular acceleration α of the object when the torque τ acts on it is, from Eq. (8–20),

Acceleration of torsion pendulum

$$\alpha = \frac{\tau}{I} = -\frac{K}{I}\theta \tag{11–16}$$

Comparing this formula with the equivalent result for a harmonic oscillator,

$$a = -\frac{k}{m}s \tag{11–4}$$

suggests that the period of a torsion pendulum can be given by the general formula for the period of a harmonic oscillator,

Acceleration of harmonic oscillator

$$T = 2\pi\sqrt{-\frac{s}{a}} \tag{11–5}$$

with θ/α replacing s/a. This idea turns out to be correct, and we have

$$T = 2\pi\sqrt{-\frac{\theta}{\alpha}} = 2\pi\sqrt{\frac{I}{K}} \qquad \textit{Torsion pendulum} \quad (11–17)$$

The simplest procedure for finding the moment of inertia of an irregular object is often to suspend it by a wire, measure the torsion constant K and the period of oscillation T, and then use Eq. (11–17) to find I.

Example A grindstone is suspended by a wire from its center. When a torque of 0.12 N·m is applied to the grindstone, it turns through 8°, and when it is released, it oscillates with a period of 1.0 s. Find the moment of inertia of the grindstone.

Solution Since $1° = 0.01745$ rad, $\theta = 0.14$ rad, and the torsion constant of the wire is

$$K = \frac{\tau}{\theta} = \frac{0.12\,\text{N·m}}{0.14\,\text{rad}} = 0.86\,\text{N·m/rad}$$

From Eq. (11–17),

$$I = \frac{KT^2}{4\pi^2} = \frac{(0.86\,\text{N·m/rad})(1.0\,\text{s})^2}{4\pi^2} = 0.022\,\text{kg·m}^2$$

FIG. 11–12 A torsion pendulum. Twisting the wire leads to a restoring torque.

11–7   THE PHYSICAL PENDULUM

Any swinging object can be treated as simple pendulum if L is properly interpreted

An object of any shape will oscillate back and forth when it is pivoted at some point other than its center of gravity and given an initial displacement to one side. Such an object is called a *physical pendulum*. The formula for the period of a simple pendulum also applies to a physical pendulum provided that the length L is properly interpreted.

Figure 11–13 shows a physical pendulum pivoted about a horizontal axis at O. The pendulum is displaced so that the line from O to its center of gravity is at the angle θ from the vertical. The pendulum's weight mg acts from the center of gravity and produces the restoring torque

$$\tau = -mgx$$

where x is the horizontal distance between O and the center of gravity. The minus sign reflects the fact that the restoring torque is always opposite in direction to the angular displacement θ of the pendulum.

When θ is small, the chord x is very nearly equal to the arc s. Assuming them to be equal is the same approximation made in analyzing the simple pendulum (see Fig. 11–11). Since h is the distance between the pivot point O and the center of gravity CG in Fig. 11–13, what we have for small θ is

$$x = s = h\theta$$

This means that

$$\tau = -mgh\theta$$

which is the condition for simple harmonic motion since the restoring torque is proportional to $-\theta$. If I is the moment of inertia of the pendulum about O and α is the angular acceleration produced by the restoring torque, then

$$\alpha = \frac{\tau}{I} = -\frac{mhg}{I}\theta$$

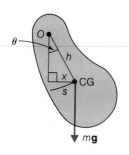

$$\tau = -(mg)(h\sin\theta)$$

FIG. 11–13 A physical pendulum pivoted at O. The center of gravity is marked CG.

Period of physical pendulum

By the same reasoning used in the preceding section, the period of the physical pendulum is

$$T = 2\pi\sqrt{-\frac{\theta}{\alpha}} = 2\pi\sqrt{\frac{I}{mgh}} \qquad\qquad \textit{Physical pendulum} \quad (11–18)$$

The simple pendulum whose period is the same as that of a given physical pendulum may be found by setting equal the formulas for their respective periods of oscillation:

$$2\pi\sqrt{\frac{L}{g}} = 2\pi\sqrt{\frac{I}{mgh}}$$

$$L = \frac{I}{mh} \qquad\qquad\qquad\qquad (11–19)$$

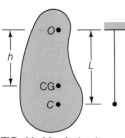

FIG. 11–14 A simple pendulum of length L = I/mh has the same period as that of a physical pendulum. The point C is the center of oscillation.

Thus the mass of a physical pendulum can be regarded as concentrated at a point C the distance $L = I/mh$ from the point O. This point is called the *center of oscillation* (Fig. 11–14) and has two interesting properties:

1. If the pendulum is pivoted at C instead of at O, it will oscillate with the same period as before and O will be the new center of oscillation.

2. If the pendulum is struck along a line of action through C, there will be no reaction force on the pivot at O. A baseball that strikes a bat at the latter's center of oscillation does not produce a sting in the batter's hands, for example (Fig. 11–15). The center of oscillation is often called the *center of percussion* for this reason. The concept of center of percussion plays an important part in the design of many mehanical devices.

Striking a physical pendulum at its center of percussion produces no reaction force on its pivot

Example Analyze the process of walking by considering the leg as a physical pendulum.

The leg as a physical pendulum

Solution A leg L long may be crudely approximated by a thin rod hinged at one end. From Fig. 8–9 the moment of inertia of such a rod is

$$I = \tfrac{1}{3}mL^2$$

where m is its mass. If the center of gravity of the leg is at its middle, $h = L/2$, and the natural period of oscillation of the leg is

$$T = 2\pi \sqrt{\frac{I}{mgh}} = 2\pi \sqrt{\frac{mL^2/3}{mgL/2}} = 2\pi \sqrt{\frac{2L}{3g}}$$

The period of a leg 1.0 m long is

$$T = 2\pi \sqrt{\frac{2(1.0\,\text{m})}{3(9.8\,\text{m/s}^2)}} = 1.6\,\text{s}$$

Each step represents only half a complete cycle; hence if the leg is swinging freely, it takes 0.8 s and the rate of walking is 75 steps/min. Walking at a slower or a faster rate than this involves more effort. With a stride 80 cm long, 75 steps/min means a speed of 60 m/min, which is 3.6 km/hr (2.24 mi/hr).

The longer the legs of an animal, the longer its stride, but T is increased as well. The natural speed of walking varies as L/T, and since T is proportional to $\sqrt{L}$, this speed is proportional to $L/\sqrt{L} = \sqrt{L}$. The larger an animal, then, the faster it walks, although it takes an increase by a factor of 4 in leg length to double the natural speed. Running is quite a different matter because the muscular strength and the mass of an animal are involved as well as its size. As we saw in Sec. 6–5, all animals have roughly similar running speeds. When running, an animal's legs are bent while moving forward, which reduces their lengths and hence their natural periods of oscillation. This reduction makes it easier for the legs to be moved rapidly. ∎

FIG. 11–15 When a physical pendulum is struck at its center of percussion, its pivot experiences no reaction force.

11–8 DAMPED HARMONIC OSCILLATOR

Once set in motion, an ideal harmonic oscillator should continue to oscillate indefinitely. Actual harmonic oscillators do not behave in this way; while in some cases they may oscillate for a long time, eventually the amplitude decreases and the motion comes to a stop. The latter phenomenon is called *damping*. If the amount of damping is very

Actual harmonic oscillators are damped

FIG. 11–16 (a) Damped harmonic oscillator. (b) An overdamped oscillator returns to its equilibrium position very slowly. (c) A critically damped oscillator returns to its equilibrium position rapidly, but not so rapidly that it overshoots and begins to oscillate.

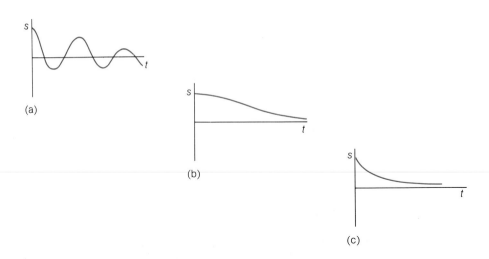

large, as it is for a car whose springs are damped with good shock absorbers, no oscillation at all occurs; after being displaced, the object gradually returns to its equilibrium position and goes no further.

The frequency of a damped oscillator is less than that of a comparable simple oscillator

In general, damping is caused by frictional forces. The potential energy of a stretched spring is not completely converted into kinetic energy at $s = 0$, since some of it goes into work done against the frictional forces that are acting. At the end of each vibration the loss in energy leads to a shorter extension of the spring, until finally the amplitude becomes infinitesimal. Damping also acts to reduce the frequency of the motion—the greater the damping, the slower the oscillations.

Figure 11–16 shows the effects of different amounts of damping on a harmonic oscillator. In (a) the damping is small, and the oscillations steadily decrease in ampli-

Damping may prevent oscillations from occurring

tude. In (b) the damping is so great that the displaced object never oscillates but returns to its equilibrium position very slowly; such an oscillator is said to be *overdamped*. A *critically damped* oscillator, as in (c), falls right on the line between the other two situations; it returns to its equilibrium position as rapidly as possible without overshooting, which would mean oscillating. Critical damping is desirable in many situations, such as that of the suspension of a car.

IMPORTANT TERMS

A body under stress possesses **elastic potential energy,** which is equal to the work done in deforming it.

Simple harmonic motion is an oscillatory motion that occurs whenever a force acts on a body in the opposite direction to its displacement from its equilibrium position, with the magnitude of the force proportional to the magnitude of the displacement. Thus the force on a body in simple harmonic motion always tends to return it to its equilibrium position.

The **period** T of a body undergoing simple harmonic motion is the time required for it to make one complete oscillation. The **frequency** f of such a body is the number of complete oscillations it makes per unit time.

The **amplitude** of a body undergoing simple harmonic motion is its maximum displacement on either side of its equilibrium position. The period of the motion is independent of the amplitude.

The **center of percussion** of a pivoted object is that point at which it can be struck without producing a reaction force on its pivot.

In a **damped harmonic oscillator,** friction progressively reduces the amplitude of the vibrations.

IMPORTANT FORMULAS

Elastic potential energy: $PE = \frac{1}{2}ks^2$

Harmonic oscillator $T = \dfrac{1}{f} = 2\pi\sqrt{\dfrac{m}{k}}$

Simple pendulum: $T = 2\pi\sqrt{\dfrac{L}{g}}$

Torsion pendulum: $T = 2\pi\sqrt{\dfrac{I}{K}}$

Physical pendulum: $T = 2\pi\sqrt{\dfrac{I}{mgh}}$

MULTIPLE CHOICE

1. A spring whose force constant is k is cut in half. Each of the new springs has a force constant of

(a) $\frac{1}{2}k$.   (b) k.

(c) $2k$.   (d) $4k$.

2. A force of 0.2 N is needed to compress a certain spring by 2 cm. Its potential energy when compressed is

(a) 2×10^{-3} J.   (b) 2×10^{-5} J.
(c) 4×10^{-5} J.   (d) 8×10^{-5} J.

3. The product of the period and the frequency of a harmonic oscillator is always equal to

(a) 1.

(b) π.

(c) 2π.

(d) the amplitude of the motion.

4. The period of a simple harmonic oscillator is independent of its

(a) frequency.

(b) amplitude.

(c) force constant.

(d) mass.

5. An object undergoes simple harmonic motion. Its maximum speed occurs when its displacement from its equilibrium position is

(a) zero.

(b) a maximum.

(c) half its maximum value.

(d) none of the above.

6. In simple harmonic motion, there is always a constant ratio between the displacement of the mass and its

(a) velocity.   (b) acceleration.

(c) period.   (d) mass.

7. An object attached to a horizontal spring executes simple harmonic motion on a frictionless surface. The ratio between its kinetic energy when it passes through the equilibrium position and its potential energy when the spring is fully extended is

(a) less than 1.

(b) equal to 1.

(c) more than 1.

(d) equal to the ratio between its mass and the spring constant.

8. The amplitude of an object undergoing harmonic motion is

(a) its total range of motion.

(b) its maximum displacement on either side of the equilibrium.position.

(c) its minimum displacement on either side of the equilibrium position.

(d) the number of cycles per second it describes.

9. The amplitude of a simple harmonic oscillator is doubled. Which of the following is also doubled?

(a) its frequency

(b) its period

(c) its maximum speed

(d) its total energy

10. The period of a simple pendulum depends on its

(a) mass.

(b) length.

(c) total energy.

(d) maximum speed.

11. A pendulum executes simple harmonic motion provided that

(a) its bob is not too heavy.

(b) the supporting string is not too long.

(c) the arc through which it swings is not too small.

(d) the arc through which it swings is not too large.

12. The period of a harmonic oscillator of mass m is proportional to

(a) $\sqrt{m}$.   (b) $1/\sqrt{m}$.
(c) m^2.   (d) $1/m^2$.

13. When a 1-kg mass is suspended from a spring, the spring stretches by 5 cm. The force constant of the spring is

(a) 0.2 N/m.   (b) 1.96 N/m.
(c) 49 N/m.   (d) 196 N/m.

14. If the suspended mass of question 13 oscillates up and down, its period will be approximately

(a) 0.032 s. (b) 0.071 s.

(c) 0.45 s. (d) 4.5 s.

15. The maximum speed of a particle that undergoes simple harmonic motion with a period of 0.5 s and an amplitude of 2 cm is

(a) π cm/s. (b) 2π cm/s.

(c) 4π cm/s. (d) 8π cm/s.

16. In a refrigeration compressor, a 1-kg piston undergoes 20 cycles/s in which its total travel is 14 cm. The maximum force on the piston

(a) is 1105 N.

(b) is 1548 N.

(c) is 2210 N.

(d) cannot be calculated from the given data.

17. A boy swings from a rope 4.9 m long. His approximate period of oscillation is

(a) 0.5 s. (b) 3.1 s.

(c) 4.4 s. (d) 12 s.

18. A lead sphere is suspended by a wire and set into rotational oscillation. If the sphere were flattened into a horizontal disk, the period of the oscillations would be

(a) shorter.

(b) the same.

(c) longer.

(d) any of the above, depending on the radius of the disk.

EXERCISES

1. Must a spring obey Hooke's law in order to oscillate?

2. At what point or points in its motion is the energy of a harmonic oscillator entirely potential? At what point or points is its energy entirely kinetic?

3. Upon what, if anything, does the ratio between the maximum kinetic energy and maximum potential energy of a harmonic oscillator depend?

4. A body pivoted at some point is given an initial displacement and then released. Under what circumstances will it oscillate back and forth? Under what circumstances will the oscillations be simple harmonic in character? Under what circumstances will it behave like a simple pendulum?

5. A wooden object is floating in a bathtub. It is pressed down and then released. Under what circumstances will its oscillations be simple harmonic in nature?

6. At what displacement relative to the amplitude is the kinetic energy of a harmonic oscillator three times the potential energy?

7. A loudspeaker generates sound waves that correspond in frequency and amplitude to the electrical oscillations that reach it from the output of an audio amplifier. Why is it desirable for the cone of a loudspeaker to be damped?

8. What is the frequency of a pendulum whose normal period is T when it is in an elevator in free fall? When it is in an elevator descending at constant velocity? When it is in an elevator ascending at constant velocity?

9. A toy rifle employs a spring whose force constant is 200 N/m. In use, the spring is compressed 5 cm, and when released, it propels a 5-g rubber ball. What is the ball's speed when it leaves the rifle?

10. A 5-kg object is dropped on a vertical spring from a height of 2 m. If the maximum compression of the spring is 40 cm, what is its force constant?

11. When a 1-kg mass is suspended from a spring, the spring stretches by 6 cm. If the mass oscillates up and down, what is its period? What is its frequency?

12. A body whose mass is 0.4 kg is suspended from a spring and oscillates with a period of 2 s. By how much will the spring contract when the body is removed?

13. A 70-kg gymnast jumps on a trampoline from a height of 60 cm. The trampoline sags 40 cm when the gymnast strikes it, and he then bounces up and down. If the motion is simple harmonic, find its period.

14. The outer end of a lightweight diving board is depressed by 30 cm when a person stands on it. If the board then vibrates, what is the approximate period of oscillation?

15. A harmonic oscillator has a period of 0.2 s and an amplitude of 10 cm. Find the speed of the moving object when it passes through the equilibrium position.

16. Atoms in a crystalline solid are in constant vibration at room temperature, where their motion has amplitudes in the neighborhood of 10^{-11} m. If the frequency of oscillation of one of the atoms in an iron bar is 2.5×10^{12} Hz, find its maximum speed and acceleration.

17. A chandelier is suspended from a high ceiling with a cable 6 m long. What is its period of oscillation?

18. A pendulum whose length is 1.53 m oscillates 24 times per minute in a particular location. What is the acceleration of gravity there?

19. Find the period of the rotational oscillations of an aluminum sphere 8 cm in diameter whose mass is 725 g which is suspended by a wire whose torsion constant is 0.1 N·m/rad.

PROBLEMS

1. A spring has a 1-s period of oscillation when a 20-N weight is suspended from it. Find the elongation of the spring when a 50-N weight is suspended from it.

2. An object whose mass is 1 kg hangs from a spring. When the object is pulled down 5 cm from its equilibrium position and released, it oscillates once per second. (a) What is the force constant of the spring? (b) What is the object's speed when it passes through its equilibrium position? (c) What is the maximum acceleration of the object?

3. An object whose mass is 0.005 kg is in simple harmonic motion with a period of 0.04 s and an amplitude of 0.01 m. (a) What is its maximum acceleration? (b) What is the maximum force on the object? (c) What is its acceleration when it is 0.005 m from its equilibrium position? (d) What is the force on it at that point?

4. The prongs of a tuning fork vibrate in simple harmonic motion at a frequency of 660 Hz and with an amplitude of 1 mm at their tips. Find the maximum speed and acceleration of the prong tips. Express the acceleration in terms of g.

5. A pendulum has a length of 50 cm. Find its period when it is suspended in (a) a stationary elevator; (b) an elevator falling at the constant velocity of 5 m/s; (c) an elevator falling at the constant acceleration of 2 m/s²; (d) an elevator rising at the constant velocity of 5 m/s; (e) an elevator rising at the constant acceleration of 2 m/s².

6. A 200-g brass hoop 40 cm in diameter is suspended on a knife edge on which it rocks back and forth. If the period of the oscillations is 1.27 s, find the moment of inertia of the hoop about an axis through its circumference perpendicular to its plane.

7. An iron bar 80 cm long is suspended from one end. What is the period of its oscillations? What would be the length of a simple pendulum with the same period?

8. A spring of force constant k_1 is connected end to end to a spring of force constant k_2. (a) What is the force constant k of the combination? (b) If $k_1 = 10$ N/m and $k_2 = 30$ N/m, find k.

9. Two springs with the same force constant k are connected to an object of mass m as in the figures below. Find the period of oscillation in each case in the absence of friction.

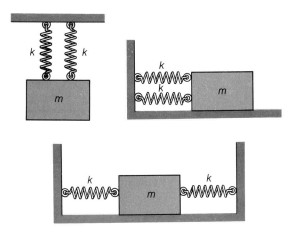

10. A wooden cube of density d that is L long on each edge floats in a liquid of density d' so that its upper and lower faces are horizontal. The cube is pushed down and released. Verify that the cube then oscillates up and down in simple harmonic motion with a period of $2\pi\sqrt{Ld/gd'}$.

11. A hole is bored through the earth along a diameter. Inside the hole the acceleration of gravity varies as rg/R, where r is the distance from the center of the earth, R is the earth's radius of 6.4×10^6 m, and g is the acceleration of gravity at the earth's surface. A stone is dropped into the hole and executes simple harmonic motion about the center of the earth. Why? Find the period of this motion.

ANSWERS TO MULTIPLE CHOICE

1. a	**4.** b	**7.** b	**10.** b	**13.** d	**16.** a
2. c	**5.** a	**8.** b	**11.** d	**14.** c	**17.** c
3. a	**6.** b	**9.** c	**12.** a	**15.** d	**18.** c

WAVES

12

The properties of a vibrating system and those of a wave traveling through a medium are similar in a number of essential respects. In both, a certain characteristic motion recurs at regular intervals; in both, the motions that occur represent the continuous conversion of potential energy into kinetic energy and back; in both, the properties of matter play important roles in providing restoring forces. The chief difference is that the energy in a vibrating system remains localized in space, whereas waves carry energy from one place to another without any actual transport of matter.

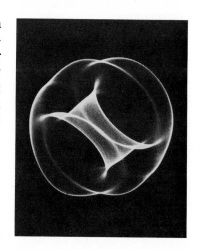

12–1   WAVE MOTION

Energy can be transmitted from one place to another in a variety of ways. Suppose we wish to supply energy to a boat in the center of a lake from a position on the shore, with the provision that the precise form in which the energy arrives does not matter. The most obvious thing to do is to throw a stone at the boat, thereby providing it with kinetic energy. Another method is to pour hot water into the lake, thereby providing the boat with thermal energy. Or we can simply drop a stone in the water near the shore; the waves that are produced transfer energy to the boat by causing it to move up and down (Fig. 12–1). When the stone strikes the water, a deformation of the water surface begins to spread. The energy that reaches the boat arrives as a periodic deformation that contains both kinetic and potential energy. Energy propagation by means of the motion of a change in a medium is called *wave motion,* and it occurs in many forms in nature.

FIG. 12–1   Waves transmit energy from one place to another through the motion of a change in a medium.

12–2   PULSES IN A STRING

If we give one end of a stretched string a quick shake, a kink or *pulse* travels down the string at some speed v (Fig. 12–2). If the string is uniform and completely flexible, the pulse keeps the same shape as it moves. It is worth examining the behavior of pulses in a string both because this is the simplest kind of wave phenomenon and because it is easy to visualize what is going on.

The speed v of a pulse depends upon the properties of the string—how heavy it is and how tightly it is stretched—rather than upon the shape of the pulse or upon exactly how it is produced. Pulses move slowly down a slack, heavy rope; they move rapidly down a taut, light string. By "heavy" and "light" are meant the mass per unit length of a string, not its total weight; a pulse has the same speed in a long string under a given tension as in a short string of the same kind under the same tension. When the mass per unit length of a string is high, the pulse speed is low because the inertia of each segment of the string is large and it therefore responds slowly to the forces acting on it. When the string is tightly stretched, the pulse speed is high because the tendency of the string to straighten out is greater.

Pulse speed in a string depends on the properties of the string, not on the properties of the pulse

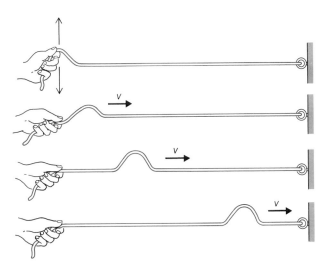

FIG. 12–2   A pulse moves along a stretched string with a constant speed v.

The above observations are reflected in the formula

$$v = \sqrt{\frac{T}{m/L}}$$

Waves in a string (12–1)

for the speed of waves in a stretched string of mass m and length L that is under the tension T.

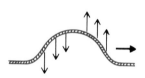

FIG. 12–3 The forward part of this traveling pulse is moving upward and the rear part is moving downward.

The energy content of a moving pulse is partly kinetic and partly potential. As the pulse travels, its forward part is moving upward and its rear part is moving downward; because the string has mass, there is a certain amount of kinetic energy associated with these up-and-down motions (Fig. 12–3). The potential energy is due to the tension in the string. Work had to be done in order to produce the pulse by pulling against the tension, and the deformed string accordingly possesses elastic potential energy.

A moving pulse has both KE and PE

When a pulse reaches the end of a string, it may be reflected and travel back toward its starting point. Depending upon how the end of the string is held in place, the reflected pulse may be inverted (upside down) or erect (right side up). Under just the right conditions, of course, the energy of the pulse may all be absorbed by the support and the pulse will then disappear.

Reflection from a fixed end

Suppose the end of the string is held firmly in place. When the pulse arrives there, the string exerts an upward force on the support (assuming the pulse is upward, as in Fig. 12–4). By the third law of motion, the support then exerts an equal and opposite reaction force on the string. The effect of this reaction force is to produce a pulse whose displacement is opposite to that of the original pulse but otherwise with the same shape.

FIG. 12–4 A pulse reaching a fixed end of the string is inverted upon reflection.

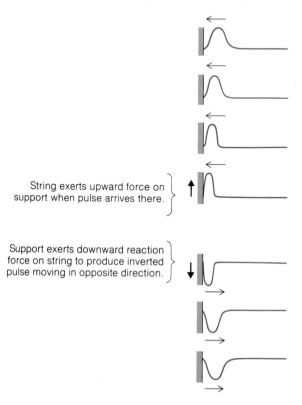

String exerts upward force on support when pulse arrives there.

Support exerts downward reaction force on string to produce inverted pulse moving in opposite direction.

The new inverted pulse proceeds back along the string in the reverse direction of that of the original pulse. Thus an upward pulse becomes a downward one upon reflection, and vice versa.

If the end of the string is not held firmly in place, however, the reflected pulse is not inverted. Figure 12–5 shows the end of a string attached to a ring free to move up and down a frictionless rod. When the pulse arrives at this end, the string moves upward until its kinetic energy is completely converted into elastic potential energy, whereupon the end of the string moves downward again to send out a pulse that is reversed in direction but otherwise the same as the original one. If the end of the string is held in a manner exactly in between complete rigidity and complete freedom, then the pulse will not be reflected at all but will disappear when it reaches the end.

A little experimentation with pulses in an actual string will show that, whereas it is very easy to hold the far end of the string so that the reflected pulse is smaller than the original one, it is not easy at all to keep some reflection from taking place. Similar difficulty is experienced in trying to construct surfaces that completely absorb sound, light, or water waves.

So far we have been considering pulses in a uniform string. Now let us connect two different strings together, one of them light (that is, with a low mass per unit length) and the other heavy (high mass per unit length). One end of the combination is fastened to something, and the other is given a shake to produce a pulse. Not surprisingly, the pulse passes from the first string to the second at the junction between them: the pulse is *transmitted*. But the transmission is not complete, since a reflected pulse also appears at the junction and proceeds in the opposite direction.

If the first string is the lighter one, the reflected pulse is inverted (Fig. 12–6). The greater inertia of the heavy string does not permit it to respond to the pulse as rapidly as the light string does, and an opposite reaction force occurs that causes the reflected pulse to be inverted, as though the junction were a rigid support. The energy of the original pulse is then split between the reflected and transmitted pulses. Since both strings have the same tension, the pulse travels more slowly in the heavy string. The length of the reflected pulse is the same as that of the original one, though its height is smaller since it has less energy. The transmitted pulse, however, is shorter, because its speed is less than that of the original pulse while the time interval in which it comes into being is the same as that of the original pulse.

On the other hand, if the first string is the heavy one, the reflected pulse is erect (Fig. 12–7). The smaller inertia of the light string permits it to follow the movements of the heavy one readily, and the situation is like that of a string whose end is able to

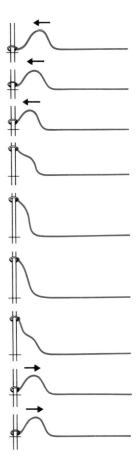

FIG. 12–5 A pulse reaching a free end of the string is not inverted on reflection.

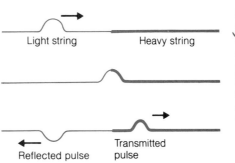

Light string   Heavy string

Reflected pulse   Transmitted pulse

FIG. 12–6 When a pulse passes from a light to a heavy string, reflection occurs with the reflected pulse being inverted. The transmitted pulse is right side up. Since both strings have the same tension, the pulse travels more slowly in the heavy string.

FIG. 12–7 When a pulse passes from a heavy to a light string, the reflected pulse stays right side up. The pulse speed is again less in the heavy string.

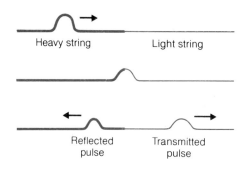

move up and down freely. However, the light string does have some inertia, and so a reflected pulse again comes into being as well as a transmitted one. The pulse speed is higher in the light string, and in consequence the pulse length is longer there than in the heavy one.

All waves are reflected at junctions

All types of waves, not just pulses in a stretched string, exhibit reflection and transmission at junctions between different media. For example, light waves are partially reflected and partially transmitted when they pass from air to glass, which is why we can see our images in a clear pane of glass (such as a shop window) even though the glass is transparent to light. The inversion of a pulse when it is reflected at a junction with a medium in which its speed is smaller also has a counterpart in the behavior of light waves, as we shall find in Chap. 25.

12–3   PRINCIPLE OF SUPERPOSITION

What happens when two pulses meet

Each end of a stretched string is given an upward shake, and the pulses thus produced move along the string toward each other. What happens when they meet? The result is a larger pulse at the moment the pulses come together, and then the separate pulses reappear and continue unchanged in their original directions of motion. Each pulse proceeds as though the other does not exist (Fig. 12–8).

The *principle of superposition* is a statement of the above behavior. This principle can be phrased as follows:

Principle of superposition for pulses

When two pulses travel past a point in a string at the same time, the displacement of the string at that point is the sum of the displacements each pulse would produce there by itself.

FIG. 12–8 Two pulses moving in opposite directions along a stretched string. The pulses are unaffected by their crossing.

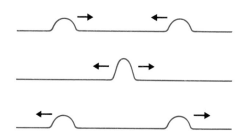

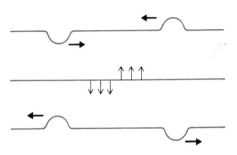

FIG. 12–9 Complete cancellation occurs when two identical pulses with opposite displacements meet. At the instant of complete cancellation, the total energy of both pulses resides in the kinetic energy of the string segment where the cancellation occurs.

What if one of the pulses is inverted relative to the other? According to the superposition principle, if the pulses have the same sizes and shapes, their displacements ought to cancel out when they meet, only to reappear later on after they have passed the crossing point. Such behavior is indeed observed in practice. At the instant of complete cancellation, the total energy of both pulses resides in the kinetic energy of the string segment where the cancellation occurs (Fig. 12–9).

Where the wave energy goes during complete cancellation

12–4 PERIODIC WAVES

In a periodic wave, one pulse follows another in regular succession. Sound waves, water waves, and light waves are almost always periodic, although in each case a different quantity varies as the wave passes.

Sinusoidal waves

In periodic waves, a certain waveform—the shape of the individual waves—is repeated at regular intervals. Periodic waves of all kinds usually have *sinusoidal* waveforms; a stretched string down which such waves move presents exactly the same appearance as a graph of sin x (or cos x) versus x that is moved along the x axis with the wave speed v (Fig. 12–10).

Sinusoidal waves are common because the particles of matter in a medium that waves can travel through undergo simple harmonic motion when momentarily displaced

Why periodic waves are usually sinusoidal

FIG. 12–10 Most periodic waves have sinusoidal waveforms.

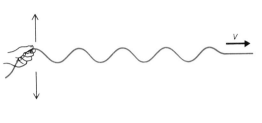

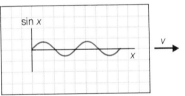

FIG. 12–11 Each particle in the path of a sinusoidal wave executes simple harmonic motion perpendicular to the wave direction.

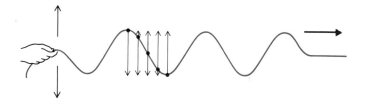

from their equilibrium positions. The passage of a wave sets up coupled harmonic oscillations in the medium, with each particle behaving like a harmonic oscillator that has begun its cycle just a trifle later than the particle behind it (Fig. 12–11). The result in the case of waves in a stretched string is a waveform that is in essence a graph of how the position of a harmonic oscillator varies with time, which is a sine curve. (Light waves are also sinusoidal in character even though their existence does not involve the motion of material particles and they can travel through empty space.)

Three related quantities are useful in describing periodic waves:

Speed, wavelength, and frequency

1. The *wave speed v,* which is the distance through which each wave moves per second;
2. The *wavelength* λ (Greek letter *lambda*), which is the distance between adjacent crests or troughs;
3. The *frequency f,* which is the number of waves that pass a given point per second.

The hertz is the unit of frequency

The unit of frequency is the *cycle/s,* or the *hertz* (Hz) after Heinrich Hertz, one of the pioneers in the study of electromagnetic waves. Multiples of the cycle/s and of the hertz are used for high frequencies:

$$1 \text{ kilocycle/s (kc/s)} = 1 \text{ kilohertz (kHz)} = 10^3 \text{ cycles/s}$$

$$1 \text{ megacycle/s (mc/s)} = 1 \text{ megahertz (MHz)} = 10^6 \text{ cycles/s}$$

Thus a frequency of 50 MHz is equal to

$$(50 \text{ MHz})\left(10^6 \frac{\text{Hz}}{\text{MHz}}\right) = 5 \times 10^7 \text{ Hz} = 5 \times 10^7 \text{ cycles/s}$$

How the speed, wavelength, and frequency of a wave are related

The wave speed, wavelength, and frequency of a train of waves are not independent of one another. In every second, *f* waves (by definition) go past a particular point, with each wave occupying a distance of λ (Fig. 12–12). Therefore a wave travels a total distance of *f*λ per second, which is the wave speed *v.* Thus

$$v = f\lambda \qquad\qquad\qquad\qquad \textit{Wave speed} \quad (12\text{–}2)$$

Wave speed = frequency × wavelength

which is a basic formula that applies to all periodic waves, sinusoidal or not.

Example　A marine radar operating at a frequency of 9400 MHz emits groups of radio waves 0.08 μs in duration. (The time needed for reflections of these groups to return indicates the distance of the target.) Radio waves, like light waves, are electromagnetic in nature and travel at 3.00×10^8 m/s. Find (a) the wavelength of these waves; (b) the length of each wave group, which is indicative of the precision with which the radar can measure distance; and (c) the number of waves in the group.

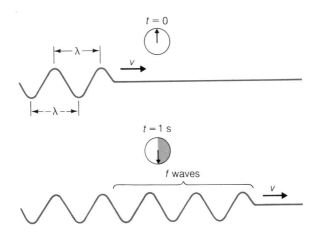

v = wave speed
= distance traveled per second
= (number of waves passing a point per second) × (length of each wave)
= $f\lambda$

Solution (a) From Eq. (12–2) the wavelength is

$$\lambda = \frac{v}{f} = \frac{3.00 \times 10^8\,\text{m/s}}{9.4 \times 10^9\,\text{Hz}} = 3.19 \times 10^{-2}\,\text{m} = 3.19\,\text{cm}$$

(b) The length s of each wave group is simply

$$s = vt = (3.00 \times 10^8\,\text{m/s})(8 \times 10^{-8}\,\text{s}) = 24\,\text{m}$$

(c) We can find the number n of waves in each group in either of these ways:

$$n = ft = (9.4 \times 10^9\,\text{Hz})(8 \times 10^{-8}\,\text{s}) = 752\,\text{cycles} = 752\,\text{waves}$$

$$n = \frac{s}{\lambda} = \frac{24\,\text{m}}{3.19 \times 10^{-2}\,\text{m}} = 752\,\text{wavelengths} = 752\,\text{waves} \qquad ■$$

Sometimes it is more useful to consider the *period T* of a wave, which is the time   **Wave period** required for one complete wave to pass a given point (Fig. 12–13). Since f waves pass by per second, the period of each wave is

$$T = \frac{1}{f} \qquad\qquad\qquad\qquad \textit{Wave period} \quad (12–3)$$

If there are five waves per second passing by, for example, each wave has a period of $\frac{1}{5}$ s. In terms of period T, the formula for wave velocity is

$$v = \frac{\lambda}{T} \qquad\qquad\qquad\qquad\qquad\qquad (12–4)$$

Example An anchored boat is observed to rise and fall once every 4 s as waves whose crests are 25 m apart pass by it. Find the frequency and speed of the waves.

FIG. 12–13 The period of a wave is the time required for one complete wave to pass by a given point.

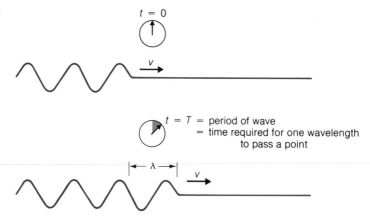

$t = 0$

v

$t = T$ = period of wave
= time required for one wavelength
to pass a point

$|\leftarrow \lambda \rightarrow|$

v

Solution   The frequency of the waves is

$$f = \frac{1}{T} = \frac{1}{4\,\text{s}} = 0.25\,\text{Hz}$$

and their speed is

$$v = \frac{\lambda}{T} = \frac{25\,\text{m}}{4\,\text{s}} = 6.25\,\text{m/s} \qquad \blacksquare$$

Wave amplitude

The *amplitude A* of a wave refers to the maximum displacement from their normal positions of the particles, which oscillate back and forth as the wave travels by (Fig. 12–14). The amplitude of a wave in a stretched string is the height of the crests above the original line of the string (or the depth of the troughs below the original line). The speed, frequency, and wavelength of a wave are independent of its amplitude, just as the period of a harmonic oscillator is independent of its amplitude.

Fourier's theorem

We have been considering sinusoidal waves thus far. However, in a medium whose properties do not vary with wave frequency, everything that is true for a sinusoidal wave is also true for all other periodic waves. An interesting and important theorem by Fourier shows why this should be so. What Fourier proved is that *any* periodic wave of frequency f, regardless of its waveform, can be thought of as a superposition of sinusoidal waves whose frequencies are f, $2f$, $3f$, and so on. The amplitudes of the various component waves depend on the precise character of the composite wave, and a mathematical procedure exists for finding these amplitudes. Figure 12–15 shows how just three waves of appropriate frequency and amplitude add up to give an approximation of a square wave; the more the waves that are included, the better the approximation becomes. Even an isolated pulse can be represented by a superposition of sinusoidal waves, although here the frequencies must be very close to each other instead of being multiples of f, and a great many waves are needed.

FIG. 12–14 The quantity A is the amplitude of the wave.

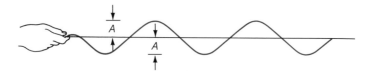

A

A

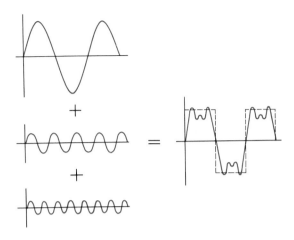

FIG. 12–15 Fourier synthesis of a square wave.

The correspondence between harmonic and wave motion permits us to obtain an interesting result. As we found in Sec. 11–4, a particle of mass m that undergoes simple harmonic motion of frequency f and amplitude A has a total energy of

Energy content of wave motion

$$E = 2\pi^2 m f^2 A^2$$

This dependence of energy on f^2 and on A^2 is also true for mechanical waves of all kinds. (A mechanical wave is one that involves moving matter, in contrast to, say, an electromagnetic wave.) Waves in a string are an example: the energy per unit length in a string due to waves of frequency f and amplitude A is $2\pi^2 (m/L)f^2A^2$, where m/L is the mass per unit length of the string.

12–5   TYPES OF WAVES

Waves in a stretched string are *transverse waves* since the individual segments of the string vibrate perpendicular to the direction in which the waves travel, that is, from side to side. *Longitudinal waves* occur when the individual particles of a medium vibrate back and forth in the direction in which the waves travel (Fig. 12–16). Longitudinal waves are easy to produce in a long coil spring; each portion of the spring is alternately compressed and extended as the waves pass by. Longitudinal waves, then, are essentially density fluctuations.

Transverse and longitudinal waves

Waves on the surface of a body of water (or other liquid) are a combination of longitudinal and transverse waves. If we were somehow to tag individual water molecules and follow them when a train of waves passes by, we would find that their paths are like those shown in Fig. 12–17. Each molecule describes a circular orbit with a

Water waves combine transverse and longitudinal wave motions

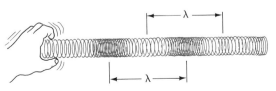

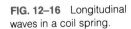

FIG. 12–16 Longitudinal waves in a coil spring.

FIG. 12–17 Water molecules move in circular orbits about their original positions when a typical deep-water wave passes by. At the crest of a wave the molecules are moving in the direction the wave is traveling, while in the trough the molecules are moving in the opposite direction. There is no net motion of water involved in the motion of such a wave.

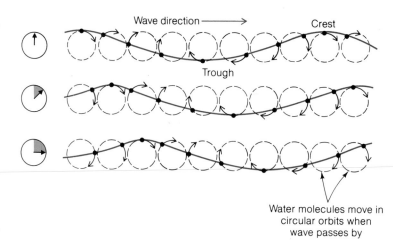

Water molecules move in circular orbits when wave passes by

period equal to the period of the wave, and does not undergo a permanent displacement. Because successive molecules reach the tops of their orbits at slightly different times, the water surface takes the form of a series of crests and troughs. At the crest of a wave the molecules move in the direction the wave is traveling, while in a trough the molecules are moving in the opposite direction. The passage of a wave across the surface of a body of water, like the passage of a wave through any medium, involves the motion of a pattern: energy is transported by virtue of the changing pattern, but there is no net transport of matter.

Example The water waves in the previous example have an amplitude of 60 cm. Find the speed of an individual molecule of water on the surface.

Solution The water molecules on the surface are moving in circles of radius 60 cm, so that as each wave passes by, the molecules travel a distance s equal to the circumference $2\pi r$ of the circle (Fig. 12–18). Hence

$$s = 2\pi r = 2\pi(0.6 \text{ m}) = 3.8 \text{ m}$$

Each wave takes $T = 4$ s to go past a given point, which means that the molecules must cover the 3.8-m circumference of their orbits in 4 s. The speed of each molecule is therefore

$$V = \frac{s}{T} = \frac{3.8 \text{ m}}{4 \text{ s}} = 0.95 \text{ m/s}$$

FIG. 12–18

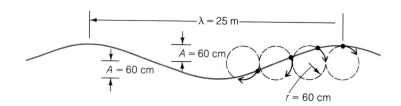

The speed of the *wave*, however, is 6.25 m/s, more than six times greater. Thus the motion of the pattern that constitutes a wave in a medium can be much more rapid than the motions of the individual particles of the medium, and energy can be transported by wave motion faster than might be possible through the net transport of matter. ∎

Wave speed can exceed particle speed in wave motion

All three kinds of waves—transverse, longitudinal, and surface—are sent out by an earthquake, and can be detected many thousands of kilometers away if the quake is a major one. The longitudinal waves are the fastest and usually have periods of 2 to 3 s; the slower transverse waves have periods of 10 to 15 s; and the still slower surface waves have periods of 10 to 60 s. The longitudinal and transverse earthquake waves travel through the earth's interior, and by analyzing the waves that arrive at the various observatories around the world after an earthquake, it is possible to determine the structure of the earth's interior (see Fig. 5–17). In particular, the inability of transverse earthquake waves to go through the central part of the earth whereas longitudinal waves can do so indicates that this region must be liquid, since longitudinal waves can occur in a liquid whereas transverse waves cannot.

Earthquake waves and the earth's interior

12–6 STANDING WAVES

When we pluck a string whose ends are fixed in place, the string starts to vibrate in one or more loops (Fig. 12–19). These *standing waves* may be thought of as the result of waves that travel down the string in both directions, are reflected at the ends, proceed across to the opposite ends and are again reflected, and so on.

The vibrations of a string fixed at both ends constitute standing waves

In order to understand how standing waves come into being, we must transfer to the case of waves our knowledge of how pulses in a string are reflected and of what happens when two pulses traveling in opposite directions meet. When a pulse in a string is reflected at a rigid support, the reflected pulse is inverted; a similar inversion occurs for periodic waves which means that, although their waveform and wavelength stay the same, the wave train is in effect shifted by $\frac{1}{2}\lambda$ so that a crest arriving at the end of the string is reflected as a trough and vice versa.

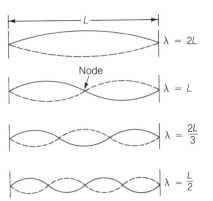

FIG. 12–19 Standing waves in a stretched string.

FIG. 12–20   (a) Constructive interference. (b) Destructive interference.

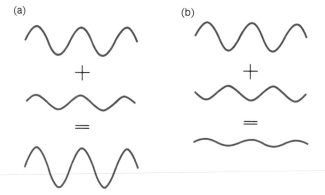

Principle of superposition for waves

Now let us look into the manner in which two waves in the same string interact. Applied to waves, the principle of superposition states that

> **When two or more waves of the same nature travel past a point at the same time, the displacement at that point is the sum of the instantaneous displacements of the individual waves.**

The principle of superposition holds for all types of waves, including waves in a stretched string, sound waves, water waves, and light waves.

What the principle of superposition signifies is that every wave train proceeds independently of any others that may also be present. Should two waves with the same wavelength come together in such a way that crest meets crest and trough meets trough, the resulting composite wave will have an amplitude greater than that of either of the original waves. When this occurs the waves are said to *interfere constructively* with each other. Should the waves come together in such a way that crest meets trough and trough meets crest, the composite wave will have an amplitude less than that of the larger of the original waves. When this occurs the waves are said to *interfere destructively* with each other (Fig. 12–20).

Constructive and destructive interference

Let us apply these ideas to a stretched string whose ends are fixed in place. When the string is plucked, waves move back and forth between its ends, undergoing an inversion each time they are reflected, and they interfere with each other in such a way that destructive interference always occurs at the ends. If the string is L long, destructive interference will occur at the ends when the waves have wavelengths of $\lambda = 2L$, L, $2L/3$, $L/2$, and so on (Fig. 12–19). In the case of the shorter wavelengths, other points, called *nodes*, are present at intermediate positions where no motion of the string takes place.

A node is a place where no vibration occurs

Figure 12–21 shows how a standing wave pattern comes into being: the dotted curve represents a wave moving to the right, and the dashed curve a wave of the same wavelength and amplitude moving to the left. The sum of these waves is found by adding together their displacements at each point on the string, and is shown as a solid line. The addition is performed in Fig. 12–21 for four successive instants one-eighth of a period apart. We note that the dotted and dashed curves do not always have displacements equal to zero at the ends of the string. There is no contradiction here

Origin of standing wave

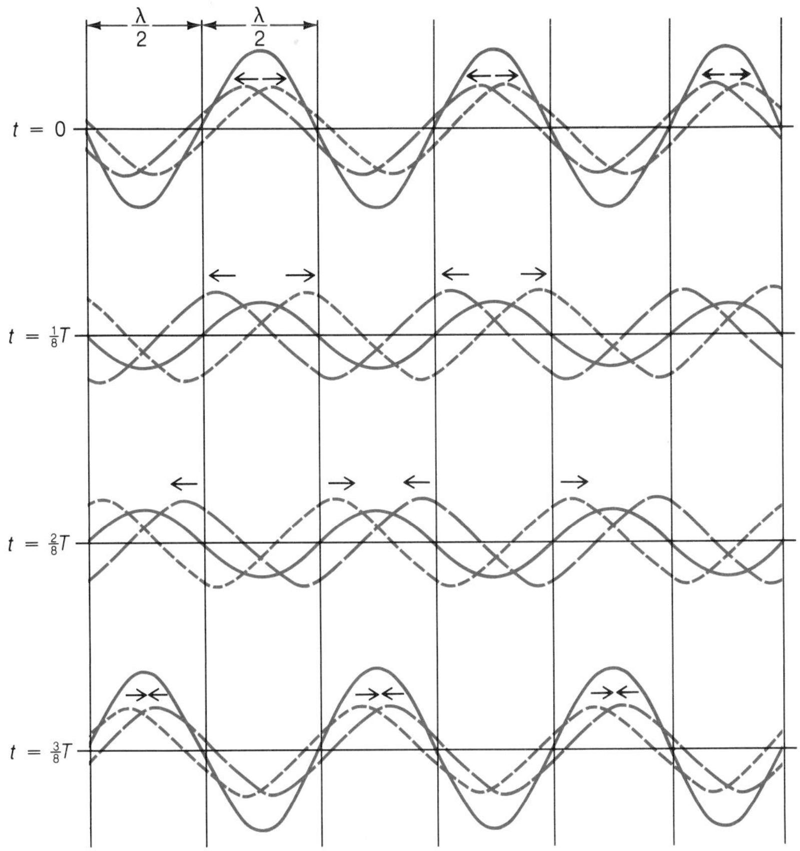

FIG. 12-21 The origin of a standing wave.

since it is their resultant, the solid curve, that corresponds to reality, and this curve exhibits the proper behavior.

Any type of wave can occur as a standing wave between suitable reflectors. The vibrating air columns in wind instruments and organ pipes—and in the throat, mouth, and nose of a person speaking—are standing sound waves, for instance. Standing light waves play an important role in the operation of lasers, as we shall find later. And standing waves of a rather remarkable kind are the key to understanding the structure of the atom.

Standing waves are a common phenomenon

12-7 RESONANCE

The condition that nodes occur at each end of the string restricts the possible wavelengths of standing waves to

$$\lambda = \frac{2L}{n} \qquad n = 1, 2, 3, \ldots \qquad \qquad \textit{Standing waves} \quad (12\text{-}5)$$

The standing waves of a system can have only certain wavelengths

The lowest possible frequency of oscillation f_1 of a stretched string corresponds to the longest wavelength, $\lambda = 2L$. Thus

$$f_1 = \frac{v}{\lambda} = \frac{v}{2L} \qquad\qquad (12\text{--}6)$$

Higher frequencies correspond to shorter wavelengths. From Eq. (12–5) we see that these higher frequencies can be represented in terms of f_1 by

$$f_n = nf_1 \qquad n = 2,\ 3,\ 4,\ \ldots \qquad\qquad (12\text{--}7)$$

Fundamental frequency of a stretched string

The frequency f_1 is called the *fundamental frequency* of the string, and the higher frequencies f_2, f_3, and so on, are called *overtones*.

Since the wave velocity v is $\sqrt{T/(m/L)}$ according to Eq. (12–1), we can express the fundamental frequency of a stretched string in terms of its length, linear density, and tension by the formula

$$f_1 = \frac{1}{2L}\sqrt{\frac{T}{m/L}} \qquad\qquad (12\text{--}8)$$

This formula is the basis for the design of stringed musical instruments such as pianos and violins. A short, light, taut string means a high fundamental frequency of vibration, while a long, heavy, slack string means a low fundamental frequency. The tuning of a stringed instrument involves changing the tensions in the various strings until their fundamental frequencies are correct.

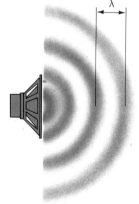

Example   The A-string of a violin has a linear density of 0.6 g/m and an effective length of 330 mm. (a) Find the tension required for its fundamental frequency to be 440 Hz. (b) If the string is under this tension, how far from one end should it be pressed against the fingerboard in order to have it vibrate at a fundamental frequency of 495 Hz, which corresponds to the note B?

Solution   (a) First we square both sides of Eq. (12–8) and then solve for T:

$$f_1{}^2 = \frac{T}{4L^2(m/L)}$$

$$T = 4L^2 f_1{}^2 (m/L) = 4(0.33\,\text{m})^2 (440\,\text{Hz})^2 (6 \times 10^{-4}\,\text{kg/m}) = 51\,\text{N}$$

This is about 11 lb.

(b) According to Eq. (12–8), $f_A/f_B = L_B/L_A$, so here

$$L_B = L_A\left(\frac{f_A}{f_B}\right) = (330\,\text{mm})\left(\frac{440\,\text{Hz}}{495\,\text{Hz}}\right) = 293\,\text{mm}$$

Hence the string should be pressed $(330 - 293)$ mm $= 37$ mm from one end.   ■

FIG. 12–22 Sound consists of longitudinal waves, representing condensations and rarefactions in the air in its path.

The 88 "strings" of a piano are metal wires whose fundamental frequencies range from 27 Hz for the lowest bass note to 4186 Hz for the highest treble note. This span is so great that the density as well as the length and tension of the wires must be varied in order that the instrument be of reasonable size: the treble strings are thin steel wires, and the bass strings are wound with copper wire to increase their mass density m/L.

The piano

The total tension of all the strings in a piano is about 20 tons and is borne by a heavy cast iron frame.

The fundamental frequency and overtones of a stretched string are its natural frequencies of oscillation: if the string is plucked, one or more of these frequencies will be excited. Eventually internal friction in the string will cause the various vibrations to die out. However, we can cause vibrations to persist by applying a periodic force to the string whose frequency is exactly the same as that of one of its natural frequencies. When this is done, the standing waves continue as long as the periodic force supplies energy to the string. If the energy provided exceeds that dissipated by internal friction, the amplitude of the standing wave will increase until the string may rupture. (A column of soldiers can destroy a flimsy bridge by marching across it in step with one of its natural frequencies, although the bridge may be capable of safely holding the static load of the soldiers.) This phenomenon is called *resonance*.

> **Resonance occurs when the frequency of an applied force equals one of the natural frequencies of an oscillator**

When periodic impulses are given to a string at frequencies other than those of its fundamental frequency and overtones, hardly any response occurs: the situation then is like pushing a child's swing at a frequency different from its natural one, which produces oscillations of negligible amplitude. All rigid structures possess characteristic natural frequencies of oscillation, even though their vibrations may be more complex in character than those of a stretched string, and these vibrations can be excited by a stimulus of the proper frequency. The traditional example of a goblet shattering when a violin is played with just the right frequency is an illustration of resonance.

> **All rigid structures have resonant frequencies**

12–8 SOUND

Sound waves are longitudinal and consist of pressure fluctuations. The air (or other medium) in the path of a sound wave becomes alternately denser and rarer; the resulting changes in pressure cause our eardrums to vibrate with the same frequency, which produces the physiological sensation of sound.

> **Sound waves are pressure fluctuations**

Most sounds are produced by vibrating objects. An example is the diaphragm of a loudspeaker (Fig. 12–22). When it moves outward, it pushes the air molecules directly in front of it closer together to form a region of high pressure that spreads out in front of the loudspeaker. The diaphragm then moves backward, thereby expanding the volume available to nearby air molecules. Air molecules now flow toward the diaphragm, and consequently a region of low pressure spreads out directly behind the high-pressure region. The continued vibrations of the diaphragm thus send out successive layers of condensation and rarefaction.

> **How a loudspeaker produces sound**

The speed of sound in air at sea level and at 20°C is 343 m/s, which is 1126 ft/s. This speed increases at about 0.6 m/s per °C because the random speeds of air molecules increase with temperature and so make the passage of pressure fluctuations more rapid.

> **Speed of sound in air**

The musical note A, the frequency of which is 440 Hz, represents a wavelength λ in air of

$$\lambda = \frac{v}{f} = \frac{343 \, \text{m/s}}{440 \, \text{Hz}} = 0.78 \, \text{m}$$

which is about 31 in. A normal ear responds to sound waves with frequencies from about 20 Hz to about 20,000 Hz, which correspond to wavelengths in air from 17 m

> **Audible sound and ultrasound**

FIG. 12–23 Echo sounding permits water depth to be rapidly and continuously determined from a boat. (a) A pulse of high-frequency sound waves is sent out by a suitable device on the boat's hull. (b) The time T needed for the pulse to return after being reflected by the sea floor is a measure of the water depth d.

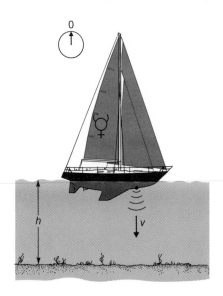

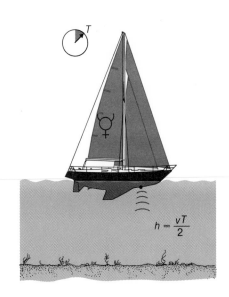

(54 ft) to 17 mm (0.65 in.). Ordinary speech is concentrated in a relatively narrow frequency band in the vicinity of 1000 Hz. Sound waves whose frequencies are above 20,000 Hz are called *ultrasonic* and can be detected by appropriate electromechanical devices. Ultrasonic waves are audible to many animals whose hearing organs are smaller than those of human beings and so are better suited to cope with short wavelengths. Figure 12–23 shows how pulses of ultrasound are used to determine the depth of the water under a boat or ship.

What the speed of sound depends upon

Sound waves are transmitted by solids, liquids, and gases. In general, the stiffer the material the faster the waves travel, which is reasonable when we reflect that stiffness implies particles that are tightly coupled together and therefore more immediately responsive to one another's motions. As in the case of waves in a stretched string, the less dense the material the less the inertia, and the higher the speed of the waves. A detailed analysis shows that the speed of sound in a fluid medium is given by

$$v = \sqrt{\frac{B}{d}}$$

where B is the bulk modulus of the fluid and d its density. The speed of sound in a solid is

$$v = \sqrt{\frac{Y}{d}}$$

where Y is the Young's modulus of the material. Table 12–1 lists the speed of sound in various materials.

The frequency remains unchanged when a wave goes from one medium to another

The wavelength of the musical note A in seawater is

$$\lambda = \frac{v}{f} = \frac{1531\,\text{m/s}}{440\,\text{Hz}} = 3.5\,\text{m}$$

which is nearly five times the wavelength of the same note in air. When a sound wave produced in one medium enters another in which its speed is different, the frequency

Medium	Speed, m/s
Gases (0°C)	
Carbon dioxide	259
Air	331
Air, 20°C	343
Helium	965
Liquids (25°C)	
Ethyl alcohol	1207
Water, pure	1498
Water, sea	1531
Solids	
Lead	1200
Wood, mahogany	~4300
Iron and steel	~5000
Aluminum	5100
Glass, pyrex	5170
Granite	6000

TABLE 12–1
Speed of sound

of the wave remains the same while the wavelength changes. This is in accord with the behavior of pulses that pass from one stretched string to another, as discussed earlier.

As an object moves through a fluid—for instance, an airplane through the atmosphere—it disturbs the fluid, and the resulting pressure waves spread out in spherical shells at the speed of sound. If the airplane itself is moving at the speed of sound, the waves pile up in front of it, as in Fig. 12–24(b), to create a wall of high pressure—the "sound barrier." For the airplane to go faster than sound, a considerable amount of force is needed to push it through this barrier, more than the force needed to give a comparable acceleration at speeds either less than (subsonic) or more than (supersonic) that of sound.

The "sound barrier"

At supersonic speeds, an airplane outdistances the waves it produces, which gives rise to the pattern of crests shown in Fig. 12–24(c). Where successive crests overlap, constructive interference takes place, and the result is a conical shell of high pressure with the airplane at its apex. This shell is called a *shock wave* because the pressure increases sharply when it passes, and it is responsible for the sonic boom heard after a supersonic airplane has passed overhead. Because the rise in pressure is so sudden, such a shock wave can do physical damage to structures in its path even though the pressure change itself may not be very great. On a smaller scale, the crack of a whip is a sonic boom that occurs when the tip of the whip moves faster than sound.

A shock wave is produced by supersonic motion

12–9 INTERFERENCE OF SOUND WAVES

Interference occurs in longitudinal as well as in transverse waves, as we just saw in the case of the shock waves produced in supersonic motion. Another example is the combination of sinusoidal sound waves to give the actual sounds we hear. A graph of pressure versus time at a certain place (or of pressure versus distance at a certain time) for a sound wave that contains a single frequency is sinusoidal in shape. A person

Pure sound waves are sinusoidal

FIG. 12–24 Waves produced by an object moving at (a) subsonic, (b) sonic, and (c) supersonic speeds.

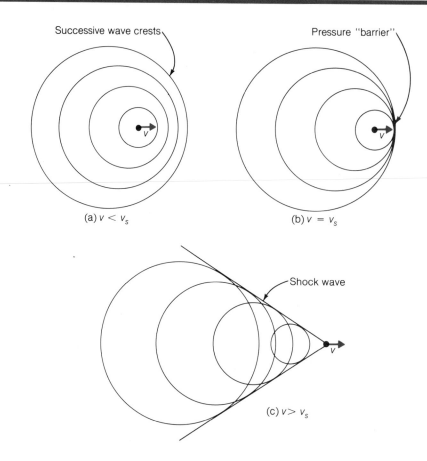

Successive wave crests

Pressure "barrier"

(a) $v < v_s$

(b) $v = v_s$

Shock wave

(c) $v > v_s$

singing the musical note "middle A" produces 440-Hz pressure waves whose amplitude might be 1 N/m², which is 0.001% of sea-level atmospheric pressure. The pressure in the path of such waves increases to 1 N/m² above normal and decreases to 1 N/m² below normal 440 times per second. The graphs of Fig. 12–25 show the pressure variations of such waves under the assumption that they are purely sinusoidal.

Actual sounds have complex waveforms

Natural sounds are never as regular as that shown in Fig. 12–25 but consist of mixtures of different frequencies that combine to give complex waveforms. The waveforms of sounds can be observed with the help of an oscilloscope, a device that displays electrical signals, such as those produced by a microphone in response to a sound, on the screen of a tube like a television picture tube. (The oscilloscope is described in Chapter 17.) Figure 12–26 shows some waveforms of musical notes, which are mixtures of a fundamental frequency and some of its overtones. The precise mixture determines the quality or timbre of the note, which is different for each instrument and for different human voices.

Musical sounds

Certain mixtures of frequencies are pleasing to the ear, and sounds that incorporate them are considered "musical." For instance, a tone combined with its first overtone, whose frequency is twice as great, appears harmonious to a listener. In music, such an

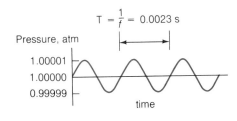

$$T = \frac{1}{f} = 0.0023 \text{ s}$$

Pressure, atm

1.00001
1.00000
0.99999

time

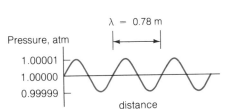

$\lambda = 0.78 \text{ m}$

Pressure, atm

1.00001
1.00000
0.99999

distance

FIG. 12–25 The pressure variations that constitute the musical note A in air. The amplitude is approximately that which a singer would produce.

interval is called an *octave* because it includes eight notes. Another harmonious combination is a tone together with another whose frequency is 50% higher, so the frequencies are in the ratio 2:3. An example is C (264 Hz) plus G (396 Hz); such an interval includes five notes (here C, D, E, F, G) and so is called a *fifth*. Somewhat less agreeable are two notes whose frequencies are in the ratio 4:5, for instance C (264 Hz) and E (330 Hz), whose interval spans three notes and is called a *third*. The larger the numbers needed to express the ratio of frequencies, the less attractive the resulting sound is. Thus C and D are in the ratio 8:9 and seem discordant, and E and F, whose ratio is 15:16, are more discordant still. Ordinary sounds are mixtures of frequencies that have no special relationships with one another, and if the mixture seems particularly harsh, it qualifies as noise.

Let us look into what happens when two sounds of different frequency are combined. If two tuning forks (or other sources of single-frequency sound waves) whose frequencies are slightly different are struck at the same time, the sound that we hear

Beats result from the interference of sound waves

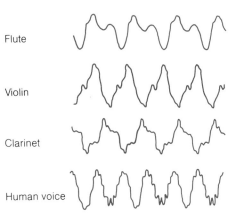

FIG. 12–26 Waveforms of some musical sounds.

Flute

Violin

Clarinet

Human voice

FIG. 12–27 *Top,* the origin of beats in sound waves. *Bottom,* beats produced by waves of 20 and 21 Hz.

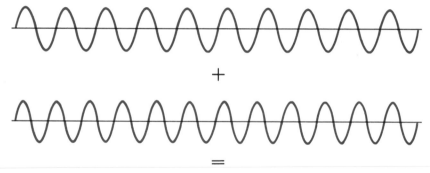

+

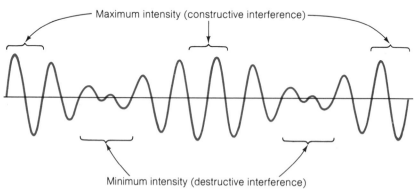

=

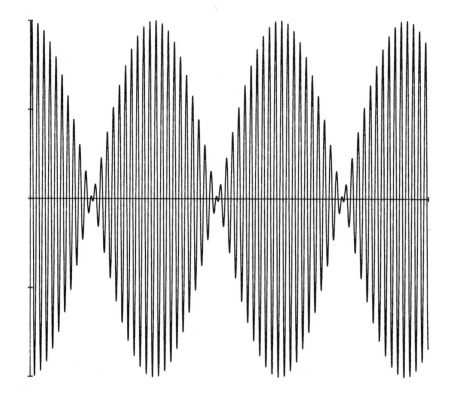

Maximum intensity (constructive interference)

Minimum intensity (destructive interference)

fluctuates in intensity. At one instant we hear a loud tone, then virtual silence, then the loud tone again, then virtual silence, and so on. The origin of this behavior is shown schematically in Fig. 12–27; the loud periods occur when the waves from the two forks interfere constructively, thus reinforcing one another, and the quiet periods occur when the waves interfere destructively, thus partially or wholly canceling one another out. These regular loudness pulsations are called *beats*.

A piano tuner can tell whether a piano string is in tune by listening for beats between the sound of that string and the sound of a tuning fork; if beats occur, he adjusts the tension of the string until they disappear. The sound we hear when beats are produced has a frequency that is the average of the two original frequencies, and the number of beats per second equals the difference between the two original frequencies. Thus the simultaneous vibrations of a 440-Hz tuning fork and an out-of-tune A-string that vibrates at 444 Hz will produce a 442-Hz tone whose amplitude rises and falls four times per second.

> The beat frequency is the difference between the two original frequencies

Beats are hard to distinguish as such when the two frequencies differ by more than perhaps 10 Hz. When the frequencies are far enough apart, however, they give rise to an audible "difference tone" responsible for the lack of harmony of certain combinations of musical notes that was mentioned above. For instance, the notes E and F have the respective frequencies 330 Hz and 352 Hz, and when they are emitted together, the result is an additional 22 Hz difference tone that makes the combination seem dissonant.

12–10 DOPPLER EFFECT

If we stand beside a road when a police car goes by with its siren blowing, we cannot help but notice that the pitch of the siren drops suddenly as the car passes by. What happens is that the pitch of the siren as the car approaches is *higher* than the pitch when the car is stationary, and the pitch when the car recedes is *lower* than its normal one. We also observe a change of pitch if the siren is at rest and we go past it in a rapidly moving car: we find a higher pitch than usual as we approach the siren, and a lower one as we recede. The change in frequency of a sound brought about by relative motion between source and listener is called the *Doppler effect*.

> Relative motion between an observer and a sound source changes the perceived frequency

The origin of the Doppler effect is straightforward. As a moving source emits sound waves, it is tending to overtake those traveling in the same direction (Fig. 12–28(a)). Hence the distance between successive waves is smaller than usual; since this distance is the wavelength of the sound, the corresponding frequency is higher than usual. At the same time the source is moving away from those of its waves that travel in the opposite direction, increasing the distance between successive waves behind it and thereby reducing their frequency.

> The Doppler effect in sound

If the source is stationary, a listener moving toward it intercepts more sound waves per unit time than if he were at rest, and accordingly he hears a higher frequency (Fig. 12–28(b)). When the observer moves away from the source, fewer of the waves catch up with him per unit time, and he hears a lower frequency.

The relationship between the frequency f_L the listener hears and the frequency f_S produced by the source is

$$f_L = f_S \left(\frac{v + v_L}{v - v_S} \right)$$

Doppler effect in sound (12–9)

FIG. 12–28 The Doppler effect occurs when there is relative motion between a source of sound and a listener.

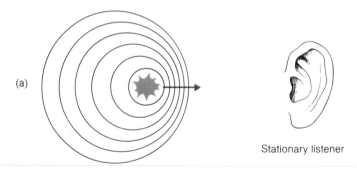

(a)

Stationary listener

Moving source

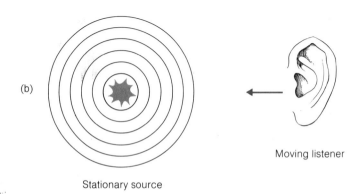

(b)

Moving listener

Stationary source

Here v is the speed of sound in air, v_L is the speed of the listener (reckoned as $+$ if he moves toward the source, as $-$ if he moves away from the source), and v_S is the speed of the source (reckoned as $+$ if it moves toward the listener, as $-$ if it moves away from the listener). If the listener is stationary, $v_L = 0$, while if the source is stationary, $v_S = 0$.

It is not difficult to derive Eq. (12–9). Let us consider first the situation shown in Fig. 12–28(a) of a sound source approaching a stationary listener with the speed v_S. The wavelength λ_S of the sound waves produced by the source when it is at rest is $\lambda_S = vT_S$, where $T_S = 1/f_S$ is the period of the waves. In the time T_S, the source moves toward the listener the distance $\Delta = v_S T_S$, so that the wavelength λ_L in the direction of motion is

$$\lambda_L = \lambda_S - \Delta = (v - v_S)T_S = \frac{v - v_S}{f_S}$$

Since the speed of sound is a property of the medium in which the sound travels and does not depend on the speed of the source, the frequency f_L the listener perceives is related to λ_L by $f_L \lambda_L = v$. Hence

$$f_L = \frac{v}{\lambda_L} = \frac{vf_S}{v - v_S} = f_S\left(\frac{v}{v - v_S}\right) \qquad \textit{Moving source} \quad (12\text{–}9a)$$

If the source is moving away from the listener, the same reasoning gives $\lambda_L = \lambda_S + \Delta$ and $f_L = f_S v/(v + v_S)$. We can use Eq. (12–9a) for this case also simply by reckoning v_S as a negative quantity.

Next we consider a listener approaching a stationary sound source, as in Fig. 12–28(b). The listener's speed relative to the air is v_L, which means that it is $v + v_L$ relative to the waves. Thus the period of the waves appears to the listener to be $T_L = \lambda_S/(v + v_L)$ and their frequency to be

$$f_L = \frac{1}{T_L} = \frac{v + v_L}{\lambda_S} = f_S\left(\frac{v + v_L}{v}\right) \qquad\qquad \textit{Moving listener} \quad (12\text{–}9b)$$

If the listener is moving away from the sound source, the listener's speed relative to the waves is $v - v_L$, which we can include in Eq. (12–9b) by reckoning v_L as negative in such a case. If both source and listener are in motion toward or away from each other, the source frequency f_S to be used in Eq. (12–9b) is the frequency f_L of the moving source from Eq. (12–9a), which gives Eq. (12–9):

$$f_L = f_S\left(\frac{v}{v - v_S}\right)\left(\frac{v + v_L}{v}\right) = f_S\left(\frac{v + v_L}{v - v_S}\right)$$

When $v_L = 0$, this formula becomes Eq. (12–9a), and when $v_S = 0$, it becomes Eq. (12–9b).

Example   The frequency of a train's whistle is 1000 Hz. (a) The train is approaching a stationary man at 40 m/s. What frequency does the man hear? (b) The train is stationary and the man is driving toward it in a car whose speed is 40 m/s. What frequency does the man hear now? (Unless otherwise specified, a temperature of 20°C is to be assumed.)

Solution   (a) Here $f_S = 1000$ Hz, $v = 343$ m/s, $v_S = 40$ m/s, and $v_L = 0$. Hence the apparent frequency of the whistle is

$$f_L = f_S\left(\frac{v + v_L}{v - v_S}\right) = (1000\,\text{Hz})\frac{343\,\text{m/s}}{(343 - 40)\,\text{m/s}} = 1132\,\text{Hz}$$

(b) Again $f_S = 1000$ Hz and $v = 343$ m/s, but now $v_S = 0$ and $v_L = 40$ m/s. The apparent frequency of the whistle is therefore

$$f_L = f_S\left(\frac{v + v_L}{v - v_S}\right) = (1000\,\text{Hz})\frac{(343 + 40)\,\text{m/s}}{343\,\text{m/s}} = 1117\,\text{Hz} \qquad\blacksquare$$

Frequency measurements can be made so accurately that relatively small speeds can be detected by means of the Doppler effect. An example is the speed of blood in an artery, which does not exceed about 0.4 m/s even in the aorta. When an ultrasound beam is directed at an artery, the waves reflected from the moving blood cells exhibit a Doppler shift because the cells then act as moving wave sources, and from this shift the speed of the blood can be found (see Problem 10).

The Doppler effect is an important tool of the astronomer, who is able to determine the speed of approach or recession of stars and galaxies from shifts in the characteristic frequencies of the light they emit. (These characteristic frequencies are discussed in

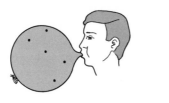 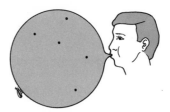

FIG. 12–29 Two-dimensional analogy of the expanding universe. As the balloon is inflated, the spots on it become increasingly distant from one another. A bug on the balloon would find that the farther away a spot is from it, the faster the spot seems to be receding; this is true regardless of where the bug is located. In the case of the universe, the more distant a galaxy is from us, the faster it is moving away (as revealed by the Doppler effect), which means that the universe is expanding uniformly.

Doppler shifts in the light from distant galaxies led to the discovery of the expanding universe

Chapter 28.) This is the method by which the expansion of the universe was detected. Throughout the sky, characteristic frequencies in the light from distant galaxies are lower than normal, a phenomenon called the "red shift" because red light is at the low-frequency end of the visible spectrum. The magnitude of the frequency change increases with distance from the earth, which suggests that the entire universe is expanding so that all the objects in it recede from one another (Fig. 12–29). The expansion apparently began 10 to 20 billion years ago in the explosion of a condensed mass of primeval matter, an event usually called the "big bang." The matter soon turned into the electrons, protons, and neutrons of which the present universe is composed, and as it expanded individual aggregates formed that became the galaxies. Gravitational forces are slowing the expansion down, and it is possible (present data are insufficient to decide) that the expansion will eventually stop. If this happens, the universe must then contract, and its collapse will be followed by another big bang. Otherwise the current expansion will continue forever.

Doppler effect in light

The Doppler effect in light differs from that in sound because light, which does not depend upon a material medium for its transmission, has the same relative speed c to all observers regardless of their state of motion (Chapter 26). The Doppler effect in light obeys the formula

$$f = f_s \sqrt{\frac{1 + v/c}{1 - v/c}} \qquad \textit{Doppler effect in light} \quad (12\text{--}10)$$

where f is the observed frequency, f_s is the frequency of the source, and v is the relative speed between source and observer. If source and observer are approaching each other, v is reckoned as $+$, and if they are receding from each other, v is reckoned as $-$. In a vacuum, the speed of light is $c = 3.00 \times 10^8$ m/s; its value in air is very close to this. Unlike the case of sound, the Doppler effect in light cannot be used to distinguish between motion of a source and motion of an observer.

Light consists of electromagnetic waves in a certain frequency interval to which the eye responds. Other electromagnetic waves, such as those used in radar and in radio communication, also exhibit the Doppler effect in accord with Eq. (12–10). Doppler shifts in radar waves are widely used by police to determine vehicle speeds, and such shifts in the radio waves emitted by a set of five earth satellites form the basis of the highly accurate Transit system of marine navigation.

12–11 SOUND INTENSITY

The rate at which a wave of any kind transports energy per unit cross-sectional area is called its *intensity I*. The unit of intensity is the watt/m^2. One joule of energy per second flows through a 1-m^2 surface perpendicular to the path of a wave whose intensity is 1 W/m^2 (Fig. 12–30). The minimum intensity a sound wave must have in order to be audible is about 10^{-12} W/m^2; at the other extreme, sound waves whose intensities exceed about 1 W/m^2 damage the ear.

The decibel is the unit of sound intensity level

The human ear does not respond linearly to sound intensity; doubling the intensity of a particular sound produces the sensation of a somewhat louder sound, but one that seems far less than twice as loud. For this reason the scale customarily used to measure the intensity level of a sound is logarithmic, which is a reasonable approximation of the actual response of the human ear. The unit is the *decibel* (dB), with a sound that is barely audible ($I_0 = 10^{-12}$ W/m^2) being assigned a value of 0 dB. A 10-dB sound is, by definition, 10 times more intense than a 0-dB sound; a 20-dB sound is 10^2 (or 100) times more intense; a 30-dB sound is 10^3 (or 1000) times more intense; and so on. Formally, the *intensity level* β, in decibels, of a sound wave whose intensity is W/m^2 is I is given by

$$\beta = 10 \log \frac{I}{I_0} \qquad \textit{Sound intensity level} \quad (12–11)$$

(If $a = 10^b$, then b is the logarithm of a. Thus 3 is the logarithm of 1000 since 1000 $= 10^3$. Logarithms are discussed in Appendix B-5.)

Ordinary conversation is usually about 60 dB, which is a sound intensity a million times greater than the faintest sound that can be heard. City traffic noise is about 80 dB, a rock band using amplifiers produces an intensity of as much as 125 dB, and the noise of a jet airplane is about 140 dB at a distance of 100 ft (Fig. 12–31). An extended exposure to sound intensities of over 85 dB usually leads to permanent hearing damage.

Example Five trumpets are being played, each at an average intensity level of 70 dB. What is the resulting intensity level?

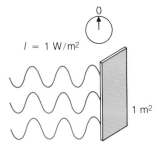

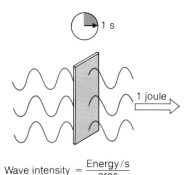

Wave intensity $= \dfrac{\text{Energy/s}}{\text{area}}$

FIG. 12–30 The intensity of a wave is a measure of the rate at which it transports energy.

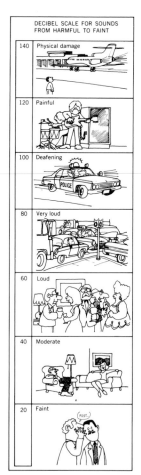

DECIBEL SCALE FOR SOUNDS
FROM HARMFUL TO FAINT

140	Physical damage
120	Painful
100	Deafening
80	Very loud
60	Loud
40	Moderate
20	Faint

FIG. 12–31 Decibel scale.

Solution The intensity of one trumpet is I_1, for an intensity level of

$$\beta_1 = 10 \log \frac{I_1}{I_0} = 70 \, \text{dB}$$

The intensity of five trumpets is $5I_1$. Since $\log xy = \log x + \log y$, the intensity level of the five trumpets is

$$\beta_5 = 10 \log \frac{5I_1}{I_0} = 10 \log 5 + 10 \log \frac{I_1}{I_0}$$

The logarithm of 5 can be found with a calculator or a table of logarithms. The result is $\log 5 = 0.70$, so that

$$\beta_5 = 10(0.70) \, \text{dB} + \beta_1 = 7 \, \text{db} + 70 \, \text{dB} = 77 \, \text{dB}$$

In this particular case, increasing the intensity fivefold only leads to a 10% increase in the intensity level. ∎

Example A change in sound intensity level of 1 dB is about the minimum that can be detected by a person with good hearing; usually the change must be 2 or 3 dB to be readily apparent. What is the actual intensity ratio between two sounds that are 3 dB apart?

Solution If the respective intensities and intensity levels of the two sounds are I_1, β_1 and I_2, β_2, then

$$\beta_2 - \beta_1 = 10 \left(\log \frac{I_2}{I_0} - \log \frac{I_1}{I_0} \right)$$

Since $\log x - \log y = \log x/y$,

$$\log \frac{I_2}{I_0} - \log \frac{I_1}{I_0} = \log \left(\frac{I_2/I_0}{I_1/I_0} \right) = \log \frac{I_2}{I_1}$$

and so

$$\beta_2 - \beta_1 = 10 \log \frac{I_2}{I_1}$$

This relationship is plotted in Fig. 12–32. Here $\beta_2 - \beta_1 = 3$, hence

$$3 = 10 \log \frac{I_2}{I_1} \quad \text{or} \quad 0.3 = \log \frac{I_2}{I_1}$$

Now we take the antilogarithm of both sides:

$$\log^{-1} 0.3 = \log^{-1} \left(\log \frac{I_2}{I_1} \right) = \frac{I_2}{I_1}$$

To find $\log^{-1} 0.3$ with a calculator, enter 0.3 and press the 10^x key (Inv Log on some calculators). To find $\log^{-1} 0.3$ from a table of logarithms, look up 0.3 in the body of

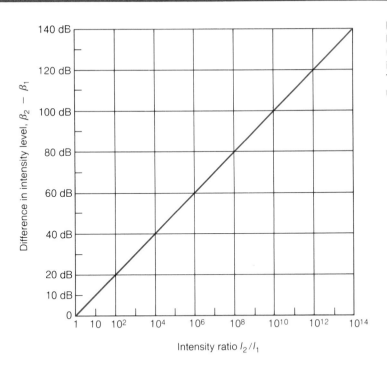

FIG. 12–32 Relationship between sound intensity ratio and difference in intensity level in decibels. The intensity ratio scale is not linear.

the table and read across to obtain the number whose logarithm is 0.3. The result is

$$\frac{I_2}{I_1} = 2$$

A 3 dB difference in intensity level corresponds to a factor of two in sound intensity. ∎

How sound intensity varies with distance

Because sound waves spread out as they move away from their source, their intensity decreases with distance. Let us consider the sound waves from a source whose power output is P; that is, P joules of energy flow from the source per second. At the distance r from the source, the total power P is distributed over the $4\pi r^2$ area of a sphere of radius r. Hence the intensity I of the sound at this distance is

$$I = \frac{P}{4\pi r^2} \qquad \textit{Intensity and distance} \quad (12\text{–}12)$$

The sound intensity is inversely proportional to the square of the distance from the source. If we go from 5 m away from a source to 10 m away, the intensity drops to $\frac{1}{4}$ its former value. This inverse-square law holds for the intensity of all waves that spread out freely in three dimensions—the intensity of the light from a lamp also varies as $1/r^2$, for example. The purpose of the concave mirror in a searchlight is to avoid the $1/r^2$ decrease in intensity by concentrating the light waves from a lamp into as nearly parallel a beam as possible. Cupping one's hands around one's mouth similarly helps one's voice to carry further.

Example The minimum sound intensity level needed for reasonable audibility is 20 dB. If a certain person speaking normally produces an intensity level of 40 dB at a distance of 1 m, what is the maximum distance at which the sound can be heard?

Solution A difference of 20 dB is equivalent to an intensity ratio of $10^2 = 100$. Since $I_2/I_1 = r_1{}^2/r_2{}^2$,

$$r_2 = r_1 \sqrt{\frac{I_1}{I_2}} = 1\,\text{m} \times \sqrt{100} = 10\,\text{m}$$ ■

Decibel scale for power gain

The decibel scale is widely used in electronics. If the power input to an amplifier or other signal-processing device is P_{in} and the power output is P_{out}, the *power gain* G in decibels is defined as

$$G = 10 \log \frac{P_{out}}{P_{in}}$$ *Power gain*

Thus the power gain of an amplifier whose power input is 0.1 W and whose power output is 50 W is

$$G = 10 \log \frac{P_{out}}{P_{in}} = 10 \log \frac{50\,W}{0.1\,W} = 10 \log 500 = 10 \times 2.7 = 27\,\text{dB}$$

As mentioned earlier, the logarithm of a product is equal to the sum of the logarithms of the factors:

$$\log xyz = \log x + \log y + \log z$$

Since power gains are logarithmic quantities, the overall gain in decibels of a system of several devices is just the sum of their separate gains expressed in decibels:

$$G\,(\text{overall}) = G_1 + G_2 + G_3 + \cdots$$

For example, an audio system might consist of a 35-dB preamplifier, a -10-dB attenuator (a negative dB rating means a negative power gain, so the output is less than the input), and a 70-dB amplifier. The overall gain of the system is

$$G\,(\text{overall}) = +35\,\text{dB} - 10\,\text{dB} + 70\,\text{dB} = +95\,\text{dB}$$

12–12 THE EAR

Sensitivity of the ear

The ear is not equally sensitive to sounds of different frequencies. Maximum response occurs for sounds between 3000 and 4000 Hz, when the threshold for hearing is somewhat less than 0 dB. A 100-Hz sound, however, must have an intensity level of at least 40 dB to be heard. Sounds whose frequencies are below about 20 Hz *(infrasound)* and above about 20,000 Hz *(ultrasound)* are inaudible to almost everybody regardless of intensity. The intensity level at which a sound produces a feeling of discomfort in the ear is relatively constant at approximately 120 dB for all frequencies. Figure 12–33 shows how the thresholds of hearing and of discomfort vary with frequency. Hearing deteriorates with age, most notably at the high-frequency end of the spectrum. A typical person 60 years old has a threshold of hearing about 10 dB higher than the lowest curve

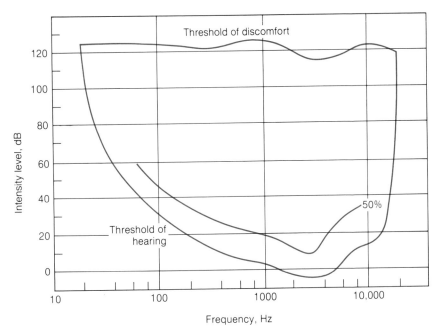

FIG. 12–33 The response of the ear to sound varies with frequency. Only 1% of the U.S. population can hear sounds with intensity levels that fall below the lower curve; 50% can hear sound with intensity levels that fall below the curve above it. Hearing acuity decreases with age.

of Fig. 12–33 for 2000-Hz tones, nearly 30 dB higher for 8000-Hz tones, and nearly 70 dB higher for 12,000-Hz tones.

The structure of the human ear is shown in Fig. 12–34. The *eardrum* vibrates when sound waves reach it from the ear canal. Inside the eardrum is the *middle ear,* a small air-filled cavity that contains three linked bones called the *hammer,* the *anvil,* and the *stirrup* because of their respective shapes. The *Eustachian tube,* which opens when a person swallows, permits the middle ear to stay at atmospheric pressure. The vibrations of the eardrum pass in succession from the hammer to the anvil to the stirrup and finally to the membrane over the *oval window* between the middle and inner ears. The liquid-filled *inner ear* consists of the *semicircular canals,* whose function is to act as a reference system in providing a sense of balance, and the *cochlea,* a snail-shaped organ in which pressure waves in the inner-ear liquid excite nerve impulses that pass along the *auditory nerve* to the brain.

Structure of the ear

The ear amplifies sound waves in three ways. The ear canal itself acts as a resonant cavity—that is, standing waves are set up in it—for sound frequencies in the neighborhood of 3000–4000 Hz, and consequently the pressure fluctuations at the eardrum may be twice as great as those outside the ear. This is the reason why the ear is most sensitive to such frequencies. Second, the linked bones in the middle ear have a mechanical advantage of two to three, which correspondingly increases the forces exerted on the oval-window membrane over those exerted on the eardrum. Third, the area of the eardrum is typically 20 times greater than that of the oval window, so that (by analogy with the hydraulic press) there is a further mechanical advantage of this amount. The total amplification of pressure changes may thus be more than 100, and the amplification of changes in sound intensity may be more than 10,000.

The ear as an amplifier

The actual mechanism by which the ear discriminates among sound frequencies depends upon the physical properties of the *basilar membrane* in the cochlea. This

FIG. 12–34   The ear.

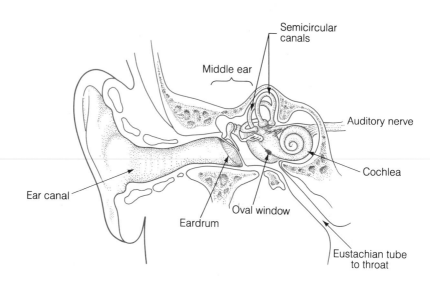

membrane varies in thickness and tension along its length, with a different portion resonating at each frequency. Thousands of tiny fibers distributed along the basilar membrane transmit disturbances in the membrane to nerve endings, and the brain interprets the impulses it receives as sound of a certain pitch according to the position of the stimulated nerve ending. The amplitude of the disturbance of the basilar membrane is registered as the loudness of the sound that produced it.

IMPORTANT TERMS

Wave motion is characterized by the propagation of a change in a medium. Waves transport energy from one place to another.

The **frequency** of a series of periodic waves is the number of waves that pass a particular point per unit time, while their **wavelength** is the distance between adjacent crests or troughs. The **period** is the time required for one complete wave to pass a particular point. The **amplitude** of a wave is the maximum displacement of a particle of the medium on either side of its normal position when the wave passes.

Longitudinal waves occur when the individual particles of a medium vibrate back and forth in the direction in which the waves travel. **Transverse waves** occur when the individual particles of a medium vibrate from side to side perpendicular to the direction in which the waves travel. The vibrations of a stretched string are transverse waves.

The **principle of superposition** states that when two or more waves of the same nature travel past a given point at the same time, the amplitude at the point is the sum of the amplitudes of the individual waves. The interaction of different wave trains is called **interference: constructive interference** occurs when the resulting composite wave has an amplitude greater than that of either of the original waves, and **destructive interference** occurs when the resulting composite wave has an amplitude less than that of either of the original waves.

Resonance occurs when periodic impulses are given to an object at a frequency equal to one of its natural frequencies of oscillation.

Sound is a longitudinal wave phenomenon which results in periodic pressure variations. Sound intensity level is measured in **decibels.**

A **shock wave** is a shell of high pressure produced by the motion of an object whose speed exceeds that of sound.

The **Doppler effect** refers to the change in frequency of a wave when there is relative motion between its source and an observer.

IMPORTANT FORMULAS

Waves in a string: $v = \sqrt{\dfrac{T}{m/L}}$

Wave motion: $v = f\lambda = \dfrac{\lambda}{T}$

Sound intensity level: $\beta \, (dB) = 10 \log \dfrac{I}{I_0}$

MULTIPLE CHOICE

1. The speed of waves in a stretched string depends upon
(a) the tension in the string.
(b) the amplitude of the waves.
(c) the wavelength of the waves.
(d) the acceleration of gravity.

2. The higher the frequency of a wave,
(a) the lower its speed.
(b) the shorter its wavelength.
(c) the greater its amplitude.
(d) the longer its period.

3. Of the following properties of a wave, the one that is independent of the others is its
(a) amplitude.   (b) speed.
(c) wavelength.   (d) frequency.

4. Waves transmit from one place to another
(a) mass.   (b) amplitude.
(c) wavelength.   (d) energy.

5. In a transverse wave, the individual particles of the medium
(a) move in circles.
(b) move in ellipses.
(c) move parallel to the direction of travel.
(d) move perpendicular to the direction of travel.

6. Sound waves are
(a) longitudinal.
(b) transverse.
(c) partly longitudinal and partly transverse.
(d) sometimes longitudinal and sometimes transverse.

7. Two waves meet at a time when one has the instantaneous amplitude A and the other has the instantaneous amplitude B. Their combined amplitude at this time is
(a) $A + B$.
(b) $A - B$.
(c) between $A + B$ and $A - B$.
(d) indeterminate.

8. Sound waves do not travel through
(a) solids.   (b) liquids.
(c) gases.   (d) a vacuum.

9. The amplitude of a sound wave determines its
(a) pitch.   (b) loudness.
(c) overtones.   (d) resonance.

10. When a sound wave goes from air into water, the quantity that remains unchanged is its
(a) speed.   (b) amplitude.
(c) frequency.   (d) wavelength.

11. A pure musical tone causes a thin wooden panel to vibrate. This is an example of
(a) an overtone.   (b) harmonics.
(c) resonance.   (d) interference.

12. A sonic boom is heard after an airplane has passed overhead. This means that the airplane
(a) is accelerating.
(b) is climbing.
(c) was just then passing through the sound barrier.
(d) is traveling faster than sound.

13. Beats are the result of
(a) diffraction.
(b) constructive interference.
(c) destructive interference.
(d) both constructive and destructive interference.

14. Relative to the radio signals sent out by a spacecraft headed away from the earth, the signals that are received on the earth
(a) have a lower speed.
(b) have a lower frequency.
(c) have a shorter wavelength.
(d) have all of the above characteristics.

15. Sound waves whose frequency is 300 Hz have a speed relative to sound waves in the same medium whose frequency is 600 Hz that is
(a) half as great.
(b) the same.
(c) twice as great.
(d) four times as great.

16. Sound waves whose frequency is 300 Hz have a wavelength relative to sound waves in the same medium whose frequency is 600 Hz that is
(a) half as great.
(b) the same.
(c) twice as great.
(d) four times as great.

17. Radio amateurs are permitted to communicate on the "10-meter band." What frequency of radio waves corresponds to a wavelength of 10 meters? The speed of radio waves is 3×10^8 m/s.

(a) 3.3×10^{-8} Hz (b) 3.0×10^7 Hz
(c) 3.3×10^7 Hz   (d) 3.0×10^9 Hz

18. A boat at anchor is rocked by waves whose crests are 40 m apart and whose speed is 10 m/s. These waves reach the boat once every

(a) 400 s.   (b) 30 s.
(c) 4 s.   (d) 0.25 s.

19. Two tuning forks of frequencies 310 and 316 Hz vibrate simultaneously. The number of times the resulting sound pulsates per second is

(a) 0.   (b) 6.
(c) 313.   (d) 626.

20. How many times more intense is a 90-dB sound than a 40-dB sound?

(a) 5   (b) 50
(c) 500   (d) 10^5

EXERCISES

(Assume that the speed of sound in sea-level air is 343 m/s, which corresponds to a temperature of 20°C.)

1. In general, in what state of matter does sound travel fastest? Why?

2. Verify that v has the dimensions of length/time in the formulas $v = \sqrt{T/(m/L)}$, $v = \sqrt{B/d}$, and $v = \sqrt{Y/d}$, which respectively give the speed of waves in a stretched string, in a fluid, and in a solid.

3. A pulse sent down a long string eventually dies away and disappears. What happens to its energy?

4. The amplitude of a wave is doubled. If nothing else is changed, how is the flow of energy affected?

5. The faintest audible sound has an amplitude of about 2×10^{-5} N/m². What percentage of sea-level atmospheric pressure is this?

6. A wave of frequency f and wavelength λ passes from a medium in which its speed is v to another medium in which its speed is $2v$. What are the frequency and wavelength of the wave in the second medium? (*Hint:* Consider what happens at the interface between the two media when the wave passes through to decide whether the frequency or the wavelength or both change.)

7. How can constructive and destructive interference be reconciled with the principle of energy conservation?

8. As a clarinet is played, the temperature of the air column inside it rises. What effect does this have on the frequencies of the notes it produces?

9. A man has two tuning forks, one marked "256 Hz" and the other of unknown frequency. He strikes them simultaneously, and hears 10 beats per second. "Aha," he says, "the other tuning fork has a frequency of 266 Hz." What is wrong with his conclusion?

10. The amplitude and wavelength of the wiggles in the grooves of a phonograph record correspond to the same quantities in the sound waves they represent. Inspection of a certain record reveals grooves whose wiggles have the same amplitude but one has a wavelength three times shorter than the other. (a) If the audio system is linear, what will be the difference in the sound intensity the two sets of wiggles produce? (b) Approximately how will the difference in apparent loudness correspond to the difference in intensity?

11. What is the wavelength of sound waves of frequency 8000 Hz in (a) air, (b) seawater, (c) steel?

12. A violin string is set in vibration at a frequency of 440 Hz. How many vibrations does it make while its sound travels 200 m in air?

13. A certain groove in a phonograph record moves past the needle at a speed of 0.3 m/s. If the wiggles in the groove are 0.1 mm apart, what is the frequency of the sound that is produced?

14. A certain groove in a phonograph record moves past the needle at a speed of 0.4 m/s. The sound produced has a frequency of 3000 Hz. What is the wavelength of the wiggles in the groove?

15. Water waves are observed approaching a lighthouse at a speed of 6 m/s. There is a distance of 7 m between adjacent crests. (a) What is the frequency of the waves? (b) What is their period?

16. The stainless steel forestay of a racing sailboat is 20 m long and 1 cm in diameter, and its mass is 10 kg. In order to determine its tension, it is struck with a hammer at the lower end and the return of the pulse is timed. If the time interval is 0.2 s, what is the tension in the stay?

17. A stretched wire 1 m long has a fundamental frequency of 300 Hz. (a) What is the speed of the waves in the wire? (b) What are the frequencies of the first three overtones?

18. A steel wire 1 m long whose mass is 10 g is under a tension of 400 N. (a) What is the wavelength of its fundamental mode of vibration? (b) What is the frequency of this mode? (c) What is the wavelength of the sound waves produced when the string vibrates in this mode?

19. A workman strikes a steel rail with a hammer, and the sound reaches an observer 0.5 km away both through the air and through the rail. How much time separates the two sounds?

20. A mine explodes at sea, and there is an interval of 5 s between the arrival of the sound through the water and its arrival through the air at a nearby ship. How far away is the ship?

21. A tuning fork vibrating at 440 Hz is placed in distilled water. (a) What are the frequency and wavelength of the waves produced within the water? (b) What are the frequency and wavelength of the waves produced in the surrounding air when the water waves reach the surface?

22. A fire engine has a siren whose frequency is 500 Hz. What frequency is heard by a stationary observer when the engine moves toward him at 12 m/s? When it moves away from him at 12 m/s?

23. A latecomer to a concert hurries down the aisle toward his seat so fast that the note middle C (256 Hz) appears 1 Hz higher in frequency. How fast is he going?

PROBLEMS

1. The G and A strings of a violin have the respective fundamental frequencies of 196 Hz and 440 Hz. If the G string has a linear density of 3 g/m, what should the linear density of the A string be if both are to be under the same tension?

2. The vibrating part of the G string of a certain violin is 330 mm long and has a fundamental frequency of 196 Hz when under a tension of 50 N. (a) Find the linear density of the string. (b) Where should the string be pressed in order for it to vibrate at 220 Hz?

3. The vibrating part of the E string of a violin is 330 mm long and has a fundamental frequency of 659 Hz. What is its fundamental frequency when the string is pressed against the fingerboard at a point 60 mm from its end? What are the first and second overtones of the string under these circumstances?

4. A heavy rope hanging from the ceiling is given a transverse shake at its lower end. Show that the speed with which the resulting pulse travels up the rope increases with height. (*Hint:* The tension at a given height equals the weight of the rope below that height.)

5. The ratio between the speed of an airplane (or other moving object) and the speed of sound is called the *Mach number* after the Austrian physicist Ernst Mach. Find the angle between the shock wave created by an airplane traveling at Mach 1.3 and the direction of the airplane.

6. The speed of surface waves in shallow water h deep is $\sqrt{gh}$. How fast is a boat moving in water 1.5 m deep if the total angular width of its wake is 50°?

7. Two identical steel wires have fundamental frequencies of vibration of 400 Hz. The tension in one of the wires is increased by 2%, and both wires are plucked. How many beats per second occur?

8. Train A is heading east at the speed v_A and train B is heading west at the speed v_B on an adjacent track. The locomotive on train A is blowing its whistle continuously. A passenger on train B observes the frequency of the sound from the whistle to be 307 Hz when the trains approach each other, 256 Hz when they are abreast, and 213 Hz when they recede from each other. From these figures find the speed of each train relative to the ground.

9. A moving reflector approaches a stationary source of sound of frequency f with the speed u. A listener nearby hears both the original sound waves and the reflected sound waves. Obtain a formula for the number of beats per second the listener hears.

10. The speed with which blood flows through an artery can be determined from the Doppler shift in high-frequency sound waves sent into the artery and detected after reflection from the moving blood cells. What is done is to aim the source and receiver of the sound so that the reflections occur from cells moving away from them. Thus the sound waves striking a cell are Doppler shifted because of its motion away from the source, and the waves that reach the receiver are further shifted because they come from a reflector moving away from it. (a) Derive a formula for the frequency f_r of the waves that reach the receiver in terms of the source frequency f_s, the speed v of sound in blood (which is 1570 m/s), and the speed v_b of the blood. (b) Solve this formula for v_b. (c) Find the blood speed when the source frequency is 10^6 Hz and the Doppler shift is 40 Hz.

11. A motorist goes through a red light and, when he is arrested, claims that the color he actually saw was green ($\lambda = 5.4 \times 10^{-7}$ m) and not red ($\lambda = 6.2 \times 10^{-7}$ m) because of the Doppler effect. The judge accepts this explanation and instead fines him for speeding at the rate of $1.00 for each km/h he exceeded the speed limit of 80 km/h. What was his fine?

12. The characteristic frequencies in the light from a distant galaxy of stars are found to be two-thirds as great as similar frequencies in the light from nearby stars. Find the recession speed of the distant galaxy.

13. A galaxy in the constellation Ursa Major is receding from the earth at 15,000 km/s. If one of the characteristic wavelengths of the light the galaxy emits is 5.50×10^{-7} m, what is the corresponding wavelength measured by astronomers on the earth?

14. What is the equivalent in decibels of a sound whose intensity is 5×10^{-6} W/m^2?

15. A siren produces a 120-dB sound. What is its intensity in W/m^2?

16. How many times more intense is a 60-dB sound than a 50-dB sound? Than a 40-dB sound? Than a 20-dB sound?

17. According to government regulations, the maximum permitted daily exposure time in a workplace to noise of 90 dB is 8 h; to noise of 95 dB, 4 h; to noise of 100 dB, 2 h; and so on. Thus each increase of 5 dB means halving the permitted exposure time. What intensity ratio corresponds to a 5-dB change in intensity level?

18. A riveting gun in a shipyard is producing 95 dB of noise. What is the intensity level when a second riveting gun begins to operate?

19. A chorus of 20 voices is singing at an intensity level of 70 dB. If the voices all have the same intensity level, what is it?

20. At a party, the intensity level of the conversation is 65 dB when a record player is switched on and set to an intensity level of 70 dB. What is the intensity level in the room now?

21. An amplifier has a power gain of 40 dB. What is the actual ratio between its output and input powers?

22. A microphone with an output of 10^{-9} W is connected to a 20-dB preamplifier and a 90-dB amplifier. What is the power output of the system.?

ANSWERS TO MULTIPLE CHOICE

1. a	**5.** d	**9.** b	**13.** d	**17.** b
2. b	**6.** a	**10.** c	**14.** b	**18.** c
3. a	**7.** a	**11.** c	**15.** b	**19.** b
4. d	**8.** d	**12.** d	**16.** c	**20.** d

THERMAL PROPERTIES OF MATTER

13

Nearly all substances expand when they are heated. Why does this happen? The fact that a phenomenon is familiar does not mean that its explanation need be obvious, and, as we shall see in this chapter, it is necessary to inquire deeply into the structure of matter in order to understand its thermal behavior. Interestingly enough, the details of why a gas tends to expand when heated turn out to be rather different from the reasons a solid or liquid tends to expand, although in both cases the effect is ultimately due to the close relationship between molecular energy and temperature.

13–1   TEMPERATURE

Temperature, like force, is a key concept in physics that requires an elaborate definition in order to be specified precisely, even though we have a clear idea

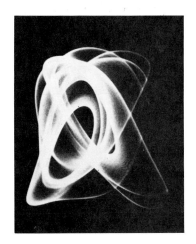

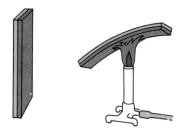

(a) Two strips of different metals that are joined together bend to one side with a change in temperature owing to different rates of expansion in the two metals, a fact employed in constructing household oven thermometers. The higher the temperature, the greater the deflection. When cooled, such a bimetallic strip bends in the opposite direction.

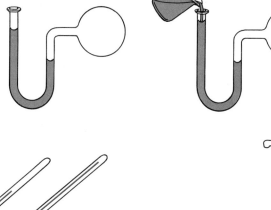

(b) In a constant-volume gas thermometer, which is a very sensitive laboratory instrument, the height of the mercury column at the left is adjusted until the mercury column at the right is just below the gas bulb. The difference in heights of the two mercury columns is a measure of the pressure needed to maintain the gas in a fixed volume, and hence a measure of the temperature.

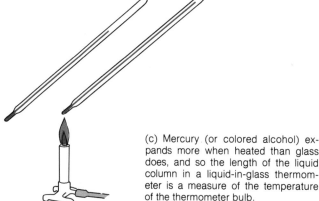

(c) Mercury (or colored alcohol) expands more when heated than glass does, and so the length of the liquid column in a liquid-in-glass thermometer is a measure of the temperature of the thermometer bulb.

FIG. 13-1   Three types of thermometer.

The color of a hot object varies with its temperature

Thermal expansion

of its meaning in terms of our sense impressions. We do not need such precision yet, and so we shall dodge the issue for the moment and consider temperature merely as that which is responsible for sensations of hot and cold.

There are a number of properties of matter that vary with temperature, and these can be used to construct *thermometers*, devices for measuring temperature. For example, when an object is heated sufficiently, it glows—at first a dull red, then bright red, and finally, at a high enough temperature, it becomes "white hot." By measuring the color of the light it gives off, we can accurately determine the temperature of an object. This method can only be used at rather high temperatures, however.

Of wider application is the fact that matter usually expands when its temperature is increased and contracts when its temperature is decreased. Concrete roads must be laid with regular gaps to allow for expansion in the summer; heated air above a radiator rises as it expands and becomes lighter than the surrounding air; a column of mercury in a glass tube changes length with a change in temperature. All three of these observations have resulted in practical thermometers (Fig. 13–1).

Before we can use any of these or other thermal properties of matter to construct a practical thermometer, we must begin by specifying a temperature scale and the method by which we shall calibrate the thermometer in terms of this scale. Water is a readily available liquid that freezes into a solid, ice, and vaporizes into a gas, steam, at definite temperatures at sea level atmospheric pressure. We can establish a temperature scale by defining the freezing point of water at 1 atm pressure (or, more exactly, the point at which a mixture of ice and water is in equilibrium, with exactly as much ice melting as water freezing) as 0° and the boiling point of water at 1 atm pressure (or, more exactly, the point at which a mixture of steam and water is in equilibrium) as 100°. This scale is called the *Celsius* scale, and temperatures measured in it are written, for example, "40°C." In the United States, the Celsius scale is sometimes called the *centigrade* scale.

To calibrate a thermometer, say an ordinary mercury thermometer, we first plunge it into a mixture of ice and water. When the mercury column has come to rest we mark the position of its top 0°C on the glass (Fig. 13–2). Then we plunge it into a mixture of steam and water, and when the mercury column has again come to rest, we mark the new position of the top of the mercury column 100°C. Finally we divide the interval between the 0°C and 100°C markings into 100 equal parts, each representing a change in temperature of 1°C, and extend the scale with divisions of the same length beyond 0°C and 100°C as far as is convenient. In doing this we have, of course, assumed that changes in the length of the mercury column are always directly proportional to the changes in temperature that brought them about.

The Celsius (centigrade) temperature scale

Calibrating a thermometer

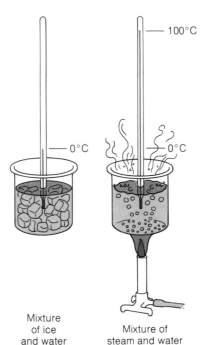

FIG. 13–2 Calibrating a thermometer on the Celsius scale. A mixture of ice and water at atmospheric pressure is, by definition, at 0°C, and a mixture of steam and water at atmospheric pressure is, again by definition, at 100°C.

100°C

0°C   0°C

Mixture of ice and water

Mixture of steam and water

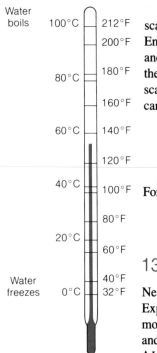

Water
boils   100°C   212°F
                  200°F

         80°C   180°F

                  160°F

         60°C   140°F

                  120°F

         40°C   100°F

                  80°F

         20°C

                  60°F

Water   40°F
freezes   0°C   32°F

FIG. 13–3 The
Fahrenheit and Celsius
temperature scales.

Although the Celsius scale is used in most of the world, a different temperature scale called the *Fahrenheit* scale is commonly used for nonscientific purposes in some English-speaking countries. In the Fahrenheit scale the freezing point of water is 32°F and the boiling point of water is 212°F (Fig. 13–3). This means that 180°F separates the freezing and boiling points of water, whereas 100°C separates them in the Celsius scale. Therefore fahrenheit degrees are 100/180 or 5/9 as large as Celsius degrees. We can convert temperatures from one scale to the other with the help of the formulas

$$T_F = \tfrac{9}{5}T_C + 32°$$
$$T_C = \tfrac{5}{9}(T_F - 32°) \hspace{3cm} (13\text{–}1)$$

For instance, the Celsius equivalent of 70°F is

$$\tfrac{5}{9}(70° - 32°) = 21°C$$

13–2 THERMAL EXPANSION

Nearly all substances expand when they are heated and contract when they are cooled. Experiments show that, to a good approximation, a change in temperature of ΔT causes most solids to change in length by an amount proportional both to their original lengths and to ΔT. If the original length of a rod of a certain material is L_0, its change in length ΔL after its temperature changes by ΔT is

$$\Delta L = aL_0\,\Delta T \hspace{2.5cm} \textit{Thermal expansion} \hspace{0.5cm} (13\text{–}2)$$

Change in length = a × original length × temperature change

The quantity a, called the *coefficient of linear expansion*, is a constant whose value depends upon the nature of the material. Different substances expand (and contract) to different extents; a lead rod, for example, changes in length by 60 times as much as a quartz rod of the same initial length when both are heated or cooled through the same temperature interval. Table 13–1 lists coefficients of linear expansion for various substances.

**Coefficient of linear
expansion**

TABLE 13–1
Coefficients of linear
expansion

Substance	Coefficient, $\times 10^{-5}/°C$
Aluminum	2.4
Brass	1.8
Concrete	0.7–1.2
Copper	1.7
Iron	1.2
Lead	3.0
Quartz	0.05
Silver	2.0
Steel	1.2

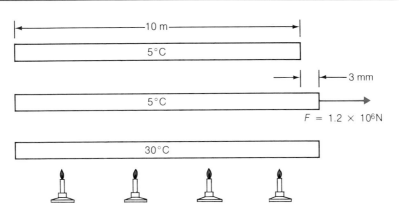

Example   What is the increase in length of a steel girder that is 10 m long at 5°C when its temperature rises to 30°C?

Solution   The coefficient of linear expansion of steel is $1.2 \times 10^{-5}/°C$, and so, from Eq. (13-2),

$$\Delta L = aL_0 \, \Delta T = (1.2 \times 10^{-5}/°C)(10\,\text{m})(25°C) = 3 \times 10^{-3}\,\text{m} = 3\,\text{mm}$$

which is about $\frac{1}{8}$ in. (Fig. 13-4).   ∎

Example   How much force is associated with the expansion of the girder if its cross-sectional area is 200 cm²?

Solution   Since the girder increases in length by 3 mm, the force is the same as that required to stretch it by 3 mm. Equation (9-4) gives the change in length ΔL of a rod that is subjected to a tension or compression force F as

$$\Delta L = \frac{L_0}{Y} \frac{F}{A}$$

where A is the cross-sectional area of the rod and Y is Young's modulus for the material of the rod. Since the girder has $A = 200\,\text{cm}^2 = 0.02\,\text{m}^2$ and, from Table 9-2, Young's modulus for steel is $2 \times 10^{11}\,\text{N/m}^2$,

$$F = \frac{YA \, \Delta L}{L_0} = \frac{(2 \times 10^{11}\,\text{N/m}^2)(0.02\,\text{m}^2)(3 \times 10^{-3}\,\text{m})}{10\,\text{m}} = 1.2 \times 10^6\,\text{N}$$

A force of 1.2 million newtons, which is equivalent to 135 tons, is associated with the expansion of the girder. Clearly, thermal expansion can involve quite considerable forces.   ∎

A formula similar to Eq. (13-2) holds for the changes in volume, ΔV, of a solid or liquid whose temperature changes by an amount ΔT. Here we have   **Volume expansion**

$$\Delta V = bV_0 \, \Delta T \qquad\qquad \textit{Volume expansion} \quad (13\text{-}3)$$

where V_0 is the original volume and b is the *coefficient of volume expansion*. Table

TABLE 13–2
Coefficients of volume expansion

Substance	Coefficient, $\times 10^{-4}/°C$
Ethyl alcohol	7.5
Glass (average)	0.2
Glycerin	5.1
Ice	0.5
Mercury	1.8
Pyrex glass	0.09
Water	2.1

Coefficient of volume expansion

13–2 is a list of coefficients of volume expansion for various substances. In general, the coefficients of linear and volume expansion are related by

$$b = 3a$$

so that we can readily determine the values of b for the materials of Table 13–1.

Example Calculate the volume of water that overflows when a Pyrex beaker filled to the brim with 250 cm^3 of water at 20°C is heated to 60°C.

Solution First, we note that a cavity in a body expands or contracts by precisely as much as a solid object having the same volume as the cavity and having the composition of the body (Fig. 13–5). This means that for the increase in capacity of the beaker we can write

$$\Delta V_b = b_p V_b \,\Delta T = \left(0.09 \times \frac{10^{-4}}{°C}\right)(250\,\text{cm}^3)(40°C) = 0.09\,\text{cm}^3$$

The increase in the volume of the water is

$$\Delta V_w = b_w V_w \,\Delta T = \left(2.1 \times \frac{10^{-4}}{°C}\right)(250\,\text{cm}^3)(40°C) = 2.1\,\text{cm}^3$$

and so the volume of water that overflows is

$$\Delta V_w - \Delta V_b = 2.0\,\text{cm}^3 \qquad\blacksquare$$

The thermal expansion of water is unusual. Above 4°C, water expands when heated, just as most other substances do. From 0°C to 4°C, however, the volume of a

FIG. 13–5 A cavity in a body expands or contracts with a change in temperature precisely as a solid object of the same size, shape, and composition would.

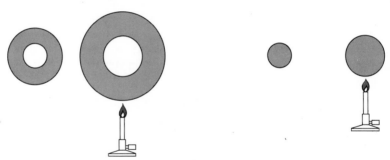

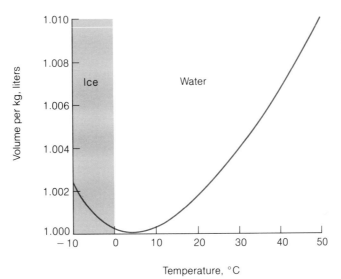

FIG. 13–6 Water has its greatest density at 4°C. Because water expands when it freezes, ice floats.

water sample *decreases* with increasing temperature (Fig. 13–6); thus water has its maximum density at 4°C. Equally unusual is the expansion of water when it freezes, so that ice floats and exposed water pipes may burst in a severe winter. (The reason for this behavior is discussed in Sec. 30–3.) Because ice floats, a body of water freezes in winter from the top down, not from the bottom up as it would if ice were denser than water and cold water denser than warm water at all temperatures. Ice is a much poorer conductor of heat than water (the difference in heat conductivity is nearly a factor of 4); hence the layer of ice that forms initially on the surface of a body of water impedes further freezing. Ice forming at the bottom would not have this effect. Many lakes, rivers, and arms of the sea are therefore able to escape total freezing in winter, which enables their plant and animal life to survive.

Thermal expansion of water

13–3 BOYLE'S LAW

A peculiar difficulty arises when we attempt to measure the coefficient of volume expansion of a gas. Unlike solids and liquids, gases do not have specific volumes at a particular temperature, but expand to fill their containers. The only way to change the volume of a gas is to change the capacity of its container. However, even though its volume may remain the same, another property of a confined gas varies with its temperature, namely the pressure it exerts on the container walls. The air pressure in an automobile tire drops in cold weather and increases in warm, an illustration of this property.

At constant volume, the pressure of a gas increases with temperature

When the temperature of a sample of gas is held constant, the absolute pressure it exerts on its container is very nearly inversely proportional to the volume of the container. Expanding the container lowers the pressure; shrinking the container raises the pressure. Conversely, increasing the pressure on a gas sample reduces its volume;

Boyle's law holds at constant temperature

FIG. 13–7 Boyle's law states that the volume of a gas sample is inversely proportional to its pressure at constant temperature. Thus $p_1V_1 = p_2V_2 = p_3V_3$ as shown.

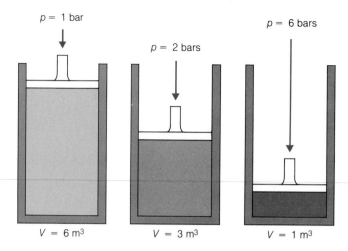

decreasing the pressure increases its volume (Fig. 13–7). This relationship is called *Boyle's law* after its discoverer, Robert Boyle (1627–1691). Though not exact, Boyle's law is an excellent approximation over a wide range of temperatures and pressures.

Boyle's law can be expressed in the form

$$p_1V_1 = p_2V_2 \qquad (T = \text{constant}) \hspace{3cm} \textit{Boyle's law} \quad (13\text{–}4)$$

where p_1 is the absolute gas pressure when the volume of the gas is V_1, and p_2 is the gas pressure when the volume is V_2.

Example   A scuba diver's 12-liter tank is filled with air at a gauge pressure of 150 bars. If the diver uses 30 liters of air per minute at the same pressure as the water pressure at her depth below the surface, how long can she remain at a depth of 15 m in seawater?

Solution   Since atmospheric pressure is ~1.0 bar, the absolute pressure of the air in the tank is

$$p_1 = p_{\text{atm}} + p_{\text{gauge}} = 1 \text{ bar} + 150 \text{ bars} = 151 \text{ bars}$$

The absolute pressure at a depth of 15 m is, according to the first example in Sec. 10–3,

$$p_2 = p_{\text{atm}} + dgh = 1.0 \text{ bar} + 1.5 \text{ bars} = 2.5 \text{ bars}$$

From Eq. (13–4) we have for the volume of air available at a pressure of 2.5 bars

$$V_2 = \frac{p_1V_1}{p_2} = \frac{(151 \text{ bars})(12 \text{ liters})}{2.5 \text{ bars}} = 725 \text{ liters}$$

However, 12 liters of air remain in the tank, so 713 liters are usable and

$$t = \frac{713 \text{ liters}}{30 \text{ liters/min}} = 24 \text{ min}$$

∎

13–4 CHARLES'S LAW

Now let us see what happens to a gas when its temperature is changed. As mentioned earlier, if the volume of a gas is held constant, the pressure it exerts on its container depends upon its temperature. According to Boyle's law, then, if we hold the gas pressure constant, its volume should vary with temperature. When this prediction is experimentally tested, which was first done over 150 years ago by Charles and Gay-Lussac, it is found that the change in volume ΔV of a gas sample is in fact related to a change ΔT in its temperature by the same formula, Eq. (13–3),

$$\Delta V = b V_0 \, \Delta T$$

that holds for solids and liquids.

    The significant thing about gases at constant pressure is that they *all* have very nearly the same coefficient of volume expansion b; by contrast, as Tables 13–1 and 13–2 indicate, the thermal coefficients for solids and liquids may have markedly different values for different substances. At 0°C the coefficient of volume expansion b_0 for all gases is very close to

$$b_0 = \frac{0.0037}{°C} = \frac{1/273}{°C}$$

If we vary the temperature of a gas sample while holding its pressure constant, its volume changes by 1/273 of its volume at 0°C for each 1°C temperature change. A child's large balloon filled with air whose volume at 0°C is 1.000 m^3 has a volume of 1.037 m^3 at 10°C and 0.963 m^3 at -10°C (Fig. 13–8).

    What happens when the balloon is cooled to -273°C? At that temperature the air in the balloon should have lost 273/273 of its volume at 0°C, and therefore should have vanished entirely! Actually, all gases condense into liquids at temperatures above -273°C, so the question has no physical meaning. But -273°C is still an important temperature. Let us set up a new temperature scale, the *absolute temperature scale,* and designate -273°C as the zero point (Fig. 13–9). Temperatures in the absolute scale are expressed in *kelvins,* denoted K, after Lord Kelvin (1824–1907), a noted British physicist. To convert temperatures from one scale to the other we note that

At constant pressure, the volume of a gas increases with temperature

All gases have the same coefficient of volume expansion at constant pressure

Absolute temperature scale

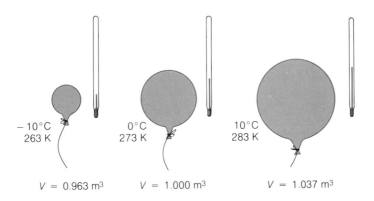

FIG. 13–8 Charles's law states that the volume of a gas sample is directly proportional to its absolute temperature at constant pressure. Thus $V_1/T_1 = V_2/T_2 = V_3/T_3$ as shown.

FIG. 13–9 The absolute temperature scale.

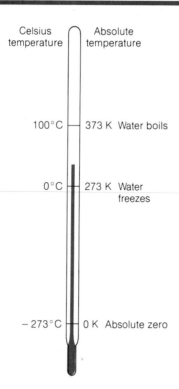

$$T_K = T_C + 273° \qquad T_C = T_K - 273° \qquad \textit{Absolute Temperature Scale} \quad (13\text{–}5)$$

Charles's law holds at constant pressure

The reason for setting up the absolute temperature scale is that, provided the pressure is constant, *the volume of a gas sample is directly proportional to its absolute temperature* (Fig. 13–10). This relationship is called *Charles's law*. Like Boyle's law, Charles's law is not a basic physical principle but deviations from it are usually quite small.

We can express Charles's law in the form

$$\frac{V_1}{T_1} = \frac{V_2}{T_2} \qquad (p = \text{constant}) \qquad\qquad \textit{Charles's law} \quad (13\text{–}6)$$

where V_1 is the volume of a gas sample at the absolute temperature T_1 and V_2 is its volume at the absolute temperature T_2.

Absolute zero

If there were a gas that did not liquify before reaching 0 K, then at 0 K its volume would shrink to zero. Since a negative volume has no meaning, it is natural to think of 0 K as *absolute zero*. Actually, 0 K is indeed the lower limit to temperatures capable of being attained, but on the basis of a stronger argument than one based on imaginary gases. This argument is discussed later in this chapter. To five significant figures the Celsius equivalent of absolute zero is $-273.16°C$.

13–5   IDEAL GAS LAW

Boyle's law and Charles's law can be combined in a single formula called the *ideal gas law*:

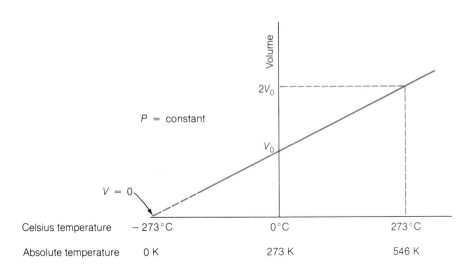

FIG. 13–10 The volume
of a gas at constant
pressure is directly
proportional to its absolute
temperature.

$$\frac{p_1 V_1}{T_1} = \frac{p_2 V_2}{T_2}$$

Ideal gas law (13–7)

**The ideal gas law
combines Boyle's law
and Charles's law**

When $T_1 = T_2$, the ideal gas law becomes Boyle's law,

$$p_1 V_1 = p_2 V_2 \qquad (\text{T} = \text{constant})$$

and when $p_1 = p_2$ it becomes Charles's law,

$$\frac{V_1}{T_1} = \frac{V_2}{T_2} \qquad (p = \text{constant})$$

The ideal gas law is obeyed approximately by all gases. The significant thing is **Ideal gas**
not that the agreement with experiment is never perfect, but that *all* gases, no matter
what kind, behave almost identically. An *ideal gas* is defined as one that obeys Eq.
(13–7) exactly. While no ideal gases actually exist, they do provide a target for theories
of the gaseous state to aim at. It is reasonable to suppose that the ideal gas law is a
consequence of the essential nature of gases. Hence the next step is to account for this
law and only afterward to seek reasons for its failure to be completely correct.

Example (a) A tank with a capacity of 1 m^3 contains helium gas at 27°C under a
pressure of 20 atm. The helium is used to fill a balloon. When the balloon is filled, the
gas pressure inside it is 1 atm, and its temperature has dropped to -33°C. (The gas
has done work in expanding at the expense of its internal energy, and the cooling reflects
this loss of internal energy.) What is the volume of the balloon at this time? (b) After
a while the helium in the balloon absorbs heat from the atmosphere and returns to its
original temperature of 27°C, and it expands further to maintain its pressure at 1 atm.
What is the final volume of the balloon? (The gas pressure in the balloon is actually
slightly greater than 1 atm to balance the tendency of the rubber to contract, but this
is ignored here for convenience.)

Solution (a) The equivalents of 27°C and -33°C on the absolute scale are 300 K
and 240 K respectively. Applying the ideal gas law to the initial expansion, we obtain

FIG. 13–11

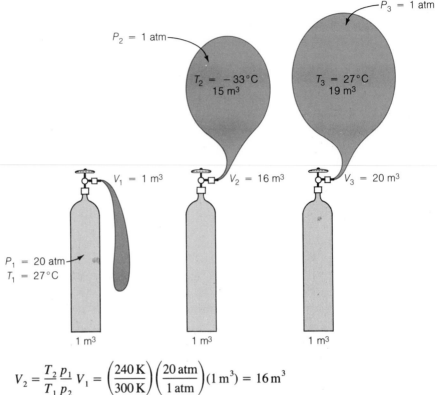

$$V_2 = \frac{T_2 \, P_1}{T_1 \, P_2} V_1 = \left(\frac{240\,\text{K}}{300\,\text{K}}\right)\left(\frac{20\,\text{atm}}{1\,\text{atm}}\right)(1\,\text{m}^3) = 16\,\text{m}^3$$

Because the tank's capacity is 1 m³, the balloon's volume after the initial expansion is 15 m³.

(b) When the helium has reached the outside air temperature of 27°C, which we shall call state 3, then $T_1 = T_3$. Hence we need only apply Boyle's law to states 1 and 3 to obtain the eventual volume of the helium:

$$V_3 = \frac{P_1}{P_3} V_1 = \left(\frac{20\,\text{atm}}{1\,\text{atm}}\right)(1\,\text{m}^3) = 20\,\text{m}^3$$

Again we subtract the 1 m³ volume of the tank to find the volume of the balloon itself, which is 19 m³ (Fig. 13–11). ∎

13–6   STRUCTURE OF MATTER

Before we go on to see how the ideal gas law is accounted for, it is appropriate to review the notions of element, compound, and solution, which apply to bulk matter, and those of atom and molecule, which apply to matter on the microscopic level.

Liquids and gases are almost always *homogeneous*, which means that every portion of a particular sample is exactly like every other portion. Solids may be either homogeneous or *heterogeneous;* if the latter, some portions of a particular sample may be different from others. A bar of gold, for example, is a homogeneous solid, while a

Homogenous and heterogeneous substances

piece of wood is a heterogeneous one. A heterogeneous solid is not always easy to recognize as such, and instruments such as the microscope (or even more sophisticated devices) may be required for definite identification.

Homogeneous substances may be further classified into *elements, compounds,* and *solutions. Elements* are the simplest substances we encounter in bulk; they cannot be decomposed or transformed into one another by ordinary chemical or physical means. There are over 100 known elements, listed in Appendix C together with their symbols and certain of their properties, of which 92 have been found in nature and the rest artificially prepared. At room temperature and sea-level atmospheric pressure, 10 elements are in the gaseous state, namely argon, chlorine, fluorine, helium, hydrogen, krypton, nitrogen, oxygen, radon, and xenon, and two are in the liquid state, namely bromine and mercury. The rest are in the solid state, the majority being metals.

Elements

Two or more elements may combine chemically to form a *compound,* a new substance whose properties are different from those of the elements that compose it. Each constituent of a *solution,* in contrast, retains its characteristic properties (except, of course, for the mechanical properties of solids and gases dissolved in liquids), and may be separated from the other constituents by relatively simple procedures. Boiling and freezing are examples of such procedures, since the temperatures at which these changes of state occur have specific values for each element or compound. Air, for instance, is a solution of several gases, chiefly nitrogen and oxygen. Oxygen boils at − 183°C while nitrogen boils at − 196°C, 13° lower; hence if we heat a sample of liquid air to a temperature over − 196°C but under − 183°C, the nitrogen will vaporize and we will be left ideally with oxygen alone. Under certain circumstances, nitrogen and oxygen unite to form the compound nitric oxide; the boiling point of nitric oxide is − 152°C, and heating a sample of liquid nitric oxide to this temperature will result in the vaporization of the entire sample. The constituents of a solution may be elements or compounds or both.

Compounds and solutions

Another distinction between compounds and solutions is that the elements in a compound are present in certain definite proportions, always the same for a particular compound, while the constituents of a solution may be present in a wide range of proportions. At sea level the mass ratio of the nitrogen and oxygen in the atmosphere varies slightly about an average of 3.2:1, and is several percent greater at high elevations; the mass ratio of the nitrogen and oxygen in nitric oxide is invariably 0.88:1. If there is an excess of either nitrogen or oxygen when nitric oxide is being prepared, the additional amount will not combine but will be left over and can be separated out at an appropriate temperature (Fig. 13–12).

Law of definite proportions

The idea that matter is not infinitely divisible, that all substances are composed of characteristic individual particles, is an ancient one. The ultimate particles of many compounds are called *molecules.* (Later chapters discuss the structure of compounds more completely.) Although molecules may be further broken down, when this happens they no longer are representative of the original substance. The molecules of a compound consist of the *atoms* of its constituent elements joined together in a definite ratio. Thus each molecule of water contains two hydrogen atoms and one oxygen atom. While the ultimate particles of elements are atoms, many elemental gases consist of molecules rather than atoms. Oxygen molecules, for instance, contain two oxygen atoms each. The molecules of other gases, such as helium and argon, are single atoms. Figure 13–13 shows schematically the composition of some common molecules.

Atoms and molecules

FIG. 13–12 The law of definite proportions.

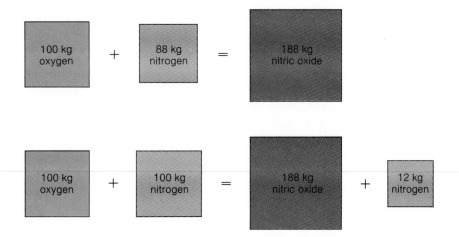

The masses of atoms and molecules are usually expressed in *atomic mass units* (u), where

Atomic mass unit

$$1 \text{ atomic mass unit} = 1 \text{ u} = 1.66 \times 10^{-27} \text{ kg}$$

A list of the atomic masses of the elements is given in Appendix C; if we know the composition of a compound, we can calculate the corresponding molecular mass.

Example How many H_2O molecules are present in 1 g of water? The atomic mass of hydrogen is 1.008 u and that of oxygen is 16.00 u.

FIG. 13–13 Molecular structures of several common substances.

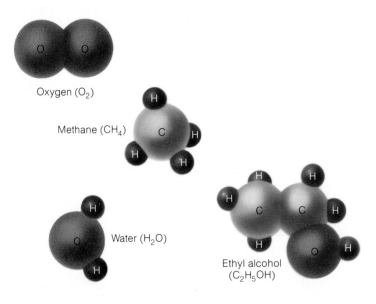

Oxygen (O_2)

Methane (CH_4)

Water (H_2O)

Ethyl alcohol (C_2H_5OH)

Solution   We begin by finding the mass of the H_2O molecule in u:

$$2H = 2 \times 1.008\,u = 2.02\,u$$
$$O = 1 \times 16.00\,u = \underline{16.00\,u}$$
$$18.02\,u$$

The mass of the H_2O molecule in kg is therefore

$$m = (18.02\,u)(1.66 \times 10^{-27}\,kg/u) = 2.99 \times 10^{-26}\,kg$$

and so the number of H_2O molecules in $1\,g = 10^{-3}\,kg$ of water is

$$\text{Molecules of }H_2O = \frac{\text{mass of }H_2O}{\text{mass of }H_2O\text{ molecule}} = \frac{10^{-3}\,kg}{2.99 \times 10^{-26}\,kg}$$

$$= 3.34 \times 10^{22}\,\text{molecules} \qquad \blacksquare$$

A considerable amount of experimentation and ingenious reasoning had to be carried out before the reality of atoms and molecules became definitely established. Although the full story of the kinetic-molecular theory of matter, a large part of which involves chemistry, will not be gone into here, it is easy to show that it can account for the ideal gas law. We will also see how the physical properties of solids and liquids, and the deviations of actual gases from the ideal gas, fit into the kinetic-molecular theory.

13–7   KINETIC THEORY OF GASES

According to the assumptions of the *kinetic theory of gases,* a gas consists of a great many tiny individual molecules that do not interact with one another except when collisions occur. The molecules are supposed to be far apart compared with their dimensions and to be in constant motion, incessantly hurtling to and fro as in Fig. 13–14, being kept from escaping into space only by the solid walls of a container (or, in the case of the earth's atmosphere, by gravity). A natural consequence of the random motion and large molecular separation is the tendency of a gas to completely fill its container and to be readily compressed or expanded.

In a solid, on the other hand, the constituent particles are close together, and mutual attractive and repulsive forces hold them in place to provide the solid with its characteristic rigidity. In a liquid the intermolecular forces are sufficient to keep the volume of a sample constant; however, they are not strong enough to prevent adjacent molecules from sliding past one another, which results in the ability of liquids to flow.

Boyle's law follows directly from the picture of a gas as a group of randomly moving molecules. The pressure the gas exerts originates in the impacts of its molecules; the vast number of molecules in even a tiny gas sample means that their separate blows appear as a continuous force to our senses and measuring instruments. Figure 13–15 shows a simplified model of a gas confined to a box. Although the molecules are actually traveling about in all directions, the effects of their collisions with the walls

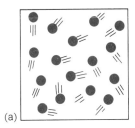
(a)

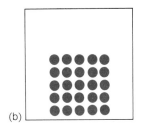
(b)

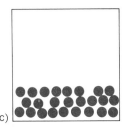
(c)

FIG. 13–14   (a) The molecules of a gas are in constant, random motion. (b) The constituent particles of a solid are also in motion, but oscillate about definite equilibrium positions. (c) The molecules of a liquid keep a more or less constant distance apart, but move about freely.

Structures of solids and liquids

Origin of Boyle's law

FIG. 13–15 A simplified model of a gas.

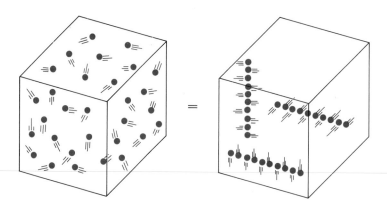

of the box are the same as if one-third of them were moving back and forth between each pair of opposite walls.

When a cylinder containing a gas is doubled in volume, as in Fig. 13–16, those molecules moving up and down have twice as far to go between impacts. Since their speed is unchanged, the time between impacts is also doubled, and the pressure they exert on the top and bottom of the cylinder falls to half its original value. The expansion of the cylinder also means that the molecules moving horizontally are now spread over twice their former area, and the pressure on the sides of the cylinder accordingly falls to half its original value as well. Thus doubling the volume means halving the pressure, which is Boyle's law. Similar reasoning accounts for a rise in pressure when the volume is reduced.

Charles's law follows from the kinetic theory of gases when a further assumption is made:

Temperature is a measure of molecular kinetic energy

The average kinetic energy of the random translational motions of the molecules of a gas is proportional to the absolute temperature of the gas.

FIG. 13–16 The origin of Boyle's law according to the kinetic theory of gases.

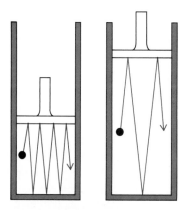

Pressure falls on top and bottom of expanded cylinder because molecules spend more time in transit between collisions

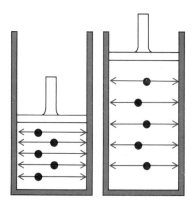

Pressure falls on sides of expanded cylinder because molecules spread their impacts over a larger area

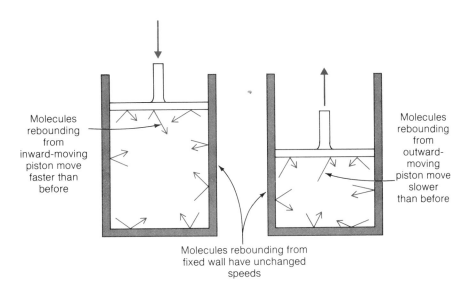

FIG. 13–17 The temperature of a gas increases when it is compressed because the average energy of its molecules increases; the temperature of a gas decreases when it is expanded because the average energy of its molecules decreases.

Molecules rebounding from inward-moving piston move faster than before

Molecules rebounding from outward-moving piston move slower than before

Molecules rebounding from fixed wall have unchanged speeds

This assumption is reasonable, since we observe that compressing a gas quickly (so no heat can enter or leave the container) raises its temperature, and such a compression must increase the average energy of the molecules because they bounce off the inward-moving piston more rapidly than they approach it (Fig. 13–17). A familiar example of the latter effect is a baseball rebounding with greater speed when struck by a bat. On the other hand, expanding a gas lowers its temperature, and such an expansion reduces molecular energies since molecules lose speed in bouncing off an outward-moving piston. The association between molecular energy and temperature is in accord with experience.

The interpretation of absolute zero in terms of the elementary kinetic theory of gases is a simple one: it is that temperature at which all molecular translational movement in a gas ceases (Fig. 13–18). A more advanced analysis shows that complete cessation of movement is impossible, but the difference is not important for the discussion here. See Sec. 27–7 for the origin of the *zero-point energy* that particles have at 0 K.

Absolute zero

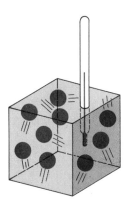 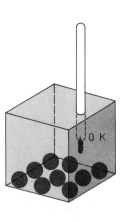

FIG. 13–18 At absolute zero, the kinetic theory of gases predicts that molecular translational motion in a gas will cease. In reality, at absolute zero the molecules would retain a small minimum amount of kinetic energy.

Molecular kinetic energy

The precise relationship between the average molecular kinetic energy KE_{av} and absolute temperature T is found to be

$$KE_{av} = \tfrac{3}{2}kT \qquad\qquad \textit{Molecular energy} \quad (13–8)$$

where k, known as Boltzmann's constant, has the value

$$k = 1.38 \times 10^{-23} \text{ J/K}$$

Boltzmann's constant

Equation (13–8) holds for the molecules of all gases regardless of the masses of their molecules and has been verified by direct measurements of molecular velocities; it is derived in Sec. 13–12.

Thus we have an interpretation of temperature in terms of molecular motion that is much more precise and definite than simply describing temperature as that which is responsible for sensations of hot and cold.

13–8   MOLECULAR SPEEDS

Rms molecular speed and temperature

We can use Eq. (13–8) to compute the average speed of gas molecules whose mass is m:

$$\tfrac{1}{2}m\overline{v^2} = \tfrac{3}{2}kT$$

$$v_{rms} = \sqrt{\overline{v^2}} = \sqrt{\frac{3kT}{m}} \qquad\qquad \textit{Rms molecular speed} \quad (13–9)$$

The above speed is denoted v_{rms} because it is the square root of the mean of the squared molecular speeds—the "root-mean-square" speed—and therefore different from the simple arithmetic average speed $\overline{v}$. To emphasize their difference with a simple example, we can evaluate both kinds of average for an assembly of two molecules, one with a speed of 1 m/s and the other with a speed of 3 m/s. We find that

$$\overline{v} = \frac{v_1 + v_2}{2} = \frac{(1 + 3)\,\text{m/s}}{2} = 2\,\text{m/s}$$

whereas

$$v_{rms} = \sqrt{\frac{v_1^2 + v_2^2}{2}} = \sqrt{\frac{1^2 + 3^2}{2}}\,\text{m/s} = \sqrt{5}\,\text{m/s} = 2.24\,\text{m/s}$$

Clearly v_{rms} and $\overline{v}$ are not at all the same. The relationship bewteen v_{rms} and $\overline{v}$ depends upon the specific variation in molecular speeds being considered. For the distribution of molecular speeds found in a gas,

$$v_{rms} \approx 1.09\overline{v}$$

so the root-mean-square speed of Eq. (13–9) is about 9 percent greater than the arithmetic average $\overline{v}$.

Example   Find the rms speed of oxygen molecules at 0°C.

Solution   Oxygen molecules are composed of two oxygen atoms each. Since the atomic mass of oxygen is 16.00 u, the molecular mass is 32.00 u, and an O_2 molecule has a mass in kg of

$$m = (32.00 \text{ u})(1.66 \times 10^{-27} \text{ kg/u}) = 5.31 \times 10^{-26} \text{ kg}$$

At an absolute temperature of 273 K (which corresponds to 0°C), the rms speed of oxygen molecules is therefore

$$v_{rms} = \sqrt{\frac{3kT}{m}} = \sqrt{\frac{3(1.38 \times 10^{-23} \text{ J/K})(273 \text{ K})}{5.31 \times 10^{-26} \text{ kg}}} = 461 \text{ m/s}$$

which is a little over 1000 mi/hr! Evidently molecular speeds are very large compared with those of the macroscopic bodies familiar to us.   ■

It is important to keep in mind that actual molecular speeds vary considerably on either side of v_{rms}. The graph in Fig. 13–19 shows the distribution of molecular speeds in oxygen at 273 K and in hydrogen at 273 K. The mass of an O_2 molecule is 16 times that of an H_2 molecule. Rms molecular speed decreases with molecular mass; hence at the same temperature molecular speeds in hydrogen are on the average greater than in oxygen. At the same temperature the average molecular *energy* is the same for all gases, however.

Variation of molecular speed

In Fig. 13–20 we see the distributions of molecular speeds in oxygen at 73 K and at 273 K. The average molecular speed increases with temperature, as predicted. The curves of Figs. 13–19 and 13–20 are not symmetrical because the lower limit to v is fixed at $v = 0$ whereas there is, in principle, no upper limit; actually, as the curves show, the likelihood of speeds many times greater than v_{rms} is small.

The distribution of molecular speeds in a gas has an interesting astronomical consequence. The higher the surface temperature of a planet, the faster the molecules of its atmosphere move, and the greater the chance they may exceed the escape speed

Planetary atmospheres

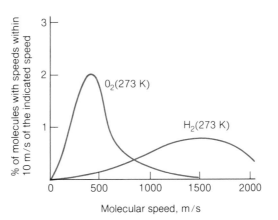

FIG. 13–19 Molecular speeds in oxygen and hydrogen at 273 K (0°C). The smaller masses of H_2 molecules means that they have higher average speeds than O_2 molecules at the same temperature, since the average kinetic energy depends only on temperature.

FIG. 13–20 Molecular speeds in oxygen at 73 K (−200°C) and 273 K (0°C). The higher the temperature, the greater the average kinetic energy.

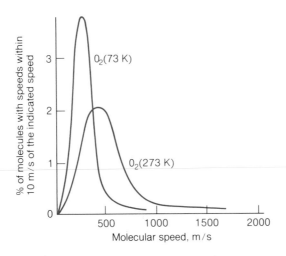

and disappear into space. The smaller a planet, the lower its escape speed; and the closer a planet is to the sun, the warmer it is. Thus it is not surprising that Mercury, small and hot, has no atmosphere, while the giant outer planets of Jupiter, Saturn, Uranus, and Neptune have extremely dense atmospheres.

Why the ideal gas law is only approximate

The kinetic theory of gases leads directly to the ideal gas law; but, as mentioned earlier, the ideal gas law is only a good approximation of reality. If we examine the initial assumptions that are made, it is easy to see why we should expect discrepancies between theory and experiment. For instance, it is assumed that gas molecules have volumes so small as to be entirely negligible; that they exert no forces upon one another except in actual collisions; and that these collisions conserve kinetic energy of translational motion. (The last assumption means that the molecules are supposed to have no internal energy of their own, such as energy of rotation or vibration, whereas in fact they often do; see Fig. 13–21.) When the kinetic theory is worked out from more realistic assumptions, the results are in excellent agreement with observational data.

13–9   MOLECULAR MOTION IN LIQUIDS

The elementary kinetic theory of matter is not as successful when applied to the liquid and solid states as it is when applied to gases; the Newtonian laws of mechanics that gas molecules obey in their translational motion are not adequate to describe the behav-

FIG. 13–21 The molecules of a gas may have energies of vibration and rotation as well as of translation, but only the kinetic energy of their translational motions affects the temperature of the gas.

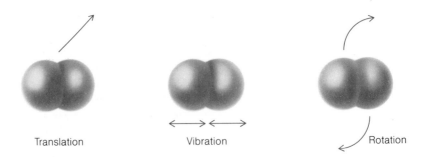

Translation   Vibration   Rotation

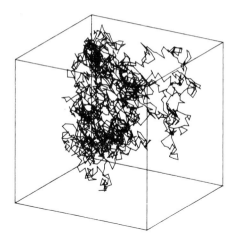

FIG. 13-22 Computer-generated representation of Brownian motion. Successive line segments correspond to the movement of the particle in sucessive equal time intervals.

ior of the molecules in liquids and solids. However, the concept that the internal energy of a substance resides at least in part in the kinetic energies of its molecules helps in understanding a variety of phenomena characteristic of liquids and solids.

The random motion of water molecules led to an important event in the history of science. In 1827 the British botanist Robert Brown noticed that pollen grains in water are in continual, agitated movement. Similar *Brownian motion* is apparent whenever very small particles are suspended in a fluid medium, for example smoke particles in air (Fig. 13–22). According to kinetic theory, Brownian motion originates in the bombardment of the particles by molecules of the fluid. This bombardment is completely random, with successive molecular impacts coming from different directions and contributing different impulses to the particles. Albert Einstein, in 1905, found that he could account for Brownian motion quantitatively by assuming that, as a result of continual collisions with fluid molecules, the particles themselves have the same average kinetic energy as the molecules. Surprising as it may seem, this was the first direct verification of the reality of molecules, and it convinced many distinguished scientists who had previously been reluctant to believe that such things actually exist.

Brownian motion

Another kinetic-molecular phenomenon characteristic of the liquid state is evaporation. A dish of water well below its boiling point of 100°C will nevertheless gradually turn into vapor, growing colder as it does so. The faster the evaporation, the more pronounced the cooling effect; alcohol and ether chill the skin upon contact because of their extreme volatility. This behavior follows from the distribution of molecular speeds in a liquid. Though not identical with that found in a gas, this distribution resembles those shown in Figs. 13–19 and 13–20 in that a certain fraction of the molecules in any sample have much greater and much smaller speeds than the average. The fastest molecules have enough energy to escape through the liquid surface despite the attractive forces of the other molecules. The molecules left behind redistribute the available energy in collisions among themselves, but, because the most energetic ones escape, the average energy that remains is less than before and the liquid is now at a lower temperature (Fig. 13–23). Boiling occurs when the average molecular energy in a liquid is equal to the work needed to pull the molecules apart against the forces that hold them together.

Evaporation and boiling on the basis of kinetic theory

FIG. 13–23 After evaporation, the remaining liquid is cooler than before.

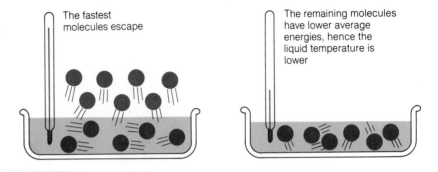

FIG. 13–23 After evaporation, the remaining liquid is cooler than before.

The fastest molecules escape

The remaining molecules have lower average energies, hence the liquid temperature is lower

Saturation vapor density increases with temperature

When molecules from the vapor above a liquid surface impinge on the surface, they may be trapped there, so that a constant two-way traffic of molecules to and from the liquid occurs. If the density of the vapor above the liquid is sufficiently great, as many molecules return as leave it at any time, a situation that is described by saying that the region is *saturated* with the substance. The higher the temperature, the greater the maximum vapor density: at 0°C the density of water vapor at saturation is 5 g/m^3, at 20°C it is 17 g/m^3, at 100°C it is 598 g/m^3, and at 300°C it is all the way up to 45.6 kg/m^3 (Fig. 13–24). If for any reason (such as a sudden drop in temperature) the vapor density exceeds the saturation value, condensation will be more rapid than evaporation until equilibrium is reestablished. It is for this reason that on a hot day moisture condenses on the outside of a glass that contains a cold drink.

Relative humidity is water vapor density relative to saturation density

The *relative humidity* of a volume of air describes its degree of saturation with water vapor. Relative humidities of 0, 50%, and 100% mean respectively that no water vapor is present, that the air contains half as much moisture as the maximum possible, and that the air is saturated. On a hot day the evaporation of sweat from the skin is the chief means by which the human body dissipates heat, and a high relative humidity is

FIG. 13–24 The variation with temperature of water vapor density in air for various relative humidities. The curve for 100% relative humidity represents the maximum vapor density.

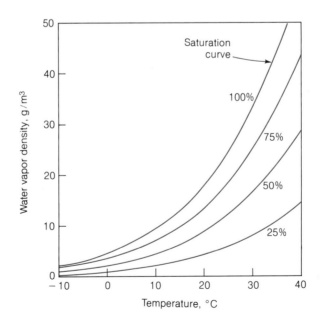

Saturation curve

100%

75%

50%

25%

Water vapor density, g /m^3

Temperature, °C

uncomfortable because it impedes the process. A low relative humidity is also undesirable because it leads to the drying of the skin and mucous membranes. The regulation
of relative humidity is as important a function of a heating or of an air-conditioning
system as the regulation of temperature.

From Fig. 13–24 we can see to what extent heating air decreases its relative
humidity and cooling air increases its relative humidity. For instance, between 10°C
and 20°C the saturated vapor density (which corresponds to 100% relative humidity)
just about doubles. This means that if outside air at 10°C whose relative humidity is,
say, 70% is taken inside a house and heated to 20°C, the relative humidity indoors will
only be 35% since the actual vapor density stays the same. A way to humidify heated
air in winter is clearly desirable. If the outside air is at 30°C with 70% relative humidity,
then cooling it down to 24°C is enough to bring it to saturation, which is 100% relative
humidity; further cooling will cause water to condense out. An air-conditioning system
therefore should incorporate means to remove water vapor from the air being cooled.

**Changing the
temperature of a body
of air also changes its
relative humidity**

13–10   THERMAL EXPANSION IN SOLIDS

Thermal expansion in a solid has a straightforward explanation in terms of kinetic
theory. Most solids are crystalline in nature, which means that the various atoms that
compose them form a regular arrangement in space. (In some crystalline solids the
basic constituents are whole molecules rather than individual atoms, but we shall refer
to them as atoms here for convenience.) The atoms behave as though they are joined
together by tiny springs, thereby accounting for Hooke's law, and constantly oscillate
about their equilibrium positions.

**Atoms in a crystal
behave as if joined by
tiny springs**

Figure 13–25 shows how the atomic potential energy of a solid varies with interatomic spacing. The spacing at room temperature a corresponds to the lower portion
of the curve, where the energy per atom is least. The amplitude of the vibrations is
determined by the width of the curve: when the atomic separation is a minimum or a
maximum, the energy of a pair of adjacent atoms is wholly potential, as in the case of
a harmonic oscillator at each end of its path, while in the middle their energy is wholly
kinetic. The average interatomic spacing is what determines a and hence the dimensions
of the solid.

**Potential energy curve
of atom in a solid**

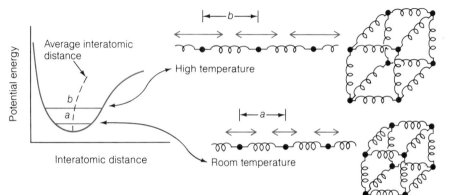

FIG. 13–25 The atomic
potential energy of a
crystal as a function of the
spacing of its constituent
atoms. The increase in the
average interatomic
distance with energy is
the cause of thermal
expansion in a solid.

**Why solids expand
when heated**

When the energies of the atoms increase, the atomic spacing alternates through a wider range than before. If the potential energy curve were symmetrical about the distance a, no change in the dimensions of the solid would occur, since a is always halfway between the two parts of the curve. However, the attractive and repulsive forces between atoms vary with distance in different ways, with the repulsive force increasing more rapidly as the atoms move closer together than the attractive force increases as the atoms move farther apart. Consequently, as shown in Fig. 13–25, the average interatomic spacing increases to b as the internal energy of the solid increases and the amplitudes of the atomic vibrations increase. As it happens, changes in the average interatomic spacing are very nearly proportional to changes in temperature, leading to the linear thermal expansion formula of Eq. (13–2).

13–11   THE MOLE

**The mole is a unit of
number of molecules**

A *mole* of any substance is that quantity of it whose mass equals its molecular mass (or atomic mass if it is an elemental substance that consists of individual atoms) expressed in grams instead of in atomic mass units. A mole of water has a mass of 18.02 g because a water molecule has a mass of 18.02 u. The utility of the mole arises from the fact that a mole of any substance contains exactly as many molecules as a mole of any other substance. The mole is widely used in chemistry in place of the gram or kilogram as a measure of quantity because the primary interest of the chemist is usually in the relative numbers of atoms and molecules that react together.

The number of molecules in a mole is a universal constant known as *Avogadro's number,* whose value is

**Avogadro's number is
the number of molecules
in a mole**

$$N_0 = 6.023 \times 10^{23} \text{ molecules/mole} \qquad\qquad \textit{Avogadro's number}$$

The number of molecules in a sample of a substance is equal to the number of moles n it contains multiplied by N_0.

Under identical conditions of pressure and temperature, equal volumes of gases contain equal numbers of molecules. The reason for this is that the molecules in a sample of a gas have negligible volumes compared with the volume of the sample itself. Under the standard conditions of 0°C and atmospheric pressure, usually referred to by the abbreviation STP, one mole of any gas is found to occupy a volume of 22.4 liters. This observation makes it easy to deal with gas volumes in chemical processes. If a certain reaction produces 5 moles of a gas, for example, at STP the volume of the gas will be 5 moles $\times$ 22.4 liters/mole $=$ 112 liters.

**At STP a mole of any
gas has a volume of 22.4
liters**

Example   Find the density of ammonia, NH_3, at STP.

Solution   At STP one mole of any gas occupies 22.4 liters. The molecular mass of NH_3 is

$$
\begin{aligned}
1\,N &= 1 \times 14.01\text{ u} = 14.01\text{ u}\\
3\,H &= 3 \times 1.008\text{ u} = \underline{3.02\text{ u}}\\
&\qquad\qquad\quad 17.03\text{ u} = 17.03\text{ g/mole}
\end{aligned}
$$

One mole of NH_3 therefore has a mass of 17.03 g at STP, and since 1 liter $=$ 10^{-3} m^3, its density is

$$d = \frac{m}{V} = \frac{17.03\,g}{22.4\,liters} = 0.76\,g/liter = 0.76\,kg/m^3$$ ■

According to the ideal gas law, the pressure, temperature, and volume of a particular gas sample obey the relationship

$$\frac{pV}{T} = \text{constant}$$

at all times. We can find the value of this constant for a sample that contains n moles by noting that, at STP, its volume must be $V = n \times 22.4$ liters/mole, its temperature must be $T = 273$ K, and its pressure must be $p = 1$ atm. Hence

$$\frac{pV}{T} = \frac{n \times 1\,atm \times 22.4\,liters/mole}{273\,K} = nR$$

where R, the *universal gas constant*, has the value 0.0821 atm·liter/mole·K. In SI units, in which p is expressed in Pa and V in m^3, the value of R is 8.31 J/mole·K. The ideal gas law is often written

$$pV = nRT \qquad\qquad\qquad \textit{Ideal gas law} \quad (13\text{–}10)$$

Example At an altitude of 10 km the temperature of the atmosphere is about $-50°$C and its pressure is about 2.6×10^4 Pa. Find the mass of hydrogen, H_2, needed to fill a balloon whose volume at this altitude is to be 4000 m^3.

Solution The first step is to find the number of moles of H_2 needed from Eq. (13–10). Since $-50°$C $= 223$ K and 1 Pa $= 1$ N/m^2,

$$n = \frac{pV}{RT} = \frac{(2.6 \times 10^4\,N/m^2)(4 \times 10^3\,m^3)}{(8.31\,J/mole \cdot K)(223\,K)} = 5.6 \times 10^4\,moles$$

The molecular mass of H_2 is 2×1.008 u $= 2.016$ u $= 2.016$ g/mole, and so the mass of H_2 required is

$$m = (5.6 \times 10^4\,moles)(2.016\,g/mole) = 11.3 \times 10^4\,g = 113\,kg$$ ■

Although at first glance it is just another fact to add to our collection, the presence of 6.023×10^{23} molecules in 22.4 liters of any gas at STP is surely remarkable. If a cubic centimeter of air at STP—a thimbleful—were to be divided equally among all the four billion people on the earth, each would receive nearly 7 billion molecules! There are about as many molecules in an average breath of air as there are breaths in the entire atmosphere, so that, as James Jeans has said, "if we assume that the last breath of, say, Julius Caesar, has by now become thoroughly scattered through the atmosphere, then the chances are that each of us inhales one molecule of it with every breath we take."

13–12   TEMPERATURE AND MOLECULAR MOTION

As discussed in Sec. 13–7, what we call temperature is associated on the microscopic level with the random translational motions of molecules. The precise relationship in the case of the molecules of an ideal gas is given by Eq. (13–8) as

$$\mathrm{KE}_{av} = \tfrac{3}{2}kT$$

in which KE_{av} is the average molecular kinetic energy and k is Boltzmann's constant. Let us see how this formula can be obtained.

We begin with a model situation: a cubical box L long on each side is filled with N identical molecules of mass m. While the molecules are actually moving about randomly, the effects of their collisions with the walls of the box are the same as if one-third of them bounce back and forth between the top and bottom of the box, one-third between the front and rear walls, and one-third between the right- and left-hand walls (see Fig. 13–15).

Now we consider what happens when a molecule strikes one of the walls. As in Fig. 13–26, the molecule undergoes a momentum change of

$$\Delta(m\mathbf{v}) = m\mathbf{v}_2 - m\mathbf{v}_1 = m\mathbf{v} - (-m\mathbf{v}) = 2m\mathbf{v}$$

as it bounces off the wall. The molecule needs the time

$$\Delta t = \frac{2L}{v}$$

to make a round trip from that wall to the opposite one and back again. According to Eq. (7–4), the momentum change $\Delta(mv)$ is related to the impulse $\mathbf{F}\Delta t$ given to the wall by

$$\mathbf{F}\Delta t = \Delta(mv)$$

FIG. 13–26 Each molecular impact on a wall of the box means a momentum change of $2mv$. The time between successive impacts is $2L/v$.

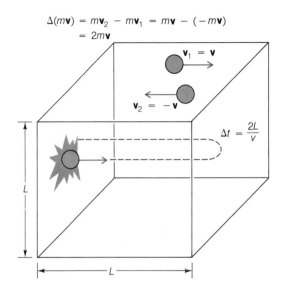

$$\Delta(m\mathbf{v}) = m\mathbf{v}_2 - m\mathbf{v}_1 = m\mathbf{v} - (-m\mathbf{v})$$
$$= 2m\mathbf{v}$$

$\mathbf{v}_1 = \mathbf{v}$

$\mathbf{v}_2 = -\mathbf{v}$

$\Delta t = \dfrac{2L}{v}$

where $\mathbf{F}$ is the average force on the wall during the time interval Δt. The magnitude F of the average force on the wall caused by the successive impacts of the molecule is therefore

$$F = \frac{\Delta(mv)}{\Delta t} = \frac{2mv}{2L/v} = \frac{mv^2}{L}$$

The next step is to find the pressure on the wall due to the impacts of all the $N/3$ molecules that strike it. If the average value of v^2 for the various molecules is $\overline{v^2}$, the sum ΣF of the individual forces they exert is

$$\Sigma F = \frac{N}{3} \times F = \frac{Nm\overline{v^2}}{3L}$$

The pressure on the wall is this total force divided by the area L^2 of the wall:

$$p = \frac{\Sigma F}{L^2} = \frac{Nn\overline{v^2}}{3L^3}$$

But L^3 is the volume V of the box! Hence

$$p = \frac{Nm\overline{v^2}}{3V}$$

which can be rewritten as

$$pV = \tfrac{1}{3}Nm\overline{v^2} \tag{13–11}$$

The significance of Eq. (13–11) becomes clear when it is expressed in the form

$$pV = \tfrac{2}{3}N\left(\tfrac{1}{2}m\overline{v^2}\right) = \tfrac{2}{3}N(\mathrm{KE}_{av}) \tag{13–12}$$

where $\mathrm{KE}_{av} = \tfrac{1}{2}m\overline{v^2}$ is the average kinetic energy per gas molecule. This result agrees with the ideal gas law of Eq. (13–10),

$$pV = nRT$$

provided that

$$\tfrac{2}{3}N(\mathrm{KE}_{av}) = nRT$$

$$\mathrm{KE}_{av} = \frac{3}{2}\frac{R}{(N/n)}T \tag{13–13}$$

Here is where the link is made between the microscopic picture of a gas as a collection of randomly moving molecules and its macroscopic picture as a fluid that obeys the ideal gas law. Since N is the number of molecules and n is the number of moles in the gas sample, their ratio N/n is just Avogadro's number:

$$\frac{N}{n} = \frac{\text{molecules}}{\text{moles}} = N_0 = \text{Avogadro's number.}$$

As we recall, R is the universal gas constant, so that $R/(N/n)$ in Eq. (13–13) has the value

$$k = \frac{R}{N/n} = \frac{R}{N_0} = \frac{8.31 \text{ J/mole} \cdot \text{K}}{6.023 \times 10^{23} \text{ molecules/mole}}$$

$$= 1.38 \times 10^{-23} \text{ J/molecule} \cdot \text{K}$$

This quantity we recognize as Boltzmann's constant, and in terms of k, Eq. (13–13) becomes

$$\text{KE}_{av} = \tfrac{3}{2}kT \qquad\qquad\qquad (13\text{–}8)$$

which is what we set out to demonstrate.

IMPORTANT TERMS

The **temperature** of a body of matter is a measure of the average kinetic energy of random translational motion of its constituent particles.

A **thermometer** is a device for measuring temperature. The two temperature scales in common use are the **Celsius** (centigrade) scale, in which the freezing point of water is assigned the value 0°C and its boiling point the value 100°C, and the **Fahrenheit** scale, in which these points are assigned the values 32°F and 212°F, respectively.

The **coefficient of linear expansion** is the ratio between the change in length of a solid rod of a particular material and its original length per 1° change in temperature. The **coefficient of volume expansion** is the ratio between the change in volume of a sample of a particular solid or liquid and its original volume per 1° change in temperature.

Boyle's law states that, at constant temperature, the absolute pressure of a sample of a gas is inversely proportional to its volume, so that $pV = $ constant at that temperature regardless of changes in either p or V individually.

Charles's law states that, at constant pressure, the volume of a sample of a gas is directly proportional to its absolute temperature, so that $V/T = $ constant at that pressure regardless of changes in either V or T individually.

The **Kelvin absolute temperature scale** has its zero point at -273°C; temperatures in this scale are designated K. **Absolute zero** is 0 K $= -273$°C.

The equation $pV/T = $ constant, a combination of Boyle's and Charles's laws, is called the **ideal gas law** and is obeyed approximately by all gases.

According to the **kinetic theory of gases,** a gas consists of a great many tiny individual molecules that do not interact with one another except when collisions occur. The molecules are far apart compared with their dimensions and are in constant random motion. The ideal gas law may be derived from the kinetic theory of gases. The average kinetic energy of gas molecules is proportional to the absolute temperature of the gas. At absolute zero, gas molecules would have virtually no kinetic energy of translational motion.

The **relative humidity** of a volume of air is the ratio between the amount of water vapor it contains and the amount that would be present at saturation.

IMPORTANT FORMULAS

Celsius and Fahrenheit scales:
$$T_F = \tfrac{9}{5}T_C + 32°$$
$$T_C = \tfrac{5}{9}(T_F - 32°)$$

Thermal expansion: $\quad \Delta L = aL_0 \, \Delta T$
$$\Delta V = bV_0 \, \Delta T$$

Boyle's law:
$$pV = \text{constant} \qquad (T = \text{constant})$$

Absolute temperature scale: $\quad T_K = T_C + 273°$

Charles's law:
$$\frac{V}{T} = \text{constant} \qquad (p = \text{constant})$$

Ideal gas law:
$$\frac{pV}{T} = \text{constant} = nR$$

Molecular kinetic energy: $\quad \text{KE}_{av} = \tfrac{3}{2}kT$

MULTIPLE CHOICE

1. Two thermometers, one calibrated in the Celsius scale and the other in the Fahrenheit scale, are used to measure the same temperature. The numerical reading on the Fahrenheit thermometer

    (a) is proportional to that on the Celsius thermometer.

    (b) is greater than that on the Celsius thermometer.

    (c) is less than that on the Celsius thermometer.

    (d) may be greater or less than that on the Celsius thermometer.

2. Two elements *cannot* be combined chemically to make

    (a) a compound.   (b) another element.

    (c) a gas.   (d) a liquid.

3. The relative proportions of the elements in a compound

    (a) may vary considerably.

    (b) may vary only slightly.

    (c) do not vary.

    (d) may or may not vary, depending on the compound.

4. Which of the following statements is not correct?

    (a) Matter is composed of tiny particles called molecules.

    (b) These molecules are in constant motion.

    (c) All molecules have the same size and mass.

    (d) The differences between the solid, liquid, and gaseous states can be attributed to the relative freedom of motion of their respective molecules.

5. The volume of a gas sample is proportional to its

    (a) Fahrenheit temperature.

    (b) Celsius temperature.

    (c) absolute temperature.

    (d) pressure.

6. Absolute zero may be regarded as that temperature at which

    (a) water freezes.

    (b) all gases become liquids.

    (c) all substances are solid.

    (d) molecular motion in a gas would be the minimum possible.

7. The kinetic-molecular theory of gases predicts that, at a given temperature,

    (a) all of the molecules in a gas have the same average speed.

    (b) all of the molecules in a gas have the same average energy.

    (c) light gas molecules have lower average energies than heavy gas molecules.

    (d) light gas molecules have higher average energies than heavy gas molecules.

8. The volume of a gas is held constant while its temperature is raised. The pressure the gas exerts on the walls of its container increases because

    (a) the masses of the molecules increase.

    (b) each molecule loses more kinetic energy when it strikes the wall.

    (c) the molecules are in contact with the wall for a shorter time.

    (d) the molecules have higher average speeds and strike the wall more often.

9. The temperature of a gas is held constant while its volume is reduced. The pressure the gas exerts on the walls of its container increases because its molecules

    (a) strike the container walls more often.

    (b) strike the container walls with higher speeds.

    (c) strike the container walls with greater force.

    (d) have more energy.

10. Oxygen boils at $-183°C$. This temperature is

    (a) $-215°F$.   (b) $-297°F$.

    (c) $-329°F$.   (d) $-361°F$.

11. A temperature of 100°F is almost exactly

    (a) 38°C.   (b) 56°C.

    (c) 122°C.   (d) 212°C.

12. A copper bar is 1 m long at 20°C. At what temperature will it be shorter by 1 mm?

    (a) $-17°C$ (b) $-39°C$

    (c) $-59°C$ (d) $-79°C$

13. An absolute temperature of 100 K is the same as a Celsius temperature of

    (a) $-173°C$.   (b) 32°C.

    (c) 212°C.   (d) 373°C.

14. A certain container holds 1 kg of air at atmospheric pressure. When an additional kg of air is pumped into the container, the new pressure is

    (a) $\frac{1}{2}$ atm.   (b) 1 atm.

    (c) 2 atm.   (d) 4 atm.

15. If the absolute pressure on 10 m^3 of air is increased from 30 bars to 120 bars, the new volume of the air will be

    (a) 2.5 m^3.   (b) 5 m^3.

    (c) 40 m^3.   (d) 900 m^3.

16. A sample of hydrogen gas is compressed to half its original volume while its temperature is held constant. If the average velocity of the hydrogen molecules was originally v, their new average speed is

    (a) $4v$.   (b) $2v$.

    (c) v.   (d) $\frac{1}{2}v$.

17. At which of the following temperatures would the molecules of a gas have twice the average kinetic energy they have at room temperature, 20°C?

    (a) 40°C   (b) 80°C

    (c) 313°C   (d) 586°C

18. The mass of a nitrogen molecule is 14 times greater than that of a hydrogen molecule. The temperature of a sample of hydrogen whose average molecular energy is equal to that in a sample of nitrogen at 300 K is

    (a) 6.5 K.   (b) 21 K.

    (c) 300 K.   (d) 4200 K.

19. A mole of helium atoms and a mole of iron atoms at 20°C have exactly the same

    (a) numbers of atoms.

    (b) rms speeds per atom.

    (c) volumes.

    (d) densities.

EXERCISES

1. In the construction of a light bulb, wires are led through the glass at the base by means of airtight seals. If the wires were made of copper, what would happen when the light is turned on and the bulb heats up? What must be true for a wire to be successfully used for this purpose?

2. Verify that the force associated with the thermal expansion or contraction of a solid object depends upon its cross-sectional area but not upon its length.

3. Starting from the ideal gas law, obtain an equation relating the pressure and temperature of a gas at constant volume.

4. Actual molecules attract one another slightly. Does this tend to increase or decrease gas pressures from values computed from the ideal gas law? Why?

5. Is it meaningful to say that an object at a temperature of 200°C is twice as hot as one at 100°C?

6. At absolute zero, an ideal gas sample would occupy zero volume. Why would an actual gas not occupy zero volume at absolute zero?

7. Molecular speeds are comparable with those of rifle bullets, yet a gas with a strong odor, such as ammonia, takes a few minutes to diffuse through a room. Why?

8. According to the kinetic theory of gases, molecular motion virtually ceases only at absolute zero. How can this be reconciled with the definite shape and volume of a solid at temperatures well above absolute zero?

9. Why does the air in a heated room tend to be dry?

10. The air in a closed container is saturated with water vapor at 20°C. (a) What is the relative humidity? (b) What happens to the relative humidity if the temperature is reduced to 10°C? (c) If the temperature is increased to 30°C?

11. The melting point of lead is 330°C and its boiling point is 1170°C. Express these temperatures on the Fahrenheit scale.

12. The normal temperature of the human body is 98.6°F. What is this temperature on the Celsius scale?

13. At what temperature would Celsius and Fahrenheit thermometers give the same reading?

14. Mercury freezes at -40°C. What is this temperature on the Fahrenheit scale?

15. Dry ice (solid carbon dioxide) vaporizes at -112°F. What is this temperature on the Celsius scale?

16. How large a gap should be left between steel rails that are 10 m long when laid at 20°C if they are to just barely touch at 30°C?

17. A rod 2 m long expands by 1 mm when heated from 8°C to 70°C. What is the coefficient of linear expansion of the material from which the rod is made?

18. The outside diameter of a wheel is 1.000 m. An iron tire for this wheel has an inside diameter of 0.992 m at 20°C. To what temperature must the tire be heated in order for it to fit over the wheel?

19. A steel tape measure is calibrated at 22°C. A reading of 40.000 m is obtained when it is used to determine the width of a building at -10°C. What is the true width of the building at -10°C?

20. A Pyrex flask holds 500 cm^3 of mercury at 0°C. How much mercury will run out when it is heated to 80°C?

21. A Pyrex beaker is filled to the brim with 250 cm^3 of glycerin at 15°C. How much glycerin overflows at 25°C?

22. A concrete swimming pool 12 m × 6 m × 2.5 m is filled with water to within 6 mm of the top when the temperature is 10°C. The coefficient of linear expansion of the concrete used is 0.9×10^{-5}/°C. What will happen to the water level as the temperature increases? If it rises, at what temperature will the water begin to overflow?

23. An aluminum mast whose cross-sectional area is 50 cm^2 is 20 m long at 12°C. (a) By how much does it increase in length when its temperature rises to 35°C? (b) How much force is associated with the expansion?

24. A flat tire contains 5 liters of air at atmospheric pressure, which means an absolute pressure of very nearly 1 bar. How many strokes of a pump whose capacity is 0.5 liter are needed to raise the pressure in the tire to a gauge pressure of 2 bars? Assume the temperature and volume of the tire do not change in the process.

25. To what Celsius temperature must a gas sample initially at 20°C be heated if its volume is to double while its pressure remains the same?

26. An air tank used for scuba diving has a safety valve set to open at an absolute pressure of 280 bars. The normal absolute pressure of the full tank at 20°C is 200 bars. If the tank is heated after being filled to the latter pressure, at what temperature will the safety valve open?

27. The tires of a stationary car contain air at a gauge pressure of 1.8 bars and a temperature of 10°C. When the car is driven for a while, the temperature of the tires increases to 50°C. If the tire volumes remain unchanged, find their new gauge pressure.

28. A sample of gas occupies 100 cm^3 at 0°C and 1 atm pressure. What is its volume (a) at 50°C and 1 atm pressure; (b) at 0°C and 2.2 atm pressure; (c) at 50°C and 2.2 atm pressure?

29. A sample of gas occupies 2 m^3 at 300 K and an absolute pressure of 2 × 10^5 Pa. (a) What is its pressure at the same temperature when it has been compressed to a volume of 1 m^3? (b) What is its volume at the same temperature when its pressure has been decreased to 1.5 × 10^5 Pa? (c) What is its volume at a temperature of 400 K and a pressure of 2 × 10^5 Pa?

30. Find the mass of a molecule of ethyl alcohol, C_2H_6O.

31. Find the mass of a molecule of glucose, $C_6H_{12}O_6$.

32. How many lead atoms are present in 50 g of lead?

33. To what temperature must a gas sample initially at 27°C be raised in order for the average energy of its molecules to double? For their average speed to double?

34. What is the average kinetic energy of the molecules of a gas (a) at 0°C? (b) at 100°C?

35. The rms speed of a hydrogen molecule at room temperature is about 1 mi/s. What is the rms speed of an oxygen molecule, whose mass is 16 times greater, at this temperature?

36. Consider the following gases: CO_2, UF_6, H_2, He, Xe, NH_3. (a) Which has the highest rms molecular speed at a given temperature? (b) Which has the lowest rms molecular speed?

PROBLEMS

1. The density of lead is 11.0 g/cm^3 at 20°C. Find its density at 200°C.

2. A sign is suspended from the middle of a steel cable attached to poles 20.00 m apart. When the temperature is 30°C, the cable sags so that the sign is 2.00 m below a horizontal line between the ends of the cable. Find the amount of sag on a winter day when the temperature is −10°C.

3. Vodka that is "100 proof" is a mixture of half ethyl alcohol and half water. How much profit per liter will a merchant make if he buys vodka at $10.00 per liter at 0°C and sells it at $10.00 per liter at 25°C?

4. A load of 4000 kg is placed on a vertical steel column 3 m long and cross-sectional area 50 cm^2 when the temperature is 20°C. To what should the temperature be increased if the length of the column is to be the same after the load is applied as it was originally?

5. A diver blows an air bubble 1 cm in diameter at a depth of 10 m in a freshwater lake where the temperature is 5°C. What is the diameter of the bubble when it reaches the surface of the lake where the temperature is 20°C?

6. The density of air is 1.293 kg/m^3 at 0°C and 1 atm pressure. Find its density at 100°C and 2 atm pressure.

7. Hydrogen and helium are the most abundant elements in the universe, yet nitrogen and oxygen are far more abundant in the earth's atmosphere. To see why, find the rms speeds of these molecules in the upper atmosphere, whose temperature is roughly 10^3 K, and compare them with the 11.2 km/s escape speed from the earth.

8. Two vessels of the same size are at the same temperature. One of them holds 1 kg of H_2 gas and the other holds 1 kg of N_2 gas. (a) Which vessel contains more molecules? How many times more? (b) Which vessel is under the greater pressure? How many times greater? (c) In which vessel are the rms molecular speeds greater? How many times greater?

9. (a) Find the rms speed of carbon dioxide (CO_2) molecules at 0°C. (b) At what temperature would this speed be doubled?

10. Silver is a vapor at 1500 K. What is the rms speed of silver atoms in a vapor at this temperature?

11. The rms speed of air molecules is roughly 4 × 10^2 m/s, and the average distance an air molecule goes between collisions with other molecules is about 10^{-7} m. What is the average number of collisions an air molecule makes per second?

12. One of the assumptions of the kinetic theory is that the average distance between molecules is much greater than the dimensions of the molecules themselves. Oxygen and nitrogen molecules are roughly 2 × 10^{-10} m in diameter, and there are 2.7 × 10^{25} molecules in a cubic meter of air at room temperature and atmospheric pressure. (a) On the average, how far apart are the molecules in air? (b) How many molecular diameters is their average separation?

13. A 1-mg mass of oleic acid placed on a water surface spreads out to cover an area of about 1.4 m^2. The density of oleic acid is 895 kg/m^3 and its molecular mass is 282.5 u. On the assumptions that the film of oleic acid is one molecule thick and that the molecules are symmetrical, use the above information to obtain a value for Avogadro's number.

14. (a) Find the mass of 75 moles of ethylene, C_2H_4. (b) How many carbon atoms are present?

15. Find the number of moles and the number of molecules in 1000 liters of acetylene, C_2H_2, at STP.

16. The density of sulfur dioxide gas at STP is 2.86 g/liter. Find its molecular mass.

17. Find the volume of 10 g of carbon dioxide, CO_2, at STP.

18. (a) What volume does 1 g of ammonia, NH_3, occupy at STP? (b) What volume does it occupy at 100°C and a pressure of 2 atm?

19. Find the mass of 12 liters of nitrogen, N_2, at 200°C and a pressure of 3 × 10^5 Pa.

20. Find the density of oxygen, O_2, at 20°C and a pressure of 5 atm.

ANSWERS TO MULTIPLE CHOICE

1. d	**5.** c	**9.** a	**13.** a	**17.** c
2. b	**6.** d	**10.** b	**14.** c	**18.** c
3. c	**7.** b	**11.** a	**15.** a	**19.** a
4. c	**8.** d	**12.** b	**16.** c	

HEAT

14

The higher the temperature of a body of matter, the faster its constituent particles move in a random manner, and the more internal energy it contains. Heat may be thought of as internal energy in transit: when heat is added to a body, its temperature generally increases, and when heat is removed, its temperature generally decreases. The temperature change that accompanies the addition or removal of a given amount of heat varies with the mass and nature of the body. Changes of state—from solid to liquid, liquid to gas, solid to gas, and their reverses—involve changes in internal energy without changes in temperature as the atoms or molecules of the substance alter their relationships with one another. The last topic of this chapter concerns the mechanisms by which heat can be transferred from one body of matter to another.

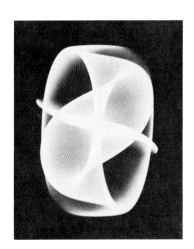

14–1   INTERNAL ENERGY AND HEAT

Every body of matter contains a certain amount of internal energy in addition to any kinetic or potential energy it may possess by virtue of its motion or

FIG. 14–1 The internal energy of a body depends upon its temperature and mass. However, the direction of internal energy flow depends only on the temperatures of the bodies involved.

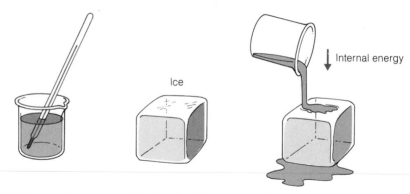

High temperature but small mass, hence little internal energy

Low temperature but large mass, hence much internal energy

Internal energy always flows from hot body to cold body, regardless of internal energy content

Temperature difference determines direction of internal energy flow

position. This internal energy resides in the random translational motions of the atoms or molecules of which the body is composed, and in their rotations and vibrations as well. The total amount of internal energy a body contains depends upon its temperature, upon its composition, upon its mass, and upon its physical state (solid, liquid, or gas). However, the temperature of the body alone is what determines whether internal energy will be transferred from it to another body with which it is in contact, or vice versa. A large block of ice at 0°C has far more internal energy than a cup of hot water, yet when the water is poured on the ice some of the ice melts and the water becomes cooler, which signifies that energy has passed from the water to the ice (Fig. 14–1).

Heat is internal energy in transit between two bodies of matter

When the temperature of a body increases, it is customary to say that *heat* has been added to it; when the temperature of a body decreases, it is customary to say that heat has been removed from it. We can therefore define heat as follows:

Heat is internal energy in transit from one body of matter to another by virtue of a temperature difference between them.

Changes of state (for instance, from ice to water or from water to steam) involve the transfer of heat to or from a body of matter without any change in its temperature. We should keep in mind that heat transfer is not the only way to change the temperature of a body of matter. A body that has work done on it may become hotter as a result, and a body that does work on something else may become cooler.

Temperature is a measure of molecular kinetic energy

The term "heat" remains in the vocabulary of physics partly because of convenience and partly because of tradition. Temperature, on the other hand, is a unique concept both in a macroscopic sense as an indicator of the direction of internal energy flow and in a microscopic sense as a measure of average molecular kinetic energy.

Since heat is a form of energy, the correct SI unit of heat is the joule. However, two older units, the *kilocalorie* (kcal) and the *British thermal unit* (Btu) are still widely used, and for the time being it is necessary to be familiar with them. The kilocalorie is the amount of heat required to raise the temperature of 1 kg of water through 1°C. Similarly, 1 kcal of heat must be removed from 1 kg of water to reduce its temperature

FIG. 14–2

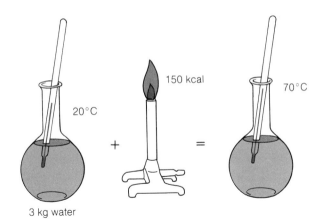

by 1°C. Because this amount of heat actually varies slightly with temperature, the **The kilocalorie** kilocalorie is formally defined as the amount of heat involved in changing the temperature of 1 kg of water from 14.5°C to 15.5°C; the difference is insignificant for most purposes, however. The relationship between the joule and the kilocalorie is as follows:

$$1\,J = 2.39 \times 10^{-4}\,kcal$$
$$1\,kcal = 4185\,J$$

Example How much heat is required to raise the temperature of 3 kg of water from 20°C to 70°C?

Solution The temperature of the water must be raised by 50°C. Since 1 kcal of heat raises the temperature of 1 kg of water by 1°C, 50 kcal is required for each kg of water. There are 3 kg of water in all, so $3 \times 50\,kcal = 150\,kcal$ is needed (Fig. 14–2). ∎

Example Water flows through a 3.5-kW instantaneous-type water heater at a rate of 2 liters/min. If the initial water temperature is 15°C and all the heat produced is added to the water, find the temperature of the hot water.

Solution Since 3.5 kW = 3500 J/s, the heat furnished in 1 min is

$$\text{Heat gained} = \frac{3500\,J/s \times 60\,s}{4185\,J/kcal} = 50\,kcal$$

The mass of water is 2 kg since the mass of 1 liter of water is 1 kg, and so

$$\text{Change in temperature} = \frac{\text{heat gained}}{\text{mass of water}}$$

$$\Delta T = \frac{50\,kcal}{2\,kg} = 25°C$$

The final water temperature is 15°C + 25°C = 40°C. ∎

The *calorie* that dieticians use is the same as the kilocalorie. Thus the energy **The dietician's calorie is** content of a 150-calorie cupcake is really 150 kcal. The carbohydrate content of foods **the kcal** averages 4.1 kcal/g, the protein content 4.2 kcal/g, and the fat content 9.3 kcal/g; pure ethanol (ethyl alcohol) has an energy content of 7.1 kcal/g. (A heat unit once widely used is also called the calorie. This is equal to the heat needed to raise the temperature of 1 g of water by 1°C, so that 1000 cal = 1 kcal. The dietician's calorie is sometimes written "Calorie" to distinguish it from the ordinary, smaller calorie.)

The Btu

The British thermal unit is the amount of heat needed to raise the temperature of 1 lb of water by 1°F; when 1 Btu of heat is removed from 1 lb of water, its temperature falls by 1°F. The Btu is related to the other units of energy and heat as follows:

$$1 \text{ Btu} = 1054 \text{ J} = 0.252 \text{ kcal} = 778 \text{ ft·lb}$$

We note that weight rather than mass is specified in defining the Btu. In practice this is not an important difference because what is being considered is the quantity of matter without regard to its dynamical properties. All the formulas and procedures that follow can be used with British units by expressing temperature in degrees Fahrenheit and heat in Btu, and letting m represent weight in pounds.

14–2   ANIMAL METABOLISM

Energy conversion in animals

The conversion of the metabolic energy of an animal into mechanical work varies in efficiency with the type of muscle and how fast it contracts—the efficiency is least at high and low speeds. An efficiency of 10 to 20% is usual. Most of the energy liberated by an animal's metabolic processes thus ends up as heat, the greater part of which escapes through the animal's skin. The ability of an animal to dissipate the heat its body produces accordingly is proportional to its surface area, which suggests that the maximum metabolic rate of an animal, and hence its power output, depends upon this area.

Basal metabolic rate depends on size

If a representative linear dimension (such as its length) of an animal is L, its surface area varies as L^2. Because the animal's mass is proportional to its volume and hence varies as L^3, the metabolic rate per kilogram ought to depend, at least approximately, upon $L^2/L^3 = 1/L$ by the above reasoning. Thus small animals should have higher metabolic rates per kilogram than large ones, which is indeed the case in nature. Typical basal metabolic rates (which correspond to an animal resting) are 5.2 W/kg for a pigeon, 1.3 W/kg for a dog, 1.2 W/kg for a person, and 0.67 W/kg for a cow. African elephants partly overcome the limitation of the small surface/mass ratio of their bodies by their enormous ears, which help them get rid of metabolic heat. It is clear why birds are small: past a certain size, a bird's weight outstrips its ability to perform the work needed for it to fly. (See also the discussion in Sec. 10–7 concerning wing area.)

We can now appreciate why natural selection has led to large sizes for whales and porpoises, which are air-breathing mammals that live in the sea. About 0.07 g of oxygen is needed for each kJ of energy released by the metabolism of food in an animal's body. The amount of oxygen an animal can store varies with its volume and hence with L^3. The larger an animal, the more reserve oxygen it has relative to its total metabolic rate (which varies as L^2), so a large aquatic mammal has the advantage over a small one of being able to remain submerged for a longer period.

The metabolic rate of an animal when active may greatly exceed its basal rate. A 70-kg person, for instance, might have a basal metabolic rate of 80 W. When the person is reading or doing light work while sitting, the metabolic rate will be perhaps 125 W, and it will be in the neighborhood of 300 W while the person is walking and as much as 1200 W while he or she is running hard. If the intake of energy from food exceeds a person's metabolic requirements, the excess is used to create additional tissue: muscle if sufficient physical activity is being carried out, otherwise fat. The energy stored in fat is available for use by the body if metabolic requirements exceed the food supply at a later time.

14–3 SPECIFIC HEAT CAPACITY

Samples of other substances respond to the addition or removal of a given amount of heat with temperature changes greater than that of an equal mass of water. One kg of water increases in temperature by 1°C when 1 kcal of heat is added to it, but 1 kcal of heat increases the temperature of 1 kg of helium (its volume held constant) by 1.3°C, of 1 kg of ice by 2°C, and of 1 kg of gold by 33°C (Fig. 14–3).

Water changes least in temperature when heat is added or removed

The *specific heat capacity* (symbol c) of a substance refers to what we might think of as its thermal inertia:

> **The specific heat capacity of a substance is the amount of heat that must be added or removed from a unit mass of it to change its temperature by 1°.**

Specific heat capacity

A high specific heat capacity means a relatively small change in temperature for a given change in internal energy content, just as a large inertial mass means a relatively small acceleration when a given force is applied.

When the kcal is being used as the heat unit, the corresponding unit of c is the kcal/kg·°C. Thus the specific heat capacity of water is $c = 1.00$ kcal/kg·°C since 1 kcal of heat is involved when 1 kg of water changes in temperature by 1°C. In the British system, the unit of c is the Btu/lb·°F. Because of the way the kcal and the Btu are defined, the numerical value of c for a substance is the same whether the kcal/kg·°C or the Btu/lb·°F is the unit; thus the specific heat capacity of water is 1.00 Btu/lb·°F.

Units of specific heat capacity

In the SI system, the unit of c is, of course, the J/kg·°C, and in this system the specific heat capacity of water is 4185 J/kg·°C. Usually it is more convenient to use the *kilojoule* (kJ) in connection with heat, in which case $c = 4.19$ kJ/kg·°C. Table

The kilojoule is a convenient SI heat unit

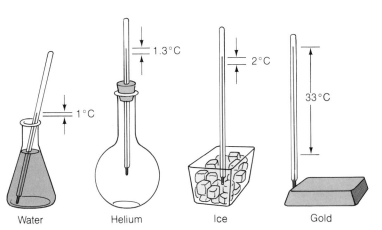

FIG. 14–3 When 1 kcal of heat is added to 1 kg of each of the substances shown, their respective rises in temperature differ considerably.

1.3°C

2°C

33°C

1°C

Water   Helium   Ice   Gold

Substance	Specific Heat Capacity (kcal/kg·°C)	(KJ/kg·°C)
Alcohol (ethyl)	0.58	2.43
Aluminum	0.22	0.92
Concrete	0.7	2.9
Copper	0.093	0.39
Glass	0.20	0.84
Gold	0.030	0.13
Granite	0.19	0.80
Human body	0.83	3.47
Ice	0.50	2.09
Iron	0.11	0.46
Lead	0.030	0.13
Mercury	0.033	0.14
Silver	0.056	0.23
Steam	0.48	2.01
Water	1.00	4.19
Wood	0.42	1.76

14–1 is a list of specific heats for some common substances. The actual values vary somewhat with temperature, and the ones given in the table represent averages.

With the help of specific heat capacity we can write a formula for the quantity of heat Q involved when a quantity m of a substance undergoes a change in temperature of ΔT. This formula is simply

Quantity of heat
$$Q = mc\,\Delta T \qquad\qquad Quantity\ of\ heat \quad (14\text{–}1)$$

Heat transferred = mass × specific heat capacity × temperature change

Example   How much heat in kilocalories must be removed from 1.4 kg of aluminum in order to cool it from 80°C to 15°C?

Solution   From Table 14–1 the specific heat capacity of aluminum is 0.22 kcal/kg·°C. Since $\Delta T = -65°C$ here,

$$Q = mc\,\Delta T = (1.4\,\text{kg})\left(0.22\frac{\text{kcal}}{\text{kg·°C}}\right)(-65°C) = -20\,\text{kcal}$$

The minus sign means that this quantity of heat is to be removed to achieve the temperature change of $-65°C$. ■

Example   If 0.20 kg of coffee at 90°C is poured into a 0.30-kg cup at 20°C, and we assume that no heat is transferred to or from the outside, what is the final temperature of the coffee?

Solution   We shall take the specific heat of coffee to be that of water and the specific heat of the cup to be that of glass. To solve the problem, we begin by noting that

$$Q_{\text{gained}} = Q_{\text{lost}}$$

Heat gained by cup = heat lost by coffee

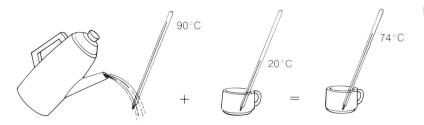

FIG. 14–4

If the final temperature of both coffee and cup is T, then the heat gained by the cup is

$$Q_{\text{gained}} = m_{\text{cup}} c_{\text{cup}} (T - 20°\text{C}) = (0.30\,\text{kg})\left(0.20\,\frac{\text{kcal}}{\text{kg}\cdot°\text{C}}\right)(T - 20°\text{C})$$

$$= (0.06\,T - 1.2)\ \text{kcal}$$

and the heat lost by the coffee is

$$Q_{\text{lost}} = m_{\text{coffee}} c_{\text{coffee}} (90°\text{C} - T)$$

$$= (0.20\,\text{kg})\left(1.0\,\frac{\text{kcal}}{\text{kg}\cdot°\text{C}}\right)(90°\text{C} - T) = (18 - 0.20\,T)\ \text{kcal}$$

Now we set the heat gained by the cup equal to the heat lost by the coffee and solve for T:

$$Q_{\text{gained}} = Q_{\text{lost}}$$

$$0.06T - 1.2 = 18 - 0.20\,T$$

$$0.26\,T = 19.2$$

$$T = 74°\text{C}$$

The temperature of the coffee drops by 16°C as it warms the cup (Fig. 14–4). Evidently it is necessary to preheat the cup if one wants really hot coffee. ∎

Example   Radiant energy from the sun arrives at the earth at the rate of about 1.4 kW per square meter of surface perpendicular to the sun's rays. On a clear day a reflector 1 m² in area is used to concentrate sunlight on a 1-kg lead bar initially at 20°C (Fig. 14–5). How long will it take the lead to reach its melting point of 330°C under the assumption that it absorbs energy at the rate of 0.5 kW?

Solution   The energy absorbed by the lead in the time t is Pt and its corresponding internal energy gain is $mc\,\Delta T$. Since $c = 0.13$ kJ/kg·°C for lead and $\Delta T = 330°\text{C} - 20°\text{C} = 310°\text{C}$, the required time is

$$t = \frac{mc\,\Delta T}{P} = \frac{(1\,\text{kg})(0.13\,\text{kJ/kg}\cdot°\text{C})(310°\text{C})}{0.5\,\text{kW}} = 81\ \text{s} \quad ∎$$

14–4 CHANGE OF STATE

Not always does the addition or removal of heat from a sample of matter lead to a change in its temperature. Instead the sample may change its state from solid to liquid

FIG. 14–5

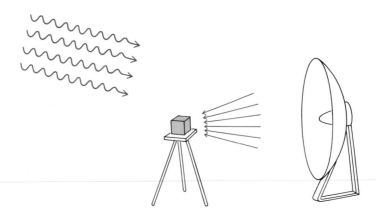

Changes of state involve changes in internal energy

or from liquid to gas when heat is added, or it may change from gas to liquid or from liquid to solid when heat is taken away. Such changes of state take place at definite temperatures for most substances at any given pressure, but for a few (glass or wax, for instance) there is only a gradual softening or hardening over a range of temperatures. Substances of the latter kind are not true solids, however; their structures are really those of liquids, and their hardness at room temperature is really a kind of exaggerated viscosity.

Figure 14–6 shows what happens when we add heat at a constant rate of 100 kcal/min to 1 kg of ice that is initially at $-50°C$. The specific heat capacity of ice is 0.5 kcal/kg·°C, and so 25 kcal is needed to bring the ice to 0°C.

The melting of ice

At 0°C the ice begins to melt. The temperature remains constant until all the ice has melted, which requires a total of 80 kcal. Thus 80 kcal/kg is the *heat of fusion* of water: the amount of heat needed to convert 1 kg of ice into 1 kg of water at its melting point of 0°C.

When all the ice has turned to water, the temperature goes up once more as further heat is supplied. Since the specific heat of water is 1 kcal/kg·°C, there is now a rise of 1°C per kcal of heat. This rate of change is less than that of ice, since the specific heat of water is greater than that of ice, and so the slope of the graph is less steep.

The boiling of water

When 100°C is reached, the water begins to turn into steam. The temperature stays constant until a total of 540 kcal is added, at which time all the water has become steam. Thus 540 kcal/kg is the *heat of vaporization* of water: the amount of heat needed to convert 1 kg of water into 1 kg of steam at its boiling point of 100°C (at atmospheric pressure).

After the water has become steam, its temperature rises again. The specific heat of steam is 0.48 kcal/kg·°C, so the temperature increase is 2.1°C per kcal of heat, and the slope of the graph is therefore steeper than it was for ice or water.

Heat of fusion

In general, the *heat of fusion* of a substance is the amount of heat that must be supplied to change 1 kg of the substance at its melting point from the solid to the liquid state; the same amount of heat must be removed from a unit amount of the substance in the liquid state at its melting point to change it to a solid. The usual symbol for heat of fusion is L_f.

Heat of vaporization

The *heat of vaporization* of a substance is the amount of heat that must be supplied to change 1 kg of the substance at its boiling point from the liquid to the gaseous (or

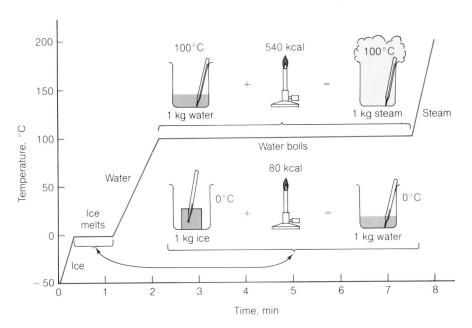

FIG. 14–6 A graph of temperature versus time for 1 kg of water, initially ice at −50°C, to which heat is being added at the constant rate of 100 kcal/min.

vapor) state; the same amount of heat must be removed from 1 kg of the substance in the gaseous state at its boiling point to change it into a liquid. The usual symbol for heat of vaporization is L_v. The heats of fusion and vaporization for a number of substances are listed in Table 14–2 together with their melting and boiling points. To convert a heat of fusion or heat of vaporization expressed in kcal/kg to its equivalent in Btu/lb, we note that 1 kcal/kg = 1.80 Btu/lb.

Example A man is sitting in the shade in an ambient air temperature of 37°C, which is the same as his body temperature. Under these circumstances the chief way his body gets rid of the 120 W his metabolic processes liberate is through the evaporation of sweat. How much sweat per hour is required? At 37°C the heat of vaporization of water is 2430 kJ/kg. (In Sec. 13–9 we saw how it is possible for a liquid to evaporate at a temperature below its boiling point.)

Substance	Melting Point, °C	L_F (kcal/kg)	(kJ/kg)	Boiling Point, °C	L_v (kcal/kg)	(kJ/kg)
Alcohol (ethyl)	−114	25	105	78	204	854
Lead	330	5.9	25	1170	208	870
Mercury	−39	2.8	12	358	71	297
Nitrogen	−210	6.1	26	−196	48	201
Oxygen	−219	3.3	14	−183	51	213
Silver	961	21	88	2193	558	2335
Tungsten	3410	44	184	5900	1150	4813
Water	0	80	335	100	540	2260

TABLE 14–2
Heats of fusion and vaporization and melting and boiling points of various substances at atmospheric pressure

Solution   If the rate at which sweat is evaporated is m/t, the rate at which heat is being lost is $P = mL_v/t$. Since $P = 120$ W $= 0.12$ kW $= 0.12$ kJ/s,

$$\frac{m}{t} = \frac{P}{L_v} = \frac{0.12\,\text{kJ/s}}{2430\,\text{kJ/kg}} = 4.94 \times 10^{-5}\,\text{kg/s}$$

Because 1 h $= 3600$ s, the mass of sweat per hour is

$$m = (4.94 \times 10^{-5}\,\text{kg/s})(3600\,\text{s}) = 0.178\,\text{kg} = 178\,\text{g} \qquad\blacksquare$$

Example   What is the minimum amount of ice at $-10°C$ that must be added to 0.50 kg of water at 20°C in order to bring the temperature of the water down to 0°C?

Solution   We begin, as before, with the statement of energy conservation,

$$Q_{\text{gained}} = Q_{\text{lost}}$$

Heat gained by ice = heat lost by water

In a problem like this, it does not matter whether we use the kcal or the kJ as the unit of heat. If we use the kcal, then the heat Q_1 absorbed by the unknown mass of ice in going from $-10°C$ to its melting point of 0°C is

$$Q_1 = m_{\text{ice}} c_{\text{ice}}\,\Delta T_{\text{ice}} = m_{\text{ice}}\left(0.50\frac{\text{kcal}}{\text{kg}\cdot°\text{C}}\right)(10°\text{C}) = m_{\text{ice}}\,(5\,\text{kcal/kg})$$

and the heat Q_2 absorbed by the ice in melting at 0°C is

$$Q_2 = m_{\text{ice}} L_{f\,\text{ice}} = (m_{\text{ice}})(80\,\text{kcal/kg})$$

Hence the total heat absorbed by the ice is

$$Q_{\text{gained}} = Q_1 + Q_2 = (5 + 80)m_{\text{ice}}\,\text{kcal/kg}$$

The heat lost by the water in cooling from 20°C to 0°C is

$$Q_{\text{lost}} = m_{\text{water}} c_{\text{water}}\,\Delta T_{\text{water}} = (0.50\,\text{kg})\left(1.0\frac{\text{kcal}}{\text{kg}\cdot°\text{C}}\right)(20°\text{C}) = 10\,\text{kcal}$$

Equating the heat gained with the heat lost and then solving for m_{ice} yields (Fig. 14–7)

$$Q_{\text{gained}} = Q_{\text{lost}}$$

$$85 m_{\text{ice}}\,\frac{\text{kcal}}{\text{kg}} = 10\,\text{kcal}$$

$$m_{\text{ice}} = 0.12\,\text{kg} \qquad\blacksquare$$

Sublimation refers to direct change from solid to vapor or from vapor to solid

Under certain circumstances most substances can change directly from the solid to the vapor state, or vice versa. Both processes are called *sublimation*. For example, "dry ice" (solid carbon dioxide) evaporates directly to gaseous carbon dioxide at temperatures above $-78.5°C$, and does not pass through the liquid state. With the exception of carbon dioxide and a few other substances, however, sublimation does not occur except at pressures well below that of the atmosphere.

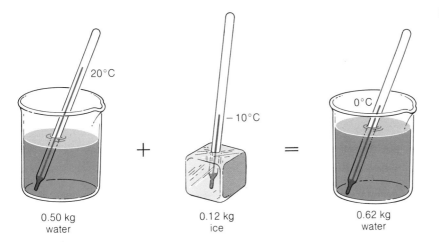

FIG. 14–7

20°C

0.50 kg
water

+

−10°C

0.12 kg
ice

=

0°C

0.62 kg
water

14–5   THE TRIPLE POINT

Changes of state are affected by pressure. Figure 14–8 shows how the boiling point of water varies with pressure: the higher the pressure, the higher the boiling point. This is the principle that underlies the pressure cooker. By heating water in a sealed container, the pressure can be raised well above 1 atm, and the temperature at which the water inside boils will be correspondingly higher than 100°C. In this way food can be cooked more rapidly than in an open pan. At a pressure of 2 atm, for instance, water boils at 120°C. The converse is also true: lowering the pressure below 1 atm reduces the boiling point below 100°C. Atmospheric pressure decreases with altitude, so the boiling point of water is lower than 100°C in regions located above sea level. Water boils at 96°C in Denver, for instance.

> **The boiling point of water increases with pressure**

The upper limit of the *vaporization curve* of Fig. 14–8, which occurs at a temperature of 374°C and a pressure of 218 atm, is known as the *critical point*. A substance cannot exist in the liquid state at a temperature above that of its critical point, regardless of how great the pressure may be. The gas becomes more and more dense with increasing pressure but does not condense into a liquid with a definite interface between liquid

> **Critical point**

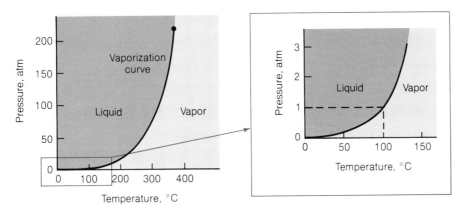

> **FIG. 14–8** The vaporization curve of water, showing how the boiling point of water varies with pressure. Above the temperature of its critical point a substance cannot exist in the liquid state.

FIG. 14–9 The fusion
curve of water, showing
how the melting point of
ice varies with pressure.

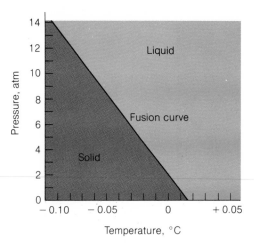

and gas. Helium has the lowest critical temperature, $-268°C$, and is therefore a gas
at all temperatures above that. The critical temperature for water is $374°C$.

The melting points of solids also depend upon pressure, although to a smaller
extent than the boiling points. The variation of the melting point of ice with pressure
is shown in the *fusion curve* of Fig. 14–9. Ice, together with gallium and bismuth, is
unusual in that its melting point *decreases* with increasing pressure; the melting points
of all other substances increase with increasing pressure. Hence it is possible to melt
ice by applying pressure to it as well as by heating it. An ice skater makes use of this
fact in an interesting way. His entire weight is supported by skate blades of very small
area, and the resulting pressure on the ice may exceed 1000 atm. The ice under the
blades melts because of the great pressure, which creates a thin film of water that acts
as an efficient lubricant. On unusually cold days even such pressures may not be
sufficient to melt the ice, and skating then becomes impossible.

**Why ice skating is
possible**

The fusion and vaporization curves of water intersect at a temperature of $0.01°C$
and a pressure of 4.6 torr, as shown on the combined plot of Fig. 14–10. Along the

FIG. 14–10 Triple-point
diagram of water. The
solid, liquid, and vapor
phases of water can exist
simultaneously at the
temperature and pressure
of the triple point (1 torr =
1 mm Hg).

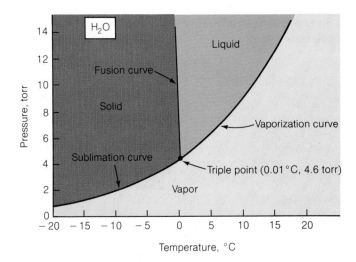

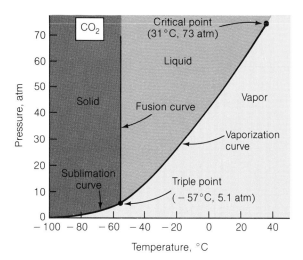

FIG. 14–11   Triple-point diagram of carbon dioxide.

fusion curve both ice and water can simultaneously exist, and along the vaporization curve both water and water vapor can simultaneously exist; hence under conditions corresponding to those of the intersection of the two curves, the solid, liquid, and vapor states of water can all exist together. This intersection is accordingly called the *triple point* of water.

At its triple point, the solid, liquid, and vapor states of a substance can exist together

At pressures below that of its triple point, no substance can exist as a liquid. The dividing line on a pressure-temperature graph between the solid and vapor states is called the *sublimation curve,* since it represents the conditions required for a solid to vaporize directly or a vapor to solidify directly. At atmospheric pressure the addition of heat causes ordinary ice to melt, since the triple point of water lies well below 1 atm, but the addition of heat causes solid carbon dioxide to sublime, since its triple point lies above 1 atm (Fig. 14–11).

Sublimation curve

At a temperature below that of its triple point, a substance passes directly from the solid state to the vapor state if the pressure is sufficiently low. This phenomenon is the basis of the *freeze-drying* process widely used for the preservation of foods, blood plasma, and biological samples. The usual procedure is to cool the material below the triple point of water in a gas-tight chamber, and then to evacuate the chamber with a vacuum pump. Freeze drying affects the structure of a material of biological origin less than other methods of dehydration do.

Freeze drying

14–6   HEAT CONDUCTION

The three mechanisms of heat transfer are illustrated in Fig. 14–12. When we place one end of an iron rod in a fire, the other end becomes warm as a result of the conduction of heat through the iron. Conduction is a very slow process in air; a stove warms a room chiefly through the actual movement of heated air, a process called convection. Neither conduction nor convection can take place appreciably in the virtual void of interplanetary space. Instead, the heat the earth receives from the sun arrives in the form of radiation. These mechanisms all embody a fundamental fact: the natural direction of heat flow is from hot bodies to cold ones.

Conduction, convection, and radiation

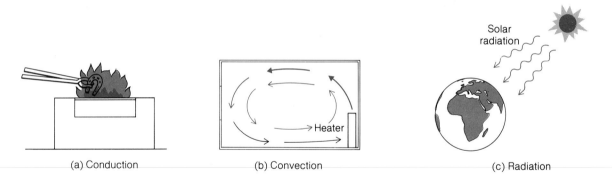

(a) Conduction   (b) Convection   (c) Radiation

FIG. 14–12 Mechanisms of heat transfer.

In most materials, conduction is a simple consequence of the kinetic behavior of matter. Molecules (or atoms, depending upon the nature of the rod) at the hot end of a rod vibrate faster and faster as the temperature there increases. When these molecules collide with their less energetic neighbors, some of their kinetic energy is transferred to the latter (Fig. 14–13). Through successive molecular collisions energy travels down the rod, and, since we perceive random molecular motion as heat, we equally well describe the situation by saying that heat travels down the rod. The average positions of the molecules themselves do not change in conduction.

Mechanism of heat conduction

For heat to be conducted through a body, its ends must be at different temperatures. If the entire body is at the same temperature, all its molecules have the same average energy, and any molecule has as much chance of losing energy in a collision with a nearby molecule as it has of gaining energy. Thus here there is no flow of energy from a region of rapidly moving molecules to an adjacent one of slowly moving molecules. In fact, we can use the necessity of a temperature difference for heat transport to define temperature: one body has a higher temperature than another if, when they are placed in contact, heat flows from the former to the latter.

There are wide differences in the ability of various substances to conduct heat. Gases are poor conductors, because their molecules are relatively far apart and collisions between them correspondingly infrequent. The molecules of liquids and non-metallic solids are closer together, leading to somewhat higher thermal conductivities. Metals exhibit by far the greatest ability to conduct heat; this is why saucepans are made of metal but have handles of wood or plastic.

Metals are the best conductors of heat

The reason for the exceptional thermal conductivity of metals is the same as for their exceptional electrical conductivity: a significant number of electrons are able to move about freely instead of being bound permanently to particular atoms. Acquiring kinetic energy at the hot end of a metal object, the free electrons can travel past many atoms before giving up their energy in collisions, and thereby can speed up the rate of

FIG. 14–13 Molecules at the hot end of a rod vibrate faster than those at the cold end.

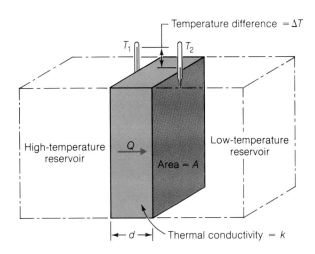

energy transport toward the cold end of the object. Heat conduction by free electrons in a metal compares with heat conduction by molecular interactions as travel by express train compares with travel by local train. We shall further explore the important subject of free electrons in metals in a later chapter.

Electrons in a metal are responsible for its ability to conduct heat and electricity well

The rate at which heat flows through a slab of some material depends upon four quantities (Fig. 14–14):

1. The temperature difference ΔT between the faces of the slab; the greater the temperature difference, the faster the heat transfer.
2. The thickness d of the slab; the thicker the slab, the slower the heat transfer.
3. The area A of the slab; the greater the area, the faster the heat transfer.
4. The *thermal conductivity* k of the material, which is a measure of its ability to conduct heat; the larger the value of k, the faster the heat transfer.

Factors that determine rate of heat conduction

The amount of heat Q that passes through the slab in the period of time t is given by

$$Q = \frac{kAt\,\Delta T}{d} \qquad\qquad Heat\ conduction \quad (14\text{–}2)$$

In SI units, Q is expressed in joules and k in W/m·°C. When Q is expressed in kcal, the corresponding unit for k is the kcal/m·s·°C. Table 14–3 lists the thermal conductivities of a number of materials.

Example   A house window consists of a pane of glass 1 m wide, 1.2 m high, and 5 mm thick. If the inner and outer faces of the glass are at temperatures of 15°C and 5°C, respectively, at what rate is heat lost by conduction through the window?

Solution   From Table 14–3 the thermal conductivity of glass is 0.80 W/m·°C. Since $A = 1.2\ \text{m}^2$, $\Delta T = 10°C$, and $d = 5 \times 10^{-3}$ m, the rate of heat transfer is

$$\frac{Q}{t} = \frac{kA\,\Delta T}{d} = \frac{(0.8\ \text{W/m}\cdot°\text{C})(1.2\ \text{m}^2)(10°\text{C})}{5 \times 10^{-3}\ \text{m}} = 1920\ \text{W}$$

This is equivalent to 1652 kcal per hour.

TABLE 14-3
Thermal conductivities

Material	$k, \dfrac{\text{kcal}}{\text{m} \cdot \text{s} \cdot {}^\circ\text{C}}$	$k, \dfrac{\text{W}}{\text{m} \cdot {}^\circ\text{C}}$
Metals		
Aluminum	4.9×10^{-2}	205
Brass	2.6×10^{-2}	109
Copper	9.2×10^{-2}	385
Iron and steel	1.2×10^{-2}	50
Silver	9.7×10^{-2}	406
Other Solids		
Body fat	0.4×10^{-4}	0.17
Brick	1.5×10^{-4}	0.6
Concrete	2×10^{-4}	0.8
Glass	1.9×10^{-4}	0.80
Ice	4×10^{-4}	1.6
Wood (pine)	0.3×10^{-4}	0.13
Snow	2×10^{-4}	0.8
Insulating materials		
Cork	0.1×10^{-4}	0.04
Glass wool	0.1×10^{-4}	0.04
Down	0.06×10^{-4}	0.02
Kapok	0.08×10^{-4}	0.03
Gases		
Hydrogen	0.3×10^{-4}	0.13
Air	0.06×10^{-4}	0.024

It is worth noting that the statement of the problem gives the temperature of the inner and outer faces of the glass pane. If instead we were merely given the room and outside air temperatures, we would not know by just how much the insulating effect of the air layers on either side of the window affects the actual temperatures at the glass surfaces. This insulating effect is not a minor one, and may reduce the heat flow through the window considerably. Only if there is a very vigorous circulation of air past both sides of the pane will the temperatures of the sides of the pane equal those of the room and outside air. ∎

Choosing insulation thickness

Because the rate of heat flow Q/t through an insulating layer is inversely proportional to its thickness d, in any particular application it usually does not make sense to increase the insulation thickness beyond a certain value. Consider a freezer that absorbs 2000 kcal/day through 5 cm of insulation. Doubling the thickness to 10 cm will decrease the heat flow to 1000 kcal/day. A further doubling to 20 cm of insulation will reduce the heat flow to 500 kcal/day. As the thickness increases, the rate of change of heat flow goes down: going from 5 cm to 10 cm cuts the heat flow by 1000 kcal/day, but going from 10 cm to 20 cm only brings it down by an additional 500 kcal/day. Clearly there is a point at which the cost of more insulation averaged over an appropriate period of time will exceed the savings in operating costs. Just what this point is depends, of course, on the relative costs involved, including that of the space occupied by the insulation.

14–7 CONVECTION

Convection is a much simpler physical process than conduction, since it consists of the actual motion of a volume of hot fluid from one place to another. The hot fluid displaces cold fluid in its path, thereby setting up a *convection current*. Convection is the chief mechanism of heat transfer in fluids under most circumstances.

Convection involves the motion of a heated fluid

Convection may be either *natural* or *forced*. In natural convection, the buoyancy of a heated fluid leads to its motion; when a portion of a fluid (either gas or liquid) is heated, it expands to become of lower density than the surrounding, cooler fluid and hence rises upward (Fig. 14–15). A steam or hot-water heating system employs radiators in each room which heat the rooms with the help of the convection currents they set up.

Natural convection

In forced convection, a blower or pump directs the heated fluid to its destination. In the cooling system of most cars, water is circulated between the engine block and the radiator by a pump; in the radiator, heat is transferred to the atmosphere by conduction through the thin-walled metal tubes of which it is constructed. A hot-air household heating system employs a fan to blow the air from the furnace through ducts to outlets in each room. In an animal's body, excess metabolic heat is carried by the blood to the skin, through which it passes by conduction. This heat is then dissipated by the evaporation of moisture from the skin, by convection to the atmosphere, and by radiation.

Forced convection

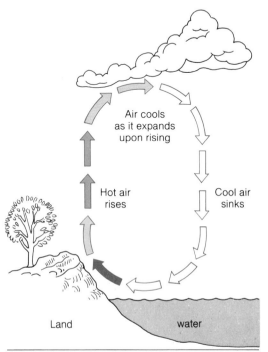

FIG. 14–15 The temperature of a land surface rises more rapidly than that of a water surface in sunlight. The resulting convection is responsible for the sea breeze experienced on sunny days near the shore of a body of water.

Air cools as it expands upon rising

Hot air rises

Cool air sinks

Land

water

Rate of heat transfer by convection

The rate Q/t at which a hot object transfers heat to a surrounding fluid by convection is approximately proportional to the area A of the object in contact with the fluid and to the temperature difference ΔT between them:

$$\frac{Q}{t} = hA \, \Delta T \qquad\qquad\qquad Convection \quad (14\text{–}3)$$

The convection coefficient h depends on the form and orientation of the object (it is nearly twice as great for the upper surface of a horizontal plate as it is for the lower surface) and on the properties and state of motion of the fluid (in the case of convective loss from the skin to air, h is twice as great when the air is moving at 1 m/s than it is in quiet air).

Example   A photographic enlarger has as its light source a 25-W lamp in a metal hood that keeps light from escaping except through the enlarger lens. When the darkroom temperature is 20°C, the metal hood is at a temperature of 50°C. If the 25-W lamp is replaced by a 50-W lamp, what is the new temperature of the hood?

Solution   Since h and A are each the same in both cases,

$$\frac{(Q/t)_1}{(Q/t)_2} = \frac{\Delta T_1}{\Delta T_2}$$

When $(Q/t)_1 = 25$ W, the temperature difference ΔT_1 between the hood and the surrounding air is 50°C − 20°C = 30°C. When $(Q/t)_2 = 50$ W,

$$\Delta T_2 = \Delta T_1 \frac{(Q/t)_2}{(Q/t)_1} = (30°C)\left(\frac{50 \text{ W}}{25 \text{ W}}\right) = 60°C$$

Hence the new hood temperature is 20°C + 60°C = 80°C.   ■

14–8   RADIATION

Electromagnetic waves

In the process of radiation, energy is transported by means of *electromagnetic waves*. These waves travel at the speed of light (3×10^8 m/s = 186,000 mi/s), and require no material medium for their passage. Radio and radar waves, light waves, and X and gamma rays are all electromagnetic waves; they differ only in their wavelength. Figure 14–16 shows the classification of electromagnetic waves according to wavelength.

All objects radiate electromagnetic waves

Every object radiates electromagnetic waves of all wavelengths, though the intensities of the different wavelengths vary considerably. We are all familiar with the glow of a hot piece of metal, where enough radiation is emitted as visible light for our eyes to respond, but other wavelengths are also given off. An object need not be so hot that it gives off visible light for it to be radiating electromagnetic energy—the radiation from an object at room temperature, for instance, is mainly in the infrared part of the spectrum to which the eye is not sensitive.

The ability of an object to emit radiation is proportional to its ability to absorb radiation: a good absorber is a good emitter, and vice versa. This conclusion follows from the fact that something at the same temperature as its environment must be

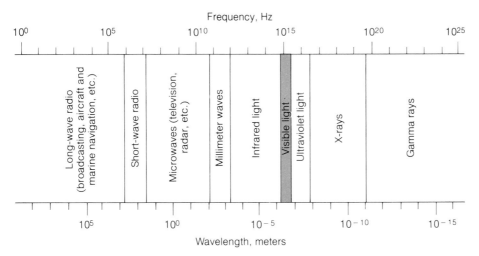

absorbing and emitting radiation at exactly the same rates. When an object is warmer than its environment, it emits more radiation than it absorbs, and it is this difference we perceive. A perfect absorber is called a *blackbody,* and it is accordingly the best possible radiator as well.

A good absorber is a good radiator

The rate R (in W/m^2) at which an object of surface area A and absolute temperature T emits radiation is given by the *Stefan-Boltzmann law* as

Radiation rate varies as T^4

$$R = \frac{P}{A} = e\sigma T^4 \qquad\qquad Stefan\text{-}Boltzmann\ law \quad (14\text{–}4)$$

The value of the constant σ (Greek letter *sigma*) is

$$\sigma = 5.67 \times 10^{-8}\ W/m^2 \cdot K^4$$

The *emissivity e* depends on the nature of the radiating surface and ranges from 0, for a perfect reflector that does not radiate at all, to 1, for a blackbody. As an example, polished copper has an emissivity of 0.3.

Figure 14–17 shows how the intensity of the radiation emitted by an object varies with wavelength at various temperatures. The total emitted radiation (which is proportional to the area under each curve) increases with increasing temperature, while the wavelength corresponding to the peak of each curve decreases. An object that glows red is not as hot as one that glows bluish-white, since red light has the longer wavelength (see Chapter 23).

The higher the temperature of an object, the shorter the predominant wavelength of its radiation

Example   A *thermograph* is a device that measures the amount of infrared radiation each small portion of a person's skin emits and presents this information in pictorial form by different shades of gray or different colors in a *thermogram*. The skin over a tumor is warmer than elsewhere (perhaps because of increased blood flow or a higher rate of metabolism), and thus a thermogram is a valuable diagnostic aid for detecting such maladies as breast and thyroid cancer. To verify that a small difference in skin temperature leads to a significant difference in radiation rate, calculate the percentage difference between the radiation from skin at 34°C and at 35°C.

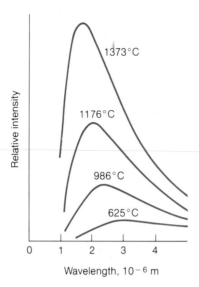

FIG. 14–17 The intensity of the electromagnetic radiation emitted by a heated object at various temperatures, as a function of wavelength.

Solution The emissivity of the skin is the same at both temperatures, whose values on the absolute scale are $T_1 = 34°C + 273 = 307$ K and $T_2 = 35°C + 273 = 308$ K. Since R_1 is proportional to $T_1^{\,4}$ and R_2 is proportional to $T_2^{\,4}$,

$$\frac{R_2 - R_1}{R_1} = \frac{T_2^{\,4} - T_1^{\,4}}{T_1^{\,4}} = \frac{(308\,\text{K})^4 - (307\,\text{K})^4}{(307\,\text{K})^4} = 0.013$$

which is 1.3 percent.   ∎

Example Radiant energy from the sun arrives at the earth's atmosphere at a rate of about 1.4 kW per m^2 of area perpendicular to the sun's rays. The average radius of the earth's orbit is 1.5×10^{11} m and the radius of the sun is 7.0×10^8 m. From these figures find the surface temperature of the sun under the assumption that it radiates like a blackbody (which is approximately true).

Solution We begin by finding the total power P radiated by the sun. The area of a sphere whose radius R is that of the earth's orbit is $4\pi R^2$. Since radiant energy falls on this sphere at a rate of $P/A = 1.4$ kW/m^2,

$$P = (P/A)(4\pi R^2) = (1.4 \times 10^3\,\text{W/m}^2)(4\pi)(1.5 \times 10^{11}\,\text{m})^2$$
$$= 3.96 \times 10^{26}\,\text{W}$$

Next we find the radiation rate R of the sun, which is its power output per square meter of surface. If r is the sun's radius, its surface area is $4\pi r^2$ and

$$R = \frac{\text{power output}}{\text{surface area}} = \frac{P}{4\pi r^2} = \frac{3.96 \times 10^{26}\,\text{W}}{(4\pi)(7.0 \times 10^8\,\text{m})^2} = 6.43 \times 10^7\,\text{W/m}^2$$

The emissivity of a blackbody is $e = 1$, so from the Stefan-Boltzmann formula $R = e\sigma T^4$, we have

$$T^4 = \frac{R}{e\sigma}$$

$$T = \left(\frac{R}{e\sigma}\right)^{1/4} = \left(\frac{6.43 \times 10^7\,\text{W/m}^2}{(1)(5.67 \times 10^{-8}\,\text{W/m}^2 \cdot \text{K}^4)}\right)^{1/4} = (1.134 \times 10^{15})^{1/4}\ \text{K}$$

To evaluate this quantity, it is necessary to have the power of 10 divisible by 4. Accordingly, we proceed as follows:

$$(1.134 \times 10^{15})^{1/4}\ \text{K} = (1134 \times 10^{12})^{1/4}\,\text{K} = (1134)^{1/4}(10^{12})^{1/4} = 5.8 \times 10^3\,\text{K}$$

The surface temperature of the sun is approximately 5800 K, based on this calculation. To find $(1134)^{1/4}$ with a calculator, enter 1134 and press the LOG key. Then divide by 4 and press the INV LOG (or 10^x) key. See Appendixes B3 and B5. ∎

IMPORTANT TERMS

Heat is internal energy in transit from one body of matter to another by virtue of a temperature difference between them. If a body of matter does not change state during the addition or removal of heat, and neither does work nor has work done on it, the change in its internal energy results in a corresponding change in temperature. The SI unit of heat is the joule. Another unit of heat in common use is the **kilocalorie** (kcal), which is the amount of heat required to change the temperature of 1 kg of water by 1°C.

The **specific heat capacity** of a substance is the amount of heat required to change the temperature of 1 kg of it by 1°C.

The **heat of fusion** of a substance is the amount of heat that must be supplied to change 1 kg of it at its melting point from the solid to the liquid state; the same amount of heat must be removed from 1 kg of the substance in the liquid state at its melting point to change it to a solid.

The **heat of vaporization** of a substance is the amount of heat that must be supplied to change 1 kg of it at its boiling point from the liquid to the gaseous (or vapor) state; the same amount of heat must be removed from the substance at its boiling point to change it into a liquid.

Sublimation is the direct conversion of a substance from the solid to the vapor state, or vice versa, without it first becoming a liquid.

The **critical point** is the upper limit of the vaporization curve of a substance; a substance cannot exist in the liquid state at a temperature above that of its critical point.

The **triple point** is the intersection of the vaporization, fusion, and sublimation curves of a substance. All three states of a substance may exist in equilibrium at the temperature and pressure of its triple point.

Heat can be transferred from one place to another by means of **conduction, convection,** or **radiation.** In conduction heat is transported by successive molecular collisions, and in convection by the motion of a volume of hot fluid from one place to another. Heat transfer by radiation takes place by means of **electromagnetic waves,** which require no material medium for their passage.

IMPORTANT FORMULAS

Heat and temperature change: $\quad Q = mc\,\Delta T$

Heat conduction: $\quad Q = \dfrac{kAt\,\Delta T}{d}$

Convection: $\quad \dfrac{Q}{t} = hA\,\Delta T$

Stefan-Boltzmann Law: $\quad R = e\sigma T^4$

MULTIPLE CHOICE

1. Two blocks of lead, one twice as heavy as the other, are both at 50°C. The ratio of the internal energy of the heavier block to that of the lighter block is

(a) $\frac{1}{2}$.

(b) 1.

(c) 2.

(d) 4.

2. Of the following substances, the one that requires the greatest amount of heat per kilogram for a given increase in temperature is

(a) ice.
(b) water.
(c) steam.
(d) copper.

3. A cup of hot coffee can be cooled by placing a cold spoon in it. A spoon of which of the following materials would be most effective for this purpose, assuming the spoons all have the same mass?

(a) aluminum.
(b) copper.
(c) iron.
(d) silver.

4. Body A is at a higher temperature than body B. When they are placed in contact, heat will flow from A to B

(a) only if A has the greater internal energy content.
(b) only if both are fluids.
(c) only if A is on top of B.
(d) until both have the same temperature.

5. When a vapor condenses into a liquid,

(a) it absorbs heat.
(b) it evolves heat.
(c) its temperature rises.
(d) its temperature drops.

6. The freezing point of a substance is always lower than its

(a) melting point.
(b) boiling point.
(c) heat of fusion.
(c) heat of vaporization.

7. The heat of vaporization of a substance is

(a) less than its heat of fusion.
(b) equal to its heat of fusion.
(c) greater than its heat of fusion.
(d) any of the above, depending on the nature of the substance.

8. Sublimation refers to

(a) the vaporization of a solid without first becoming a liquid.
(b) the melting of a solid.
(c) the vaporization of a liquid.
(d) the condensation of a gas into a liquid.

9. Under conditions corresponding to its triple point, a substance

(a) is in the solid state.
(b) is in the liquid state.
(c) is in the gaseous state.
(d) may be in any or all of the above states.

10. The natural direction of the heat flow between two reservoirs depends on

(a) their temperatures.
(b) their internal energy contents.
(c) their pressures.
(d) whether they are in the solid, liquid, or gaseous state.

11. Metals are good conductors of heat because

(a) they contain free electrons.
(b) their atoms are relatively far apart.
(c) their atoms collide infrequently.
(d) they have reflecting surfaces.

12. The materials with the highest heat conductivities are the

(a) gases.
(b) liquids.
(c) woods.
(d) metals.

13. In natural convection, a heated portion of a fluid moves because

(a) its molecular motions become aligned.
(b) of molecular collisions within it.
(c) its density is less than that of the surrounding fluid.
(d) of currents in the surrounding fluid.

14. Four pieces of iron are heated in a furnace to different temperatures. The one at the highest temperature appears

(a) white.
(b) yellow.
(c) orange.
(d) red.

15. Electromagnetic radiation is emitted

(a) only by radio and television antennas.
(b) only by bodies at higher temperatures than their surroundings.
(c) only by bodies at lower temperatures than their surroundings.
(d) by all bodies.

16. If 2 kg of punch of specific heat capacity 1.7 kJ/kg·°C at a temperature of 5°C is poured into a 1-kg glass punch bowl that is at 20°C, the final temperature of the punch is

(a) 6°C.
(b) 8°C.
(c) 10°C.
(d) 12°C.

17. If 10 kg of ice at 0°C is added to 2 kg of steam at 100°C, the temperature of the resulting mixture is

(a) 0°C.
(b) 23°C.
(c) 28°C.
(d) 40°C.

18. A 1-kg lead bar at 80°C is placed in 2 kg of water at 20°C. The final temperature of the lead bar is

(a) 22°C.
(b) 28°C.
(c) 40°C.
(d) 50°C.

19. A brass rod at a temperature of 150°C radiates energy at a rate of 20 W. If its temperature is increased to 300°C, at what rate does it radiate now?

(a) 27 W   (b) 40 W

(c) 67 W   (d) 320 W

EXERCISES

1. A jar of water is shaken vigorously. What becomes of the work that is done?

2. Why will the engine of a car whose cooling system is filled with an alcohol antifreeze be more likely to overheat in summer than one whose cooling system is filled with water?

3. Which is more effective in cooling a drink, 10 g of water at 0°C or 10 g of ice at 0°C?

4. How does perspiration give the body a means of cooling itself?

5. Why does turning the flame higher under a pan of boiling water not reduce the time needed to cook an egg in the water?

6. When a certain quantity of a vapor condenses into a liquid, what happens to its internal energy content and to its temperature?

7. What condition is necessary for heat to flow through an object?

8. In the winter, why does the steel blade of a shovel seem colder than its wooden handle?

9. A Thermos bottle consists of two glass vessels, one inside the other, with the space between them evacuated. The vessels are both coated with thin films of silver. Why is this device so effective in keeping the contents of the bottle at a constant temperature?

10. By what mechanism or mechanisms does a person seated in front of a fire receive heat from it? A person seated in front of a radiator through which hot water is circulated?

11. Under what circumstances does an object radiate electromagnetic waves? How is the predominant wavelength in the radiation related to the temperature of the object?

12. Why is it advantageous to install the heating element of an electric hot water heater near the bottom of the tank?

13. If 200 g of water is to be heated from 15°C to 100°C to make a cup of tea, how much heat is needed?

14. How much heat must be added to 1 g of silver to raise its temperature from −5°C to 65°C?

15. If 10^4 kcal of heat is removed from a metric ton (10^3 kg) of iron at 300°C, what is its final temperature?

16. A 25-kg storage battery has an average specific heat of 0.84 kJ/kg·°C. When fully charged the battery contains 1.4 MJ of electrical energy. If all this energy were dissipated within the battery, find the increase in its temperature.

17. In an effort to lose weight, a person runs 5 km per day at a speed of 4 m/s. While running, the person's body processes consume energy at a rate of 1.4 kW. Fat has an energy content of about 40 kJ/g. How much fat is metabolized during each run?

18. A 60-kg woman does 1.2 MJ of work on a certain day at an average efficiency of 15%. (a) How many kcal of energy must be provided by her food if she is to neither store energy nor use stored energy on that day? (b) If she were so well insulated that only 90% of the waste heat is lost to the outside world, what would her temperature be at the end of the day? Normal body temperature is 37°C.

19. A man drinks a bottle of beer and proposes to work off its 460 kJ by exercising with a 20-kg barbell. If each lift of the barbell from chest height to over his head is through 60 cm and the efficiency of his body is 10% under these circumstances, how many times must he lift the barbell?

20. A bathtub contains 70 kg of water at 26°C. If 10 kg of water at 90°C is poured in, what is the final temperature of the mixture?

21. A 0.6-kg copper container holds 1.5 kg of water at 20°C. A 0.1-kg iron ball at 120°C is dropped into the water. What is the final temperature of the water?

22. A 0.1-kg piece of silver is taken from a bath of hot oil and placed in a 0.08-kg glass jar containing 0.2 kg of water at 15°C. The temperature of the water increases by 8°C. What was the temperature of the oil?

23. Carbon dioxide is usually shipped in tanks under a pressure of approximately 70 atm. At 20°C, is the carbon dioxide a solid, a liquid, or a gas?

24. The pressure and temperature of the atmosphere at 35,000 m are 4 torr and −23°C respectively. What is the state of water under those conditions? Of carbon dioxide?

25. What thicknesses of (a) concrete, (b) brick, and (c) pine wood have the same insulating ability as 10 cm of glass wool?

26. A pine wood door 60 mm thick is to be removed and the opening bricked up. How thick should the brick wall be if the rate of heat transfer through this part of the building is to be unchanged?

27. The steel hull of a boat is 6 mm thick and its underwater area is 120 m^2. If the water temperature is 26°C and the temperature of the boat's interior is 20°C, how much heat per day enters the boat through the hull?

28. The neoprene foam material of a scuba diver's wet suit has a thermal conductivity of 0.075 W/m·°C. (a) Find the rate of heat loss of a diver whose skin area is 1.7 m^2 wearing a suit of such material 6 mm thick who is diving in 10°C water. Assume a skin temperature of 32°C. (b) How much heat does the diver lose in a dive lasting 1 h? (The suit material when submerged is squeezed by water pressure to a smaller thickness than at the surface. At a depth of 20 m the thickness is about half that at the surface, for instance. The figure quoted above refers to the actual thickness at the diver's depth.)

29. A double boiler has an aluminum upper pan whose bottom is 20 cm in diameter and 0.9 mm thick. If the lower pan contains boiling water and the upper one milk at 5° C, compute the rate of heat transfer to the milk. What happens to this rate as the milk warms up?

30. An insulated copper rod 5 mm on a side and 1 m long has one end in a steam bath and the other in contact with a block of ice at 0°C. How much ice melts per hour?

31. A roast turkey cools from 65°C to 60°C in 5 min when it is in a room whose temperature is 20°C. How long will it take to cool from 45°C to 40°C in the same room?

32. A glass of beer warms from 7°C to 10°C in 3 min when the air temperature is 35°C. How long will it take to warm from 12°C to 15°C?

33. A small hole in a cavity behaves like a blackbody because any radiation that falls on it is trapped inside by multiple reflections until it is absorbed. At what rate does radiation escape from a hole 10 cm^2 in area in the wall of a furnace whose interior is at a temperature of 700°C?

34. A copper sphere 5 cm in diameter whose emissivity is 0.3 is heated in a furance to 400°C. At what rate does it radiate energy?

35. An object is at a temperature of 400°C. At what temperature would it radiate energy twice as fast?

PROBLEMS

1. A person decides to lose weight by eating only cold food. A 100-gram piece of apple pie yields about 350 kcal of energy when eaten. If its specific heat capacity is 0.4 kcal/kg·°C, how much greater is its energy content at 50°C than at 20°C? What percentage difference is this?

2. What is the difference in temperature between the water at the top and at the bottom of a waterfall of height *h?*

3. If all the heat lost by 1 kg of water at 0°C when it turns into ice at 0°C could be turned into kinetic energy, what would the speed of the ice be?

4. A 1-kg block of ice at 0°C falls into a lake whose water is also at 0°C, and 0.01 kg of ice melts. What was the minimum altitude from which the ice fell?

5. A lead bullet at 100°C strikes a steel plate and melts. What was its minimum speed?

6. A 20-g lead bullet whose temperature is 50°C and whose speed is 400 m/s strikes a large block of ice at 0°C and stops inside it. How much ice melts?

7. The minimum speed an artificial earth satellite can have is 7.9 km/s. Aluminum melts at 660°C. If an aluminum satellite reenters the earth's atmosphere when its temperature is 0°C, can it be brought to rest rapidly by air resistance without melting? If not, how do actual spacecraft avoid this dilemma?

8. A 1200-kg car whose transmission is in neutral is coasting down a 10° hill with its brakes being used to maintain a constant speed. If the total mass of the car's iron brake drums is 20 kg, find the increase in their temperature per 100 m of travel. Assume that the only friction acting occurs in the brakes and that all the heat produced is absorbed by the drums, so the resulting figure is a maximum.

9. A 50-kg block of ice at 0°C is pushed across a wooden floor also at 0°C for a distance of 20 m. A total of 25 g of ice melts as a result of the friction of the block on the floor. What is the minimum coefficient of friction in this case?

10. The 30-kg copper tank of a 5-kW water heater has a capacity of 100 kg of water. The water temperature in the tank is 15°C when it is switched on and is 50°C an hour later. What percentage of the energy input was lost?

11. A 300-g aluminum pot containing 0.5 liter of water is placed on the 1-kg iron hot plate of an electric stove. The heating element of the hot plate is rated at 1.5 kW. If the hot plate, the pot, and the water are all at 20°C when the heating element is switched on, how long will it take until the water begins to boil? Neglect heat losses to the room. (One liter of water has a mass of 1 kg.)

12. Air is blown past the 2-kW element of an electric heater at a rate of 0.05 m^3/s. If the density of air is 1.3 kg/m^3 and its specific heat capacity is 1 kJ/kg·°C, what is the increase in temperature of the air that passes through the heater?

13. How much ice at −10°C is required to cool a mixture of 0.1 kg ethyl alcohol and 0.1 kg water from 20°C to 5°C?

14. If 6 kg of ice at $-10°C$ is added to 6 kg of water at $+10°C$, find the temperature of the resulting mixture.

15. By mistake, 0.2 kg of water at 0°C is poured into a vessel containing liquid nitrogen at $-196°C$. How much nitrogen vaporizes?

16. A 5-kg iron bar is taken from a forge at a temperature of 1000°C and plunged into a pail containing 10 kg of water at 60°C. How much steam is produced?

17. If 10 kg of steam at 104°C is passed through 10 kg of water at 93°C, what is the temperature and physical state of the mixture afterward?

18. How much steam at 120°C is required to melt 5 kg of ice at 0°C?

19. The layer of ice on a frozen lake is 5 cm thick at 5 P.M. During the night the air temperature is $-2°C$ and the temperature of the water beneath the ice is 0°C. How thick is the ice at 8 A.M. the next day?

20. An igloo of compacted snow has a surface area of 24 m^2 and a wall thickness of 40 cm. The thermal conductivity of the snow is 2×10^{-4} kcal/m·s·°C. If the temperature inside the igloo is 20°C and the outside temperature is $-10°C$, how much blubber (heat of combustion = 9500 kcal/kg) must be burned per hour to keep the interior temperature constant? Neglect the melting of the igloo wall and the heat evolved by the igloo's inhabitants.

21. A woman is taking a sunbath with 0.5 m^2 of her skin exposed to solar radiation, which reaches her at the rate of 1.2 kW/m^2. Her skin temperature is the same as the air temperature. She has light skin whose emissivity is 0.7, so that 0.7 of the radiation reaching her is absorbed. Her metabolic processes result in the liberation of heat at a rate of 80 W. How much sweat must she evaporate per hour in order to get rid of the metabolic heat and absorbed solar energy? The heat of vaporization of sweat is 2430 kJ/kg.

22. A naked person whose skin area is 1.5 m^2 and whose skin temperature is 33°C is sitting in a room at 25°C. Find the rates at which the person loses heat by convection and by radiation. The convection coefficient h for such a person is about 7 W/m^2·°C, and at such temperatures the emissivity of human skin is very nearly 1 (the radiation is chiefly in the infrared part of the spectrum).

ANSWERS TO MULTIPLE CHOICE

1. c	**5.** b	**9.** d	**13.** c	**17.** d
2. b	**6.** b	**10.** a	**14.** a	**18.** a
3. a	**7.** c	**11.** a	**15.** d	**19.** c
4. d	**8.** a	**12.** d	**16.** b	

THERMODYNAMICS

15

Thermodynamics has as its basic concern the transformation of heat into mechanical energy. Thermodynamics thus plays a central role in technology, since almost all the "raw" energy available for our use is liberated in the form of heat. A device or system that converts heat into mechanical energy is called a heat engine, and the principles that govern its operation are the same whether it is an automobile engine whose heat source is the burning of gasoline, a steam turbine whose heat source is a nuclear reactor, the earth's atmosphere whose heat source is the sun, or a human being whose heat source is the food he or she eats.

15–1   FIRST LAW OF THERMODYNAMICS

Three characteristic processes take place in all heat engines:

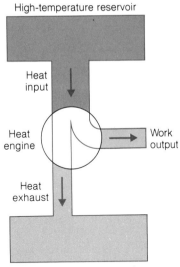

High-temperature reservoir

Heat
input

Heat
engine

Work
output

Heat
exhaust

Low-temperature reservoir

FIG. 15–1 The work output of a heat engine is the difference between the amount of heat it takes in from a high-temperature reservoir and the heat it exhausts to a low-temperature reservoir.

1. Heat is absorbed from a source at a high temperature.
2. Mechanical work is done.
3. Heat is given off at a lower temperature.

Heat engine

Different heat engines carry out these processes in different ways, but the general pattern of operation is always the same (Fig. 15–1).

Two general principles have been found to apply to all heat engines. The *first law of thermodynamics* expresses the conservation of energy: Energy cannot be created or destroyed, but may be converted from one form to another. In terms of a heat engine, this law states that

First law of thermodynamics

Net heat input = work output + change in internal energy of engine

If the engine operates in a cycle, energy may be alternately stored and released from storage, but the engine does not experience a net change in its internal energy. In this case,

Net heat input = work output

The net heat input equals the amount of heat the engine takes in from a reservoir at high temperature minus the amount of heat it exhausts to a reservoir at low temperature. In a steam engine the high-temperature reservoir is the boiler, and the low-temperature reservoir is the escaping steam; in a gasoline engine the high-temperature reservoir is the exploding mixture of air and gasoline vapor in each cylinder, and the low-temperature reservoir is the exhaust gas; in the earth's atmosphere the ultimate high-temperature reservoir is the sun, and the ultimate low-temperature reservoir is the rest of the universe.

A *refrigerator* is a heat engine operating in reverse. Ordinarily a heat engine absorbs heat from a high-temperature reservoir and exhausts it to a low-temperature one, with an output of mechanical work being provided in the process. In a refrigerator, on the other hand, heat is transferred from a low-temperature reservoir (typically a

A refrigerator reverses the natural flow of heat

FIG. 15–2 A refrigerator takes heat from a low-temperature reservoir and transfers it to a high-temperature one. Energy must be supplied to permit this reverse flow of heat to occur.

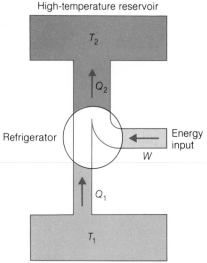

High-temperature reservoir

T_2

Q_2

Refrigerator

Energy input

W

Q_1

T_1

Low-temperature reservoir

storage chamber) to a high-temperature one (the outside world), and mechanical energy must be supplied in order to do this. In effect, heat must be "pushed uphill" if it is to go from a cold region to a warm one (Fig. 15–2).

It is important to keep in mind that a refrigerator does not "produce cold," since cold is a relative deficiency of internal energy and not something in its own right, as heat is. What a refrigerator does is to remove internal energy from a specific region and transport it elsewhere.

15–2 SECOND LAW OF THERMODYNAMICS

Heat is disordered energy

Heat is the easiest and cheapest form of energy to obtain, since all we need do to liberate it is to burn a fuel such as wood, coal, or oil. The real problem is to turn heat into mechanical energy so it can power cars, ships, airplanes, electric generators, and machines of all kinds. To appreciate the problem, we recall that heat consists of the kinetic energies of moving atoms and molecules. In order to change heat into a more usable form, we must extract some of the energy of the random motions of atoms and molecules and convert it into regular motions of a piston or a wheel. Such conversions cannot take place efficiently, for the same reason that it is easier to shatter a wineglass than to reassemble the fragments: the natural tendency of all physical systems is toward increasing disorder. The *second law of thermodynamics* is an expression of this tendency, whose role in the evolution of the universe is quite as central as are those of the various conservation principles.

Many processes permitted by the conservation laws nevertheless do not occur

The first law of thermodynamics prohibits an engine from operating without a source of energy, but it does not tell us anything about the character of possible sources of energy. For instance, there is an immense amount of internal energy in the atmosphere, yet it is impossible to run a car by taking in air at ambient temperature, extracting some of its internal energy, and then exhausting liquid air. Or, to give an even more extreme case, it is energetically possible for a puddle of water to rise spontaneously

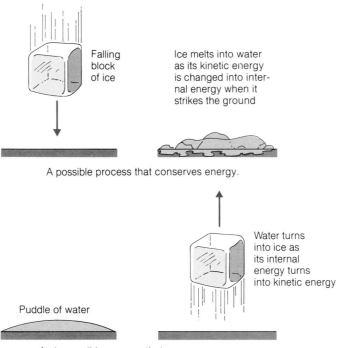

Falling block of ice

Ice melts into water as its kinetic energy is changed into internal energy when it strikes the ground

A possible process that conserves energy.

Water turns into ice as its internal energy turns into kinetic energy

Puddle of water

An impossible process that conserves energy.

FIG. 15–3 The second law of thermodynamics provides a way to identify impossible processes that are in accord with all other physical principles.

into the air, cooling and freezing into ice as its internal energy changes into potential energy (Fig. 15–3). After all, a block of ice dropped from a sufficient height melts when it strikes the ground, its initial potential energy first being converted to kinetic energy and then into heat. Needless to say, water does not rise upward of its own accord, and we must find an appropriate way of expressing this conclusion.

The second law of thermodynamics is the physical principle, independent of the first law and not derivable from it, that supplements the first law in limiting our choice of heat sources for our engines. It can be stated in a number of equivalent ways, a common one being as follows:

> **It is impossible to construct an engine, operating in a cycle (that is, continuously), which does nothing other than take heat from a source and perform an equivalent amount of work.**

Second law of thermodynamics

According to the second law of thermodynamics, then, no engine can be completely efficient—some of its heat input *must* be ejected. As we shall see, the greatest efficiency any heat engine is capable of depends upon the temperatures of its heat source and of the reservoir to which it exhausts heat. The greater the difference between these temperatures, the more efficient the engine. The second law is a consequence of the empirical fact we have already noted:

A heat engine extracts energy from the flow of heat through it

> **The natural direction of heat flow is from a reservoir of internal energy at a high temperature to a reservoir of internal energy at a low temperature, regardless of the total energy content of each reservoir.**

FIG. 15–4 The second
law of thermodynamics.

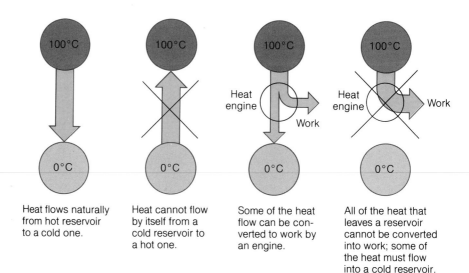

Heat flows naturally
from hot reservoir
to a cold one.

Heat cannot flow
by itself from a
cold reservoir to
a hot one.

Some of the heat
flow can be con-
verted to work by
an engine.

All of the heat that
leaves a reservoir
cannot be converted
into work; some of
the heat must flow
into a cold reservoir.

The latter statement, in fact, may be regarded as an alternative expression of the second law (Fig. 15–4).

A low-temperature reservoir is essential for a heat engine

If we are to utilize the internal energy content of the atmosphere or the oceans, we must first provide a reservoir at a lower temperature than theirs in order to extract heat from them. There is no reservoir in nature suitable for this purpose, for if there were, heat would flow into it until its temperature reached that of its surroundings. To establish a low-temperature reservoir, we must employ a refrigerator (which is a heat engine running in reverse by using up energy to extract heat), and in so doing we will perform more work than we can successfully obtain from the heat of the atmosphere or oceans.

The laws of thermodynamics can be summarized by saying that the first law prohibits us from getting something for nothing, and the second law prohibits us from doing as well as breaking even.

15–3 CARNOT ENGINE

Every heat engine behaves in the same general way: it absorbs heat at a certain temperature, converts some of the heat into work, and exhausts the rest at a lower temperature. Because of the second law of thermodynamics, we cannot expect a heat engine to be 100% efficient in turning heat into work. But suppose we have an engine that is not subject to friction, to the loss of stored heat by conduction or radiation, or to other modes of energy dissipation that can, in principle, be reduced as much as we like by careful construction. What is the maximum efficiency of such an idealized engine?

Reversible engines are the most efficient

It is clear that maximum efficiency can be attained only if all the processes that occur in the engine's operation do not involve permanent changes in the engine, because any such changes must mean the discard of energy in a nonproductive way. That is, every process in the engine must be *reversible*. Only isothermal and adiabatic processes meet this specification. An *isothermal* process is one in which the temperature of the substance that undergoes a change of some kind remains constant. If the substance is

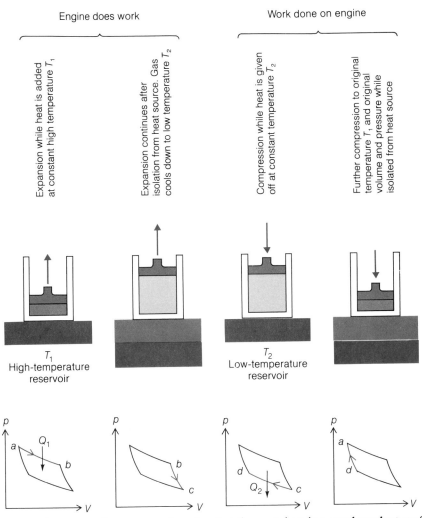

FIG. 15–5 The Carnot cycle.

a gas confined to a container, we can imagine that the container is a good conductor of heat and is immersed in a constant-temperature reservoir of heat. No process that occurs in the real world is ever wholly isothermal, just as no actual gas exactly resembles an ideal gas, but many processes are quite close to being isothermal.

An isothermal process takes place at a constant temperature

At the other extreme from being in such intimate contact with a heat reservoir that its temperature never changes, a system might be so completely isolated from its surroundings that heat can neither enter nor leave it. Any process undergone by a system in this situation is called *adiabatic*. Most rapid thermodynamic processes are approximately adiabatic, since heat transfer takes time and the process may be completed before an appreciable amount of heat has passed through the walls of the system.

No heat enters or leaves a system during an adiabatic process

A heat engine that employs only isothermal and adiabatic processes was devised in 1824 by the French engineer Sadi Carnot. A *Carnot engine* consists of a cylinder that is filled with an ideal gas and has a movable piston at one end. The four stages in its operating cycle are shown in Fig. 15–5, together with a graph of each stage on a pressure-volume diagram. These stages are as follows:

Operating cycle of a Carnot engine

1. An amount of heat Q_1 is added to the gas, which expands isothermally at its initial temperature T_1. The heat added equals exactly the work done by the gas, which is why its temperature does not change.

2. The heat source is removed, and the expansion is allowed to continue. The second expansion is adiabatic and takes place at the expense of the energy stored in the gas, and so the gas temperature falls from T_1 to T_2. During expansions 1 and 2 the piston exerts a force on whatever it is attached to, and thereby performs work.

3. Having done work in pushing the piston outward, the engine must now be returned to its initial state in order for it to be able to do further work. The third stage involves an isothermal compression of the gas at the constant temperature T_2 during which an amount of heat Q_2 is given off. The heat given off exactly equals the work done on the gas by the piston, which is why its temperature does not change.

4. The gas is returned to its initial temperature, pressure, and volume by an adiabatic compression in which heat is neither added to it nor removed from it. Work is done on the gas in this compression, which is why its temperature rises.

15–4 ENGINE EFFICIENCY

In each cycle a Carnot engine performs some net amount of work W, which is the difference between the work it does during the two expansions and the work done on it during the two compressions. It has taken in the heat Q_1 and ejected the heat Q_2; we note that the heat Q_2 *must* be ejected in order that the engine return to its initial state from which it can begin another cycle (Fig. 15–6). According to the first law of thermodynamics,

$$W = Q_1 - Q_2 \tag{15–1}$$

The efficiency of the engine is the ratio between its work output W and its heat input , Q_1, so that

FIG. 15–6 The efficiency of a Carnot engine depends upon the ratio between Q_2 and Q_1.

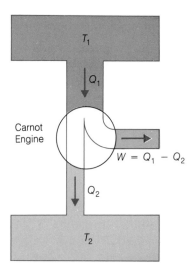

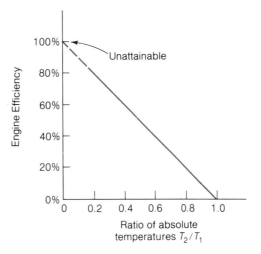

FIG. 15–7 The efficiency of a Carnot engine is equal to $1 - T_2/T_1$.

$$\text{Eff} = \frac{W}{Q_1} = \frac{Q_1 - Q_2}{Q_1} = 1 - \frac{Q_2}{Q_1} \qquad (15\text{-}2)$$

The smaller the ratio of the ejected heat Q_2 to the absorbed heat Q_1, the more efficient the engine.

As it happens, the heat Q transferred to or from a Carnot engine is directly proportional to the absolute temperature T of the reservoir with which it is in contact. That is, for a given Carnot engine,

Heat transfer to and from a Carnot engine is proportional to temperature

$$\frac{Q}{T} = \text{constant} \qquad \qquad \textit{Carnot engine} \quad (15\text{-}3)$$

(The argument that leads to this conclusion is rather lengthy, although it does not involve any physical principles we have not yet encountered, and so is omitted here.) According to Eq. (15–3), the ratio Q_2/Q_1 between the amounts of heat ejected and absorbed per cycle by a Carnot engine is equal to the ratio T_2/T_1 between the temperatures of the respective reservoirs. The efficiency of such an engine is therefore

$$\text{Eff} = 1 - \frac{T_2}{T_1} \qquad \qquad \textit{Carnot efficiency} \quad (15\text{-}4)$$

The smaller the ratio between the absolute temperatures T_2 and T_1, the more efficient the engine (Fig. 15–7). No engine can be 100% efficient because no reservoir can have an absolute temperature of 0 K. (Even if such a reservoir could somehow be created, the exhaust of heat to it by the engine would raise its temperature above 0 K immediately.)

Ideal engine efficiency depends upon temperatures at which heat is absorbed and exhausted

Example   Steam enters a certain steam turbine (Fig. 15–8) at a temperature of 570°C and emerges into a partial vacuum at a temperature of 95°C. What is the upper limit to the efficiency of this engine?

Solution   The absolute temperatures equivalent to 570°C and 95°C are respectively

FIG. 15–8 A primitive steam turbine. In a modern turbine, steam flows past a dozen or more sets of blades on the same shaft to extract as much power as possible. Stationary blades are interleaved between the moving blades to direct the flow of steam in the most advantageous way.

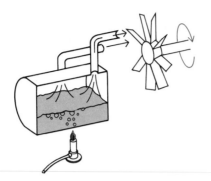

843 K and 368 K. The efficiency of a Carnot engine operating between these two absolute temperatures is

$$\text{Eff} = 1 - \frac{T_2}{T_1} = 1 - \frac{368\,\text{K}}{843\,\text{K}} = 0.56$$

which is 56%. The efficiency of a Carnot engine is the maximum possible for an engine operating between a given pair of temperatures. An actual steam turbine operating between 570°C and 95°C would have an efficiency of no more than about 40% because of the inevitable presence of friction and heat losses to the atmosphere. ■

All reversible engines have the same efficiency

As mentioned earlier, the significance of the Carnot engine is that, being reversible, it has the highest efficiency permitted by the laws of thermodynamics. All other reversible engines must have the same efficiency when operated between the same two internal-energy reservoirs. A Carnot engine with an ideal gas as its working substance is thus just one representative of the whole class of reversible engines. For example, analogs of the simple Carnot cycle can be devised based upon the electrochemical changes that occur in a storage battery or upon the magnetic changes that occur in a paramagnetic substance. A real engine is never exactly reversible because of such irreversible transformations as those involved in friction and in heat losses through the engine walls, and its efficiency is less than that of a Carnot engine.

15–5 INTERNAL COMBUSTION ENGINES

Diesel engines are more efficient than gasoline engines

An internal combustion engine is able to achieve a relatively high operating efficiency by generating the input heat within the engine itself. In a gasoline engine a mixture of air and gasoline vapor is ignited in each cylinder by a spark plug, and the evolved heat is converted into mechanical energy by the pressure of the hot gases on a piston. The greater the ratio between the initial and final volumes of the expanding gases, the greater the engine efficiency. In a gasoline engine this ratio is limited to about 8 to 1, since the gasoline-air mixture in the cylinder will otherwise spontaneously ignite during its compression before the end of the stroke is reached, which leads to a Carnot efficiency of about 55%. The more efficient diesel engine circumvents this difficulty by compressing only air and injecting fuel oil into the hot, compressed air at the instant

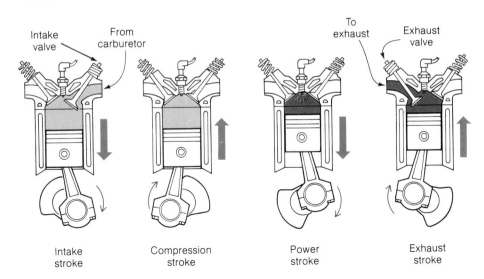

FIG. 15–9 The operating cycle of a four-stroke gasoline engine.

| Intake valve | From carburetor | | To exhaust | Exhaust valve |

| Intake stroke | Compression stroke | Power stroke | Exhaust stroke |

the piston has reached the top of its travel. No spark plug is required. The compression ratio in a diesel engine is typically 20 to 1, for a Carnot efficiency of as much as 70%.

Gasoline engine cycle

Figure 15–9 shows the operating cycle of a typical four-stroke gasoline engine. In the intake stroke a mixture of gasoline vapor and air from the carburetor is drawn into the cylinder through the intake valve by the suction of the downward-moving piston. In the compression stroke both valves are closed and the upward-moving piston compresses the fuel-air mixture. At the top of the stroke the spark plug is fired, which ignites the fuel-air mixture. The burning fuel expands and forces the piston down in the power stroke. At the end of the power stroke the exhaust valve opens and the upward-moving piston expels the waste gases.

Each cylinder in a four-stroke engine has one power stroke in every two shaft revolutions. When a lightweight engine is necessary a two-stroke cycle can be used, which increases the power output by permitting a power stroke in each cylinder in every shaft revolution. Usually a two-stroke cycle means reduced efficiency in the case of a gasoline engine, which is unimportant in such applications as outboard motors for boats, and greater complexity in the case of a diesel engine, which is a fair price to pay for decreased weight in an engine that must be heavily built to withstand 20-to-1 compressions.

Diesel engine cycle

Figure 15–10 illustrates the operation of a two-stroke diesel engine. When the piston is at the bottom of its path, air from a blower enters to flush waste gases from the previous power stroke out through the exhaust valve. As the piston moves upward it compresses the fresh air to a fraction of its initial volume. At the top of the compression stroke the air temperature is perhaps 550°C, and fuel oil sprayed in by the injector is ignited at once. The burning fuel presses down on the piston during the ensuing power stroke.

Gas turbine

The outstanding efficiency of reciprocating gasoline and diesel engines (nearly as high as those of steam turbines) has to some extent retarded the development of the still more efficient gas turbine. A gas turbine is similar to a steam turbine except that

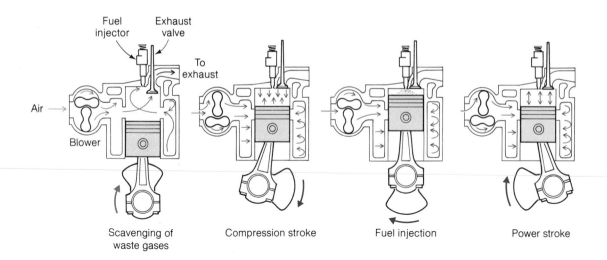

Fuel injector

Exhaust valve

To exhaust

Air

Blower

Scavenging of waste gases

Compression stroke

Fuel injection

Power stroke

FIG. 15–10 The operating cycle of a two-stroke diesel engine.

hot gases from the burning fuel pass through its sets of blades instead of steam. A gas turbine is lighter in weight and has fewer moving parts than a reciprocating internal combustion engine, but the high temperature and high rotational speeds at which it operates present difficulties in manufacture. Gas turbines are nevertheless coming into wider and wider use: "turboprop" aircraft engines are gas turbines, for instance, and a number of ships are already powered by gas turbines.

Jet engine

The rapidly rotating shaft of a turboprop engine is coupled to a propeller through a reduction gear. In a jet engine the propeller is eliminated and the hot gases from the burning fuel are ejected at high speed from the rear of the engine to furnish a reaction force that pushes the aircraft forward. The energy liberated by the burning fuel is thus converted directly into propulsion with no moving parts intervening (except for a turbine that powers the necessary air compressor). Rocket motors are jet engines in which the required oxygen or other oxidizing agent for fuel combustion comes from an internal reservoir instead of from the atmosphere. In a solid-fuel rocket, the ultimate in simplicity, both the components required for combustion are combined in a stable mixture whose reaction rate when ignited is relatively slow and steady rather than explosive.

Heat of combustion

The *heat of combustion* of a substance is the amount of heat liberated when 1 kg of it is completely burned. Table 15–1 lists heats of combustion for a number of common fuels. If the fuel consumption rate of an engine is known at a certain power output, this table can be used to calculate its efficiency.

Example The diesel engine of a boat uses 9.3 kg/h of fuel when its power output is 35 kW. Find its efficiency.

Solution The power input to the engine is

$$P_{\text{input}} = \left(9.3\,\frac{\text{kg}}{\text{h}}\right)\left(44.8 \times 10^6\,\frac{\text{J}}{\text{kg}}\right)\left(\frac{1}{3600\,\text{s/h}}\right) = 1.16 \times 10^5\,\text{W} = 116\,\text{kW}$$

Substance	Heat of Combustion	
Solids	*kcal/kg*	*MJ/kg*
Charcoal	8,100	33.9
Coal	7,800	32.6
Wood	4,500	18.8
Liquids	*kcal/kg*	*MJ/kg*
Diesel oil	10,700	44.8
Domestic fuel oil	10,800	45.2
Ethyl alcohol	7,800	32.6
Gasoline	11,300	47.3
Kerosene	11,000	46.0
Gases*	*kcal/m^3*	*MJ/m^3*
Acetylene	12,900	54.0
Coal gas	4,300	18.0
Hydrogen	2,445	10.2
Natural gas	8,000–17,000	33–71
Propane	20,600	86.2

TABLE 15–1
Typical heats of combustion for common fuels

*Volumes at 0°C and atmospheric pressure.

The engine's efficiency is therefore

$$\text{Eff} = \frac{P_{\text{output}}}{P_{\text{input}}} = \frac{35\,\text{kW}}{116\,\text{kW}} = 0.30 = 30\%$$

which is less than half its Carnot efficiency.

15–6 REFRIGERATORS AND HEAT PUMPS

A refrigerator absorbs heat at a low temperature and exhausts it at a high temperature. According to the second law of thermodynamics, such a cycle cannot take place without an external supply of energy to force heat to flow in the reverse of its usual direction (Fig. 15–2). In a typical refrigerator, two to three times as much heat is extracted from the storage chamber as the amount of external energy provided.

Energy is needed to reverse the natural flow of heat

In nearly all refrigerators the working substance (or *refrigerant*) is a gas that is readily liquified. Common refrigerants are ammonia, Freon 12, methyl chloride, and sulfur dioxide. Figure 15–11 shows the vaporization curve of Freon 12. Under conditions of temperature and pressure corresponding to points above the curve only liquid Freon is present, while under conditions corresponding to points below the curve only Freon vapor is present. Other refrigerants have different vaporization curves. At atmospheric pressure Freon 12 boils at −30°C, while at this pressure ammonia boils at −33°C, methyl chloride at −24°C, and sulfur dioxide at −10°C. The choice of a refrigerant depends upon the precise kind of refrigerator involved and the temperatures between which it is to operate.

FIG. 15–11   Vaporization curve of Freon 12, a common refrigerant. Freon 12 is a trade name for dichlorodifluoromethane, CCl_2F_2.

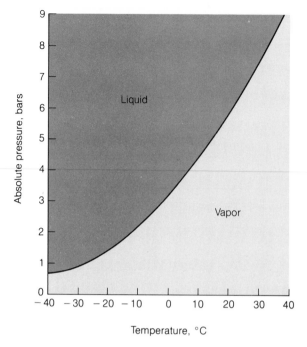

Typical refrigeration system

Let us examine a refrigeration system that uses Freon 12. As in Fig. 15–12, this system consists of a *compressor,* a *condenser,* an *expansion valve,* and an *evaporator.* When the piston of the compressor moves downward, Freon vapor at 1.4 bars (gauge pressure) and approximately room temperature is sucked into the cylinder. As the piston reaches the bottom of its stroke and begins to move upward, the intake valve closes and the discharge valve opens. The compressed Freon, which emerges at a pressure of

FIG. 15–12   A refrigeration system using Freon 12. Gauge pressures are shown.

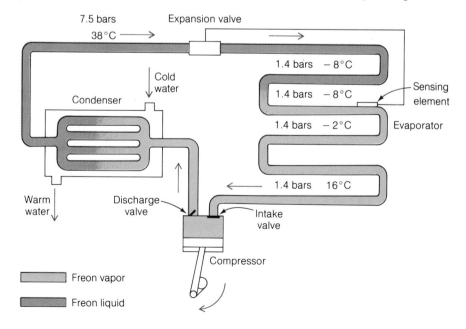

7.5 bars and a high temperature, passes into the condenser in which it is cooled until it liquefies. The condenser may be water-cooled, as shown in the sketch, or air-cooled. It is in this stage that the heat extracted from the refrigerated space is dissipated.

The liquid Freon then goes into the expansion valve from which it emerges at a lower pressure (1.4 bars) and temperature (−8°C). The amount of Freon supplied by the valve is regulated by a sensing element placed in the refrigerated space; this may be simply a gas-filled bulb which responds to temperature changes by pressure changes that actuate a bellows in the valve. As the cold liquid Freon flows through the evaporator tubes it absorbs heat from the region being cooled. From Fig. 15–11 we see that, at an absolute pressure of 2.4 bars, the boiling point of Freon 12 is −8°C, so that the heat absorbed by the liquid Freon in the evaporator causes it to vaporize. Farther along in the evaporator the Freon vapor itself absorbs heat, rising in temperature to perhaps −2°C. The amount of this temperature rise (called *super heat* by refrigeration engineers) is critical for the efficiency of the system, and is usually 6°C or less. Finally the Freon vapor leaves the evaporator and enters the compressor to begin another cycle.

A *heat pump* is a refrigeration system that can extract heat from the cold outdoors in winter and deliver it to the interior of a house. In summer, the same system can serve as an air conditioner to take heat from the house and exhaust it to the warmer outdoors (Fig. 15–13). The concept of the heat pump is interesting because more heat is transferred than the work done. For example, a heat pump with an energy input of 5 kW might be able to "pump" an additional 15 kW of heat to provide a total of 20 kW. An ordinary furnace would have to burn fuel at a rate of 20 kW and so would require four times as much energy input as the heat pump. However, heat pumps have a cost problem to overcome: energy in the form of fossil fuel (coal, oil, gas) is still quite a bit cheaper than the electrical energy needed to operate a heat pump, and in addition a heat pump is a more expensive installation than a conventional heating plant. Only when a heat

Heat pump

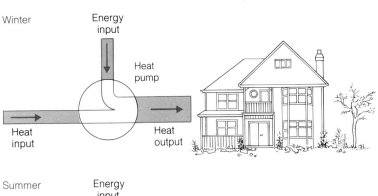

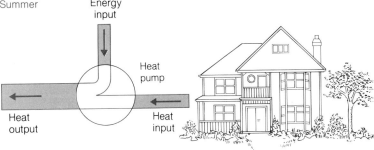

FIG. 15–13 A heat pump is a refrigeration system that can be used to heat a house in winter and cool it in summer.

pump is needed for air conditioning as well as for space heating can it at present be a practical proposition.

15–7　STATISTICAL MECHANICS

According to the second law of thermodynamics, it is impossible to convert heat into any other form of energy completely. Some of the heat input to an engine *must* be lost. Why? The reason lies in the nature of heat, which is molecular kinetic energy; the temperature of a body is a measure of the average kinetic energy of each of its constituent molecules. Let us see how the microscopic picture of matter as consisting of molecules in motion leads to the second law of thermodynamics. Although the argument will be based on molecules in a gas, the essential ideas hold for matter in any state.

Molecular energy distribution at equilibrium

The molecules of a gas are in constant random motion and undergo frequent collisions with one another. Although we cannot hope to follow an individual gas molecule in its wanderings, it is possible to predict on the basis of statistical arguments what fraction of the time it will have any specified amount of kinetic energy. Hence we can calculate the distribution of molecular energies in a gas sample at a particular temperature. This distribution, which has been confirmed by experiment, has the form shown in Fig. 15–14, and holds for all equilibrium conditions in which each molecule has the same average energy over a period of time. A molecule that moves more swiftly than usual at one instant will move less swiftly at a later instant after a number of collisions have taken place.

Equilibrium corresponds to the most probable energy distribution

Equilibrium is the most probable condition according to *statistical mechanics*, a branch of physics that mathematically deduces the behavior of assemblies of so many particles that deviations from statistically probable behavior are not significant. If we toss two coins, it is unlikely that one will come up heads and the other tails; but if we toss a million coins, the percentage deviation from an equal number of heads and tails will be extremely small. We can appreciate why departures from the equilibrium distribution of molecular energies for more than the briefest instant are so unlikely if we look at a cubic centimeter—a thimbleful—of air at atmospheric pressure and room temperature. There are 2.7×10^{19} molecules in the cubic centimeter, and each mol-

FIG. 15–14 The distribution of molecular energies in a gas at a particular temperature.

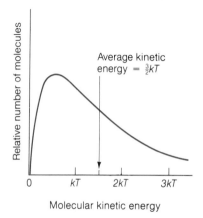

Molecular kinetic energy

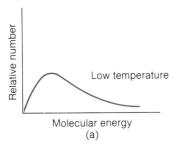

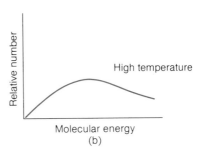

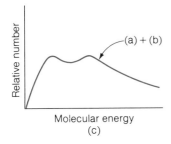

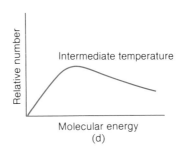

FIG. 15–15 The molecular systems (c) and (d) have the same total energy, but the distribution of energies in (d) has the greater probability.

ecule undergoes an average of 4×10^9 collisions per second (equivalent to every person in the world colliding with every other person, one at a time, in each second).

Let us consider a heat reservoir at a high temperature and a heat reservoir at a low temperature. The molecules of each are in equilibrium and have the molecular energy distributions shown in Fig. 15–15(a) and (b). If we consider the two reservoirs as a single system, the molecular energy distribution in the system is like that of (c). This distribution is, in a statistical sense, very improbable; if they were mixed together, the molecules of the two reservoirs would soon blend their energies in collisions to attain the equilibrium distribution of (d), which corresponds to a temperature intermediate between the initial ones of the two reservoirs.

We note the important fact that the total energy contents of the distributions of Fig. 15–15(c) and (d) are identical; the only distinction between them is the manner in which the energy is allotted to the molecules on the average. However, it is just this distinction with which the second law of thermodynamics is concerned, because this law states that a system of two heat reservoirs at different temperatures can be made to yield a net work output, while a single heat reservoir, no matter how much energy it contains, cannot be made to perform any net work. A system of molecules whose energies are distributed in the most probable way is "dead" thermodynamically, while a system having a different distribution of molecular energies is capable of doing mechanical work as it progresses to an equilibrium state (Fig. 15–16).

The universe may be thought of as a single system of molecules, and its evolution is powered by the flow of energy from high-temperature reservoirs (the stars) to low-temperature reservoirs (everything else). Ultimately the entire universe will be at the same temperature and all its constituent particles will have the same average energy, a condition sometimes called the "heat death" of the universe.

The second law of thermodynamics is evidently an unusual kind of physical principle. It does not apply to the interactions of individual particles, only to trends in the

Heat reservoirs at different temperatures constitute a thermodynamically improbable system

Probability and the second law of thermodynamics

Heat death of the universe

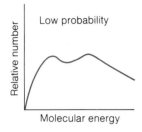

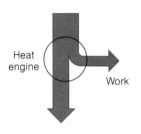

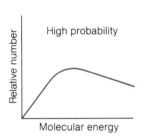

FIG. 15–16 Work can be done by a system whose distribution of molecular energies is statistically improbable.

The sun is the source of most of the energy available on the earth's surface

Fossil fuels furnish most energy today but are being rapidly depleted

evolution of assemblies of many particles. In fact, the second law is hardly a basic principle in the usual sense, since it is the result of combining the laws of mechanics with the theory of probability, and cannot be used to predict anything specific except in the sense that it establishes upper limits to the efficiencies of various processes, shows that certain events have negligible likelihoods of occurrence, and so forth.

However, the second law has the unique property of being correlated with the direction of time. Events that involve individual particles are always reversible—the same laws of motion apply to the billiard balls seen in the film of a game whether the film is run forward or backward. But events that involve systems of large numbers of particles are not always reversible—the film of an egg being dropped makes no sense at all when run backward. The transformation of a broken egg into a whole egg is not totally impossible, it is simply exceedingly unlikely.

The second law of thermodynamics is thus a statement of probability: as time goes on, order becomes disorder in an isolated system. The sequence can be reversed in parts of the universe now and then—after all, heat engines do turn heat into work, simple forms of life evolve into highly complex ones—but in the universe as a whole, which is an isolated system, increasing disorder is inevitable.

15–8 ENERGY AND CIVILIZATION

The development of modern civilization has been paralleled by a steady increase in the world's use of energy. This is no coincidence: all of our activities require energy, and the more energy that is readily available in convenient form, the more effectively we can satisfy our desires for food, clothing, shelter, warmth, illumination, transport, communication, and manufactured goods. Our earliest ancestors had only food as an energy source, and utilized a total of perhaps 8 MJ/day; the mastery of fire and the harnessing of domestic animals yielded an approximately sixfold increase in this figure; the flowering of the industrial revolution a century ago meant a rise in energy consumption to about 300 MJ/day in the more advanced countries; and today in the United States each person uses an average of nearly 1,000 MJ/day. Energy production involved 23% of the country's industrial plant in 1979.

A single source has provided almost all the energy available to us today—the sun. Light and heat arrive directly from the sun; food and wood owe their energy contents to photosynthesis; water power exists because solar heat evaporates water from the oceans to fall later as rain on high ground; wind power is a consequence of convective motions in the atmosphere whose energy source is solar radiation. The fossil fuels coal, oil, and natural gas were formed from plants and animals that lived and stored energy derived from sunlight millions of years ago.

Figure 15–17 shows how rapidly the world's use of energy has increased in recent times and also indicates the relative importance of the chief energy sources. Existing trends have been extrapolated to the year 2000, but it is not possible to go beyond that with any confidence for a number of reasons. An obvious one is that fossil fuels, which today furnish nearly 98% of our energy, cannot last much longer at current rates of consumption. Natural gas, the least polluting of them, will be the first to run out, perhaps soon after the turn of the century. Oil will be next, some decades later: our

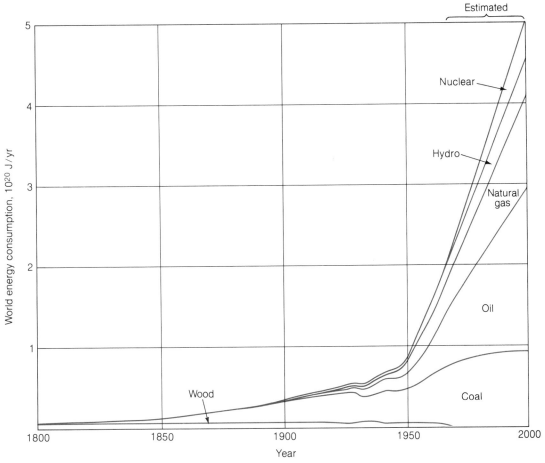

FIG. 15–17 World energy sources and consumption.

descendants will view with disbelief the exhaustion by mere burning of these magnificent feedstocks for synthetic materials of all kinds. Even though the coal currently consumed each year took about two million years to accumulate, enough remains to last another two or three centuries. But coal is far from being a desirable fuel. Not only does its mining leave large tracts of land unfit for further use, but the air pollution due to coal burning is responsible for impairing the health of millions of people through cancer and respiratory diseases. Coal-burning power plants actually expose their surrounding populations to more radioactivity than do normally operating nuclear plants because of the radon gas they liberate.

Nuclear fuel reserves exceed those of fossil fuels, especially since "breeder" reactors already have been built that convert the normally unusable ^{238}U form of uranium into plutonium, which can serve as a reactor fuel (see Sec. 31–10). The energy content of the U.S. stockpile of refined ^{238}U already about equals that in all the country's coal reserves. To be sure, nuclear energy has serious drawbacks. Although the safety record of nuclear installations has been good on an overall basis, even a single malfunction has the potential of doing enormous harm. A power station reactor produces several tons of wastes each year whose radioactivity will remain high for thousands of years,

Nuclear fuel reserves are large but nuclear power also has large drawbacks

and their safe disposal remains a major question. And the widespread production and use of plutonium as a fuel cannot ignore its suitability for nuclear weapons, in many people's eyes a sufficient reason to avoid this technology.

Nuclear fusion is a long-term prospect

In the long run, practical ways to utilize fusion energy seem certain to be developed. As described in Sec. 31–11, a fusion reactor will obtain its fuel from the sea, will be safe and nonpolluting, and cannot be adapted for military purposes. But nobody can predict when this ultimate source of energy will become an everyday reality or how expensive it will be.

Coal is plentiful but is not an ideal fuel

The big question today is how to bridge the transition to fusion energy, a transition whose time scale can only be guessed at. Expanded use of coal can help, but only at great cost—either in terms of environmental damage and human suffering, or in terms of the expense of minimizing them. And any acceleration in the rate of burning fossil fuels will bring with it an increase in the amount of carbon dioxide in the atmosphere, which may well have serious consequences for the world's weather because of the role of this gas in absorbing heat reradiated from the earth's surface. Since 1880 the carbon dioxide content of the atmosphere has gone up by 12%, and currently about 12 billion tons of carbon dioxide pour from our chimneys and exhaust pipes per year. Like coal, nuclear energy can fill the impending energy gap, but also not without hazards that would best be avoided.

Other energy sources

What about the energy of trees and plants, of winds and tides, of the sun's radiation, and of the heat of the earth's interior? The technology already exists, and no resources are depleted. A brief look shows that such energy sources are unlikely to provide more than a modest fraction of future energy needs. In the case of burning wood and vegetable matter in general ("biomass"), the pollution produced (especially of carcinogens) would exceed that due to burning coal of equal energy content, and the diversion of agricultural land from food production in the required quantity would be catastrophic in a time of soaring population. The land needed to provide enough alcohol by conversion of plant sugar or starch to operate one car could provide enough food for 8 to 16 people. Windmills to generate electricity certainly represent a better approach, but locations with sufficiently strong and reliable winds are few and seldom near the locations where the electricity is needed most. The ebb and flow of tides in the Rance River estuary in northern France drive generators of 240 MW capacity, and a few other sites in the world (such as the Bay of Fundy between Maine and Nova Scotia) promise the capture of as much or more tidal energy, but again the potential contribution on a global basis is not impressive. Geothermal plants have operated for some time in California, in Italy, and in New Zealand, but the geological conditions needed for the successful extraction of the earth's heat are not very common.

Solar energy

The direct use of solar radiation is more promising, but problems of cost and scale exist here, too. Even if a substantial fraction of American homes were to be fitted for solar space heating, the energy saving relative to the total energy consumed in the country would be only a percent or two—most desirable, but hardly a solution to the overall problem. The economic production of electricity from sunlight needs more efficient photovoltaic cells and better methods of storing electric energy than now exist, but even if they are developed, the size of the required investment in technology and land area means that only a gradual shift to their use is to be expected.

Clearly there is no magic solution possible in the near future to the problem of safe, cheap, and abundant energy. The only sensible course is to try to get the best from each available technology while proceeding as rapidly as possible to perfect fusion energy. But will the advent of fusion energy solve the energy problem forever? The ultimate limit to world energy production is set by the intrinsic inefficiency of the means by which thermal energy is converted into mechanical energy and thence into electrical energy—in other words, by the second law of thermodynamics. The best of today's power stations have overall efficiencies of only about 35%, and this figure is unlikely to improve by much in the foreseeable future. The waste heat must go somewhere, and even today heavily industrialized countries find its disposal difficult if environmental damage is to be avoided. In the United States about 10% of the flow of all rivers and streams is already being used to provide cooling water for generating plants. The biological consequences of large-scale heating of inland waters are considerable. The oceans can absorb vast amounts of waste heat with minimal side effects, but if most power plants were located on their shores the transmission of electricity inland would then be a major problem. More and more power plants are discharging waste heat into the atmosphere through cooling towers, but here too there are snags in the long run, since local heating of the atmosphere alters the weather and climate of a region, not necessarily for the better.

As energy consumption increases, the disposal of waste heat becomes more difficult

The situation is far from hopeless with respect to both resources and environmental damage, provided the world's total energy requirements do not continue to increase much longer at the present rate. It is precisely here that social rather than technical considerations enter the picture, because to taper off the growth of energy consumption without at the same time limiting population growth (which is by far the most critical problem of the modern world) means a decrease in average living standards, which are too low already in most of the world. No comfortable resolution is in sight to the fundamental conflict between a rising demand for energy from an exploding population and the inability of our planet to provide an unlimited supply.

The real problem is population growth

IMPORTANT TERMS

A **heat engine** is any device that converts heat into mechanical energy or work.

The **first law of thermodynamics** states that the work output of a heat engine is equal to its net heat input.

The **second law of thermodynamics** states that it is impossible to construct an engine, operating in a repeatable cycle, which does nothing other than take energy from a source and perform an equivalent amount of work.

A **Carnot engine** is an idealized engine which is not subject to such practical difficulties as friction or heat losses by conduction or radiation but which obeys all physical laws. No engine operating between the same two temperatures can be more efficient than a Carnot engine operating between them.

IMPORTANT FORMULA

Carnot efficiency: $\text{Eff} = 1 - \dfrac{T_2}{T_1}$

MULTIPLE CHOICE

1. A heat engine operates by taking in heat at a particular temperature and
  (a) converting it all into work.
  (b) converting some of it into work and exhausting the rest at a lower temperature.
  (c) converting some of it into work and exhausting the rest at the same temperature.
  (d) converting some of it into work and exhausting the rest at a higher temperature.

2. The natural direction of heat flow is from a high-temperature reservoir to a low-temperature reservoir, regardless of their respective heat contents. This fact is incorporated in the
(a) first law of thermodynamics.
(b) second law of thermodynamics.
(c) law of conservation of energy.
(d) principle of superposition.

3. A refrigerator
(a) produces cold.
(b) causes heat to vanish.
(c) removes heat from a region and transports it elsewhere.
(d) changes heat to cold.

4. A refrigerator exhausts
(a) less heat than it absorbs from its contents.
(b) the same amount of heat it absorbs from its contents.
(c) more heat than it absorbs from its contents.
(d) any of the above, depending on the circumstances.

5. The work output of every heat engine
(a) equals the difference between its heat intake and heat exhaust.
(b) equals that of a Carnot engine with the same intake and exhaust temperatures.
(c) depends only upon its intake temperature.
(d) depends only upon its exhaust temperature.

6. A Carnot engine turns heat into work
(a) with 100% efficiency.
(b) with 0% efficiency.
(c) without itself undergoing a permanent change.
(d) with the help of expanding steam.

7. In any process, the maximum amount of heat that can be converted to mechanical energy
(a) depends on the amount of friction present.
(b) depends on the intake and exhaust temperatures.
(c) depends on whether kinetic or potential energy is involved.
(d) is 100%.

8. In any process, the maximum amount of mechanical energy that can be converted to heat
(a) depends on the amount of friction present.
(b) depends on the intake and exhaust temperatures.
(c) depends on whether kinetic or potential energy is involved.
(d) is 100%.

9. An adiabatic process in a system is one in which
(a) no heat enters or leaves the system.
(b) the system does no work nor is work done on it.
(c) the temperature of the system remains constant.
(d) the pressure of the system remains constant.

10. A frictionless heat engine can be 100% efficient only if its exhaust temperature is
(a) equal to its input temperature.
(b) less than its input temperature.
(c) 0°C.
(d) 0 K.

11. When a gas is in equilibrium, its molecules
(a) all have the same energy.
(b) have different energies which remain constant.
(c) have a certain constant average energy.
(d) do not collide with one another.

12. A system of molecules whose energies are distributed in the most probable way
(a) can perform an amount of mechanical work equal to its total energy content.
(b) can perform an amount of mechanical work that depends on its absolute temperature.
(c) cannot perform any mechanical work.
(d) is a Carnot engine.

13. The chief source of energy in the world today is
(a) coal.   (b) oil.
(c) natural gas.   (d) uranium.

14. The source of energy whose reserves are greatest is
(a) coal.   (b) oil.
(c) natural gas.   (d) uranium.

15. A Carnot engine absorbs heat at a temperature of 127°C and exhausts heat at a temperature of 77°C. Its efficiency is
(a) 13%.   (b) 39%.
(c) 61%.   (d) 88%.

16. If a heat engine exhausting heat at 100°C is to have an efficiency of 33%, it must take in heat at
(a) 149°C.   (b) 284°C
(c) 300°C.   (d) 557°C.

EXERCISES

1. Two identical watches, one wound and the other unwound, are dropped into beakers of acid and completely dissolved. Is there any difference between the two reactions? Justify your answer using physical principles.

2. The sun's corona is a very dilute gas at a temperature of about 10^6 K that is believed to extend into interplanetary space at least as far as the earth's orbit. Why can we not use the corona as the high-temperature reservoir of a heat engine in an earth satellite?

3. An attempt is made to cool a kitchen in the summer by switching on an electric fan and closing the kitchen door and windows. What will happen?

4. In another attempt to cool the kitchen, the refrigerator door is left open, again with the kitchen door and windows closed. Now what will happen?

5. There are four parts to each cycle of a gasoline engine: (1) a gasoline-air mixture enters the cylinder from the carburetor and is compressed adiabatically by the piston; (2) the mixture is detonated by the spark plug at the instant of maximum compression, and the pressure rises sharply before the piston begins to move outward; (3) as the piston moves out, the burnt gases expand adiabatically; and (4) most of the gases leave the cylinder through the exhaust valves when the piston is at the bottom of its stroke, so this is essentially a constant-volume process. Plot the entire cycle on a *p-V* diagram and indicate in what parts of the cycle heat is absorbed, heat is rejected, and work is done on the outside world.

6. The operation of a steam engine proceeds approximately as follows: (1) water is heated to the boiling point; (2) the water turns into steam and expands at constant pressure at its boiling point; (3) the steam enters the cylinder of the engine and expands adiabatically against the piston; (4) the spent steam condenses into water and is returned to the boiler. Plot the entire cycle on a *p–V* diagram and indicate in what parts of the cycle heat is absorbed, heat is rejected, and work is done on the outside world.

7. An engine operating between 300°C and 50°C is 15% efficient. What would its efficiency be if it were a Carnot engine?

8. A Carnot engine takes in 1 MJ of heat from a reservoir at 327°C and exhausts heat to a reservoir at 127°C. How much work does it do?

9. One of the most efficient engines ever developed operates between about 2000 K and 700 K. Its actual efficiency is 40%. What percentage of its maximum possible efficiency is this?

10. An engine is proposed which is to operate between 200°C and 50°C with an efficiency of 35%. Will the engine perform as predicted? If not, what would its maximum efficiency be?

11. A Carnot engine absorbs 200 kcal of heat at 500 K and exhausts 150 kcal. What is the exhaust temperature?

12. In a certain power station coal is consumed at the rate of 0.4 kg/h for each kilowatt of electrical output. Find the overall efficiency of the power station.

13. A typical turboprop aircraft engine consumes 250 g/h of kerosene for each horsepower it develops. Find the efficiency of the engine.

14. How many cubic meters of propane must be burned to heat 50 liters of water from 5°C to 90°C? Assume that 25% of the heat is wasted.

PROBLEMS

1. A Carnot engine whose efficiency is 35% takes in heat at 500°C. What must the intake temperature be if the efficiency is to be 50% with the same exhaust temperature?

2. Three designs for a heat engine to operate between 450 K and 300 K are proposed. Design *A* is claimed to require a heat input of 0.2 kcal for each 1000 J of work output, design *B* a heat input of 0.6 kcal, and design *C* a heat input of 0.8 kcal. Which design would you choose and why?

3. The total drop of the Wollomombi Falls in Australia is 482 m. What would be the Carnot efficiency of an engine operating between the top and bottom of the falls if the water temperature at the top were 10°C and all the potential energy of the water at the top were converted to heat at the bottom?

4. A coal-fired electric power plant of 400 MW output operates at an efficiency of 39%, which is about three-quarters of its Carnot efficiency. Water from a river flowing at 50 m^3/s is used to absorb the waste heat. By how much does the river's temperature rise as a result?

5. Starting from the definition of work, show that the amount of work done by a gas that expands by ΔV at the constant pressure *p* is $W = p \, \Delta V$.

6. An adiabatic process is one in which heat neither enters nor leaves the system in which the process occurs. A gas sample expands from V_1 to V_2. Does it perform the most work when the expansion takes place at constant pressure, at constant temperature, or adiabatically? In which process does the gas perform the least work?

7. Use the result of Problem 5 to find the percentage of the heat of vaporization of water that represents the work involved in expanding water into steam against the pressure of the atmosphere. At 100°C and atmospheric pressure the density of steam is 0.6 kg/m^3.

8. The power output of each cylinder of a gasoline or diesel engine depends upon four factors: the average pressure *p* on the piston during a power stroke; the length *L* of piston travel;

the area A of the piston; and the number N of power strokes per second. Show that the power output is given by $P = pLAN$.

9. A four-cylinder, two-stroke diesel engine has pistons 108 mm in diameter whose travel is 127 mm. The engine develops 100 kW at 33 rev/s. Find the average pressure on the pistons during the power stroke.

10. An ideal refrigerator is a Carnot engine operating backwards. If an ideal refrigerator extracts heat from a storage chamber at the absolute temperature T_1 and ejects heat to the outside world at the absolute temperature T_2, show that the ratio between the work done on the refrigerator and heat extracted is $T_2/T_1 - 1$.

11. A Carnot refrigerator extracts heat from a freezer at $-5°C$ and exhausts it at $25°C$. How much work per joule of heat extracted is required?

12. A Carnot refrigerator is used to make 1 kg of ice at $-10°C$ from 1 kg of water at $20°C$, which is also the temperature of the kitchen. How many joules of work must be done?

13. Three designs for a refrigerator to operate between $-20°C$ and $40°C$ are proposed. Design A is claimed to require 300 J of work for each kcal of heat extracted, design B to require 950 J, and design C to require 2000 J. Which design would you choose and why?

ANSWERS TO MULTIPLE CHOICE

1. b	**5.** a	**9.** a	**13.** b
2. b	**6.** c	**10.** d	**14.** d
3. c	**7.** b	**11.** c	**15.** a
4. c	**8.** d	**12.** c	**16.** b

ELECTRICITY

16

The success of the laws of motion, of the law of gravitation, and of the kinetic-molecular theory of matter might tempt us into thinking that we now have, at least in outline, a complete picture of the workings of the physical universe. To dispel this notion all we need do is perform a simple experiment: on a dry day, we run a hard rubber comb through our hair, and find that the comb is now able to pick up small bits of paper and lint. The attraction is surely not due to gravity, because the gravitational force between comb and paper is far too small and should not, in any event, depend upon whether the comb is run through our hair or not. What has been revealed by this experiment is an electrical phenomenon, so called after *elektron,* the Greek word for amber, a substance used in the earliest studies of electricity.

16–1   ELECTRIC CHARGE

Electricity is familiar to all of us as the name for that which causes our light bulbs to glow, many of our motors to turn, our telephones and radios to

FIG. 16–1 A rubber rod stroked with fur becomes negatively charged; two negatively charged objects repel each other.

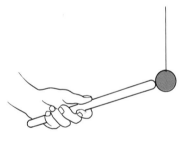

A pith ball suspended by a fine string is touched by a hard rubber rod. Nothing happens.

The rubber rod is stroked with a piece of fur.

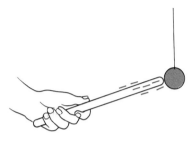

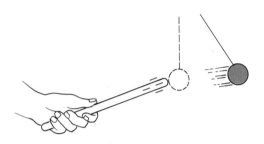

The pith ball is again touched by the rubber rod.

After the touch, the ball flies away from the rod.

communicate sounds, our television screens to communicate pictures. But there is more to electricity than its technological uses. Electrical forces bind electrons to nuclei to form atoms, and they hold atoms together to form molecules, solids, and liquids. All of the chief properties of matter in bulk—with the notable exception of mass—can be traced to the electrical nature of its constituent particles.

Let us begin our study of electricity by examining three basic experiments. The first experiment is shown in Fig. 16–1. By convention, we call whatever it is that a rubber rod possesses by virtue of having been stroked with a piece of fur *negative electric charge*. Part of the negative charge on the rubber rod of Fig. 16–1 flowed to the pith ball when it was touched, and the fact that the ball then flew away from the rod suggests that negative electric charges repel each other.

The next experiment, shown in Fig. 16–2, is very similar. By convention, we call whatever it is that a glass rod possesses by virtue of having been stroked with a silk cloth *positive electric charge*. Part of the positive charge on the glass rod of Fig. 16–2 flowed to the pith ball when it was touched, and the fact that the ball then flew away from the rod suggests that positive electric charges repel each other.

Why is it assumed that the electric charge on the glass rod is different from that on the rubber rod? The reason lies in the result of the third experiment, which is shown

Negative charges repel each other

Positive charges repel each other

Unlike charges attract each other

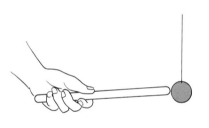

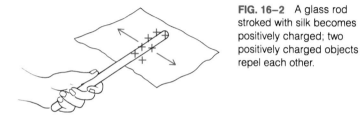

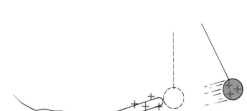

FIG. 16–2 A glass rod stroked with silk becomes positively charged; two positively charged objects repel each other.

A second pith ball is touched by a glass rod. Nothing happens.

The glass rod is stroked with a silk cloth.

The pith ball is again touched by the glass rod.

After the touch the ball flies away from the rod.

in Fig. 16–3. The attraction of the two pith balls means (1) that the charges they carry are different, since like charges have already been observed to repel, and (2) that unlike charges attract each other.

The preceding results can be summarized very simply:

Like charges repel; unlike charges attract.   **Behavior of charges**

Where do the charges come from when one substance is stroked with another? When we charge one pith ball with a rubber rod and another with the fur the rod was stroked with, we find (Fig. 16–4) that the two balls attract; since the rubber rod is

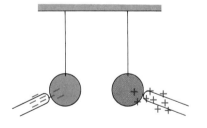

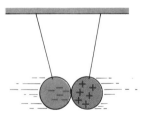

FIG. 16–3 Objects with unlike charges attract each other.

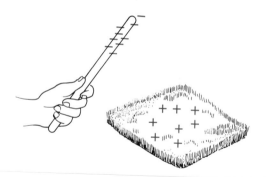

FIG. 16–4 The process of stroking a rubber rod with a piece of fur serves to separate charges so that the rod becomes negatively charged and the fur positively charged.

negatively charged, this experiment indicates that the fur is positively charged. A similar experiment with a glass rod and a silk cloth indicates that the cloth acquires a negative charge during the stroking. Evidently the process of stroking serves to *separate* charges. We might infer that rubber has an affinity of some kind for negative charges and fur an affinity for positive charges, so that, when rubbed together, each tends to acquire a different kind of charge.

A great many experiments with a variety of substances have shown that there are only the two kinds of electric charge, positive and negative, that we have spoken of. All electrical phenomena involve either or both kinds of charge. An "uncharged" body of matter actually possesses equal amounts of positive and negative charge, so that appropriate treatment—mere rubbing is sufficient for some substances—can leave an excess of either kind on the body and thereby cause it to exhibit electrical effects.

Only two kinds of charge exist

What is electric charge? All that can be said is that charge, like rest mass, is a fundamental property of certain of the elementary particles of which all matter is composed. Three types of particle are found in atoms, the positively charged *proton,* the negatively charged *electron,* and the neutral (that is, uncharged) *neutron.* The proton and electron have exactly equal amounts of charge, though of opposite sign. Neutrons and protons have nearly the same mass, which is almost 2000 times greater than the electron mass (Table 16–1). An atom normally contains equal numbers of protons and electrons, so it is electrically neutral unless disrupted in some way.

Ordinary matter consists of neutrons, protons, and electrons

TABLE 16–1
Neutrons, protons, and electrons are the constituents of atoms. An atom contains equal numbers of protons and electrons, so it is electrically neutral. The neutrons and protons are responsible for most of an atom's mass.

	Neutron	Proton	Electron
Charge	0	$+e = 1.60 \times 10^{-19}$ C	$-e = 1.60 \times 10^{-19}$ C
	Uncharged	Equal in magnitude, opposite in sign	
Mass	1.675×10^{-27} kg	1.673×10^{-27} kg	9.11×10^{-31} kg
	Nearly equal		Electron mass is 1/1836 of proton mass

The *principle of conservation of charge* states that

The net electric charge in an isolated system remains constant.

By "net charge" is meant the algebraic sum of the charges present—the total positive charge minus the total negative charge. Net charge can be positive, negative, or zero.

Every known physical process in the universe conserves electric charge. Separating or bringing together charges does not affect their magnitudes, so such rearrangements leave the net charge unaffected. Under certain circumstances matter can be created from energy, but whenever this happens, the number of positively charged particles created is always exactly the same as the number of negatively charged particles created. Under other circumstances matter can be completely converted into energy, and in such events the number of positively charged particles that disappear is again always exactly the same as the number of negatively charged particles. Unlike rest mass, charge is invariably conserved.

Electric charge is the fourth quantity we have studied which is conserved in every physical process. The others are energy, linear momentum, and angular momentum, and their conservation can be traced to fundamental symmetry properties of nature. These properties are, respectively, the independence of physical laws to shifts in time, in space, and in orientation. Conservation of charge is also associated with a symmetry property of nature, although this property is too abstract to be described here.

16–2 COULOMB'S LAW

In order to arrive at the law of gravitation

$$F = G\,\frac{m_A m_B}{r^2}$$

Newton had to make use of astronomical data and an indirect argument, because gravitational forces are appreciable only when the masses involved are very large. However, the law that electrical forces obey can be readily determined in the laboratory, because these forces are so much greater in magnitude than gravitational ones.

The law of force between charges was first published by the eighteenth century French scientist Charles Coulomb, and is called Coulomb's law in his honor. If we use the symbol Q for electric charge, Coulomb's law for the magnitude F of the force $\mathbf{F}$ between two charges Q_A and Q_B the distance r apart states that

$$F = k\,\frac{Q_A Q_B}{r^2} \qquad\qquad \textit{Coulomb's law} \quad (16\text{–}1)$$

The force between two charges is proportional to both of the charges and is inversely proportional to the square of the distance between them. The quantity k is a constant whose value depends upon the units employed and upon the medium (air, vacuum, oil, and so forth) in which the charges are located.

Electric force is, of course, a vector quantity, and the formula above gives only its magnitude. The direction of $\mathbf{F}$ is always along the line joining Q_A and Q_B, and the

FIG. 16–5 The electric force one charge exerts on another is always along the line joining the two charges. The force is attractive if the charges have opposite signs, repulsive if they have the same sign.

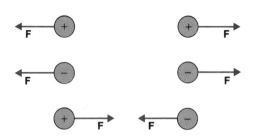

The coulomb is the unit of charge

force is attractive if Q_A and Q_B have opposite signs and repulsive if they have the same signs (Fig. 16–5).

The unit of electric charge is the *coulomb* (abbreviated C). The formal definition of the coulomb, given in Chapter 19, is expressed in terms of magnetic forces. A more realistic way to think of the coulomb is in terms of the number of individual elementary charges that add up to this amount of charge. All charges, both positive and negative, occur only in multiples of 1.60×10^{-19} C. No elementary particle has ever been found with a charge of other than $\pm 1.60 \times 10^{-19}$ C or 0. The electron has a charge of -1.60×10^{-19} C; the proton has a charge of $+1.60 \times 10^{-19}$ C. If 6.25×10^{18} electrons were assembled, the total charge would be -1 C; if 6.25×10^{18} protons were assembled, the total charge would be $+1$ C (Fig. 16–6).

Because electric charge always occurs in multiples of $\pm 1.60 \times 10^{-19}$ C, this amount of charge has been given a special name, the electron (or electronic) charge, and a special symbol, e:

$$e = 1.60 \times 10^{-19} \text{ C} \qquad \qquad \textit{Electron charge}$$

Thus a charge of $+1.60 \times 10^{-19}$ C is abbreviated $+e$, and one of -3.20×10^{-19} C is abbreviated $-2e$.

In most processes that lead to a net charge on some object, electrons are either added to it or removed from it. Hence we can think of an object whose charge is -1 C as having 6.25×10^{18} electrons more than its normal number, and of an object whose charge is $+1$ C as having 6.25×10^{18} electrons less than its normal number. (By "normal number" is meant a number of electrons equal to the number of protons present, so that the object has no net charge.)

When Q_A and Q_B in Coulomb's law are expressed in coulombs and r in meters, the constant k has the value in vacuum of

$$k = 9.0 \times 10^9 \text{ N·m}^2/\text{C}^2$$

The value of k in air is very slightly greater.

The constant k is often written

$$k = \frac{1}{4\pi \varepsilon_0}$$

where ε_0, called the permittivity of free space, is equal to

$$\varepsilon_0 = 8.85 \times 10^{-12} \text{ C}^2/\text{N·m}^2$$

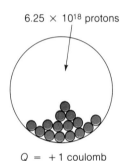

6.25 × 10¹⁸ protons

$Q = +1$ coulomb

6.25 × 10¹⁸ electrons

$Q = -1$ coulomb

FIG. 16–6

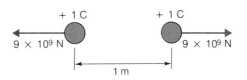

FIG. 16–7

Example   Find the force between two charges of 1 C each that are 1 m apart (Fig. 16–7).

Solution   From Coulomb's law,

$$F = k\,\frac{Q_A Q_B}{r^2} = \left(9 \times 10^9 \frac{\text{N·m}^2}{\text{C}^2}\right)\left(\frac{1\,\text{C} \times 1\,\text{C}}{1\,\text{m}^2}\right) = 9 \times 10^9\,\text{N}$$

This force is equal to about 2 billion lb. Evidently even the most highly charged objects that can be produced seldom contain more than a minute fraction of a coulomb of net charge of either sign. ■

The coulomb is a very large unit

At the beginning of this chapter a familiar observation was noted: a hard rubber comb that has been charged by being passed through someone's hair on a dry day is able to attract small bits of paper. Since the paper bits were originally uncharged, how could the comb exert a force on them?

The explanation depends upon Coulomb's law (Fig. 16–8). When the negatively charged comb is brought near the paper, some of the negative charges in the paper that are not tightly bound in place move as far away as they can from the comb, while some of the positive charges that are not tightly bound move toward the comb. Because electrical forces vary inversely with distance, the attraction between the comb and the closer positive charges is greater than the repulsion between the comb and the farther negative charges, and so the paper moves toward the comb. Only a small amount of charge separation actually occurs, and so, with little force available, only very light objects can be attracted in this way.

How a charge attracts an uncharged object

16–3 MULTIPLE CHARGES

When more than two charges are in the same region, the force on any one of them may be calculated by adding vectorially the forces exerted on it by each of the others.

Vector addition of electric forces

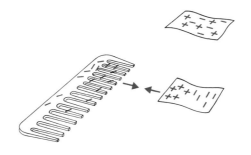

FIG. 16–8   How a charged object attracts an uncharged one.

FIG. 16–9

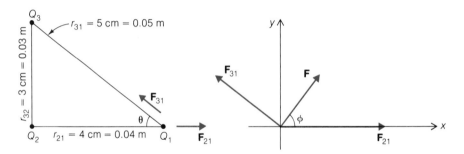

Usually the component method of vector addition provides the most straightforward means of carrying out the calculation.

Example   Three charges, $Q_1 = +2 \times 10^{-9}$ C, $Q_2 = +4 \times 10^{-9}$ C, and $Q_3 = -5 \times 10^{-9}$ C, are located as shown in Fig. 16–9. Find the magnitude and direction of the net force on Q_1.

Solution   If we call $\mathbf{F}_{21}$ the repulsive force exerted by Q_2 on Q_1 and $\mathbf{F}_{31}$ the attractive force exerted by Q_3 on Q_1, Coulomb's law yields, for the magnitudes of these forces,

$$F_{21} = \frac{kQ_2Q_1}{r_{21}^2} = \frac{(9 \times 10^9 \, \text{N·m}^2/\text{C}^2)(4 \times 10^{-9}\,\text{C})(2 \times 10^{-9}\,\text{C})}{(4 \times 10^{-2}\,\text{m})^2} = 4.5 \times 10^{-5}\,\text{N}$$

$$F_{31} = \frac{kQ_3Q_1}{r_{31}^2} = \frac{(9 \times 10^9 \, \text{N·m}^2/\text{C}^2)(5 \times 10^{-9}\,\text{C})(2 \times 10^{-9}\,\text{C})}{(5 \times 10^{-2}\,\text{m})^2} = 3.6 \times 10^{-5}\,\text{N}$$

The directions of $\mathbf{F}_{21}$ and $\mathbf{F}_{31}$ are parallel to the 4-cm and 5-cm sides of the triangle.

In order to determine $\mathbf{F}$, the net force on Q_1, we first resolve $\mathbf{F}_{21}$ and $\mathbf{F}_{31}$ into components. We have, since $\sin \theta = \frac{3}{5}$ and $\cos \theta = \frac{4}{5}$,

$$F_{21x} = F_{21} = 4.5 \times 10^{-5}\,\text{N}$$

$$F_{21y} = 0$$

$$F_{31x} = F_{31} \cos \theta = \tfrac{4}{5} F_{31} = -2.9 \times 10^{-5}\,\text{N}$$

$$F_{31y} = F_{31} \sin \theta = \tfrac{3}{5} F_{31} = 2.1 \times 10^{-5}\,\text{N}$$

Hence the components of $\mathbf{F}$ are

$$F_x = F_{21x} + F_{31x} = 1.6 \times 10^{-5}\,\text{N}$$

$$F_y = F_{21y} + F_{31y} = 2.1 \times 10^{-5}\,\text{N}$$

and the magnitude of $\mathbf{F}$ is accordingly

$$F = \sqrt{F_x^2 + F_y^2} = 2.6 \times 10^{-5}\,\text{N}$$

The direction of $\mathbf{F}$ can be specified in various ways. If ϕ is the angle between $\mathbf{F}$ and the $+x$-axis, then

$$\phi = \tan^{-1} \frac{F_y}{F_x} = \tan^{-1} 1.31 = 53°$$

■

16–4 ELECTRICITY AND MATTER

Coulomb's law for the electric force between charges is very similar to Newton's law for the gravitational force between masses. The most striking difference is that electric forces may be either attractive or repulsive, whereas gravitational forces are always attractive. The latter fact means that matter in the universe tends to come together to form large bodies, such as stars and planets, and these bodies are always found in groups, such as galaxies of stars and families of planets.

Gravitational forces dominate on a large scale

There is no comparable tendency for electric charges of either sign to come together; quite the contrary. Unlike charges attract strongly, which makes it hard to separate neutral matter into portions of opposite signs. Furthermore, like charges repel, so it becomes harder and harder to add further charge to an already charged object. Hence the large-scale structure of the universe is largely governed by gravitational forces.

Electric forces dominate on a small scale

On an atomic scale, though, the relative importance of gravity and electricity is reversed. Elementary particles are so tiny that the gravitational forces between them are insignificant, whereas their electric charges are sufficiently great for electric forces to govern the structures of atoms, molecules, liquids, and solids.

Example The hydrogen atom has the simplest structure of all atoms, consisting of a proton and an electron whose average separation is 5.3×10^{-11} m. (For the time being we can think of the electron as circling the proton much as the moon circles the earth, as in Fig. 16–10. A more realistic model of the hydrogen atom—but one that is harder to visualize—will be given later.) Compare the electrical and gravitational forces between the proton and the electron in a hydrogen atom.

The hydrogen atom consists of a single proton and a single electron

Solution The electrical force between the electron and proton is

$$F_e = k\frac{Q_e Q_p}{r^2} = \frac{(9.0 \times 10^9 \,\text{N}\cdot\text{m}^2/\text{C}^2)(1.6 \times 10^{-19}\,\text{C})^2}{(5.3 \times 10^{-11}\,\text{m})^2}$$
$$= 8.2 \times 10^{-8}\,\text{N}$$

while the gravitational force between them is

$$F_g = G\frac{m_e m_p}{r^2} = \frac{(6.7 \times 10^{-11}\,\text{N}\cdot\text{m}^2/\text{kg}^2)(9.1 \times 10^{-31}\,\text{kg})(1.7 \times 10^{-27}\,\text{kg})}{(5.3 \times 10^{-11}\,\text{m})^2}$$
$$= 3.7 \times 10^{-47}\,\text{N}$$

The electrical force is over 10^{39} times greater than the gravitational force. Clearly the electrical forces that subatomic particles exert upon one another are so much stronger than their mutual gravitational ones that the latter can be neglected completely. ∎

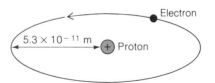

FIG. 16–10 A simple model of the hydrogen atom.

16–5   ATOMIC STRUCTURE

By the beginning of this century a substantial body of evidence had been accumulated in support of the idea that the chemical elements consist of atoms. The nature of the atoms themselves, however, was still a mystery, although a significant clue had been discovered. This clue was the fact that electrons are constituents of atoms, which suggests that electrial forces are involved in atomic phenomena. J. J. Thomson, whose work had led to the identification of the electron, proposed in 1898 that atoms are spheres of positively charged matter that contain embedded electrons, much as a fruit cake is studded with raisins. The reality turned out to be very different.

The Rutherford experiment established the structure of the atom

The most direct way to find out what is inside a fruitcake is simply to plunge a finger into it. In essence this is the classic experiment performed in 1911 by Geiger and Marsden at the suggestion of Ernest Rutherford. The probes they used were fast *alpha particles* spontaneously emitted by certain radioactive elements. For the present all we need to know about alpha particles is that they consist of two neutrons and two protons held tightly together, so that each one has a charge of $+2e$.

Geiger and Marsden placed a sample of an alpha-emitting substance behind a lead screen with a small hole in it, so that a narrow beam of alpha particles was produced. On the other side of a thin metal foil in the path of the beam they placed a zinc sulfide screen that gave off a flash of light when struck by an alpha particle, thus indicating the extent to which the alpha particles were scattered from their original direction of motion (Fig. 16–11).

Geiger and Marsden expected to find that most of the alpha particles go through the foil without being affected by it, with the remainder receiving only slight deflections. If the positive and negative charges within an atom are spread more or less evenly throughout its volume, only weak electric forces would be exerted on alpha particles passing through a thin foil, and their momenta would be enough to carry them through with only minor departures—at most 1° or so—from their original paths.

What Geiger and Marsden actually found was that, while most of the alpha particles indeed emerged unaffected from the foil, the others underwent deflections through very large angles, in some cases even being scattered in the backward direction. As Rutherford remarked, "It was almost as incredible as if you fired a 15-inch shell at a piece of tissue paper and it came back and hit you." Since alpha particles are relatively heavy (almost 8000 times more massive than electrons) and have fairly high initial speeds (typically 2×10^7 m/s), it was clear that strong forces had to be exerted upon them to cause such marked deflections. The only atomic model able to account for such forces is one that consists of a tiny *nucleus* in which its positive charge and nearly all of its mass are concentrated, with the electrons some distance away (Fig. 16–12).

Only the nuclear model of the atom can explain alpha-particle scattering

With the atom largely empty space, it is easy to see why most alpha particles proceed right through a thin foil. On the other hand, an alpha particle that happens to come near a nucleus experiences a strong electric force and is likely to be scattered through a large angle. (The atomic electrons, being very light, are readily knocked out of the way by alpha particles, while the situation is reversed for the nuclei, which are heavier than alpha particles.) Rutherford was able to obtain a formula for the scattering of alpha particles by thin foils on the basis of his hypothesis that agreed with the experimental results. He is therefore credited with the discovery of the nucleus.

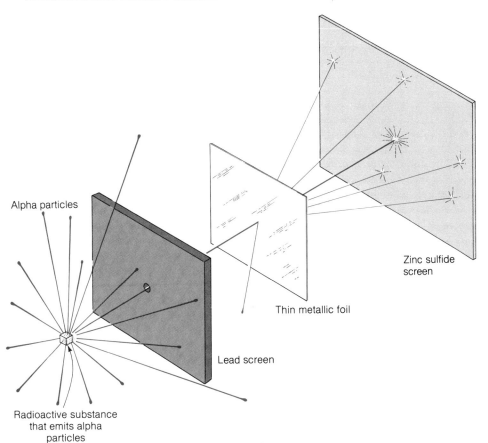

FIG. 16–11 Diagram of the Rutherford experiment.

Alpha particles

Zinc sulfide screen

Thin metallic foil

Lead screen

Radioactive substance that emits alpha particles

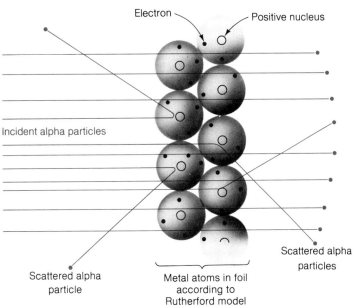

Electron

Positive nucleus

Incident alpha particles

Scattered alpha particle

Metal atoms in foil according to Rutherford model

Scattered alpha particles

FIG. 16–12 According to the Rutherford model of the atom, positive charge is concentrated in a tiny nucleus at its center with electrons some distance away. Strong electric forces can occur within atoms on the basis of this model, and it accordingly predicts considerable deflections of some of the alpha particles striking a thin foil. This prediction agrees with experiment.

16–6   ELECTRICAL CONDUCTION

Most substances conduct electricity either very well or very badly

An electric current is a flow of charge from one place to another. Nearly all substances fall into two categories: *conductors,* through which charge can flow easily, and *insulators,* through which charge can flow only with great difficulty. Metals, many liquids, and plasmas (gases whose molecules are charged) are conductors, whereas nonmetallic solids, certain liquids, and gases whose molecules are electrically neutral are insulators. Several substances, called *semiconductors,* are intermediate in their ability to conduct charge.

Conduction in a metal

In a solid metal, each atom gives up one or more electrons to a common "gas" of freely moving electrons that pervades the entire metal. These electrons can migrate quite readily through the crystal structure of the metal, so if one end of a metal wire is given a positive charge and the other end a negative charge, electrons will flow through the wire from the negative to the positive end. This flow, of course, constitutes an electric current. By supplying new electrons to the negative end of the wire and removing electrons from the positive end as they arrive there—which can be done by connecting the wire to a battery or to a generator—a constant current can be maintained in the wire.

Insulators

In nonmetallic solids, such as salt, glass, rubber, minerals, wood, and plastics, all the atomic electrons are bound to particular atoms or groups of atoms and cannot move from place to place. Such solids are accordingly classed as insulators. Actually, nonmetallic solids do conduct very small amounts of current, but their abilities to do this are vastly inferior to those of metals. For instance, when identical bars of copper and sulfur are connected to the same battery, about 10^{23} times more current flows in the copper bar.

Semiconductors

As mentioned earlier, there are a few substances called semiconductors through which current flows more readily than through insulators but still with distinctly more difficulty than through conductors. Thus about 10^7 times more current flows in a germanium bar connected to a battery than in a sulfur bar of the same size, but this is still about 10^{16} times less current than in a copper bar. The electrical conductivity of solids is discussed in more detail in Chapter 30.

Superconductivity is a low-temperature phenomenon

At temperatures near absolute zero (0 K, which is $-273°C$) certain metals, alloys, and chemical compounds lose all of their resistance to the flow of electric current. This phenomenon, called *superconductivity,* was discovered by Kamerlingh Onnes in Holland in 1911. For example, aluminum is superconducting at temperatures under 1.20 K, lead at temperatures under 7.22 K, and CuS (copper sulfide) at temperatures under 1.6 K. If a current is set up in a closed wire loop at room temperature, it will die out in less than a second even if the wire is made of a good conductor such as copper or silver, whereas if the wire is made of a superconducting material and is kept cold enough, the current will continue indefinitely. Currents have persisted in superconducting loops with no apparent diminution for several years.

Superconductivity is of immense potential importance for the transmission of electric energy and in applications where strong magnetic fields are required. Already laboratory electromagnets with superconducting coils are in use, and an experimental electric motor whose windings are superconducting has been built in England. There is no fundamental reason why superconducting magnets cannot be used to levitate trains and thereby both increase their speeds and reduce their power requirements. The

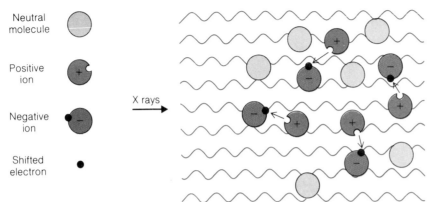

Neutral molecule

Positive ion

X rays

Negative ion

Shifted electron

FIG. 16-13 Schematic representation of the ionization of air by X rays. A molecule losing an electron becomes a positive ion; a molecule gaining an electron becomes a negative ion.

immediate problem is that the best materials for the purpose thus far discovered exhibit superconductivity only at temperatures under about 20 K, although it is possible that in time superconductors will be developed that function at more practical temperatures.

The mechanism of electrical conduction in liquids and gases is different from that in metals. The current in a metal consists of a flow of electrons past the stationary atoms in its structure. The current in a fluid medium other than a liquid metal, however, consists of a flow of entire atoms or molecules that are electrically charged. An atom or molecule that carries a net charge is called an *ion,* and both positive and negative ions participate in the conduction process in liquids and gases.

Conduction in liquids and gases

An atom or molecule becomes a positive ion when it loses one or more of its electrons; if it gains one or more electrons in addition to its usual complement, it becomes a negative ion. The fundamental positive charges in matter are protons, which are so tightly bound in the nucleus of every atom that they can be dislodged only under exceptional circumstances. Atomic electrons, however, are held more loosely, and one or two of them can be detached from an atom with relative ease. Thus the oxygen and nitrogen gases in ordinary air become ionized when a spark occurs, in the presence of a flame, and by the passage of x-rays or even ultraviolet light. These processes so disturb the air molecules that some electrons are dislodged, leaving behind positive ions. The liberated electrons almost at once become attached to other nearby molecules to create negative ions (Fig. 16-13).

A positive ion has a deficiency of electrons; a negative ion has a surplus of electrons

The electrical attraction between positive and negative charges in time brings the ions together, and the extra electrons on the negative ions become reattached to the positive ions. The gas molecules are then neutral, as they were originally. This *recombination* is rapid at normal atmospheric pressure and temperature.

Recombination

In the upper atmosphere, where air molecules are so far apart that the recombination of ions is a slow process, the continual bombardment of X rays and ultraviolet light from the sun maintains a perceptible proportion of ions at all times. The layers of ions in the upper atmosphere constitute the *ionosphere,* and they make possible long-range radio communication by their ability to reflect radio waves (see Fig. 23-9). The ionosphere is an example of a *plasma,* which is a gas whose constituent particles are electrically charged. The behavior of a plasma, unlike that of an ordinary gas, is strongly influenced by electric and magnetic forces. Most of the universe is in the plasma state.

Origin of the ionosphere

16–7   IONS IN SOLUTION

Many liquids contain positive and negative ions at all times and hence are able to conduct electricity. Let us look into how the ions in a liquid come into being and how they are able to resist the recombination that occurs so readily in a gas.

Polar and nonpolar molecules

When atoms join together to form a molecule, their electrons are shifted in such a manner that electric forces hold the atoms together. We shall consider the details of the binding process in Chapter 30, but for the moment it is sufficient for us to note that certain molecules have asymmetrical (nonsymmetrical) distributions of charge and behave as though negatively charged at one end and positively charged at the other. A molecule of this kind is called a *polar molecule*; the water molecule is an example (Fig. 16–14). A *nonpolar molecule,* on the other hand, has a uniform distribution of charge. All molecules are normally electrically neutral, and the distinction between the polar and nonpolar varieties lies solely in the way their electrons are arranged.

Water is a polar liquid; gasoline is a nonpolar liquid

The fact that polar molecules exist helps to explain a number of familiar phenomena. The behavior of compounds in solution is a good example. Water readily dissolves such compounds as salt and sugar, but cannot dissolve fats or oils. Gasoline readily dissolves fats and oils, but cannot dissolve salt or sugar. The key to these differences lies in the strongly polar nature of water molecules and the nonpolar nature of gasoline molecules. Water molecules tend to form aggregates under the influence of the electric forces between the ends of adjacent molecules, as shown in Fig. 16–15.

Polar molecules of other substances, such as sugar, can join in the aggregates of water molecules, and are therefore easily dissolved by water (Fig. 16–16). The nonpolar molecules of fats and oils, however, do not interact with water molecules. If samples

FIG. 16–14   The end of a water molecule where the hydrogen atoms are attached behaves as if positively charged, and the opposite end behaves as if negatively charged. The water molecule is therefore polar.

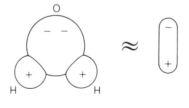

FIG. 16–15   Because water molecules are polar, they tend to clump together under the influence of electric forces.

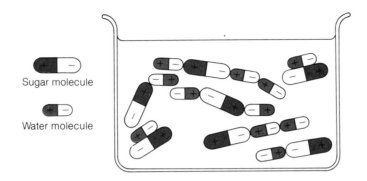

FIG. 16–16 Polar components such as sugar dissolve in water because their molecules can link up with water molecules.

of oil and water are mixed together, the attraction of water molecules for one another acts to squeeze out the oil molecules, and the mixture soon separates into layers of each substance. Fat and oil molecules dissolve only in liquids whose molecules are similar to theirs, which is why gasoline is a solvent for these compounds (Fig. 16–17). In general, then, "like dissolves like."

Like dissolves like

Many solid compounds have structures that consist of ions rather than of neutral atoms. Thus the sodium chloride (NaCl) of ordinary salt consists of Na^+ and Cl^- ions in the regular geometrical array shown in Fig. 16–18. (The symbol Na^+ refers to a sodium atom that has lost an electron to leave it with a net charge of $+e$, and the symbol Cl^- refers to a cholrine atom that has gained an electron to give it a net charge of $-e$.)

When a crystal of an ionic compound such as NaCl is placed in water, the water molecules cluster around the crystal's ions with their positive ends toward negative ions and their negative ends toward positive ions. The attraction of several water molecules is usually sufficient to pull an ion from the rest of the crystal, and it moves away surrounded by water molecules (Fig. 16–19). The resulting solution contains ions rather than molecules of the dissolved compound.

How water dissolves an ionic solid

Substances that separate into free ions when dissolved in water are called *electrolytes* since they are able to conduct electric current by the migration of positive and negative ions. All ionic compounds soluble in water and certain other soluble compounds, such as HCl, are electrolytes. Still other compounds, such as sugar, are nonelectrolytes even though they are soluble in water.

FIG. 16–17 Nonpolar compounds dissolve only in nonpolar liquids. Thus oil dissolves in gasoline but not in water.

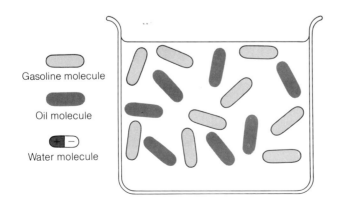

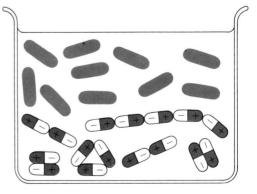

FIG. 16–18 A sodium chloride crystal consists of NA⁺ and Cl⁻ ions in a regular geometrical arrangement.

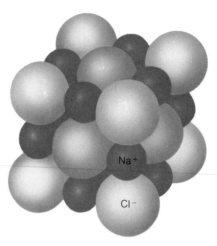

FIG. 16–19 The solution of solid NaCl.

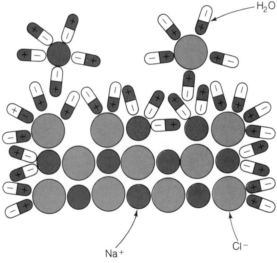

Ions in solution have their own characteristic properties

Since the outer electron structure of an ion may be very different from that of a neutral atom of the same species, it is not surprising that the ions of an element may behave very differently from its atoms or molecules. Thus gaseous chlorine is greenish in color, has a strong, irritating taste, and is very active chemically, while a solution of chlorine ions is colorless, has a mild, pleasant taste, and is only feebly active.

16–8   ELECTROLYSIS

Let us inquire into the effect of passing an electric current through a liquid containing ions. Because water itself may participate in the events that occur in a solution, for simplicity we shall first consider molten NaCl rather than a NaCl solution. When a

Cathode

Anode

−

+

Molten sodium
chloride

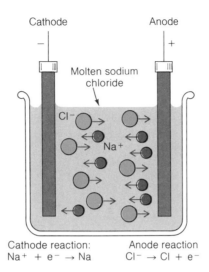

Cl^-

Na^+

Cathode reaction:
$Na^+ + e^- \rightarrow Na$

Anode reaction
$Cl^- \rightarrow Cl + e^-$

FIG. 16–20 The
electrolysis of molten
sodium chloride.

current flows through a bath of molten NaCl, as in Fig. 16–20, metallic sodium is observed to deposit out at the negative electrode (or *cathode*) and gaseous chlorine to bubble up from the positive electrode (or *anode*). (At the temperature of molten NaCl, metallic sodium is also in the liquid state.) These results are not hard to understand in view of the presence of free Na^+ and Cl^- ions in the bath. The negative electrode attracts Na^+ ions and, when they arrive, neutralizes them by transferring an electron to each one:

$$Na^+ + e^- \rightarrow Na$$

The resulting sodium atoms, unlike sodium ions, are not soluble and appear at this electrode as ordinary metallic sodium. The positive electrode at the same time attracts Cl^- ions and neutralizes them by absorbing an electron from each one:

$$Cl^- \rightarrow Cl + e^-$$

The insoluble chlorine atoms are evolved as chlorine gas. The entire phenomenon is an example of *electrolysis,* the process by which free elements are liberated from a liquid by the passage of an electric current.

A widely used application of electrolysis is the depositing, or *plating,* of a thin layer of one metal on an object made of another metal. Sometimes this is done because the plating metal is expensive, for instance, gold or silver; in other cases the reason is to protect the base metal from corrosion, as in the chromium plating of steel. Non-metallic items can be plated by first coating them with a conducting substance such as graphite. Figure 16–21 shows an arrangement for silver-plating a spoon. The bath is a solution of silver nitrate, which dissociates into Ag^+ and NO_3^- (nitrate) ions. Silver atoms enter the solution as Ag^+ ions at the anode, and these ions are attracted to the spoon where they gain electrons to become silver atoms once more. Because an Ag atom loses an electron to become Ag^+ more readily than an NO_3^- ion loses its odd electron, the NO_3^- ions stay in solution and do not participate in the plating process.

In electrolysis, the passage of an electric current through a liquid liberates a free element

Electroplating is an example of electrolysis

FIG. 16–21   Silver plating.

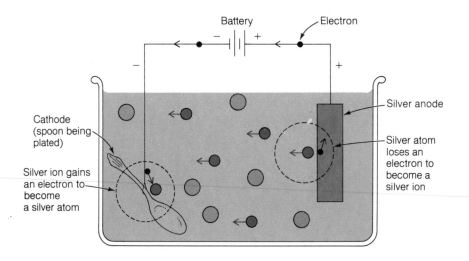

Battery   Electron

Cathode (spoon being plated)

Silver ion gains an electron to become a silver atom

Silver anode

Silver atom loses an electron to become a silver ion

16–9   BATTERIES AND FUEL CELLS

In a battery, chemical energy becomes electric energy

A chemical reaction that involves the transfer of electrons from one substance to another can be used to produce an electric current if matters are so arranged that the electrons do not move directly between the reacting substances but instead pass through an external circuit. The dry cell of a flashlight, the storage battery of a car, and the fuel cell of a spacecraft are all based on reactions of this kind and are collectively called *electrochemical cells*.

A storage battery is "charged" by the electrolytic reactions that occur when a current is passed through it from an outside source; when it acts as a battery, the same reactions occur backwards to produce a current in the reverse direction as the ingredients of the cell restore themselves to their initial state. In principle, all batteries can be reversed to charge them, but in practice the construction of certain types of cells prevents this, and such cells must be discarded when they are exhausted.

Each cell of the lead-acid storage battery of a car in its charged state has positive plates of lead dioxide, PbO_2, interleaved with negative plates of pure lead. The electrolyte is sulfuric acid, H_2SO_4, which in solution dissociates into H^+ and SO_4^- (sulfate) ions. When the cell provides current, electrons leave the negative plates and travel through the external circuit to the positive plates. The reactions that occur at each set of plates are shown in Fig. 16–22. The effect of drawing current from the cell is evidently to deposit lead sulfate, which is virtually insoluble, on both sets of plates. When no more Pb and PbO_2 are accessible to the electrolyte, the cell is "dead" and can produce no further current. To recharge the cell, a current must be passed through it in the opposite direction, which reverses the original reactions and restores the plates to their original compositions.

Specific gravity of a battery's electrolyte

Since the hydrogen and sulfate ions of the electrolyte become part of the lead sulfate deposits on plates as the cell is discharged, the sulfuric acid content of the electrolyte falls and its density decreases. We recall that the specific gravity of a substance is its density relative to that of water. A fully charged cell has a specific gravity of about 1.26, which falls to about 1.19 when the cell is half discharged and to 1.11 when it is completely discharged.

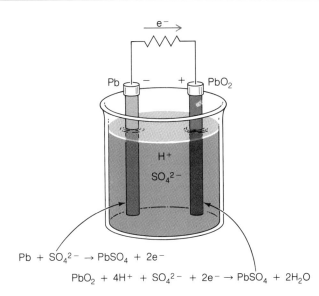

FIG. 16-22 The lead-acid storage battery. The reactions that occur at each electrode when the battery provides current are shown. To charge the battery, a current is passed through it in the opposite direction, which reverses these reactions.

$$Pb + SO_4^{2-} \rightarrow PbSO_4 + 2e^-$$

$$PbO_2 + 4H^+ + SO_4^{2-} + 2e^- \rightarrow PbSO_4 + 2H_2O$$

As discussed in Chapter 17, the electromotive force (emf) of a battery is the electrical equivalent of pressure in a liquid. A fully charged lead-acid cell has an emf of 2.1 volts; the "12-volt" batteries used in cars consist of six cells connected together. One of the major advantages of lead-acid batteries is their low internal resistances, which permit large currents to be drawn for short periods to operate the starting motors of gasoline and diesel engines. The internal resistance to the flow of charge increases at low temperatures largely because ions move very slowly in a cold electrolyte. As a result, the electric current available for starting an engine in freezing weather may be less than half that available on a warm day.

Lead-acid batteries can provide large currents

Other types of storage cell have been developed for special purposes. An example is the nickel-cadmium cell, which consists of nickel and cadmium electrodes in an electrolyte of potassium hydroxide. Such cells have good low-temperature performance and keep their charge for long periods, but are expensive. A nickel-cadmium battery may retain 50% of its initial charge after a year of inactivity, whereas a lead-acid battery may lose as much as 1% of its charge per day. Several new types of storage cells with high energy densities have been proposed for powering electric cars and as adjuncts of electric power stations for peak loads, but they are still in the experimental stage.

A *fuel cell* is a type of battery in which a continuous flow of the initial reactant chemicals is possible, which means that the cell need never be exhausted (like a dry battery) or need recharging (like a storage battery). The efficiency with which chemical energy is converted to electrical energy in a fuel cell is not limited by the second law of thermodynamics, as would be the case if the same fuel were burned to power an internal combustion engine or a steam turbine that drives an electric generator. The absence of moving parts is another advantage. The only mechanism of energy loss in a fuel cell is heat production by electric currents within the cell (described in Chap. 18), which is minor compared with the energy losses in conventional generating plants. Fuel cells are already used to supply electricity in space vehicles, where their high power/weight ratio is important, and are being developed as power sources for electric cars and as self-contained units to furnish electricity to homes.

A fuel cell is a battery that can operate indefinitely

FIG. 16–23   A hydrogen-oxygen fuel cell. The reactions in the electrolyte are promoted by catalysts. The reaction at the negative electrode must occur twice for each time the reaction at the positive electrode occurs. The net result is the formation of water and the liberation of energy.

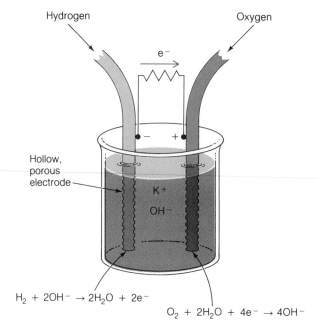

The hydrogen-oxygen cell (Fig. 16–23) is convenient to illustrate how a fuel cell works. The hollow electrodes are made of inert conducting materials with microscopic pores that permit the gases to come in contact with the electrolyte at a gradual rate. The electrolyte of the cell is a potassium hydroxide (KOH) solution that contains K^+ and OH^- ions. The net effect of the reactions at the electrodes is the combination of hydrogen and oxygen to form water, together with the flow of electrons between the electrodes:

$$2H_2 + O_2 \rightarrow 2H_2O + \text{flow of 4 electrons}$$

If two volumes of hydrogen gas are added to one of oxygen and the mixture ignited, a violent explosion occurs, with water as the product. In the hydrogen-oxygen fuel cell the same chemical process occurs, but the liberated energy appears in the form of electric current.

IMPORTANT TERMS

Electric charge, like rest mass, is a fundamental property of certain of the elementary particles of which all matter is composed. There are two kinds of electric charge, **positive charge** and **negative charge**; charges of like sign repel, unlike charges attract. The unit of charge is the **coulomb.** All charges, of either sign, occur in multiples of the fundamental **electron charge** of 1.6×10^{-19} coulomb.

The principle of **conservation of charge** states that the net electric charge in an isolated system remains constant.

Coulomb's law states that the force one charge exerts upon another is directly proportional to the magnitudes of the charges and inversely proportional to the square of the distance between them.

An **atom** consists of a tiny, positively charged nucleus surrounded at some distance by electrons. The nucleus consists of protons and neutrons held tightly together by nuclear forces, and the number of electrons equals the number of protons so the atom as a whole is electrically neutral.

An **ion** is an atom or group of atoms that carries a net electric charge. An atom or group of atoms becomes a negative ion when it picks up one or more electrons in addition to its normal number, and becomes a positive ion when it loses one or more of its usual number.

A **polar molecule** is one whose charge distribution is not uniform, so that one end is positive and the other negative even though the molecule as a whole is electrically neutral.

A substance that separates into free ions when dissolved in water is called an **electrolyte** since the resulting solution is able to conduct electric current.

Electrolysis is the process by which free elements are liberated from a liquid by the passage of an electric current.

A **battery** is a device in which chemical reactions produce an electric current. A **fuel cell** is a type of battery in which a continuous supply of the initial reactant chemicals is possible.

IMPORTANT FORMULA

Coulomb's law: $\quad F = k \dfrac{Q_A Q_B}{r^2}$

MULTIPLE CHOICE

1. Electric charge
   (a) is a continuous quantity that can be subdivided indefinitely.
   (b) is a continuous quantity but it cannot be subdivided into smaller parcels than $\pm 1.6 \times 10^{-19}$ C.
   (c) occurs only in separate parcels, each of $\pm 1.6 \times 10^{-19}$ C.
   (d) occurs only in separate parcels, each of ± 1 C.

2. An object has a positive electric charge whenever
   (a) it has an excess of electrons.
   (b) it has a deficiency of electrons.
   (c) the nuclei of its atoms are positively charged.
   (d) the electrons of its atoms are positively charged.

3. Coulomb's law belongs in the same general category as
   (a) the law of gravitation.
   (b) the laws of motion.
   (c) the laws of thermodynamics.
   (d) the conservation principles of mechanics.

4. A negative electric charge
   (a) interacts only with positive charges.
   (b) interacts only with negative charges.
   (c) interacts with both positive and negative charges.
   (d) may interact with either positive or negative charges, depending on circumstances.

5. According to the Rutherford model of the atom, the positive charge in an atom is
   (a) concentrated at its center.
   (b) spread uniformly throughout its volume.
   (c) in the form of positive electrons at some distance from its center.
   (d) readily deflected by an incident alpha particle.

6. The nucleus of an atom cannot be said to
   (a) contain most of the atom's mass.
   (b) be small in size.
   (c) be electrically neutral.
   (d) deflect incident alpha particles.

7. All molecules are normally
   (a) neutral.   (b) charged.
   (c) polar.   (d) nonpolar.

8. A molecule to which an electron is added becomes
   (a) a negative ion.   (b) a positive ion.
   (c) a polar molecule.   (d) an electrolyte.

9. A molecule whose charge distribution is not perfectly symmetrical is called
   (a) a polar molecule.   (b) a nonpolar molecule.
   (c) an electrolyte.   (d) an organic molecule.

10. Water is an excellent solvent because its molecules are
    (a) neutral.   (b) highly polar.
    (c) nonpolar.   (d) anodes.

11. Oil does not mix with water because
    (a) their respective molecules are different in mass.
    (b) their respective molecules are different in size.
    (c) water molecules are polar whereas oil molecules are nonpolar.
    (d) oil molecules are polar whereas water molecules are nonpolar.

12. Substances that separate into free ions when dissolved in water are called
    (a) polar.   (b) nonpolar.
    (c) anodes.   (d) electrolytes.

13. Crystalline solids such as NaCl that consist of ions dissolve only in liquids that are
    (a) polar.   (b) nonpolar.
    (c) ionized.   (d) oily.

14. The atoms of an element and its ions always have the same
    (a) nuclei.   (b) electron structures.
    (c) color.   (d) chemical behavior.

15. A fuel cell does not require
    (a) an electrolyte.   (b) an anode.
    (c) a cathode.   (d) recharging.

16. In the formula $F = kQ_AQ_B/r^2$, the value of the constant k

(a) is the same under all circumstances.
(b) depends upon the medium the charges are located in.
(c) is different for positive and negative charges.
(d) has the numerical value 1.6×10^{-19}.

17. Two charges of $+Q$ are 1 cm apart. If one of the charges is replaced by a charge of $-Q$, the magnitude of the force between them is

(a) zero.   (b) smaller.
(c) the same.   (d) larger.

18. A charge of $+q$ is placed 2 cm from a charge of $-Q$. A second charge of $+q$ is then placed next to the first. The force on the charge of $-Q$

(a) decreases to half its former magnitude.
(b) remains the same.
(c) increases to twice its former magnitude.
(d) increases to four times its former magnitude.

19. Two charges repel each other with a force of 10^{-6} N when they are 10 cm apart. When they are brought closer together until they are 2 cm apart, the force between them becomes

(a) 4×10^{-8} N.   (b) 5×10^{-6} N.
(c) 8×10^{-6} N.   (d) 2.5×10^{-5} N.

20. Two charges, one positive and the other negative, are initially 2 cm apart and are then pulled away from each other until they are 6 cm apart. The force between them is now smaller by a factor of

(a) $\sqrt{3}$.   (b) 3.
(c) 9.   (d) 27.

21. If 10,000 electrons are removed from a neutral pith ball, its charge is now

(a) $+1.6 \times 10^{-15}$ C.   (b) $+1.6 \times 10^{-23}$ C.
(c) -1.6×10^{-15} C.   (d) -1.6×10^{-23} C.

22. The force between two charges of -3×10^{-9} C that are 5 cm apart is

(a) 1.8×10^{-16} N.   (b) 3.6×10^{-15} N.
(c) 1.6×10^{-6} N.   (d) 3.2×10^{-5} N.

EXERCISES

1. Electricity was once regarded as a weightless fluid, an excess of which was "positive" and a deficiency of which was "negative." What phenomena can this hypothesis still explain? What phenomena can it not explain?

2. (a) When two objects attract each other electrically, must both of them be charged? (b) When two objects repel each other electrically, must both of them be charged?

3. How do we know that the inverse square force holding the earth in its orbit around the sun is not an electrical force?

4. An insulating rod has a charge of $+Q$ at one end and a charge of $-Q$ at the other. How will the rod behave when it is placed near a fixed positive charge that is initially equidistant from the ends of the rod?

5. How can the principle of charge conservation be reconciled with the fact that a rubber rod can be charged by stroking it with a piece of fur?

6. What reasons might there be for the universal belief among scientists that there are only two kinds of electric charge?

7. Nearly all the mass of an atom is concentrated in its nucleus. Where is its charge located?

8. In what ways do the Thomson and Rutherford models of the atom agree? In what ways do they disagree?

9. Most alpha particles pass through gases and thin metal foils with no deflection. To what conclusion regarding atomic structure does this observation lead?

10. What property of the electrons in a metal enables it to conduct electric current readily? What property of the electrons in an insulator prevents it from conducting electric current readily?

11. How does electrical conduction in a metal differ from that in an ionized gas?

12. List several good conductors of electricity and several good insulators. How well do these substances conduct heat? What general relationship between an ability to conduct electricity and an ability to conduct heat can you infer?

13. What aspect of superconductivity has prevented its large-scale application thus far?

14. How could you experimentally distinguish between a solution of an electrolyte and one of a nonelectrolyte?

15. Distinguish between a molecular ion and a polar molecule.

16. In what fundamental way or ways is a fuel cell different from a battery?

17. Give an example of a polar liquid and one of a nonpolar liquid, and state several substances soluble in each but not the other.

18. A calcium atom has 20 protons in its nucleus. (a) How many electrons does this atom contain? (b) How many protons are in the nucleus of a Ca^{++} ion? (c) How many electrons does this ion contain?

19. An oxygen atom has eight protons in its nucleus. (a) How many electrons does this atom contain? (b) How many protons are in the nucleus of an O^{--} ion? (c) How many electrons does this ion contain?

20. Two charges attract each other with a force of 4×10^{-6} N when they are 4 mm apart. Find the force between them when their separation is increased to 5 mm.

21. Two electric charges originally 8 cm apart are brought closer together until the force between them is greater by a factor of 16. How far apart are they now?

22. Two charges of unknown magnitude and sign are observed to repel one another with a force of 0.1 N when they are 5 cm apart. What will the force be when they are (a) 10 cm apart? (b) 50 cm apart? (c) 1 cm apart?

23. The nucleus of a hydrogen atom is a single proton. Find the force between the two protons in a hydrogen molecule, H_2, that are 7.42×10^{-11} m apart. (The two electrons in the molecule spend more time between the protons than outside them, which leads to attractive forces that balance the repulsion of the protons and permit a stable H_2 molecule; see Chapter 30.)

24. A charge of -5×10^{-7} C is 10 cm from a charge of $+6 \times 10^{-6}$ C. Find the magnitude and direction of the force on each charge.

25. A charge of $+5 \times 10^{-9}$ C is attracted by a charge of -3×10^{-7} C with a force of 0.135 N. How far apart are they?

26. Two metal spheres, one with a charge of $+2 \times 10^{-5}$ C and the other with a charge of -1×10^{-5} C, are 10 cm apart. (a) What is the force between them? (b) The two spheres are brought into contact, and then separated again by 10 cm. What is the force between them now?

PROBLEMS

1. How far apart should two electrons be if the force each exerts on the other is to equal the weight of an electron?

2. At what distance apart (if any) are the electric and gravitational forces between two electrons equal in magnitude? Between two protons? Between an electron and a proton?

3. According to one model of the hydrogen atom, it consists of a proton circled by an electron whose orbit has a radius of 5.3×10^{-11} m. How fast must the electron be moving

if the required centripetal force is provided by the electric force exerted by the proton?

4. As mentioned in Sec. 16–2, the permittivity of free space is $\varepsilon_0 = 8.85 \times 10^{-12}$ C^2/N·m^2. The permittivity of air is only a trifle greater than this, but it is considerably greater for some other materials. For example, the permittivity of water is $\varepsilon = 80\varepsilon_0$. Find the force, in air and in water, between two charges of $+2 \times 10^{-10}$ C that are 1 mm apart.

5. A particle carrying a charge of $+6 \times 10^{-7}$ C is located halfway between two other charges, one of $+1 \times 10^{-6}$ C and the other of -1×10^{-6} C, that are 40 cm apart. All three charges lie on the same straight line. What is the magnitude and direction of the force on the $+6 \times 10^{-7}$ C charge?

6. A test charge of -5×10^{-7} C is placed between two other charges so that it is 5 cm from a charge of -3×10^{-7} C and 10 cm from a charge of -6×10^{-7} C. The three charges lie along a straight line. What is the magnitude and direction of the force on the test charge?

7. Two charges, one of -1×10^{-6} C and the other of -3×10^{-6} C are 0.4 m apart. (a) Where should a charge of -1×10^{-7} C be placed on the line between the other charges in order that there be no resultant force on it? (b) Where should a charge of $+1 \times 10^{-7}$ C be placed in order that there be no resultant force on it?

8. Two charges, one of $+2 \times 10^{-8}$ C and the other of $+1 \times 10^{-8}$ C are 0.2 m apart. (a) Where should a charge of -1×10^{-7} C be placed in order that there be no resultant force on it? (b) Where should a charge of $+1 \times 10^{-7}$ C be placed in order that there be no resultant force on it?

9. Three charges, $+Q$, $+Q$, and $-Q$, are at the vertexes of an equilateral triangle a long on each side. Find the magnitude and direction of the force on one of the positive charges.

10. Four charges of $+1 \times 10^{-8}$ C are at the corners of a square 0.2 m on each side. Find the magnitude and direction of the force on one of them.

ANSWERS TO MULTIPLE CHOICE

1. c	**6.** c	**11.** c	**16.** b	**21.** a
2. b	**7.** a	**12.** d	**17.** c	**22.** d
3. a	**8.** a	**13.** a	**18.** c	
4. c	**9.** a	**14.** a	**19.** d	
5. a	**10.** b	**15.** d	**20.** c	

ELECTRIC FIELD

17

Electric forces, like gravitational forces, act between objects that may be widely separated. An appropriate way to regard such forces involves the concept of a force field. When a charge is present somewhere, the properties of space in its vicinity can be considered to be so altered that another charge brought to this region will experience a force there. The "alteration in space" caused by a charge at rest is called its electric field, and any other charge is thought of as interacting with the field and not directly with the charge responsible for it. All forces, not just electric ones, can be interpreted as arising through the intermediacy of a force field of one kind or another. Thus the sun is regarded as being surrounded by a gravitational field, and it is the forces exerted by this field on the planets that are the centripetal forces that hold them in their orbits. In the case of the "direct contact" forces between solid objects in everyday life, the force fields involved are electric fields.

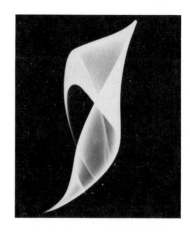

17–1 WHAT IS A FIELD?

A force field is a model devised to provide a framework for understanding how forces are transmitted from one object to another across empty space. A successful model does more than just organize all the information we have on a certain phenomenon into a unified picture; it also enables us to predict hitherto unsuspected effects and relationships, which of course must then be verified by experiment. The creative role of a scientific model is beautifully illustrated by the electromagnetic field model proposed by James Clerk Maxwell a century ago. With the help of this model, Maxwell predicted in 1864 that electromagnetic waves should exist; he calculated that their velocity should be $c = 3 \times 10^8$ m/s; and he suggested that light consists of such waves. In 1887, after Maxwell's death, Heinrich Hertz confirmed Maxwell's hypothesis in the laboratory.

A successful scientific model predicts new effects as well as organizing existing knowledge

A field—in the sense that physicists use the word—is a region of space in which a certain quantity has a definite value at every point. Thus it is appropriate to speak of the temperature field in a room, of the velocity field of the water in a river, and of the gravitational field around the earth. But it is not appropriate to speak of the "chair field" in a room, or of the "rowboat field" in a river, or of the "airplane field" around the earth because chairs, rowboats, and airplanes are separate objects whose presence somewhere is not a property of a region of space that varies throughout that region in the way that temperature, water velocity, and gravitational force vary.

A field is a region of space with a continuous distribution of a certain property

To determine the temperature field in a room, we must measure the temperature at a great many points with a thermometer. The results might be displayed by a series of cross-sectional maps of the room with temperature values written in at each point of measurement, as in Fig. 17–1.

A better way to picture the temperature field is to draw a series of lines on each map that connect points having the same temperature (Fig. 17–2). These lines are called *isotherms,* and might be drawn, for example, for temperatures that are 2°C apart. Naturally the actual measurements are not necessarily 16°, 18°, 20°, and so on; we must interpolate between the actual measurements to find the contours of the various isotherms.

Temperature is a scalar quantity because it involves a magnitude only. To say that the temperature somewhere is 20°C describes it completely. The field of a scalar quantity is a *scalar field,* and it can always be pictured by a plot of isolines (lines joining sets of identical values) as in the case of a temperature field.

Scalar field

A vector quantity has direction as well as magnitude, so picturing a *vector field* is not a simple matter. One method is to draw lines on a map of the region occupied by the field so that the lines always point in the direction of the field quantity at each point. For example, the velocity field in a river can be shown with the help of streamlines that represent the paths of successive particles of water (Fig. 17–3). Several rubber balls thrown into a river and photographed with a movie camera would permit us to make a plot of the velocity field at the surface of the river, since each ball would follow a streamline as it moves with the current.

Vector field

17–2 ELECTRIC FIELD

An *electric field* is a force field that exists wherever an electric force acts on a charge. The source of the field may be a single charge or many charges. We would like to

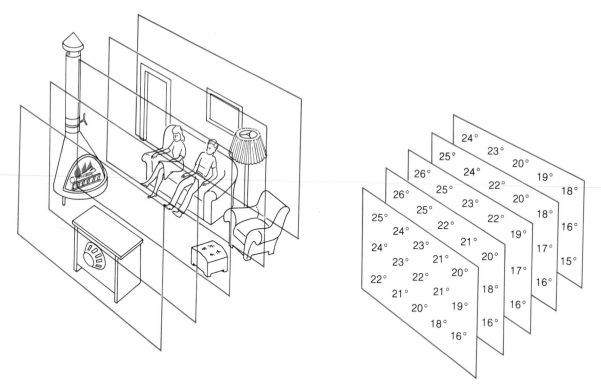

FIG. 17–1 The temperature field in a room is a scalar field, since the temperature at a point has no direction associated with it. The temperatures here are expressed in degrees Celsius.

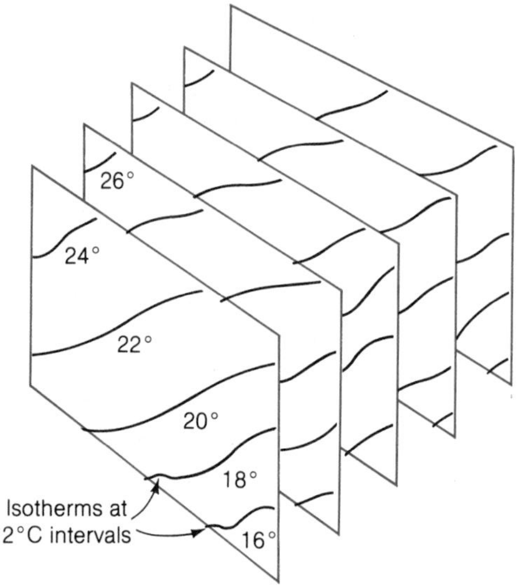

FIG. 17–2 An isotherm is a line that joins points whose temperature is the same. Here the temperature field in the room of Fig. 17–1 is displayed by means of isotherms.

Isotherms at 2°C intervals

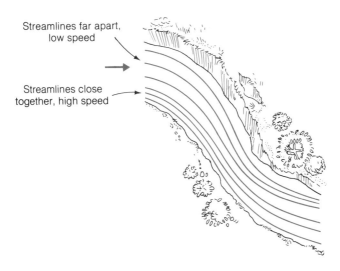

FIG. 17-3 The velocity field in a river can be visualized to some extent with the help of streamlines that represent the paths of successive particles of water.

Streamlines far apart, low speed

Streamlines close together, high speed

specify electric field in such a way that it will be possible to determine the force acting on any arbitrary charge present at any point in the field. Accordingly the electric field **E** at a point in space is defined as the ratio between the force **F** on a charge Q (assumed positive) at that point and the magnitude of Q:

Electric field is a vector quantity

$$\mathbf{E} = \frac{\mathbf{F}}{Q} \qquad\qquad \textit{Electric field} \quad (17\text{--}1)$$

The units of **E** are newtons per coulomb. Electric field is a vector quantity that possesses both magnitude and direction.

Once we know what the electric field **E** is at some point, from the definition we see that the force that the field exerts on a charge Q at that point is

An electric force on a charge may be regarded as due to the action of an electric field on the charge

$$\mathbf{F} = Q\mathbf{E} \qquad\qquad\qquad\qquad (17\text{--}2)$$

Force = charge × electric field

Example The electric field between the electrodes of the gas discharge tube used in a certain neon sign has the magnitude 5.0×10^4 N/C. Find the acceleration of a neon ion of mass 3.3×10^{-26} kg in the tube if it carries a charge of $+e$ (Fig. 17–4).

Solution Since $e = 1.6 \times 10^{-19}$ C, the force on the ion has the magnitude

$$F = QE = (1.6 \times 10^{-19}\ \text{C})(5.0 \times 10^4\ \text{N/C}) = 8.0 \times 10^{-15}\ \text{N}$$

$E = 5.0 \times 10^4$ N/C

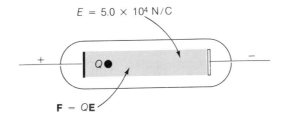

$\mathbf{F} = Q\mathbf{E}$

FIG. 17-4 A charge Q in an electric field **E** experiences a force equal to $Q\mathbf{E}$.

Hence the acceleration of the ion is

$$a = \frac{F}{m} = \frac{8.0 \times 10^{-15}\,\text{N}}{3.3 \times 10^{-26}\,\text{kg}} = 2.4 \times 10^{11}\,\text{m/s}^2$$

which is 25 billion times greater than the acceleration of gravity. The speed such an ion actually attains is limited to a relatively modest figure by frequent collisions with the neon atoms in its path. ■

Electric field of a single charge

We can use Coulomb's law to determine the magnitude of the electric field around a single charge Q. First we find the force F that Q exerts upon a test charge q at the distance r away, which is

$$F = k\frac{Qq}{r^2}$$

Since $E = F/q$ by definition, we have

$$E = \frac{F}{q} = k\frac{Q}{r^2} \qquad\qquad \textit{Electric field of a charge} \quad (17\text{–}3)$$

This formula tells us that the electric field magnitude the distance r from a point charge is directly proportional to the magnitude Q of the charge and is inversely proportional to the square of r.

Electric field of many charges

When more than one charge contributes to the electric field at a point P, the net field $\mathbf{E}$ is the *vector sum* of the fields of the individual charges. That is,

$$\mathbf{E} = \mathbf{E}_1 + \mathbf{E}_2 + \mathbf{E}_3 + \cdots = \Sigma\,\mathbf{E}_n \qquad\qquad (17\text{–}4)$$

Example   Find the electric field intensity at the position of the charge Q_1 due to the other charges in the situation shown in Fig. 16–9.

Solution   Making use of our previous results, we obtain

$$E = \frac{F}{Q_1} = \frac{2.6 \times 10^{-5}\,\text{N}}{2 \times 10^{-9}\,\text{C}} = 1.3 \times 10^4\,\text{N/C}$$

The direction of $\mathbf{E}$ is the same as that of $\mathbf{F}$ since Q_1 is a positive charge; if Q_1 were negative, $\mathbf{E}$ would be in the opposite direction. The usefulness of knowing $\mathbf{E}$ here is that we can determine from it what the force on *any* charge at the position of Q_1 is by simply multiplying $\mathbf{E}$ and Q. ■

17–3   ELECTRIC LINES OF FORCE

An electric field is a region of space in which an object by virtue of its electric charge is acted upon by a force. To visualize an electric field, we may use *lines of force*. These are constructed as follows. At several points in space we draw arrows in the direction a positive charge there would experience a force (Fig. 17–5(a)). Then these arrows are connected to form smooth curves called lines of force (Fig. 17–5(b)). Two rules should be kept in mind:

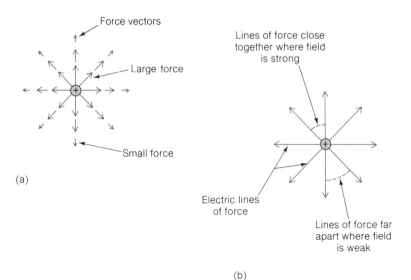

FIG. 17–5 How electric lines of force around a positive charge are drawn. In (a) the forces whose vectors are shown are those exerted by the charge on a positive test charge.

1. Lines of force leave positive charges and enter negative ones.
2. The spacing of lines of force is such that they are close together where the field is strong and far apart where the field is weak.

Rules for lines of force of electric field

If we place a positive charge in an electric field, then, it will experience a force in the direction of the line of force it is on, and the magnitude of the force will be proportional to the concentration of lines of force in its vicinity. Figure 17–6 shows the patterns of lines of force around a positive and a negative charge and Fig. 17–7 shows the patterns around pairs of charges.

It is important to keep in mind that lines of force do not actually exist as threads in space; they are simply a device for giving intuitive form to our thinking about force fields. The use of lines of force is not limited to electric fields. Gravitational and magnetic fields, for instance, are often described in this manner.

Lines of force are imaginary

The notion of lines of force is often helpful as a guide to our thinking when electric fields of charge distributions are concerned. For example, let us inquire into the electric field around a spherical distribution of charge whose total amount is Q. Such a distribution is quite easy to arrange; all we need do is to add the charge Q to a metal sphere.

Electric field around a charged sphere

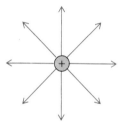

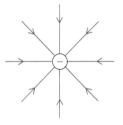

(a)   (b)

FIG. 17–6 Lines of force around a positive and a negative charge.

FIG. 17–7 Lines of force near two equal charges (a) of the same sign and (b) of opposite sign. At large distances relative to the separation of the charges, the field in (a) approaches that of a single charge.

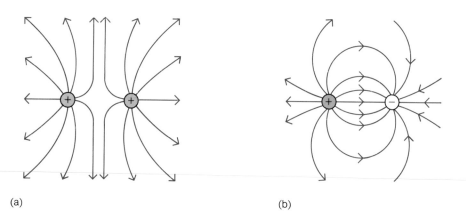

(a)   (b)

Since metals are good conductors of electricity, the mutual repulsion of the individual charges that make up Q causes them to spread out uniformly over the surface of the sphere; in this way the charges are as far apart as possible. Evidently it doesn't matter whether the sphere is hollow or solid.

Whatever the number of lines of force we choose to represent the electric field around a point charge Q, the same number must emerge from a body that carries the same net charge Q. In the case of a charged sphere, the lines of force are symmetrically arranged, and therefore the pattern of lines of force outside the sphere is identical with that around a point charge of the same magnitude (Fig. 17–8). The same is true for the electric field E itself, of course.

The greater the curvature, the stronger the field

When a nonspherical metal body is charged, the individual charges do not distribute themselves uniformly on its surface. As a general rule, the more highly curved parts have greater concentrations of charge than the gently curved parts, and the electric field is accordingly most intense near the highly curved parts (Fig. 17–9).

At a point, the electric field intensity may become so great that it causes a separation of charge in nearby air molecules. The resulting electrical "discharge" is visible as a luminous glow. If two oppositely charged pointed rods are brought close together, charge flows between them via disrupted air molecules and a "spark" occurs. If more gently curved charged bodies, such as large spheres, are similarly brought near each other, their separation must be much smaller before a spark can occur.

FIG. 17–8 (a) Electric lines of force around a uniformly charged sphere. (b) Electric lines of force around a point charge.

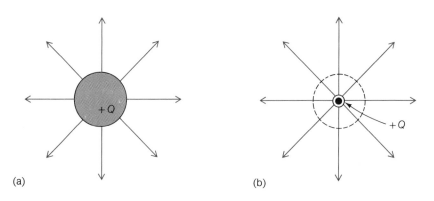

(a)   (b)

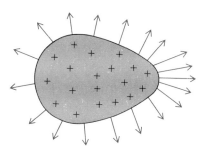

FIG. 17–9 The electric field near a charged asymmetric metal object is strongest near the parts of the object that have the smallest radii of curvature.

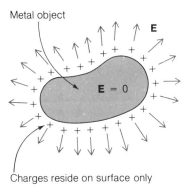

Metal object

Charges reside on surface only

FIG. 17–10 There is no electric field inside a charged conducting object.

Inside a charged conducting object of any shape the electric field is 0 everywhere; the electric fields due to the individual charges on the surface all cancel out in the interior, although outside the object they do not. We can see why this must be true even without a formal calculation. Suppose there *were* an electric field **E** in the interior of the object. Then the charges inside the object that are free to move (electrons in the case of a metal) would do so under the influence of the field **E.** But no currents are observed in a charged conducting object except for a moment after the charge is placed on it; since energy is needed to maintain an electric current, a supply of energy would be needed for currents to persist in such an object (unless it were superconducting). The only conclusion is that the interior of an isolated conducting object is always free of electric field (Fig. 17–10).

E = 0 inside a charged conductor

17–4 ELECTRIC POTENTIAL ENERGY

In our study of mechanics we found the related concepts of work and potential energy to be useful in analyzing a wide variety of situations. These concepts are equally useful in the study of electrical phenomena, in particular electric current. Instead of potential energy itself, a related quantity called potential difference turns out to be especially appropriate in electrical problems.

Let us examine the potential energy of a charge in a uniform electric field together with a gravitational analogy. At the left in Fig. 17–11(a) is a uniform electric field **E** between two parallel, uniformly charged plates *A* and *B,* and at the right is a region near the earth's surface in which the gravitational field is also uniform. Now we place

FIG. 17–11 Analogy between electric and gravitational potential energy.

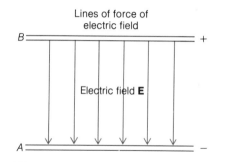

Lines of force of electric field

Electric field **E**

B +

A −

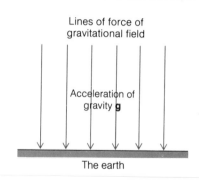

Lines of force of gravitational field

Acceleration of gravity **g**

The earth

(a)

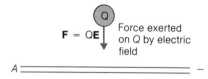

B +

$\mathbf{F} = Q\mathbf{E}$ Force exerted on Q by electric field

A −

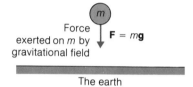

Force exerted on m by gravitational field $\mathbf{F} = m\mathbf{g}$

The earth

(b)

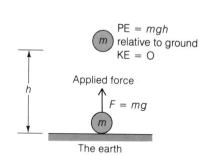

B +

$PE = QEs$ relative to A
$KE = O$

Applied force

$F = QE$

s

A −

PE = mgh relative to ground
$KE = O$

Applied force

$F = mg$

h

The earth

(c)

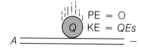

B +

$PE = O$
$KE = QEs$

A −

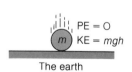

$PE = O$
$KE = mgh$

The earth

(d)

a particle of charge Q in the electric field and a particle of mass m in the gravitational field. As shown in Fig. 17–11(b) the charge is acted upon by the electric force

$$\mathbf{F}_{elec} = Q\mathbf{E}$$

Electric force on charge

and the mass is acted upon by the gravitational force

$$\mathbf{F}_{grav} = m\mathbf{g}$$

Gravitational force on mass

If the charge is on plate A and we want to move it to plate B, we must apply a force of magnitude QE to it because we have to push against a force of this magnitude (and opposite direction) exerted by the electric field. When the charge is at B, we will have performed the amount of work

Work = force × distance

$$W = QEs$$

Work done on charge

on it, where s is the distance the charge has moved (Fig. 17–11(c)). Similarly, to raise the mass from the ground to a height h, we must apply a force of magnitude mg to it, and the work we do is

$$W = mgh$$

Work done on mass

At plate B the charge has the potential energy

$$PE = QEs \qquad\qquad \textit{PE of charge in uniform electric field} \quad (17\text{–}5)$$

with respect to A. If we let it go, the potential energy will become kinetic energy as the electric field $\mathbf{E}$ accelerates the charge, and when the charge is back at A it will have a kinetic energy equal to QEs (Fig. 17–11(d)). In the same way, the mass has potential energy *with respect to the ground* in its new location. This potential energy is equal to the work done in raising it through the height h, and is

$$PE = mgh$$

If we let the mass go, it will fall to the ground with a final kinetic energy of mgh.

To summarize: The amount of work which must be performed to move a charge Q from A to B, the distance s apart, in a uniform electric field $\mathbf{E}$ is QEs. At B the charge accordingly has the potential energy QEs. If the charge is released at B, the force QE acting on it produces an acceleration such that the charge has the kinetic energy QEs when it is back at A. Thus the work done in moving the charge in the field becomes potential energy, which in turn becomes kinetic energy when it is released.

Moving a charge perpendicular to an electric field does not affect its energy

There is no change in the energy of a charge moved perpendicular to an electric field, just as there is no change in the energy of a mass moved perpendicular to a gravitational field (for instance, along the earth's surface).

17–5 POTENTIAL ENERGY OF TWO CHARGES

A particle of charge Q_A that is located the distance r from another particle of charge Q_B has electrical potential energy because there is a force exerted on it by the electric

FIG. 17–12 A charge has potential energy when it is in the electric field of another charge. Actually, the potential energy belongs to the system of the two charges. Here Q_B is assumed to be fixed in place, so the PE of the system becomes kinetic energy of Q_A when Q_A is released.

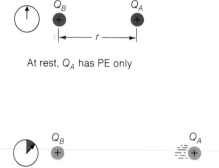

At rest, Q_A has PE only

When released, Q_A gains KE at the expense of its original PE

field of Q_B. When Q_A and Q_B have the same sign, the force is repulsive; when the charges have opposite signs, the force is attractive. In either case, when Q_A is released, it will begin to move and acquire kinetic energy at the expense of its original potential energy (Fig. 17–12).

We have been assuming that Q_B is somehow fixed in place. If instead Q_A is fixed in place, then we can speak of the potential energy of Q_B in the electric field of Q_A. This potential energy is exactly the same as before, since by Newton's third law of motion the force one object exerts on another is equal in magnitude to the force the second object exerts on the first.

PE of two charges is a property of the system of both of them

In reality, the potential energy belongs to the *system* of the two particles. If the particles are both released, *both* of them begin to move, and they share the original potential energy between them. The relative velocities of the two particles will be such as to conserve linear momentum, so the lighter particle will move faster than the heavier one. (The total amount of energy available depends upon the charges of the particles, and the division of the energy depends upon their masses.) If one of the particles has a much greater mass than the other, its velocity will be very small, and it may then be legitimate to regard it as being fixed in place. For example, the mass of a proton is 1836 times that of an electron, so in a hydrogen atom, which consists of a proton and an electron, we may think of the proton as being stationary and of the electron as possessing the potential energy of the system.

A potential energy of any kind must be specified relative to a reference location. Near the earth, for example, the gravitational potential energy of an object is usually given with the earth's surface as the reference location for PE = 0. In the case of

PE = 0 for two charges infinitely far apart

individual charges interacting with one another, the reference location is chosen to be infinity, since the electric field of a charge falls to zero an infinite distance away.

Because the electric field **E** of the charge Q_B (which we assume to be fixed in place) is not uniform, it is not easy to calculate the potential energy of Q_A when it is the distance r away. What we must do is start with Q_A at infinity and imagine we move it a short distance toward Q_B. In this short distance we can consider **E** as having a constant magnitude, so we can find the work needed for this step from the formula $W = QEs$. Then we move Q_A through another short interval closer to Q_B; again we can consider **E** as having a constant magnitude, though a greater one than before, and we find the work needed for the second step. By continuing this process until Q_A is r away

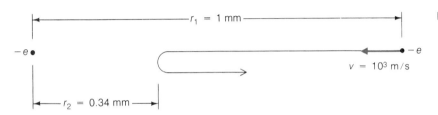

FIG. 17–13

from Q_B and adding up all the amounts of work needed, we can determine the potential energy of Q_A. With the help of calculus this addition yields the result

> **PE relative to infinite separation**

$$PE = k \frac{Q_A Q_B}{r} \qquad \text{\textit{Potential energy of system of two charges}} \quad (17\text{–}6)$$

If the charges have the same sign, their potential energy is positive; thus a positive potential energy corresponds to a repulsive force. If the charges have opposite signs, their potential energy is negative; thus a negative potential energy corresponds to an attractive force. The potential energy of a charge decreases as it moves away from another charge of the same sign and increases as it moves away from another charge of opposite sign.

> **PE > 0 for a repulsive force; PE < 0 for an attractive force**

Example An electron of initial speed 10^3 m/s is aimed at another electron, whose position is fixed, from a distance of 1 mm. How close to the stationary electron will the other one approach before it comes to a stop and reverses its direction?

Solution Figure 17–13 illustrates the situation. Since $Q_A = Q_B = -e$ here, the potential energies of the moving electron at the initial distance $r_1 (= 10^{-3}$ m) and the final distance r_2 are respectively

$$PE_1 = \frac{ke^2}{r_1} \qquad PE_2 = \frac{ke^2}{r_2}$$

The potential energies are both positive, corresponding to a repulsive force between the electrons. The difference between the two potential energy values is equal to the initial kinetic energy $\frac{1}{2}mv^2$ of the moving electron, and therefore

$$KE = PE_2 - PE_1$$

$$\frac{1}{2}mv^2 = ke^2 \left(\frac{1}{r_2} - \frac{1}{r_1} \right)$$

$$\frac{1}{r_2} = \frac{mv^2}{2ke^2} + \frac{1}{r_1}$$

$$r_2 = \frac{1}{(mv^2/2ke^2) + (1/r_1)} = 3.4 \times 10^{-4} \, \text{m} = 0.34 \, \text{mm} \qquad \blacksquare$$

17–6 POTENTIAL DIFFERENCE

The quantity *potential difference* is introduced to describe the situation of a charge in an electric field in an especially convenient way. The potential difference V_{AB} between

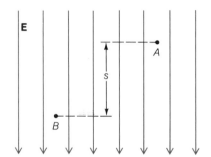

FIG. 17–14 The potential difference between two points in a uniform electric field **E** is equal to *Es*, where *s* is the component parallel to **E** of the distance between the points.

two points A and B is defined as the ratio between the work that must be done to take charge Q from A to B and the value of Q:

$$V_{AB} = \frac{W_{AB}}{Q}$$

Potential difference   (17–7)

Potential difference = work per unit charge

The volt is the unit of potential difference

The unit of potential difference is the joule per coulomb. Because this quantity is so frequently used, its unit has been given a name of its own, the *volt* (V). Thus

1 volt = 1 joule/coulomb

In a uniform electric field, $W_{AB} = QEs$, with the result that the potential difference between A and B is QEs/Q or

$$V_{AB} = Es$$

Potential difference in a uniform electric field   (17–8)

In a uniform electric field, the potential difference between two points is the product of the field magnitude E and the separation s of the two points in a direction parallel to that of **E** (Fig. 17–14).

Positive and negative potential differences

A positive potential difference means that the energy of the charge is *greater* at B than at A; a negative potential difference means that its energy is *less* at B than at A. If V_{AB} is positive, then a charge at B tends to return to A, whereas if V_{AB} is negative, the charge tends to move further away from A.

Batteries and generators are sources of potential difference

One advantage of specifying the potential difference between two points in an electric field, rather than the magnitude of the field between them, is that an electric field is normally created by imposing a difference of potential between two points in space. A battery is a device that uses chemical means to produce a potential difference between two terminals (Fig. 17–15). A "6-volt" battery is one that has a potential difference of 6 V between its terminals. A generator is another device for producing a potential difference. Batteries and generators are to electric charges what elevators are to masses: all of them increase the potential energy of what they act upon.

When a charge Q goes from one terminal of a battery whose potential difference is V to the other, Eq. (17–7) tells us that the work

$$W = QV$$

Work done on a charge   (17–9)

is done on it regardless of the path taken by the charge and regardless of whether the actual electric field that caused the motion of the charge is strong or weak. Given V we can find W at once, no matter what the details of the process are. Hence the notion of

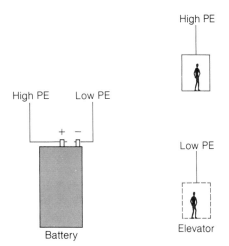

FIG. 17–15 A battery is to an electric charge as an elevator is to a mass.

potential difference permits us to simplify our analyses of electrical phenomena, just as the notion of potential energy permitted us to simplify our analysis of mechanical phenomena.

Example Figure 17–16 shows a tube that has a source of electrons at one end and a metal plate at the other. A 100-V battery is connected between the electron source and the plate, so that there is a potential difference of 100 V between them. The negative terminal of the battery is connected to the electron source. What is the speed of the electrons when they arrive at the metal plate? (The tube is evacuated to prevent collisions between the electrons and air molecules.)

Solution To find the kinetic energy, we note that the work done by the electric field within the tube on an electron is $W = QV = eV$. Since the KE of the electron is equal to the work done on it, its KE after passing through the entire field is eV. Hence

$$\text{KE} = W = eV = \tfrac{1}{2}mv^2$$

$$v = \sqrt{\frac{2eV}{m}} = \sqrt{\frac{2(1.6 \times 10^{-19}\,\text{C})(100\,\text{V})}{9.1 \times 10^{-31}\,\text{kg}}} = 5.9 \times 10^6\,\text{m/s}$$ ∎

Example The electron source and the positive electrode are 40 cm apart in the tube of the preceding problem. What is the magnitude of the electric field (assumed uniform)

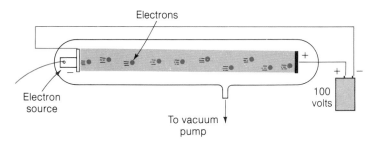

FIG. 17–16

between them? If this distance is reduced to 20 cm, what is the new magnitude of the electric field and what effect does this change have on the final speed of the electrons?

The electric field magnitude in a uniform field is given by $E = V/s$ from the definition of potential difference in an electric field. Hence

$$E = \frac{100\,\text{V}}{0.4\,\text{m}} = 250\,\text{V/m}$$

If s is reduced to 20 cm, the field magnitude increases to

$$E = \frac{100\,\text{V}}{0.2\,\text{m}} = 500\,\text{V/m}$$

The final electron speed depends on the potential difference, not on the details of the field

The energy given to an electron by this field when it travels through the entire 20 cm is the same as the energy given to an electron that travels through the 40-cm length of the previous, weaker field, since in both cases QEs is the same. We reach the same conclusion by noting that $KE = QV$ and the potential difference V in this problem is independent of the spacing of the electrodes. The final electron speed is therefore unchanged. ∎

17–7   CATHODE-RAY TUBE

In a cathode-ray tube, an electron beam traces out an image on a fluorescent screen

Before their nature was understood, beams of electrons were known as "cathode rays." Even today, a tube in which an electron beam can be controlled in such a way as to trace a desired pattern on a fluorescent screen is called a *cathode-ray tube*. Figure 17–17 shows how such a tube is constructed. Electrons emitted from the heated cathode are accelerated through a potential difference of 5 kV to 50 kV between the cathode and the anode and emerge from a hole in the latter. Near the cathode is an electrode whose potential difference relative to the cathode controls the number of electrons in

FIG. 17–17 Cathode-ray tube.

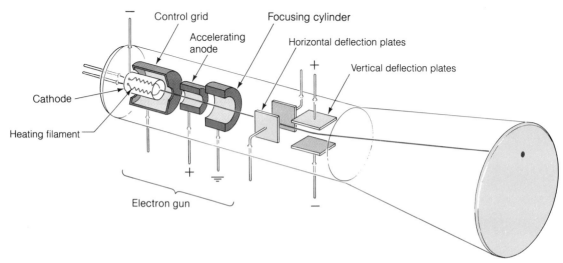

Fluorescent screen

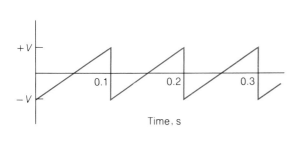

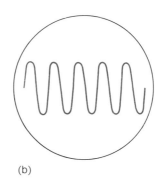

FIG. 17–18 (a) A sawtooth voltage with a period of 0.1 s. (b) Oscilloscope trace of a 50-Hz alternating signal applied to the vertical deflection plates when the sawtooth voltage of (a) is applied to the horizontal deflection plates. Each cycle takes 0.02 s, so five appear on the screen.

the beam and therefore the brightness of the spot on the screen. The entire assembly, which also incorporates another electrode to focus the beam, is called an *electron gun*.

The electron beam then passes through a horizontal electric field set up between a pair of parallel metal plates. The direction of the field determines whether the beam is deflected to the right or to the left, and its magnitude (which is proportional to the voltage applied to the plates) determines the amount of the deflection. The vertical electric field between the next set of plates similarly governs the vertical position of the electron beam when it reaches the screen, which is coated with a substance called a *phosphor* that emits light when struck by the electrons. The screen continues to glow for a short time after the electron beam has moved away from each point, which permits a picture to be built up that lasts long enough to be perceived as a whole.

An *oscilloscope* uses a cathode-ray tube to display electrical signals that vary rapidly, since the small mass of the electron permits a beam of them to respond at once to a changing voltage applied to a pair of deflecting plates. The horizontal deflection plates are connected to a source of sawtooth pulses (Fig. 17–18(a)), which rise steadily from a certain voltage $-V$ through 0 to $+V$ and then return rapidly to $-V$ to start a new cycle. In the figure the pulses take 0.1 s to go from $-V$ to $+V$, where V is chosen to correspond to a full horizontal deflection in a particular tube. Thus each sawtooth pulse sweeps the electron beam from left to right across the screen in 0.1 s and then returns the beam to the left to start another sweep. If we connect the vertical deflection plates to a voltage source that alternates in polarity 50 times per second (in other words, whose frequency is 50 Hz), what we will see are five complete cycles on the oscilloscope screen, as in Fig. 17–18(b), because each cycle involves 1/50 s = 0.02 s. A wide range of sweep times is provided on an oscilloscope, and the input signal can be amplified as necessary to give a trace large enough to be inspected and measured.

An oscilloscope displays electrical signals for visual inspection

The picture tube of a television set is a cathode-ray tube, but in this application it is more convenient to use magnetic rather than electric fields to deflect the electron beam; magnetic fields are considered in Chapter 19. The electron beam is moved horizontally across the screen just as in an oscilloscope, but after each sweep the beam is shifted downward in order to cover a new line on the screen (Fig. 17–19). In the United States, a complete image is divided into 525 lines, but in many other countries a larger number of lines is used to give better picture quality. A complete image is built up by two scans of the screen, each covering alternate lines, a process that takes place 30 times per second. During the scans the beam intensity is varied to produce the variations in brightness that constitute the image. In color television separate signals

A TV picture tube is a cathode-ray tube

FIG. 17–19 The electron beam of a television picture tube covers the screen in a pattern of horizontal lines starting at the upper left. The returns from right to left take about one-tenth as long as the sweeps from left to right that produce the image.

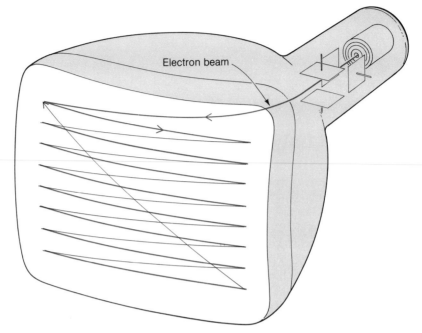

are transmitted that correspond to the red, green, and blue contents of the image. The receiver screen consists of red, green, and blue phosphor dots that flash accordingly when the electron beam strikes them.

17–8   THE ELECTRON VOLT

The electron volt is an energy unit

The electron volt (abbreviated eV) is a widely used energy unit in atomic and nuclear physics. By definition, 1 eV is the energy acquired by an electron that has been accelerated through a potential difference of 1 volt. Hence

$$W = QV$$
$$1\,\text{eV} = (1.60 \times 10^{-19}\,\text{C})(1.00\,\text{V})$$

and so

$$1\,\text{eV} = 1.60 \times 10^{-19}\,\text{J} \qquad\qquad\qquad \textit{Electron volt} \quad (17\text{–}9)$$

Typical quantities expressed in electron volts are the ionization energy of an atom (which is the work needed to remove one of its electrons) and the binding energy of a molecule (which is the work needed to break it apart into separate atoms). Thus the ionization energy of nitrogen is usually stated to be 14.5 eV and the binding energy of the hydrogen molecule, which consists of two hydrogen atoms, is usually stated to be 4.5 eV.

The eV is too small a unit for nuclear physics, where its multiples the MeV (10^6 eV) and the GeV (10^9 eV) are commonly used. The M and G respectively signify *mega* ($= 10^6$) and *giga* ($=10^9$) and are used in connection with other units as well, for

MeV and GeV

instance the megabuck (10^6) and the gigawatt (10^9 watts). The GeV was formerly called the BeV, where the B stood for "billion," but this proved confusing since in Europe a billion is 10^{12} whereas in the United States a billion is 10^9.

A typical quantity expressed in MeV is the energy liberated when the nucleus of a uranium atom splits into two parts. Such *fission* of a uranium nucleus releases an average of 200 MeV; this is the process that powers nuclear reactors and atomic bombs.

Example A hydrogen atom consists of a proton and an electron an average of 5.3 $\times$ 10^{-11} m apart. Find the potential energy of the electron in eV.

Solution Here the two charges are $-e$ and $+e$, where $e = 1.60 \times 10^{-19}$ C. Hence

$$\text{PE} = k\frac{Q_A Q_B}{r} = k\frac{(-e)(e)}{r} = \left(9.0 \times 10^9 \frac{\text{N}\cdot\text{m}^2}{\text{C}^2}\right) \times \frac{-(1.60 \times 10^{-19}\,\text{C})^2}{5.3 \times 10^{-11}\,\text{m}}$$

$$= -4.35 \times 10^{-18}\,\text{J}$$

Since $1\,\text{eV} = 1.60 \times 10^{-19}$ J,

$$\text{PE} = \frac{-4.35 \times 10^{-18}\,\text{J}}{1.60 \times 10^{-19}\,\text{J/eV}} = -27.2\,\text{eV}$$

The minus sign signifies that the force on the electron is directed toward the proton. (The KE of the electron in a hydrogen atom is 13.6 eV, so the total energy of the electron is

$$\text{PE} + \text{KE} = -27.2\,\text{eV} + 13.6\,\text{eV} = -13.6\,\text{eV}$$

The negative total energy means that the hydrogen atom is a stable system, since work must be done by an outside agency to liberate the electron from the proton.) ■

17–9 THE ACTION POTENTIAL

Animal cells have an excess of negative charge inside their membranes and an excess of positive charge on the outside. The potential difference across the membrane of a nerve or muscle cell is normally 90 mV (0.090 V). This potential difference is a consequence of the different permeability of the membrane to different ions. To see how the process works, let us consider two solutions of potassium chloride, KCl, separated by a membrane that permits K^+ ions to pass through it but hinders Cl^- ions from doing so (Fig. 17–20). The solution at the left has the higher concentration of KCl. K^+ ions from both solutions diffuse through the membrane, but because the solution on the left has more of them per unit volume, the number going to the right at first exceeds the number going to the left. As a result the solution on the left soon has a deficiency of K^+ ions relative to its Cl^- ions and hence a net negative charge, while the solution on the right has an excess of K^+ ions and hence a net positive charge. Now an electric field exists across the membrane which favors the diffusion of K^+ ions from right to left. When the influence of the electric field exactly balances the opposite influence of the difference in concentration, as many K^+ ions pass through the membrane in one direction as in the other.

Potential differences occur across cell membranes

FIG. 17–20 (a) Initially the KCl concentration is higher in the solution to the left of the semipermeable membrane, but in each solution the numbers of K^+ and Cl^- ions are equal. Because the concentration of K^+ ions is greater in the solution at left, more K^+ ions diffuse through the membrane to the right than to the left. (b) At equilibrium, the solution at left has an excess of Cl^- ions, and the one at right has an excess of K^+ ions. The rate of diffusion of K^+ ions is now the same in both directions, and there is a potential difference across the membrane.

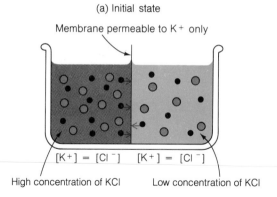

(a) Initial state

Membrane permeable to K^+ only

$[K^+] = [Cl^-]$ $[K^+] = [Cl^-]$

High concentration of KCl   Low concentration of KCl

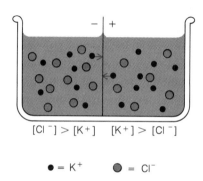

(b) Equilibrium state

$[Cl^-] > [K^+]$ $[K^+] > [Cl^-]$

$\bullet = K^+$ $\bigcirc = Cl^-$

Actual cells do have higher concentrations of K^+ ions inside their membranes compared with the concentration in the fluid around them, and the observed difference is consistent with a potential difference of 90 mV. Sodium ions, Na^+, are also present in living tissue, but their concentration is higher in the extracellular fluid, and cell membranes are ordinarily impermeable to them. When a nerve or muscle cell is stimulated, its membrane momentarily becomes permeable to Na^+ ions, which flow into the cell to neutralize the excess negative charge there. (Cl^- is not the only negative ion involved.) The membrane, in effect, is short-circuited by the Na^+ ions. Because the concentration of Na^+ ions is greater outside the cell (the opposite of the case of K^+ ions), in a short time the potential difference across the cell membrane becomes reversed with the inside at a small positive potential relative to the outside (Fig. 17–21). Then the cell membranes again become impermeable to Na^+ ions, and K^+ diffusion restores the potential difference to its normal value. The entire process involved in producing such an *action potential* takes about a millisecond (0.001 s). A mechanism that is not completely understood then "pumps" the excess Na^+ ions out of the cell into the fluid around it.

Origin of action potential in nerve and muscle cells

Action potentials of the above kind are propagated from nerve cell to nerve cell at speeds of about 30 m/s. When a nerve impulse arrives at a muscle cell, an action potential triggered in it travels along the muscle fiber and causes it to contract. The

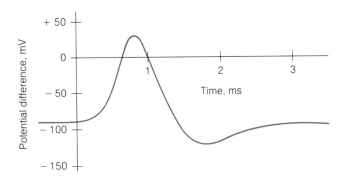

action potentials in heart muscle are especially large and are easily detected by electrodes placed on the chest; an *electrocardiogram* is a record of these potentials.

Electrocardiogram

Some fish, notably the giant electric eel, contain specialized cells called *electroplaques* whose chief purpose is to develop action potentials when stimulated. Large numbers of such cells are stacked in the same manner as the cells of a high-voltage battery, and although each cell is limited to a pulse of not much more than 0.1 V, the total can amount to several hundred volts. Such fish use these potentials to stun their prey. Over two-thirds of the mass of an electric eel consists of 4000 rows of electroplaques, which can produce a 600-V pulse.

Electric eel

IMPORTANT TERMS

A **force field** is a region of space at every point of which an appropriate test object would experience a force.

An **electric field** exists wherever an electric force acts on a charged particle. The magnitude of an electric field at a point is equal to the force that would act on a charge of +1 C placed there; the direction of the field is the direction of the force on the charge. The unit of electric field is the V/m, which is equal to 1 N/C.

The electric **potential difference** between two points is the work that must be done to take a charge of 1 C from one of the points to the other. The unit of potential difference is the **volt,** which is equal to 1 J/C.

In a **cathode-ray tube,** an electron beam controlled by electric (or magnetic) fields traces out an image on a fluorescent screen.

The **electron volt** is the energy acquired by an electron that has been accelerated by a potential difference of 1 V. It is equal to 1.6×10^{-19} J.

IMPORTANT FORMULAS

Electric field: $\mathbf{E} = \dfrac{\mathbf{F}}{Q}$

Electric field of a charge: $E = k \dfrac{Q}{r^2}$

Potential energy of a charge in uniform field: $PE = QEs$

Potential energy of two charges: $PE = k \dfrac{Q_A Q_B}{r}$

Potential difference: $V_{AB} = \dfrac{W_{AB}}{Q}$

$\qquad\qquad = Es$ (in uniform electric field)

MULTIPLE CHOICE

1. Of the following quantities, the one that is vector in character is electric

(a) charge.
(b) field.
(c) energy.
(d) potential difference.

2. Electric lines of force

(a) exist everywhere.
(b) exist only in the immediate vicinity of electric charges.
(c) exist only when both positive and negative charges are near one another.
(d) are imaginary.

3. The electric field at a point in space is equal in magnitude to

(a) the potential difference there.
(b) the electric charge there.
√(c) the force a charge of one coulomb would experience there.
(d) the force an electron would experience there.

4. The magnitude of the electric field in the region between two parallel oppositely-charged metal plates is

(a) zero.
(b) uniform throughout the region.
(c) greatest near the positive plate.
(d) greatest near the negative plate.

5. Ten million electrons are placed on a solid copper sphere. The electrons become

(a) uniformly distributed on the sphere's surface.
(b) uniformly distributed in the sphere's interior.
(c) concentrated at the center of the sphere.
(d) concentrated at the bottom of the sphere.

6. A system of two charges has a positive potential energy. This signifies that

(a) both charges are positive.
(b) both charges are negative.
(c) both charges are positive or both are negative.
(d) one charge is positive and the other is negative.

7. From its definition, the unit of electric field E is the N/C. An equivalent unit of E is the

(a) V·m.   (b) V·m^2.
(c) V/m.   (d) V/m^2.

8. The electron volt is a unit of

(a) charge.
(b) potential difference.
(c) energy.
(d) momentum.

9. The force on an electron in an electric field of 200 N/C is

(a) 8×10^{-22} N.   (b) 3.2×10^{-21} N.
(c) 3.2×10^{-17} N.   (d) 6.4×10^{-15} N.

10. The electric field 2 cm from a certain charge has a magnitude of 10^5 N/C. The value of E 1 cm from the charge is

(a) 2.5×10^4 N/C.   (b) 5×10^4 N/C.
(c) 2×10^5 N/C.   (d) 4×10^5 N/C.

11. An electric field of magnitude 200 N/C can be produced by applying a potential difference of 10 V to a pair of parallel metal plates separated by

(a) 2 cm.   (b) 5 cm.
(c) 20 m.   (d) 2000 m.

12. A charge of 10^{-10} C between two parallel metal plates 1 cm apart experiences a force of 10^{-5} N. The potential difference between the plates is

(a) 10^{-5} V.   (b) 10 V.
(c) 10^3 V.   (d) 10^5 V.

13. In charging a certain storage battery, a total of 2×10^5 C is transferred from one set of electrodes to another. The potential difference between the electrodes is 12 V. The energy stored in the battery is

(a) 1.7×10^4 J.   (b) 2.4×10^6 J.
(c) 2.4×10^7 J.   (d) 2.9×10^7 J.

14. The rest energy of the proton is 938 MeV, which is equal to

(a) 1.67×10^{-27} J.   (b) 1.50×10^{-10} J.
(c) 1.04×10^{-8} J.   (d) 1.50×10^{-8} J.

15. An electron whose speed is 10^7 M/s has an energy of

(a) 4.6×10^{-17} eV.   (b) 160 eV.
(c) 284 eV.   (d) 568 eV.

EXERCISES

1. How could you distinguish between an electric field and a gravitational field?

2. What can you tell about the force a charged object would experience at a given point in an electric field by looking at a sketch of the lines of force of the field?

3. Sketch the pattern of electric lines of force in the neighborhood of two charges, $+Q$ and $+2Q$, that are a short distance apart.

4. Sketch the pattern of electric lines of force in the neighborhood of two charges, $-Q$ and $+2Q$, that are a short distance apart.

5. Can lines of force ever intersect in space? Explain.

6. A charge of $+Q$ is placed on a cubical copper box 20 cm on an edge. What is the magnitude and direction of the electric field at the center of the box?

7. An insulating rod has a charge of $+Q$ at one end and a charge of $-Q$ at the other. How will the rod behave when it is placed in a uniform electric field whose direction is (a) parallel to the rod; (b) perpendicular to the rod?

8. The electron in a hydrogen atom averages 5.3×10^{-11} m away from the proton that is the nucleus of this atom. What is the electric field the electron experiences?

9. Find the electric field 40 cm from a charge of $+ 7 \times 10^{-5}$ C.

10. The potential difference between two parallel metal

plates which are 0.5 cm apart is 10^4 V. Find the force on an electron located between the plates.

11. A potential difference of 50 V is applied across two parallel metal plates and an electric field of 10^4 V/m is produced. How far apart are the plates?

12. Four charges of $+1 \times 10^{-8}$ C are at the corners of a square which measures 1 m on each side. Find the electric field at the center of the square.

13. Two parallel metal plates are 4 cm apart. If the force on an electron between the plates is to be 10^{-14} N, what should the potential difference between them be?

14. How strong an electric field is needed to support a proton against gravity at sea level?

15. A cloud is at a potential of 8×10^6 V relative to the ground. A charge of 40 C is transferred in a lightning stroke between the cloud and the ground. Find the energy dissipated.

16. A 12-V storage battery is being charged at the rate of 10 C/s. (a) How much power is needed to charge the battery at this rate? (b) How much energy is stored in the battery if it is charged for 1 h?

17. What is the energy in electron volts of a potassium atom of mass 6.5×10^{-26} kg whose speed is 10^6 m/s?

18. What is the energy in electron volts of an electron whose speed is 10^6 m/s?

19. What is the speed of an electron whose energy is 50 eV?

20. What is the speed of a neutron whose kinetic energy is 50 eV? The neutron mass is 1.67×10^{-27} kg.

PROBLEMS

1. Two charges, one of 1.5×10^{-6} C and the other of 3×10^{-6} C, are 0.2 m apart. Where is the electric field in their vicinity equal to zero?

2. Two charges of $+4 \times 10^{-6}$ C and $+8 \times 10^{-6}$ C are 2 m apart. What is the electric field halfway between them?

3. A particle carrying a charge of 10^{-5} C starts moving from rest in a uniform electric field whose intensity is 50 V/m. (a) What is the force on the particle? (b) How much kinetic energy will the particle have after it has moved 1 m?

4. The electrodes in a neon sign are 1.2 m apart and the potential difference across them is 8000 V. (a) Find the acceleration of a neon ion of mass 19.9 u and charge $+e$ in the field. (b) If the ion starts at the positive electrode of the

sign and moves unimpeded to its negative electrode, how much energy (in joules and in electron volts) would it gain? (c) Why is it extremely unlikely that the ion would actually acquire this much energy?

5. A potential difference of 10 V is applied across two parallel metal plates 2 cm apart. An electron is projected at a speed of 10^7 m/s halfway between the plates and parallel to them. How far will the electron travel before striking the positive plate?

6. What is the electric field at one vertex of an equilateral triangle whose sides are 1 m long if there are charges of $+2 \times 10^{-5}$ C at the other vertexes?

7. In charging a certain 20-kg storage battery, a total of 2×10^5 C is transferred from one set of electrodes to another. The potential difference between the electrodes is 12 V. (a) How much energy is stored in the battery? (b) If this energy were used to raise the battery above the ground, how high would it go? (c) If this energy were used to provide the battery with kinetic energy, what would its speed be?

8. Find the rest energy of the electron in MeV.

9. Typical chemical reactions absorb or release energy at the rate of several eV per molecular change. What change in mass is associated with the absorption or release of 1 eV?

10. The electron gun of a television picture tube has an accelerating potential difference of 15 kV and a power rating of 25 W. How many electrons reach the screen per second? At what speed?

11. An electric dipole consists of the charges $+Q$ and $-Q$ a distance A apart. Find the electric field produced by a dipole at a distance R from the line joining the charges along the perpendicular bisector of that line, and show that $E = kQA/R^3$ when $R >> A$.

12. Find the electric field produced by an electric dipole at a distance R from the center of the line joining the two charges along an extension of that line, and show that $E = 2kQA/R^3$ when $R >> A$. How does the direction of E in this case compare with its direction in Problem 11?

ELECTRIC CURRENT

18

A flow of charge from one place to another constitutes an electric current. Currents and not static charges are involved in nearly all the practical applications of electricity. In an electric circuit, current is the means by which energy is transferred from a source such as a battery or generator to a load, which may be a lamp, a motor, or any other device that absorbs electrical energy and converts it into some other form of energy or into work. In this chapter we shall consider the chief factors that govern direct currents in simple circuits, and in later chapters the magnetic effects of currents and how they are exploited in technology will be examined. The important topic of alternating current is the subject of Chapter 22.

18–1 ELECTRIC CURRENT

The magnitude of an electric current, denoted I, is the rate at which charge passes a given point. If the net charge Q goes past in the time interval t, then the average current is

$$I = \frac{Q}{t}$$
Electric current (18–1)

$$\text{Current} = \frac{\text{charge}}{\text{time interval}}$$

The unit of electric current is the *ampere* (A), where

1 ampere = 1 coulomb/second

The direction of a current is, by convention, taken as that in which *positive* charges would have to move in order to produce the same effects as the observed current. Thus a current is always assumed to proceed from the positive terminal of a battery or generator to its negative terminal in an external circuit (Fig. 18–1).

Despite the above convention, actual electric currents in metals consist of flows of electrons, which carry negative charges. However, a current that consists of negative particles moving in one direction is electrically the same as a current that consists of positive particles moving the other way. Since there is no overwhelming reason to prefer one way of designating current to the other, we shall follow the usual practice of considering current as a flow of positive electric charge.

Two conditions must be met in order for an electric current to exist between two points. These are:

1. There must be a path between the two points along which charge can flow. As was discussed earlier, metals, many liquids, and plasmas (gases whose molecules are charged) allow charge to pass through them readily and are classed as conductors. Nonmetallic solids, certain liquids, and gases whose molecules are electrically neutral allow charge to pass through them only with great difficulty and are classed as insulators. A few substances have an intermediate ability to permit the flow of charge and are classed as semiconductors.

2. There must be a difference of potential between the two points. (A superconductor is an exception to this requirement.) Just as the rate of flow of water between the ends of a pipe depends upon the difference of pressure between them, so the rate of flow of charge between two points depends upon the difference of potential between them. A large potential difference means a large "push" given to each charge.

The analogy between electric current and water flow is a close one. The rate of flow of water in a pipe may be increased by having the water fall through a greater height, which increases the pressure in the pipe and thereby leads to a greater force on each parcel of water. Similarly the current in a wire may be increased by increasing the potential difference across it, which means a stronger electric field in the wire and thus more force on the moving charges that constitute the current (Fig. 18–2).

A particular conducting path—for instance, a copper wire, a light bulb, an electric heater, a transistor—is usually called a conductor, even though this is also the name of the class of substances through which current flows readily. The *resistance* of a conductor is the ratio between the potential difference V across it and the resulting current I that flows:

$$R = \frac{V}{I}$$
Resistance (18–2)

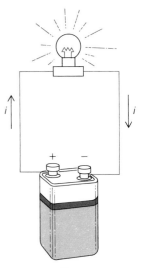

FIG. 18–1 By convention, an electric current is assumed to flow from the positive terminal of a battery or generator to its negative terminal in an external circuit. Actual currents in metals consist of electrons that move in the opposite direction.

A conducting path is needed for a current to occur

A potential difference is also needed

Potential difference is analogous to water pressure

Definition of resistance

FIG. 18–2 The role of potential difference in producing an electric current is analogous to the role of height in producing a flow of water.

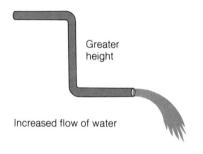

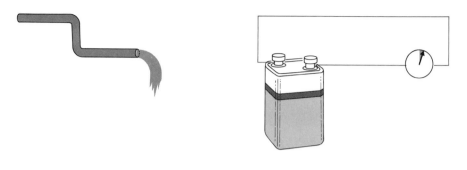

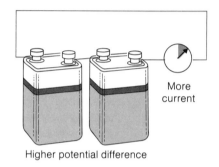

Greater height

Increased flow of water

More current

Higher potential difference

$$\text{Resistance} = \frac{\text{potential difference}}{\text{current}}$$

The unit of resistance is the *ohm* (Ω), where

1 ohm = 1 volt/ampere

A conductor in which there is a current of 1 A when a potential difference of 1 V exists across it has a resistance of 1 Ω.

Example A 120-V electric heater draws a current of 15 A. Find its resistance.

Solution From the definition of resistance,

$$R = \frac{V}{I} = \frac{120\,\text{V}}{15\,\text{A}} = 8\,\Omega$$

18–2 OHM'S LAW

The resistance of a conductor depends in general both upon its properties—its nature and its dimensions—and upon the potential difference applied across it. In some conductors R increases when V increases, in others R decreases when V increases, and in still others R depends upon the direction of the current (Fig. 18–3).

Ohm's law holds for metallic conductors

Metallic conductors usually have constant resistances (at constant temperature), so that I is directly proportional to V in them. This relationship is called *Ohm's law*, since it was first verified experimentally by the German physicist Georg Ohm (1787–1854). Ohm's law states that

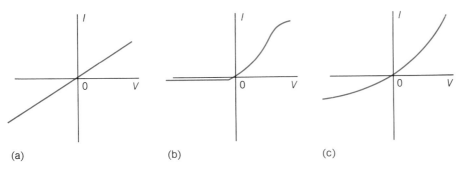

FIG. 18-3 The relationship between current and voltage for (a) a metal, (b) a vacuum tube, and (c) a semiconductor diode. Only in (a) is *I* proportional to *V*, a relationship known as Ohm's law.

(a)   (b)   (c)

$$I = \frac{V}{R} \quad (R = \text{constant}) \qquad \qquad \textit{Ohm's law} \quad (18\text{--}3)$$

Despite its name, Ohm's law is not a fundamental physical principle, but rather an empirical relationship obeyed by most metals under a wide range of circumstances.

Ohm's law must be distinguished from the definition of resistance,

$$R = \frac{V}{I}$$

Ohm's law only holds for conductors in which the ratio V/I is constant regardless of the values of V and I. The resistance of a certain conductor is given by $R = V/I$ whether or not it is constant as the voltage is changed.

Example A light bulb has a resistance of 240 Ω. Find the current in it when it is placed in a 120-V circuit (Fig. 18–4).

Solution From Ohm's law,

$$I = \frac{V}{R} = \frac{120\,\text{V}}{240\,\Omega} = 0.5\,\text{A}$$ ∎

Example The current in the coil of an 8-Ω loudspeaker is 0.5 A. Find the voltage across its terminals.

Solution We rewrite Ohm's law in the form $V = IR$ and obtain

$$V = IR = 0.5\,\text{A} \times 8\,\Omega = 4\,\text{V}$$ ∎

FIG. 18-4

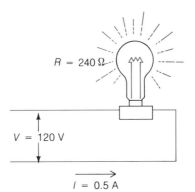

$R = 240\,\Omega$

$V = 120\,\text{V}$

$I = 0.5\,\text{A}$

Electrical hazards

An electric current in body tissue affects it both by stimulating nerves and muscles and through the heat produced. Tissue is a fairly good conductor because of the ions in solution it contains. Dry skin has a higher resistance and so is able to protect to some extent the rest of the body in the event of accidental exposure to a high potential difference, protection that disappears when the skin is wet. A current of as little as 0.5 mA (that is, 0.0005 ampere) is perceptible to most people, one of 5 mA is painful, and one of 10 mA or more causes muscle contractions. Such contractions may prevent the person involved from letting go of the source of the current, and breathing becomes impossible when the current exceeds about 18 mA.

Since a closed conducting path is necessary for a current to occur, touching a single "live" conductor has no effect if the body is isolated. However, if a person at the same time is in contact with a water pipe, or is standing on wet soil, or otherwise is grounded, a current will pass through his body. The resistance of the human body itself is of the order of magnitude of 1000 Ω, so contact via wet skin with a 120-V line will lead to a current in the neighborhood of $I = 120$ V/1000 $\Omega = 0.12$ A = 120 mA. Such a current is extremely dangerous because it is likely to cause the heart muscles to contract rapidly and irregularly, and, if allowed to persist, will cause death.

Because the water in a bathtub is grounded via the tub's drainpipe, a person in the tub is at risk if he or she touches any electrical device, even a switch; the moisture on a wet finger may be sufficient to provide a conducting path to the interior of the device. Another situation of great potential danger occurs in hospitals where electrical appliances are often attached to patients to monitor various functions, or to control them as in the case of cardiac pacemakers. Even a minor malfunction of such an appliance, or the failure to properly ground it, may have fatal consequences because the conducting path in the body is then short compared with what it is when the contact is made via a finger.

18–3   RESISTIVITY

Factors that govern resistance

The resistance of a conductor that obeys Ohm's law depends upon three factors:

1. The material of which it is composed; the ability to carry an electric current varies more than almost any other physical property of matter.
2. Its length L; the longer the conductor, the greater its resistance.
3. Its cross-sectional area A; the thicker the conductor, the less its resistance.

Once again we note the correspondence to water flowing through a pipe: the longer the pipe, the more chance friction against the pipe wall has to slow down the water, and the wider the pipe, the larger the volume of water that can pass through per second when everything else is the same (Fig. 18–5).

The simple formula

$$R = \rho \frac{L}{A} \qquad \textit{Resistance of ohmic conductor} \quad (18\text{–}4)$$

Resistivity

has been found to hold for the resistance of a conductor that obeys Ohm's law. The quantity ρ (the Greek letter *rho*) is called the *resistivity* of the material from which the conductor is made. Table 18–1 lists the resistivities of various substances at room

Short, wide pipe

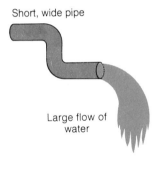

Large flow of
water

Short, thick wire

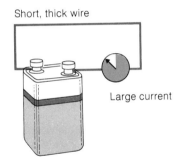

Large current

FIG. 18–5 The way in which the dimensions of a conductor affect the flow of charge in it is analogous to the way in which the dimensions of a pipe affect the flow of water in it.

Long, narrow pipe

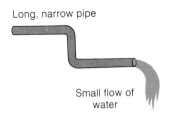

Small flow of
water

Long, thin wire

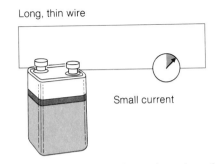

Small current

temperature (20°C). Given the nature of a conductor and its dimensions, the value of R can be calculated at once. In the SI system, lengths are measured in meters and areas in square meters, and the unit of resistivity is accordingly the ohm-meter ($\Omega \cdot m$). Metric wire sizes are usually specified by their cross-sectional areas in square millimeters, so a particular wire might be referred to as 2.5 mm^2, for instance, instead of having its diameter of 1.8 mm quoted.

Example What length of copper wire whose area is 0.1 mm^2 is needed to provide a resistance of 3 Ω?

Solution Since 1 mm $= 10^{-3}$ m, 1 mm$^2 = 10^{-6}$ m^2 and

$$L = \frac{RA}{\rho} = \frac{(3\,\Omega)(0.1\,mm^2)(10^{-6}\,m^2/mm^2)}{1.7 \times 10^{-8}\,\Omega \cdot m} = 18\,m$$ ∎

The resistivities of nearly all substances vary with temperature. In general, metals increase in resistivity with an increase in temperature while nonmetals decrease in resistivity. If R is the resistance of a conductor at a particular temperature, then the change ΔR in its resistance when the temperature changes by ΔT is approximately proportional to both R and ΔT, and therefore

Temperature variation of resistivity

$$\Delta R = \alpha R\,\Delta T \tag{18–5}$$

The quantity α is the temperature coefficient of resistivity of the material. In Table 18–1 the temperature coefficient of carbon is labeled negative because its resistivity decreases with increasing temperature.

TABLE 18–1
Approximate resistivities (at 20°C) and their temperature coefficients

Substance	ρ ($\Omega \cdot$ m)	α (per °C)
Conductors		
Aluminum	2.6×10^{-8}	0.0039
Constantan (60% Cu, 40% Ni)	49×10^{-8}	0.000002
Copper	1.7×10^{-8}	0.0039
Iron	12×10^{-8}	0.0050
Lead	21×10^{-8}	0.0043
Manganin (84% Cu, 12% Mn, 4% Ni)	44×10^{-8}	0.000000
Mercury	98×10^{-8}	0.00088
Platinum	11×10^{-8}	0.0036
Silver	1.6×10^{-8}	0.0038
Semiconductors		
Carbon	3.5×10^{-5}	−0.0005
Germanium	0.5	
Copper oxide (CuO)	1×10^{3}	
Insulators		
Glass	$10^{10} - 10^{14}$	
Quartz	7.5×10^{17}	
Sulfur	10^{15}	

Example   A *resistance thermometer* makes use of the temperature variation of resistivity. When a coil of platinum wire whose resistance at 20°C is 11 Ω is placed in a furnace, its resistance triples. What is the temperature of the furnace, assuming that α remains constant?

Solution   Since $R = 11\ \Omega$ and $\Delta R = 33\ \Omega - 11\ \Omega = 22\ \Omega$,

$$\Delta T = \frac{\Delta R}{\alpha R} = \frac{22\ \Omega}{(0.0036/°\text{C})(11\ \Omega)} = 556°\text{C}$$

The temperature of the furnace is $T + \Delta T = 20°\text{C} + 556°\text{C} = 576°\text{C}$ ∎

18–4   DETERMINING WIRE SIZE

The choice of the wire size for a particular application is determined by either or both of two factors:

1. The maximum current the wire may have to carry.
2. The maximum permissible voltage drop in the wire.

The procedure is to find the smallest wire that each factor permits, and then to use the larger of the two.

The maximum current a wire may safely carry depends both upon how its temperature varies with current and upon the nature of its insulation: a wire that gets too hot may be a fire hazard, and its insulation may melt to produce an electrical hazard as well. The larger the diameter of a wire, the greater its current-carrying capacity, partly because its resistance is lower and partly because it has more surface area to dissipate heat. Tables have been compiled that give the allowable currents for copper wires of standard sizes. For example, asbestos-covered No. 14 wire (diameter 1.63 mm) can safely carry 45 A in the open air, according to the National Electrical Code, whereas rubber-covered No. 14 wire is limited to 20 A.

Current-carrying capacity is sometimes called ampacity

For the second factor, it is necessary to begin by calculating the highest resistance the wire can have in order that the voltage drop not exceed the specified limit. Given the resistance and the length of the wire, the wire size can be found in the usual way.

Example   A water heater draws 30 A from a 120-V power source 10 m away. What is the minimum cross-section of the wire that can be used if the voltage is not to be lower than 115 V at the heater?

Solution   The permissible voltage drop is 5 V, and the resistance that corresponds to this drop when the current is 30 A is

$$R = \frac{V}{I} = \frac{5\,\text{V}}{30\,\text{A}} = 0.167\,\Omega$$

The total length of wire involved is twice the distance between the source and heater, so $L = 2 \times 10\,\text{m} = 20\,\text{m}$. From Eq. (18–4) we have

$$A = \frac{\rho L}{R} = \frac{(1.7 \times 10^{-8}\,\Omega\cdot\text{m})(20\,\text{m})}{0.167\,\Omega} = 2.04 \times 10^{-6}\,\text{m}^2 = 2.04\,\text{mm}^2 \qquad \blacksquare$$

In the United States, wire is manufactured in the standard sizes specified by the American Wire Gage (AWG) system. The largest wire in this system is AWG No. 0000, which is 11.7 mm in diameter, and the smallest is the hair-thin No. 40, which is 0.079 mm in diameter. The sequence of diameters is such that every third gage number means a cross-sectional area half as great and so a resistance twice as great for the same wire length. Thus No. 17 copper wire has a resistance of 1.57 Ω per 100 m, double the 0.786 Ω/100 m resistance of No. 14 wire. Even sizes are normally employed for wiring purposes, with the odd sizes finding use in coils of various kinds, such as the windings of motors and transformers. No. 14 wire is the smallest permitted by the National Electrical Code for residential, farm, and industrial wiring.

American Wire Gage system

18–5 ELECTRIC POWER

Electrical energy in the form of electric current is converted into heat in an electric stove, into radiant energy in a light bulb, into chemical energy when a storage battery is charged, and into mechanical energy in an electric motor. The widespread use of electrical energy is due as much to the ease with which it can be transformed into other kinds of energy as to the ease with which it can be transported through wires.

Electric energy is readily transformed into other forms of energy

The work that must be done to take a charge Q through the potential difference V is, by definition,

$$W = QV$$

Since a current I carries the amount of charge $Q = It$ in the time t, the work done is

Work done by a current

$$W = IVt \qquad\qquad \textit{Electric work} \quad (18–6)$$

The energy input to a device of any kind through which the current I flows when the potential difference V is placed across it is equal to the product of the current, the potential difference, and the time span.

We recall from Chapter 6 that *power* is the term given to the rate at which work is being done, so that

$$P = \frac{W}{t} \tag{6-4}$$

$$\text{Power} = \frac{\text{work done}}{\text{time interval}}$$

When the work is done by an electric current, $W = IVt$, and so

$$P = IV \qquad\qquad\qquad \textit{Electric power} \quad (18\text{-}7)$$

$$\text{Electric power} = \text{current} \times \text{potential difference}$$

Watts = amperes × volts

The unit of power is the watt, and when I and V are in amperes and volts, respectively, P will be in watts.

Example   A solar cell 10 cm in diameter produces a current of 2.15 A at 0.45 V in bright sunlight whose intensity is 0.1 W/cm². Find the efficiency of the cell.

Solution   The area of the cell is $A = \pi d^2/4 = 78.5$ cm², so it receives solar energy at the rate of

$$P_{\text{input}} = (0.1\text{ W/cm}^2)(78.5\text{ cm}^2) = 7.85\text{ W}$$

The power output of the cell is

$$P_{\text{output}} = IV = (2.15\text{ A})(0.45\text{ V}) = 0.97\text{ W}$$

The efficiency of the cell is therefore

$$\text{Eff} = \frac{P_{\text{output}}}{P_{\text{input}}} = \frac{0.97\text{ W}}{7.85\text{ W}} = 0.12 = 12\% \qquad \blacksquare$$

Example   How much current is drawn by a $\frac{1}{2}$-hp electric motor operated from a 120-V source of electricity? Assume that all of the electrical energy absorbed by the motor is turned into mechanical work.

Solution   Since 1 hp = 746 W, we find from $P = IV$ that

$$I = \frac{P}{V} = \frac{(\frac{1}{2}\text{ hp})(746\text{ W/hp})}{120\text{ V}} = 3.1\text{ A} \qquad \blacksquare$$

Formulas for electric power

The power consumed by a resistance that obeys Ohm's law ($I = V/R$) through which current passes may be expressed in the alternative forms

$$P = IV \tag{18-7}$$

$$P = I^2 R \qquad\qquad \textit{Electric power} \quad (18\text{-}8)$$

$$P = \frac{V^2}{R} \tag{18-9}$$

Equation (18–7) holds regardless of the nature of the current-carrying device. Depending upon which quantities are known in a specific case, any of the above expressions for P may be used.

Example   Find the power consumed by a 240-Ω light bulb when the current through it is 0.5 A.

Solution   The formula $P = I^2R$ is easiest to use here. We have

$$P = I^2R = (0.5\,\text{A})^2(240\,\Omega) = 60\,\text{W} \qquad \blacksquare$$

Owing to the resistance that all conductors offer to the flow of charge through them, electric power is dissipated whenever a current is maintained regardless of whether the current also supplies energy that is converted to some other form. Electrical resistance is analogous to friction, and so the power consumed in causing a current to flow is dissipated as heat. If too much current flows in a particular wire, it becomes so hot that it may start a fire or even melt. To prevent this from happening, nearly all electric circuits are protected by fuses or circuit breakers, which interrupt the current when I exceeds a safe value. For example, a 15-A fuse in a 120-V power line means that the maximum power that can be carried is

Fuses and circuit breakers protect circuits from overloads

$$P = IV = (15\,\text{A})(120\,\text{V}) = 1800\,\text{W}$$

Table 18–2 summarizes the various formulas for potential difference V, current I, resistance R, and power P that follow from Ohm's law, $I = V/R$, and from the power formula, $P = IV$.

18–6   RESISTORS IN SERIES

In analyzing a direct-current circuit, it is convenient to group its various components into individual *resistors* that are imagined to be joined together by resistanceless wires. The *equivalent resistance* of a set of interconnected resistors is the value of the single resistor that can be substituted for the entire set without affecting the current in the rest of the circuit.

Equivalent resistance

TABLE 18–2

	V and I	I and R	V and R	P and I	P and V	P and R
$V =$		IR		$\dfrac{P}{I}$		$\sqrt{PR}$
$I =$			$\dfrac{V}{R}$		$\dfrac{P}{V}$	$\sqrt{\dfrac{P}{R}}$
$R =$	$\dfrac{V}{I}$			$\dfrac{P}{I^2}$	$\dfrac{V^2}{P}$	
$P =$	IV	I^2R	$\dfrac{V^2}{R}$			

FIG. 18–6

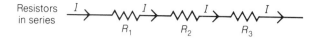

Resistors in series

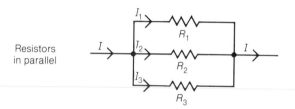

Resistors in parallel

Resistors in series and parallel

　　The symbol for a resistor, $\wedge\!\wedge\!\wedge$, represents any circuit component that has electrical resistance and obeys Ohm's law. Although there are an unlimited number of ways in which resistors can be put together, many of them are merely combinations of the basic *series* and *parallel* arrangements (Fig. 18–6). Resistors in series are connected consecutively so the same current is present in all of them. On the other hand, resistors in parallel have their ends connected together, so the total current is split up among them.

　　The potential difference V across the ends of a series set of resistors is the sum of the potential differences V_1, V_2, V_3, . . . , across each one. This statement follows from the principle of conservation of energy: if V is the work done per coulomb in "pushing" a charge through the set of resistors, then it must equal the sum $V_1 + V_2 + V_3 + \cdots$ of the amounts of work done per coulomb in pushing the charge through each resistor in turn. In the case of the three resistors in series of Fig. 18–7,

$$V = V_1 + V_2 + V_3$$

Resistors in series have the same current

　　Since the same current I passes through all the resistors, the individual potential drops are

$$V_1 = IR_1 \qquad V_2 = IR_2 \qquad V_3 = IR_3$$

If we denote the equivalent resistance of the set by R, the potential difference across it is

$$V = IR$$

We therefore have

$$V = V_1 + V_2 + V_3$$
$$IR = IR_1 + IR_2 + IR_3$$

Dividing through by the current I yields

$$R = R_1 + R_2 + R_3$$

Equivalent resistance of series resistors

　　In general, *the equivalent resistance of a set of resistors connected in series is equal to the sum of the individual resistances:*

$$R = R_1 + R_2 + R_3 + \cdots \qquad\qquad \textit{Resistors in series} \quad (18\text{–}10)$$

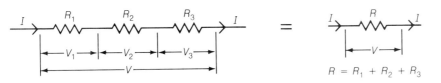

FIG. 18–7 Resistors in series. The same current I passes through each resistor.

$$R = R_1 + R_2 + R_3$$

Example A 5-Ω resistor and a 20-Ω resistor are connected in series and a potential difference of 100 volts is applied across them by means of a generator. Find (a) the equivalent resistance of the circuit, (b) the current that flows in it, (c) the potential difference across each resistor, (d) the power dissipated by each resistor, and (e) the power dissipated by the entire circuit.

Solution Successive stages in the solution of this problem are shown in Fig. 18–8.
  (a) The equivalent resistance of the two resistors is

$$R = R_1 + R_2 = 5\ \Omega + 20\ \Omega = 25\ \Omega$$

This resistance is more than that of either of the resistors.
  (b) The current in the circuit is, from Ohm's law,

$$I = \frac{V}{R} = \frac{100\ \text{V}}{25\ \Omega} = 4\ \text{A}$$

  (c) The potential difference across R_1 is

$$V_1 = IR_1 = (4\ \text{A})(5\ \Omega) = 20\ \text{V}$$

and the potential difference across R_2 is

$$V_2 = IR_2 = (4\ \text{A})(20\ \Omega) = 80\ \text{V}$$

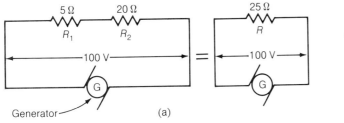

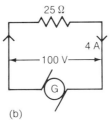

FIG. 18–8

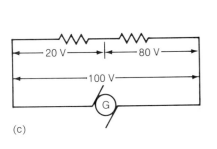

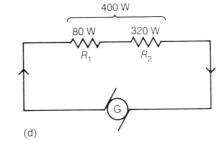

We note that the sum of V_1 and V_2 is 100 V, the same as the impressed potential difference, as of course it must be.

(d) The power dissipated by R_1 is

$$P_1 = IV_1 = (4 \text{ A})(20 \text{ V}) = 80 \text{ W}$$

We can also find P_1 from the alternative formulas $P = I^2R$ and $P = V^2/R$. In each case we get the same answer:

$$P_1 = I^2R_1 = (4 \text{ A})^2(5 \text{ }\Omega) = 80 \text{ W} \qquad P_1 = \frac{V^2}{R} = \frac{(20 \text{ V})^2}{5 \text{ }\Omega} = 80 \text{ W}$$

The power dissipated by R_2 is

$$P_2 = IV_2 = (4 \text{ A})(80 \text{ V}) = 320 \text{ W}$$

(e) The power dissipated by the entire circuit is

$$P = IV = (4 \text{ A})(100 \text{ V}) = 400 \text{ W}$$

and is equal to $P_1 + P_2$. The dissipated power appears as heat.   ■

Example   A 20-Ω load whose maximum power rating is 5 W is to be connected to a 24-V battery. What is the minimum resistance of the series resistor (Fig. 18–9) that is required?

Solution   The maximum allowable current in the resistor R_1 may be found from the formula $P = I^2R_1$ to be

$$I = \sqrt{\frac{P}{R_1}} = \sqrt{\frac{5 \text{ W}}{20 \text{ }\Omega}} = 0.5 \text{ A}$$

For a current of 0.5 A, the equivalent resistance R of the circuit should be

$$R = \frac{V}{I} = \frac{24 \text{ V}}{0.5 \text{ A}} = 48 \text{ }\Omega$$

Since $R = R_1 + R_2$, the minimum resistance R_2 of the required series resistor is

$$R_2 = R - R_1 = 48 \text{ }\Omega - 20 \text{ }\Omega = 28 \text{ }\Omega$$ ■

18–7   RESISTORS IN PARALLEL

When two or more resistors are connected in parallel, the same potential difference is applied across all of them. As we shall find, the current that passes through each resistor is inversely proportional to its resistance: the less the resistance, the more the current.

Resistors in parallel have the same voltage across them

Let us consider three resistors, R_1, R_2, and R_3, that are connected in parallel, as in Fig. 18–10. The total current I through the set is equal to the sum of the currents through the separate resistors, so that

$$I = I_1 + I_2 + I_3$$

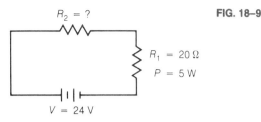

FIG. 18–9

The potential difference V is the same across all the resistors, and by applying Ohm's law to each of them in turn we find that

$$I_1 = \frac{V}{R_1} \qquad I_2 = \frac{V}{R_2} \qquad I_3 = \frac{V}{R_3}$$

The smaller the resistance, the greater the proportion of the total current that flows through it.

The total current flowing through the set of three resistors is given in terms of their equivalent resistance R by

$$I = \frac{V}{R}$$

Hence we have

$$I = I_1 + I_2 + I_3$$
$$\frac{V}{R} = \frac{V}{R_1} + \frac{V}{R_2} + \frac{V}{R_3}$$

The final step is to divide through by V, which yields

$$\frac{1}{R} = \frac{1}{R_1} + \frac{1}{R_2} + \frac{1}{R_3}$$

In general, *the reciprocal of the equivalent resistance of a set of resistors connected in parallel is equal to the sum of the reciprocals of the individual resistances*:

$$\frac{1}{R} = \frac{1}{R_1} + \frac{1}{R_2} + \frac{1}{R_3} + \cdots \qquad \text{Resistors in parallel} \quad (18\text{–}11)$$

Equivalent resistance of parallel resistors

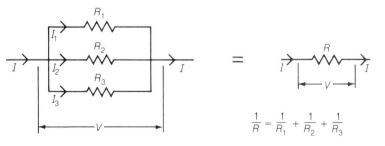

FIG. 18–10 Resistors in parallel. The same potential difference V exists across each resistor.

The formulas for the equivalent resistances of series and parallel arrangements of resistors are in accord with the basic formula

$$R = \rho \frac{L}{A}$$

for the resistance of a conductor. When several resistors are connected in series, the effect is the same as increasing the length L, whereas when they are connected in parallel, the effect is the same as increasing the cross-sectional area A. Thus a series set of resistors lets through less current than any of the individual resistors, and a parallel set of resistors lets through more current than any of the individual resistors, in each case assuming the same potential difference.

As a check after making a calculation, it is worth recalling that in a series circuit, the equivalent resistance is always *greater* than any of the individual resistances. In a parallel circuit, on the other hand, the equivalent resistance is always *smaller* than any of the individual resistances.

Two resistors in parallel   In the case of two resistors in parallel, the equivalent resistance is

$$\frac{1}{R} = \frac{1}{R_1} + \frac{1}{R_2}$$

The lowest common denominator of the right-hand side of this formula is $R_1 R_2$, which enables us to write

$$\frac{1}{R} = \frac{R_1 + R_2}{R_1 R_2}$$

Taking the reciprocal of both sides of this equation yields the convenient result

$$R = \frac{R_1 R_2}{R_1 + R_2} \qquad\qquad \textit{Two resistors in parallel} \quad (18\text{--}12)$$

When using a calculator that has a reciprocal [1/X] key, it is easier to use Eq. (18–11) directly, as shown in part (a) of the example below.

Example   A 5-Ω resistor and a 20-Ω resistor are connected in parallel and a potential difference of 100 volts is applied across them by means of a generator. Find (a) the equivalent resistance of the circuit, (b) the current that flows in each resistor and in the circuit as a whole, (c) the power dissipated by each resistor, and (d) the power dissipated by the entire circuit.

Solution   Successive stages in the solution of this problem are shown in Fig. 18–11.

(a) The equivalent resistance of the resistors is

$$R = \frac{R_1 R_2}{R_1 + R_2} = \frac{(5\,\Omega)(20\,\Omega)}{5\,\Omega + 20\,\Omega} = 4\,\Omega$$

Alternatively, we can find R from

$$\frac{1}{R} = \frac{1}{5\,\Omega} + \frac{1}{20\,\Omega}$$

FIG. 18–11

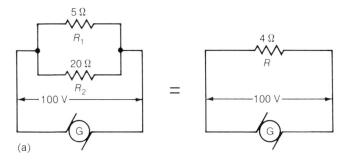

(a)

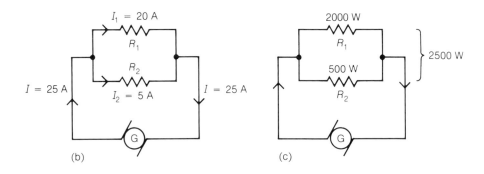

(b)   (c)

with the following key sequence on a calculator: $[5] [1/X] [+] [20] [1/X] [=] [1/X]$.
The generalization to three or more resistors is obvious. For three resistors, for instance,
the sequence would be: $[R_1][1/X][+][R_2][1/X][+][R_3][1/X][=][1/X]$.

(b) The current that flows in R_1 is

$$I_1 = \frac{V}{R_1} = \frac{100\text{ V}}{5\ \Omega} = 20\text{ A}$$

and the current that flows in R_2 is

$$I_2 = \frac{V}{R_2} = \frac{100\text{ V}}{20\ \Omega} = 5\text{ A}$$

The total current in the circuit is

$$I = \frac{V}{R} = \frac{100\text{ V}}{4\ \Omega} = 25\text{ A}$$

and is equal to $I_1 + I_2$.

(c) The power dissipated by R_1 is

$$P_1 = I_1 V = (20\text{ A})(100\text{ V}) = 2000\text{ W}$$

and that dissipated by R_2 is

$$P_2 = I_2 V = (5\text{ A})(100\text{ V}) = 500\text{ W}$$

FIG. 18–12

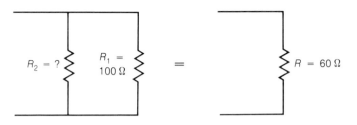

(d) The power dissipated by the entire circuit is

$$P = IV = (25\,\text{A})(100\,\text{V}) = 2500\,\text{W}$$

and is equal to $P_1 + P_2$.

We recall from the previous section that, when a 100-V potential difference is applied across the same two resistors connected in series, the power dissipated is only 400 W. The difference is due to the lower equivalent resistance of the parallel combination (4 Ω instead of 25 Ω), which permits more current to flow. Since $P = IV$ and V is the same in both cases, the greater current in the parallel circuit leads to a greater power dissipation.   ■

Example   A circuit has a resistance of 100 Ω. How can it be reduced to 60 Ω?

Solution   In order to obtain an equivalent resistance of $R = 60\,\Omega$, a resistor R_2 must be connected in parallel with the circuit of $R_1 = 100\,\Omega$ (Fig. 18–12). To find the value of R_2 we proceed as follows:

$$\frac{1}{R} = \frac{1}{R_1} + \frac{1}{R_2}$$

$$\frac{1}{R_2} = \frac{1}{R} - \frac{1}{R_1} = \frac{R_1 - R}{R_1 R}$$

$$R_2 = \frac{R_1 R}{R_1 - R} = \frac{(100\,\Omega)(60\,\Omega)}{100\,\Omega - 60\,\Omega} = 150\,\Omega$$

With a calculator, we would work from

$$\frac{1}{R} = \frac{1}{60\,\Omega} - \frac{1}{100\,\Omega}$$

and use this key sequence:

$$[60][1/X][-][100][1/X][=][1/X]$$ ■

Example   Find the equivalent resistance of the set of resistors shown in Fig. 18–13.

Solution   The procedure to follow here is shown in Fig. 18–14, where the original complex circuit is decomposed into its series and parallel elements. The first step is to find the equivalent resistance R' of the parallel resistors R_2 and R_3.

$$R' = \frac{R_2 R_3}{R_2 + R_3} = \frac{(12\,\Omega)(8\,\Omega)}{12\,\Omega + 8\,\Omega} = 4.8\,\Omega$$

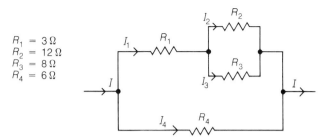

FIG. 18-13

$R_1 = 3\,\Omega$
$R_2 = 12\,\Omega$
$R_3 = 8\,\Omega$
$R_4 = 6\,\Omega$

FIG. 18-14 Successive steps in determining the equivalent resistance R of the resistor network shown in Fig. 18-13.

This pair of resistors is in series with R_1, and the equivalent resistance R'' of the upper branch of the circuit is therefore

$$R'' = R' + R_1 = 4.8\,\Omega + 3\,\Omega = 7.8\,\Omega$$

The upper and lower branches of the circuit are in parallel, which means that the equivalent resistance R of the entire circuit is

$$R = \frac{R''R_4}{R'' + R_4} = \frac{7.8\,\Omega \times 6\,\Omega}{7.8\,\Omega + 6\,\Omega} = 3.4\,\Omega \qquad \blacksquare$$

Example A potential difference of 12 V is applied across the set of resistors shown in Fig. 18-13. Find the current that flows through each resistor.

Solution The full potential difference is applied across R_4, and so, by Ohm's law, the current I_4 in it is

$$I_4 = \frac{V}{R_4} = \frac{12\,\text{V}}{6\,\Omega} = 2.0\,\text{A}$$

The current I_1 that flows through R_1 also flows through the entire upper branch of the circuit. Since the equivalent resistance of this branch is R'',

$$I_1 = \frac{V}{R''} = \frac{12\,\text{V}}{7.8\,\Omega} = 1.54\,\text{A}$$

The potential difference V' across the parallel resistors R_2 and R_3 is equal to the current I_1 through the equivalent resistance R' multiplied by R':

$$V' = I_1 R' = (1.54\ \text{A})(4.8\ \Omega) = 7.4\ \text{V}$$

Another way to find V' is to subtract the potential difference

$$V_1 = I_1 R_1 = (1.54\ \text{A})(3\ \Omega) = 4.6\ \text{V}$$

from the total of 12 volts to obtain

$$V' = V - V_1 = 12\ \text{V} - 4.6\ \text{V} = 7.4\ \text{V}$$

Hence the currents I_2 and I_3 through resistors R_2 and R_3 are respectively

$$I_2 = \frac{V'}{R_2} = \frac{7.4\,\text{V}}{12\,\Omega} = 0.62\,\text{A} \qquad I_3 = \frac{V'}{R_3} = \frac{7.4\,\text{V}}{8\,\Omega} = 0.92\,\text{A} \qquad \blacksquare$$

18–8   ELECTROMOTIVE FORCE

Emf is no-load voltage of a source

The potential difference that exists across a battery, generator, or other source of electrical energy when it is not connected to any external circuit is called its *electromotive force*. Electromotive force is usually referred to simply as emf, and its symbol is $\mathscr{E}$.

As charges pass through a source of electrical energy, work is done on them, and the emf of the source is the work done per coulomb on the charges. The emf of an automobile storage battery is 12 V, which means that 12 J of work are done on each coulomb of charge that passes through the battery. In the case of a battery, chemical energy is converted into electrical energy by means of the work done on the charges in transit through it; in a generator, mechanical energy is converted into electrical energy; in a thermocouple, heat energy is converted into electrical energy; and so on. The emf of an electrical source bears a relationship to its power output analogous to that of applied force to mechanical power in a machine, which is the reason for its name.

Internal resistance of an emf source

When a source of electrical energy is part of a complete circuit, a current I flows, and the potential difference across the terminals of the source is always *less* than its emf owing to its *internal resistance*. Every electrical source has a certain amount of internal resistance r, which means that a potential drop Ir occurs *within* the source. Hence the actual terminal voltage V across a source of emf $\mathscr{E}$ and internal resistance r is

$$V = \mathscr{E} - Ir \qquad\qquad \textit{Terminal voltage} \quad (18\text{–}13)$$

Terminal voltage

Terminal voltage = emf − potential drop within source

If the source is disconnected, no current flows, and $V = \mathscr{E}$; the existence of a current lowers the value of V by an amount proportional to I.

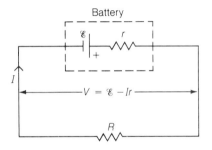

Figure 18–15 shows a battery of emf $\mathscr{E}$ connected to an external circuit whose equivalent resistance is R. The total resistance in the entire circuit is R plus the internal resistance r of the battery, so that the current I that flows is

$$I = \frac{\mathscr{E}}{R + r} \tag{18–14}$$

The actual potential difference across the battery is given by Eq. (18–13).

The internal resistance of a battery governs the maximum current it can supply. As mentioned in Sec. 16–9, this is one of the reasons why it is so hard to start a car on a cold day. The lower the temperature of the electrolyte in a battery, the more slowly its ions move and the higher the internal resistance. As a result the current available for the starting motor in freezing weather may be less than half that available on a warm day.

Example A "D" cell of emf 1.5 V and internal resistance 0.3 Ω is connected to a flashlight bulb whose resistance is 3.0 Ω. Find the current in the circuit and the terminal voltage of the cell.

Solution The current in the circuit is

$$I = \frac{\mathscr{E}}{R + r} = \frac{1.5\,\text{V}}{3.0\,\Omega + 0.3\,\Omega} = 0.45\,\text{A}$$

The terminal voltage of the cell is

$$V = \mathscr{E} - Ir = 1.5\,\text{V} - (0.45\,\text{A})(0.3\,\Omega) = 1.37\,\text{V} \qquad\blacksquare$$

Example A storage battery whose emf is 12 V and whose internal resistance is 0.2 Ω is to be charged at a rate of 20 A. What applied voltage is required?

Solution The applied voltage V must exceed the battery's emf $\mathscr{E}$ by the amount Ir to provide the charging current I. Hence

$$V = \mathscr{E} + Ir = 12\,\text{V} + (20\,\text{A})(0.2\,\Omega) = 16\,\text{V} \qquad\blacksquare$$

Example A battery of emf $\mathscr{E}$ and internal resistance r is connected to an external resistance R. What should the value of R be in order that it dissipate the maximum amount of power?

Solution The current in the circuit is $I = \mathscr{E}/(R + r)$ as in Eq. (18–14). The power dissipated in the external resistance is therefore

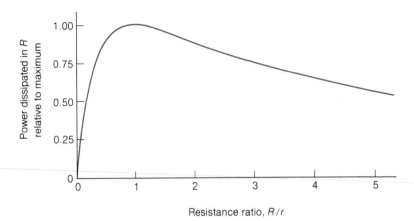

FIG. 18–16   The power transferred to an external load of resistance R is a maximum when R equals the internal resistance r of the source.

$$P = I^2R = \frac{\mathcal{E}^2R}{(R + r)^2}$$

From Fig. 18–16, which is based on this formula, it is clear that P has its maximum value when $R = r$. ■

Impedance matching is important in energy transfer

The conclusion reached in the preceding example is an example of the phenomenon of *impedance matching*: when energy is being transferred from one system to another (here from the battery to the external resistance), the efficiency is greatest when both systems have the same impedance, which is a general term for resistance to the flow of energy in whatever form it may take in a particular case. We encountered impedance matching earlier, the first time in Sec. 7–6. There we saw that when a moving object strikes a stationary one, the maximum transfer of energy takes place when both have the same mass. In this situation the inertia of each object, as measured by its mass, represents its impedance. The curve of relative energy transfer versus mass ratio in Fig. 7–15 is identical in form to that of Fig. 18–16.

Another example of impedance matching occurs when a pulse or wave moves down a stretched string, as discussed in Sec. 12–2. When two strings having the same mass per unit length are joined together, a pulse in one passes to the other with no reflection. However, if the second string has either a greater or a smaller mass per unit length, reflection occurs at the junction and not all the energy of the pulse is transmitted (Figs. 12–6 and 12–7). Though best known for its application to electric circuits, impedance matching is an important concept in many other branches of physics and engineering as well.

18–9   KIRCHHOFF'S RULES

It is often difficult or impossible to determine the currents that flow in the various branches of a complex network merely by computing equivalent resistances. Two rules formulated by Gustav Kirchhoff (1824–1887) make it possible to find the current in each part of a direct-current circuit, no matter how complicated, if we are given the emf's of the sources of potential difference and the resistances of the various circuit

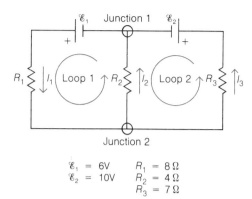

FIG. 18–17 A *junction* is a point where three or more wires come together; a *loop* is any closed conducting path in the network. Here the internal resistances of the batteries are included in R_1 and R_3. The directions of the currents I_1, I_2, and I_3 are chosen arbitrarily; the value of I_3 turns out to be negative, meaning that its actual direction is opposite to the one shown.

elements. These rules apply to *junctions,* which are points where three or more wires come together, and to *loops,* which are closed conducting paths that are part of the circuit (Fig. 18–17).

 Kirchhoff's first rule follows from the conservation of electric charge. Charge is never found to accumulate at any point in a circuit, nor can it be created there, and so there cannot be any net current into or out of a junction. Hence the first rule:

1. The sum of the currents flowing into a junction is equal to the sum of the currents flowing out of the junction.

 The second rule is a consequence of the conservation of energy. The sum of the emf's in a loop equals the amount of work done per coulomb by the sources of emf *on* a charge that moves once around the loop. The work done per coulomb *by* the charge as it moves around the loop equals the sum of the *IR* potential drops along the way. To conserve energy, the work done on the charge must be the same as the work done by the charge, and so we have Kirchhoff's second rule:

2. The sum of the emf's around a loop is equal to the sum of the *IR* potential drops around the loop.

 A definite procedure must be followed when Kirchhoff's rules are applied to a network. First, a direction is assumed for the current through each resistor. If we have guessed correctly, the solution of the problem will give a positive value for the current; if we have guessed wrong, a negative value will indicate that the actual current is in the reverse direction. Second, in going around a loop to apply the second rule, we must follow a consistent path, either clockwise or counterclockwise. An emf is reckoned positive if we meet the negative terminal of its source first; if instead we meet the positive terminal first, the emf is reckoned negative. An *IR* drop is considered positive if the current in the resistor is in the same direction as the path we are following, and negative if its direction is opposite to that of our path.

Example Find the current in each of the resistors in the network shown in Fig. 18–17.

Junctions and loops

Kirchhoff's first rule

Kirchhoff's second rule

How the second rule is applied

Solution The assumed directions of the unknown currents I_1, I_2, and I_3 are shown in the figure. Applying Kirchoff's first rule to junction 1, we obtain

Junction 1
$$I_1 = I_2 + I_3$$

This rule applied to junction 2 yields the same result.

In applying Kirchhoff's second rule to the two loops we shall follow counterclockwise routes. For loop 1,

Loop 1
$$\mathscr{E}_1 = I_1 R_1 + I_2 R_2$$

and for loop 2,

Loop 2
$$-\mathscr{E}_2 = -I_2 R_2 + I_3 R_3$$

In loop 2 we consider $\mathscr{E}_2$ as negative because we encounter its positive terminal first, and I_2 as negative because its direction is opposite to our counterclockwise path.

There is also a third loop, namely the outside one in Fig. 18–17, which similarly must obey Kirchhoff's second rule. Proceeding counterclockwise yields

Outside loop
$$-\mathscr{E}_2 + \mathscr{E}_1 = I_1 R_1 + I_3 R_3$$

We note that this last equation is just the sum of the two preceding loop equations. Thus we may use the junction equation and any two of the loop equations to solve for the unknown currents; nothing will be gained by using all three loop equations.

There are three unknown currents, and we have three independent equations relating them, which means that the problem can be solved by routine algebra. As an example of how this might be done, we shall start by substituting $I_2 + I_3$ for I_1 in the first loop equation to obtain

$$\mathscr{E}_1 = I_2 R_1 + I_3 R_1 + I_2 R_2$$

We now solve both this equation and the second loop equation for I_3:

$$I_3 = \frac{\mathscr{E}_1 - I_2 R_1 - I_2 R_2}{R_1} \qquad I_3 = \frac{-\mathscr{E}_2 + I_2 R_2}{R_3}$$

Setting the two expressions for I_3 equal enables us to solve for I_2 as follows:

$$\frac{\mathscr{E}_1 - I_2 R_1 - I_2 R_2}{R_1} = \frac{-\mathscr{E}_2 + I_2 R_2}{R_3}$$

$$\mathscr{E}_1 R_3 - I_2 R_1 R_3 - I_2 R_2 R_3 = -\mathscr{E}_2 R_1 + I_2 R_1 R_2$$

$$I_2 (R_1 R_3 + R_2 R_3 + R_1 R_2) = \mathscr{E}_2 R_1 + \mathscr{E}_1 R_3$$

$$I_2 = \frac{\mathscr{E}_2 R_1 + \mathscr{E}_1 R_3}{R_1 R_3 + R_2 R_3 + R_1 R_2}$$

Finally we insert the values of $\mathscr{E}_1$, $\mathscr{E}_2$, R_1, R_2, and R_3 given in Fig. 18–17 to obtain

$$I_2 = 1.05 \text{ A}$$

With the value of I_2 known we can find the value of I_3 from the second loop equation:

$$I_3 = \frac{-\mathscr{E}_2 + I_2 R_2}{R_3} = -0.83 \text{ A}$$

The minus sign means that the direction of I_3 is opposite to that shown in Fig. 18–17. The junction equation finally provides us with the value of I_1.

$$I_1 = I_2 + I_3 = 0.22\,\text{A}$$

The same results would have been obtained had we chosen other directions for the currents or taken other routes around the loops in applying Kirchhoff's second rule. The important thing is not what choice of directions is made but to follow that choice consistently in working out the problem.

In the above calculation symbols for the various quantities were used until the very end to make clear the operations involved. However, in working out problems of this kind it is usually faster to substitute numerical values from the start. ∎

18–10 AMMETERS, VOLTMETERS, AND OHMMETERS

The traditional instrument for measuring electrical quantities is the *galvanometer*, whose principle of operation is described in Chapter 19. The deflection of a galvanometer's pointer is proportional to the current that passes through it. A galvanometer can be used to measure potential difference as well as current. It has a certain resistance, and according to Ohm's law the current in it is directly proportional to the potential difference across its terminals. Hence the meter's scale can be calibrated equally well in volts as in amperes. A meter that has a full-scale reading of, say, 1 mA (10^{-3} A) and a resistance of 50 Ω can also be used to measure potential differences of up to

$$V_{max} = I_{max}R_{coil} = (10^{-3}\,\text{A})(50\,\Omega) = 0.05\,\text{V}$$

A galvanometer employed to measure the current in a circuit is called an *ammeter* and is inserted in series in the circuit. The inherent range of a galvanometer (0 to 1 mA in the case of the above meter) can be extended by connecting a low resistance shunt in parallel with it to carry the bulk of the current, leaving a known fraction of the current to be registered by the meter itself.

Ammeter

As an example, let us convert the above galvanometer to measure currents of up to 1 A. What we must do is place a resistor in parallel with the meter that will divert

$$1.000\,\text{A} - 0.001\,\text{A} = 0.999\,\text{A}$$

of the maximum current, leaving 0.001 A to give a full-scale reading on the meter (Fig. 18–18). To obtain the value of R_{shunt} we note that, since the potential difference V is the same across each branch of a parallel circuit,

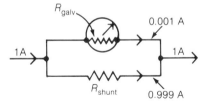

FIG. 18–18 The range of current a galvanometer can measure can be extended by placing a shunt resistor in parallel with it. The equivalent resistance of such an ammeter should be very low.

FIG. 18–19 The range of voltage a galvanometer can measure can be extended by placing a resistor in series with it. The equivalent resistance of such a voltmeter should be very high.

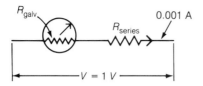

$$V = I_{shunt}R_{shunt} = I_{galv}R_{galv}$$

and so

$$R_{shunt} = R_{galv}\frac{I_{galv}}{I_{shunt}} = (50\,\Omega)\left(\frac{0.001\,A}{0.999\,A}\right) = 0.050\,\Omega$$

An ammeter should have a low resistance

An ideal ammeter has a very low equivalent resistance so that its presence in any circuit alters the properties of the circuit as little as possible. The above ammeter has an equivalent resistance of very nearly 0.050 Ω; unless the equivalent resistance of a circuit in which it is inserted is large compared with 0.050 Ω, the meter reading will not be an accurate measure of the current in the absence of the ammeter.

Voltmeter

A galvanometer employed to measure potential difference is called a *voltmeter*. The inherent voltage range of a galvanometer can be extended by inserting a resistor in series with it in order to hold the maximum current down to whatever figure is appropriate for the meter's movement.

We might wish to convert the above 0 to 1 mA galvanometer to a voltmeter with a full-scale reading of 1 V. This means that when a potential difference of 1 V is across the voltmeter's terminals, a current of 1 mA flows in the meter's coil. The equivalent resistance of the voltmeter must therefore be

$$R = R_{series} + R_{galv} = \frac{V}{I}$$

and so

$$R_{series} = \frac{V}{I} - R_{galv} = \frac{1\,V}{0.001\,A} - 50\,\Omega = 950\,\Omega$$

A 950-Ω resistor connected in series with the galvanometer converts it to a voltmeter whose range is 0 to 1 V (Fig. 18–19).

A voltmeter should have a high resistance

Placing a voltmeter across a circuit element is the same thing as connecting a resistor in parallel with it, which changes the properties of the circuit and so leads to an incorrect potential difference measurement. For this reason an ideal voltmeter has a very high resistance. A voltmeter rated at 10,000 Ω/V, which is typical of inexpensive combination volt-ohm-milliammeters, has a resistance of 10,000 Ω on the 0–1-V scale, 100,000 Ω on the 0–10-V scale, and so forth. Better voltmeters of the conventional type have ratings of 50,000 Ω/V, and electronic voltmeters may go as high as 10^7 Ω/V, which means that they have essentially no effect on a circuit they are connected to.

Ohmmeter

A galvanometer can be used to measure an unknown resistance R with the help of a battery and a series resistor, as in Fig. 18–20. Such a device is called an *ohmmeter*. The meter's scale is calibrated backwards, so that a full-scale deflection corresponds

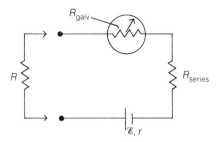

FIG. 18–20 How a galvanometer can be connected to measure resistance.

to $R = 0$, when the current is a maximum; no deflection corresponds to $R = \infty$, when the current is 0. The purpose of the series resistor is to have $I = I_{max}$ for the galvanometer when the only resistance in the circuit is that of the resistor, that of the meter, and that of the battery. In the case of a 0–1-mA, 50-Ω galvanometer used with a 1.5-V, 0.05-Ω battery,

$$R_{series} + R_{galv} + r_{bat} = \frac{\mathscr{E}}{I}$$

and so

$$R_{series} = \frac{\mathscr{E}}{I} - R_{galv} - r_{bat} = \frac{1.5\,\text{V}}{0.001\,\text{A}} - 50\,\Omega - 0.05\,\Omega = 1450\,\Omega$$

An ohmmeter is not a precision instrument, but is handy and often sufficient. A Wheatstone bridge (Problem 19) offers greater accuracy.

IMPORTANT TERMS

A flow of electric charge from one place to another is called an **electric current.** The unit of electric current is the **ampere,** which is equal to a flow of 1 coulomb/second.

Ohm's law states that the current in a metallic conductor is proportional to the potential difference between its ends. The **resistance** of a conductor is the ratio between the potential difference across its ends and the current that flows. The unit of resistance is the **ohm,** which is equal to 1 volt/ampere. The resistance of a conductor is proportional to its length and to the **resistivity** ρ of the material of which it is made, and inversely proportional to its cross-sectional area.

The **equivalent resistance** of a set of interconnected resistors is the value of the single resistor that can be substituted for the entire set without affecting the current that flows in the rest of any circuit of which it is a part. Resistors in **series** are connected consecutively so that the same current flows through all of them, whereas resistors in **parallel** have their terminals connected together so that the total current is split up among them.

The **electromotive force** (emf) of a battery, generator, or other source of electrical energy is the potential difference across its terminals when no current flows. When a current is flowing, the terminal voltage is less than the emf owing to the potential drop in the **internal resistance** of the source.

Kirchhoff's rules for network analysis are: (1) The sum of the currents flowing into a junction of three or more wires is equal to the sum of the currents flowing out of the junction; (2) the sum of the emf's around a closed conducting loop is equal to the sum of the IR potential drops around the loop.

IMPORTANT FORMULAS

Electric current: $\quad I = \dfrac{Q}{t}$

Resistance: $\quad R = \dfrac{V}{I}$

Ohm's law: $\quad I = \dfrac{V}{R} \quad$ *(holds for most metals)*

Resistance of ohmic conductor: $R = \rho \dfrac{L}{A}$

Power: $P = IV$

$$= I^2 R = \dfrac{V^2}{R} \qquad \text{(ohmic conductor)}$$

Resistors in series: $R = R_1 + R_2 + R_3 + \cdots$

Resistors in parallel:

$$\dfrac{1}{R} = \dfrac{1}{R_1} + \dfrac{1}{R_2} + \dfrac{1}{R_3} + \cdots$$

$$R = \dfrac{R_1 R_2}{R_1 + R_2} \qquad \text{(two resistors)}$$

Emf and terminal voltage: $V = \mathscr{E} - IR$

MULTIPLE CHOICE

1. The resistance of a conductor does not depend on its
(a) mass.
(b) length.
(c) cross-sectional area.
(d) resistivity.

2. A certain wire has a resistance R. The resistance of another wire, identical with the first except for having twice its diameter, is
(a) $\frac{1}{4} R$.
(b) $\frac{1}{2} R$.
(c) $2 R$.
(d) $4 R$.

3. A certain piece of copper is to be shaped into a conductor of minimum resistance. Its length and cross-sectional area
(a) should be, respectively, L and A.
(b) should be, respectively, $2 L$ and $\frac{1}{2} A$.
(c) should be, respectively, $\frac{1}{2} L$ and $2 A$.
(d) do not matter, since the volume of copper remains the same.

4. The temperature of a copper wire is raised. Its resistance
(a) decreases.
(b) remains the same.
(c) increases.
(d) any of the above, depending upon the temperatures involved.

5. Of the following combinations of units, the one that is not equal to the watt is the
(a) J/s.
(b) AV.
(c) $A^2 \Omega$.
(d) Ω^2 / V.

6. The unit of emf is the
(a) ohm.
(b) ampere.
(c) volt.
(d) watt.

7. Which of the following is neither a basic physical law nor derivable from one?
(a) Coulomb's law
(b) Ohm's law
(c) Kirchhoff's first law
(d) Kirchhoff's second law

8. A battery is connected to an external circuit. The potential drop within the battery is proportional to
(a) the emf of the battery.
(b) the equivalent resistance of the circuit.
(c) the current in the circuit.
(d) the power dissipated in the circuit.

9. A resistor R_1 dissipates the power P when connected to a certain generator. If a resistor R_2 is inserted in series with R_1, the power dissipated by R_1
(a) decreases.
(b) increases.
(c) remains the same.
(d) may do any of the above, depending on the values of R_1 and R_2.

10. If the resistor R_2 of question 9 is placed in parallel with R_1, the power dissipated by R_1
(a) decreases.
(b) increases.
(c) remains the same.
(d) may do any of the above, depending on the values of R_1 and R_2.

11. A battery of emf $\mathscr{E}$ and internal resistance r is connected to an external circuit of equivalent resistance R. If $R = r$,
(a) the current in the circuit will be a minimum.
(b) the current in the circuit will be a maximum.
(c) the power dissipated in the circuit will be a minimum.
(d) the power dissipated in the circuit will be a maximum.

12. If the wrong direction is assumed for a current I in a network being analyzed by Kirchhoff's rules, the value of the current that is obtained will be
(a) I.
(b) $-I$.
(c) incorrect.
(d) 0.

13. An electric iron draws a current of 15 A when connected to a 120-V power source. Its resistance is
(a) 0.125 Ω.
(b) 8 Ω.
(c) 16 Ω.
(d) 1800 Ω.

14. The power rating of an electric motor which draws a current of 3 A when operated at 120 V is
(a) 40 W.
(b) 360 W.
(c) 540 W.
(d) 1080 W.

15. When a 100-W, 240-V light bulb is operated at 200 V, the current that flows in it is

(a) 0.35 A.   (b) 0.42 A.

(c) 0.50 A.   (d) 0.58 A.

16. A 20-V potential difference is applied across a series combination of a 10-Ω resistor and a 30-Ω resistor. The current in the 10-Ω resistor is

(a) 0.5 A.   (b) 0.67 A.

(c) 1 A.   (d) 2 A.

17. The potential difference across the 10-Ω resistor of question 16 is

(a) 5 V.   (b) 10 V.

(c) 15 V.   (d) 20 V.

18. The equivalent resistance of a 10-Ω resistor and a 30-Ω resistor connected in parallel is

(a) 0.13 Ω.   (b) 7.5 Ω.

(c) 20 Ω.   (d) 40 Ω.

19. A 20-V potential difference is applied across the resistors of question 18. The current in the 10-Ω resistor is

(a) 0.5 A.   (b) 1 A.

(c) 2 A.   (d) 2.67 A.

20. A 50-V battery is connected across a 10-Ω resistor and a current of 4.5 A flows. The internal resistance of the battery is

(a) 0.   (b) 0.5 Ω.

(c) 1.1 Ω.   (d) 5 Ω.

21. The equivalent resistance of a network of three 2-Ω resistors cannot be

(a) 0.67 Ω.   (b) 1.5 Ω.

(c) 3 Ω.   (d) 6 Ω.

22. A 12-V potential difference is applied across a series combination of four 6-Ω resistors. The current in each resistor is

(a) 0.5 A.   (b) 2 A.

(c) 8 A.   (d) 18 A.

23. A 12-V potential difference is applied across a parallel combination of four 6-Ω resistors. The current in each resistor is

(a) 0.5 A.   (b) 2 A.

(c) 8 A.   (d) 18 A.

24. A 0–10-mA galvanometer with a coil resistance of 20 Ω is converted to a 0–10-A ammeter by using a

(a) 0.02-Ω shunt resistor.

(b) 0.2-Ω shunt resistor.

(c) 9.99-Ω shunt resistor.

(d) 9.99-Ω series resistor.

25. A 0–10-mA galvanometer with a coil resistance of 20 Ω is converted to a 0–50-V voltmeter by using a

(a) 50-Ω shunt resistor.

(b) 4980-Ω series resistor.

(c) 5000-Ω series resistor.

(d) 5020-Ω series resistor.

EXERCISES

1. It is sometimes said that an electrical appliance "uses up" electricity. What does such an appliance actually use in its operation?

2. Why are two wires used to carry electric current instead of a single one?

3. How should two identical batteries be connected in order to obtain the maximum emf from the combination? Why?

4. Why is it undesirable to connect cells of different emf in parallel?

5. The light bulbs in the circuit of the figure below are identical. Which bulb gives off the most light? The least light?

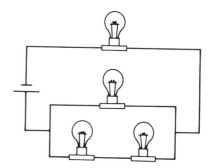

6. Alice and Fred are discussing whether an electric heater with a large or with a small resistance will yield the greater heat output. Alice favors a small value of R because $P = V^2/R$; Fred favors a large value of R because $P = I^2R$. What is your conclusion?

7. When a metal object is heated, both its dimensions and its resistivity increase. Is the increase in resistivity likely to be a consequence of the increase in length?

8. How many electrons flow through the filament of a 120-V, 60-W electric light bulb per second?

9. Currents of 3 A flow through two wires, one that has a potential difference of 60 V across its ends and another that has a potential difference of 120 V across its ends. Compare the rates at which charge and energy pass through each wire.

10. What potential difference must be applied across a 1500-Ω resistor in order that the resulting current be 50 mA?

11. Find the resistance of a 120-V electric toaster that draws a current of 8 A.

12. A metal rod 1 m long and 1 cm in diameter is drawn out into a wire 1 mm in diameter. (a) What is the length of the wire? (b) Compare the resistance of the rod with the resistance of the wire.

13. AWG 000 wire has a cross-sectional area of 85 mm^2. Find the resistance of 10 km of copper wire of this size at 20°C.

14. A 40-Ω resistor is to be wound from platinum wire 0.1 mm in diameter. How much wire is needed?

15. A silver wire 2 m long is to have a resistance of 0.5 Ω. What should its diameter be?

16. A copper wire 1 mm in diameter carries a current of 12 A. Find the potential difference between two points in the wire that are 100 m apart.

17. The resistance of a copper wire is 100 Ω at 20°C. Find its resistance at 0°C and at 80°C.

18. Motors, generators, transformers, and other electrical devices should not be operated above certain temperatures, whose values depend upon their construction. The temperature in the interior of such a device can be found by measuring the resistance of one of its windings before it is run and after it has been run for some time. If the copper field windings of an electric motor have resistances of 100 Ω at 20°C and of 115 Ω when the motor is operating, what is their temperature under the latter conditions?

19. An iron wire has a resistance of 2.00 Ω at 0°C and a resistance of 2.46 Ω at 45°C. Find the temperature coefficient of resistivity of the wire.

20. Find the maximum current in a 50-Ω, 10-W resistor if its power rating is not to be exceeded.

21. An electric drill rated at 400 W is connected to a 240-V power line. How much current does it draw?

22. An electric motor whose power output is 0.5 hp draws 4 A at 120 V. What is its efficiency?

23. A 240-V clothes dryer draws a current of 15 A. How much energy, in kilowatthours and in joules, does it use in 45 min of operation?

24. A certain 12-V storage battery is rated at 80 A·h, which means that when it is fully charged it can deliver a current of 1 A for 80 h, 2 A for 40 h, 80 A for 1 h, and so forth. (a) How many coulombs of charge can this battery deliver? (b) How much energy is stored in it?

25. A mercury cell with a capacity of 1.5 A·h and an emf of 1.35 V is to be used to power a cardiac pacemaker. If the power required is 0.1 mW, how long will the cell last? What will be the average current?

26. When a certain 1.5-V battery is used to power a 3-W flashlight bulb, it is exhausted after an hour's use. (a) How much charge has passed through the bulb in this period of time? (b) If the battery costs $0.50, find the cost of a kilowatthour of electric energy obtained in this way. How does this compare with the cost of the electric energy supplied to your home?

27. A 200-Ω and a 500-Ω resistor are in series as part of a larger circuit. If the voltage across the 200-Ω resistor is 2 V, find the voltage across the 500-Ω resistor.

28. It is desired to limit the current in an 80-Ω resistor to 0.5 A when it is connected to a 50-V power source. What is the value of the series resistor that is needed?

29. How can the resistance of a 20-Ω circuit be reduced to 5 Ω?

30. In a certain bus, forty 15-W, 30-V light bulbs are connected in parallel to a 30-V power source. (a) What is the current provided by the source? (b) What is the current in each bulb? (c) How much power is provided by the source?

31. A set of Christmas tree lights consists of 12 bulbs connected in series to a 120-V power source. Each bulb has a resistance of 5 Ω. (a) What is the current in the circuit? (b) How much power is dissipated in the circuit?

32. A 5-Ω light bulb and a 10-Ω light bulb are connected in series with a 12-V battery. (a) What is the current in each bulb? (b) What is the voltage across each bulb? (c) What is the power dissipated by each bulb and the total power dissipated by the circuit?

33. The light bulbs of Exercise 32 are connected in parallel across the same battery. Answer the same questions for this arrangement.

34. List the resistances that can be obtained by combining three 100-Ω resistors in all possible ways.

35. A source of what potential difference is needed to charge a battery of emf 24 V and internal resistance 0.1 Ω at a rate of 70 A?

36. A 12-V battery of internal resistance 1.5 Ω is connected to an 8-Ω resistor. Find the total power produced by the battery and the percentage of this power dissipated as heat within it.

37. A battery having an emf of 24 V is connected to a 10-Ω load and a current of 2.2 A flows. Find the internal resistance of the battery and its terminal voltage.

38. A generator has an emf of 240 V and an internal resistance of 0.3 Ω. When the generator is supplying a current of 20 A, find (a) its terminal voltage, (b) the power supplied to the load, and (c) the power dissipated in the generator itself.

39. The brightness of a light bulb depends upon the power dissipated by its filament. As a dry cell ages, its internal resistance increases while its emf remains approximately unchanged at 1.5 V. A fresh No. 6 dry cell might have an internal resistance of 0.05 Ω and an old one an internal resistance of 0.20 Ω. Find the ratio between the powers dissipated in a 0.25-Ω bulb when it is connected to a fresh and to an old dry cell.

PROBLEMS

1. Approximately 10^{20} electrons/cm participate in conducting electric current in a certain wire. (That is, 10^{20} electrons in each centimeter of the wire are in motion when a current is being carried by the wire.) What is the average speed of the electrons when there is a current of 1 A in the wire?

2. Aluminum wires are sometimes used to transmit electric power instead of copper wires. What is the ratio between the masses of an aluminum and a copper wire of the same length whose cross-sectional areas are such that they have the same resistance? (The density of aluminum is 2.70×10^3 kg/m^3 and that of copper is 8.89×10^3 kg/m^3.)

3. In a Van de Graaff generator an insulating belt is used to carry charges to a large metal sphere. In a typical generator of this kind, the potential difference between the sphere and the source of the charges is 5×10^6 V. (a) If the belt carries charge to the sphere at a rate of 10^{-3} A, how much power is required? (b) How much energy in eV will an electron have if it is accelerated by such a potential difference? (c) Express the answer to (b) in joules.

4. A certain 16-cell "32-volt" storage battery has an emf of 33.5 V when charged to 75% of its 250 ampere-hour capacity. The internal resistance of the battery is 0.1 Ω. It is desired to charge the battery to its full capacity, when its emf will be 34.3 V. (a) What potential difference must be applied to the battery if it is to be charged at the initial rate of 40 A? (b) If this potential difference is held constant, what will be the rate of charge at the end of the process? (c) The emf of the battery arises from the conversion of chemical to electrical energy; the higher potential difference is required for charging in order to pass a current through the battery and thereby produce chemical changes that store energy. Find the proportion of the power supplied during the

charging process that is stored as chemical energy and the proportion that is dissipated as heat. (Assume an average emf during charging of 33.9 V.)

5. A 12-volt storage battery with an internal resistance of 0.012 Ω delivers 80 A when used to crank a gasoline engine. If the battery mass is 20 kg and it has an average specific heat of 0.2 kcal/kg·°C, what is its rise in temperature during 1 min of cranking the engine?

6. Each of the resistors in the circuit of the figure below can safely dissipate 10 watts. What is the maximum power the entire circuit can dissipate?

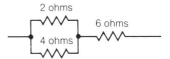

7. Three identical 1.5-V dry cells with internal resistances of 0.15 Ω are connected in parallel with an external 0.5-Ω resistor. How much current flows through the resistor?

8. (a) Find the equivalent resistance of the circuit of the figure below. (b) What is the current in the 12-Ω resistor when a potential difference of 100 V is applied to the circuit?

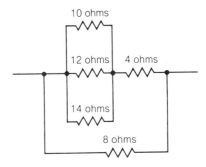

9. (a) Find the equivalent resistance of the circuit of the figure below. (b) What is the current in the 8-Ω resistor when a potential difference of 12 V is applied to the circuit?

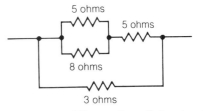

10. A 60-V potential difference is applied to the circuit of the figure on page 446. Find the current in the 10-Ω resistor. (*Hint:* Redraw the circuit to bring out the series and parallel combinations of resistors more clearly.)

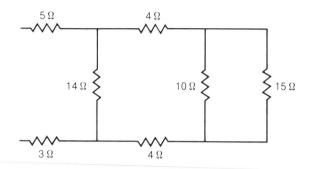

11. A 5-Ω and a 10-Ω resistor are connected in parallel. This combination is connected in series with another pair of parallel resistors whose resistances are both 8 Ω. (a) What is the equivalent resistance of the network? (b) The network is connected to a 24-V battery whose internal resistance is 1.5 Ω. Find the current in each of the resistors.

12. Find the values of R_1 and R_2 in the circuit below.

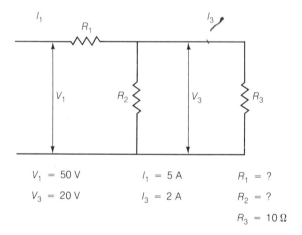

$V_1 = 50$ V	$I_1 = 5$ A	$R_1 = ?$
$V_3 = 20$ V	$I_3 = 2$ A	$R_2 = ?$
		$R_3 = 10$ Ω

13. Find the current in the 20-Ω resistor in the circuit shown in the figure below.

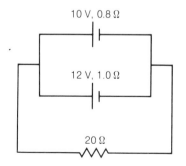

14. Find the currents in the three resistors in the circuit shown in the figure below. The internal resistances of the batteries are included in the values of the resistors.

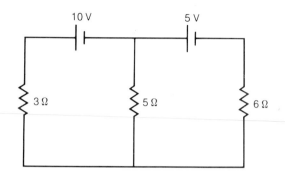

15. Find the currents in each of the resistors of the circuit shown in the figure below. The internal resistances of the batteries are included in the values of the resistors.

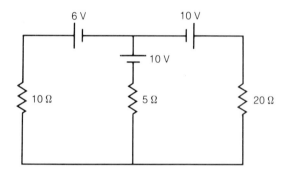

16. (a) Find the current in the 5-Ω resistor in the circuit shown below. (b) Find the potential difference between the points a and b.

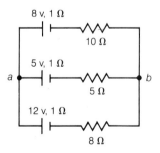

17. Find the potential differences between a and b and between a and c in the circuit shown on page 447.

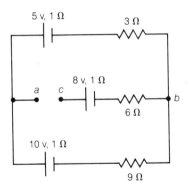

18. If a and c in the circuit shown above are connected, find the potential difference between a and b.

19. A Wheatstone bridge (shown below) provides a convenient means for measuring an unknown resistance R in terms of the known resistances A and B and the calibrated variable resistance C. The resistance of C is varied until no current flows through the galvanometer, in which case the bridge is said to be *balanced*. Show that $R = AC/B$ when the bridge is balanced.

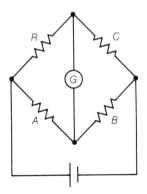

20. A galvanometer whose coil has a resistance of 12 Ω requires a current of 1.7 mA for full-scale deflection.

(a) What shunt resistance is needed to convert the meter to a 0–10-mA ammeter? (b) What series resistance is needed to convert the meter to a 0–10-V voltmeter?

21. A galvanometer whose coil has a resistance of 60 Ω requires a current of 0.02 mA for full-scale deflection. (a) What shunt resistance is needed to convert the meter to a 0–1-A ammeter? (b) What series resistance is needed to convert the meter to a 0–5-V voltmeter?

22. An ammeter has a full-scale reading of 100 mA. The potential difference across the meter when it reads 50 mA is 0.02 V. What must be done to convert the meter to have a full-scale reading of 1.0 A?

23. A voltmeter has a full-scale reading of 10 V. The current through the meter when it reads 10 V is 0.06 mA. What must be done to convert the meter to have a full-scale reading of 100 V?

24. The coil of an ammeter has a resistance of 4 Ω and its shunt has a resistance of 0.02 Ω. What is the current in the coil when the meter reads 15 A?

25. A voltmeter whose resistance is 2000 Ω is placed across a resistor of unknown resistance, and the combination is connected in series with an ammeter. The ammeter reads 0.040 A when the voltmeter reads 12 V. Find the unknown resistance.

ANSWERS TO MULTIPLE CHOICE

1. a	**6.** c	**11.** d	**16.** a	**21.** b
2. a	**7.** b	**12.** b	**17.** a	**22.** a
3. c	**8.** c	**13.** b	**18.** b	**23.** b
4. c	**9.** a	**14.** b	**19.** c	**24.** a
5. d	**10.** c	**15.** a	**20.** c	**25.** b

MAGNETIC FIELD

19

Magnetism and electricity are both manifestations of the same basic interaction between electric charges. Charges at rest relative to an observer appear to him or her to exert only electric forces upon one another. When the charges are in motion relative to the observer, however, the forces acting between them seem different from before, and these differences are traditionally attributed to "magnetic" forces. In reality, magnetic forces represent the modifications to electric forces that arise because of the motions of the charges involved. It is convenient to consider magnetic and electric forces separately, and to think in terms of separate magnetic and electric fields, but we should keep in mind that these distinctions are artificial.

19–1   THE NATURE OF MAGNETISM

Electric charges in motion exert forces upon one another quite different from those they exert while at rest. For instance, if we place a current-carrying wire

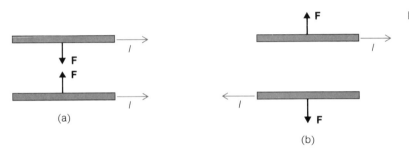

parallel to another current-carrying wire, with the currents in the same direction, we find that the wires attract each other (Fig. 19–1). If the currents are in opposite directions, the forces on the wires are repulsive.

These observations cannot be accounted for unless the motion of the charges is taken into account. Gravitational forces cannot be responsible since they are never repulsive, and the electric forces discussed earlier cannot be responsible since there is no net charge on a wire when a current is present in it. The forces that come into being when electric currents interact are called *magnetic forces*. All magnetic effects can ultimately be traced to currents or, more exactly, to moving electric charges. Of course, the word "magnetic" suggests ordinary magnets and their familiar attraction for iron objects, but, as we shall see, this is but one aspect of the whole subject of magnetism.

> **Magnetic forces arise from the interactions of moving charges**

The gravitational force between two masses and the electrical force between two charges at rest are both *fundamental forces* in the sense that they cannot be accounted for in terms of anything else. On the other hand, the force a bat exerts on a ball is not fundamental because it can be traced to the electric forces between the atomic electrons of the bat and the atomic electrons of the ball.

What about magnetic forces? It is an important fact that whatever it is in nature that manifests itself as an electric force between stationary charges *must,* according to the theory of relativity, also manifest itself as a magnetic force between moving charges. It is impossible to have one without the other. There is only a single interaction between charges, the *electromagnetic interaction*. The distinction we make between electric and magnetic forces is an artificial one for the sake of convenience only.

> **The electromagnetic interaction**

The theory of relativity is taken up in Chapter 26, and the nature of magnetic forces is further discussed there. In the meantime, it is sufficient for us to note that it is always possible to separate the force on a charge into an electric part, which is independent of its motion, and a magnetic part, which is proportional to its speed relative to the observer. These forces are additive and hence, for example, it is entirely possible for a magnetic force on a moving charge to exactly cancel an electric force on it under the proper circumstances. Because fields are always defined in terms of the forces they exert, the additive character of electric and magnetic forces permits a similar separation of an electromagnetic field into electric and magnetic parts.

> **The electromagnetic field can be separated into electric and magnetic parts**

19–2   MAGNETIC FIELD

We recall from Chapter 17 that the electric field **E** at a given location is defined in terms of the force **F** the field exerts on a stationary positive charge Q placed there. Because the electric force on a charge is always found to be proportional to Q, the magnitude of $\mathbf{E}$ is appropriately specified by the ratio

$$E = \frac{F}{Q}$$ *Electric field magnitude*

The direction of $\mathbf{E}$ is taken as the same as the direction of $\mathbf{F}$. Once we know $\mathbf{E}$, we can readily find the magnitude and direction of the electric force on *any* charge at that location.

Magnetic field is defined in terms of the force on a moving charge

The symbol for *magnetic field* is $\mathbf{B}$. Because magnetic forces only act on moving charges, $\mathbf{B}$ is defined in terms of the magnetic force $\mathbf{F}$ exerted on a positive charge Q whose velocity is $\mathbf{v}$. Experiment and theory both show that the magnetic force on a moving charge is proportional to two factors. One is the product Qv: the larger the charge and the faster it moves, the greater the magnetic force.

No force acts on a charge moving in the direction of B

The other factor concerns direction. When a charge is in a magnetic field, there is always a certain line along which it can move with no magnetic force acting on it. The direction of $\mathbf{B}$ is taken to lie along this line. When $\mathbf{v}$ is at the angle θ with respect to $\mathbf{B}$, the magnetic force on the charge is found to be proportional to $\sin\theta$, so that F is a maximum at $\theta = 90°$; that is, the maximum force occurs for motion perpendicular to $\mathbf{B}$. Because F depends on $Qv\sin\theta$, the magnitude B of the magnetic field responsible for $\mathbf{F}$ is defined, by analogy with the definition of electric field $E = F/Q$, as

$$B = \frac{F}{Qv\sin\theta}$$ *Magnetic field magnitude*   (19–1)

Although $\mathbf{B}$ lies along the line where $F = 0$, two opposite directions are possible along that line, so the direction of $\mathbf{B}$ needs further specification. Let us consider a charge moving perpendicular to that line. The direction of $\mathbf{B}$ is, by convention, given by a right-hand rule (Fig. 19–2):

Right-hand rule for direction of B given F and v.

Open your right hand so that the fingers are together and the thumb sticks out. When your thumb is in the direction of v and your palm faces in the direction of F, your fingers are in the direction of B.

(An easy way to remember this rule is to associate the outstretched thumb with hitch-hiking and so with velocity, the palm with pushing on something and so with force, and the parallel fingers with magnetic lines of force.)

FIG. 19–2   (a) Right-hand rule for the direction of the magnetic field **B** that exerts the force **F** on a positively charged particle of velocity **v.** If the particle has a negative charge, the force is in the opposite direction. (b) The magnitude of **B** is defined in terms of F, Q, and v.

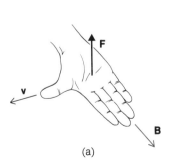

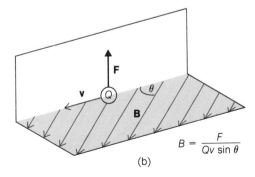

$$B = \frac{F}{Qv\sin\theta}$$

(a)   (b)

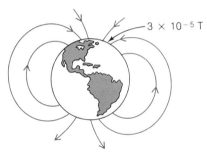

The magnitude of the earth's magnetic field at sea level is about 3×10^{-5} T.

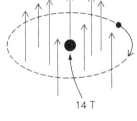

14 T

The magnetic field produced at the nucleus of a hydrogen atom by the electron circling around it is about 14 T.

FIG. 19–3 Some representative values of magnetic field.

The magnetic field near a strong permanent magnet is about 0.1 T.

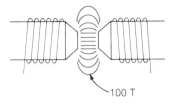

100 T

The most powerful magnetic fields achieved in the laboratory have magnitudes in the neighborhood of 100 T.

This completes an operational definition of **B,** since we now have an unambiguous way to establish **B** in any region of space by performing suitable experiments. Such an experiment might use a cathode-ray tube (Sec. 17–7) in which the electron beam is deflected by the magnetic field as well as by an electric field, as described in the illustrative problem at the end of this section.

The unit of magnetic field is, from the above definition, the newton/ampere-meter (N/A·m), since the units of Qv are coulomb-meter/second = ampere-meter. The name *tesla,* abbreviated T, has been given to this unit:

The tesla is the unit of magnetic field

$$1 \text{ T} = 1 \text{ N/A·m}$$

Thus a force of 1 N will be exerted on a charge of 1 C when it is moving at 1 m/s perpendicular to a magnetic field whose magnitude is 1 T. (When **v** is perpendicular to **B,** $\theta = 90°$ and $\sin \theta = 1$.)

The tesla is also referred to as the *weber/m^2*. Another unit of **B** in common use is the *gauss*, where

$$1 \text{ gauss} = 10^{-4} \text{ T} \quad \text{or} \quad 1 \text{ T} = 10^{4} \text{ gauss}$$

Figure 19–3 contains some representative values of magnetic field that may help in acquiring a feeling for the magnitude of the tesla.

Example The electrons in the beam of the cathode-ray tube shown in Fig. 19–4 are accelerated through a potential difference V of 500 V. A magnetic field **B** applied to

FIG. 19–4 Experimental arrangement to determine **B.** The electric field between the plates is adjusted until the electron beam is undeflected. The same procedure can be used with a known magnetic field to determine the charge-to-mass ratio e/m of the electron.

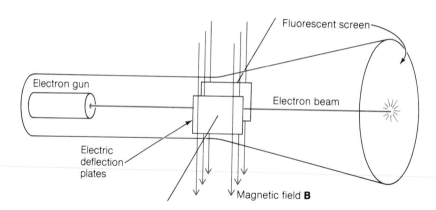

Electron gun

Fluorescent screen

Electron beam

Electric deflection plates

Magnetic field **B**

the region between the deflection plates bends the beam into the paper, but this effect is canceled out when the electric field E between the plates is 10^4 V/m. Find the direction and magnitude of **B**.

Solution   Since electrons have negative charges, the direction of **B** must be opposite to that given by the right-hand rule, so it is upward. To find B, we start with the speed v of the electrons. Since the electron kinetic energy $\frac{1}{2}mv^2$ is equal to the energy eV they gain in the electron gun,

$$\tfrac{1}{2}mv^2 = eV$$

Electron speed

$$v = \sqrt{\frac{2eV}{m}}$$

When the electric force eE on the electron balances the magnetic force $evB \sin \theta = evB$ (since $\theta = 90°$ here),

$$eE = evB$$

Force balance

$$B = \frac{E}{v}$$

Substituting for v gives

$$B = \frac{E}{v} = E\sqrt{\frac{m}{2eV}} = 10^4 \,\text{V/m} \times \sqrt{\frac{9.1 \times 10^{-31}\,\text{kg}}{2(1.6 \times 10^{-19}\,\text{C})(10^3\,\text{V})}} = 5.3 \times 10^{-4}\,\text{T}$$

(We must be careful not to confuse eV, the product of the electron charge e and the potential difference V, with eV, the abbreviation for the electron volt, which is a unit of energy.)

The same experimental arrangement can be used with a known B to find the ratio e/m between the charge and mass of the electron. This was first done in 1897 by the English physicist J. J. Thomson, whose finding that e/m is always the same provided the first definite evidence that "cathode rays" are actually streams of particles. Later work by R. A. Millikan in the United States showed that electrons all have the same charge $-e = 1.60 \times 10^{-9}$ C, a discovery which permitted the electron mass to be established. ∎

19–3 MAGNETIC FIELD OF A CURRENT

Now that both the direction and magnitude of magnetic field **B** have been defined in terms of procedures for finding them, we can go on to the nature of the magnetic field produced by various electric currents.

Unfortunately the calculations needed to determine **B** in a given situation are usually fairly difficult. To see why, let us consider a short length of wire ΔL long that carries the current I, as in Fig. 19–5. (We ignore for the moment the rest of the circuit.) The field $\Delta \mathbf{B}$ due to the current element at a point P a distance r away has its magnitude given by *Biot's law*:

Biot's law enables the magnetic field of a current to be calculated

$$\Delta B = \frac{\mu I \, \Delta L \sin \theta}{4 \pi r^2} \qquad \qquad \textit{Biot's law} \quad (19\text{–}2)$$

Here θ is the angle between the direction of the current and that of the vector **r**.

The constant μ is called the *permeability* of the medium in which the magnetic field exists. In free space,

Magnetic permeability

$$\mu_0 = 4\pi \times 10^{-7} \text{ T·m/A} = 1.257 \times 10^{-6} \text{ T·m/A}$$

so that

$$\frac{\mu_0}{4\pi} = 10^{-7} \text{ T·m/A}$$

The value of μ in air is very close to μ_0; they will be assumed to be the same here. The unit of permeability is sometimes expressed in other ways, for instance as N/A^2.

The direction of $\Delta \mathbf{B}$ is, as shown in Fig. 19–5, perpendicular to the plane formed by $\Delta \mathbf{L}$ and **r**. Another right-hand rule specifies the sense of $\Delta \mathbf{B}$:

Grasp the wire with the right hand so that the thumb points in the direction of the current; the curled fingers of that hand point in the direction of the magnetic field.

Direction of B around straight current

To apply Biot's law in an actual situation, we must compute the value of $\Delta \mathbf{B}$ at

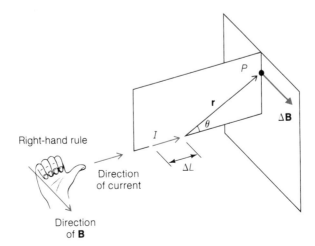

FIG. 19–5 The direction of the magnetic field of a current element is given by the right-hand rule; its magnitude is given by Biot's law.

FIG. 19–6 The lines of
force of the magnetic field
around a long, straight
current consist of
concentric circles. The
sense of the field is given
by the right-hand rule.

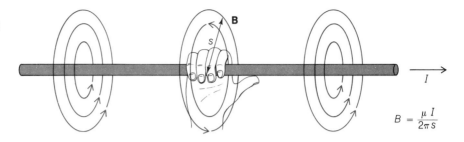

$$B = \frac{\mu\,I}{2\pi\,s}$$

some point P for each of the tiny lengths of wire $\Delta\mathbf{L}$ that make up the complete circuit, and then add up the results to find the total field $\mathbf{B}$ at P. It is necessary to take the directions of the various $\Delta\mathbf{B}$ contributions into account, since they generally are not parallel to one another. With few exceptions such calculations involve calculus, so only the results in the most important cases will be given here.

**Field around long,
straight current**

Figure 19–6 shows the configuration of the magnetic field around a long, straight wire that carries the current I. The lines of force take the form of a series of concentric circles with the current at the center. The magnitude of the field a distance s from the wire is given by

$$B = \frac{\mu I}{2\pi s} \qquad\qquad \textit{Long, straight current} \quad (19\text{–}3)$$

The greater the current and the closer one is to it, the stronger the magnetic field.

Example   Find the magnetic field in air 1 cm from a wire that carries a current of 1 A.

Solution   Since 1 cm $= 10^{-2}$m, we have (Fig. 19–7)

$$B = \frac{\mu_0 I}{2\pi s} = \frac{(4\pi \times 10^{-7}\,\text{T·m/A})(1\,\text{A})}{(2\pi)(10^{-2}\,\text{m})} = 2 \times 10^{-5}\,\text{T}$$

This is only a little smaller than the magnitude of the earth's magnetic field. For this reason great care is taken aboard ships to keep current-carrying wires away from magnetic compasses. ∎

FIG. 19–7

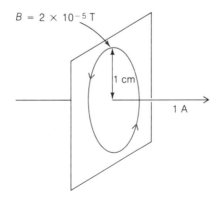

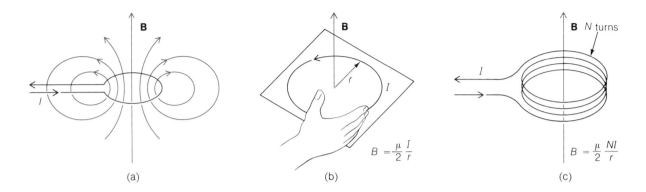

(a)   (b)   (c)

The magnetic field around a circular current loop has the configuration shown in Fig. 19–8. At the center of the loop **B** is perpendicular to the plane of the loop and has the magnitude

$$B = \frac{\mu I}{2r} \qquad \text{Center of current loop} \quad (19\text{–}4)$$

where I is the current in the loop and r is its radius. The direction of **B** is given by still another right-hand rule:

Grasp the loop so that the curled fingers of the right hand point in the direction of the current; the thumb of that hand then points in the direction of B.

In the case of a flat coil of more than one loop, as in Fig. 19–8(c), the magnetic fields of each individual loop add up to give a proportionately stronger field. If there are N turns, then,

$$B = \frac{\mu N I}{2r} \qquad \text{Center of flat coil} \quad (19\text{–}5)$$

A *solenoid* is a coil of wire in the form of a helix (Fig. 19–9). If the turns are close together and the solenoid is long relative to its diameter, then the magnetic field within it is uniform and parallel to its axis except near the ends. The direction of the field inside a solenoid is given by the same right-hand rule that gives the direction of **B** inside a current loop. The magnetic field in the interior of a solenoid l long that has N turns of wire and carries the current I has the magnitude

$$B = \mu \frac{N}{l} I \qquad \text{Interior of solenoid} \quad (19\text{–}6)$$

The diameter of the solenoid does not matter, provided it is small compared with the length l.

Example A solenoid 20 cm long and 4 cm in diameter with an air core is wound with a total of 200 turns of wire. The solenoid is aligned with its axis parallel to the earth's magnetic field at a place where the latter is 3×10^{-5} T in magnitude. What should

FIG. 19–8 (a) The magnetic field around a circular current loop. (b) At the center of the loop, **B** is perpendicular to the plane of the loop and its direction is given by the right-hand rule shown. (c) If there are N loops, the magnetic fields of the individual loops add up to give a field N times as strong as each one produces by itself.

The field inside a solenoid is uniform

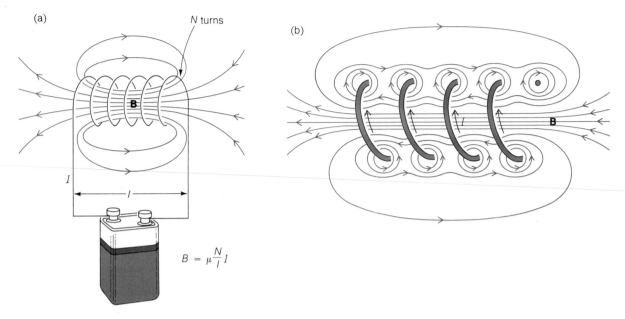

(a)

N turns

B

I

l

(b)

B

$$B = \mu \frac{N}{l} I$$

FIG. 19–9 (a) The magnetic field inside a solenoid is uniform except near its ends if the solenoid is long relative to its diameter and if its turns are close together. (b) An expanded view of a solenoid showing how the magnetic fields of the individual turns add together to yield a uniform field inside it.

the current in the solenoid be in order for its field to exactly cancel the earth's field inside the solenoid?

Solution   In air $\mu = \mu_0$, and so, since $l = 0.2$ m here, the required current is

$$I = \frac{Bl}{\mu_0 N} = \frac{(3 \times 10^{-5}\,\text{T})(0.2\,\text{m})}{(4\pi \times 10^{-7}\,\text{T·m/A})(2 \times 10^2)} = 0.024\,\text{A} = 24\,\text{mA}$$

The solenoid diameter here has no significance except as a check that the solenoid is long relative to its diameter.   ∎

The magnetic field of a bar magnet is identical with that of a solenoid, which is not surprising since all permanent magnets owe their character to an alignment of atomic current loops no different in principle from the alignment of the current loops in a solenoid (Fig. 19–10). The behavior of permanent magnets is discussed in Sec. 19–11.

Field of a bar magnet

19–4   MAGNETIC PROPERTIES OF MATTER

Ferromagnetism increases B greatly

The magnetic field produced by a current-carrying solenoid is changed in strength when a rod of almost any material is inserted in it. Some materials increase *B* (for instance,

FIG. 19–10 The magnetic fields of a bar magnet and of a solenoid are the same.

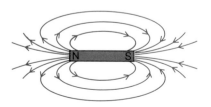

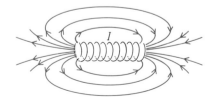

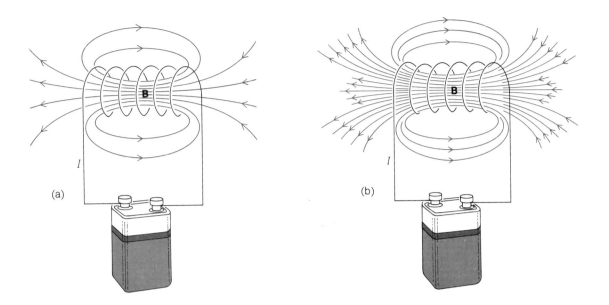

oxygen and aluminum), others decrease B (for instance, mercury and bismuth), but in almost all cases the difference is very small. However, a few substances yield a dramatic increase in B when placed in a solenoid—the new field may be hundreds or thousands of times greater in magnitude than before. Such substances are called *ferromagnetic*; iron is the most familiar example (Fig. 19–11).

FIG. 19–11 (a) Solenoid with no core. (b) Solenoid with ferromagnetic core.

The magnetic properties of matter can be traced almost entirely to the electrons every atom contains; the contribution of the nucleus is very minor. There are two sources of the magnetic behavior of an atomic electron. First, an electron in certain respects resembles a spinning charged sphere, which we may imagine as a series of ultraminute current loops (Fig. 19–12(a)). Hence every electron has the magnetic field of a tiny bar magnet due to its spin. (Although this model permits us to visualize the origin of the

An electron behaves like a tiny bar magnet

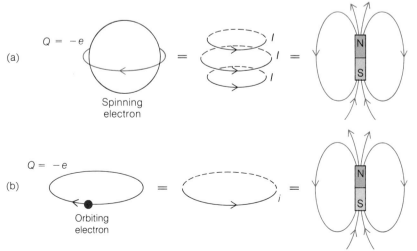

FIG. 19–12 Sources of atomic magnetism.

electron's magnetic behavior, experiments show that electrons are actually point particles with no measurable dimensions; see Chapter 32.)

An atomic electron also behaves like a current loop

Second, when it is part of an atom, an electron may be thought of as revolving around the nucleus much like a planet revolving around the sun. (This is also a crude approximation of the actual situation, but it is adequate for many purposes.) An electron circulating in an orbit is a current loop, and so the result is again the magnetic field of a tiny bar magnet (Fig. 19–12(b)).

Diamagnetism decreases B slightly

When any material whatever is placed in a magnetic field, the orbital motions of the electrons are affected by the field and, by Lenz's law (described in Sec. 20–3) the result is that the alterations in the orbital magnetic fields tend to oppose the external magnetic field. This effect is called *diamagnetism*. In most substances whose atoms or molecules have an even number of electrons, the various magnetic fields of the electrons cancel each other out in pairs when there is no external magnetic field. (If we shake an even number of small bar magnets together in a box, they will also end up paired off with opposite poles together.) In an external field, the orbital magnetic fields of the electrons will then reduce the magnetic field inside the material.

Paramagnetism increases B slightly

In other substances one or more electrons per atom or molecule have spin magnetic fields that are not canceled out. In an external magnetic field the fields of these electrons tend to line up to enhance the external field. This effect is called *paramagnetism*. The increase in B due to paramagnetism (when it is present) is always greater than the decrease due to diamagnetism (which is always present), so the net result is an increase in B. The increase is usually small, however, because the constant thermal agitation of the atoms or molecules prevents complete alignment of the spin magnetic fields with the external magnetic field. At low temperatures, as we might expect, there is less random thermal motion, so the alignment of the electron spins is more complete and the increase in B is correspondingly greater. The diamagnetic decrease in B described in the preceding paragraph is not affected by temperature.

Magnetic domains

In a ferromagnetic material the unpaired electrons in each atom interact strongly with their counterparts in adjacent atoms, which causes the unpaired spin magnetic fields in all the atoms to be locked together. Atoms in a ferromagnetic material are accordingly grouped together in assemblies called *domains,* each about 5×10^{-5} m across and just visible in a microscope. In an unmagnetized sample, the directions of magnetization of the domains are randomly oriented, although within each domain the unpaired electron spins are parallel. When such a sample is placed in an external magnetic field, either the spins within the domains turn to line up with the field or, in pure and homogeneous materials, the domain walls change so that those domains already lined up with the field grow at the expense of the others (Fig. 19–13). The former process requires stronger fields to take place than the latter. Hence good "permanent" magnets are irregular in structure—for instance, steel rather than pure iron—and once magnetized, cannot change their magnetization by the easy process of domain wall motion. When all the unpaired spins in a ferromagnetic sample are lined up, no further increase in B is possible, and the sample is said to be *saturated*.

Ferromagnetism disappears at high temperatures

Above a certain temperature (770°C in the case of iron), the atoms in a domain acquire enough kinetic energy to overcome the interatomic forces that hold their spins in alignment. The spins and their magnetic fields then become randomly oriented, and the ferromagnetic material loses its special magnetic properties. Thus heating a "permanent magnet" sufficiently will cause it to become demagnetized. Hammering a

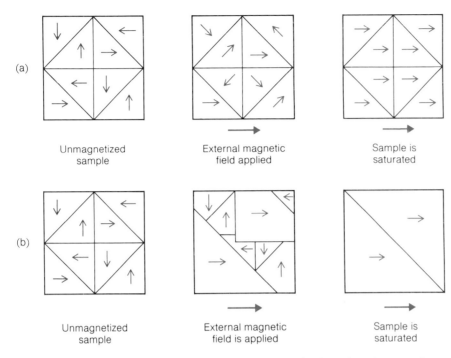

FIG. 19–13 (a) Magnetization of a ferromagnetic material by domain alignment. (b) Magnetization by domain growth.

permanent magnet also tends to disturb the alignment of spins, though some ferromagnetic materials are able to retain their magnetization despite almost any mechanical disturbance.

Iron, unlike steel (which is an alloy, or mixture, of iron with carbon and other elements), tends to lose its magnetization when an external magnetic field is removed. Hence if an iron rod is placed inside a solenoid, we have a very strong magnet that can be turned on and off just by switching the current in the solenoid on and off. Such an *electromagnet* is much stronger than the solenoid itself and can be stronger than a permanent magnet as well; also, unlike a permanent magnet, its field can be controlled at will by adjusting the current in the solenoid. Electromagnets are among the most widely used electrical devices. They range in size from the tiny one in a telephone receiver that causes a steel plate to vibrate and thus produce the sounds we hear to the giant electromagnets used to pick up automobiles in scrap yards.

Why electromagnets have iron cores

19–5 HYSTERESIS

The magnetic field inside a solenoid with no core is, according to Eq. (19–6), $B_0 = \mu_0(N/l)I$. When an iron core is inside the solenoid, the magnetic field increases to $B = \mu(N/l)I$, where μ is the permeability of the iron. Let us examine the magnetic behavior of annealed (soft) iron by comparing B and B_0 as the current I is increased. Figure 19–14 is a plot of B (the total field, including the contribution of the iron core) against B_0 (the field due to the solenoid itself, without the core). Because B is so much greater than B_0, the respective scales on the graph are different. At first, B_0 is too weak to align the domains in the iron to any great extent, so B increases slowly with increasing

Magnetic behavior of annealed iron

FIG. 19–14   The magnetization curve of annealed iron. B_0 is the magnetic field without the iron present, and B is the total field including the effect of the iron.

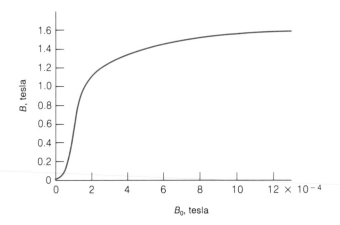

B_0. When B_0 is made stronger, it is more effective in aligning the domains, and B rises rapidly until it is over 5500 times more than B_0. Eventually, when the domains are virtually all aligned with B_0, B levels off to a nearly constant saturated value. Clearly μ, the permeability of the iron, is not a fixed quantity but varies with the magnetizing field B_0.

Hysteresis refers to the magnetic "memory" of a ferromagnetic material

A ferromagnetic material tends to retain a degree of magnetization even when the magnetic field that originally aligned its domains is removed. Hence μ does not even have a fixed value at a given B_0 but may take on different values depending on its past history. This phenomenon is called *hysteresis*.

Suppose that we place a sample of unmagnetized iron in a coil and vary the current from zero to a maximum in one direction, down through zero to a maximum in the other direction, back through zero to the first maximum, and so on. Figure 19–15 is a plot of B versus B_0 for this cycle. When we first turn on the current, the B-B_0 curve is the same as that of Fig. 19–14. At the point b we reduce the current, so that B_0 drops, but now B does *not* retrace its original path. From b to c the values of B are higher than they were from a to b at corresponding values of B_0 owing to the magnetic "memory" of ferromagnetic materials. At c there is no current in the coil and $B_0 = 0$, yet B nevertheless has the magnitude B_c: the iron sample is now permanently magnetized.

Reversing the direction of B_0 does not at first reverse B but merely reduces it, until finally, at d, B_0 is sufficiently negative to bring B to $B = 0$. A further increase in $-B_0$ takes the B-B_0 curve to e, where B and B_0 are equal in magnitude and opposite in sign

FIG. 19–15   Hysteresis loop.

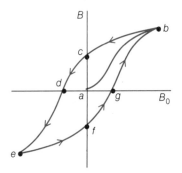

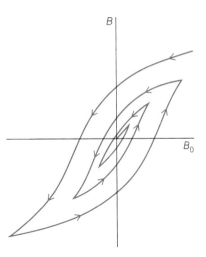

FIG. 19–16 Successive hysteresis loops during the demagnetization of a ferromagnetic sample.

to what they were at b. When B_0 is again brought to $B_0 = 0$, B is at f, where $B_f = -B_c$. Increasing B_0 in the positive sense returns B_0 to point b, but along a curve on which B is always less than it was from a to b. Further cycles simply retrace the curve $bcdefgb$.

From Fig. 19–15 it would seem that, once a ferromagnetic material is magnetized, it cannot be demagnetized (without heating) to make $B = 0$ when $B_0 = 0$. There is a procedure that can accomplish this, however. What is done is to carry a magnetized sample through a succession of hysteresis curves, each with a smaller $B_{0_{max}}$. Figure 19–16 shows that, as $B_{0_{max}}$ is brought closer and closer to zero, the curve approaches the origin where $B_0 = B = 0$. The method by which jewelers demagnetize wristwatches is based upon this procedure: the watch is placed in a coil connected to a source of alternating current, and the current is gradually decreased to zero.

Demagnetization

The shape of the hysteresis loop of a particular material provides important information about its magnetic behavior. For instance, what properties should we look for in choosing a material for a permanent magnet? Obviously we want the residual magnetization to be a maximum, which means that the point c in Fig. 19–15 should represent as high a value of B as possible. The value of B corresponding to this point is called the *retentivity* of the material involved. Thus we might regard an alloy of 98% Fe, 0.86% C, and 0.9% Mn as ideal, since its retentivity of 0.95 T is quite high. However, a permanent magnet should also be able to keep its magnetization despite stray magnetic fields from nearby currents, a factor as significant as its retentivity. The quantity in a hysteresis curve that is a measure of the ability of a material to resist changes in its magnetization is the value of B_0 at the point d, which is known as the *coercive force*. The steel alloy mentioned above happens to have a coercive force of only about 0.0045 T, which is not particularly large. The alloy Alnico 2 (55% Fe, 10% Al, 17% Ni, 12% Co, 6% Cu) has a retentivity of 0.76 T, somewhat lower than that of the steel alloy above, but its coercive force is about 0.053 T, more than ten times greater. Which alloy is best in a particular permanent magnet application must be decided on the basis of a comparison of both retentivity and coercive force.

Retentivity and coercive force

Under other circumstances the area enclosed by the hysteresis loop is of interest, since this area is proportional to the energy dissipated as heat when a sample is carried

The area of hysteresis loop represents dissipated energy

through an entire magnetization cycle. The heat may be thought of as arising from the work done by the minute magnetic elements within the material as they shift their directions. In a transformer, for instance, a current whose direction is periodically reversed passes through a coil with a ferromagnetic core. A core with a large hysteresis loop becomes very hot in a transformer, evidence of inefficiency since it is electrical energy that is being wasted. Hence a narrow hysteresis loop is desirable in selecting alloys for transformer cores.

19–6   FORCE ON A MOVING CHARGE

The defining property of a magnetic field is its ability to exert a force on an electric current, whether it is a current in a wire, a moving charged particle, or an atomic current as in an iron bar. This property has been exploited both technologically, as in the electric motor, and scientifically, as in such research tools as the mass spectrometer and various kinds of particle accelerators. The law that governs the magnetic force on a current element in a magnetic field is a straightforward one, and in the remainder of this chapter we shall see how it is applied in a number of situations.

According to Eq. (19–1), the force on a particle of charge Q and velocity $\mathbf{v}$ in the magnetic field $\mathbf{B}$ has the magnitude

$$F = QvB \sin \theta \qquad\qquad\qquad \textit{Force on moving charge} \quad (19\text{–}7)$$

where θ is the angle between $\mathbf{v}$ and $\mathbf{B}$ (Fig. 19–2(b)). The direction of the force $\mathbf{F}$ is given by the right-hand rule shown in Fig. 19–2(a).

A constant magnetic field does no work on a charged particle

The work done by a force on a body upon which it acts depends upon the component of the force in the direction the body moves. Because the force on a charged particle in a magnetic field is perpendicular to its direction of motion, the force does no work on it. Hence the particle keeps the same speed v and energy it had when it entered the field, even though it is deflected. On the other hand, the speed and energy of a charged particle in an *electric* field are always affected by the interaction between the field and the particle, unless $\mathbf{v}$ is perpendicular to $\mathbf{E}$.

A particle of charge Q and velocity $\mathbf{v}$ that is moving in a uniform magnetic field so that $\mathbf{v}$ is perpendicular to $\mathbf{B}$ experiences a force of magnitude

A charged particle moving perpendicular to a magnetic field follows a circular path

$$F = QvB \qquad (\mathbf{v} \perp \mathbf{B}) \tag{19–8}$$

since $\sin 90° = 1$. This force is directly perpendicular to both $\mathbf{v}$ and $\mathbf{B}$, so the particle travels in a circular path (Fig. 19–17).

To find the radius R of the circular path of the charged particle, we note that the magnetic force QvB provides the particle with the centripetal force mv^2/R that keeps it moving in a circle. Equating the magnetic and centripetal forces yields

$$F_{\text{magnetic}} = F_{\text{centripetal}}$$
$$QvB = \frac{mv^2}{R}$$

and so, solving for R, we obtain

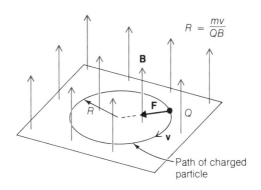

$$R = \frac{mv}{QB}$$

FIG. 19–17 The path of a charged particle moving perpendicular to a uniform magnetic field is a circle.

$$R = \frac{mv}{QB}$$

Orbit radius (19–9)

The radius of a charged particle's orbit in a uniform magnetic field is directly proportional to its momentum mv and inversely proportional to its charge and to the magnitude of the field. The greater the momentum, the larger the circle, and the stronger the field, the smaller the circle.

A charged particle moving parallel to a magnetic field experiences no force and is not deflected; the same particle moving perpendicular to the field follows a circular path. Hence a charged particle whose direction of motion is oblique with respect to **B** follows a helical (corkscrew) path. If we call $v_\parallel$ the component of the particle's velocity **v** that is parallel to **B** and $v_\perp$ the component of **v** perpendicular to **B,** then the motion of the particle is the resultant of a forward motion at the velocity $v_\parallel$ and a circular motion perpendicular to this whose radius is $mv_\perp/QB$ (Fig. 19–18).

An extremely interesting phenomenon occurs when a charged particle moving in a magnetic field approaches a region where the field becomes stronger. The magnetic lines of force that describe such a field converge, since their spacing is always proportional to the magnitude of the field they describe. The force the particle experiences now has a backward component as well as the inward component that leads to its helical path, as shown in Fig. 19–19. The backward force may be strong enough and extend over a long enough distance to reverse the particle's direction of motion. A converging magnetic field can thus act as a *magnetic mirror*.

Magnetic mirrors are found both in the laboratory and in nature. In the laboratory a pair of them can be used as a "magnetic bottle," as in Fig. 19–20, to contain a hot plasma (highly ionized gas) in research on thermonuclear fusion. If a solid container

Magnetic mirror

FIG. 19–18 A charged particle that has velocity components both parallel and perpendicular to a magnetic field follows a helical path in the field.

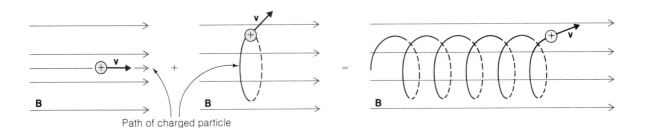

Path of charged particle

FIG. 19–19 The principle of the magnetic mirror.

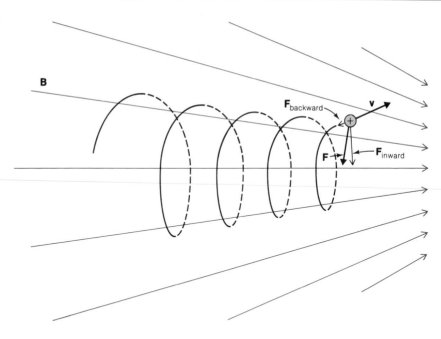

Magnetic bottle

were used, contact with its walls would contaminate the plasma and also cool it so that the ions would not have enough energy to interact. Magnetic bottles of this kind are somewhat leaky, because ions moving along the axis of a magnetic mirror experience no backward force and hence are able to escape.

Magnetosphere

The earth's magnetic field traps electrons and protons from space in the *magnetosphere*, a giant doughnut-shaped magnetic bottle that surrounds the earth and extends from about 1000 km above the equator out to perhaps 65,000 km. The magnetosphere contains large numbers of particles with relatively high energies (100 MeV, for instance). Figure 19–21 shows a typical particle trajectory in the magnetosphere.

19–7 THE MASS SPECTROMETER

A mass spectrometer measures atomic masses

The mass of an atom is one of its most characteristic properties and an accurate knowledge of atomic masses provides considerable insight into nuclear phenomena. A variety

FIG. 19–20 A "magnetic bottle."

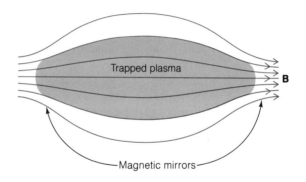

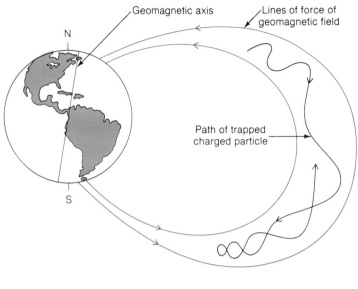

FIG. 19–21 Protons and electrons are trapped by the earth's magnetic field.

Geomagnetic axis

Lines of force of geomagnetic field

N

S

Path of trapped charged particle

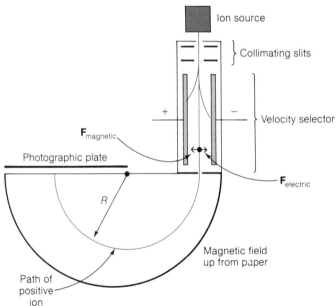

FIG. 19–22 A simple mass spectrometer. Modern instruments use electrical ion detectors and more complicated fields.

Ion source

Collimating slits

+   −   Velocity selector

$\mathbf{F}_{magnetic}$

Photographic plate

$\mathbf{F}_{electric}$

R

Magnetic field up from paper

Path of positive ion

of instruments with the generic name of *mass spectrometers* have been devised to measure atomic masses, and we shall consider the operating principles of the particularly simple one shown in Fig. 19–22.

The first step in the operation of this spectrometer is to produce ions of the substance under study. If the substance is a gas, ions can be formed by electron bombardment; if it is a solid, it is often convenient to incorporate it into an electrode that is used as one terminal of an electric arc discharge. The ions emerge from their source through a slit with the charge $+e$ and are then accelerated by an electric field. (Ions with other charges are sometimes present but are easily taken into account.)

Velocity selector

When the ions enter the spectrometer, as a rule they are traveling in slightly different directions with slightly different speeds. A pair of slits serves to collimate the beam, that is, to eliminate those ions not moving in the desired direction. Then the beam passes through a *velocity selector*. The velocity selector consists of uniform electric and magnetic fields that are perpendicular to each other and to the beam of ions. The electric field **E** exerts the force $F_{electric} = eE$ on the ions to the right, whereas the magnetic field **B** exerts the force $F_{magnetic} = evB$ on those to the left. In order for an ion to reach the slit at the far end of the velocity selector it must suffer no deflection inside the selector, which means that the condition for escape is

$$F_{electric} = F_{magnetic}$$

$$eE = evB$$

Hence the ions that escape all have the speed

$$v = \frac{E}{B}$$

Once past the velocity selector the ions enter a uniform magnetic field and follow circular paths whose radius is given by Eq. (19–9). Since v, e, and B are known, a measurement of R yields a value for m, the ion mass.

Example The velocity selector of a mass spectrometer consists of an electric field of $E = 40,000$ V/m perpendicular to a magnetic field of $B = 0.0800$ T. The same magnetic field is used to deflect the ions that have passed through the velocity selector. Ions of a certain isotope of lithium are found to have radii of curvature in the magnetic field of 39.0 cm. What is their mass?

Solution The speed of the ions is

$$v = \frac{E}{B} = \frac{4.00 \times 10^4 \, \text{V/m}}{8.00 \times 10^{-2} \, \text{T}} = 5.00 \times 10^5 \, \text{m/s}$$

From Eq. (19–9) we obtain

$$m = \frac{QBR}{v} = \frac{(1.60 \times 10^{-19} \, \text{C})(8.00 \times 10^{-2} \, \text{T})(0.390 \, \text{m})}{5.00 \times 10^5 \, \text{m/s}} = 9.98 \times 10^{-27} \, \text{kg} \quad \blacksquare$$

19–8 FORCE ON A CURRENT

Since an electric current is a flow of charge, we would expect a current-carrying wire to be affected by a magnetic field in a manner similar to that of a moving charged particle. According to Eq. (19–1), the force on a charge Q whose velocity is **v** when it is in the magnetic field **B** has the magnitude

$$F = QvB \sin \theta \qquad (19\text{–}1)$$

where θ is the angle between **v** and **B.** What we must do to find an expression for the force on a current is to replace the Qv of the above formula with the quantity appropriate for a current.

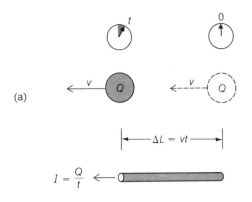

FIG. 19–23 The force on a charge Q moving with the speed v in a magnetic field is the same as that on a wire ΔL long carrying the current I, where $I\,\Delta L = Qv$.

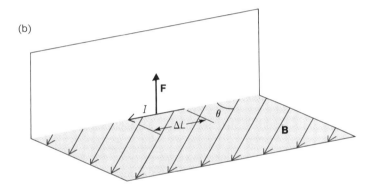

Figure 19–23 (a) shows a particle of charge Q and speed v. In the time t the particle travels the distance

Equivalence of current element and moving charge

$$\Delta L = vt$$

and while it does so it is equivalent to a current of

$$I = \frac{Q}{t}$$

Hence

$$v = \frac{\Delta L}{t} \qquad \text{and} \qquad Q = It$$

so that

$$Qv = I\,\Delta L$$

We conclude that the force on an element ΔL long of current I when it is in a magnetic field **B** has the magnitude

$$F = I\,\Delta LB \sin\theta \qquad\qquad \textit{Force on current element} \quad (19\text{–}10)$$

where θ is the angle between the direction of I and that of **B** (Fig. 19–23(b)).

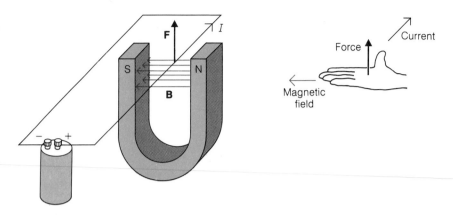

Right-hand rule for force on a current

There are two simple ways to determine the direction of the force on a current element in a magnetic field. Both give the same result, of course, and deciding which one to use in a particular case is largely a matter of personal preference. The first is essentially the same as the right-hand rule used for a moving charge in Sec. 19–1, except that now the thumb points in the direction of the current. This version of the rule is illustrated in Fig. 19–24.

Lines-of-force rule for force on a current

Another method for finding the direction of **F** is based upon the pattern of lines of force around a current in a magnetic field. Figure 19–25(a) shows the lines of force of a uniform field **B** in the absence of a current, and (b) shows the lines of force around a wire carrying a current I in the absence of a magnetic field of external origin. Since the current is into the paper, the lines of force are concentric circles in the clockwise sense. When field (a) is added vectorially to field (b), the resulting pattern of lines of force is like that shown in (c): the lines are closer together in the region above the wire where the field of the current is in the same direction as **B** and farther apart under the wire where the field of the current is opposite to **B**. *The direction of the force on the current element is from the region of strong field to the region of weak field,* as though the lines of force were rubber bands that try to straighten out when distorted by the presence of the current.

While the pictorial method for establishing the direction of **F** is easy to use and appeals to the intuition, it must be kept in mind that lines of force do not in fact exist but are only a device for visualizing the magnitude and direction of a force field. It is

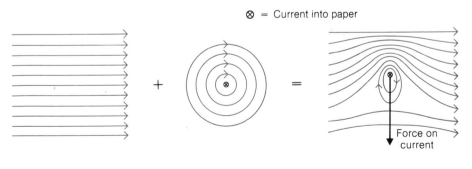

(a) Uniform field **B**   (b) Field of current   (c) Resultant field

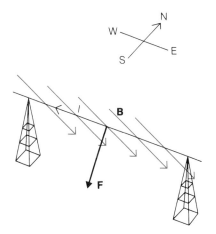

FIG. 19–26

the field itself that exists in space as a continuous property of the region it occupies, not a series of strings. However, despite the artificial nature of lines of force, they can be extremely convenient in representing various aspects of the interaction between magnetic fields and electric currents, and we can make use of them for this purpose whenever appropriate.

Example   A wire carrying a current of 100 A due west is suspended between two towers 50 m apart. The lines of force of the earth's magnetic field enter the ground there in a northerly direction at a 45° angle; the magnitude of the field at that location is 5×10^{-5} T. Find the force on the wire exerted by the earth's field.

Solution   The wire is perpendicular to **B**, and so the magnitude of the force is

$$F = I\ \Delta LB \sin \theta = (100\,\text{A})(50\,\text{m})(5 \times 10^{-5}\,\text{T})(\sin 90°) = 0.25\,\text{N}$$

Since 1 N = 0.225 lb, the force is 0.056 lb, about an ounce. By either of the methods described above, the force acts downward at a 45° angle with the ground toward the south (Fig. 19–26).

19–9   FORCE BETWEEN TWO CURRENTS

Every current is surrounded by a magnetic field, and because of this nearby currents exert forces upon one another. The forces are magnetic in origin; a current-carrying wire has no net electric charge, and hence cannot interact electrically with another such wire.

Figure 19–27 shows two parallel wires a distance s apart that carry the currents I_1 and I_2 respectively. The magnetic fields a distance s from each of the wires are, from Eq. (19–3),

$$B_1 = \frac{\mu I_1}{2\pi s} \qquad B_2 = \frac{\mu I_2}{2\pi s}$$

Magnetic field of the currents

The fields are perpendicular to the wires, which means that $\theta = 90°$ and $\sin \theta = 1$,

FIG. 19–27 Equal and opposite forces are exerted by parallel currents on each other. The forces are attractive when the currents are in the same direction, repulsive when they are in opposite directions.

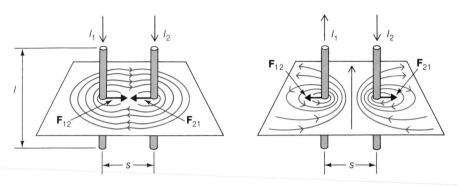

and therefore the force F_{12} on a length L of current 1 exerted by the magnetic field of current 2 is

Force on current 1 due to field of current 2

$$F_{12} = I\,\Delta LB\sin\theta = \frac{\mu I_1 I_2}{2\pi s}\,L \qquad (19\text{--}11)$$

The force F_{21} on current 2 that is exerted by the magnetic field of current 1 has exactly the same magnitude, and we may express both of them in the alternative form

$$\frac{F}{L} = \frac{\mu I_1 I_2}{2\pi s} \qquad\qquad \textit{Force between parallel currents} \quad (19\text{--}12)$$

Parallel currents attract, antiparallel currents repel

where F/L is the *force per unit length* each wire exerts on the other by virtue of its magnetic field. From the pattern of lines of force around each wire, it is clear that the forces are always opposite in direction and hence obey Newton's third law of motion, as they must. The forces are attractive when the currents are in the same direction and repulsive when they are in opposite directions.

Example The cables that connect the starting motor of a car with its battery are 1 cm apart for a distance of 40 cm. Find the forces between the cables when the current in them is 300 A.

Solution The currents in the cables are opposite in direction, and hence the forces are repulsive. Their magnitudes, taking $\mu = \mu_0$, are

$$F = \frac{\mu_0 I^2}{2\pi s}L = \frac{(4\pi \times 10^{-7}\,\text{T·m/A})(300\,\text{A})^2(0.4\,\text{m})}{2\pi \times 10^{-2}\,\text{m}} = 0.72\,\text{N}$$

which is 0.16 lb, a perceptible amount. ∎

Definition of the ampere

Equation (19–12) is used to define the ampere: An ampere is that current in each of two parallel wires 1 m apart in free space which produces a force on each wire of exactly 2×10^{-7} N per meter of length. (Thus $\mu_0 = 4\pi \times 10^{-7}$ *T·*m/A.) In turn, the coulomb is defined in terms of the ampere as that amount of charge transferred per second by a current of 1 A. The ampere is chosen as the primary electrical unit instead of the coulomb because it can be defined in terms of a more direct and unambiguous experiment than would be possible with the coulomb.

19–10   TORQUE ON A CURRENT LOOP

A straight current-carrying wire is acted upon by a force when it is in a magnetic field, provided that it is not parallel to the direction of **B.** A loop of current in a uniform magnetic field experiences no net force, but instead a torque occurs that tends to rotate the loop to bring its plane perpendicular to **B.** This is the principle that underlies the operation of all electric motors, from the tiniest one in a clock to the many-thousand-horsepower giant in a locomotive.

Forces on a current loop in a magnetic field

Let us examine the forces on each side of a rectangular current-carrying wire loop whose plane is parallel to a uniform magnetic field **B,** as in Fig. 19–28(a). The sides A and C of the loop are parallel to **B** and so there is no magnetic force on them. Sides B and D are perpendicular to **B,** however, and each therefore experiences a force. To find the directions of the forces on B and D we can use the right-hand rule: with the fingers of the right hand in line with **B** and the outstretched thumb in line with **I,** the palm faces the same way as **F.** What we find is that $\mathbf{F}_B$ is opposite in direction to $\mathbf{F}_D$. The same conclusion can be obtained by examining the pattern of lines of force.

The forces $\mathbf{F}_B$ and $\mathbf{F}_D$ are the same in magnitude, so there is no net force on the current loop. But $\mathbf{F}_B$ and $\mathbf{F}_D$ do not act along the same line, and hence they exert a torque on the loop that tends to turn it. This is a perfectly general conclusion that holds for a current loop of any shape in a magnetic field.

If the plane of the loop is perpendicular to the magnetic field instead of parallel to it, there is neither a net force nor a net torque on it. This is easy to verify from Fig. 19–28(b), bearing in mind the right-hand rule for the direction of the force on each side of the loop. Evidently $\mathbf{F}_A$ and $\mathbf{F}_C$ cancel each other out, and $\mathbf{F}_B$ and $\mathbf{F}_D$ also cancel each other out. There is no torque now because $\mathbf{F}_A$ and $\mathbf{F}_C$ have the same line of action, and $\mathbf{F}_B$ and $\mathbf{F}_D$ have the same line of action.

FIG. 19–28   (a) A current-carrying wire loop whose plane is parallel to a magnetic field experiences a torque. (b) If the plane of the loop is perpendicular to the magnetic field, there is no torque on the loop. In both cases there is no net force on the loop.

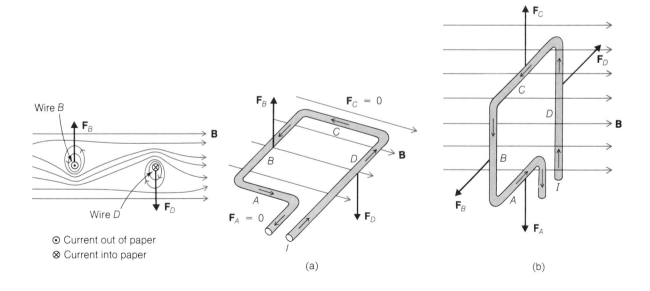

⊙ Current out of paper
⊗ Current into paper

(a)   (b)

FIG. 19–29 The construction of a common type of galvanometer.

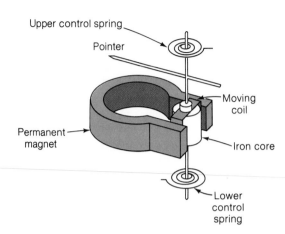

The above results can be summarized by saying that

Current loop in magnetic field

A current loop in a magnetic field always tends to turn so that its plane becomes perpendicular to the field.

Galvanometer

We have seen that a current-carrying wire loop tends to rotate in a magnetic field. The *galvanometer* capitalizes upon this behavior to furnish a means for measuring current. Figure 19–29 shows the basic construction of a galvanometer. A U-shaped permanent magnet is used to provide a magnetic field, and between its poles is a small coil wound on an iron core to enhance the torque developed when the unknown current is passed through it. The coil assembly is held in place by two bearings that permit it to rotate, and a pair of hairsprings keeps the pointer at 0 when there is no current in the coil.

When a current flows, there is a torque on the coil because of the interaction between the current and the magnetic field, and the coil rotates as far as it can against the opposing torque of the springs. The more the current, the stronger the torque, and the farther the coil turns. The restoring torque of the hairsprings is proportional to the angle through which they are twisted, and as a result the deflection of the pointer is directly proportional to the current I in the coil.

Galvanometers of the above type can be constructed that are able to respond to currents of as little as 0.1 microampere (10^{-7} A), though ordinary commercial meters are less sensitive. Even greater sensitivity can be attained if the moving coil is suspended by a thin wire to which a small mirror is attached: bearing friction is avoided in this way, and the mirror deflects a light beam so that the "pointer" may be a meter or more long instead of a few centimeters. Laboratory galvanometers like this can be used to measure currents of 10^{-10} A.

The torque which a magnetic field exerts on a current loop disappears when the loop turns so that its plane is perpendicular to the field direction. If the loop swings past this position, the torque on it will be in the opposite sense and will return the loop to the perpendicular orientation. In order to construct a motor capable of continuous rotation, then, the current in the loop must be automatically reversed each time it turns through 180°. The method by which this reversal is accomplished is shown in Fig. 19–30. The current is led to the loop by means of graphite rods called *brushes* which

How an electric motor achieves continuous rotation

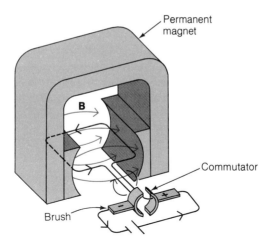

FIG. 19–30 A simple dc electric motor. The commutator automatically reverses the current in the rotating loop twice per rotation so that the torque will stay in the same direction.

press against a split ring called a *commutator.* As the loop rotates, the current is reversed twice per turn as the commutator segments make contact alternately with the brushes. The torque is always in the same direction, except at the moments of switching when it is zero because the loop is perpendicular to the field. However, the angular momentum of the loop carries it past this point, and it can continue to turn indefinitely.

While actual direct-current electric motors, such as the starter motor of a car, are the same in principle as the simple device of Fig. 19–30, they employ a number of stratagems to increase the available torque. Electromagnets rather than permanent magnets provide the field, and there are six or more different coils with many turns each on a slotted iron core called an *armature,* instead of a single loop (Fig. 19–31). A commutator with a pair of segments for each coil is provided so that only those coils approximately parallel to the magnetic field receive current at any time, which means that maximum torque is developed continuously.

Electrical energy for industrial and domestic purposes is usually transmitted by *alternating current* (ac) whose direction periodically reverses itself. In the United States the frequency of ordinary ac is 60 Hz, which means that the current changes direction 120 times per second; this frequency is 50 Hz in much of the rest of the world. In ac electric motors, commutators and brushes are not needed because the current itself does

Alternating-current motors

FIG. 19–31 Actual dc electric motors employ various means to increase the available torque.

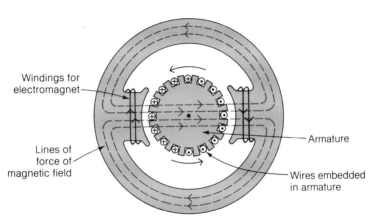

the required reversing, which makes such motors easier to build and more reliable than dc motors. In order to start the armature of an ac electric motor turning, and to ensure that it begins to turn the desired way, an auxiliary stationary winding is used that, together with the operating winding, creates a magnetic field whose direction rotates about the motor's axis. This rotating field pulls the armature around when the motor is switched on. As the armature approaches its normal running speed, the alternating magnetic fields of the operating windings are sufficient to keep it going, and (depending on the motor design) the starting winding may then be cut out of the circuit.

19–11 MAGNETIC POLES

It may seem strange that there has been no mention until now of the "magnetic poles" that figure so prominently in naive discussions of magnetism. The reason is that all magnetic fields, including those of permanent magnets, orignate in electric currents (or, more precisely, in moving charges); and all magnetic forces arise from interactions between currents and magnetic fields. To understand electromagnetic phenomena of any kind, it is necessary to start directly from these fundamental concepts.

The field of a bar magnet is like that of a solenoid

The magnetic field of a bar magnet is identical with that of a solenoid, as we saw in Fig. 19–10, because in a permanent magnet atomic current loops are aligned by their mutual interactions. Hence we can use the ideas of this chapter to understand the behavior of permanent magnets.

Because the external magnetic field of a bar magnet seems to originate in its ends, these are by custom called its *poles*. At one time it was believed that the poles were "magnetic charges" analogous to electric charges, and that the field of the magnet was due to these poles. This belief was reinforced by the repulsion of like poles and the attraction of unlike ones, phenomena that have their true explanation in the forces between parallel and antiparallel currents (Fig. 19–32).

FIG. 19–32 Interactions between magnets can be traced to interactions between current loops.

An important difference between magnetic poles and electric charges is that the former invariably occur in pairs of equal strength and opposite polarity: if a magnet is

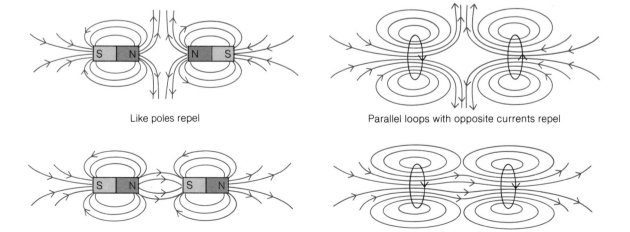

Like poles repel

Parallel loops with opposite currents repel

Unlike poles attract

Parallel loops with similar currents attract

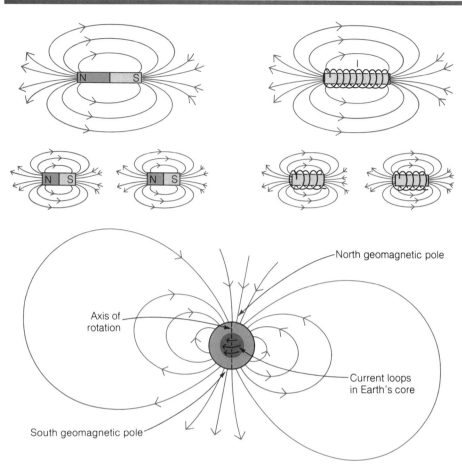

FIG. 19–33 Cutting a magnet in half produces two new magnets.

FIG. 19–34 The earth's magnetic field originates in currents in its core of molten iron. The magnetic axis is titled by 11° from the axis of rotation.

North geomagnetic pole

Axis of rotation

Current loops in Earth's core

South geomagnetic pole

sawed in half, the poles are not separated but instead two new magnets are created, as in Fig. 19–33. Magnetic poles are therefore only superficially like electric charges, and all effects that can be attributed to them can be explained in terms of the behavior of current-carrying solenoids.

Magnetic poles differ profoundly from electric charges

Measurements of the earth's magnetic field show that it is very much like the field that would be produced by a powerful current loop whose center is a few hundred km from the earth's center and whose plane is tilted by 11° from the plane of the earth's equator (Fig. 19–34). On the basis of geological evidence the earth is thought to have a core of molten iron 3470 km (2160 mi) in radius, a little over half the earth's radius, and there is no doubt today that electric currents in this core are responsible for the observed geomagnetic field; the details of how these currents came into being and how they are maintained are still uncertain, however.

As we saw, a current loop tends to rotate in a magnetic field until its axis is parallel to the field. A bar magnet, too, tends to rotate in a magnetic field until it is aligned with the field direction. Because of the earth's magnetic field, a magnet suspended by a string turns so as to line up in an approximately north-south direction. A compass consists of a pivoted magnetized iron needle together with a card that permits directions relative to magnetic north to be determined (Fig. 19–35). The end of a freely swinging

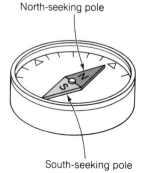

North-seeking pole

South-seeking pole

FIG. 19–35 A magnetic compass.

FIG. 19–36 How a permanent magnet attracts a ferromagnetic object.

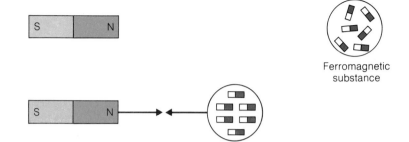

Ferromagnetic substance

magnet that points toward the north is called its north-seeking pole, usually shortened to just *north pole,* and the other end is its south-seeking pole, or *south pole.* Magnetic lines of force leave the north pole of a magnet and enter its south pole. (The north geomagnetic pole is thus in reality a south pole, and the south geomagnetic pole is a north pole; this has been a source of confusion for several hundred years.)

The mechanism by which a magnet or a solenoid attracts an iron object (or an object of any other ferromagnetic material such as cobalt or nickel) is very similar to the way in which an electric charge attracts an uncharged object. First the presence of the magnet causes the atomic magnets in the iron object to line up with its field by one of the mechanisms shown in Fig. 19–13, and then the attraction of opposite poles leads to a force on the object that draws it toward the magnet, as in Fig. 19–36.

IMPORTANT TERMS

Charged particles in motion relative to an observer exert forces upon one another that are different from the electric forces they exert when at rest. These differences are by custom said to arise from **magnetic forces.** In reality, magnetic forces represent modifications of electric forces due to the motion of the charges involved, as predicted by the theory of relativity.

A **magnetic field** exists wherever a magnetic force would act on a moving charged particle. The direction of a magnetic field **B** at a point is such that a charged particle would experience no force if it moves in that direction at the point. The magnitude of **B** is numerically equal to the force that would act on a charge of 1 C moving at 1 m/s perpendicular to **B**. The unit of magnetic field is the **tesla** (T), equal to 1 N/A·m.

When different substances are inserted in a current-carrying wire coil, the magnetic field in its vicinity changes. Those substances that lead to a slight increase in B are called **paramagnetic,** those that lead to a great increase in B are called **ferromagnetic,** and those that lead to a slight decrease in B are called **diamagnetic.** A **permanent magnet** is an object composed of ferromagnetic material whose

atomic current loops have been aligned by an external current.

The **permeability** of a medium is a measure of its magnetic properties. Paramagnetic and ferromagnetic substances have higher permeabilities than free space, and diamagnetic ones have lower permeabilities. The permeability of a ferromagnetic material in a given magnetizing field B_0 depends upon its past history as well as upon B_0, a phenomenon called **hysteresis.**

IMPORTANT FORMULAS

Magnetic field: $B = \dfrac{F}{Qv\sin\theta}$

Field around long, straight current: $B = \dfrac{\mu I}{2\pi s}$

Field at center of flat coil: $B = \dfrac{\mu NI}{2r}$

Field in interior of solenoid: $B = \mu \dfrac{N}{l} I$

Force on moving charge: $F = QvB \sin \theta$

Force on current element: $F = I \, \Delta L \, B \sin \theta$

Force between parallel currents: $\dfrac{F}{L} = \dfrac{\mu I_1 I_2}{2\pi s}$

MULTIPLE CHOICE

1. All magnetic fields originate in
(a) iron atoms.
(b) permanent magnets.
(c) magnetic domains.
(d) moving electric charges.

2. An observer moves past a stationary electron. His instruments measure
(a) an electric field only.
(b) a magnetic field only.
(c) both electric and magnetic fields.
(d) any of the above, depending upon his speed.

3. Magnetic fields do not interact with
(a) stationary electric charges.
(b) moving electric charges.
(c) stationary permanent magnets.
(d) moving permanent magnets.

4. A drawing of the lines of force of a magnetic field provides information on
(a) the direction of the field only.
(b) the magnitude of the field only.
(c) both the direction and magnitude of the field.
(d) the source of the field.

5. In a drawing of magnetic lines of force, the stronger the field is,
(a) the closer together the lines of force are.
(b) the farther apart the lines of force are.
(c) the more nearly parallel the lines of force are.
(d) the more divergent the lines of force are.

6. The magnetic field near a strong permanent magnet might be
(a) 10^{-9} T.
(b) 10^{-5} T.
(c) 0.1 T.
(d) 100 T.

7. A typical value for the earth's magnetic field at sea level is
(a) 3×10^{-9} T.
(b) 3×10^{-5} T.
(c) 3×10^{5} T.
(d) 3×10^{9} T.

8. When a paramagnetic substance is inserted in a current-carrying solenoid, the magnetic field is
(a) slightly decreased.
(b) greatly decreased.
(c) slightly increased.
(d) greatly increased.

9. When a diamagnetic substance is inserted in a current-carrying solenoid, the magnetic field is
(a) slightly decreased.
(b) greatly decreased.
(c) slightly increased.
(d) greatly increased.

10. When a ferromagnetic substance is inserted in a current-carrying solenoid, the magnetic field is
(a) slightly decreased.
(b) greatly decreased.
(c) slightly increased.
(d) greatly increased.

11. Permanent magnets are made from
(a) diamagnetic substances.
(b) paramagnetic substances.
(c) ferromagnetic substances.
(d) any of the above.

12. The magnetic field a distance d from a long, straight wire is proportional to
(a) d.
(b) d^2.
(c) $1/d$.
(d) $1/d^2$.

13. Inside a solenoid the magnetic field
(a) is zero.
(b) is uniform.
(c) increases with distance from the axis.
(d) decreases with distance from the axis.

14. A current is flowing east along a power line. If we neglect the earth's field, the direction of the magnetic field below it is
(a) north.
(b) east.
(c) south.
(d) west.

15. A charged particle moves perpendicularly through a magnetic field. The effect of the field is to change the particle's
(a) charge.
(b) mass.
(c) velocity.
(d) energy.

16. The right-hand rule for the direction of the force on a charged particle in a magnetic field applies
(a) only to positive charges.
(b) only to negative charges.
(c) to both positive and negative charges.
(d) only when the particle is moving parallel to the field.

17. A current-carrying wire is in a uniform magnetic field with the direction of the current the same as that of the field.
(a) There is a force on the wire that tends to move it parallel to the field.
(b) There is a force on the wire that tends to move it perpendicular to the field.
(c) There is a torque on the wire that tends to rotate it until it is perpendicular to the field.
(d) There is neither a force nor a torque on the wire.

18. A current-carrying loop in a magnetic field always tends to rotate until the plane of the loop is
  (a) parallel to the field.
  (b) perpendicular to the field.
  (c) either parallel or perpendicular to the field, depending on the direction of the current.
  (d) at a 45° angle with the field.

19. The magnetic field 2 cm from a long, straight wire is 10^{-6} T. The current in the wire is
  (a) 0.01 A.   (b) 0.1 A.
  (c) 1 A.   (d) 10 A.

20. The magnetic field inside a 100-turn solenoid 2 cm long that carries a 10-A current is
  (a) 0.00063 T.   (b) 0.00126 T.
  (c) 0.063 T.   (d) 0.126 T.

21. A "magnetic mirror" that can reflect approaching charged particles is a
  (a) sheet of ferromagnetic material.
  (b) uniform magnetic field.
  (c) magnetic field whose lines of force converge.
  (d) magnetic field whose lines of force diverge.

22. Two parallel wires in free space are 10 cm apart and carry currents of 10 A each in the same direction. The force each wire exerts on the other per meter of length is
  (a) 2×10^{-7} N, attractive.
  (b) 2×10^{-7} N, repulsive.
  (c) 2×10^{-4} N, attractive.
  (d) 2×10^{-4} N, repulsive.

23. The magnetic field of a bar magnet most closely resembles the magnetic field of
  (a) a straight current-carrying wire.
  (b) a stream of electrons moving parallel to one another.
  (c) a current-carrying wire loop.
  (d) a horseshoe magnet.

24. The needle of a magnetic compass
  (a) is affected only by permanent magnets.
  (b) rotates continuously in the magnetic field of an electric current.
  (c) aligns itself parallel to a magnetic field.
  (d) aligns itself perpendicular to a magnetic field.

25. At different places on the earth's surface, the earth's magnetic field
  (a) is the same in direction and magnitude.
  (b) may be different in direction but not in magnitude.
  (c) may be different in magnitude but not in direction.
  (d) may be different in both magnitude and direction.

EXERCISES

1. A current is flowing north along a power line. What is the direction of the magnetic field above it? Below it?

2. A current is flowing vertically upward when it enters a magnetic field directed to the east. In what direction is the force on the current?

3. An electron is moving vertically upward when it enters a magnetic field directed to the east. In what direction is the force on the electron?

4. A current is passed through a loop of highly flexible wire. What shape does the loop assume? Why?

5. A current-carrying wire is in a magnetic field. (a) What angle should the wire make with **B** for the force on it to be zero? (b) What should the angle be for the force to be a maximum?

6. A beam of protons, initially moving slowly, is accelerated to higher and higher speeds. What happens to the diameter of the beam during this process?

7. A stream of protons is moving parallel to a stream of electrons. Is the force between the two streams necessarily attractive? Explain.

8. The electron beam in a cathode-ray tube is aimed parallel to a nearby wire carrying a current in the same direction. Is the electron beam deflected? If so, in what direction?

9. A current-carrying wire loop is in a uniform magnetic field. Under what circumstances, if any, will there be no torque on the loop? No net force? Neither torque nor net force?

10. An alternating current whose variation with time follows a sine curve is sent through an iron-core solenoid. Does the resulting magnetic field also vary sinusoidally with time?

11. Would you expect a compass to be more accurate near the equator or in the polar regions? Why?

12. Why is a piece of iron attracted by *either* pole of a magnet?

13. The alloy Alnico 2 has a retentivity of 0.76 T and a coercive force of 0.053 T. The corresponding figures for a certain cobalt steel alloy are 0.95 T and 0.023 T. (a) Which material would make the initially stronger permanent magnet? (b) Which would be more likely to retain its magnetization?

14. A power line 10 m above the ground carries a current of 5000 A. Find the magnetic field of the current on the ground directly under the cable.

15. At what distance from a long, straight wire carrying a current of 12 A does the magnetic field equal that of the earth, approximately 3×10^{-5} T?

16. A 10-turn circular coil of radius 2 cm carries a current of 0.5 A. Find the magnetic field at its center.

17. What should the current be in a wire loop 1 cm in diameter if the magnetic field at the center of the loop is to be 0.001 T?

18. A solenoid 10 cm long is meant to have a magnetic field of 0.002 T inside it when the current is 3 A. How many turns are needed?

19. A long solenoid is wound with 30 turns/cm. What is the magnetic field inside the solenoid when the current is 3.8 A?

20. A vertical wire 2 m long is in a 10^{-2} T magnetic field whose direction is northeast. (a) What is the magnitude and direction of the force on the wire when a 5-A current flows upward in it? (b) When the same current flows downward in it?

21. A horizontal north-south wire 5 m long is in a 0.02 T magnetic field whose direction is northeast. (a) What is the magnitude and direction of the force on the wire when a 4-A current flows north in it? (b) When the same current flows south in it?

22. A copper wire whose linear density is 10 g/m is stretched horizontally perpendicular to the direction of the horizontal component of the earth's magnetic field at a place where the magnitude of that component is 2×10^{-5} T. What must the current in the wire be for its weight to be supported by the magnetic force on it? What do you think would happen to such a wire if this current were passed through it?

23. What is the radius of the path of an electron whose speed is 10^7 m/s in a magnetic field of 0.02 T when the electron's path is perpendicular to the field?

24. Compute the maximum radius of curvature in the earth's magnetic field at sea level, assuming $B = 3 \times 10^{-5}$ T, of (a) a proton whose speed is 2×10^7 m/s, and (b) an electron of the same speed.

25. An electron in a television picture tube travels at 3×10^7 m/s and is acted upon both by gravity and by the earth's magnetic field. Which exerts the greater force on the electron?

26. A certain electric transmission line consists of two wires 4 m apart that carry currents of 10^4 A. If the towers sup-

porting the wires are 200 m apart, how much force does each current exert on the other between the towers?

27. The parallel wires in a lamp cord are 2.0 mm apart. What is the force per meter between them when the cord is used to supply power to a 120-V, 200-W light bulb?

PROBLEMS

1. Two parallel wires 10 cm apart carry currents in the same direction of 8 A. Find the magnetic field midway between the wires.

2. Two parallel wires 10 cm apart carry currents in opposite directions of 8 A. Find the magnetic field midway between the wires.

3. Two parallel wires 20 cm apart carry currents in the same direction of 5 A. Find the magnetic field between the wires 5 cm from one of them and 15 cm from the other.

4. Two parallel wires 20 cm apart carry currents in opposite directions of 5 A. Find the magnetic field between the wires 5 cm from one of them and 15 cm from the other.

5. A tube 20 cm long is wound with 400 turns of wire in one layer and with 200 turns in a second layer in the opposite sense to the first. A current of 0.1 A is passed through the coils. What is the magnetic field in the interior of the tube?

6. A solenoid is wound with 18 turns/cm and carries a current of 1.2 A. Another layer of turns is wound over the solenoid with 10 turns/cm and a current of 5.0 A is passed through the new coil, opposite to the direction of the current in the solenoid. What is the magnetic field in the interior of the solenoid?

7. A current of 0.5 A is passed through a solenoid wound with 20 turns/cm. A single loop of wire 1 cm in radius is bent around the middle of the solenoid. What should the current in this loop be in order for the magnetic field at its center to cancel out the field of the solenoid there?

8. Two long, parallel wires a distance s apart each carry the current I in opposite directions. Verify that the magnetic field at a point equidistant from the wires and x away from the plane in which they lie is given by $2\mu_0 Is/\pi(4x^2 + s^2)$.

9. The table on page 480 lists corresponding values of B_0 and B for a type of carbon steel, where B_0 is the magnetic field of a current-carrying coil with no core and B is the magnetic field of the same coil with a carbon steel core. (a) Plot a graph of B versus B_0 for this material. (b) Plot a graph of the relative permeability μ/μ_0 of this material versus B_0.

Approximately what is the maximum value of μ/μ_0, and at approximately what value of B_0 is this maximum reached?

B_0	B
0.0×10^{-5} T	0.0 T
4.2	0.2
6.3	0.4
7.7	0.6
9.0	0.8
12	1.0
20	1.2
36	1.4
75	1.6

10. A charge of $+10^{-6}$ C is moving at 500 m/s along a path parallel to a long, straight wire and 0.1 m from it. The wire carries a current of 2 A in the same direction as that of the charge. What is the magnitude and direction of the force on the charge?

11. A long, straight wire carries a current of 100 A. (a) What is the force on an electron traveling parallel to the wire, in the opposite direction to the current, at a speed of 10^7 m/s when it is 10 cm from the wire? (b) Find the force on the electron under the above circumstances when it is traveling perpendicularly toward the wire.

12. A charge of $+2 \times 10^{-6}$ C is moving at 10^3 m/s at a distance of 12 cm away from a straight wire carrying a current of 4 A. Find the magnitude and direction of the force on the charge when it is moving parallel to the wire (a) in the same direction as the current, and (b) in the opposite direction.

13. An electron is moving at 6×10^7 m/s at a distance of 5 cm from a long, straight wire carrying a current of 40 A. Find the magnitude and direction of the force on the electron when it is moving parallel to the wire (a) in the same direction as the current, and (b) in the opposite direction to the current.

14. The charge of Problem 12 is moving in a direction perpendicular to the same wire. Find the magnitude and direction of the force on the charge when it is moving (a) toward the wire, and (b) away from the wire.

15. The electron of Problem 13 is moving in a direction perpendicular to the same wire. Find the magnitude and direction of the force on the electron when it is moving (a) toward the wire, and (b) away from the wire.

16. An electron moving through an electric field of 500 V/m and a magnetic field of 0.1 T experiences no force. The two fields and the electron's direction of motion are all mutually perpendicular. What is the speed of the electron?

17. A velocity selector uses a magnet to produce a 0.05-T magnetic field and a pair of parallel metal plates 1 cm apart to produce a perpendicular electric field. What potential difference should be applied to the plates to permit singly charged ions of speed 5×10^6 m/s to pass through the selector?

18. A mass spectrometer employs a velocity selector consisting of a magnetic field of 0.0400 T perpendicular to an electric field of 50,000 V/m. The same magnetic field is then used to deflect the ions. Find the radius of curvature of singly charged lithium ions of mass 1.16×10^{-26} kg in this spectrometer.

19. Prove that the time required for a charged particle in a magnetic field to make a complete revolution is independent of its speed and the radius of its orbit.

20. In a typical loudspeaker, a permanent magnet creates a radial magnetic field in which a wire coil attached to the apex of a paper cone can move perpendicular to the field. An alternating current in the coil causes the cone to oscillate and thereby produce sound waves. If the 40-turn coil of a certain loudspeaker is 8 mm in radius and is located in a 0.4-T magnetic field, find the force on the coil when the current in it is 0.05 A.

21. Use Biot's law, Eq. (19–2), to obtain the formula $B = \mu I/2r$ for the magnetic field at the center of a circular current loop of radius r. To do this, first find the field element ΔB due to a small segment ΔL of the loop. Since all the segments of the loop are at the same distance r from the center and the angle θ is 90° for all of them, B can then be found simply by replacing ΔL by the circumference of the loop.

ANSWERS TO MULTIPLE CHOICE

1. d	6. c	11. c	16. a	21. c
2. c	7. b	12. c	17. d	22. c
3. a	8. c	13. b	18. b	23. c
4. c	9. a	14. a	19. b	24. c
5. a	10. d	15. c	20. c	25. d

ELECTROMAGNETIC INDUCTION

20

We have seen that, by causing a current to flow, an electric field is able to produce a magnetic field. Is there any way in which a magnetic field can produce an electric field? This problem was unsuccessfully tackled by many of the early workers in electricity and magnetism. Finally, in 1831, Michael Faraday in England and Joseph Henry in the United States independently discovered the phenomenon of electromagnetic induction. Two familiar applications of electromagnetic induction are the electric generator, which is the source of most electric power, and the transformer, which enables the emf of an alternating current to be increased or decreased easily.

20–1   ELECTROMAGNETIC INDUCTION

There is no current in a stationary wire in a static magnetic field when the wire is not connected to a source of emf. What Faraday and Henry found is that

FIG. 20–1 (a) There is no current in a stationary wire in a magnetic field. (b) When the wire is moved across the field, a current is produced. Reversing the direction of motion also reverses the direction of the current.

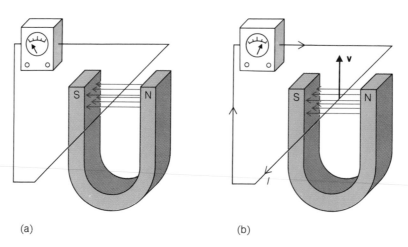

(a)   (b)

moving the wire produces a current (Fig. 20–1). It does not matter whether the wire or the source of the magnetic field is being moved, provided that a component of the motion is perpendicular to the field; the origin of the current lies in the *relative motion* between a conductor and a magnetic field. This effect is known as *electromagnetic induction*.

Faraday generalized his observations with the help of the notion of lines of force:

Electromagnetic induction

An electromotive force is produced in a conductor whenever it cuts across magnetic lines of force.

It is this electromotive force that leads to the current that flows whenever there is relative motion between a conductor and a magnetic field. In fact, it is not even necessary for there to be actual motion of either a wire or of a source of magnetic field, because a magnetic field that changes in strength has moving lines of force associated with it.

No emf arises from motion parallel to a magnetic field

Moving a wire *parallel* to a magnetic field does not give rise to a current: electromagnetic induction occurs only when there is a component of the wire's velocity perpendicular to the lines of force.

Origin of electromagnetic induction

Electromagnetic induction has a straightforward explanation in terms of the force exerted by a magnetic field on a moving charge. As we saw in the previous chapter, a wire carrying a current is pushed sideways in a magnetic field because of the forces exerted on the moving electrons. In Faraday's and Henry's experiments electrons are also moved through a magnetic field, but now by shifting the entire wire. As before, the electrons are pushed to the side, and in consequence move along the wire. The motion of these electrons along the wire is the electric current we measure.

Direction of induced current

The right-hand rule for the force on a moving positive charge in a magnetic field can also be used to give the direction of the induced current in a wire moving across a magnetic field. Hold your right hand so that the fingers point in the direction of **B** and the outstretched thumb is in the direction of **v.** The palm then faces in the same direction as the force on the positive charges in the wire, and hence in the direction of the conventional current (Fig. 20–2).

In the arrangement of Fig. 20–1, a current will also flow if the magnet is moved past the wire, instead of the wire being moved past the magnet. This is the basis of the

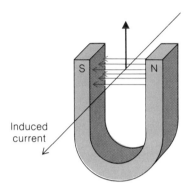

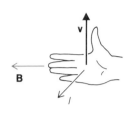

FIG. 20–2 Right-hand rule for the direction of an induced current.

playing head of a tape recorder. The tape used is a plastic film with a ferromagnetic coating, and it is magnetized by being passed close to an electromagnet whose current varies in accordance with the signal to be recorded. For instance, the recording head would be connected via an amplifier to a microphone if music or other sound is to be registered on the tape. The playing head contains a small coil, and the current induced in it by the changing magnetic fields of the moving tape is amplified and fed into a loudspeaker to reconstruct the original sound.

How a tape recorder works

20–2 MOVING WIRE IN A MAGNETIC FIELD

When a wire moves through a magnetic field **B** with a velocity **v** that is not parallel to **B,** an electromotive force $\mathcal{E}$ comes into being between the ends of the wire. It is not hard on the basis of what we already know to determine the magnitude of $\mathcal{E}$ in terms of **B, v,** and the length L of the wire. We shall assume that the wire, its direction of motion, and the magnetic field are all perpendicular to one another as in the previous diagrams.

According to Eq. (19–8) the force on a charge Q in a moving wire when the wire has the velocity component $\mathbf{v}_\perp$ perpendicular to a magnetic field **B** has the magnitude

$$F = Qv_\perp B$$

The direction of this force is along the wire, as in Fig. 20–3.

Let us consider a charge Q that moves from one end of the wire to the other under the influence of the magnetic force $Qv_\perp B$. The work done on the charge by the agent that moves the wire is

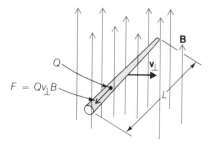

FIG. 20–3 The force on a charge Q within a conductor when the conductor is moved through a magnetic field is $Qv_\perp B$, where $v_\perp$ is the component of the conductor's velocity perpendicular to **B.**

FIG. 20–4 The magnitude of the motional emf induced in a conductor moving through a magnetic field depends upon the magnetic field strength, upon the length and speed of the conductor, and upon the direction of **v** relative to **B**.

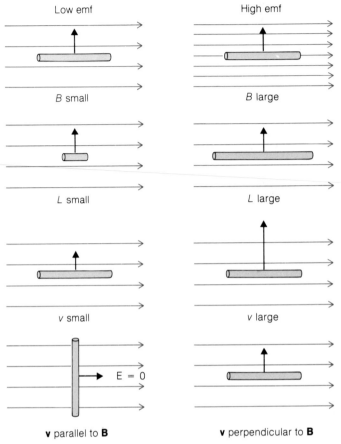

Voltage across ends of wire moving in a magnetic field

$$\text{Work} = \text{force} \times \text{distance}$$
$$W = FL = Qv_{\perp}BL$$

since the wire is L long. By definition the potential difference between two points is the work done in moving a unit charge between these points, so the potential difference between the ends of the wire is W/Q or

$$\mathcal{E} = BLv_{\perp} \qquad\qquad\qquad \textit{Motional emf} \quad (20\text{–}1)$$

This potential difference is the emf induced in the wire by its motion through the magnetic field (Fig. 20–4).

How large will the current that flows in the wire be? The answer is given by Ohm's law, since an induced emf is like any other emf in its ability to cause a current to flow. If the total resistance in the circuit is R, the resulting current will be equal to $I = \mathcal{E}/R$ as long as the emf exists.

Example   Find the potential difference between the wing tips of a jet airplane induced by its motion through the earth's magnetic field. The total wing span of the airplane is 40 m and its speed is 300 m/s in a region where the vertical component of the earth's field has the magnitude 3×10^{-5} T (Fig. 20–5).

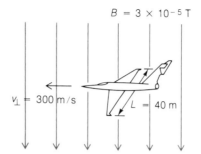

$B = 3 \times 10^{-5}$ T   **FIG. 20–5**

$v_\perp = 300$ m/s

$L = 40$ m

Solution   Substituting the given values of $v_\perp$, L, and B in Eq. (20–1) for the induced emf yields

$$\mathscr{E} = BLv_\perp = (3 \times 10^{-5}\,\text{T})(40\,\text{m})\left(300\,\frac{\text{m}}{\text{s}}\right) = 0.36\,\text{V}$$

20–3   FARADAY'S LAW

A convenient approach to electromagnetic induction makes use of the notion of *magnetic flux*. If the area of a wire loop is A and it is perpendicular to a magnetic field **B**, the total magnetic flux Φ (Greek letter *phi*) through the loop is defined as

Magnetic flux

$$\Phi = BA \qquad\qquad\qquad \textit{Magnetic flux} \quad (20\text{–}2)$$

The loop can have any shape, but its area must be taken from its projection on a plane perpendicular to **B**; see Fig. 20–6. The unit of flux is the *weber* (Wb), where 1 Wb = 1 T·m².

Faraday found that the electromotive force $\mathscr{E}$ in such a wire loop is *equal to the rate of change of the flux through it*. That is,

Faraday's law of electromagnetic induction

$$\mathscr{E} = -\frac{\Delta\Phi}{\Delta t} \qquad\qquad \textit{Induced emf} \quad (20\text{–}3)$$

where $\Delta\Phi$ is the change in the flux Φ that takes place during a period of time Δt. The values of the magnetic field B and the loop area A are, in themselves, irrelevant; only the rate at which either or both of them changes is important. Equation (20–3) is

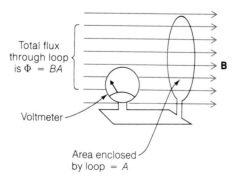

Total flux through loop is $\Phi = BA$

Voltmeter

Area enclosed by loop = A

FIG. 20–6   The emf induced in the wire loop is equal to the rate at which the flux Φ it encloses is changing.

called Faraday's law of electromagnetic induction. The reason for the minus sign is given below.

The induced emf in a wire loop is equal to the potential difference that would be found between the ends of an identical *open* wire loop. The notion of emf is useful here because it is meaningless to speak of the potential difference around a closed circuit; the emf in a closed circuit is the potential difference that would be found between the ends of the circuit if it were cut anywhere.

If a coil of N turns replaces the single loop of Fig. 20–6, the induced emf's in the turns are added together, and the total emf in the entire coil is

$$\mathcal{E} = - N\frac{\Delta\Phi}{\Delta t} \tag{20–4}$$

The minus sign in Eqs. (20–3) and (20–4) is a consequence of the law of conservation of energy. If the sign of $\mathcal{E}$ were the same as that of $\Delta\Phi/\Delta t$, the induced electric current would be in such a direction that its own magnetic field would *add* to that of the external field **B**; this additional changing field would then augment the existing rate of change of the flux Φ, and more and more current would flow even if the external contribution to Φ were to stay constant. But energy is associated with every current, and no current can increase without external energy being supplied to it. The only possibility, then, is that $\mathcal{E}$ be opposite in sign to $\Delta\Phi/\Delta t$, which means that

Lenz's law follows from conservation of energy

The direction of an induced current is always such that its own magnetic field opposes the changes in flux responsible for producing it.

This observation is known as *Lenz's law*. An example of Lenz's law is illustrated in Fig. 20–7.

Example   A 12-turn coil 10 cm in diameter has its axis parallel to a magnetic field of 0.5 T which is produced by a nearby electromagnet. The current in the electromagnet is cut off, and as the field collapses an average emf of 8 V is induced in the coil. What was the length of time required for the field to disappear?

Solution   The change in the flux through the coil is

$$\Delta\Phi = BA = B\pi r^2 = (-0.5 \text{ T})(\pi)(0.05 \text{ m})^2 = -0.0039 \text{ Wb}$$

If we assume that the flux drops to zero at a uniform rate,

$$\Delta t = -\frac{N\,\Delta\Phi}{\mathcal{E}} = -\frac{(12)(-0.0039 \text{ Wb})}{8\text{V}} = 0.0059\text{ s} \qquad \blacksquare$$

FIG. 20–7 The current induced in the wire loop is such that its own magnetic field is opposite to the field of the permanent magnet, in accordance with Lenz's law.

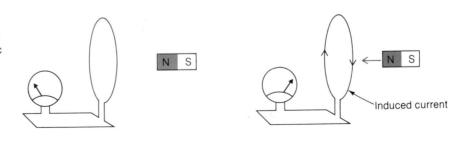

Induced current

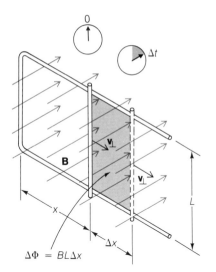

FIG. 20–8

$\Delta\Phi = BL\Delta x$

It is not hard to derive Faraday's law of electromagnetic induction from the formula for the emf induced in a wire moved across a magnetic field. Figure 20–8 shows a moving wire that slides across the legs of a ∪-shaped metal frame, so that the frame completes the loop. At the moment (say $t = 0$) when the moving wire is the distance x from the closed end of the frame, the flux enclosed by the loop is

Derivation of Faraday's law

$$\Phi = BA = BLx$$

At the later time $t = \Delta t$ the moving wire is $x + \Delta x$ from the closed end of the frame, so the flux now enclosed by the loop is

$$\Phi + \Delta\Phi = BLx + BL\,\Delta x$$

Hence the increase in the enclosed flux during the time Δt is

$$\Delta\Phi = BL\,\Delta x$$

The speed of the moving wire is $v_\perp$, and so

$$\Delta x = v_\perp\,\Delta t$$

and

$$\Delta\Phi = BLv_\perp\,\Delta t$$
$$\frac{\Delta\Phi}{\Delta t} = BLv_\perp$$

But $BLv_\perp$ is the emf induced in the moving wire. Hence we have, inserting a minus sign to remind ourselves of Lenz's law,

$$\mathscr{E} = -\frac{\Delta\Phi}{\Delta t}$$

which is Faraday's law.

20–4   THE BETATRON

An interesting application of electromagnetic induction is the betatron. In a betatron (Fig. 20–9) electrons are accelerated to high speeds through the action of an increasing magnetic field. As the magnetic field **B** increases, the associated electric field contributes to the energy of an electron moving in a circular path of area A by an amount per revolution equal to its increase in energy when it moves through an emf of

$$\mathscr{E} = -\frac{\Delta \Phi}{\Delta t} = -A\frac{\Delta B}{\Delta t}$$

That is, we can regard a circular path of this kind exactly as though it were a loop of wire, insofar as electromagnetic induction is concerned. As the electron goes faster and faster, gaining the energy $Q\mathscr{E}$ in each revolution, it will require a stronger magnetic field **B** if it is to stay in the same orbit; it is not difficult to arrange matters so that the very increase in **B** that accelerates the electron in the first place exactly keeps pace with its increasing speed, thereby maintaining a constant orbit radius.

A typical betatron that accelerates electrons to 100 MeV might have an orbit radius of 1 m and a magnetic field changing at the rate of 100 T/s. The area enclosed by the orbit is

$$A = \pi R^2 = 3.1 \text{ m}^2$$

and the emf per revolution is, omitting the minus sign,

$$\mathscr{E} = A\frac{\Delta B}{\Delta t} = (3.1 \text{ m}^2)(100 \text{ Ts}) = 310 \text{ V}$$

An electron in this betatron acquires 310 eV of additional energy each time it makes a complete circle; to acquire 100 MeV it must make

$$\frac{(100)(10^6 \text{ eV})}{310 \text{ eV/rev}} = 3.23 \times 10^5 \text{ revolutions}$$

20–5   THE GENERATOR

Electromagnetic induction underlies the operation of the generator

Electromagnetic induction is of immense practical importance since it is the means whereby nearly all the world's electric power is produced. In a generator a coil of wire is rotated in a magnetic field so that the flux through the coil changes constantly. The resulting potential difference across the ends of the coil causes a current to flow in an external circuit, and this current can be transmitted by a suitable system of wires for long distances from its origin.

Despite its name, a generator does not *create* electrical energy; what it does is *convert* mechanical energy into electrical energy, just as a battery converts chemical energy into electrical energy.

The construction of an idealized simple generator is shown in Fig. 20–10. A wire loop is rotated in a magnetic field, and the ends of the loop are connected to two *slip rings* on the shaft. Brushes pressing against the slip rings permit the loop to be connected to an external circuit. Only sides B and D of the loop contribute to the induced emf.

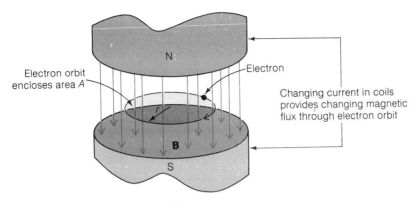

FIG. 20–9 Schematic diagram of a betatron.

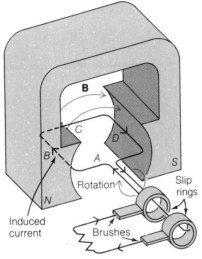

FIG. 20–10 A simple ac generator.

Because the sides of the loop reverse their direction of motion through the magnetic field twice per rotation, the induced current is also reversed twice per rotation and for this reason is called an *alternating current*. The variation with time of the emf of this generator is shown in Fig. 20–11. The shape of the curve is that of a sine (or cosine) curve. In a complete cycle, which corresponds to one turn of the loop, the emf increases to a maximum, falls to zero, continues to a negative maximum, and then returns to zero. If the loop rotates f times per second, the generator output will go through f cycles per second. An ac generator is often called an *alternator*.

Alternating current generator

The ac output of a simple generator can be changed to dc by substituting a commutator for the slip rings that connect the rotating coil with the external circuit (Fig. 20–12). The emf of such a dc generator varies with time as shown in Fig. 20–13; the emf is always in the same direction, but it rises to a maximum and drops to zero twice per complete rotation. The resemblance between the dc generator of Fig. 20–12 and the dc motor of Fig. 19–30 is not accidental: one is the inverse of the other. If the output terminals of a dc generator are connected to a battery, it will run as a motor.

Direct current generator

The emf of Fig. 20–13, though unidirectional, pulsates too much for most applications. A steadier emf can be obtained by using a pair of coils on the armature set

FIG. 20–11 The variation with time of the emf of a simple ac generator.

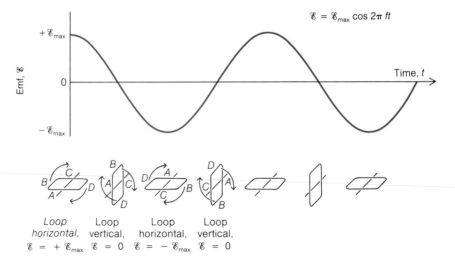

$$\mathscr{E} = \mathscr{E}_{max} \cos 2\pi \, ft$$

$+\mathscr{E}_{max}$

Emf, $\mathscr{E}$

0

Time, t

$-\mathscr{E}_{max}$

Loop horizontal, $\mathscr{E} = +\mathscr{E}_{max}$ Loop vertical, $\mathscr{E} = 0$ Loop horizontal, $\mathscr{E} = -\mathscr{E}_{max}$ Loop vertical, $\mathscr{E} = 0$

FIG. 20–12 A simple dc generator. The current in the brushes is always in the same direction.

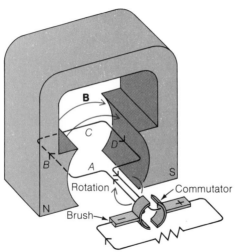

perpendicular to each other. When one of the coils has its maximum emf, the other has an emf of zero. Figure 20–14 shows the emf's of each coil and also what happens when the outputs of the two are connected in series: the resulting emf has only a moderate ripple. Commercial generators have many coils in their armatures to produce very nearly constant emf's.

A rectifier converts ac to dc

Another approach is to use a *rectifier* in the output circuit of an alternator. A rectifier is a device that permits current to pass through it in only one direction; the operation of semiconductor rectifiers is described in Chapter 30. Because alternators are simpler to construct than dc generators and are more reliable, they are often used in combination with semiconductor rectifiers to produce the direct current needed in cars and other vehicles.

In nearly all actual generators, electromagnets produce the magnetic field. (If a permanent magnet is used, the generator is called a *magneto*.) In an alternator, the

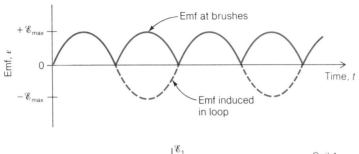

FIG. 20–13 The variation with time of the emf of a simple dc generator.

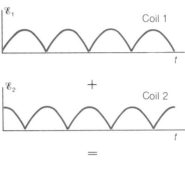

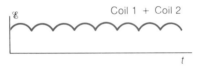

FIG. 20–14 The output of a dc generator whose armature contains two perpendicular coils has only a slight ripple.

Alternator construction

armature and field windings are often exchanged so that, in effect, the field rotates inside a stationary armature. This makes no difference to the physics involved since the essential condition is that there be relative movement of a conductor and a magnetic field. The advantage of a stationary armature is that a high-voltage output can be taken directly from it without any moving electrical contacts, since the field current can be supplied at a modest voltage.

20–6 BACK EMF

Induced emf's occur in all electric motors

In an electric motor a torque is produced through the interaction of the current in its armature with a magnetic field. As we have just learned, a coil rotated in a magnetic field is a source of emf, which implies that there is an emf generated in the rotating armature of an electric motor. This induced emf exists simply because wires are being moved through a magnetic field: the presence of a current of external origin in the armature at the same time, which is what causes it to turn in the first place, does not alter the situation. Every electric motor is also a generator.

Back emf reduces the armature current

By Lenz's law all induced emf's are such as to oppose the changes that bring them into being, which means that the induced emf in a motor armature is opposite in direction to the external voltage applied to the armature. The effect of this *back emf* (or *counter emf*) is to reduce the current in the armature to

$$I = \frac{V - \mathscr{E}_b}{R} \qquad \text{Armature current} \quad (20\text{–}5)$$

where V is the external voltage, $\mathscr{E}_b$ the back emf, and R the resistance of the armature coils.

Back emf and motor speed

Because the back emf is proportional to the speed of the armature, it is zero when the motor is turned on and increases as the speed increases. The current is therefore a maximum when the motor starts and drops afterward. The existence of back emf in a motor helps to keep its speed constant. At the normal speed,

$$V = \mathscr{E}_b + IR \qquad (20\text{–}6)$$

Impressed voltage = back emf + potential drop in armature

If the speed falls, $\mathscr{E}_b$ decreases and I goes up, and the greater current means more torque on the armature to restore the speed to its original value. If the speed rises above normal, $\mathscr{E}_b$ increases and I goes down, with the result that the torque is reduced and the motor loses speed until it assumes its normal value.

Shunt and series wound motors

Example   A 120-V dc electric motor has its armature and field windings connected in parallel. (Such a motor is said to be "shunt wound." When armature and field are in series, the motor is said to be "series wound.") The armature resistance is 2.0 Ω and the field resistance is 200 Ω. When the motor is at its operating speed of 1800 rpm, the total current is 3.0 A. Find (a) the back emf of the motor at its operating speed, (b) the power output and efficiency of the motor, and (c) the current in the motor at the moment its switch is turned on.

Solution   (a) The current in the field winding is

$$I_f = \frac{V}{R_f} = \frac{120\text{ V}}{200\ \Omega} = 0.6\text{ A}$$

Since the total current is 3.0 A, the armature current is

$$I_a = I - I_f = 3.0\text{ A} - 0.6\text{ A} = 2.4\text{ A}$$

The magnitude of the back emf is therefore

$$\mathscr{E}_b = V - I_a R_a = 120\text{ V} - (2.4\text{ A})(2.0\ \Omega) = 115\text{ V}$$

The direction of the back emf is, of course, opposite to that of the impressed potential difference V.

Power output of motor is product of back emf and armature current

(b) The mechanical power output of the motor (ignoring friction) is equal to the product $\mathscr{E}_b I_a$ of the back emf and the armature current. Hence

$$P_{\text{output}} = \mathscr{E}_b I_a = (115\text{ V})(2.4\text{ A}) = 276\text{ W}$$

The electrical power input of the motor is equal to IV, the total current multiplied by the applied potential difference:

$$P_{\text{input}} = IV = (3.0\text{ A})(120\text{ V}) = 360\text{ W}$$

Hence the motor's efficiency is

$$\text{Eff} = \frac{\text{power output}}{\text{power input}} = \frac{276 \text{ W}}{360 \text{ W}} = 0.77 = 77\%$$

Thus 23% of the power input is dissipated as heat in the armature and field windings.

(c) At the moment the motor is started, there is no back emf because the armature is still stationary. The total resistance of the armature and field windings is

$$R = \frac{R_a R_f}{R_a + R_f} = \frac{(2.0\,\Omega)(200\,\Omega)}{2.0\,\Omega + 200\,\Omega} = 1.98\,\Omega$$

The current is therefore

$$I = \frac{V}{R} = \frac{120 \text{ V}}{1.98\,\Omega} = 61 \text{ A}$$

The starting current is more than 20 times greater than the operating current! ◼

20–7 THE TRANSFORMER

Earlier it was mentioned that a current can be induced in a wire or other conductor by a changing magnetic field as well as by relative motion between a wire and a constant magnetic field. The essential condition for an induced emf is that magnetic lines of force cut across the wire (or vice versa), and it does not matter exactly how this comes about.

A change in the current in a wire loop is accompanied by a corresponding change in its magnetic field, and if there is another wire loop nearby, an emf will therefore be induced in it. If an alternating current flows in the first loop, the magnetic field around it will vary periodically and an alternating emf will be induced in the second loop. A *transformer* is a device based upon this effect which is used to produce an emf in a secondary ac circuit larger or smaller than the emf in the primary circuit. A transformer thus permits electric energy to be transferred from one ac circuit to another without a direct connection between them.

The output emf of a transformer may be greater or smaller than its input emf

With the help of Fig. 20–15 we can see how a changing current in a wire loop induces a changing current in another wire loop not connected to it.

(a) The switch connecting loop (1) with the battery is open, and no current is present in either loop.

(b) The switch has just been closed, and the expanding lines of force resulting from the increasing current in loop (1) cut across loop (2), inducing a current in it. The current in loop (2) is in the *opposite* direction to that in loop (1) because of Lenz's law: the direction of an induced current is always such that its own magnetic field opposes the change in flux that is inducing it.

(c) Now there is a constant current in loop (1), and since the flux through it does not change, no current is induced in loop (2).

(d) The switch has just been opened, and the contracting lines of force resulting from the decreasing current in loop (1) cut across loop (2), inducing a current in it. The current in loop (2) is now in the *same* direction as that in loop (1),

FIG. 20–15 When the current in a wire loop is turned on or off, the magnetic field around it changes. The changing field can induce a current in another nearby loop, even though both loops are stationary.

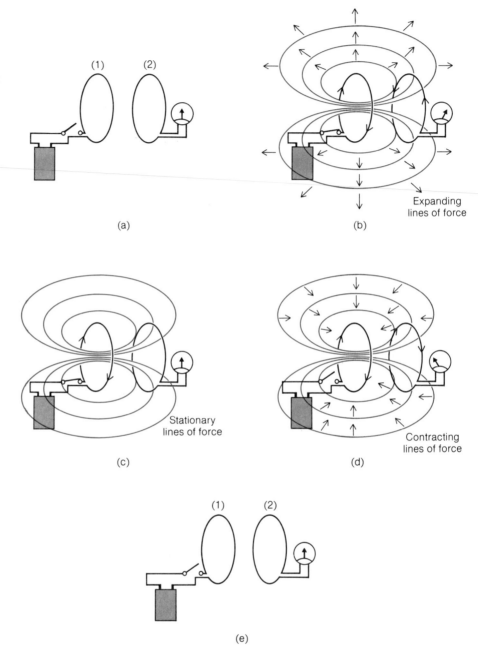

since it is the decreasing current in (1) that leads to the current in (2) and Lenz's law requires that an induced current oppose the change in flux that brings it about.

(e) Finally the current in loop (1) disappears, and no current is present in either loop.

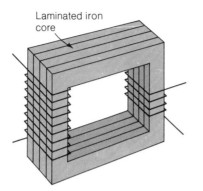

Laminated iron core

FIG. 20–16 A simple transformer.

Instead of being switched on and off, the current input to a transformer is alternating current which reverses its direction regularly. The construction of a simple transformer is shown in Fig. 20–16. Coils are used instead of single loops, and they are wound on a common iron core so that the alternating magnetic flux set up by an alternating current in one coil links with the other coil. This alternating flux induces an alternating emf in the latter coil. The *primary winding* of a transformer is the coil which is fed with an alternating current, and the *secondary winding* is the coil to which power is transferred via the changing magnetic flux. The designations of primary and secondary depend only upon how the transformer is connected; either winding may be the primary, and the other is then the secondary. A transformer may have more than one secondary winding so that it can provide several different secondary emf's.

A transformer operates on ac

Let us consider an ideal transformer in which there is no "leakage" of flux outside the iron core and no losses within it, so that $\Delta\Phi/\Delta t$ is the same for each of the turns of both windings. The emf *per turn* is therefore the same in both windings, and the total emf in each winding is proportional to the number of turns it contains. Hence

$$\frac{\mathscr{E}_1}{\mathscr{E}_2} = \frac{N_1}{N_2} \tag{20-7}$$

$$\frac{\text{Primary emf}}{\text{Secondary emf}} = \frac{\text{total primary turns}}{\text{total secondary turns}}$$

When the secondary winding has more turns than the primary winding, $\mathscr{E}_2$ is greater than $\mathscr{E}_1$ and the result is a *step-up transformer*; when the reverse is the case and $\mathscr{E}_2$ is less than $\mathscr{E}_1$, the result is a *step-down transformer*.

Step-up and step-down transformers

In an ideal transformer the power output is equal to the power input. If the primary current is I_1 and the second current is I_2, then

$$\mathscr{E}_1 I_1 = \mathscr{E}_2 I_2$$

Power input = power output

and

$$\frac{I_1}{I_2} = \frac{\mathscr{E}_2}{\mathscr{E}_1} = \frac{N_2}{N_1} \qquad\qquad \textit{Transformer} \quad (20-8)$$

Increasing the voltage with a step-up transformer means a proportionate decrease in the current, while dropping the voltage leads to a greater current.

Example   A transformer connected to a 120-V ac power line has 200 turns in its primary winding and 50 turns in its secondary winding. The secondary is connected to a 100-Ω light bulb. How much current is drawn from the 120-V power line?

Solution   The voltage across the secondary of the transformer is

$$V_2 = \frac{N_2}{N_1}V_1 = \left(\frac{50\,\text{turns}}{200\,\text{turns}}\right)(120\,\text{V}) = 30\,\text{V}$$

and so the current in the secondary circuit is

$$I_2 = \frac{V_2}{R} = \frac{30\,\text{V}}{100\,\Omega} = 0.3\,\text{A}$$

Hence the current in the primary circuit is

$$I_1 = \frac{N_2}{N_1}I_2 = \left(\frac{50\,\text{turns}}{200\,\text{turns}}\right)(0.3\,\text{A}) = 0.075\,\text{A}\qquad\blacksquare$$

Eddy currents

A transformer core is itself an electrical conductor, and the changing magnetic flux within it leads to internal currents called *eddy currents*. Eddy currents are wasteful partly because of the power lost as heat and partly because the flux they establish interferes with the proper operation of the transformer. To minimize eddy currents, actual transformer cores are usually laminated, with many thin sheets arranged parallel to the flux. The laminations are insulated from each other by natural oxide layers on their surfaces or by varnish coatings; only very weak eddy currents are induced in a core of this kind. In small transformers an alternative technique is to use a core of compressed powdered iron, so that each grain of iron is fairly well insulated from the others, although magnetically the core behaves as if it were solid.

There are still other sources of power loss in a transformer. The windings themselves have a certain amount of resistance, and part of the energy that flows through them is dissipated as heat. Also, in all magnetic materials some energy is lost as heat per hysteresis cycle in the atomic rearrangements that accompany flux changes (see Sec. 19–5). Leakage of flux from the core of Fig. 20–16 between the separated primary and secondary windings is more easily remedied. Instead, transformers normally have concentric windings as in Fig. 20–17, in which case the flux through each winding is virtually the same. Efficiencies as high as 99% are achieved in large transformers.

High-voltage transmission minimizes power loss

Alternating current owes its wide use largely to the ability of transformers to change the voltage at which it is transmitted from one value to another. Electricity generated at perhaps 11,000 volts is stepped up by transformers at the power station to as much as 750,000 volts (or even more) for transmission, and subsequently is stepped down by transformers at substations near the point of consumption to less than a thousand volts. Local transformers then step this down further to 240 or 120 volts. High voltages are desirable because, since $P = IV$, the higher the voltage, the smaller the current for a given amount of power; and since power is dissipated as heat at the rate I^2R, the smaller the current, the less the power lost due to resistance in the transmission line.

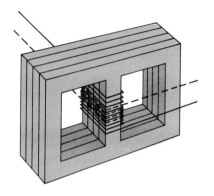

FIG. 20–17 Most transformers have concentric windings to minimize flux leakage.

Example   Find the power lost as heat when a 10 Ω cable is used to transmit 1 kW of electricity at 240 V and at 240,000 V.

Solution   The current in the cable in each case is

$$I_a = \frac{P}{V_a} = \frac{1000\,\text{W}}{240\,\text{V}} = 4.17\,\text{A}$$

$$I_b = \frac{P}{V_b} = \frac{1000\,\text{W}}{240,000\,\text{V}} = 4.17 \times 10^{-3}\,\text{A}$$

The respective rates of heat production per kilowatt are therefore

$$I_a^2 R = (4.17\,\text{A})^2 (10\,\Omega) = 174\,\text{W}$$

$$I_b^2 R = (4.17 \times 10^{-3}\,\text{A})^2 (10\,\Omega) = 1.74 \times 10^{-4}\,\text{W}$$

a difference of a factor of 10^6. Transmission at 240 V therefore means a million times more power lost as heat than does transmission at 240,000 V.   ∎

IMPORTANT TERMS

Electromagnetic induction refers to the production of an electric field in a conductor whenever magnetic lines of force move across it.

Lenz's law states that the direction of an induced current must be such that its own magnetic field opposes the changes in flux that are inducing it.

A **generator** is a device that converts mechanical energy into electrical energy.

The direction of an **alternating current** reverses itself periodically.

A **back** (or **counter**) **emf** is induced in the rotating coils of an electric motor and is opposite in direction to the external voltage applied to them.

An alternating current flowing in the primary coil of a **transformer** induces another alternating current in the secondary coil. The ratio of the emf's is proportional to the ratio of turns in the coils.

IMPORTANT FORMULAS

Motional emf: $\mathcal{E} = BLv$

Magnetic flux: $\Phi = BA$

Induced emf: $\mathcal{E} = -N\dfrac{\Delta\Phi}{\Delta t}$

Transformer: $\dfrac{I_1}{I_2} = \dfrac{\mathcal{E}_2}{\mathcal{E}_1} = \dfrac{N_2}{N_1}$

MULTIPLE CHOICE

1. Electromotive force is most closely related to
- (a) electric field.
- (b) magnetic field.
- (c) potential difference.
- (d) mechanical force.

2. The emf produced in a wire by its motion across a magnetic field does *not* depend on
- (a) the length of the wire.
- (b) the diameter of the wire.
- (c) the orientation of the wire.
- (d) the flux density of the field.

3. The magnetic flux through a wire loop in a magnetic field **B** does *not* depend upon
- (a) the area of the loop.
- (b) the shape of the loop.
- (c) the angle between the plane of the loop and the direction of **B**.
- (d) the magnitude B of the field.

4. A wire loop is moved parallel to a uniform magnetic field. The induced emf in the loop
- (a) depends upon the area of the loop.
- (b) depends upon the shape of the loop.
- (c) depends upon the magnitude of the field.
- (d) is 0.

5. Lenz's law is a consequence of the law of conservation of
- (a) charge.
- (b) momentum.
- (c) lines of force.
- (d) energy.

6. The unit of magnetic flux is the weber, where 1 Wb =
- (a) 1 T·m^2.
- (b) 1 T/m^2.
- (c) 1 A·m^2.
- (d) 1 A/m^2.

7. When a wire loop is rotated in a magnetic field, the direction of the induced emf changes once in each
- (a) $\frac{1}{4}$ revolution.
- (b) $\frac{1}{2}$ revolution.
- (c) 1 revolution.
- (d) 2 revolutions.

8. The emf produced by an ac-generator varies between $\mathscr{E}_{max}$ and
- (a) 0.
- (b) $\frac{1}{2}\mathscr{E}_{max}$.
- (c) $-\mathscr{E}_{max}$.
- (d) the back emf.

9. When the speed of a dc motor drops because of an increased load, there is also a drop in the
- (a) impressed voltage.
- (b) back emf.
- (c) current.
- (d) armature resistance.

10. The ac-output of a simple generator can be changed to dc by connecting the output of the rotating coil to the external circuit with
- (a) slip rings.
- (b) a commutator.
- (c) a resistor.
- (d) a transformer.

11. The alternating current in the secondary coil of a transformer is induced by
- (a) a varying electric field.
- (b) a varying magnetic field.
- (c) the iron core of the transformer.
- (d) motion of the primary coil.

12. The ratio between primary and secondary currents in a transformer does *not* depend on the
- (a) ratio of turns in the two windings.
- (b) resistance of the windings.
- (c) nature of the core.
- (d) primary voltage.

13. The primary winding of a transformer has 200 turns and its secondary winding has 50 turns. If the current in the secondary winding is 40 A, the current in the primary is
- (a) 10 A.
- (b) 80 A.
- (c) 160 A.
- (d) 8000 A.

14. The primary winding of a transformer has 200 turns and its secondary winding has 50 turns. If an ac emf of 12 V is applied to the primary, the secondary emf will be
- (a) 3 V.
- (b) 6 V.
- (c) 24 V.
- (d) 48 V.

15. A 100% efficient transformer has 100 turns in its primary winding and 300 turns in its secondary. If the power input to the transformer is 60 W, the power output is
- (a) 20 W.
- (b) 60 W.
- (c) 180 W.
- (d) 540 W.

EXERCISES

1. A bar magnet held vertically with its north pole downward is dropped through a wire loop whose plane is horizontal. Is the current induced in the loop just before the magnet enters it clockwise or counterclockwise as seen by an observer above? What is the direction of the current when the magnet passes through the center of the loop? Just after it leaves the loop?

2. One end of a bar magnet is thrust into a coil, and the induced current is clockwise as seen from the side of the coil into which it was thrust. (a) Was the end of the magnet its north or south pole? (b) What will be the direction of the induced current when the magnet is withdrawn?

3. A car is traveling from New York to Florida. Which of its wheels have a positive charge and which a negative charge?

4. Why is it easy to turn the shaft of a generator when it is not connected to an outside circuit, but much harder when such a connection is made?

5. A coil is rotated 100 times per second in a magnetic field. What time interval separates successive instants when the induced emf is zero?

6. Why does an electric motor require more current when it is turned on than when it is running continuously?

7. A loop of copper wire is rotated in a magnetic field about an axis along a diameter. (a) Why does the loop resist this rotation? (b) If the loop were made of aluminum wire, would the resistance to rotation be different?

8. How is the back emf in the armature of a dc motor at a given speed of rotation related to the emf developed in the same armature when the motor is used as a generator at the same speed of rotation and with the same current in its field coils?

9. For which of the following can alternating current be used without first being rectified into direct current: an electromagnet; a light bulb; an electric heater; electroplating; charging a storage battery?

10. Why are transformer cores made from thin sheets of steel? Would there be any advantage in making permanent magnets in this way?

11. Why must alternating current be used in a transformer?

12. What would happen if the primary winding of a transformer were connected to a battery?

13. What is it whose action on the secondary winding of a transformer causes an alternating potential difference to occur across its ends even though the primary and secondary windings are not connected?

14. A train is traveling at 130 km/h in a region where the vertical component of the earth's magnetic field is 3×10^{-5} T. The railway is standard gauge with its rails 1.435 m apart. Find the potential difference between the wheels on each axle of the train's cars.

15. A car is traveling at 30 m/s on a road where the vertical component of the earth's magnetic field is 3×10^{-5} T. What is the potential difference between the ends of its axles, which are 2 m long?

16. A potential difference of 1.8 V is found between the ends of a 2-m wire moving in a direction perpendicular to a magnetic field at a speed of 12 m/s. What is the magnitude of the field?

17. A rectangular wire loop 5 cm $\times$ 10 cm in size is perpendicular to a magnetic field of 10^{-3} T. (a) What is the flux through the loop? (b) If the magnetic field drops to zero in 3 s, what is the potential difference induced between the ends of the loop during that period?

18. A wire loop is 5 cm in diameter and is oriented with its plane perpendicular to a magnetic field. How rapidly should the field change if a potential difference of 1 V is to appear across the ends of the loop?

19. The magnetic flux through a 20-turn coil drops from 0.3 weber to 0 in 1 s. What is the induced potential difference across the ends of the coil?

20. A square wire loop 10 cm on a side is oriented with its plane perpendicular to a magnetic field. The resistance of the loop is 5 Ω. How rapidly should the magnetic field change if a current of 2 A is to flow in the loop?

21. A transformer has 100 turns in its primary winding and 500 turns in its secondary winding. If the primary voltage and current are respectively 120 V and 3 A, find the secondary voltage and current. (Assume 100% efficiency in Exercises 21–24.)

22. An electric welding machine employs a current of 400 A. The device uses a transformer whose primary winding has 400 turns and which draws 4 A from a 220-V power line. (a) How many turns are there in the secondary winding of the transformer? (b) What is the potential difference across the secondary?

23. A transformer rated at a maximum power of 10 kW is used to couple a 5000-W transmission line to a 240-V circuit. (a) What is the ratio of turns in the windings of the transformer? (b) What is the maximum current in the 240-V circuit?

24. A transformer connected to a 120-V power line has 100 turns in its primary winding and 40 turns in its secondary winding. An 80-Ω light bulb is connected to the secondary. What is the current in the primary winding?

PROBLEMS

1. In a shunt-wound generator the electromagnet windings are connected in parallel with the armature windings. What happens to the emf in such a generator when the current drawn by the external circuit increases? When the current decreases? Why?

2. In a series-wound generator the magnetic field is produced by an electromagnet whose windings are connected in series with the armature windings. What happens to the emf in such a generator when the current drawn by the external circuit increases? When the external circuit current decreases? Why?

3. The armature of a dc generator has a resistance of 0.20 Ω. The terminal potential difference of the generator is 120 V with no load and 115 V on full load. How much power is delivered at full load?

4. A 1.5 kW dc generator has a full-load potential difference of 35 V and a no-load potential difference of 38 V. How much power is dissipated as heat in the armature at full load?

5. A shunt-wound 120-V, 3-kW generator has an armature resistance of 0.1 Ω and a field resistance of 120 Ω. Find its efficiency.

6. A series-wound 24-V dc electric motor has an armature resistance of 0.1 Ω and a field resistance of 0.4 Ω. The motor draws 20 A at its normal operating speed. Find the power output, power input, and efficiency of the motor.

7. A shunt-wound 120-V dc electric motor has an armature resistance of 1.0 Ω and a field resistance of 200 Ω. The motor has a power input of 420 W at 2000 rpm. (a) Find the power output and efficiency of the motor. (b) What is the starting current? (c) What series resistance is required if the starting current is to be limited to 25 A?

8. Find the power output, power input, and efficiency of the motor of Problem 7 when it is operating at 1800 rev/min.

9. The electrons in a certain betatron make 150,000 revolutions in an orbit 2 m in diameter in the course of being accelerated to 20 MeV. Find the rate of change of the magnetic field in this betatron.

ANSWERS TO MULTIPLE CHOICE

1. c	**6.** a	**11.** b
2. b	**7.** b	**12.** d
3. b	**8.** c	**13.** a
4. d	**9.** b	**14.** a
5. d	**10.** b	**15.** b

CAPACITANCE AND INDUCTANCE

21

An electric field requires the expenditure of energy to be brought into existence, and this energy is available for doing work under suitable circumstances. A capacitor is a device that stores energy in the form of an electric field. Energy is also associated with a magnetic field, and is thus present in the vicinity of every electric current. Devices called inductors are designed to exploit this property of currents. When a charged capacitor is connected to an inductor, energy flows back and forth between them in the same way that KE and PE are interchanged in a harmonic oscillator. As in the case of a harmonic oscillator, the alternations in energy flow in a given capacitor-inductor combination occur at a certain natural frequency, a property that is made use of in practically every branch of electronics.

21–1   ELECTRIC FIELD ENERGY

Work must be done to create an electric field, since work must be done to separate positive and negative charges against the Coulomb forces attracting

FIG. 21–1 As a capacitor is charged, the potential difference between its plates builds up and more work must be done to transfer each successive charge. (In an actual capacitor, the electrons are transferred by means of an electric circuit connecting the plates, not directly.)

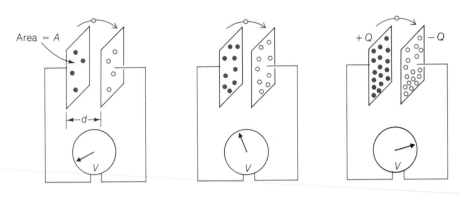

•$+e$ ○$-e$

A capacitor stores energy in the form of an electric field

them together. Where does the work go? The answer is that it is stored as potential energy, which we can think of as residing in the electric field between the charges. A system of conductors that stores energy in the form of an electric field is called a *capacitor*.

An example of a capacitor is a pair of parallel metal plates of area A that are a distance d apart, as in Fig. 21–1. We imagine that we create an electric field between them by bringing electrons from one plate to the other until a total charge of Q has been transferred. When we are finished, there will be a potential difference of V between

Charging a capacitor

the plates.

At the beginning, the potential difference between the plates is small, and little work is needed to move each electron. As the charge builds up, the potential difference becomes greater, and more work is needed per electron. The average potential difference $\overline{V}$ during the charge transfer is

$$\overline{V} = \frac{V_{final} + V_{initial}}{2} = \frac{V + 0}{2} = \tfrac{1}{2}V$$

Since the total charge transferred is Q, the work W that was done is the product of Q and the average potential difference $\overline{V}$, namely,

Energy of charged capacitor

$$W = Q\overline{V} = \tfrac{1}{2}QV \tag{21–1}$$

Because the metal plates are good conductors, the $+$ and $-$ charges on them are spread out evenly, and the field between the plates (except near the edges) is uniform with the magnitude

Electric field inside the capacitor

$$E = 4\pi k\frac{Q}{A} \tag{21–2}$$

The derivation of this formula, which follows from the considerations discussed in Chapter 17, is not given here because it involves calculus. We can rewrite Eq. (21–2) in the form

$$Q = \frac{AE}{4\pi k} \tag{21–3}$$

The potential difference between the plates is equal to

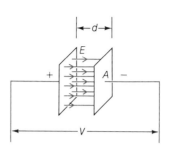

FIG. 21–2 The energy density of the electric field in a region is the total energy in the region divided by its volume.

$$V = Ed \tag{21-4}$$

Hence the electric potential energy of the system of two charged plates is

$$W = \tfrac{1}{2}QV = \frac{Ad}{8\pi k}E^2 \tag{21-5}$$

We note that Ad, the product of the area of each plate and the distance between them, is the volume occupied by the electric field $\mathbf{E}$ (Fig. 21–2). If we define the *energy density* w of an electric field as the electric potential energy per unit volume associated with it, we have here

Energy density of electric field

$$w = \frac{W}{Ad}$$

$$= \frac{E^2}{8\pi k} \qquad\qquad \textit{Electric energy density} \quad (21\text{-}6)$$

This important formula states that the energy density of an electric field is directly proportional to the square of its magnitude E. Even though we derived this formula for a special situation, it is a completely general result.

Example Dry air is an insulator provided the electric field in it does not exceed about 3×10^6 V/m. What energy density does this correspond to?

Solution Here $E = 3 \times 10^6$ V/m, and so

$$w = \frac{E^2}{8\pi k} = \frac{(3 \times 10^6 \,\text{V/m})^2}{8\pi(9 \times 10^9 \,\text{N·m}^2/\text{C}^2)} = 40 \,\text{J/m}^3$$

At sea level 1 m^3 of air has a mass of 1.3 kg, so if this amount of energy were gravitational potential energy, it would correspond to an elevation of about 3 m. ∎

21–2 CAPACITANCE

The potential difference V across a capacitor is always directly proportional to the charge Q on either of its plates (Fig. 21–3): the more the charge, the stronger the electric field between the plates, and the greater the potential difference. The ratio between Q and V is therefore a constant for any capacitor and is known as its *capacitance* (symbol C):

Capacitance

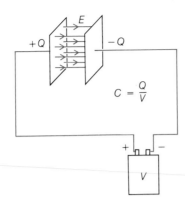

$$C = \frac{Q}{V} \qquad\qquad\qquad\qquad \textit{Capacitance} \quad (21\text{–}7)$$

$$\text{Capacitance} = \frac{\text{charge on either conductor}}{\text{potential difference between conductors}}$$

The farad, microfarad, and picofarad

The unit of capacitance is the *farad,* abbreviated F, where

1 farad = 1 coulomb/volt

The farad is so large a unit that, for practical purposes, it is usually replaced by the *microfarad* (μF) or *picofarad* (pF), whose values are

$$1\mu\text{F} = 10^{-6}\,\text{F} \qquad 1\,\text{pF} = 10^{-12}\,\text{F}$$

Parallel-plate capacitor

The capacitance of a pair of separated conductors depends solely upon their geometry and upon the material between them. In the case of a parallel-plate capacitor in vacuum, from Eqs. (21–3) and (21–4) we have

$$C = \frac{Q}{V} = \frac{AE/4\pi k}{Ed}$$
$$= \frac{1}{4\pi k}\frac{A}{d} \qquad\qquad \textit{Parallel-plate capacitor} \quad (21\text{–}8)$$

To a good degree of approximation the same formula can be used for such a capacitor in air as well.

Example   The plates of a parallel-plate capacitor are 10 cm square and 1 mm apart. Find its capacitance in air and the charge each plate will acquire when a potential difference of 100 V is applied.

Solution   Here $A = 10^{-2}\,\text{m}^2$ and $d = 10^{-3}\,\text{m}$. In air the capacitance is

$$C = \frac{1}{(4\pi)(9 \times 10^9\,\text{N·m}^2/\text{C}^2)} \times \frac{10^{-2}\,\text{m}^2}{10^{-3}\,\text{m}} = 8.85 \times 10^{-11}\,\text{F} = 88.5\,\text{pF}$$

If a potential difference of 100 V is placed across the plates of this capacitor, they will acquire charges of

$$Q = CV = 8.85 \times 10^{-11}\,\text{F} \times 100\,\text{V} = 8.85 \times 10^{-9}\,\text{C}$$

A parallel-plate capacitor can be used as a microphone if one of its plates is light and flexible enough to respond to the changing air pressure of a sound wave (Fig. 21–4). When the pressure increases, the diaphragm moves closer to the fixed plate, and C increases; when the pressure drops, the diaphragm moves outward, and C decreases. Because the voltage across the capacitor is fixed, the changes in C vary the charge on it, and the result is an electrical output whose variations match those in the incoming sound wave.

The formula $\frac{1}{2}QV$ for the potential energy of a charged capacitor derived in the previous section can be written in three equivalent ways:

Energy of charged capacitor

$$W = \tfrac{1}{2}QV \tag{21–9}$$

$$W = \tfrac{1}{2}CV^2 \qquad \text{\textit{Potential energy of charged capacitor}} \tag{21–10}$$

$$W = \frac{1}{2}\frac{Q^2}{C} \tag{21–11}$$

These equations hold for all capacitors, regardless of their construction.

Example A heart attack often leads to a condition called *fibrillation* in which the heart's actions lose their synchronization and it is unable to pump blood effectively. This condition can often be corrected by an electric shock to the heart which completely stops it for a moment; the heart may then start again spontaneously in its normal rhythm. An appropriate such shock can be provided by the discharge of a 10-μF capacitor that has been charged to a potential difference of 6000 V. (a) How much energy is released in the current pulse? (b) How much charge passes through the patient's body? (c) If the pulse lasts 5 ms, what is the average current?

Solution

(a) $W = \tfrac{1}{2}CV^2 = \tfrac{1}{2}(10^{-5}\,\text{F})(6000\,\text{V})^2 = 180\,\text{J}$

(b) $Q = CV = (10^{-5}\,\text{F})(6000\,\text{V}) = 0.06\,\text{C}$

(c) $\bar{i} = \dfrac{Q}{t} = \dfrac{0.06\,\text{C}}{5 \times 10^{-3}\,\text{s}} = 12\,\text{A}$

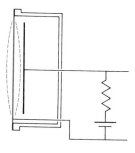

FIG. 21–4 Sound waves cause the diaphragm of a capacitor microphone to vibrate, and the resulting changes in its capacitance are transformed into electrical signals.

Table 21–1 is a summary of the various formulas for charge Q, potential difference V, capacitance C, and energy W that follow from the definition of capacitance $C = Q/V$ and from the energy formula $W = \tfrac{1}{2}QV$.

TABLE 21–1

	C and V	*C and Q*	*Q and V*	*W and C*	*W and V*	*W and Q*
$Q =$	CV			$\sqrt{2WC}$	$\dfrac{2W}{V}$	
$V =$		$\dfrac{Q}{C}$		$\sqrt{\dfrac{2W}{C}}$		$\dfrac{2W}{Q}$
$C =$			$\dfrac{Q}{V}$		$\dfrac{2W}{V^2}$	$\dfrac{Q^2}{2W}$
$W =$	$\dfrac{CV^2}{2}$	$\dfrac{Q^2}{2C}$	$\dfrac{QV}{2}$			

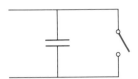

FIG. 21–5 A capacitor connected across a switch prevents arcing when the switch is opened by providing an alternate path of lower resistance than the initial small air gap.

Although capacitors have their most important application in ac circuits, as we shall find in the next chapter, they have dc uses as well, such as the ones described above. Another is to minimize the effects of arcing across switch contacts. Immediately after a switch is opened in a current-carrying circuit, the air gap between the switch contacts is so small that the voltage across them exceeds the insulating ability of the gap and an arc discharge occurs briefly. If the original current in the circuit was large, the contacts can be damaged. The problem is especially severe in dc motors and generators, whose brushes and commutators are really switches that may open and close hundreds of times per second. A capacitor connected across a switch prevents such arcing (Fig. 21–5). When the switch is closed, the capacitor is uncharged since its plates are at the same potential. When the switch is opened, the uncharged capacitor provides a path for current whose resistance is lower than that of the air gap, so the momentary current after the switch is opened goes to charge the capacitor rather than to produce an arc in the air gap.

21–3 DIELECTRIC CONSTANT

Let us now examine what happens when a slab of an insulating material is placed between the plates of a capacitor.

How polar molecules respond to electric fields

Although an insulator cannot conduct electric current, it can respond to an electric field in another way. The molecules of all substances either normally have a nonuniform distribution of electric charge within them or assume such a distribution under the influence of an electric field. As we know, a molecule of the former kind is called a polar molecule, and behaves as though one end is positively charged and the other negatively charged. In an assembly of polar molecules when there is no external electric field, the molecules are randomly oriented as in Fig. 21–6(a). When an electric field is present, it acts to align the molecules opposite to the field, as in Fig. 21–6(b).

While nonpolar molecules ordinarily have symmetric charge distributions, an electric field is able to distort their arrangements of electrons so that an effective separation of charge takes place (Fig. 21–7). Again the molecules have their charged ends aligned opposite to the external field. In either case, then, the net electric field between the plates of the capacitor is *less* than it would be with nothing between them.

How nonpolar molecules respond to electric fields

FIG. 21–6 An electric field tends to align polar molecules opposite to the field.

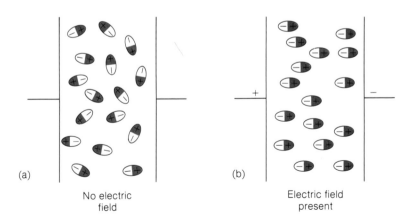

(a) No electric field

(b) Electric field present

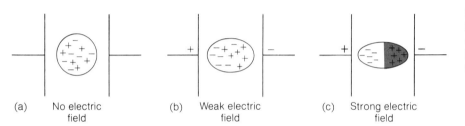

FIG. 21–7 An electric field tends to distort the charge distributions in molecules that are ordinarily nonpolar.

(a)   No electric field

(b)   Weak electric field

(c)   Strong electric field

The *dielectric constant,* symbol K, of a substance is a measure of how effective it is in reducing an electric field set up across a sample of it. For a capacitor with a given charge Q, reducing the electric field means reducing V as well, and since $C = Q/V$, this means an increase in its capacitance. If the capacitance of a capacitor is C_0 when there is a vacuum between its plates, its capacitance will be

Dielectric constant

$$C = KC_0 \tag{21–12}$$

when a substance of dielectric constant K is between the plates (Fig. 21–8). Table 21–2 is a list of dielectric constants for various substances. Water and alcohol molecules are highly polar, and the values of K for water, ice, and ethyl alcohol are accordingly high.

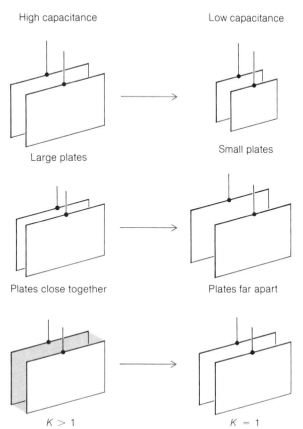

High capacitance

Low capacitance

Large plates

Small plates

Plates close together

Plates far apart

$K > 1$

$K = 1$

FIG. 21–8 The capacitance of a parallel-plate capacitor depends upon the area and spacing of its plates and upon the dielectric constant K of the medium between them.

TABLE 21–2
Dielectric constants at
20°C except where noted

Substance	K	Substance	K
Air	1.0006	Mica	2.5–7
Air, liquid ($-191°C$)	1.4	Neoprene	6.7
Alcohol, ethyl	26	Sulfur	3.9
Benzene	2.3	Teflon	2.1
Glass	5–8	Water	80
Ice ($-2°C$)	94	Waxed paper	2.2

Example   A capacitor with air between its plates is connected to a 50-V source and then disconnected. The space between the plates of the charged capacitor is filled with Teflon ($K = 2.1$). What is the potential difference across the capacitor now?

Solution   The initial charge on the capacitor is $Q = C_1 V_1$. When the Teflon dielectric is inserted, the charge remains the same but the capacitance increases to $C_2 = KC_1$. The new voltage across the capacitor is therefore

$$V_2 = \frac{Q}{C_2} = \frac{C_1 V_1}{KC_1} = \frac{V_1}{K} = \frac{50\text{ V}}{2.1} = 23.8\text{ V}$$

The ratio V_2/V_1 is independent of the original capacitance of the capacitor.   ∎

Dielectric strength

Besides its dielectric constant, an important property of a material used between the plates of a capacitor is its *dielectric strength,* which is the maximum electric field that can safely be applied to it before it breaks down and loses its insulating ability. Air has a dielectric strength of about 3×10^6 V/m; thus a voltage of 300 V will produce a spark in an air gap of 0.1 mm. Most materials used as separators in capacitors and to insulate wires have dielectric strengths that exceed 10^7 V/m.

Capacitor construction

Commercial capacitors consist of many interleaved plates to make possible a high capacitance in a small unit (Fig. 21–9). Solid dielectrics are usually used, both to increase C and to maintain a fixed distance between the plates. With sheets of a solid dielectric such as waxed paper or mica between them, the plates can be of inexpensive metal foil, without the rigidity required to prevent accidental contact if only air separated them. In an *electrolytic capacitor* the plates have extremely thin dielectric layers formed on their surfaces by chemical action. A conducting paste between these plates constitutes the other electrode of the capacitor. The very small thickness of the dielectrics in

FIG. 21–9 Most fixed capacitors are made of interleaved sheets of metal foil separated by layers of dielectric material.

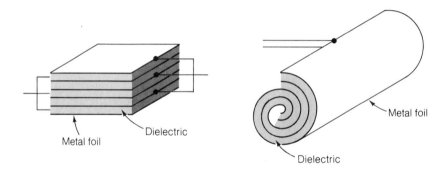

Metal foil   Dielectric   Metal foil   Dielectric

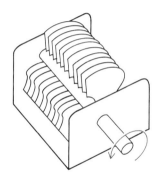

FIG. 21-10 Variable capacitors of this type are widely used to tune radio receivers.

electrolytic capacitors permits them to have capacitances of as much as 50 μF without excessive bulk.

Variable capacitors normally have two sets of rigid aluminum plates that are inter- leaved with air as the dielectric, as in Fig. 21–10. One of the sets is mounted on a shaft, and by rotating the shaft the amount of overlap between the plates can be adjusted. Since the overlapped area of the plates is the chief contributor to their capacitance, turning the shaft varies C.

Variable capacitors

21-4 CAPACITORS IN COMBINATION

The equivalent capacitance of two or more capacitors connected together can be deter- mined in a manner analogous to that used in the case of resistors in combination, though the results are different.

Figure 21–11 shows three capacitors in series. Each has charges of the same magnitude Q on its plates, in agreement with the principle of conservation of charge. Hence the potential differences across the capacitors are respectively

Capacitors in series

$$V_1 = \frac{Q}{C_1} \qquad V_2 = \frac{Q}{C_2} \qquad V_3 = \frac{Q}{C_3}$$

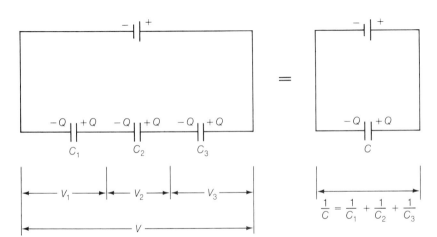

FIG. 21-11 Capacitors in series. A charge of the same magnitude Q is present on all the plates of the capacitors.

and so, if C is the equivalent capacitance of the set, we have

$$V = V_1 + V_2 + V_3$$

$$\frac{Q}{C} = \frac{Q}{C_1} + \frac{Q}{C_2} + \frac{Q}{C_3}$$

$$\frac{1}{C} = \frac{1}{C_1} + \frac{1}{C_2} + \frac{1}{C_3}$$

For any number of capacitors in series,

$$\frac{1}{C} = \frac{1}{C_1} + \frac{1}{C_2} + \frac{1}{C_3} + \cdots \qquad \textit{Capacitors in series} \quad (21\text{--}13)$$

The reciprocal of the equivalent capacitance of a series arrangement of capacitors is equal to the sum of the reciprocals of the individual capacitors. Evidently C is smaller than the capacitance of any of the individual capacitors.

If there are only two capacitors in series, Eq. (21–13) becomes

Two capacitors in series $$C = \frac{C_1 C_2}{C_1 + C_2} \qquad (21\text{--}14)$$

As in the case of the similar Eq. (18–11) for resistors in parallel, a calculator that has a reciprocal [1/X] key makes it easy to solve Eq. (21–13) for C in an actual problem. If three capacitors are in series, the key sequence would be: $[C_1][1/X][+][C_2]$ $[1/X][+][C_3][1/X][=][1/X]$.

Example Two capacitors, one of 10 μF and the other of 20 μF, are connected in series across a 12-V battery, as in Fig. 21–12. Find the equivalent capacitance of the combination, the charge on each capacitor, and the potential difference across it.

Solution The equivalent capacitance of the two capacitors is

$$C = \frac{C_1 C_2}{C_1 + C_2} = \frac{(10 \, \mu\text{F})(20 \, \mu\text{F})}{10 \, \mu\text{F} + 20 \, \mu\text{F}} = 6.7 \, \mu\text{F}$$

and so the charges on them are

$$Q_1 = Q_2 = Q = CV = (6.7 \times 10^{-6} \, \text{F})(12 \, \text{V}) = 8.0 \times 10^{-5} \, \text{C}$$

The potential differences across the capacitors are respectively

FIG. 21–12

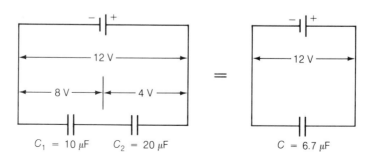

$C_1 = 10 \, \mu F \qquad C_2 = 20 \, \mu F \qquad\qquad C = 6.7 \, \mu F$

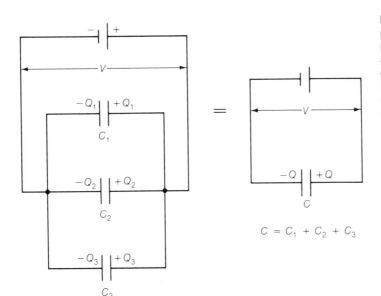

FIG. 21-13 Capacitors in parallel. The same potential difference V is across all of them, but each has a charge on its plates whose magr itude is proportional to its capacitance.

$$V_1 = \frac{Q}{C_1} = \frac{8.0 \times 10^{-5}\,\mathrm{C}}{1.0 \times 10^{-5}\,\mathrm{F}} = 8.0\,\mathrm{V}$$

$$V_2 = \frac{Q}{C_2} = \frac{8.0 \times 10^{-5}\,\mathrm{C}}{2.0 \times 10^{-5}\,\mathrm{F}} = 4.0\,\mathrm{V}$$

The potential difference is greatest across the capacitor of smaller C. The sum of V_1 and V_2 is 12 V, as it should be. ■

Figure 21-13 shows three capacitors connected in parallel. The same potential difference is across all of them, so that the charges on their plates have the respective magnitudes   **Capacitors in parallel**

$$Q_1 = C_1 V \qquad Q_2 = C_2 V \qquad Q_3 = C_3 V$$

The total charge $Q_1 + Q_2 + Q_3$ on either the positive or negative plates of the capacitors is equal to the charge Q on the corresponding plate of the equivalent capacitor, and hence

$$Q = Q_1 + Q_2 + Q_3$$
$$CV = C_1 V + C_2 V + C_3 V$$
$$C = C_1 + C_2 + C_3$$

This result can be generalized to

$$C = C_1 + C_2 + C_3 + \cdots \qquad\qquad \textit{Capacitors in parallel} \quad (21\text{-}15)$$

More energy is stored in a set of capacitors when they are connected in parallel across a given potential difference than when they are connected in series, because the electric fields in them are stronger in the former case.

Example   The capacitors of Fig. 21–12 are reconnected in parallel across the same battery. Compare the energies of the capacitors now with what they were when connected in series.

Solution   In the series connection, the capacitors had the energies

$$W_1 \text{ (series)} = \tfrac{1}{2}QV_1 = \tfrac{1}{2}(8.0 \times 10^{-5}\text{C})(8.0\text{V}) = 3.2 \times 10^{-4}\text{J}$$

$$W_2 \text{ (series)} = \tfrac{1}{2}QV_2 = \tfrac{1}{2}(8.0 \times 10^{-5}\text{C})(4.0\text{V}) = 1.6 \times 10^{-4}\text{J}$$

The capacitor of smaller C had the greater energy. In the parallel connection, the capacitors have the same potential difference $V = 12$ V, and their energies are

$$W_1(\text{parallel}) = \tfrac{1}{2}C_1V^2 = \tfrac{1}{2}(1.0 \times 10^{-5}\text{F})(12\,\text{V})^2 = 7.2 \times 10^{-4}\text{J}$$

$$W_2(\text{parallel}) = \tfrac{1}{2}C_2V^2 = \tfrac{1}{2}(2.0 \times 10^{-5}\text{F})(12\,\text{V})^2 = 14 \times 10^{-4}\text{J}$$

In this parallel connection, the stored energy is 4.4 times as great as it was in the series connection.   ∎

21–5 MAGNETIC FIELD ENERGY

A self-induced emf occurs whenever the current in a wire loop changes

Energy is associated with magnetic as well as with electric fields. To appreciate why, let us look at what happens when a wire loop is connected to a battery. As the current starts to flow, a magnetic field begins to build up. But this changing magnetic field induces an emf in the *same* wire loop whose current causes the field in the first place. By Lenz's law the self-induced emf is such as to oppose the change in flux that is responsible for it. The battery must therefore push electrons through the wire loop against the self-induced emf due to the increasing magnetic field, which means that work has to be done in order to produce the magnetic field. (The work needed to overcome the effect of the self-induced emf has nothing to do with the resistance of the wire; it is a consequence of electromagnetic induction only.)

When the current in the wire loop reaches its final value, there is no more self-induced emf. The work performed to establish the magnetic field has become magnetic energy.

Let us determine the amount of energy contained in a solenoid when a steady current I flows through it. This energy is equal to the work W that had to be done against the self-induced emf $\mathscr{E}$ in order to establish the current starting from $I = 0$ (Fig. 21–14).

If we suppose that the current rises at a uniform rate from $I = 0$ to $I = I$ in a time interval Δt, then the average current $\bar{I}$ during Δt is

$$\bar{I} = \frac{I_{\text{final}} + I_{\text{initial}}}{2} = \frac{I + 0}{2} = \tfrac{1}{2}I$$

Thus the total charge Q that passes through the coil while the current is building up to its final value I is

$$Q = \bar{I}\,\Delta t = \tfrac{1}{2}I\,\Delta t$$

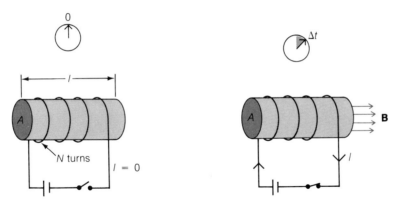

FIG. 21–14 A self-induced emf opposes the establishment of a current in a solenoid. The work done against this emf becomes magnetic energy.

The work that was done during the buildup of the current is the product of the total charge Q and the emf $\mathscr{E}$:

Work done to create field

$$W = Q\mathscr{E} = \tfrac{1}{2}I\mathscr{E}\,\Delta t \qquad (21\text{–}16)$$

To find expressions for I and $\mathscr{E}$, we begin with Eq. (19–6), $B = \mu(N/l)I$, for the magnetic field in the interior of an N-turn solenoid l long when it carries the current I. This formula can be rewritten to given an expression for I in terms of B, namely

$$I = \frac{Bl}{\mu N} \qquad (21\text{–}17)$$

The magnitude of the emf is given by Faraday's law (the minus sign is irrelevant here) as

$$\mathscr{E} = N\frac{\Delta\Phi}{\Delta t}$$

Self-induced emf

The final flux Φ through the solenoid is BA, where A is the cross-sectional area (see Fig. 21–14). Since the initial flux is $\Phi = 0$ its rate of change is

$$\frac{\Delta\Phi}{\Delta t} = \frac{BA}{\Delta t}$$

and so the emf is

$$\mathscr{E} = \frac{NAB}{\Delta t} \qquad (21\text{–}18)$$

By substituting Eqs. (21–17) and (21–18) in Eq. (21–16) we obtain for the energy contained in the magnetic field of the solenoid

$$W = \tfrac{1}{2}I\mathscr{E}\,\Delta t = \tfrac{1}{2}\left(\frac{Bl}{\mu N}\right)\left(\frac{NAB}{\Delta t}\right)(\Delta t) = \frac{lA}{2\mu}B^2 \qquad (21\text{–}19)$$

Energy of magnetic field

The quantity lA is the volume occupied by the magnetic field **B**, since it is the product of the length of the solenoid and its cross-sectional area. (Since we are considering an ideal solenoid whose length is large relative to its diameter, we can ignore the magnetic field escaping from the ends.) The *magnetic energy density* of the field is therefore

$$w = \frac{W}{lA} = \frac{B^2}{2\mu}$$ *Magnetic energy density*   (21-20)

The energy density of a magnetic field is directly proportional to the square of its magnitude B. This formula holds for all magnetic fields, not just for those inside solenoids. We recall from Eq. (21-6) that the energy density of an electric field $\mathbf{E}$ is similarly proportional to E^2.

Example   The earth's magnetic field in a certain region has the magnitude 6×10^{-5} T. Find the magnetic energy per cubic kilometer in this region.

Solution   The magnetic energy density corresponding to $B = 6 \times 10^{-5}$ T is, since here $\mu = \mu_0$,

$$w = \frac{B^2}{2\mu_0} = \frac{(6 \times 10^{-5}\,\text{T})^2}{2(1.26 \times 10^{-6}\,\text{T·m/A})} = 1.4 \times 10^{-3}\,\text{J/m}^3$$

A cubic kilometer is equal to $(10^3\,\text{m})^3 = 10^9\,\text{m}^3$, so the total magnetic energy in a cubic kilometer is

$$W = wV = (1.4 \times 10^{-3}\,\text{J/m}^3)(10^9\,\text{m}^3) = 1.4 \times 10^6\,\text{J}$$

This amount of energy is enough to raise the 365,000-ton Empire State Building by 0.42 mm. ∎

21-6   INDUCTANCE

An inductor stores energy in the form of magnetic field

A circuit element in which a self-induced emf accompanies a changing current is called an *inductor*. A solenoid is an example of an inductor.

Normally the energy content W of an inductor is most conveniently expressed in terms of the current I present in it. Since the magnetic field B in an inductor is proportional to I in the absence of ferromagnetic materials, W is proportional to I^2. The constant of proportionality is written as $L/2$, where L is called the *inductance* of the inductor. Thus

$$W = \tfrac{1}{2}LI^2$$ *Potential energy of inductor*   (21-21)

The unit of inductance is the *henry* (H), where

$$1\,\text{H} = 1\,\frac{\text{J}}{\text{A}^2} = 1\,\frac{\text{V·s}}{\text{A}}$$

The henry, millihenry, and microhenry

The henry, like the farad, is usually too large a unit for convenience. Inductances are accordingly often expressed in *millihenries* (mH) or *microhenries* (μH), where

$$1\,\text{mH} = 10^{-3}\,\text{H} \qquad 1\,\mu\text{H} = 10^{-6}\,\text{H}$$

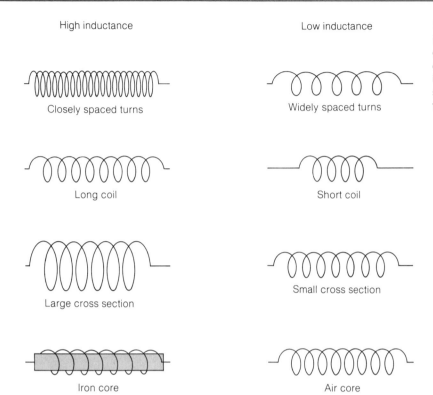

High inductance

Closely spaced turns

Long coil

Large cross section

Iron core

Low inductance

Widely spaced turns

Short coil

Small cross section

Air core

FIG. 21–15 The inductance of a coil depends upon how closely wound it is, upon its length, upon its cross-sectional area, and upon the nature of its core.

The inductance of an inductor depends upon its geometry and upon the presence of a core with magnetic properties. A coil consisting of many turns has a greater inductance than one consisting of few turns because its magnetic field is stronger for a given current (Fig. 21–15). The longer the coil and the greater its cross-sectional area, the greater its inductance because it contains a larger volume of magnetic field. A core in a coil changes its inductance by changing the flux through the coil (see Sec. 19–4). Thus a diamagnetic core decreases the inductance slightly from its value in free space or air, a paramagnetic core increases it slightly, and a ferromagnetic core increases it considerably—by a factor of as much as 10^4 in some cases—but by an amount that is not constant.

Factors that influence inductance

Let us calculate the inductance of a solenoid. From Eqs. (21–19) and (21–21), which are different ways to express the energy of a solenoid when it carries the current I, we have

Inductance of a solenoid

$$\frac{lA}{2\mu}B^2 = \tfrac{1}{2}LI^2$$

$$L = \frac{lA}{\mu I^2}B^2$$

Since the magnetic field within the solenoid is

$$B = \mu\frac{N}{l}I$$

we see that

$$L = \left(\frac{lA}{\mu I^2}\right)\left(\mu\frac{N}{l}I\right)^2$$

$$= \mu N^2 \frac{A}{l} \qquad\qquad \textit{Inductance of solenoid} \quad (21\text{-}22)$$

Example   Find the inductance in air of a 1000-turn solenoid 10 cm long that has a cross-sectional area of 20 cm². How much energy is stored in the magnetic field of the solenoid when it carries a 0.01-A current?

Solution   Since 10 cm $= 10^{-1}$ m, 20 cm² $= 2 \times 10^{-3}$ m², and $\mu = \mu_0$ in air (very nearly),

$$L = \mu_0 N^2 \frac{A}{l} = \left(1.26 \times 10^{-6}\frac{\text{T·m}}{\text{A}}\right)(10^3)^2\left(\frac{2 \times 10^{-3}\,\text{m}^2}{10^{-1}\,\text{m}}\right)$$

$$= 2.5 \times 10^{-2}\,\text{H} = 25\,\text{mH}$$

The energy stored in the solenoid's magnetic field when $I = 0.01$ A $= 10^{-2}$ A is

$$W = \tfrac{1}{2}LI^2 = \tfrac{1}{2}(2.5 \times 10^{-2}\,\text{H})(10^{-2}\,\text{A})^2 = 1.3 \times 10^{-6}\,\text{J} \qquad \blacksquare$$

21-7   SELF-INDUCED EMF

Inductance and self-induced emf

It is easy to show that the inductance L of an inductor detemines the magnitude of the self-induced emf $\mathscr{E}$ that accompanies a changing current in the inductor. From Eqs. (21-16) and (21-21) we have

$$\tfrac{1}{2}LI^2 = \tfrac{1}{2}I\mathscr{E}\,\Delta t$$

$$\mathscr{E} = L\frac{I}{\Delta t}$$

Since the current in the inductor rose from 0 to I in the time Δt, we can write $\Delta I/\Delta t$ in place of $I/\Delta t$. Also, Lenz's law requires that $\mathscr{E}$ be opposite in sign to $\Delta I/\Delta t$. Hence we have the important formula

$$\mathscr{E} = -L\frac{\Delta I}{\Delta t} \qquad\qquad \textit{Self-induced emf} \quad (21\text{-}23)$$

A high value of L means a large $\mathscr{E}$ for a given rate of change $\Delta I/\Delta t$, and a low value of L means a small $\mathscr{E}$.

Example   An average self-induced emf of -0.75 V is produced in a 25-mH coil when the current in it falls to 0 in 0.01 s. What was the original current in the coil?

Solution   From Eq. (21-33),

$$\Delta I = \frac{-\mathscr{E}\,\Delta t}{L} = \frac{-(-0.75\text{ V})(0.01\text{ s})}{2.5 \times 10^{-2}\,\text{H}} = 0.3\text{ A}$$

Since $\Delta I = I_0 - 0$, the original current was $I_0 = 0.3$ A. $\qquad \blacksquare$

21–8 ELECTRICAL OSCILLATIONS

Let us connect a charged capacitor to an inductor, as in Fig. 21–16. The following sequence of events occurs:

Oscillations of an LC circuit

(a) Initially the capacitor's plates each possess the charge Q and its electric energy is $\frac{1}{2}Q^2/C$, while the inductor has no energy since $I = 0$.

(b) The capacitor is partially discharged, and the current in the inductor leads to a magnetic field whose energy is $\frac{1}{2}Li^2$.

(c) No charge is left on the capacitor's plates, and all the energy of the circuit is magnetic energy $\frac{1}{2}LI^2$.

(d) Now charge begins to build up on the capacitor's plates with polarities the reverse of those originally present, and the electric energy of the capacitor grows at the expense of the magnetic energy of the inductor as the current drops.

(e) At the instant the capacitor is fully charged again, the current is 0 and the energy of the circuit is once more entirely electric.

(f) Next the capacitor begins to discharge, and eventually the entire circuit returns to

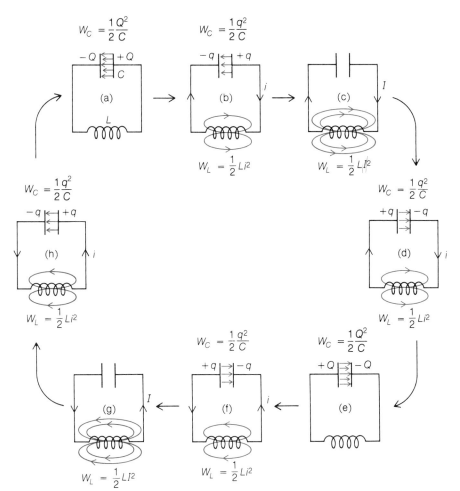

FIG. 21–16 The various stages in the oscillation of a circuit containing the inductance L and the capacitance C. The maximum charge on the capacitor is Q, and the maximum current in the circuit is I.

its original state at (a). The cycle will continue to repeat itself indefinitely if no resistance is present.

Electrical oscillator is analog of harmonic oscillator

The above sequence is wholly analogous to the behavior of a harmonic oscillator. As we learned in Chapter 11, the basic process involved in the operation of a harmonic oscillator is the continual interchange of energy between KE and PE; here the same sort of interchange occurs, and we can regard the electric energy of a charged capacitor as corresponding to PE and the magnetic energy of an inductor through which a current flows as corresponding to KE.

A harmonic oscillator whose mass is m and whose spring constant is k has the potential and kinetic energies

Harmonic oscillator

$$PE = \tfrac{1}{2}ks^2 \qquad KE = \tfrac{1}{2}mv^2$$

where s is the displacement of the object from its equilibrium position and v is its speed. The corresponding formulas for the electric and magnetic energies in an electrical oscillator are

Electrical oscillator

$$W_e = \frac{1}{2}\frac{1}{C}Q^2 \qquad W_m = \frac{1}{2}LI^2$$

We note that $1/C$ is analogous to k, since both determine the amount of energy present at the moment when nothing is moving in the respective oscillators. Similarly L is analogous to m. Both are measures of inertia—a large inductance tends to slow down changes in current, and a large m tends to slow down changes in velocity. The frequency of a harmonic oscillator is

Harmonic oscillator frequency

$$f = \frac{1}{2\pi}\sqrt{\frac{k}{m}}$$

and so it is not surpising that the frequency of an electrical oscillator is

Electrical oscillator frequency

$$f = \frac{1}{2\pi\sqrt{LC}} \qquad\qquad \textit{Oscillator frequency} \quad (21\text{--}24)$$

The larger the inductance L and the capacitance C, the lower the frequency.

The parallel between electrical and mechanical oscillations is a very close one: just as every mechanical system has certain specific natural frequencies of vibration that depend upon its properties, so every electrical circuit has a natural frequency of oscillation that depends upon L and C. Just as the energy of every actual mechanical vibration is eventually dissipated as heat owing to the inevitable presence of friction, so the energy of every electrical oscillation is eventually dissipated as heat owing to the inevitable presence of resistance in the wires of the circuit. Electrical resonance is much like mechanical resonance: if a source of potential that alternates at the natural frequency of a circuit is connected to it, energy is fed into the oscillations that can maintain or increase their amplitude. At any other frequency the energy absorbed is small.

Radio communication is made possible by the ability of an LC circuit to oscillate only at the frequency given by Eq. (21–24). Such a circuit is used to "tune" a transmitter so that only a single frequency of radio waves is produced, and another similar circuit **How radio transmitters and receivers are tuned** is used to "tune" a receiver to that same frequency. Electrical oscillations are discussed further in the next chapter.

Example   In the antenna circuit of a certain radio receiver, the inductance is fixed at 4 mH but the capacitance can be varied. When the capacitance is 10 pF, what is the frequency of the radio waves the receiver responds to?

Solution   From Eq. (21–24),

$$f = \frac{1}{2\pi\sqrt{LC}} = \frac{1}{2\pi\sqrt{(4 \times 10^{-3}\,\mathrm{H})(10 \times 10^{-12}\,\mathrm{F})}} = 7.96 \times 10^5\,\mathrm{Hz} = 796\,\mathrm{kHz}$$

■

21–9   TIME CONSTANTS

Circuits that contain resistance as well as capacitance or inductance do not respond instantaneously to changes in the applied emf. For instance, when a capacitor is connected to a battery, as in Fig. 21–17, it does not immediately become fully charged. At first the only limit to the current that flows to the capacitor is the resistance R in the circuit, so that the initial current is $I = \mathscr{E}/R$, where $\mathscr{E}$ is the emf of the battery. As the capacitor becomes charged, however, a potential difference appears across it, whose polarity is such as to tend to oppose the further flow of current. When the charge on the capacitor has built up to some value q, this opposing potential difference is $V = q/C$. Hence the net potential difference is $\mathscr{E} - q/C$, and the current is

$$I = \frac{\mathscr{E} - q/C}{R} \tag{21–25}$$

As q increases, then, its *rate* of increase drops. This gives a steadily decreasing slope in the curve of Fig. 21–18, which is a graph showing how q varies with time when a capacitor is being charged; the capacitor is connected to the battery at $t = 0$.

A mathematical analysis of Eq. (21–25) shows that after a time interval of RC (the product of the resistance R in the circuit and the capacitance C of the capacitor), the charge on the capacitor reaches 63% of its ultimate value of $Q = C\mathscr{E}$. The time RC is therefore a convenient measure of how rapidly the capacitor becomes charged and is accordingly called the *time constant* of the circuit. In principle the capacitor acquires its ultimate charge Q only after an infinite time has elapsed, but, as we can see from Fig. 21–18 this value is very nearly reached after only a few time constants; at $t = 3RC$, the charge is 95% of Q, and at $t = 4RC$ it is 98% of Q.

If a capacitor with an initial charge is discharged through a resistance, its charge decreases with time as shown in Fig. 21–19. After the time RC the charge on the capacitor is reduced to 37% of its original value, a drop of 63%.

Charging or discharging a capacitor takes time

Time constant of capacitive circuit

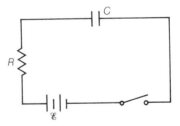

FIG. 21–17   A circuit that contains a capacitor, a resistor, and a source of emf in series. When the switch is closed, the charge on the capacitor increases gradually to its ultimate value of $C\mathscr{E}$.

FIG. 21–18 The growth of charge in a capacitor.

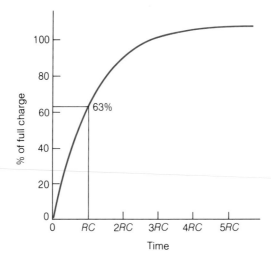

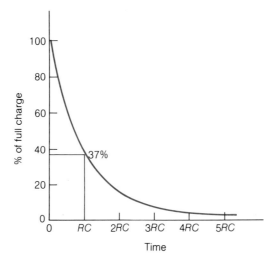

Example   A 5-μF capacitor is charged by being connected to a 1.5-V dry cell. The total resistance of the circuit is 2 Ω. (a) What is the final charge on the capacitor and how long does it take to reach 63% of this charge? (b) The battery is then disconnected from the capacitor. If the resistance of the dielectric between the capacitor plates is 10^{10} Ω, after what period of time will the charge on the capacitor drop to 37% of its initial value?

Solution   (a) The ultimate charge on the capacitor is

$$Q \doteq CV = (5 \times 10^{-6} \text{ F})(1.5 \text{ V}) = 7.5 \times 10^{-6} \text{ C}$$

The time constant of the circuit is

$$RC = (2 \ \Omega) \ (5 \times 10^{-6} \text{ F}) = 10^{-5} \text{ s}$$

and so this is the time needed for the capacitor to acquire 63% of its ultimate charge.

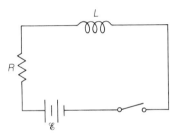

FIG. 21–20 A circuit with an inductor, a resistor, and a source of emf in series. When the switch is closed, the current increases gradually to its ultimate value of $\mathscr{E}/R$ because of the opposing self-induced emf in the inductor.

(b) When the battery is disconnected from the capacitor, charge gradually leaks through the dielectric of the latter. The time constant for discharge is

$$RC = (10^{10}\,\Omega)(5 \times 10^{-6}\,\text{F}) = 5 \times 10^4\,\text{s} = 14\,\text{h}$$

and in this period the charge on the capacitor will drop to 37% of its original value. ■

When a circuit containing inductance is connected to a battery, the current in the circuit does not rise instantly to its ultimate value $I = \mathscr{E}/R$, where $\mathscr{E}$ is the emf of the battery and R is the total resistance in the circuit. At the moment the switch in Fig. 21–20 is closed, the current i starts to grow, and as a result the induced emf $- L(\Delta i/\Delta t)$ comes into being in the opposite direction to the battery emf $\mathscr{E}$. The net emf acting to establish current in the circuit is therefore $\mathscr{E} - L(\Delta i/\Delta t)$, and the current reaches its final value of I in a gradual manner.

A current in an inductor takes time to build up or to disappear

The graph in Fig. 21–21 shows how i varies with time when a current is being established in a circuit containing inductance. A mathematical analysis shows that, after a time interval of L/R, the current reaches 63% of I. The time L/R is therefore a convenient measure of how rapidly a current rises in a circuit containing inductance, and, like RC in a circuit containing capacitance, is called the *time constant* of the circuit.

Time constant of inductive circuit

When the battery in Fig. 21–20 is short-circuited by a wire, the current i drops slowly in the manner shown in Fig. 21–22, since the induced emf now tends to maintain

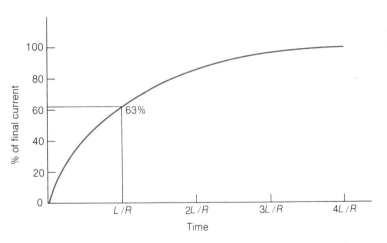

FIG. 21–21 The growth of current in a circuit containing inductance and resistance.

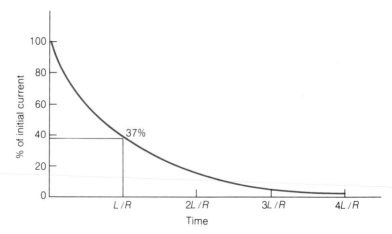

the existing current. The current falls to 37% of its original value in the time L/R after the battery is short-circuited.

A coil of inductance L has the magnetic energy of $\frac{1}{2}LI^2$ stored in it when a current I is present in it. When the potential difference across the coil that is responsible for the current is shorted out, the energy $\frac{1}{2}LI^2$ is what powers the self-induced emf that retards the drop in current. The gradual rise of current in a circuit containing inductance may be thought of as the result of the initial absorption of $\frac{1}{2}LI^2$ of potential energy by the circuit, and its gradual drop may be thought of as the result of the transfer of the inductor's potential energy to the circuit.

Example A 2-H inductor whose resistance is 100 Ω is connected to a 24-V battery of negligible internal resistance. Find the initial current in the circuit and the initial rate at which the current is increasing, the time required for the current to reach 63% of its ultimate value, and the magnitude of the final current.

Solution The initial current is 0, and the initial rate of current increase is

$$\frac{\Delta i}{\Delta t} = \frac{\mathscr{E}}{L} = \frac{24\,\text{V}}{2\,\text{H}} = 12\,\text{A/s}$$

The time needed for the current to rise to 63% of the final value is

$$\frac{L}{R} = \frac{2\,\text{H}}{100\,\Omega} = 0.02\,\text{s}$$

and the final current will be

$$I = \frac{\mathscr{E}}{R} = \frac{24\,\text{V}}{100\,\Omega} = 0.24\,\text{A}$$

IMPORTANT TERMS

The **energy density** of an electric field is the electric potential energy per unit volume associated with it. The **energy density** of a magnetic field is the magnetic energy per unit volume associated with it.

A **capacitor** is a device that stores electrical energy in the form of an electric field. The ratio between the charge on either plate of a capacitor and the potential difference between the plates is called its **capacitance.** The unit of capacitance is the **farad,** which is equal to 1 coulomb per volt.

The **dielectric constant** K of a particular material is a measure of how effective it is in reducing an electric field set up across a sample of it.

The **inductance** L of a circuit is the ratio between the magnitude of the self-induced emf $\mathscr{E}$ due to a changing current in it and the rate of change $\Delta I / \Delta t$ of the current. The unit of inductance is the **henry.**

IMPORTANT FORMULAS

Energy density of electric field: $\quad w = \dfrac{E^2}{8\pi k}$

Capacitance: $\quad C = \dfrac{Q}{V}$

Parallel-plate capacitor:

$$C = \frac{KA}{4\pi kd} \quad (K = \text{dielectric constant})$$

Potential energy of charged capacitor:

$$W = \tfrac{1}{2}QV = \tfrac{1}{2}CV^2 = \frac{Q^2}{2C}$$

Capacitors in series: $\quad \dfrac{1}{C} = \dfrac{1}{C_1} + \dfrac{1}{C_2} + \dfrac{1}{C_3} + \cdots$

Capacitors in parallel: $\quad C = C_1 + C_2 + C_3 + \cdots$

Magnetic energy density: $\quad w = \dfrac{B^2}{2\mu}$

Inductance of solenoid: $\quad L = \mu N^2 \dfrac{A}{l}$

Potential energy of inductor: $\quad W = \tfrac{1}{2}LI^2$

Self-induced emf: $\quad \mathscr{E} = -L \dfrac{\Delta I}{\Delta t}$

Oscillator frequency: $\quad f = \dfrac{1}{2\pi\sqrt{LC}}$

Time constants:

$$t = RC$$
$$t = \frac{L}{R}$$

MULTIPLE CHOICE

1. The energy content of a charged capacitor resides in its
(a) plates.   (b) potential difference.
(c) charge.   (d) electric field.

2. When a slab of insulating material is placed between the plates of a charged capacitor, the electric field there relative to what it was before is
(a) less.
(b) the same.
(c) more.
(d) any of the above, depending on the circumstances.

3. The farad is not equivalent to which of the following combination of units?
(a) C^2/J (b) C/V
(c) $C \cdot V^2$ (d) J/V^2

4. Magnetic fields invariably contain
(a) a ferromagnetic material.
(b) inductance.
(c) electric current.
(d) energy.

5. The unit of inductance is the henry, where 1 H =
(a) $1 \ J \cdot A^2$.   (b) $1 \ J/A^2$.
(c) $1 \ V \cdot A$.   (d) $1 \ V/A$.

6. A wire coil carries the current I. The potential energy of the coil does not depend upon
(a) the value of I.
(b) the number of turns in the coil.
(c) whether the coil has an iron core or not.
(d) the resistance of the coil.

7. A large increase in the inductance of a coil can be achieved by using a core that is
(a) diamagnetic.   (b) paramagnetic.
(c) ferromagnetic.   (d) polar.

8. A capacitor acquires a charge of 0.002 C when connected across a 50-V battery. Its capacitance is
(a) $1 \ \mu F$.   (b) $2 \ \mu F$.
(c) $4 \ \mu F$.   (d) $40 \ \mu F$.

9. A 50-μF capacitor has a potential difference of 8 V across it. Its charge is
(a) 4×10^{-3} C.
(b) 4×10^{-4} C.
(c) 6.25×10^{-5} C.
(d) 6.25×10^{-6} C.

10. The plates of a parallel-plate capacitor of capacitance C are brought together to one-third their original separation. The capacitance is now
(a) $\frac{1}{9}C$.
(b) $\frac{1}{3}C$.
(c) $3C$.
(d) $9C$.

11. If a 20-μF capacitor is to have an energy content of 2.5 J, it must be placed across a potential difference of
(a) 150 V.
(b) 350 V.
(c) 500 V.
(d) 250,000 V.

12. A parallel-plate capacitor has a capacitance of 50 pF in air and 110 pF when immersed in turpentine. The dielectric constant of turpentine is
(a) 0.45.
(b) 0.55.
(c) 1.1.
(d) 2.2.

13. A parallel-plate capacitor with air between its plates is charged until a potential difference of V appears across it. Another capacitor, having hard rubber (dielectric constant = 3) between its plates but otherwise identical, is also charged to the same potential difference. If the energy of the first capacitor is W, that of the second is
(a) $\frac{1}{3}W$.
(b) W.
(c) $3W$.
(d) $9W$.

14. Two 50-μF capacitors are connected in series. The equivalent capacitance of the combination is
(a) 25 μF.
(b) 50 μF.
(c) 100 μF.
(d) 200 μF.

15. The capacitor combination of question 14 is connected across a 100-V battery. The potential difference across each capacitor is
(a) 25 V.
(b) 50 V.
(c) 100 V.
(d) 200 V.

16. Two 50-μF capacitors are connected in parallel. The equivalent capacitance of the combination is
(a) 25 μF.
(b) 50 μF.
(c) 100 μF.
(d) 200 μF.

17. The capacitor combination of question 16 is connected across a 100-V battery. The potential difference across each capacitor is
(a) 25 V.
(b) 50 V.
(c) 100 V.
(d) 200 V.

18. The energy contained in a cubic meter of space in which the magnetic induction is 1 T is
(a) 6.3×10^{-7} J.
(b) 3.97×10^{-5} J.
(c) 3.97×10^{5} J.
(d) 7.94×10^{5} J.

19. The self-induced emf in a 0.1-H coil when the current in it is changing at the rate of 200 A/s is
(a) 125 V.
(b) 20 V.
(c) 8×10^{-4} V.
(d) 8×10^{-5} V.

20. The current in a circuit falls to 0 from 16 A in 0.01 s. The average emf induced in the circuit during the drop is 64 V. The inductance of the circuit is
(a) 0.032 H.
(b) 0.04 H.
(c) 0.25 H.
(d) 4 H.

21. A 2-mH coil carries a current of 10 A. The energy stored in its magnetic field is
(a) 0.05 J.
(b) 0.1 J.
(c) 1.0 J.
(d) 100 J.

22. An LC circuit has the initial capacitance C. In order to double its natural frequency of oscillation, the capacitance must be changed to
(a) $\frac{1}{4}C$.
(b) $\frac{1}{2}C$.
(c) $2C$
(d) $4C$.

23. The time constant of an RL circuit is the time needed for the current to reach which percentage of its final value?
(a) 50%.
(b) 63%.
(c) 90%.
(d) 100%.

24. The time constant of an RC circuit depends upon
(a) the applied potential difference.
(b) the value of R only.
(c) the value of C only.
(d) the values of both R and C.

EXERCISES

1. Is there any kind of material that, when inserted between the plates of a capacitor, reduces its capacitance?

2. What effect does placing a slab of a material of dielectric constant K between the plates of a charged capacitor have on the energy content of the capacitor? (The capacitor is disconnected from the charging circuit before the dielectric is inserted.) If the energy is greater than before, where does the additional energy come from? If the energy is less than before, where does the lost energy go?

3. A parallel-plate capacitor of capacitance C is given the charge Q and then disconnected from the circuit. How much work is required to pull the plates of this capacitor to twice their original separation?

4. A sheet of mica whose dielectric constant is 5 is placed between the plates of a charged, isolated parallel-plate capacitor. How is the potential difference across the capacitor affected? How is the charge on the capacitor affected?

5. The greater its capacitance, the less energy is stored in a capacitor when it is given a certain charge. On the other hand, the greater its inductance, the more energy is stored in an inductor when a certain current is present in it. Explain the difference.

6. What becomes of the work done against the back emf in an inductive circuit when a current is being established in it?

7. What is the direction of the self-induced emf in a coil when the current in its increases? When the current decreases? What is the reason in each case?

8. A potential difference of 300 V is applied across a pair of parallel metal plates 1 cm apart. What is the energy density of the electric field between the plates?

9. A 25-μF capacitor is connected to a source of potential difference of 1000 V. What is the resulting charge on the capacitor? How much energy does it contain?

10. What is the potential difference between the plates of a 20-μF capacitor whose charge is 0.01 C? How much energy does it contain?

11. The capacitance of a parallel-plate capacitor is increased from 8 μF to 50 μF when a sheet of glass is inserted between its plates. What is the dielectric constant of the glass?

12. What potential difference must be applied across a 10-μF capacitor if it is to have an energy content of 1 J?

13. The plates of a parallel-plate capacitor are 50 cm^2 in area and 1 mm apart. (a) What is its capacitance? (b) When the capacitor is connected to a 45-V battery, what is the charge on either plate? (c) What is the energy of the charged capacitor?

14. The space between the plates of the capacitor of the previous problem is filled with sulfur. Answer the same questions for this case.

15. A capacitor with air between its plates is connected to a battery and each of its plates receives a charge of 10^{-4} C. While still connected to the battery the capacitor is immersed in oil, and a further charge of 10^{-4} C is added to each plate. What is the dielectric constant of the oil?

16. Find the equivalent capacitance of a 20-μF capacitor and a 50-μF capacitor that are connected in series.

17. Find the equivalent capacitance of a 20-μF capacitor and a 50-μF capacitor that are connected in parallel.

18. List the capacitances that can be obtained by combining three 10-μF capacitors in all possible ways.

19. Find the inductance of a coil 40 cm long and 4 cm in diameter that has 1000 turns of wire.

20. A solenoid 20 cm long and 2 cm in diameter has an

inductance of 0.178 mH. How many turns of wire does it contain?

21. A 20-mH coil carries a current of 0.2 A. (a) How much energy is stored in its magnetic field? (b) What should the current be in order that it contain 1 J of energy?

22. The current in a circuit drops from 5 A to 1 A in 0.1 s. If an average emf of 2 V is induced in the circuit while this is happening, find the inductance of the circuit.

23. What is the self-induced emf in a 0.4-H coil when the current in it is changing at a rate of 500 A/s?

24. A 2-H coil carries a current of 0.5 A. (a) How much energy is stored in it? (b) In how much time should the current drop to 0 if an emf of 100 V is to be induced in it?

25. Find the natural frequency of an LC circuit in which L = 12 mH and C = 5 μF.

26. What inductance is needed in a circuit in which C = 60 μF if its natural frequency is to be 30 Hz?

27. What capacitance is needed in a circuit in which L = 2 H if its natural frequency is to be 200 Hz?

28. The frequencies used in commercial radio broadcasting range from 550 to 1600 kHz. What range of capacitance should a variable capacitor have if it is connected to a coil of inductance 1 mH in a circuit designed to respond to frequencies in this band?

PROBLEMS

1. The strongest magnetic fields that have been produced in the laboratory have been about 10^2 T. (a) How much energy is contained in 1 liter of such a field? (b) What electric field would have the same energy density?

2. The electric field near the earth's surface is about 100 V/m. How much electrical energy is stored in the lowest kilometer of the atmosphere?

3. A potential difference of 100 V is applied across a pair of parallel metal plates 5 cm square and 1 mm apart. (a) What is the force between the plates? (b) Is the force attractive or repulsive? (c) What is the energy density in the region between the plates?

4. A capacitor is charged by connecting it through a resistance to a battery. Verify that half the work done by the battery is dissipated as heat.

5. Three capacitors whose capacitances are 5, 10, and 50 μF are connected in series across a 12-V battery. Find the charge on each capacitor and the potential difference across it.

6. Three capacitors whose capacitances are 2, 4, and 5 μF are connected in series across a 100-V battery. Find the charge on each capacitor and the potential difference across it.

7. The three capacitors of problem 5 are connected in parallel across the same battery. Find the charge on each capacitor and the potential difference across it.

8. The three capacitors of problem 6 are connected in parallel across the same battery. Find the charge on each capacitor and the potential difference across it.

9. A 1-μF capacitor and a 2-μF capacitor are each charged across a potential difference of 1200 V. The capacitors are then connected with terminals of opposite sign together. What is the final charge of each capacitor?

10. An inductor consists of an iron ring 5 cm in diameter and 1 cm^2 in cross-sectional area that is wound with 1000 turns of wire. (Such an inductor is essentially a solenoid bent into a circle.) If the permeability of the iron is constant at 400 times that of free space at the magnetic fields at which the inductor will be used, find its inductance.

11. A solenoid 20 cm long and 2.4 cm in diameter is wound with 1200 turns of wire whose total resistance is 40 Ω. The solenoid is connected to a 12-V battery whose internal resistance is 2 Ω. Find the inductance of the solenoid, the final current in it, and its energy content when the final current flows.

12. The solenoid of problem 11 is immersed in liquid oxygen whose permeability is 1.0049 times greater than that of free space. Find the inductance of the solenoid, the final current in it, and its energy content when the final current flows.

13. A potential difference of 50 V is suddenly applied across a 12-mH, 8-Ω coil. Find (a) the initial current and the initial rate of change of current, (b) the current when the rate of change of current is 2000 A/s, and (c) the final current and final rate of change of current.

14. A potential difference of 100 V is suddenly impressed across a 0.5-H, 20-Ω inductor. (a) What is the initial rate of increase of current? (b) What is the rate of increase of current when the current is 3 A? (c) What is the final current? (d) What is the energy content of the inductor when the final current flows in it?

15. A 1-μF capacitor is charged by being connected to a 10-V battery. The battery is then removed and the capacitor connected to a 10-mH coil. (a) Find the frequency of the resulting oscillations. (b) Find the maximum value of the charge on the capacitor. (c) Find the maximum current that flows through the inductor.

16. A 2-μF capacitor is charged by being connected to a 24-V battery. The battery is then removed and the capacitor connected to a 50-mH coil. After 1/8 cycle, the initial energy will be equally divided between the electric field in the capacitor and the magnetic field in the coil, as in Fig. 21–16(b). (a) Find the time interval required for this to occur. (b) Find the charge on the capacitor and the current in the inductor at this moment.

17. A reusable flash bulb requires an energy of 100 J for its discharge. A 450-V battery is used to charge a capacitor for this purpose. The resistance of the charging circuit is 15 Ω. (a) What is the required capacitance? (b) What is the time constant of the circuit?

18. A 100-μF electrolytic capacitor has a leakage current of 5 μA when the potential difference across its terminals is 12 V. If the capacitor is connected to a 12-V battery and then removed, how long will it take for the charge to fall to 37% of its original value?

19. A 5-μF capacitor is connected across a 1000-V battery with wires whose resistance is a total of 5000 Ω. (a) What is the time constant of this circuit? (b) What is the initial current that flows when the battery is connected? (c) How long would it take to charge the capacitor if this current remained constant?

20. A coil 20 cm long and 3 cm in diameter is tightly wound with one layer of copper wire 1 mm in diameter. Find its time constant.

21. What is the inductance of a coil whose resistance is 14 Ω and whose time constant is 0.1 s?

22. A 50-mH coil with a resistance of 20 Ω is connected to a 90-V battery of negligible internal resistance. (a) What is the time constant of the circuit? (b) How much energy is stored in the magnetic field of the coil when the current has reached its final value?

ANSWERS TO MULTIPLE CHOICE

1. d	6. d	11. c	16. c	21. b
2. a	7. c	12. d	17. c	22. a
3. c	8. d	13. c	18. c	23. b
4. d	9. b	14. a	19. b	24. d
5. b	10. c	15. b	20. b	

ALTERNATING CURRENT

22

Nearly all the world's electrical energy is carried by alternating current. The preference for ac is chiefly due to the economy of high-voltage transmission, which minimizes I^2R heat losses; with step-up transformers at the production end and step-down transformers at the consumption end, the transmission voltage is limited only by insulation and atmospheric discharge problems. Alternating current is also preferred in industry because ac electric motors are, as a class, cheaper, more durable, and less in need of maintenance than dc motors; dc motors, however, have certain characteristics, such as better speed regulation, that make them more suitable for specialized duties. Alternating currents are involved in all aspects of modern communication: the electrical equivalent of a sound wave of a certain frequency is an alternating current of that frequency, and radio waves are produced by antennas fed with high-frequency ac. Alternating current behaves in a circuit in a very different way from direct current, and some knowledge of ac circuit behavior is necessary to understand much of modern technology.

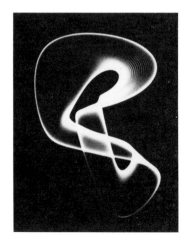

22–1   EFFECTIVE CURRENT AND VOLTAGE

A direct current is described in terms of its direction and magnitude. In a simple circuit there might be a current of 6 A that flows from the positive terminal of a battery through a resistance network to its negative terminal. An alternating current has neither a constant direction nor a constant magnitude. How shall we speak of it quantitatively?

Since alternating current flows back and forth in a circuit, it has no "direction" in the same sense as a direct current has. However, the fluctuations have a certain frequency in each case, and the value of this frequency—that is, how many times per second the curent goes through a complete cycle—forms part of the description of the current.

The effective value of an ac current is a measure of its ability to do work or produce heat

An alternating current varies with time in the manner shown in Fig. 22–1, and it would seem natural to specify its maximum value, I_{max}, as well as its frequency. The flaw in doing this is that I_{max} is not a measure of the ability of a current to do work or produce heat. A 6-A direct current is not equivalent to an alternating current in which $I_{max} = 6$ A. A better procedure is to define an *effective current* I_{eff} such that a direct current of this magnitude produces heat in a resistor at the same rate as the alternating current.

The variation of an alternating current with time obeys the formula

$$I = I_{max} \sin 2\pi ft \qquad (22\text{–}1)$$

where f is the frequency of the current. In this formula, the current is assumed to be $I = 0$ and is increasing when $t = 0$. Figure 22–1 is a graph of this formula. The rate at which heat is dissipated in a resistance R by an alternating current is, at any time t, given by

$$I^2R = I^2_{max}R \sin^2 2\pi ft$$

The average value of I^2R over a complete cycle is

$$[I^2R]_{av} = I^2_{eff}R = I^2_{max}R[\sin^2 2\pi ft]_{av} \qquad (22\text{–}2)$$

What we must find is the average value of $\sin^2 2\pi ft$ over a complete cycle. (The average value of I over a cycle is 0, since I is positive for half the cycle and negative for the other half. However, I^2 is always positive, and its average is a positive quantity. See Fig. 22–2.)

We begin with the trigonometric identity

$$\sin^2 \theta = \tfrac{1}{2}(1 - \cos 2\theta)$$

How effective current is calculated

The average value of $\sin^2 \theta$ over a complete cycle is therefore

FIG. 22–1   The variation of an alternating current with time. The frequency of the current is the number of cycles that occur per second.

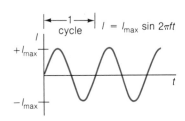

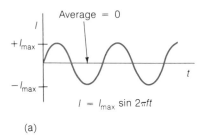

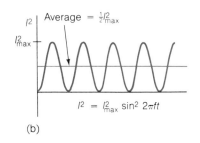

FIG. 22–2 (a) The average value of I in an ac circuit is zero. (b) The average value of I^2 in an ac circuit is $\frac{1}{2} I^2_{max}$.

$$[\sin^2 \theta]_{av} = \tfrac{1}{2}[1 - \cos 2\theta]_{av} = \tfrac{1}{2} - \tfrac{1}{2}[\cos 2\theta]_{av} = \tfrac{1}{2}$$

since, over a complete cycle, the average value of $\cos 2\theta$ is 0 by the same reasoning as in the case of $\sin \theta$. Hence we see that

$$I^2_{eff}R = I^2_{max}R\,[\sin^2 2\pi ft]_{av} = \tfrac{1}{2}I^2_{max}R$$

and

$$I_{eff} = \frac{I_{max}}{\sqrt{2}} = 0.707 I_{max} \qquad\qquad \textit{Effective current} \quad (22\text{–}3)$$

The effective magnitude of an alternating current is 70.7% of its maximum value.

In a similar way the effective voltage in an ac circuit turns out to be

$$V_{eff} = \frac{V_{max}}{\sqrt{2}} = 0.707\,V_{max} \qquad\qquad \textit{Effective voltage} \quad (22\text{–}4)$$

Effective current and voltage are 70.7% of maximum values

It is customary to express currents and voltages in ac circuits in terms of their effective values. Thus the potential difference across a "120-volt, 60-Hz" power line actually varies from

$$+V_{max} = +\frac{V_{eff}}{0.707} = +\frac{120\,\text{volts}}{0.707} = +170\,\text{volts}$$

through 0 to -170 volts and back to $+170$ volts a total of 60 times per second.

In what follows, when current, potential difference, and emf values are given for an ac circuit without other qualification, they will refer to the effective magnitudes of these quantities.

By analogy with circular motion, angular frequency ω (in radians per second) is often used instead of frequency f (in hertz) in discussing alternating currents, where

Angular frequency of alternating current

$$\omega = 2\pi f$$

In this notation, which will not be used here, the instantaneous current in an ac circuit is written

$$I = I_{max}\sin \omega t$$

What are called here "effective" values of current and voltage are elsewhere sometimes called "root-mean-square" or "rms" values from their definitions as the square roots of the average values of I^2 and V^2.

22–2 PHASORS

A phasor is a rotating vector that can represent an alternating current or voltage

A convenient way to represent ac currents and voltages is in terms of *phasors*. This scheme is based on the fact that the component in any direction of a uniformly rotating vector varies sinusoidally—that is, in the manner that sin θ varies with θ—with time. In the case of an alternating current, the length of the phasor $\mathbf{I}_{max}$ corresponds to I_{max} and we imagine it to rotate *f* times per second in a counterclockwise sense (Fig. 22–3). Since the angle θ is equal to $2\pi ft$, the vertical component of the phasor at any time *t* corresponds to the instantaneous current *I* of Eq. (22–1):

$$I = I_{max} \sin \theta = I_{max} \, 2\pi ft$$

In a similar way, a phasor $\mathbf{V}_{max}$ can be used to represent an ac voltage *V*.

Phasors are useful when *I* and *V* are out of phase in a circuit

Phasors are helpful because the current and voltage in an ac circuit or circuit element always have the same frequency but may differ in phase, so that the maxima in each quantity do not occur at the same times. Suppose, for instance, that the voltage in a certain circuit leads the current by $\frac{1}{6}$ cycle, as in Fig. 22–4(a). In a phasor diagram, such as Fig. 22–4(b), this situation is shown by having the phasor $\mathbf{V}_{max}$ at an angle of 60° counterclockwise from the phasor $\mathbf{I}_{max}$, since 360°/6 = 60°. We must imagine the two phasors rotating together, always with the same angle of 60° between them, and generating the changing values of *V* and *I* shown in Fig. 22–4(a) as times goes on.

22–3 PHASE RELATIONSHIPS

All actual electric circuits exhibit resistance, capacitance, and inductance to some degree. When direct current flows through a circuit, only its resistance is significant, but all three properties of the circuit affect the flow of alternating current.

We shall first consider a pure-resistance ac circuit, an idealized circuit whose capacitance and inductance are negligible. The instantaneous values of the voltage and

FIG. 22–3 Phasor representation of I_{max} and *I* in an alternating current of frequency *f*. The vertical component of the phasor I_{max} at any time *t* is equal to the instantaneous value *I* of the current at that time.

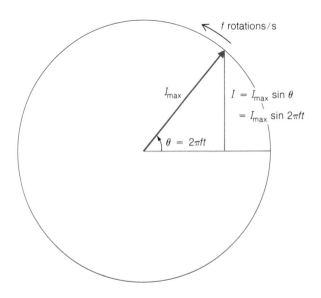

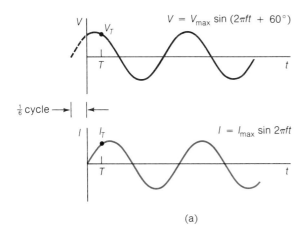

$V = V_{max} \sin (2\pi ft + 60°)$

$\frac{1}{6}$ cycle

$I = I_{max} \sin 2\pi ft$

(a)

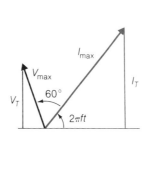

(b)

FIG. 22–4 (a) In a certain ac circuit, the oscillations in V occur earlier than those in I by $\frac{1}{6}$ cycle. (b) In a phasor diagram, the phasor V_{max} leads the phasor I_{max} by $360°/6 = 60°$. The orientations of the phasors correspond to the time T in the graphs of part (a).

current are *in phase* at all times in such a circuit: both V and I are 0 at the same time, both V and I pass through their maximum values at the same time, and so on (Fig. 22–5). The phasors for V and I therefore remain together.

    Now let us look at a pure-inductance ac circuit, an idealized circuit whose resistance and capacitance are negligible. There is no IR potential drop across the inductance, and the potential difference across it is therefore proportional to $\Delta I/\Delta t$, the rate of change of the current (see Sec. 21–7). In this situation V and I cannot be in phase with each other: I changes most rapidly when $I = 0$, so $V = \pm V_{max}$ when $I = 0$, while $\Delta I/\Delta t = 0$ at $I = I_{max}$, so $V = 0$ when $I = \pm I_{max}$. As shown in Fig. 22–6,

In a resistor, V is in phase with I

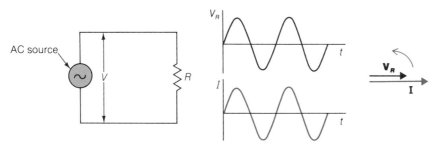

FIG. 22–5 Current and voltage are in phase in a pure resistance.

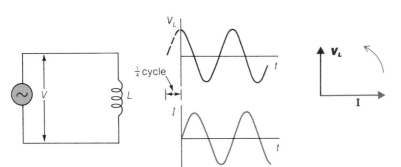

FIG. 22–6 The voltage across a pure inductance leads the current in the inductance by $\frac{1}{4}$ cycle.

FIG. 22–7 The voltage across a pure capacitor lags behind the current into and out of the capacitor by $\frac{1}{4}$ cycle.

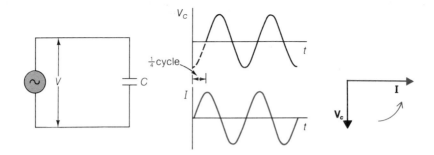

In an inductor, V leads I

The voltage across a pure inductor leads the current in the inductor by $\frac{1}{4}$ cycle.

That is, the variations in the voltage occur $\frac{1}{4}$ cycle *earlier than* the corresponding variations in the current.

In a phasor diagram, a difference in phase of $\frac{1}{4}$ cycle means that the angle between the phasors **V** and **I** is $360°/4 = 90°$. Since **V** leads **I** in a pure inductor, **V** is 90° counterclockwise from **I**, as in Fig. 22–6.

The potential difference across a capacitor depends upon the amount of charge stored on its plates. The charge is a maximum at each moment that $I = 0$, which is when the current is about to reverse direction and carry away the stored charge. The potential difference across a capacitor is

$$V = \frac{Q}{C}$$

and so $V = \pm V_{max}$ when $I = 0$. The stored charge is 0 at each moment that $I = \pm I_{max}$, because at these times the former stored charge is all gone and charge of the opposite sign is about to build up. Hence $V = 0$ when $I = \pm I_{max}$. As shown in Fig. 22–7,

In a capacitor, V lags behind I

The voltage across a pure capacitor lags behind the current into and out of the capacitor by $\frac{1}{4}$ cycle.

That is, the variations in the voltage occur $\frac{1}{4}$ cycle *later than* the corresponding variation in the current. In a phasor diagram, **V** is 90° clockwise from **I**, as in Fig. 22–7.

How ac passes through a capacitor

In a pure-capacitance circuit the phase relationship between current and voltage is different from what it is in pure-resistance and pure-inductance circuits. Although alternating current does not flow *through* a capacitor, it does flow *into one plate and out of the other* since changes in the amount of charge stored on one of the plates are mirrored by changes in the charge stored on the other plate. A flow of $+Q$ into one plate means that $+Q$ flows out of the other plate to leave the latter with a net charge of $-Q$. If direct current were involved, eventually enough charge would accumulate to stop the arrival of any more, and the current would cease. In the case of an alternating current, however, the current always stops and reverses itself periodically anyway, so the presence of a series capacitor does not prevent ac from flowing in the circuit.

22–4 INDUCTIVE REACTANCE

Resistors, inductors, and capacitors all impede the flow of alternating current in a circuit. The effect of resistance is to dissipate part of the electrical energy that passes through it into heat. In those conductors in which Ohm's law is valid for direct current, it is valid for alternating current as well, and

$$I = \frac{V_R}{R} \tag{22–5}$$

Here V_R represents the effective potential difference across the resistance R, and I is the effective current through it.

The opposition an inductor offers to the flow of alternating current arises from the self-induced back emf produced in it by the changing current. The back emf represents a potential drop across the inductor, and the current in the circuit is correspondingly reduced.

Origin of inductive reactance

The *inductive reactance* X_L of an inductor is a measure of its effect on an alternating current passing through it. The effective current I in an inductor is related to the effective potential difference V_L across it and the inductive reactance X_L by

$$I = \frac{V_L}{X_L} \tag{22–6}$$

The unit of inductive reactance is the ohm. Equation (22–6) is analogous to Ohm's law, but there is a basic distinction between reactance and resistance in that there is no power loss in an inductor whereas power is dissipated as heat in a resistor.

No power is lost in a pure inductor

The inductive reactance of an inductor is given by the formula

$$X_L = 2\pi f L \qquad\qquad \textit{Inductive reactance} \tag{22–7}$$

where f is the frequency of the current in hertz (cycles/second) and L is the inductance in henries. The direct dependence of X_L on f and L is reasonable: the self-induced back emf, which is what opposes the current, is proportional to both $\Delta I/\Delta t$ and L, and so the more rapidly the current changes and the larger the value of L, the greater the back emf (Fig. 22–8).

Example What is the current in a coil of negligible resistance and inductance 0.40 H when it is connected to a 120-V, 60-Hz power line?

Solution The reactance of the coil is

$$X_L = 2\pi f L = (2\pi)(60 \text{ Hz})(0.4 \text{ H}) = 151 \ \Omega$$

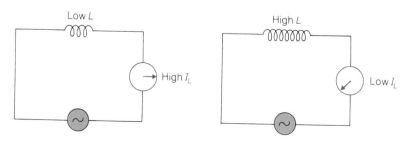

FIG. 22–8 At a given frequency, the lower the inductance L, the lower the inductive reactance X_L and the higher the current I_L.

and the current in it is accordingly

$$I = \frac{V_L}{X_L} = \frac{120 \text{ V}}{151 \, \Omega} = 0.80 \text{ A}$$

Both the 120-V and 0.80-A figures represent effective values.

22–5 CAPACITIVE REACTANCE

The extent to which a capacitor opposes the flow of alternating current depends upon its *capacitive reactance* X_C. If V_C is the effective potential difference across a capacitor whose reactance is X_C, the effective current into and out of the capacitor is

$$I = \frac{V_C}{X_C} \qquad\qquad (22\text{–}8)$$

The capacitive reactance of a capacitor is given by the formula

$$X_C = \frac{1}{2\pi f C} \qquad\qquad \textit{Capacitive reactance} \quad (22\text{–}9)$$

If f is in hertz and the capacitance C in farads, the unit of X_C is the ohm.

Origin of capacitive reactance

A capacitor impedes the flow of alternating current by virtue of the reverse potential difference that appears across it as charge builds up on its plates. Thus there is a potential drop across a capacitor in an ac circuit that affects the current just as the potential drop in a resistor does.

The inverse dependence of X_C upon f and C can be understood from the following argument. If the charge on the plates of a capacitor is changed by ΔQ in the time Δt, then the instantaneous current is $I_{inst} = \Delta Q/\Delta t$. If V_C is the change in the potential difference across the capacitor that leads to ΔQ, then $\Delta Q = C(\Delta V_C)$, and $I_{inst} = C(\Delta V_C/\Delta t)$. The higher the frequency, the greater the rate of change $\Delta V_C/\Delta t$ of the potential difference, and the greater the current. The larger C is, also, the greater the current (Fig. 22–9). Since a high current means a low reactance, increasing f and C decreases X_C.

Capacitive and inductive reactances vary differently with frequency, as Fig. 22–10 shows: X_C decreases with increasing f, whereas X_L increases with increasing f.

Example A capacitor whose reactance is 80 Ω at 50 Hz is used in a 60-Hz circuit. What is its reactance in the latter circuit?

Solution The capacitance of the capacitor, from Eq. (22–9), is

FIG. 22–9 At a given frequency, the higher the capacitance C, the lower the capacitive reactance X_C and the higher the current I_C.

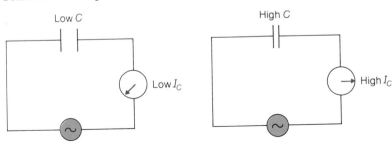

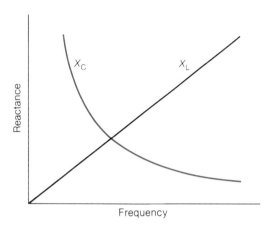

FIG. 22–10 The reactance X_L of an inductor increases with frequency, whereas the reactance X_C of a capacitor decreases with frequency. In the dc limit of $f = 0$, $X_L = 0$ and $X_C = \infty$.

$$C = \frac{1}{2\pi f X_C} = \frac{1}{(2\pi)(50\,\text{Hz})(80\,\Omega)} = 4 \times 10^{-5}\,\text{F}$$

which is 40μF. Its reactance at 60 Hz is

$$X_C = \frac{1}{2\pi f C} = \frac{1}{(2\pi)(60\,\text{Hz})(4 \times 10^{-5}\,\text{F})} = 66\,\Omega \qquad \blacksquare$$

In the limit of $f = 0$, which means direct current, $X_L = 0$ and $X_C = \infty$. When the current does not vary, there is no self-induced back emf in an inductor, and no inductive reactance to impede current. On the other hand, a capacitor completely obstructs direct current because the charge that builds up on its plates remains there instead of surging back and forth as it does when an ac potential is applied.

Only ac can pass through a capacitor

Because capacitors pass ac but not dc, they can be used to favor the transmission of signals of one kind or the other between two circuits. In Fig. 22–11(a), the capacitor

Capacitors as filters for ac or dc

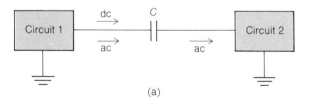

(a)

FIG. 22–11 A capacitor can be used to favor the passage of either ac or dc between two circuits, depending upon how it is connected.

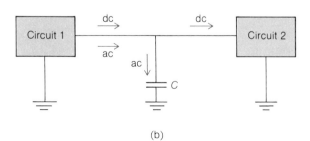

(b)

permits an ac signal to go from circuit 1 to circuit 2 while stopping any direct current. In Fig. 22–11(b), a direct current can pass between the circuits, but the capacitor provides a path to ground for alternating current, and most of the ac will take this path if the reactance is low.

22–6   IMPEDANCE

**How to add
instantaneous voltages**

A series circuit that contains resistance, inductance, and capacitance can be represented as in Fig. 22–12, where each of these circuit properties is considered as lumped into a single resistor, inductor, and capacitor. If an ac source of emf is connected to the circuit, at any instant the applied voltage V is equal to the sum of the voltage drops across the various circuit elements:

$$V = V_R + V_L + V_C \qquad\qquad \textit{Instantaneous voltage} \quad (22\text{–}10)$$

However, V_R, V_L, and V_C are *out of phase with one another*. The voltage across the resistor, V_R, is always in phase with the current I, but V_L is $\frac{1}{4}$ cycle ahead of I, and V_C is $\frac{1}{4}$ cycle behind I. This situation is shown in Fig. 22–12. While Eq. (22–10) is always correct when V, V_R, V_L, and V_C refer to the instantaneous values of the various voltages, it is *not* correct when effective (or maximum) values are involved, and we must use a vectorial approach to take the phase differences into account.

**How to add effective
voltages**

Figure 22–13 is a phasor diagram that shows the phase differences between $\mathbf{V}_R$, $\mathbf{V}_L$, and $\mathbf{V}_C$: $\mathbf{V}_L$ is 90° ahead of $\mathbf{V}_R$ (hence 90° counterclockwise from $\mathbf{V}_R$, since phasors rotate counterclockwise) and $\mathbf{V}_C$ is 90° behind $\mathbf{V}_R$. The vector sum $\mathbf{V}$ of $\mathbf{V}_R$, $\mathbf{V}_L$, and $\mathbf{V}_C$ represents the effective voltage across the terminals of the circuit, and its magnitude is

$$V = \sqrt{V_R{}^2 + (V_L - V_C)^2} \qquad\qquad \textit{Effective voltage} \quad (22\text{–}11)$$

Phase angle

The angle ϕ between $\mathbf{V}$ and $\mathbf{V}_R$ is called the *phase angle* because it is a measure of how much the voltage in the circuit leads or lags behind the current. The direction of the effective current phasor $\mathbf{I}$ is always the same as the direction of $\mathbf{V}_R$.

FIG. 22–12   The instantaneous value of the voltage V applied to a series *RLC* circuit is equal to the sum of the instantaneous values of V_R, V_L, and V_C. This relationship does not hold for the effective values of the various voltages because of the phase differences among them.

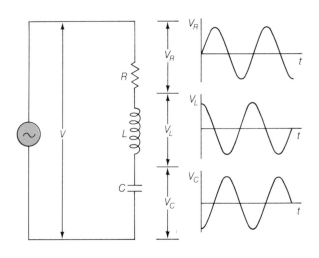

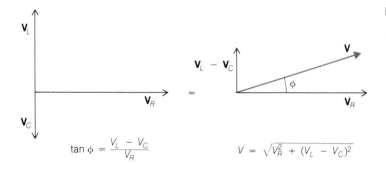

FIG. 22–13 Phasor diagram of the effective voltages across the resistor, inductor, and capacitor of Fig. 22–12. The magnitude of the vector sum of the voltage phasors equals the effective voltage across the entire circuit. The angle ϕ is the phase angle. The direction of the effective current phasor **I** is always the same as that of $\mathbf{V}_R$.

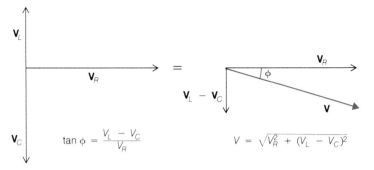

FIG. 22–14 The procedures for finding **V** and ϕ are the same whether $\mathbf{V}_L$ is greater than $\mathbf{V}_C$ or not.

In Fig. 22–13, V_L is greater than V_C, and $(V_L - V_C)$ is a positive quantity. If instead V_L is the smaller quantity, $(V_L - V_C)$ is negative and the vector **V** is below the x-axis (Fig. 22–14). However, Eq. (22–11) still applies, since $(V_L - V_C)^2 = (V_C - V_L)^2$. The phase angle in either case is specified by

$$\tan \phi = \frac{V_L - V_C}{V_R} \qquad\qquad \textit{Phase angle} \quad (22\text{–}12)$$

If $V_C > V_L$, the result will be a negative value for ϕ, which signifies that **V** lags behind $\mathbf{V}_R$ (and behind the current **I**).

Example A resistor, a capacitor, and an inductor are connected in series across an ac power source. The effective voltages across the circuit components are $V_R = 5$ V, $V_C = 10$ V, and $V_L = 12$ V. Find the effective voltage of the source and the phase angle in the circuit.

Solution (a) From Eq. (22–11),

$$V = \sqrt{V_R^{\,2} + (V_L - V_C)^2} = \sqrt{(5\text{ V})^2 + (12\text{ V} - 10\text{ V})^2} = 5.4\text{ V}$$

(b) Since

$$\tan \phi = \frac{V_L - V_C}{V_R} = \frac{12\text{ V} - 10\text{ V}}{5\text{ V}} = 0.4$$

we have for the phase angle $\phi = 22°$.

Because

$$V_R = IR \qquad V_L = IX_L \qquad \text{and} \qquad V_C = IX_C$$

we can rewrite Eq. (22–11) in the form

$$V = I\sqrt{R^2 + (X_L - X_C)^2}$$

The quantity

$$Z = \sqrt{R^2 + (X_L - X_C)^2} \hspace{3cm} \textit{Impedance} \quad (22\text{–}13)$$

Impedance is ac equivalent of resistance; its unit is the ohm

is known as the *impedance* of a series circuit containing resistance, inductance, and capacitance. The unit of impedance is evidently the ohm. Impedance in an ac circuit plays the same role that resistance does in a dc circuit, and in an ac circuit

$$I = \frac{V}{Z} \hspace{3cm} \textit{Current in ac circuit} \quad (22\text{–}14)$$

is the effective current that flows when the effective voltage V is applied. It is important to keep in mind that Z not only depends upon the circuit parameters R, L, and C but also varies with the frequency f.

Because the current I is the same in all parts of the circuit at all times and

$$Z = \frac{V}{I} \hspace{2cm} R = \frac{V_R}{I}$$

$$X_L = \frac{V_L}{I} \qquad \text{and} \qquad X_C = \frac{V_C}{I}$$

the phasor voltage diagram of Fig. 22–13 can be replaced by the phasor impedance diagram of Fig. 22–15. The phase angle ϕ is the same in both cases, of course, and can be calculated from the relation

$$\tan \phi = \frac{X_L - X_C}{R} \hspace{3cm} \textit{Phase angle} \quad (22\text{–}15)$$

Example   Analyze in detail a series circuit that consists of a 10-mH inductor, a 10-μF capacitor, and a 30-Ω resistor conneced to a source of 100-V, 400-Hz alternating current.

FIG. 22–15 The phasor impedance diagram that corresponds to the phasor voltage diagram shown in Fig. 22–13.

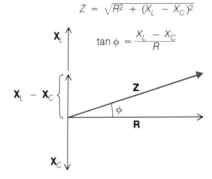

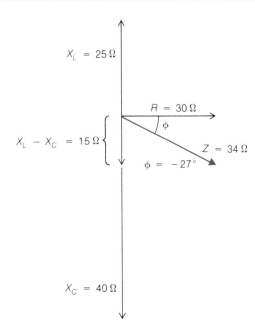

FIG. 22–16 A negative phase angle occurs when X_C is greater than X_L, and signifies that the voltage in the circuit lags behind the current.

Solution The reactances of the inductor and capacitor at 400 Hz are

$$X_L = 2\pi fL = (2\pi)(400\,\text{Hz})(0.01\,\text{H}) = 25\,\Omega$$

$$X_C = \frac{1}{2\pi fC} = \frac{1}{(2\pi)(400\,\text{Hz})(10^{-5}\,\text{F})} = 40\,\Omega$$

The vector impedance diagram for this circuit is shown in Fig. 22–16. The impedance Z is

$$Z = \sqrt{R^2 + (X_L - X_C)^2} = \sqrt{(30\,\Omega)^2 + (25\,\Omega - 40\,\Omega)^2} = 34\,\Omega$$

The phase angle ϕ is found as follows:

$$\tan \phi = \frac{X_L - X_C}{R} = -\frac{15\,\Omega}{30\,\Omega} = -0.50 \qquad \phi = -27°$$

A negative phase angle signifies that the voltage lags behind the current. Here the lag is 27°, which is $\frac{27}{360}$ or 0.075 of a complete cycle.

The current in the circuit is

$$I = \frac{V}{Z} = \frac{100\,\text{V}}{34\,\Omega} = 2.94\,\text{A}$$

The effective potential difference across each of the circuit elements is

$$V_L = IX_L = (2.94\,\text{A})(25\,\Omega) = 74\,\text{V}$$

$$V_C = IX_C = (2.94\,\text{A})(40\,\Omega) = 118\,\text{V}$$

$$V_R = IR = (2.94\,\text{A})(30\,\Omega) = 88\,\text{V}$$

Evidently the effective voltage across an inductor or capacitor in an ac circuit can exceed the effective voltage applied to the entire circuit.

The arithmetic sum of the above voltages is 280 V, but this sum means nothing because the voltages are not in phase with one another. The vector sum of the voltage, which takes into account the phase differences among them, is

$$V = \sqrt{V_R{}^2 + (V_L - V_C)^2} = \sqrt{(88 \text{ V})^2 + (74 \text{ V} - 118 \text{ V})^2} = 98 \text{ V}$$

The difference between this figure and the applied voltage of 100 V is entirely due to the rounding-off of V_L, V_C, and V_R when they were calculated. ∎

22–7 RESONANCE

Impedance is a minimum when $X_L = X_C$

When an ac voltage is applied to a series circuit, the current that flows depends upon the frequency. The greatest current flows when the impedance Z is a minimum. Since

$$Z = \sqrt{R^2 + (X_L - X_C)^2}$$

the condition for minimum impedance in a given circuit is that the frequency be such that the inductive and capacitive reactances are equal. When this is true,

$$X_L = X_C$$

$$2\pi f_0 L = \frac{1}{2\pi F_0 C}$$

and thus

$$f_0 = \frac{1}{2\pi\sqrt{LC}} \qquad\qquad \textit{Resonance frequency} \quad (22\text{--}16)$$

Resonance frequency

When the impressed voltage has this frequency, the current is a maximum and is limited only by the resistance R. The frequency f_0 is called the *resonance frequency* of the circuit, and *resonance* occurs in a series circuit when the impressed voltage oscillates with the resonance frequency.

We note that Eq. (22–16) is the same as Eq. (21–24) for the "natural" frequency at which an LC circuit will oscillate if the capacitor is initially charged. When a circuit is in resonance with an applied voltage, the energy alternately stored and discharged from the capacitor is precisely equal to the energy stored and discharged from the inductor. At other frequencies the energy contents of the capacitor and inductor are different, which interferes with the back-and-forth flow of power and gives rise to an impedance that exceeds the resistance R. The current that flows in an RLC circuit at resonance is

$$I = \frac{V}{R}$$

just as in a dc circuit.

Figure 22–17 shows how the current in a series circuit varies with frequency. The less the resistance R relative to the inductive reactance X_L, the sharper the peak in the

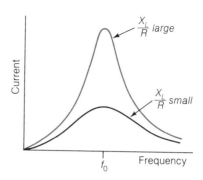

curve. The antenna circuit of a radio receiver is tuned to respond to a particular frequency of radio waves by adjusting a variable capacitor or inductor until the resonance frequency of the circuit is equal to the signal frequency. A circuit in which X_L/R is large has a narrow response curve and can separate signals from two stations very close together in frequency.

Sharpness of tune depends upon the X_L/R ratio

At resonance, $X_L = X_C$, and there is no phase difference between current and voltage. Another way to specify the condition for resonance, then, is to require that current and voltage be in phase at all times, just as they are in a pure-resistance circuit.

Current and voltage are in phase at resonance

Example The antenna circuit of a radio receiver consists of a 10-mH coil and a variable capacitor; the resistance in the circuit is 50 Ω. An 880-kHz (kilohertz) radio wave produces a potential difference of 10^{-4} V across the circuit. Find the capacitance required for resonance and the current at resonance.

Solution The capacitance required for resonance at 880 kHz is

$$C = \frac{1}{(2\pi f)^2 L} = \frac{1}{(2\pi \times 8.8 \times 10^5 \,\text{Hz})^2 (10^{-2}\,\text{H})} = 3.3 \times 10^{-12}\,\text{F} = 3.3\,\text{pF}$$

At resonance, inductive and capacitive reactances cancel each other out, and the current that flows is

$$I = \frac{V}{R} = \frac{10^{-4}\,\text{V}}{50\,\Omega} = 2 \times 10^{-6}\,\text{A} \qquad \blacksquare$$

22–8 POWER IN AC CIRCUITS

No power is consumed in a pure inductor or capacitor in an ac circuit, since these elements act merely as temporary reservoirs of energy and return whatever energy they absorb in one quarter of a cycle to the circuit in the next quarter of that cycle. No power is therefore needed to maintain an ac current in the inductive and capacitive parts of a circuit. The resistance in the circuit, however, dissipates power as heat at the rate $P = IV_R$, where, as usual, I and V_R are effective values. From Fig. 22–13 we see that

Inductors and capacitors do not absorb power

$$V_R = V \cos \phi \qquad\qquad\qquad (22\text{–}17)$$

since $V \cos \phi$ is the component of $\mathbf{V}$ that is in phase with the current phasor $\mathbf{I}$. Hence the effective power absorbed in an ac circuit is

$$P = IV \cos \phi \qquad \qquad \textit{Power in ac circuit} \quad (22\text{--}18)$$

Power factor

The quantity $\cos \phi$ is called the *power factor* of the circuit. The power factor is equal to 1 only at resonance, when current and voltage are in phase; at resonance, $\phi = 0$, and $\cos \phi = 1$. Under other circumstances the voltage is not in phase with the current, and the actual power in the circuit is less than IV.

From Fig. 22–15 we see that

$$\cos \phi = \frac{R}{Z} = \frac{R}{\sqrt{R^2 + (X_L - X_c)^2}} \qquad \qquad \textit{Power factor} \quad (22\text{--}19)$$

The power factor of a circuit is the ratio between its resistance and its impedance. Often power factors are expressed as percentages rather than as decimals or fractions. Thus a phase angle of 60° means a power factor of

$$\cos 60° = 0.50 = 50\%$$

The volt·ampere is the unit of apparent power; apparent power may be more than actual consumed power in an ac circuit

Instruments called wattmeters have been devised which respond directly to the effective product of V and I. An ac wattmeter connected in a circuit gives a lower value for the power than the product of the effective values of V and I obtained from a separate voltmeter and ammeter in the same circuit (except at resonance), because the separate meters are not affected by the phase difference between current and voltage. Alternating-current generators, transformers, and power lines are usually rated in *volt-amperes,* the product of effective voltage and current without regard to actual power, because higher values of V and/or I must be supplied to a circuit whose power factor is less than 1 than is reflected in its power consumption. It is convenient to think of the power factor as having the unit of watts per volt-ampere.

Alternating-current devices of various kinds—for example, ac electric motors and fluorescent lamps—may have net inductive or capacitive reactances and consequently have power factors of less than 100%. A power factor of 70%, for instance, means that 1 kVA (kilovolt-ampere) of apparent power must be supplied for every 700 W of power actually consumed. This is an uneconomical situation because of the additional generator capacity required as well as because of the additional heat losses in the transmission lines as the unused power circulates between the generator and the device. The remedy is to introduce capacitors or inductors into the power line to increase the power factor to an acceptable figure.

Example   An inductive load connected to a 24-V, 400-Hz power source draws a current of 2 A and dissipates 40 W. (a) Find the power factor of the load and the phase angle ϕ. (b) What series capacitance is needed to make the power factor 100%? (c) What would the current then be? (d) How much power would the load then dissipate?

Solution   (a) Since $P = IV \cos \phi$, the power factor is

$$\cos \phi = \frac{P}{IV} = \frac{40 \text{ W}}{(2 \text{ A})(24 \text{ V})} = 0.83 = 83\%$$

The phase angle is therefore $\phi = 34°$.

(b) The power factor will be 100% when $X_C = X_L$. In the original circuit, $X_C = 0$ and so, from Eq. (22–15), $X_L = R \tan \phi$. The resistance in the circuit is

$$R = \frac{P}{I^2} = \frac{40 \text{ W}}{(2 \text{ A})^2} = 10 \, \Omega$$

Hence the inductive reactance is

$$X_L = R \tan \phi = 10 \, \Omega \times \tan 34° = 6.75 \, \Omega$$

Since the required capacitive reactance X_C must equal X_L when $f = 400$ Hz,

$$C = \frac{1}{2\pi f X_C} = \frac{1}{(2\pi)(400 \text{ Hz})(6.75 \, \Omega)} = 5.9 \times 10^{-5} \text{F} = 59 \, \mu\text{F}$$

(c) When $X_C = X_L$, $Z = R$ and

$$I = \frac{V}{Z} = \frac{V}{R} = \frac{24 \text{ V}}{10 \, \Omega} = 2.4 \text{ A}$$

(d) $P = I^2 R = (2.4 \text{ A})^2 (10 \, \Omega) = 57.6$ W.   ■

22-9 IMPEDANCE MATCHING

In Sec. 18–7 we saw that the maximum power transfer between two dc circuits occurs when their resistances are equal. The same conclusion applies to ac circuits with the impedance of the circuits rather than their resistances as the quantities to be matched. An additional consideration with ac circuits is that an impedance mismatch may result in a distorted signal by altering the properties of the circuits, for instance by changing their resonance frequencies.

Power transfer is a maximum between circuits with the same impedance

Ac circuits have the advantage that a transformer can be used to correct impedance mismatches. A common use of a transformer for this purpose occurs in an audio system where the amplifier circuit might have an impedance of several thousand ohms whereas the voice coil of the loudspeaker has an impedance of only a few ohms. Connecting the voice coil directly to the amplifier would be grossly inefficient.

A transformer can be used to match impedances

Let us calculate the ratio of turns N_1/N_2 between the primary and secondary windings of a transformer needed to couple a circuit of impedance Z_1 to a circuit of impedance Z_2. If the voltages and currents in the circuits are respectively V_1, I_1 and V_2, I_2, then from Eq. (20–8)

$$\frac{V_1}{V_2} = \frac{N_1}{N_2} \quad \text{and} \quad \frac{I_2}{I_1} = \frac{N_1}{N_2}$$

Since $Z_1 = V_1/I_1$ and $Z_2 = V_2/I_2$,

$$\frac{Z_1}{Z_2} = \frac{V_1 I_2}{V_2 I_1} = \left(\frac{N_1}{N_2}\right)^2$$

and the ratio of the turns is

$$\frac{N_1}{N_2} = \sqrt{\frac{Z_1}{Z_2}} \qquad \qquad \textit{Impedance matching} \quad (22\text{–}20)$$

FIG. 22–18 A parallel *RLC* circuit. The potential differences across the circuit elements are the same.

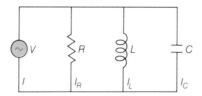

Thus if a loudspeaker whose voice coil has an impedance of 10 Ω is to be used with an amplifier whose load impedance is 9000 Ω, a transformer should be used whose ratio of turns is

$$\frac{N_1}{N_2} = \sqrt{\frac{Z_1}{Z_2}} = \sqrt{\frac{9000\ \Omega}{10\ \Omega}} = 30$$

22–10   PARALLEL AC CIRCUITS

When a resistor, an inductor, and a capacitor are connected in parallel, as in Fig. 22–18, the potential difference is the same across each circuit element:

$$V = V_R = V_L = V_C$$

The branch currents in a parallel circuit are not in phase

The total instantaneous current is the sum of the instantaneous currents in each branch, as in the case of a dc-parallel circuit, but this is not true of the total effective current because the branch currents are not in phase. The current I_R is always in phase with the voltage V, but I_C leads V by 90° and I_L lags behind V by 90°. A phasor diagram of the situation is shown in Fig. 22–19. The magnitudes of the currents in the branches are

$$I_R = \frac{V}{R} \qquad I_C = \frac{V}{X_C} \qquad I_L = \frac{V}{X_L}$$

and their vector sum is

$$I = \sqrt{I_R^2 + (I_C - I_L)^2} \qquad\qquad \textit{Effective current} \quad (22\text{–}21)$$

The phase angle ϕ specified by

$$\tan\phi = \frac{I_C - I_L}{I_R} \qquad\qquad \textit{Phase angle} \quad (22\text{–}22)$$

is the angle between the current and the voltage. A positive phase angle means that the

FIG. 22–19 A phasor diagram of the effective currents in the resistor, inductor, and capacitor of Fig. 22–18. The magnitude *I* of the vector sum of the current vectors is equal to the effective current through the entire circuit.

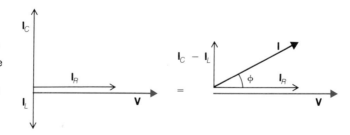

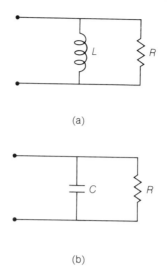

FIG. 22–20 (a) A high-pass filter. The higher the frequency, the higher the reactance X_L and the greater the proportion of the current that passes through the load R. (b) A low-pass filter. The lower the frequency, the higher the reactance X_C and the greater the proportion of the current that passes through the load R.

current leads the voltage, a negative one means that the current lags behind the voltage. The power dissipated in a parallel ac circuit is given by the same formula as in the case of a series circuit, namely $P = IV \cos \phi$.

Parallel ac circuits have various uses. For instance, an inductor connected in parallel across a resistive load, as in Fig. 22–20(a), acts as a "high-pass filter" to discriminate against low-frequency currents when more than one frequency is present. Since $X_L = 2\pi fL$, the higher the frequency, the less the current through the inductor and the greater the current through the load R. Low-frequency currents are diverted through the inductor and little of them reach the load. Similarly, a capacitor in parallel across a resistive load, as in Fig. 22–20(b), serves as a "low-pass filter." Since $X_C = \frac{1}{2}\pi fC$, the lower the frequency, the less the current through the capacitor and the greater the current through R.

High- and low-pass filters

Example Design a simple high-pass filter for a 50-Ω load that discriminates against frequencies under 20 kHz. What proportion of the total current pases through the load when the frequency of the applied voltage is 2 kHz? When it is 200 kHz?

Solution (a) An inductor is needed in parallel with the load, as in Fig. 22–20(a). If $X_L = R$ at $f = 20$ kHz, then for the same applied voltage, $I_L > I_R$ for $f < 20$ kHz and $I_L < I_R$ for $f > 20$ kHz. Hence

$$X_L = R$$

$$2\pi fL = R$$

$$L = \frac{R}{2\pi f} = \frac{50}{(2\pi)(2 \times 10^4 \, \text{Hz})} = 4 \times 10^{-4} \, \text{H} = 0.4 \, \text{mH}$$

(b) Let us assume an applied voltage of 100 V for the sake of convenience. (Any other voltage could equally well be used, or we could just call the applied voltage V.) We first find I_L and I_R for $f = 2$ kHz:

$$X_L = 2\pi fL = (2\pi)(2 \times 10^3 \, \text{Hz})(4 \times 10^{-4} \, \text{H}) = 5 \, \Omega$$

$$I_L = \frac{V}{X_L} = \frac{100 \text{ V}}{5\,\Omega} = 20 \text{ A}$$

$$I_R = \frac{V}{R} = \frac{100 \text{ V}}{50\,\Omega} = 2 \text{ A}$$

The total current I in the circuit is

$$I = \sqrt{I_R^2 + I_L^2} = \sqrt{(2\,\text{A})^2 + (20\,\text{A})^2} = 20.1 \text{ A}$$

and the proportion of I that passes through the load is therefore

$$\frac{I_R}{I} = \frac{2\,\text{A}}{20.1\,\text{A}} = 0.0995 = 9.95\%$$

Less than 10% of the total current passes through the load.

(c) When $f = 200$ kHz we proceed in exactly the same way, with the result that $I_R/I = 0.995 = 99.5\%$. Now nearly all the current passes through the load. ∎

Band-pass and band-reject filters

In a series RLC circuit, the impedance is a minimum at resonance and increases at higher and lower frequencies than f_0. At resonance, $X_L = X_C$, $Z = R$, and $I = V/R$. In a parallel RLC circuit, resonance again corresponds to $X_L = X_C$, $Z = R$, and $I = V/R$, but now the impedance is a *maximum* at f_0 since at higher and lower frequencies some current can pass through the inductor and capacitor as well as through the resistor. Thus a series circuit can be used as a selector to favor a particular frequency, and a parallel circuit with the same L and C can be used as a selector to discriminate against the same frequency. Such circuits form the basis of "band-pass" and "band-reject" filters.

Figure 22–21(a) shows one type of band-pass filter. At frequencies near f_0 (from Fig. 22–17 we can see why a band of frequencies rather than a single one is involved) the impedance of the parallel LC circuit is high, so most of the current flows through the load R. At other frequencies, current is diverted through the LC circuit and so reduces the load current. The band-reject filter of Fig. 22–21(b) acts in the opposite

FIG. 22–21 (a) A band-pass filter. At and near the resonance frequency f_0, the impedance of the LC circuit is high, and most of the current passes through the load R. (b) A band-reject filter. At and near f_0, most of the current passes through L and C instead of through R since the impedance of this branch of the circuit is then low.

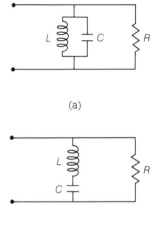

(a)

(b)

way, since the series LC circuit provides a low-impedance alternate path for current at frequencies near f_0. At other frequencies, the impedance of the LC circuit is high and most of the current therefore passes through the load. More elaborate circuits than those of Figs. 22–20 and 22–21 can be designed that provide sharper cutoffs between the passed and rejected frequency bands.

IMPORTANT TERMS

The **effective value** of an alternating current is such that a direct current of this magnitude produces heat in a resistor at the same rate as the alternating current.

A **phasor** is a rotating vector whose projection can represent either current or voltage in an ac circuit.

The **phase relationships** between the instantaneous voltage and instantaneous current in ac circuit components are as follows: the voltage across a pure resistor is in phase with the current; the voltage across a pure inductor leads the current by $\frac{1}{4}$ cycle; the voltage across a pure capacitor lags behind the current by $\frac{1}{4}$ cycle.

The **inductive reactance** X_L of an inductor is a measure of its effect on an alternating current. The **capacitive reactance** X_C of a capacitor is a measure of its effect on an alternating current. Both X_L and X_C vary with the frequency of the current.

The **impedance** Z of an ac circuit is analogous to the resistance of a dc circuit. The **resonance frequency** of an ac circuit is that frequency for which the impedance is a minimum.

The **power factor** of an ac circuit is the ratio between the power consumed in the circuit and the product of the effective current and voltage there; this ratio is equal to that between the resistance and the impedance of the circuit, and is less than 1, except at resonance. The unit of the apparent power $V_{eff} I_{eff}$ is the **volt-ampere**, as distinct from the watt, which is the unit of consumed power.

IMPORTANT FORMULAS

Effective current: $I_{eff} = \dfrac{I_{max}}{\sqrt{2}} = 0.707 I_{max}$

Effective voltage: $V_{eff} = \dfrac{V_{max}}{\sqrt{2}} = 0.707 V_{max}$

Inductive reactance: $X_L = 2\pi f L$

Capacitive reactance: $X_C = \dfrac{1}{2\pi f C}$

Impedance: $Z = \sqrt{R^2 + (X_L - X_C)^2}$

Current in ac circuit: $I = \dfrac{V}{Z}$

Phase angle:

$$\tan \phi = \frac{X_L - X_C}{R} \qquad \cos \phi = \frac{R}{Z}$$

Resonance frequency: $f_0 = \dfrac{1}{2\pi\sqrt{LC}}$

Power in ac circuit: $P = IV \cos \phi$

Impedance-matching transformer: $\dfrac{N_1}{N_2} = \sqrt{\dfrac{Z_1}{Z_2}}$

MULTIPLE CHOICE

1. The effective voltage in an ac circuit is equal to
  (a) $0.5 V_{max}$.
  (b) $0.707 V_{max}$.
  (c) $V_{max}/0.707$.
  (d) $I_{max}/0.707$.

2. The current in an ac circuit varies between
  (a) 0 and I_{eff}.
  (b) 0 and I_{max}.
  (c) $-I_{eff}$ and $+I_{eff}$.
  (d) $-I_{max}$ and $+I_{max}$

3 The reactance of a capacitor is proportional to
  (a) $\sqrt{f}$.
  (b) f.
  (c) f^2.
  (d) $1/f$.

4. The reactance of an inductor is proportional to
  (a) $\sqrt{f}$.
  (b) f.
  (c) f^2.
  (d) $1/f$.

5. The unit of inductive reactance is the
  (a) henry.
  (b) tesla.
  (c) weber.
  (d) ohm.

6. In an ac circuit, the voltage
  (a) leads the current.
  (b) lags the current.
  (c) is in phase with the current.
  (d) is any of the above, depending on the circumstances.

7. When voltage and current are in phase in an ac-circuit, the
   (a) impedance is 0.
   (b) reactance is 0.
   (c) resistance is 0.
   (d) phase angle is 90°.

8. The impedance of a circuit does not depend on
   (a) I.   (b) f.
   (c) R.   (d) C.

9. The power dissipated in an ac circuit depends on its
   (a) resistance.
   (b) inductive reactance.
   (c) capacitive reactance.
   (d) impedance.

10. A coil of inductance L has an inductive reactance of X_L in an ac circuit in which the effective current is I. The coil is made from a superconducting material and has no resistance. The rate at which power is dissipated in the coil is
    (a) 0.   (b) IX_L.
    (c) $I^2 X_L$.   (d) IX_L^2.

11. The power factor of a circuit is equal to
    (a) RZ.   (b) R/Z.
    (c) X_L/Z.   (d) X_C/Z.

12. At resonance, it is *not* true that
    (a) $R = Z$.   (b) $X_L = 1/X_C$.
    (c) $P = IV$.   (d) $I = V/R$.

13. The power factor of a certain circuit in which the voltage lags behind the current is 80%. To increase the power factor to 100%, it is necessary to add to the circuit additional
    (a) resistance.   (b) capacitance.
    (c) inductance.   (d) impedance.

14. The capacitive reactance of a 5-μF capacitor in a 20-kHz circuit is
    (a) 0.63 Ω.   (b) 1.6 Ω.
    (c) 5 Ω.   (d) 16 Ω.

15. The inductive reactance of a 1-mH coil in a 5-Hz circuit is
    (a) 3.1 Ω.   (b) 6.3 Ω.
    (c) 10 Ω.   (d) 31 Ω.

16. In a series ac circuit $R = 10$ Ω, $X_L = 8$ Ω, and $X_C = 6$ Ω when the frequency is f. The impedance at this frequency is
    (a) 10.2 Ω.   (b) 12 Ω.
    (c) 24 Ω.   (d) 104 Ω.

17. The phase angle in the circuit of question 16 is
    (a) 0.2°.   (b) 2°.
    (c) 11°.   (d) 45°.

18. The resonance frequency of the circuit of question 16 is
    (a) less than f.
    (b) equal to f.
    (c) more than f.
    (d) any of the above, depending on the applied voltage.

19. The power factor of a circuit in which $X_L = X_C$
    (a) is 0.
    (b) is 1.
    (c) depends on the ratio X_L/X_C.
    (d) depends on the value of R.

20. A voltmeter across an ac circuit reads 50 V and an ammeter in series with the circuit reads 5 A. The power consumption of the circuit
    (a) is less than or equal to 250 W.
    (b) is exactly equal to 250 W.
    (c) is equal to or more than 250 W.
    (d) may be less than, equal to, or more than 250 W.

EXERCISES

1. What properties of an ac circuit are described by its resistance, capacitive reactance, inductive reactance, and impedance? What are the similarities and differences among these quantities?

2. Should the capacitors of Fig. 22–11 have large or small values of C in order to act most efficiently as filters?

3. An alternating potential difference whose frequency is higher than the resonance frequency of a series RLC circuit is applied to it. Does the voltage in the circuit lead or lag the current?

4. The frequency of the alternating potential difference applied to a series RLC circuit is halved. What happens to the resistance, the inductive reactance, and the capacitive reactance of the circuit? What further information is needed to establish what happens to the impedance of the circuit?

5. What is the significance of the power factor of a circuit? It is independent of frequency? Under what circumstances (if any) can it be zero? Under what circumstances (if any) can it be 100%?

6. The dielectric used in a certain capacitor breaks down at a voltage of 300 V. Find the highest effective sinusoidal ac voltage that can be applied to it.

7. An ammeter in series with an ac circuit reads 10 A and a voltmeter across the circuit reads 60 V. (a) What is the maximum current in the circuit? (b) What is the maximum potential difference across the circuit? (c) Does the maxi-

mum current necessarily occur at the same moments as the maximum voltage?

8. What is the reactance of an 80-pF capacitor at 10 kHz? At 10 MHz?

9. The reactance of a capacitor is 50 Ω at 200 Hz. What is its capacitance?

10. The reactance of an inductor is 80 Ω at 500 Hz. Find its inductance.

11. What is the reactance of a 5-mH inductor at 10 Hz? At 10 kHz?

12. Find the current that flows when a 10-μF capacitor is connected to a 15-V, 5-kHz power source.

13. Find the current that flows when a 3.0-mH inductor is connected to a 15-V, 5-kHz power source.

14. The current in a resistor is 2 A when it is connected across a 240-V, 50-Hz line. How much capacitance should be connected in series with the resistor to reduce the current to 1 A?

15. How much inductance should be connected in series with the resistor of Exercise 14 to reduce the current to 1 A?

16. A current of 0.8 A flows through a 50-mH inductor of negligible resistance that is connected to a 120-V power source. What is the frequency of the source?

17. A 5-μF capacitor is connected in series with a 300-Ω resistor and a 120-V, 50-Hz potential difference is applied. Find the current in the circuit and the power dissipated.

18. A 30-mH, 60-Ω inductor is connected to a 20-V, 400-Hz power source. Find the current in the inductor and the power dissipated in it.

19. A pure capacitor, a pure inductor, and a pure resistor are connected in series across a 60-V ac power source. The potential difference across the capacitor is 60 V, and that across the inductor is also 60 V. What is the potential difference across the resistor?

20. A pure capacitor, a pure inductor, and a pure resistor are connected in series across an ac power source. A voltmeter placed in turn across each circuit element reads 15 V, 20 V, and 20 V respectively. What is the potential difference of the source?

21. A series circuit has a resistance of 20 Ω, an inductive reactance of 20 Ω, and a capacitive reactance of 20 Ω when connected to an ac source. Find (a) the impedance of the circuit, (b) the phase angle, and (c) the potential difference required for a current of 20 A to flow.

22. A series circuit has a resistance of 40 Ω, an inductive reactance of 30 Ω, and a capacitive reactance of 50 Ω when connected to a certain ac source. Find (a) the impedance of the circuit, (b) the phase angle, and (c) the potential difference required for a current of 1.5 A to flow.

23. A 10-kW electric motor has an inductive power factor of 70%. What minimum rating in kVA must the power line have? A capacitor is connected in series with the motor to increase the power factor to 100%. How does this effect the required rating of the power line?

24. A microphone whose impedance is 20 Ω is to be used with an amplifier whose input impedance is 50,000 Ω. What should the ratio of turns be in the required transformer?

PROBLEMS

1. The potential difference across a source of alternating current varies sinusoidally with time with $V_{max} = 100$ V. Find the value of the instantaneous potential difference (a) $\frac{1}{8}$ cycle, (b) $\frac{1}{4}$ cycle, (c) $\frac{3}{8}$ cycle, and (d) $\frac{1}{2}$ cycle after $V = 0$.

2. A certain 250-Hz ac power source has a maximum potential difference of 24 V. Find the value of the instantaneous potential difference (a) 0.0005 s, (b) 0.001 s, (c) 0.002 s, (d) 0.0025 s, (e) 0.004 s, and (f) 0.005 s after $V = 0$.

3. A circuit that contains inductance and resistance has an impedance of 50 Ω at 100 Hz and an impedance of 100 Ω at 500 Hz. What are the values of the inductance and the resistance of the circuit?

4. A circuit that contains capacitance and resistance has an impedance of 30 Ω at 80 Hz and an impedance of 4 Ω at 240 Hz. What are the values of the capacitance and resistance of the circuit?

5. An inductor of negligible resistance whose reactance is 120 Ω at 200 Hz is connected to a 240-V, 60-Hz power line. What is the current in the inductor?

6. A capacitor of unknown capacitance is found to have a reactance of 120 Ω at 200 Hz. The capacitor is connected to a 240-V, 60-Hz power line. (a) What is the current in the circuit? (b) Why is this current not the same as that of problem 5?

7. The voltage leads the current in a certain ac circuit by 30°. The effective current in the circuit is 5 A. (a) Is the capacitive reactance greater than or less than the inductive reactance? (b) What is the value of I when $V = 0$?

8. A capacitor of unknown capacitance is connected in series with an 80-Ω resistor. The combination is found to draw 0.5 A when connected to a 120-V, 60-Hz power source. Find (a) the capacitance of the capacitor, (b) the power dissipated in the capacitor, and (c) the power dissipated in the resistor.

9. A coil of unknown inductance and resistance is observed to draw 8 A when a 40-V dc potential difference is applied, and 5 A when a 40-V, 60 Hz ac potential difference is applied. (a) Find the inductance and resistance of the coil. (b) Find the power dissipated in the coil in each situation.

10. A coil of unknown inductance and resistance is observed to draw 50 mA when a dc potential difference of 5 V is applied. When the coil is connected to a source of 400-Hz ac, however, a potential difference of 8 V is required to yield the same current. (a) Find the inductance and resistance of the coil. (b) Find the power dissipated in the coil in each situation.

11. In a series ac circuit, $R = 20 \ \Omega$, $X_L = 10 \ \Omega$, and $X_C = 25 \ \Omega$ when the frequency is 400 Hz. Find the resonance frequency of the circuit.

12. In the antenna circuit of a radio receiver that is tuned to a certain station, $R = 5 \ \Omega$, $L = 5$ mH, and $C = 5$ pF. (a) Find the frequency of the station. (b) If the voltage applied to the circuit is 0.5 mV, find the resulting current. (c) What should the capacitance be in order to receive an 800-Hz radio signal?

13. A 60-μF capacitor, a 0.3-H inductor, and a 50-Ω resistor are connected in series with a 120-V, 60-Hz power source. Find (a) the impedance of the circuit, (b) the current in it, (c) the power factor, (d) the power dissipated, and (e) the minimum apparent-power rating of the source.

14 A 10-μF capacitor, a 30-mH inductor, and a 15-Ω resistor are connected in series with a 10-V, 250-Hz power source. Find (a) the impedance of the circuit, (b) the current in it, (c) the power factor, (d) the power dissipated, and (e) the minimum apparent-power rating of the source.

15. (a) Find the resonance frequency of the circuit of problem 13. (b) What current will flow if the circuit is connected to a 120-V power line whose frequency is equal to the resonance frequency?

16. (a) Find the resonance frequency of the circuit of problem 14. (b) What current will flow if the circuit is connected to a 10-V power line whose frequency is equal to the resonance frequency?

17. A 10-μF capacitor, a 0.10-H inductor, and a 60-Ω resistor are connected in series across a 120-V, 60-Hz power line. Find (a) the current in the circuit, (b) the power dissipated in it, and (c) the potential difference across each of the circuit elements.

18. The circuit of problem 17 is connected across a 120-V, 30-Hz power line. Answer the same questions for this situation.

19. An inductor dissipates 75 W of power when it draws 1.0 A from a 120-V, 60-Hz power line. (a) What is its power factor? (b) What capacitance should be connected in series with it to increase its power factor to 100%? (c) How much current would the circuit then draw? (d) How much power would it then dissipate? (e) What should the minimum volt-ampere rating of the power line be then?

20. A circuit consisting of a capacitor in series with a resistor draws 3.6 A from a 50-V, 100-Hz power line. The circuit dissipates 120 W. (a) What is the precise phase relationship between voltage and current in the circuit? (b) What series inductance should be inserted in the circuit if the current and voltage are to be in phase? (c) How much current would the circuit then draw? (d) How much power would it then dissipate?

21. A 10-Ω resistor and an 8-μF capacitor are connected in parallel across a 10-V, 1-kHz power source. Find the current in each component, the total current, the impedance of the circuit, the phase angle, and the power dissipated by the circuit.

22. A 10-Ω resistor and a 2-mH inductor are connected in parallel across a 10-V, 1-kHz power source. Answer the same questions for this circuit.

23. A 10-Ω resistor, an 8-μF capacitor, and a 2-mH inductor are connected in parallel across a 10-V, 1-kHz power source. Answer the same questions for this circuit.

24. Design a simple low-pass filter for a 50-Ω load that discriminates against frequencies above 20 kHz. What proportion of the total current passes through the load when the frequency of the applied voltage is 2 kHz? When it is 200 kHz?

ANSWERS TO MULTIPLE CHOICE

1. b	5. d	9. a	13. c	17. c
2. d	6. d	10. a	14. b	18. a
3. d	7. b	11. b	15. d	19. b
4. b	8. a	12. b	16. a	20. a

LIGHT

23

Among the most noteworthy achievements of nineteenth-century science was the realization that light consists of electromagnetic waves. Electromagnetic waves themselves were hypothesized by James Clerk Maxwell in 1864 on the basis of his theory of electric and magnetic fields. The speed Maxwell calculated for these waves turned out to be the same as the speed of light. He then concluded that, since both were transverse waves, they were the same phenomenon, a conclusion that has since been verified in every detail. Although every aspect of the wave behavior of light can be determined from Maxwell's electromagnetic theory of light, such calculations are quite complicated. A less comprehensive but simpler and more intuitive approach to optical phenomena was devised in 1678 by Christian Huygens; it can be applied to all kinds of waves, not just light waves. In this chapter the reflection and refraction of light at plane surfaces will be considered with the help of

Huygens' method. We should note in passing that in certain important respects light has the character of a stream of particles rather than that of a series of waves. This duality is examined in later chapters.

23–1   ELECTROMAGNETIC WAVES

A changing electric field is equivalent to a magnetic field

In electromagnetic induction, a changing magnetic field induces an emf in a nearby wire loop or other conducting path. Thus a changing magnetic field is equivalent in its effects to an electric field. The converse is also true: a changing electric field is equivalent in its effects to a magnetic field. This is true even in empty space, where electric currents cannot flow. No simple experiment can directly demonstrate the latter equivalence (unlike electromagnetic induction, which is very easy to exhibit), and it was first proposed by Maxwell on the basis of an indirect argument. Electromagnetic waves occur as a consequence of these two effects—a changing magnetic field produces an electric field, and a changing electric field produces a magnetic field. The two constantly varying fields are coupled together as they travel through space.

Weak electric fields are easier to detect than weak magnetic fields

It is not surprising that electromagnetic induction became known many years before its converse was suspected. Even a slight electric field causes a current to flow in a conductor, and, if the resistance of the conductor is low enough, the current may be sufficiently large to detect despite the feebleness of the field itself. However, electric current has no magnetic counterpart because single magnetic poles do not exist; the opposite poles of a magnet cannot be separated from one another the way opposite electric charges can. Unlike weak electric fields, weak magnetic fields are therefore hard to measure no matter where they occur; and those due to changing electric fields are seldom strong. Maxwell's notion that an electric field that varies with time gives rise to a magnetic field accordingly did not originate in an observation, but instead developed from an intuitive feeling for order in the natural world.

Em waves consist of coupled electric and magnetic fields

What emerged from Maxwell's analysis was that electromagnetic waves spread out in space from an initial disturbance in the same manner that waves spread out from a disturbance in a body of water. If we throw a stone into a pond or otherwise alter the state of the water surface at some point, oscillations occur in which energy is continually interchanged between the kinetic energy of moving water and the potential energy of water higher than its normal level. These oscillations begin where the stone lands, and spread out as waves across the surface of the pond. The wave speed depends upon the properties of the pond water, varying with temperature, impurity content, and so on, but it is independent of the wave amplitude. As we saw in Chapter 12, this is typical wave behavior. When electromagnetic waves spread out from an electric or magnetic disturbance, their energy is constantly being interchanged between the fluctuating electric field and the fluctuating magnetic field of the waves.

Formation of an em wave

Let us connect a pair of metal rods to a source of alternating emf, as in Fig. 23–1. (Such a source is called an oscillator, as we know.) For clarity we will imagine that there is only a single charge in each rod at any time.

(a) When the oscillator is switched on, a positive charge in the upper rod begins to move upward and a negative charge in the lower rod begins to move downward. The electric lines of force around the charges are indicated by the colored lines, and

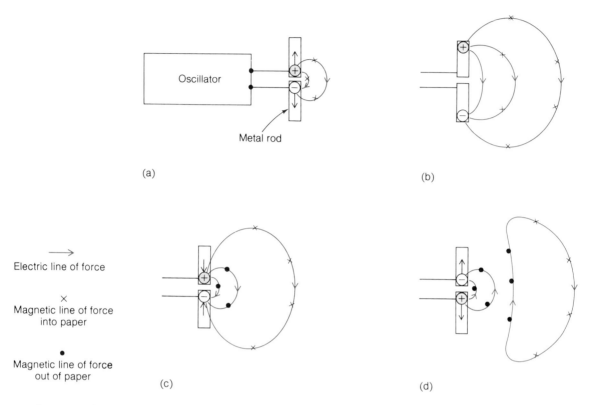

(a)

(b)

Electric line of force

× Magnetic line of force into paper

• Magnetic line of force out of paper

(c)

(d)

the magnetic lines of force due to the motion of the charges (which are concentric circles perpendicular to the paper) are indicated by crosses when their direction is into the paper and by dots when their direction is out of the paper. (The dots represent arrowheads and the crosses represent the tail feathers of arrows.)

FIG. 23–1 A pair of metal rods connected to an electrical oscillator emit electromagnetic waves.

(b) The charges have reached the limit of their motion and have stopped, so that they cease to produce a magnetic field. The outer magnetic lines of force do not disappear instantly because of the finite speed at which changes in electric and magnetic fields travel.

(c) The emf of the oscillator now begins to decrease, and the charges move toward each other. The result is a magnetic field in the opposite direction to the earlier field. The electric field is in the same direction as before.

(d) The emf has passed through 0 and begun to increase in the opposite sense. Consequently there is a negative charge in the upper rod and a positive one in the lower rod which begin to move apart. The electric field is therefore opposite in direction to the earlier field, but the magnetic field is in the same direction since magnetically a positive charge moving downward is equivalent to a negative charge moving upward.

Owing to this sequence of changes in the fields, the outermost electric and magnetic lines of force respectively form into closed loops. These loops of force, which lie in perpendicular planes, are divorced from the oscillating charges that gave rise to them and continue moving outward, constituting an electromagnetic wave. As the charges

The electric and magnetic fields of an em wave form closed loops

continue oscillating back and forth, further associated loops of electric and magnetic lines of force are emitted, forming an expanding pattern of loops.

Figure 23–2 shows the configuration of the electric and magnetic fields that spread outward from a pair of oscillating charges. The actual fields are three-dimensional, so that the magnetic lines of force form loops in planes perpendicular to the line joining the charges.

Three properties of electromagnetic waves are worth noting:

Properties of em waves

1. The variations occur simultaneously in both fields (except close to the oscillating charges), so that the electric and magnetic fields have maxima and minima at the same times and in the same places.
2. The directions of the electric and magnetic fields are perpendicular to each other and to the direction in which the waves are moving. Light waves are therefore transverse.
3. The speed of the waves depends only upon the electric and magnetic properties of the medium they travel in, and not upon the amplitudes of the field variations.

Figure 23–3 is an attempt at portraying (1) and (2) of the above in terms of lines of force of **E** and **B** a long distance from a source of electromagnetic waves. Closer to the source the lines of force are curved, as in Fig. 23–2. It is worth keeping in mind that, unlike the other types of waves considered in Chapter 12—waves in a stretched string, water waves, sound waves—nothing material moves in the path of an electromagnetic wave. The only changes are in electric and magnetic fields.

23–2   RADIATION FROM AN ACCELERATED CHARGE

All accelerated charges produce em waves

We have been considering a special kind of electromagnetic wave source. Actually, *all* accelerated charges radiate electromagnetic waves, regardless of the manner in which the acceleration occurs.

Electric field of an accelerated charge

An intuitive picture of how an electromagnetic pulse is produced when a charge is accelerated can be put together in the following way. We start with a charge $+Q$ at rest at some point A at the time $t = 0$. The electric field of the charge can be represented by lines of force like those shown in Fig. 23–4(a). The charge then undergoes an acceleration **a** for the brief interval Δt in the upward direction, and its final velocity is **v** at the point B. The charge continues to move upward at the constant velocity **v** and at the later time t it is at the point C. The electric field around the charge at this time is shown in Fig. 23–4(b). The electric field of a charge moving at constant velocity is the same as that of a stationary charge when v is much smaller than c, the speed of light, which we shall assume is the case here.

Disturbances in electric and magnetic fields are propagated with the speed of light c. Hence the electric field beyond a sphere of radius ct centered at A cannot have been affected by the charge's motion: there is no way any event involving the charge that occurs after $t = 0$ can influence the outside world farther than the distance ct away. The electric field past this sphere is therefore exactly the same as in Fig. 23–4(a). Inside the sphere the electric field has another radial pattern of lines of force centered at C except for a transition region $c\,\Delta t$ wide between the two radial fields. The effect of the charge's acceleration has thus been to put kinks in the lines of force of its electric field.

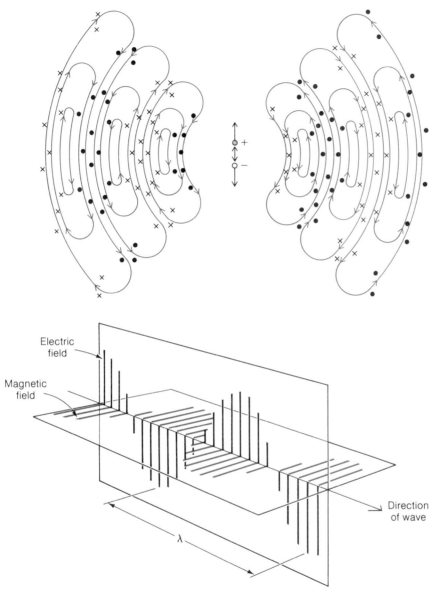

FIG. 23–2 The configuration of electric and magnetic fields that spread outward from a pair of oscillating charges.

FIG. 23–3 The electric and magnetic fields in an electromagnetic wave vary simultaneously. The field directions are perpendicular to each other and to the direction of propagation.

Electric field

Magnetic field

Direction of wave

λ

More precisely, the electric field in the transition region, which corresponds to the period of acceleration, has a transverse as well as a radial component.

The moving charge gives rise to a magnetic field as well as to an electric one, and a similar analysis shows that kinks also appear in the magnetic lines of force during the charge's acceleration. Accelerating the charge has given rise to an electromagnetic pulse that consists of mutually perpendicular electric and magnetic fields that travel at the speed c. This pulse has a life of its own, so to speak, since no subsequent motion of the charge can affect it. The pulse exactly fits the specification of an electromagnetic wave.

The acceleration produces kinks in the electric and magnetic fields of the charge

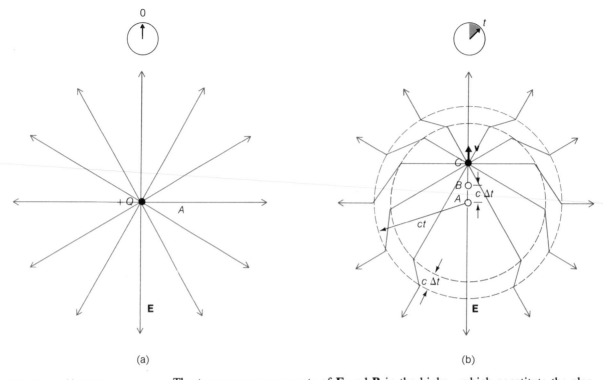

(a)   (b)

FIG. 23–4 How an accelerated charge gives rise to an electromagnetic pulse. (a) Electric field of a charge $+Q$ at rest. (b) Between A and B the charge is accelerated for the time Δt during which its electric field develops kinks. The transverse components of these kinks constitute the electric part of the electromagnetic pulse that is produced by the charge's acceleration and are accompanied by similar transverse magnetic field components.

The transverse components of **E** and **B** in the kinks—which constitute the electromagnetic pulse—have the important property that their magnitudes fall off with distance r as $1/r$. This is in constast to the **E** and **B** fields around a charge whose velocity is constant, which fall off as $1/r^2$, a much more rapid variation with distance. A charge is severely limited in the range of its interaction with other charges by the inverse-square decrease in its normal **E** and **B** fields, but no such limitation applies to the electromagnetic waves produced when it is accelerated. For this reason electromagnetic signals may be detected for quite remarkable distances from their sources—light and radio waves reach us from galaxies at the outer limit of the universe—even though static electric and magnetic fields produced by these sources would be imperceptible not very far away.

Maxwell's theory of electromagnetic waves showed that their speed c in free space depends solely upon ε_0 and μ_0, the permittivity and permeability of free space. Maxwell found that the speed c is given by

$$c = \frac{1}{\sqrt{\varepsilon_0 \mu_0}}$$

Speed of light

$$= \frac{1}{\sqrt{8.85 \times 10^{-12}\,C^2/N{\cdot}m^2 \times 1.26 \times 10^{-6}\,T{\cdot}m/A}}$$

$$= 3.00 \times 10^8\,\text{m/s} \qquad\qquad \textit{Speed of light} \quad (23–1)$$

which is the same speed that had been experimentally measured for light waves in free space! The correspondence was too great to be accidental, and, as further evidence became known, the electromagnetic nature of light found universal acceptance. To five significant figures the value of c is 2.9979×10^8 m/s.

23–3 TYPES OF ELECTROMAGNETIC WAVES

Light is not the only example of an electromagnetic wave. Although all electromagnetic waves share certain basic properties, other features of their behavior depend upon their frequencies. Light waves themselves span a brief frequency interval, from about 4.3×10^{14} Hz for red light to about 7.5×10^{14} Hz for violet light. Electromagnetic waves with frequencies between these limits are the only ones that the eye responds to, and specialized instruments of various kinds are required to detect waves with higher and lower frequencies. Figure 23–5 shows the electromagnetic wave *spectrum* from the

There are many kinds of em waves but all have the same basic nature

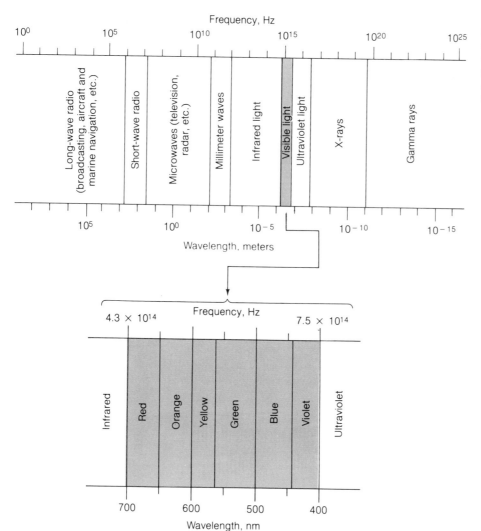

FIG. 23–5 The electromagnetic wave spectrum. The boundaries of the various categories are not sharp. (1 nm = 1 nanometer = 10^{-9} m.)

low frequencies used in radio communication to the high frequencies found in X rays and gamma rays (which are considered in later chapters). The wavelengths corresponding to the various frequencies are also shown.

As we recall, the product of the frequency f of a wave and its wavelength λ is just the wave speed, here c. Hence, given the wavelength or frequency of a particular electromagnetic wave, we can immediately find the complementary quantity.

Example Find the wavelengths of yellow light whose frequency is 5×10^{14} Hz and of radio waves whose frequency is 1 MHz. (1 MHz = 1 megahertz = 10^{6} Hz.)

Solution The wavelength of the yellow light is

$$\lambda = \frac{c}{f} = \frac{3 \times 10^{8}\,\mathrm{m/s}}{5 \times 10^{14}\,\mathrm{Hz}} = 6 \times 10^{-7}\,\mathrm{m} = 600\,\mathrm{nm}$$

which is less than 1/1000 of a millimeter. (1 nm = 1 nanometer = 10^{-9} m. The nanometer is often used for expressing the wavelengths found in light.) By the same procedure we find that a 1-MHz radio wave has a wavelength of 300 m. ■

Electromagnetic waves provide a means for transmitting information from one place to another without wires or other material links between them. There are two principal methods of incorporating information in an electromagnetic wave, *amplitude modulation* and *frequency modulation*.

Amplitude modulation

In amplitude modulation, a "carrier wave" of constant frequency is varied in amplitude, with the variations constituting the signal. The simplest example is a flashlight switched on and off to give a series of dots and dashes. Another example is the radiotelegraph. Coded sequences of dots and dashes are used to represent letters of the alphabet, numbers, or other specific information; the Morse code is an example (Fig. 23–6).

A carrier wave can also be modulated in such a way that the amplitude variations correspond to sound waves, as shown schematically in Fig. 23–7. A microphone converts sound waves to an equivalent electric signal, which is then amplified and combined with the carrier. The final wave is broadcast from an antenna, and at the receiving station the carrier is removed to leave an audio signal. The latter is then amplified and used to generate sound waves in a loudspeaker.

Frequency modulation

One difficulty with amplitude modulation is that such sources of random electromagnetic waves as electric storms and electric machinery can interfere with the broadcast waves to cause "static." In frequency modulation the frequency, not the amplitude, of the carrier is varied in accordance with the audio signal (Fig. 23–8). The information

FIG. 23–6 In radiotelegraphy, a constant-frequency electromagnetic wave is switched on and off in a coded sequence to transmit information. The letter "v," which is $\cdots$ — in Morse code, is shown.

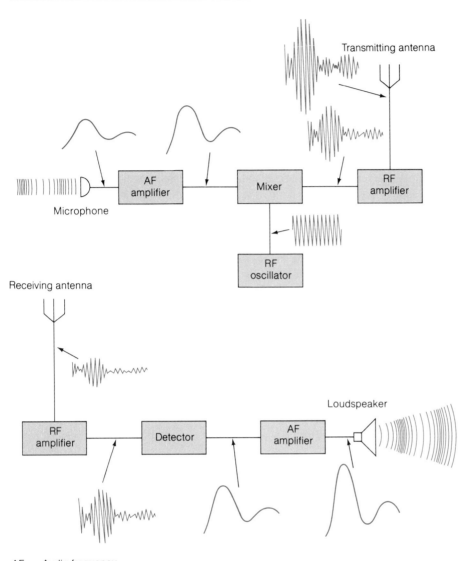

FIG. 23-7 In amplitude modulation, the amplitude of a constant-frequency carrier wave is varied in accordance with an audio signal. The transmitting and receiving systems shown are highly simplified.

AF = Audio frequency
RF = Radio frequency

content of a frequency-modulated electromagnetic wave is virtually immune to disturbance.

In television the desired scene is focussed on the light-sensitive screen of a special tube which is then scanned in a zig-zag fashion by an electron beam. The signal extracted from the screen varies with the image brightness at each successive point that is scanned, and this signal is then used to modulate a carrier wave for broadcasting. At the receiver, suitable circuits demodulate the wave, and in the picture tube the image is reproduced with the help of an electron beam that moves in synchronism with the electron beam in the camera tube.

Radio waves in the short-wave band of Fig. 23–5 have the useful property of being reflected from the earth's ionosphere (Sec. 16–6), a region high in the atmosphere

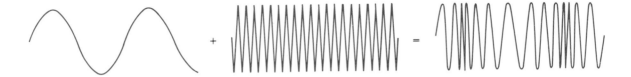

FIG. 23–8 The information content of a frequency-modulated wave resides in its frequency variations rather than in its amplitude variations.

Reflection from the ionosphere permits long-range radio communication

Radar uses a beam of short-wavelength radio waves to detect targets

where ions are relatively abundant. Without the ionosphere, radio communication would be limited to short distances since electromagnetic waves travel in straight lines and would be shielded from remote receivers by the curvature of the earth. However, since radio waves of appropriate frequencies can bounce one or more times between the ionosphere and the earth's surface, long-range transmission is possible, even to the opposite side of the earth (Fig. 23–9). Because the ion content of the ionosphere and the heights of the individual ion layers vary with time of day and are influenced by streams of fast protons and electrons emitted by the sun during periods of sunspot activity, there are no absolute rules for which frequencies are best for which distances. Generally speaking, though, the higher the frequency, the greater the range. Thus a frequency of 10 MHz would ordinarily give good results between 300 and 2400 km whereas 20 MHz would be better for 2400 to 11,000 km.

The higher-frequency waves used for FM and television broadcasting are not reflected by the ionosphere, so their reception is limited to the line of sight between transmitting and receiving antennas. When the frequencies are up around 10^{10} Hz, corresponding to wavelengths in the centimeter range, the waves can readily be focussed into narrow beams. Such beams are reflected by objects such as ships and airplanes, which is the basis of *radar*. A rotating antenna is used to send out a pulsed beam, and the distance of a particular target is established by the time needed for the echo to return to the antenna.

23–4   HUYGENS' PRINCIPLE

Most optical phenomena can be understood on the basis of the wave nature of light alone, without any direct reference to its electromagnetic character. As mentioned

FIG. 23–9 The ionosphere is a region in the upper atmosphere whose ionized layers make possible long-range radio communication by their ability to reflect radio waves.

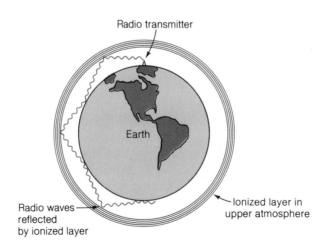

Radio transmitter

Earth

Radio waves reflected by ionized layer

Ionized layer in upper atmosphere

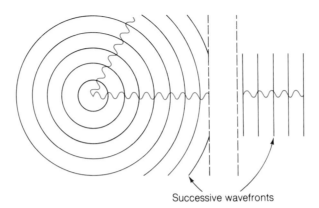

FIG. 23–10 The spherical wavefronts of a point source become plane wavefronts at a long distance from the source.

Successive wavefronts

earlier, three centuries ago Huygens devised an approach to the behavior of waves of all kinds that is particularly convenient to use. His method concerns *wavefronts*. A wavefront is an imaginary surface that joins points where all of the waves involved are in the same phase of oscillation. As in Fig. 23–10, waves from a point source spread out in a succession of spherical wavefronts. (In the case of water waves, the wavefronts that result when a stone is dropped in a lake are circular.) At a long distance from a point source, the curvature of the wavefronts is so small that they can be considered as a succession of planes.

Wavefronts

Huygens' principle states that

Every point on a wavefront can be considered as a point source of secondary wavelets which spread out in all directions with the wave speed of the medium. The wavefront at any time is the envelope of these wavelets.

Huygens' principle

Figure 23–11 shows how Huygens' principle is applied to the propagation of plane and

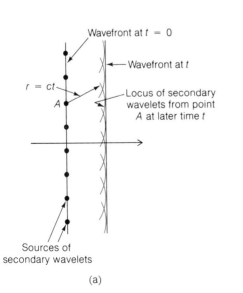

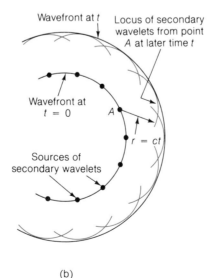

FIG. 23–11 Huygens' principle applied to the propagation of (a) plane and (b) spherical wavefronts.

(a)

(b)

spherical wavefronts in a uniform medium. It does not matter just where on the initial wavefront we imagine the sources of the secondary wavelets to be.

Despite its name, Huygens' principle is not in the same category as such fundamental principles as those of conservation of mass, energy, momentum, and electric charge, but is rather a convenient means for studying wave motion in a geometrical manner. We shall find Huygens' principle useful in understanding a variety of optical phenomena.

Rays are a convenient way to represent paths taken by light waves

Although we shall use the notion of wavefronts for the actual analysis of wave propagation, the results are commonly represented in terms of *rays*. A light ray is simply an imaginary line in the direction in which the wavefronts advance, and so it is perpendicular to the wavefronts. The picture most of us have of a light ray is a narrow pencil of light, which is perfectly legitimate. However, an approach based exclusively on rays does not reveal such characteristic wave behavior as diffraction (the bending of waves around an obstacle into the "shadow" region); hence we must remember that the motion of wavefronts is what is really significant, although it is both proper and convenient to use rays to summarize our conclusions.

23–5 REFLECTION

Diffuse and specular reflection

All real objects reflect a certain proportion of the light falling upon them, and it is this reflected light that enables us to see them. In most cases the surface of the object has irregularities that spread out an initially parallel beam of light in all directions to produce *diffuse reflection* (Fig. 23–12(a)). A surface so smooth that any irregularities in it are small relative to the wavelength of the light falling upon it behaves differently; when a parallel beam of light is directed at such a surface, it is *specularly reflected* in only one direction (Fig. 23–12(b)). Reflection from a brick wall is diffuse whereas reflection from a mirror is specular. Often a mixture of both kinds of reflection occurs, as in the case of a surface coated with varnish or glossy enamel. Our concern here is with specular reflection only.

When we see an object, what enters our eyes are light waves reflected from its surface. What we perceive as the object's color therefore depends upon two things: the kind of light falling on it, and the nature of its surface. If white light is used to illuminate an object that absorbs all colors other than red, the object will appear red. If green light is used instead, the object will appear black because it absorbs green light. A white object reflects light of all wavelengths equally well, and its apparent color is the same

Appearance of a colored object

FIG. 23–12

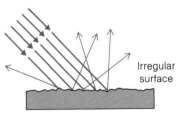

(a) Diffuse reflection

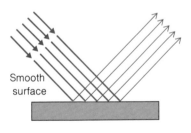

(b) Specular reflection

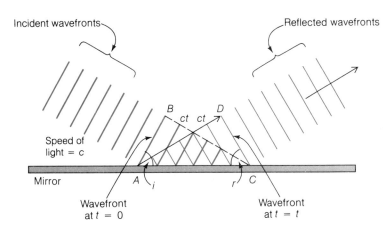

Incident wavefronts

Reflected wavefronts

Speed of light = c

Mirror

B
ct ct
D

A i

r C

Wavefront at $t = 0$

Wavefront at $t = t$

FIG. 23–13 The behavior of successive wavefronts during reflection from a plane mirror. The angle of incidence i and the angle of reflection r are equal.

as the color of the light reaching it. A black object, on the other hand, absorbs light of all wavelengths, and it appears black no matter what color light reaches it.

Let us examine the specular reflection of a series of plane wavefronts, which corresponds to a parallel beam of light. At $t = 0$ in Fig. 23–13 the wavefront AB just touches the mirror, and secondary wavelets start to spread out from A. After a time t the end B of the wavefront reaches the mirror at C. The secondary wavelets from A are now at the point D. The wavefront which was AB at $t = 0$ is therefore CD at the later time t.

The angle i between an approaching wavefront and the reflecting surface is called the *angle of incidence,* and the angle r between a receding wavefront and the reflecting surface is called the *angle of reflection.* Here

Angles of incidence and reflection

$$\sin i = \frac{BC}{AC} \quad \text{and} \quad \sin r = \frac{AD}{AC}$$

The wavelet that originates at A takes the time t to reach D. Hence $AD = ct$, where c is the velocity of light. The point B of the wavefront AB also takes the time t to reach C, and so $BC = ct$ as well. Therefore $AD = BC$ and $\sin i = \sin r$, from which we conclude that the angles i and r are equal. Thus we have the basic law of reflection:

The angle of reflection of a plane wavefront with a plane mirror is equal to the angle of incidence.

Law of reflection

In the ray model of light propagation, the angles of incidence and reflection are measured with respect to the *normal* to the reflecting surface at the point where the light strikes it (Fig. 23–14). The normal is a line drawn perpendicular to the surface at that point. In this representation, too, the angles of incidence and reflection are equal. The incident ray, the reflected ray, and the normal all lie in the same plane.

The normal to a surface is a line perpendicular to it

The image of an object we see in a plane mirror appears to be the same size as the object and as far behind the mirror as the object is in front of it. Let us investigate how this situation arises.

In Fig. 23–15 three typical light rays from a point object at A impinge on a mirror and are reflected, in each case with an angle of reflection equal to the angle of incidence. To the eye, the three diverging rays apparently come from the point A' behind the

FIG. 23–14 Ray representation of reflection. The angles of incidence and reflection are measured with respect to the normal to the reflecting surface.

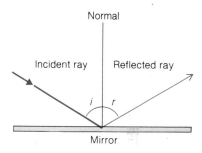

FIG. 23–15 Rays from a point object at *A* appear to come from *A'* after reflection from a plane mirror. The image is as far behind the mirror as the object is in front of it.

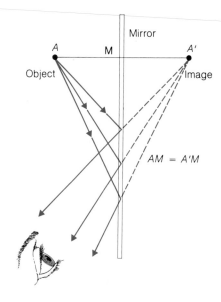

Real and virtual images

mirror. The rays that seem to come from the image do not actually pass through it, and for this reason the image is said to be *virtual*. (A *real image* is formed by light rays that pass through it.) From the geometry of the figure it is clear that $AM = A'M$, so that the object and the image are the same distance on either side of the mirror.

A plane mirror reverses left and right

In Fig. 23–16, an actual object is being reflected in a plane mirror. Again simple geometry shows that the object and its virtual image have the same dimensions. Every point in the image is directly behind the corresponding point on the object, and so the orientation of the image is the same as that of the object; the image is therefore *erect*. However, left and right are interchanged, as in Fig. 23–17: a "mirror image" of anything is the same size and shape as the original but its transverse features are reversed. A printed page appears backward in a mirror, and what seems to be one's left hand in a mirror is actually one's right hand.

23–6   REFRACTION

It is a matter of experience that a beam of light passing obliquely from one medium to another, say from air to water, is deflected at the surface between the two media. The

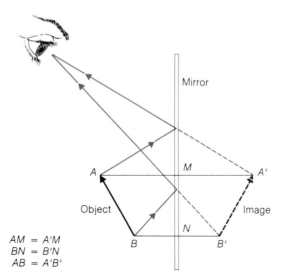

FIG. 23–16 Image formation by a plane mirror. The image is erect and is the same size and shape as the object.

AM = A'M
BN = B'N
AB = A'B'

bending of a light beam when it passes from one medium to another is called *refraction,* and it is responsible for such familiar phenomena as the apparent distortion of objects partially submerged in water.

Refraction refers to the bending of a light ray when it goes from one medium to another

Several important aspects of refraction are illustrated in Fig. 23–18. When a light beam goes from air into water along the normal to the surface between them, it simply continues along the same path, but when it enters the water at any other angle, it is bent toward the normal. The paths are reversible; thus a light beam emerging from the water is bent *away* from the normal as it enters the air.

Refraction occurs because light travels at different speeds in the two media. Let us see what happens to the plane wavefront AB in Fig. 23–19 when it passes obliquely from a medium in which its speed is v_1 to a medium in which its speed is v_2, where v_2 is less than v_1. At $t = 0$ the wavefront AB just comes in contact with the interface between the two media. After a time t, the end B of the wavefront reaches the interface at C, and the secondary wavelets from A are now at D. Since v_2 is less than v_1, the secondary wavelets generated at the interface travel a shorter distance in the same time interval than do wavelets in the first medium, and the distance AD is shorter than BC. The refracted wavefronts accordingly move in a different direction from that of the incident wavefronts.

The angle i between an approaching wavefront and the interface between two media is called the *angle of incidence* of the wavefront, and the angle r between a receding wavefront and the interface it has passed through is called the *angle of refraction.* From Fig. 23–19 we see that

$$\sin i = \frac{BC}{AC} \quad \text{and} \quad \sin r = \frac{AD}{AC}$$

Because $BC = v_1 t$ and $AD = v_2 t$,

$$\frac{\sin i}{\sin r} = \frac{BC}{AD} = \frac{v_1 t}{v_2 t}$$

FIG. 23–17 Left and right are interchanged in reflection.

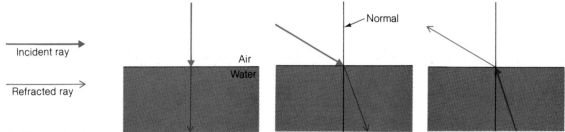

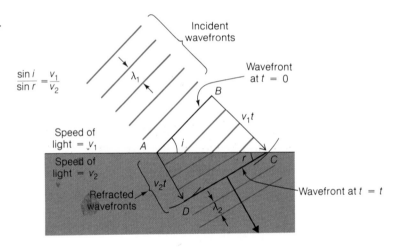

FIG. 23–18 The bending of a light beam when it passes obliquely from one medium to another is called *refraction*.

FIG. 23–19 The behavior of successive wavefronts during refraction. The speed of light v_1 in the first medium is greater than that in the second medium.

Snell's law relates angles of incidence and refraction

and so we have

$$\frac{\sin i}{\sin r} = \frac{v_1}{v_2} \tag{23-2}$$

This useful result is known as *Snell's law* after its discoverer, the seventeenth-century Dutch astronomer Willebrord Snell. It states that

The ratio of the sines of the angles of incidence and refraction is equal to the ratio of the speeds of light in the two media.

Refraction, like reflection, can be described in terms of the ray model of light. As in Fig. 23–20, the angles i and r are taken with respect to the normal to the interface at the point where the rays meet it.

Refraction is not confined to light waves. A conspicuous example of refraction can be observed in water waves that approach a sloping beach. Regardless of the direction

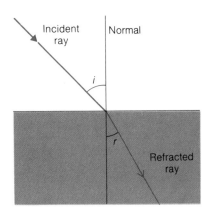

FIG. 23–20 Ray representation of refraction.

of the waves in open water, their direction becomes more and more perpendicular to the shoreline as they come nearer (Fig. 23–21). The reason is that the speed of a water wave decreases in shallow water because of friction with the bottom, so a wavefront moving obliquely shoreward is progressively affected by the change in speed and swings around until it is parallel to the shore.

23–7 INDEX OF REFRACTION

The ratio between the speed of light c in free space and its speed v in a particular medium is called the *index of refraction* of the medium. The greater the index of

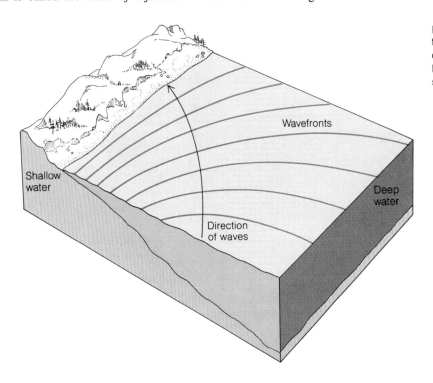

FIG. 23–21 Water waves that approach the shore obliquely are refracted because they move more slowly in shallow water.

FIG. 23–22 The greater the index of refraction *n* of a medium, the greater the deflection of a light beam on entering or leaving it.

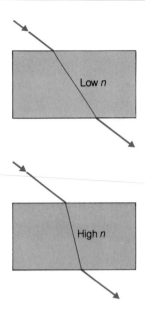

The greater the index of refraction of a medium, the slower light travels in it

refraction, the greater the extent to which a light beam is deflected upon entering or leaving the medium (Fig. 23–22). The symbol for index of refraction is *n,* so that

$$n = \frac{c}{v}$$ *Index of refraction*   (23–3)

Table 23–1 is a list of the values of *n* for a number of substances.

It is easy to rewrite Snell's law in terms of the indexes of refraction n_1 and n_2 of two successive media. In these media light has the respective speeds

$$v_1 = \frac{c}{n_1} \quad \text{and} \quad v_2 = \frac{c}{n_2}$$

and so Snell's law becomes

$$\frac{\sin i}{\sin r} = \frac{v_1}{v_2} = \frac{c/n_1}{c/n_2} = \frac{n_2}{n_1}$$

This is usually written in the form

Alternate form of Snell's law

$$n_1 \sin i = n_2 \sin r$$ *Snell's law*   (23–4)

TABLE 23–1
Indexes of refraction

Substance	n	Substance	n
Air	1.0003	Glass, flint	1.63
Benzene	1.50	Glycerin	1.47
Carbon disulfide	1.63	Ice	1.31
Diamond	2.42	Quartz	1.46
Ethyl alcohol	1.36	Water	1.33
Glass, crown	1.52	Zircon	1.92

FIG. 23–23

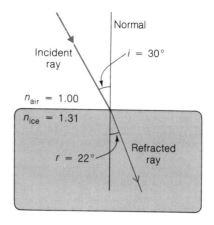

Normal

Incident ray

$i = 30°$

$n_{air} = 1.00$

$n_{ice} = 1.31$

$r = 22°$

Refracted ray

Example A beam of parallel light enters a block of ice at an angle of incidence of 30° (Fig. 23–23). What is the angle of refraction in the ice?

Solution The indexes of refraction of air and ice are respectively 1.00 and 1.31. From Snell's law,

$$\sin r = \frac{n_1}{n_2}\sin i = \frac{1.00}{1.31}\sin 30° = 0.382$$

$$r = 22°$$

Example A ray of light is incident at an angle of 45° on one side of a glass plate of index of refraction 1.6. Find the angle at which the ray emerges from the other side of the plate.

Solution The geometry of this problem is shown in Fig. 23–24. The angle of refraction r_1 at the first side of the plate is found as follows:

$$\sin r_1 = \frac{n_{air}}{n_{glass}}\sin i_1 = \frac{1.0}{1.6}\sin 45° = 0.442$$

$$r_1 = 26°$$

Because the sides of the plate are parallel, the angle of incidence i_2 at the second side

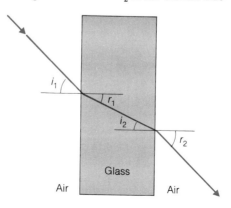

FIG. 23–24 Light entering a glass plate with parallel sides emerges parallel to its original direction but displaced to one side.

i_1

r_1

i_2

r_2

Glass

Air

Air

FIG. 23–25 Dispersion by a prism.

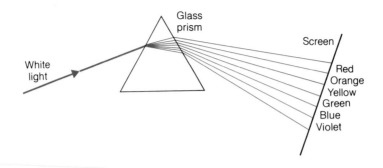

is equal to r_1. If we let r_2 be the angle of refraction at which the ray emerges from the plate,

$$\sin r_2 = \frac{n_{\text{glass}}}{n_{\text{air}}} \sin i_2 = \frac{1.6}{1.0} \sin 26° = 0.701$$

$$r_2 = 45°$$

The ray leaves the glass plate parallel to its original direction but displaced to one side. This is a general result for parallel-sided plates that holds for all angles of incidence and all indexes of refraction; of course, when $i_1 = 0$, the ray is not displaced. ∎

Dispersion by a prism separates white light into its component colors

The index of refraction of a medium depends to some extent upon the frequency of the light involved, with the highest frequencies having the highest values of n. In ordinary glass the index of refraction for violet light is about one percent greater than that for red light, for example. Since a different index of refraction means a different degree of deflection when a light beam enters or leaves a medium, a beam containing more than one frequency is split into a corresponding number of different beams when it is refracted. This effect, called *dispersion,* is illustrated in Fig. 23–25, which shows the result of directing a narrow pencil of white light at one face of a glass prism. The initial beam separates into beams of various colors, from which we conclude that white light is actually a mixture of light of these different colors. The band of colors that emerges from the prism is known as a *spectrum.*

Origin of the rainbow

Dispersion in water droplets is responsible for rainbows, which are seen when the sun is behind an observer who is facing falling rain. Figure 23–26 shows what happens when a ray of sunlight enters a raindrop, where it is first refracted, then reflected at the far surface, and finally refracted again when it emerges. Dispersion occurs at each refraction, with the result that the angles between the incoming sunlight and the violet and red ends of the spectrum are respectively 40° and 42°. Thus the light that reaches the observer comes from a ring in the sky between 40° and 42° from the direction of the sunlight, and in this ring red light is on the outside because, being deviated most, it originates in droplets farther away. Between the red outside of the rainbow and the violet inside appear all the other colors of the spectrum. A person in an airplane can see the entire ring, but from the ground only the upper part appears to give the familiar arc-shaped spectrum. Light arriving at raindrops at a larger angle than that shown in Fig. 23–26 can undergo two reflections inside each drop before emerging, which leads to an outer secondary rainbow whose colors are in the reverse order. Because some

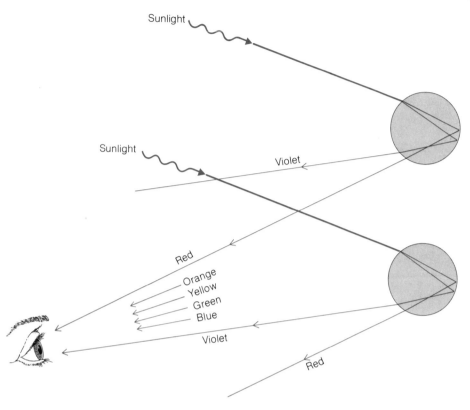

FIG. 23–26 Rainbows originate in the dispersion of sunlight by raindrops. Shown here are a drop responsible for the red end of the spectrum and one responsible for the violet end; the myriad other drops present yield the intermediate colors and produce a continuous arc in the sky.

light is refracted out of a water drop at each reflection, the double reflection makes the secondary rainbow much fainter than the primary one.

23–8 APPARENT DEPTH

A familiar example of refraction is the apparent reduction in the depth of an object submerged in water or other transparent liquid. Thus an oar dipped in a lake seems bent upward where it enters the water because its submerged portion appears to be closer to the surface than it actually is. Figure 23–27 shows how this effect comes about. Light leaving the tip of the oar at B is bent away from the normal upon entering the air. To an observer above, who instinctively interprets what he sees in terms of the straightline propagation of light, the tip of the oar is at C, and the submerged part of the oar seems to be AC and not AB.

It is not difficult to relate the apparent and actual depths of a submerged object. Figure 23–28 shows a fish at the point F, an actual depth h below the surface of a body of water, whereas to an observer in the air the fish is at F', only h' below the surface. If we restrict ourselves to rays that are nearly vertical,

Submerged objects seem closer to the surface than they actually are

$$\frac{\sin i}{\sin r} \approx \frac{\tan i}{\tan r} \tag{23-5}$$

since $\sin \theta \approx \tan \theta$ when θ is small. From Fig. 23–28 we see that

FIG. 23–27 An oar appears bent when partly immersed in water because of refraction.

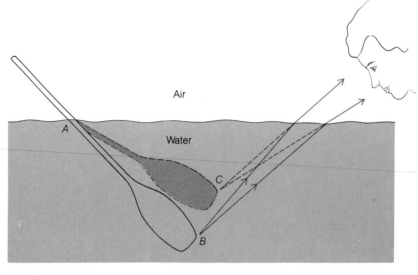

$$\tan i = \frac{x}{h} \quad \text{and} \quad \tan r = \frac{x}{h'}$$

where x is the horizontal distance between the position of the fish and the point O where the ray under consideration leaves the water. Hence

$$\frac{\tan i}{\tan r} = \frac{x/h}{x/h'} = \frac{h'}{h}$$

and so, from Eq. (23–5) and Snell's law,

$$\frac{h'}{h} = \frac{n_2}{n_1} \qquad\qquad\qquad \textit{Apparent depth} \quad (23\text{–}6)$$

Example To a sailor standing on the deck of a yacht, the water depth appears to be 3.0 m. If this estimate is correct in terms of what he sees, what is the true depth?

Solution Here, with $n_1 = 1.33$ and $n_2 = 1.00$ (the indexes of refraction of water and air respectively),

$$\frac{h'}{h} = \frac{1.00}{1.33} = 0.752$$

The apparent depth h' is only about three-quarters of the true depth h. Hence

$$h = \frac{h'}{0.752} = \frac{3.0\,\text{m}}{0.752} = 4.0\,\text{m}$$

23–9 TOTAL INTERNAL REFLECTION

An interesting phenomenon known as *total internal reflection* can occur when light passes from one medium to another which has a *lower* index of refraction, for instance

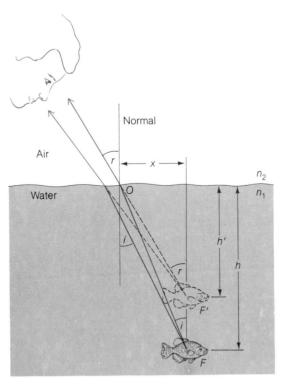

FIG. 23–28 Submerged bodies seem closer to the surface than they actually are.

from water or glass to air. In this case the angle of refraction is greater than the angle of incidence, and a light ray is bent *away* from the normal (ray 2 in Fig. 23–29). As the angle of incidence is increased, a certain *critical angle* i_c is reached for which the angle of refraction is 90°. The "refracted" ray now travels along the interface between the two media and cannot escape (ray 3). A ray approaching the boundary at an angle of incidence exceeding the critical angle is reflected back into the medium it comes from, with the angle of reflection being equal to the angle of incidence as in any other instance of reflection (ray 4).

To find the value of the critical angle, we set $i = i_c$ and $r = 90°$ in Snell's law, and we obtain

Past the critical angle, light is reflected rather than refracted

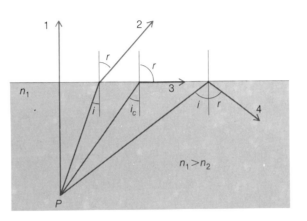

FIG. 23–29 Total internal reflection occurs when the angle of refraction of a light ray going from one medium to another of lower index of refraction equals or exceeds 90°.

FIG. 23–30   An underwater observer sees a circle of light at the surface. All the light reaching him or her is concentrated in a cone 98° wide.

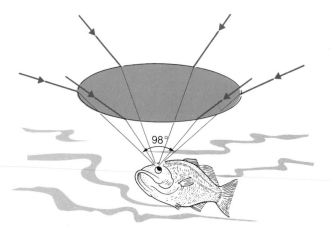

$$n_1 \sin i_c = n_2 \sin 90°$$

$$\sin i_c = \frac{n_2}{n_1} \qquad\qquad \textit{Critical angle}\quad (23\text{–}7)$$

For a ray going from water to air,

$$\sin i_c = \frac{n_{\text{air}}}{n_{\text{water}}} = \frac{1.00}{1.33} = 0.752$$

$$i_c = 49°$$

What a fish sees

What does a person (or a fish) who is underwater see when looking upward? By reversing all the rays of light in Fig. 23–29 (the paths taken by light rays are always reversible), it is clear that light from everywhere above the water's surface reaches the eyes through a circle on the surface. The rays are all concentrated in a cone whose angular width is twice i_c, which is 98°. Outside this cone is darkness (Fig. 23–30).

The sharpness and brightness of a light beam are better preserved by total internal reflection than by reflection from an ordinary mirror, and optical instruments accordingly utilize the former in preference to the latter whenever light is to be changed in direction. Figure 23–31 shows three types of totally reflecting prisms in common use.

How to pipe light

Total internal reflection makes it possible to "pipe" light from one place to another with a rod of glass or transparent plastic. As in Fig. 23–32, successive internal reflections occur at the surface of the rod, and nearly all the light entering at one end emerges at the other. If a cluster of narrow glass fibers is used instead of a single thick rod, an image can be transferred from one end to the other since each fiber carries intact a part

FIG. 23–31   Three applications of total internal reflection.

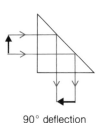

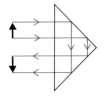

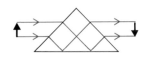

90° deflection   180° deflection   Image inversion

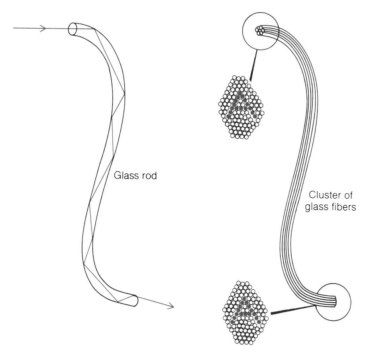

FIG. 23–32 Light can be "piped" from one place to another by means of successive internal reflections in a glass rod.

Glass rod

Cluster of glass fibers

of the image. Because a tube of glass fibers is flexible, it can be used for such purposes as examining a person's stomach by passing the tube in through the mouth. Some of the fibers are used to provide light for illumination, and the others carry the reflected light back outside for viewing.

23–10 ILLUMINATION

Visible light represents only part of the electromagnetic radiation emitted by most light sources, perhaps 10 percent in the case of an ordinary light bulb. Also, the eye does not respond with equal sensitivity to light of different colors; the sensitivity is a maximum for green light and it decreases toward both ends of the visible spectrum (see Sec. 24–6). Simply stating the power output of a light source in watts therefore does not convey much information about the illumination it actually provides, and a separate system of units has been devised for this purpose that is more closely related to the visual respose of the human eye.

Eye sensitivity varies with the color of the light

The brightness of a light source is referred to as its *luminous intensity I,* for which the *candela* (cd) is the SI unit. The candela is defined in terms of the light emitted by a small pool of platinum at its melting point. A candle has a luminous intensity of about 1 cd, and in fact the former standard of this quantity was an actual candle of specified composition and dimensions.

Luminous intensity

The *luminous flux F* emitted by a light source describes the total amount of visible light it gives off; a small bright source may give off less light than a large dim source, just as a small hot object may give off less heat than a large warm one. The unit of luminous flux is the *lumen* (lm), which equals the luminous flux that falls on each

Luminous flux

FIG. 23–33 How the
luminous flux emitted by
an isotropic light source
and the illumination it
produces are related to
the intensity of the source.

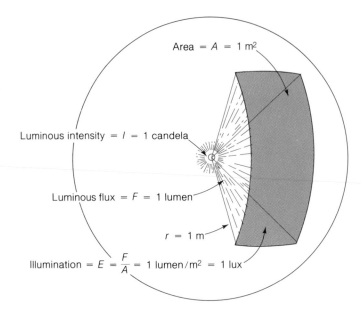

FIG. 23–33 How the luminous flux emitted by an isotropic light source and the illumination it produces are related to the intensity of the source.

square meter of a sphere 1 m in radius at whose center is a 1-candela light source that radiates equally well in all directions (Fig. 23–33). Such a source is said to be *isotropic*. A lumen of flux is equivalent to about 0.0015 W of light of wavelength 555 nm, which corresponds to the yellow-green light the eye is most sensitive to. Since the area of a sphere of radius r is $4\pi r^2$, the above sphere has a total area of 4π m^2, and the total luminous flux radiated by a 1-cd source is thus 4π lm. The luminous flux F radiated by an isotropic source whose intensity is I is accordingly given by

$$F = 4\pi I \tag{23–8}$$

Luminous flux $= 4\pi \times$ luminous intensity

Luminous efficiency   The luminous flux emitted per watt of power input by a light source is its *luminous efficiency*. Ordinary tungsten-filament lamps increase in efficiency with their power, because the higher the power, the higher the filament temperature and the greater the proportion of visible light in the total radiation. A 10-W lamp, for instance, has a luminous efficiency of about 8 lm/W, whereas it is about 22 lm/W for a 1000-W lamp. Fluorescent lamps have higher efficiencies, from 50 to 90 lm/W.

Illumination   The *illumination E* of a surface is the luminous flux per unit area that reaches it:

$$E = \frac{F}{A} \tag{23–9}$$

$$\text{Illumination} = \frac{\text{luminous flux}}{\text{area}}$$

In the SI system the unit of illumination is the lumen per square meter, or *lux*; in the British system it is the lumen per square foot, or *footcandle*. About 500 luxes is a reasonable figure for general interior lighting, with twice that better for reading.

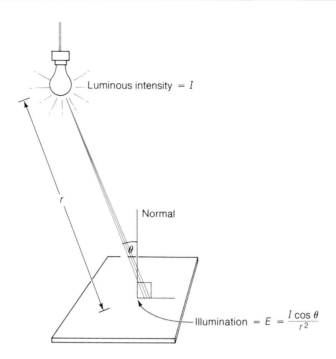

The illumination on a surface a distance r away from an isotropic light source of intensity I is given by

$$E = \frac{I \cos \theta}{r^2.} \qquad (23\text{–}10)$$

In this formula θ is the angle between the direction of the light and a normal to the surface where the light strikes it (Fig. 23–34). The illumination from an isotropic source varies as $1/r^2$, just as in the case of sound waves (Sec. 12–11). Doubling the distance from such a source reduces the illumination to $(\frac{1}{2})^2 = \frac{1}{4}$ of its original value. When the light arrives perpendicular to the surface, $\theta = 0$ and $\cos \theta = 1$, so $E = I/r^2$ in that case.

Example (a) A certain 60-W light bulb has a luminous intensity of 70 cd. A reflector concentrates all the light from the bulb into a circle whose area is 0.8 m² when the bulb is 1 m from a book that is being held perpendicular to the light rays. Find the illumination of the book. (b) If the lamp had no reflector and radiated isotropically, how far from the book would it have to be located to produce the same illumination?

Solution (a) The bulb emits a total luminous flux of

$$F = 4\pi I = 4\pi \times 70 \, \text{cd} = 880 \, \text{lm}$$

The illumination of the book is therefore

$$E = \frac{F}{A} = \frac{880 \, \text{lm}}{0.8 \, \text{m}^2} = 1100 \, \text{luxes}$$

(b) Here $\theta = 0$, $\cos \theta = 1$, and $E = I / r^2$. Hence

$$r = \sqrt{\frac{I}{E}} = \sqrt{\frac{70\,\text{cd}}{1100\,\text{luxes}}} = 0.25\,\text{m}$$

The reflection of light from the walls of the room is neglected here, but obviously it would contribute to the illumination of the book ∎

IMPORTANT TERMS

Maxwell's hypothesis is that a changing electric field produces a magnetic field.

Electromagnetic waves consist of coupled electric and magnetic field oscillations. Radio waves, microwaves, light waves, X rays, and gamma rays are all electromagnetic waves differing only in their frequency.

In **amplitude modulation,** information is contained in variations in the amplitude of a constant-frequency carrier wave. In **frequency modulation,** information is contained in variations in the frequency of the carrier wave.

A **wavefront** is an imaginary surface that joins points where all the waves from a source are in the same phase of oscillation. According to **Huygens' principle,** every point on a wavefront can be considered as a point source of secondary wavelets which spread out in all directions with the wave speed of the medium. The wavefront at any time is the envelope of these wavelets.

In **diffuse reflection** an incident beam of parallel light is spread out in many directions, while in **specular reflection** the angle of reflection is equal to the angle of incidence. A **mirror** is a specular reflecting surface of regular form that can produce an image of an object placed before it.

A **real image** of an object is formed by light rays that pass through the image; the image would therefore appear on a properly placed screen. A **virtual image** can only be seen by the eye because the light rays that seem to come from the image actually do not pass through it.

The bending of a light beam when passing from one medium to another is called **refraction.** The quantity that governs the degree to which a light beam will be deflected in entering a medium is its **index of refraction,** defined as the ratio between the speed of light in free space and its speed in the medium. **Dispersion** refers to the splitting up of a beam of light containing different frequencies by passage through a substance whose index of refraction varies with frequency.

Snell's law states that the ratio between the sine of the angle of incidence of a light ray upon an interface between two media and the sine of the angle of refraction is equal to the ratio of the speeds of light in the two media.

In **total internal reflection,** light arriving at a medium of lower index of refraction at an angle greater than the **critical angle** of incidence is reflected back into the medium it came from.

The **luminous intensity** of a light source refers to its brightness with respect to the visual response of the eye, and the **luminous flux** it emits is a measure of its light output. The **illumination** of a surface is equal to the flux reaching it per unit area.

IMPORTANT FORMULAS

Index of refraction: $\quad n = \dfrac{c}{v}$

Snell's law: $\quad n_1 \sin i = n_2 \sin r$

Apparent depth: $\quad \dfrac{h'}{h} = \dfrac{n_2}{n_1}$

Critical angle: $\quad \sin i_c = \dfrac{n_2}{n_1}$

Luminous flux: $\quad F = 4\pi I$

Illumination:

$$E = \frac{F}{A}$$

$$E = \frac{I \cos \theta}{r^2} \quad \text{(isotropic source)}$$

MULTIPLE CHOICE

1. According to Maxwell's hypothesis, a changing electric field gives rise to
(a) an electric current.
(b) an emf.
(c) a magnetic field.
(d) radiation pressure.

2. Electromagnetic waves transport
(a) wavelength.　　(b) frequency.
(c) charge.　　(d) energy.

3. Compared with the electric and magnetic fields of a charge moving at constant velocity, those of an electromagnetic wave decrease in magnitude with distance
(a) less rapidly.
(b) the same.
(c) more rapidly.
(d) any of the above, depending upon the medium.

4. The direction of the magnetic field in an electromagnetic wave is
(a) parallel to the electric field.
(b) perpendicular to the electric field.
(c) parallel to the direction of propagation.
(d) random.

5. Which of the following are *not* electromagnetic in nature?
(a) infrared rays   (b) ultraviolet rays
(c) radar waves   (d) sound waves

6. In a vacuum, the speed of an electromagnetic wave
(a) depends upon its frequency.
(b) depends upon its wavelength.
(c) depends upon its electric and magnetic fields.
(d) is a universal constant.

7. Reflection from a mirror is said to be
(a) specular.   (b) diffuse.
(c) real.   (d) virtual.

8. All real images
(a) are erect.
(b) are inverted.
(c) can appear on a screen.
(d) cannot appear on a screen.

9. When an object is reflected in a plane mirror, the image is always
(a) real.   (b) inverted.
(c) enlarged.   (d) left-right reversed.

10. When a beam of light enters one medium from another, a quantity that never changes is its
(a) direction.   (b) speed.
(c) frequency.   (d) wavelength.

11. The bending of a beam of light when it passes from one medium to another is known as
(a) refraction.   (b) reflection.
(c) diffraction.   (d) dispersion.

12. The index of refraction of a material medium
(a) is always less than 1.
(b) is always equal to 1.
(c) is always greater than 1.
(d) may be less than, equal to, or greater than 1.

13. Relative to the angle of incidence, the angle of refraction
(a) is smaller.
(b) is the same.
(c) is larger.
(d) may be any of the above.

14. A light ray enters one medium from another along the normal. The angle of refraction
(a) is 0.
(b) is 90°.
(c) equals the critical angle.
(d) depends upon the indexes of refraction of the two media.

15. Dispersion is the term used to describe
(a) the splitting of white light into its component colors in refraction.
(b) the propagation of light in straight lines.
(c) the bending of a beam of light when it goes from one medium to another.
(d) the bending of a beam of light when it strikes a mirror.

16. The index of refraction of benzene is 1.5. The speed of light in benzene is
(a) 1.5×10^8 m/s.   (b) 2×10^8 m/s.
(c) 3×10^8 m/s.   (d) 4.5×10^8 m/s.

17. According to Snell's law,
(a) $\sin i = \sin r$.   (b) $v_1 \sin i = v_2 \sin r$.
(c) $v_2 \sin i = v_1 \sin r$.   (d) $\sin i / \sin r = v_2 v_2$.

18. The depth of an object submerged in a transparent liquid
(a) always seems less than its actual depth.
(b) always seems more than its actual depth.
(c) may seem less or more than its actual depth, depending on the index of refraction of the liquid.
(d) may seem less or more than its actual depth, depending on the angle of view.

19. Total internal reflection can occur when light passes from one medium to another
(a) that has a lower index of refraction.
(b) that has a higher index of refraction.
(c) that has the same index of refraction.
(d) at less than the critical angle.

20. When a light ray approaches a glass-air interface from the glass side at the critical angle, the angle of refraction is
(a) 0.
(b) 45°.
(c) 90°.
(d) equal to the angle of incidence.

21. To double the illumination directly below it produced by an isotropic light bulb that is 1 m above a table, its height should be reduced to

(a) 0.75 m.

(b) 0.707 m.

(c) 0.5 m.

(d) 0.25 m.

EXERCISES

1. Why was electromagnetic induction discovered much earlier than its converse, the production of a magnetic field by a varying electric field?

2. Under what circumstances does a charge radiate electromagnetic waves?

3. Why are light waves able to travel through a vacuum whereas sound waves cannot?

4. Light is said to be a transverse wave phenomenon. What is it that varies at right angles to the direction in which a light wave travels?

5. In an electromagnetic wave, what is the relationship, if any, between the variations in the electric and magnetic fields?

6. A radio transmitter has a vertical antenna. Does it matter whether the receiving antenna is vertical or horizontal? Explain.

7. Do electromagnetic waves have the same speed in all transparent media? If not, how do their speeds in a transparent medium compare with their speed in free space?

8. What is a wavefront? What is the relationship between a light ray and the wavefronts whose motion it is used to describe?

9. What is the difference between a real image and a virtual image?

10. Under what circumstances do the incident ray, the reflected ray, and the normal all lie in the same plane?

11. Flint glass and carbon disulfide have almost the same index of refraction. How does this explain the fact that a flint-glass rod immersed in carbon disulfide is nearly invisible?

12. Why is a beam of white light not dispersed into its component colors when it passes perpendicularly through a pane of glass?

13. Explain why a cut diamond held in white light shows flashes of color. What would happen if it were held in red light?

14. The marine radiotelephone station on Tahiti transmits at the frequency 4390.2 kHz. What wavelength does this correspond to?

15. A nanosecond is 10^{-9} s. (a) What is the frequency of an electromagnetic wave whose period is 1 nanosecond? (b) What is its wavelength? (c) To what class of electromagnetic waves does it belong?

16. A radar sends out 0.05-μs pulses of microwaves whose wavelength is 2.5 cm. What is the frequency of these microwaves? How many waves does each pulse contain?

17. A plane mirror is mounted on the back of a truck which is traveling at 30 km/h. How fast does the image of a man standing in the road behind the truck seem to be moving away from him?

18. What is the height of the smallest mirror in which a woman 160 cm tall can see herself at full length? Does it matter how far from the mirror she stands?

19. Using the indexes of refraction given in Table 23–1 and taking the speed of light in free space as 3×10^8 m/s, find the speed of light in (a) air, (b) diamond, (c) crown glass, and (d) water.

20. What is the angle of refraction of a beam of light that enters the surface of a lake at an angle of incidence of 50°?

21. A flashlight is frozen into a block of ice. If its beam strikes the surface of the ice at an angle of incidence of 37°, what is the angle of refraction?

22. A beam of light enters a liquid of unknown composition at an angle of incidence of 30° and is deflected by 5° from its original path. Find the index of refraction of the liquid.

23. A beam of light strikes a pane of glass at an angle of incidence of 60°. If the angle of refraction is 35°, find the index of refraction of the glass.

24. The critical angle for total internal reflection in Lucite is 41°. Find its index of refraction.

25. Find the critical angle for total internal reflection in a diamond when the diamond is (a) in air, and (b) immersed in water.

PROBLEMS

1. The frequency of the light in a particular beam depends solely upon its source, whereas the wavelength depends upon the speed of light in the medium it travels through. Find the ratio between the wavelengths of a light beam that passes from one medium to another in terms of their indexes of refraction.

2. A ray of light passes through a plane boundary separating two media whose indexes of refraction are $n_1 = 1.5$ and $n_2 = 1.3$. (a) If the ray goes from medium 1 to medium 2 at an angle of incidence of 45°, what is the angle of refraction? (b) If the ray goes from medium 2 to medium 1 at the same angle of incidence, what is the angle of refraction?

3. A double-glazed window consists of two panes of glass separated by an air space for better insulation. (a) If the glass has an index of refraction of 1.5, find the angle at which a ray of sunlight enters a room if it arrives at the window at an angle of incidence of 70°. (b) The window leaks, and the space between the panes becomes filled with water. What is the angle at which the above ray enters the room now?

4. A ray of light strikes a glass plate at an angle of incidence of 55°. If the reflected and refracted rays are perpendicular to each other, find the index of refraction of the glass.

5. The index of refraction of ordinary crown glass for red light is 1.51 and for violet light is 1.53. A beam of white light falls on a cube of such glass at an incident angle of 40°. What is the difference between the angles of refraction of the red and the violet light?

6. A light bulb is on the bottom of a swimming pool 1.8 m below the water surface. A person on a diving board above the pool sees a circle of light. What is its diameter?

7. The olive in a martini cocktail ($n = 1.35$) appears to be 4 cm below the surface. What is the actual depth of the olive?

8. A glass paperweight in the form of a 6-cm cube is placed on a letter. How far underneath the top of the cube does the letter appear if the index of refraction of the glass is 1.5?

9. A barrel of ethyl alcohol is 0.80 m high. How high does it appear to somebody looking into it from above?

10. Prisms are used in optical instruments such as binocu-lars instead of mirrors because total internal reflection better preserves the sharpness and brightness of a light beam. Find the minimum index of refraction of the glass to be used in a prism that is meant to change the direction of a light beam by 90°.

11. A skating rink 30 m by 60 m is to have an average illumination at night of 300 luxes. How many 1-kW, 2500-cd lamps are needed if they are equipped with reflectors that direct ⅔ of their luminous flux on the rink?

12. A certain 100-W light bulb has a luminous intensity of 120 cd. (a) What is its luminuus efficiency? (b) If the bulb radiates isotropically and is suspended 1.2 m above a table, what is the illumination directly below it?

13. A spotlight concentrates all the light from a 200-cd bulb in a beam 1 m in diameter. Find the illumination it produces on a surface perpendicular to the beam.

ANSWERS TO MULTIPLE CHOICE

1. c	**8.** c	**15.** a
2. d	**9.** d	**16.** b
3. a	**10.** c	**17.** c
4. b	**11.** a	**18.** a
5. d	**12.** c	**19.** a
6. d	**13.** d	**20.** c
7. a	**14.** a	**21.** b

LENSES

24

A lens refracts light from an object that passes through it in such a way as to be able to form an image of the object. The image may be real or virtual, erect or inverted, and larger, smaller, or the same size as the object. In this chapter we shall investigate not only how individual lenses produce images but also how this is done by such systems of lenses as the microscope and the telescope. The discussion will be limited to lenses whose thicknesses are small relative to their diameters.

24–1 LENSES

A *lens* is a transparent object of regular form that alters the shape of wavefronts of light that pass through it. A simple way to appreciate how a lens produces its effects is to begin with a pair of prisms, as in Fig. 24–1(a). When the

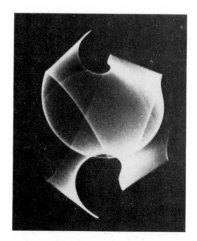

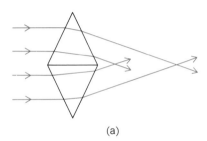

FIG. 24-1 A converging lens may be thought of as a development of two prisms placed base to base.

(a)

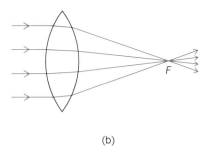

(b)

prisms have their bases together, parallel light that approaches is deviated so that the various rays intersect. However, they do not intersect at a single focal point because each prism merely changes the direction of the rays without affecting the shape of the wavefronts. If properly curved rather than flat surfaces are employed, as in Fig. 24–1(b), the object is called a *converging lens,* and it acts to bring an incoming parallel beam of light to a single focal point *F*.

A converging lens brings parallel rays to a real focal point

A *diverging lens* may similarly be thought of as a development of a pair of prisms with their apexes together (Fig. 24–2). The prisms deviate incoming parallel light outward, but the diverging rays cannot be projected back to a single point. The corresponding lens has curved surfaces so shaped that the diverging rays have a single virtual focal point *F*.

A diverging lens spreads out parallel rays as though they came from a virtual focal point

Figure 24–3 shows how lenses affect wavefronts. The speed of light in glass (or other suitable transparent material) is less than in air, and light takes more time to pass through a certain thickness of glass than through the same thickness of air. Light that

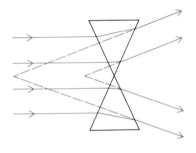

(a)

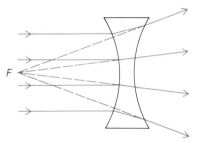

(b)

FIG. 24-2 A diverging lens may be thought of as a development of two prisms placed apex to apex.

FIG. 24–3 How converging and diverging lenses affect plane wavefronts.

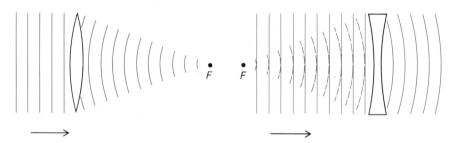

enters the center of a converging lens is accordingly retarded more than light that enters toward the edge, with the result that the wavefronts converge after passing through the lens. Exactly the opposite effect occurs with a diverging lens.

Thin lenses with spherical surfaces produce undistorted images

A lens with spherical surfaces, or one plane and one spherical surface, has the important characteristic that, provided it is relatively thin, it produces an undistorted image of something placed in front of it. Such lenses are also the easiest to manufacture. *Aspherical* (nonspherical) lens surfaces are used in special applications to avoid certain kinds of distortion that occur with spherical lenses. The various forms of simple lenses are shown in Fig. 24–4; converging lenses are always thickest in the center, whereas diverging lenses are always thinnest in the center. A meniscus lens has one concave and one convex surface.

Focal length

The distance from a lens to its focal point is called its *focal length f*. The focal length of a particular lens depends both upon the index of refraction *n* of its material relative to that of the medium it is in, and upon the radii of curvature R_1 and R_2 of its surfaces. In the case of a lens whose thickness is small compared with R_1 and R_2, these quantities are related by the *lensmaker's equation*:

$$\frac{1}{f} = (n - 1)\left(\frac{1}{R_1} + \frac{1}{R_2}\right) \qquad \textit{Lensmaker's equation} \quad (24–1)$$

In using this equation, it does not matter which surface of the lens is considered as 1 and which as 2. However, the sign given to each radius of curvature *is* important. We shall reckon a radius as + if the surface is convex (curved outward), and as − if the **Sign conventions for lensmaker's equation** surface is concave (curved inward). A plane surface has, in effect, an infinite radius of curvature, and $1/R$ for a plane surface is therefore 0. In Fig. 24–4(a), both radii are

FIG. 24–4 Some simple lenses. The focal length of a converging lens is reckoned positive, that of a diverging lens is reckoned negative.

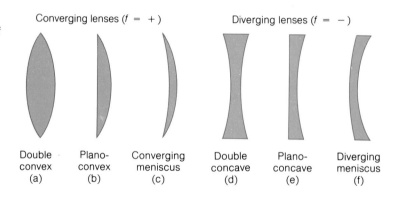

Converging lenses (*f* = +)

Diverging lenses (*f* = −)

| Double convex (a) | Plano-convex (b) | Converging meniscus (c) | Double concave (d) | Plano-concave (e) | Diverging meniscus (f) |

+; in (b), the first radius is infinite and the second is +; in (c), the first radius is −
and the second is +; in (d), both radii are −; and so on.

Depending upon its shape, a lens may have a positive or a negative focal length. A positive focal length signifies a converging lens, and a negative one signifies a diverging lens.

Focal length is + for a converging lens, − for a diverging lens

Example A meniscus has a convex surface whose radius of curvature is 25 cm and a concave surface whose radius of curvature is 15 cm. The index of refraction is 1.52. Find the focal length of the lens and whether it is converging or diverging.

Solution Here $R_1 = +25$ cm and $R_2 = -15$ cm in accord with the sign convention. From the lensmaker's equation,

$$\frac{1}{f} = (n - 1)\left(\frac{1}{R_1} + \frac{1}{R_2}\right) = (1.52 - 1.00)\left(\frac{1}{25\,\text{cm}} - \frac{1}{15\,\text{cm}}\right)$$
$$= (0.52)(0.040 - 0.067)\,\text{cm}^{-1} = -0.014\,\text{cm}^{-1}$$

and so

$$f = -71\,\text{cm}$$

The negative focal length indicates a diverging lens. ■

Example The above lens is placed in water, whose index of refraction is 1.33. Find its focal length there.

Solution The index of refraction of the glass relative to water is

$$n' = \frac{\text{index of refraction of glass}}{\text{index of refraction of water}} = \frac{1.52}{1.33} = 1.14$$

From the lensmaker's equation, since R_1 and R_2 are the same in both air and water, the ratio between the focal length f' of the lens in water and its focal length f in air is

$$\frac{f'}{f} = \frac{n - 1}{n' - 1} = \frac{1.52 - 1}{1.14 - 1} = 3.7$$

Hence

$$f' = 3.7f = (3.7)(-71\,\text{cm}) = -263\,\text{cm}$$

The focal length of *any* lens made of this glass is 3.7 times longer in water than in air. ■

24–2 IMAGE FORMATION

A scale drawing provides a convenient way to determine the size and position of an image formed by a lens. The procedure is to consider two different light rays that originate at a certain point on an object and to trace their paths until they (or their extensions) come together again after being refracted by the lens. In this connection it is worth noting that a lens has *two* focal points, one on each side of the lens the distance

Ray tracing

FIG. 24–5 The position and size of an image produced by a thin lens can be determined by tracing any two of the rays shown. In tracing the rays, any deviations produced by the lens are assumed to occur at its central plane.

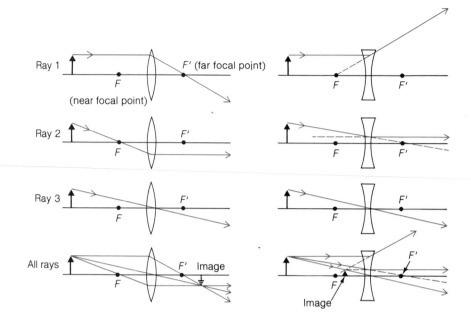

f from its center. We shall call the focal point on the side of the lens from which the light comes the *near focal point,* and the one on the other side of the lens the *far focal point.*

Three rays that are particularly easy to trace are shown in Fig. 24–5. They are:

Rays for finding an image produced by a lens

1. A ray that leaves the object parallel to the lens axis. When this ray is refracted by the lens, it passes through the far focal point of a converging lens or seems to come from the near focal point of a diverging lens.

2. A ray that leaves the object and passes through the near focal point of a converging lens or is directed toward the far focal point of a diverging lens. When this ray is refracted, it proceeds parallel to the axis.

3. A ray that leaves the object and proceeds through the center of the lens. If the lens is thin relative to the radii of curvature of its surfaces, this ray is not deviated.

Images produced by a converging lens

Ordinarily only two of these rays are required to establish the image of an object. Figure 24–6 shows how the properties of the image produced by a converging lens depend upon the position of the object. As usual, solid lines represent the actual paths taken by light rays, and dashed lines represent virtual paths. When the object is closer to the lens than F, the near focal point, the image is erect, enlarged, and virtual. The image seems to be behind the lens because the refracted rays diverge as though coming from a point behind it. (A virtual image can be seen by the eye, but it cannot appear on a screen because no light rays actually pass through such an image.) No image is formed of an object precisely at the focal point because the refracted rays are all parallel and hence never intersect; the image in this case may be said to be at infinity. An object farther from the lens than F always has an inverted real image which may be smaller than, the same size as, or larger than the object, depending upon whether the object distance is between f and 2f, equal to 2f, or greater than 2f.

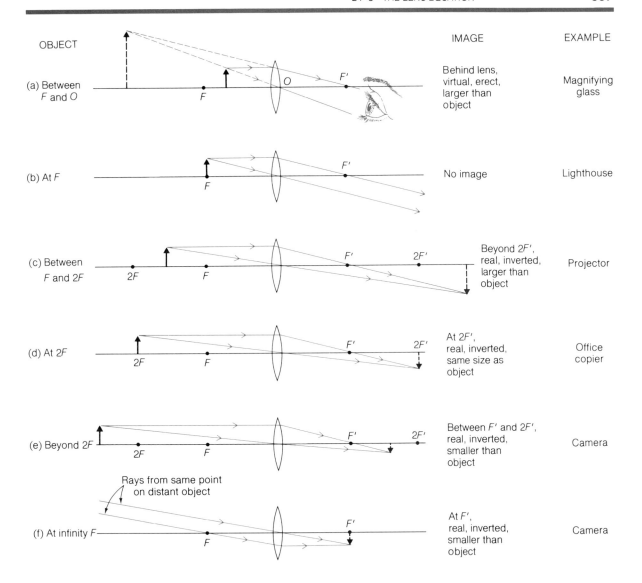

OBJECT

IMAGE

EXAMPLE

(a) Between F and O

Behind lens, virtual, erect, larger than object

Magnifying glass

(b) At F

No image

Lighthouse

(c) Between F and 2F

Beyond 2F', real, inverted, larger than object

Projector

(d) At 2F

At 2F', real, inverted, same size as object

Office copier

(e) Beyond 2F

Between F' and 2F', real, inverted, smaller than object

Camera

(f) At infinity

Rays from same point on distant object

At F', real, inverted, smaller than object

Camera

In contrast to the diversity of image sizes, natures, and locations produced by a converging lens, the image of a real object formed by a diverging lens is always virtual, erect, and smaller than the object (Fig. 24–7).

FIG. 24–6 Image formation by a converging lens.

24–3 THE LENS EQUATION

A simple formula relates the positions of the image and the object of a thin lens to the lens's focal length f. In terms of the symbols

p = distance of object from lens
q = distance of image from lens
f = focal length of lens

FIG. 24–7 The image of a real object formed by a diverging lens is always virtual, erect, and smaller than the object.

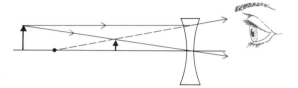

this formula is

$$\frac{1}{p} + \frac{1}{q} = \frac{1}{f}$$

Lens equation   (24–2)

When any two of the three quantities are known, the third can be calculated. The sign conventions to be observed when using the lens equation are:

Sign conventions for lens equation

1. The focal length f is considered $+$ for a converging lens, $-$ for a diverging lens.
2. Object distance p and image distance q are considered $+$ for real objects and images, $-$ for virtual objects and images.

Derivation of lens equation

The lens equation can be derived geometrically with the help of Fig. 24–8. We observe that the triangles ABO and $A'B'O$ are similar, which means that corresponding sides of these triangles are proportional. Hence

$$\frac{A'B'}{AB} = \frac{A'O}{AO} = \frac{q}{p}$$

(24–3)

The triangles $OB''F$ and $A'B'F$ are also similar, and

$$\frac{A'B'}{OB''} = \frac{A'F}{OF} = \frac{q-f}{f} = \frac{q}{f} - 1$$

(24–4)

Now, it is clear that

$$OB'' = AB$$

since the light ray BB'' is parallel to the axis. Therefore

$$\frac{A'B'}{OB''} = \frac{A'B'}{AB}$$

which permits us to set the right-hand sides of Eqs. (24–3) and (24–4) equal:

$$\frac{q}{f} - 1 = \frac{q}{p}$$

FIG. 24–8 A ray diagram for deriving the lens equation.

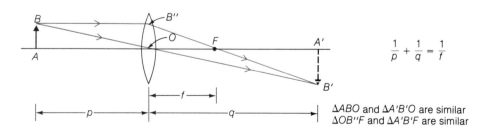

$$\frac{1}{p} + \frac{1}{q} = \frac{1}{f}$$

$\triangle ABO$ and $\triangle A'B'O$ are similar
$\triangle OB''F$ and $\triangle A'B'F$ are similar

Dividing each term of this equation by q yields

$$\frac{1}{f} - \frac{1}{q} = \frac{1}{p}$$

$d_o = p$

$d_i = q$

and, rearranging terms, we obtain

$$\frac{1}{p} + \frac{1}{q} = \frac{1}{f}$$

The above derivation involved a converging lens with an object distance greater than the focal length, but the resulting formula is valid for diverging lenses as well and for any object distance.

It is often convenient to solve the lens equation for p or q before using it in a calculation. We find that

Alternate forms of lens equation

$$\frac{1}{p} = \frac{1}{f} - \frac{1}{q} = \frac{q - f}{qf}$$

$$p = \frac{qf}{q - f} \tag{24–5}$$

and, in a similar way, that

$$q = \frac{pf}{p - f} \tag{24–6}$$

$$f = \frac{pq}{p + q} \tag{24–7}$$

If a calculator with a reciprocal [1/X] key is employed, Eq. (24–2) can be used directly to solve a numerical problem. The sequence of key strokes to find f, for instance, would be $(p)[1/X][+](q)[1/X][=][1/X]$.

Example A plano-convex lens of focal length 5 cm is used in a reading lamp to focus light from a bulb on a book. If the lens is 60 cm from the book, how far from the bulb's filament should it be?

Converging lens

Solution Here the focal length of the lens is $+5$ cm and the image distance is $+60$ cm. From the thin lens equation the object distance is

$$p = \frac{qf}{q - f} = \frac{(60\,\text{cm})(5\,\text{cm})}{60\,\text{cm} - 5\,\text{cm}} = 5.45\,\text{cm} \qquad \blacksquare$$

Example Motion picture directors often use diverging lenses for preliminary inspection of scenes to see how they would appear on the screen without having to move the camera into place first. If a double-concave lens of focal length -50 cm is used for this purpose, where does the image of a scene 10 m away seem to be located?

Diverging lens

Solution Since $f = -0.5$ m and $p = 10$ m,

$$q = \frac{pf}{p - f} = \frac{(10\,\text{m})(-0.5\,\text{m})}{10\,\text{m} - (-0.5\,\text{m})} = -0.48\,\text{m}$$

The image is located 48 cm behind the lens. The negative image distance signifies a virtual image, which is always on the same side of a lens as its object. This situation is pictured in Fig. 24–7. ■

24–4 THE CAMERA

A camera contains a converging lens that forms images on a light-sensitive photographic film in an arrangement like that of Fig. 24–9. The distance between the lens and the film is adjustable to permit bringing objects at varying distances to a sharp focus on the film. The film is exposed by opening the shutter for a fraction of a second. The shorter the exposure time, the better the ability to "freeze" a moving object; the longer the exposure time, the more light reaches the film and hence the better the ability to photograph a poorly illuminated scene.

For a given shutter speed, the amount of light reaching the film from a particular scene depends upon the area of the lens aperture, which is regulated by an iris diaphragm (Fig. 24–10). Apertures are conventionally specified in terms of the focal length f of the lens: $f/8$ means that the diameter d of the opening is $\frac{1}{8}$ of f. The "8" in this case is referred to as the "f-number" of the aperture, so that, in general,

The larger the f-number, the smaller the aperture and the fainter the light reaching the film

$$f\text{-number} = \frac{\text{focal length}}{\text{aperture diameter}} = \frac{f}{d}$$

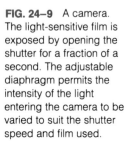

FIG. 24–9 A camera. The light-sensitive film is exposed by opening the shutter for a fraction of a second. The adjustable diaphragm permits the intensity of the light entering the camera to be varied to suit the shutter speed and film used.

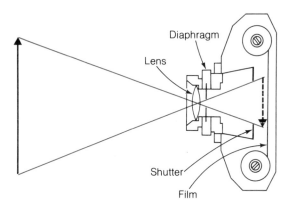

Diaphragm

Lens

Shutter

Film

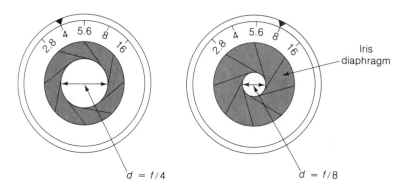

FIG. 24–10 The larger the f-number of a camera diaphragm, the smaller the aperture. The sequence of values is such that changing the aperture by one f-number changes its area and the amount of light reaching the film by a factor of 2. An aperture of $f/4$ admits four times as much light as one of $f/8$.

2.8 4 5.6 8 16

2.8 4 5.6 8 16

Iris diaphragm

$d = f/4$

$d = f/8$

The larger the *f*-number, the smaller the aperture and the less the light reaching the film. The illumination of the film is proportional to the square of *d* and inversely proportional to the square of the image distance *q*. Since *q* is usually quite close to *f* in a camera, an aperture of a given *f*-number yields essentially the same illumination whatever the focal length of the lens. This is the reason for the *f*-number scheme. The diaphragm of a camera lens is calibrated using a sequence of *f*-number values (for instance, 2.8, 4, 5.6, 8, 11, 16) such that their squares are approximately in the ratios 1:2:4:8 and so on, which means that changing the setting by one *f*-number changes the illumination of the film by the factor 2. The same amount of light reaches the film of a camera during exposures of $\frac{1}{50}$ s at *f*/8, $\frac{1}{100}$ s at *f*/5.6, and $\frac{1}{200}$ s at *f*/4.

In addition to helping determine the amount of light reaching the film, the size of a camera's lens aperture affects the *depth of field,* which is the range of object distances at which reasonably sharp images are produced. As Fig. 24–11 shows, reducing the aperture (increasing the *f*-number) leads to an increased depth of field. Thus more of a scene will appear sharp if a slow shutter speed and small aperture are used (provided motion in the scene is not a problem, of course).

The film size and the desired angle of view are what determine the appropriate focal length of a camera lens. A "normal" lens usually provides an angle of view of about 45°, which means a focal length of 50 mm for a camera using 35-mm film on which the image size is 24 × 36 mm. A lens of shorter focal length is a *wide-angle lens* because it captures more of a given scene, though at the expense of reducing the sizes of details in the scene. Typical wide-angle lenses for a 35-mm camera have focal lengths of 35 and 28 mm, which give angles of view respectively of 63° and 75°. A *telephoto lens* has a long focal length to provide larger images of distant objects, but

A small aperture provides greater depth of field

FIG. 24–11 Reducing the aperture of a lens increases the depth of field. Here arrowhead *A* is in focus on the film, whereas arrowhead *B* is in focus behind the film because it is closer to the lens. (a) With a large aperture, the image of *B* on the film is very indistinct. (b) With a small aperture, the image of *B* on the film is sharper.

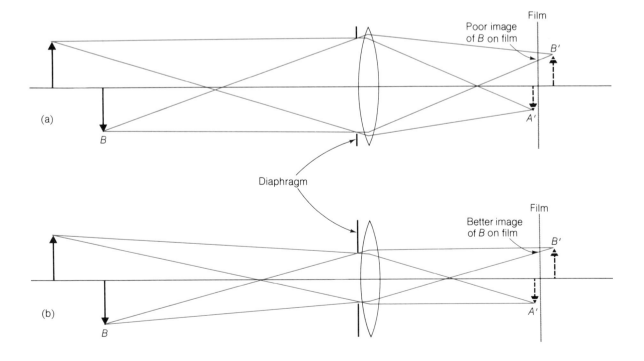

this means a reduced angle of view: a 135-mm telephoto lens for a 35-mm camera has an angle of view of 18°, and it is only 8° for a 300-mm telephoto lens.

Example   What range of motion should the 50-mm lens of a 35-mm camera have if the camera is to be capable of photographing objects as close as 50 cm from the lens? What is the aperture diameter when the diaphragm is set at $f/2.8$? At $f/16$?

Solution   (a) Here the image distance q for an object distance of $p = 50$ cm $= 500$ mm is

$$q = \frac{pf}{p - f} = \frac{(500\,\text{mm})(50\,\text{mm})}{500\,\text{mm} - 50\,\text{mm}} = 55.6\,\text{mm}$$

An object at infinity is brought to a sharp focus on the film when the image distance is equal to the focal length (see Fig. 24–6(f)), which in this case is 50 mm. Hence a range of adjustment of 5.6 mm will permit the camera to photograph objects from 50 cm away to infinity.

    (b) At $f/2.8$ the aperture diameter of the lens is

$$d = \frac{f}{2.8} = \frac{50\,\text{mm}}{2.8} = 17.9\,\text{mm}$$

and at $f/16$ it is

$$d = \frac{50\,\text{mm}}{16} = 3.13\,\text{mm} \qquad\qquad \blacksquare$$

24–5   MAGNIFICATION

Definition of linear magnification

The *linear magnification* m of an optical system of any kind is the ratio between the size of the image and that of the object. A magnification of exactly 1 means that the image and the object are the same size; a magnification of more than 1 means that the image is larger than the object; and a magnification of less than 1 means that the image is smaller than the object. By "size" is meant any transverse linear dimension, for instance height or width. Thus we can write

$$m = \frac{h'}{h}$$

$$\text{Linear magnification} = \frac{\text{height of image}}{\text{height of object}}$$

where $h = $ height of object and $h' = $ height of image.

    Figure 24–12 shows a converging lens that forms the image $A'B'$ of the object AB. The object height is $AB = h$, and that of the image is $A'B' = h'$. The image is inverted and therefore its height is considered negative. The triangles ABO and $A'B'O$ are similar, and their corresponding sides are proportional. Hence

$$\frac{A'B'}{AB} = \frac{A'O}{AO}$$

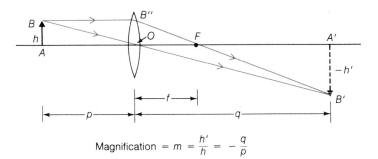

FIG. 24–12 The magnification of a lens is equal to minus the ratio of image and object distances. A positive magnification means an erect image, while a negative magnification means an inverted image.

Because $A'B' = -h'$, $AB = h$, $A'O = q$, and $AO = p$, we find that

The magnification formula holds for all lenses and object distances

$$\frac{h'}{h} = -\frac{q}{p}$$

and the magnification produced by the lens is

$$m = \frac{h'}{h} = -\frac{q}{p} \qquad \textit{Linear magnification of lens} \quad (24\text{–}8)$$

This formula is a general one that holds for diverging as well as converging lenses and for any object distance.

A useful feature of Eq. (24–8) is that it automatically indicates whether an image is erect (m positive) or inverted (m negative). Table 24–1 summarizes the various sign conventions we have been using.

Example A "magnifying glass" is a converging lens held less than its focal length from an object being examined (see Fig. 24–6(a)). How far should a double-convex lens whose focal length is 15 cm be held from an object to produce an erect image three times larger?

Magnifying glass

Solution The required magnification is $+3$ since an erect image is required. From Eq. (24–8),

$$m = -\frac{q}{p}$$

$$q = -mp = -3p$$

We also know from the thin lens equation that

$$q = \frac{pf}{p - f}$$

Quantity	Positive	Negative
Focal length f	Converging lens	Diverging lens
Object distance p	Real object	Virtual object
Image distance q	Real image	Virtual image
Magnification m	Erect image	Inverted image

TABLE 24–1
Sign conventions for lenses

FIG. 24–13 An optical projection system. The condenser causes the slide to be evenly illuminated, thereby making possible an image of uniform brightness on the screen.

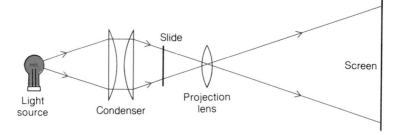

By equating these expressions for q, with $f = +15$ cm, we obtain

$$\frac{pf}{p-f} = -3p$$

$$pf = -3p(p-f)$$

$$f = -3p + 3f$$

$$p = \frac{2}{3}f = \frac{2}{3} \times 15\,\text{cm} = 10\,\text{cm}$$

When the lens is held 10 cm from the object, the image will be virtual, erect, and enlarged three times. ∎

Projector

Example   A slide projector is to be used with its lens 6 m from a screen. If a projected image 1.5 m square of a slide 5 cm square is desired, what should the focal length of the lens be?

Solution   Projectors use converging lenses to produce enlarged, inverted, real images of transparencies which are illuminated from behind (Fig. 24–13). In the present case, the image distance is given as $q = 6$ m, and the required magnification is

$$m = \frac{h'}{h} = \frac{-1.5\,\text{m}}{0.05\,\text{m}} = -30$$

The image height is reckoned negative as it is inverted. From Eq. (24–8),

$$p = -\frac{q}{m} = -\frac{6\,\text{m}}{-30} = 0.2\,\text{m}$$

From Eq. (24–7) we find that the lens should have a focal length of

$$f = \frac{pq}{p+q} = \frac{(0.2\,\text{m})(6\,\text{m})}{0.2\,\text{m} + 6\,\text{m}} = 0.194\,\text{m} = 19.4\,\text{cm} \quad ∎$$

24–6 THE EYE

Structure of the eye

The structure of the human eye is shown in Fig. 24–14. Its diameter is typically a little less than 3 cm, and its approximately spherical shape is maintained by internal pressure. Incoming light first passes through the *cornea,* the transparent outer membrane, and then enters a liquid called the *aqueous humor.* The *lens* of the eye is a jellylike assembly of minute transparent fibers that slide over one another when the shape of the lens is

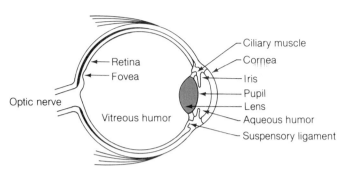

FIG. 24–14 The human eye slightly larger than life size.

altered by the *ciliary muscle* to which it is attached by ligaments. In front of the lens is the colored *iris* whose aperture is the *pupil*. Behind the lens is a cavity filled with another liquid, the *vitreous humor,* and the lining of this cavity is the *retina.*

The retina contains millions of tiny structures called *rods* and *cones* that are sensitive to light; the cones are responsible for vision in bright conditions, the rods for vision in dim conditions. Because the sensation of color is produced by the cones, colors are difficult to distinguish in a faint light. The rods and cones respond to the image formed by the lens on the retina and transmit the information to the brain through the *optic nerve.* There are no photoreceptors near the optic nerve and that region is accordingly called the "blind spot." The blind spot covers a field of view about 8° high and 6° wide. Its existence is not conspicuous for two reasons: the blind spots of the two eyes obscure different fields of vision, and the eyes are never completely at rest but are in constant scanning motion. Vision is most acute for the image focused on the *fovea,* a small region near the center of the retina, whereas the rest of the field of view appears less distinct. Since the fovea contains only cones, it is easier to see something in a dim light by looking a bit to one side instead of directly at it.

The retina

In bright light the pupil contracts to reduce the amount of light reaching the retina, and it dilates in faint light. A fully opened pupil admits about 16 times as much light as a fully contracted one, since the respective pupil diameters are about 2 and 8 mm for a ratio of 4:1. The retina itself is also able to cope with a considerable range of brightnesses. The sensitivity of the eye varies with wavelength, so light of the same intensity but of different wavelengths will give different impressions of brightness. Maximum sensitivity in bright light occurs in the middle of the visible spectrum at about $\lambda = 555$ nm, which corresponds to green (Fig. 24–15). Maximum sensitivity in dim light, when the rods rather than the cones are chiefly involved, occurs at about 510 nm, which is closer to the blue end of the spectrum and accounts for the reduced ability of the eye to respond to red light against a dark background. For this reason red illumination is provided when accommodation to darkness must be preserved, for instance on the bridge of a ship at night.

The pupil changes size according to light intensity

Refraction at the cornea is responsible for most of the focusing power of the eye; the lens is less effective because its index of refraction is not very different from those of the aqueous and vitreous humors on both sides of it. The shape of the lens is controlled by the ciliary muscle. When this relaxes, the lens brings objects at infinity to a sharp focus on the retina. To permit a closer object to be viewed, the ciliary muscle forces the lens into a more convex shape whose focal length is appropriate for the object distance involved. The image distance in this situation, which is the lens-retina distance,

The focal length of the eye's lens changes to suit object distance

FIG. 24–15 How the sensitivity of a normal human eye to bright light varies with wavelength. In dim light, the wavelength of maximum sensitivity shifts from about 555 nm (green light) to about 510 nm.

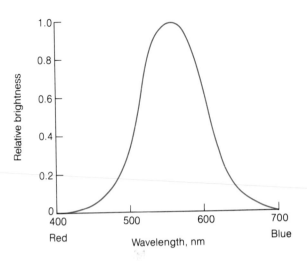

is fixed, and focusing is thus accomplished by changing the focal length of the lens. The latter process is called *accommodation*; the focal length of a normal eye can be varied between about 10 and 17 mm. In the case of a camera, on the other hand, the focal length of the lens is fixed, and focusing is done by changing the image distance.

Underwater vision

The limited range of accommodation of the lens is not sufficient to make up for the loss of refractive power that occurs when the eye is immersed in water, which is considerable since the index of refraction of the cornea (1.38) is close to that of water (1.33). Hence underwater objects cannot be brought to a sharp focus unless goggles are used to keep water away from the cornea. Fish are able to see clearly when submerged because the lenses of their eyes are spherical and have high indexes of refraction. Focusing in most fish eyes is carried out by shifting the lens closer to or farther from the retina, in the same way a camera lens is focused.

24–7 DEFECTS OF VISION

Nearsightedness and farsightedness

Two common defects of vision are *myopia* (nearsightedness) and *hyperopia* (farsightedness). In myopia, the eyeball is too long, and light from an object at infinity comes to a focus in front of the retina, as in Fig. 24–16(b). Accommodation permits nearby objects to be seen clearly, but not more distant ones. A diverging lens of the proper focal length can correct this condition. In hyperopia, the eyeball is too short, and light from an object at infinity does not come to a focus within the eyeball at all. Its power of accommodation permits a hyperopic eye to focus on distant objects, but the range of accommodation is not enough for nearby objects to be seen clearly. The correction for hyperopia is a converging lens, as in Fig. 24–16(c).

Dioptric power

Opticians often use the *power D* of a lens, expressed in *diopters*, in place of its focal length. If the focal length *f* is given in meters, then

$$D \text{ (diopters)} = \frac{1}{f} \qquad (24\text{–}9)$$

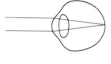

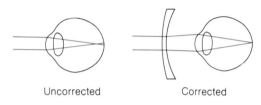

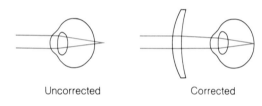

FIG. 24–16 Myopia and hyperopia are common defects of vision that can be remedied with corrective lenses.

(a) Normal eye

Uncorrected   Corrected

(b) Myopia (nearsightedness)

Uncorrected   Corrected

(c) Hyperopia (farsightedness)

Thus a converging lens of $f = +50$ cm $= +0.50$ m could also be described as having a power of $+2$ diopters.

Example A certain nearsighted eye cannot see objects distinctly when they are more than 25 cm away. Find the power in diopters of a correcting lens that will enable this eye to see distant objects clearly.   *Correcting for nearsightedness*

Solution The purpose of the lens is to form an image 25 cm in front of the eye of an object that is infinitely far away. Since the image is to be on the same side of the lens as the object, the image must be virtual, so $q = -25$ cm. The object distance is $p = \infty$; hence

$$\frac{1}{f} = \frac{1}{p} + \frac{1}{q} = \frac{1}{\infty} - \frac{1}{25\,\text{cm}} = 0 - \frac{1}{25\,\text{cm}}$$

$$f = -25\,\text{cm} = -0.25\,\text{m}$$

The minus sign means that the lens is diverging. Its power in diopters is

$$D = \frac{1}{f} = \frac{-1}{0.25\,\text{m}} = -4\,\text{diopters}$$ ∎

Correcting for farsightedness

Example   A certain farsighted eye cannot see objects distinctly when they are closer than 1 m away. Find the power in diopters of a correcting lens that will enable this eye to read a letter 25 cm away.

Solution   The purpose of the lens is to form an image 1 m in front of the eye of an object that is 25 cm away. Since the image is to be on the same side of the lens as the object, the image must be virtual, and so $q = -1$ m. The object distance is $p = 25$ cm $= 0.25$ m, hence

$$f = \frac{pq}{p+q} = \frac{(0.25 \, \text{m})(-1 \, \text{m})}{0.25 \, \text{m} + (-1 \, \text{m})} = +0.33 \, \text{m}$$

The plus sign means that the lens is converging. Its power in diopters is

$$D = \frac{1}{f} = \frac{1}{0.33 \, \text{m}} = +3 \, \text{diopters} \qquad \blacksquare$$

Presbyopia refers to the gradual loss of accommodation

The range of accommodation decreases with age as the lens hardens, a condition known as *presbyopia*. This does not affect distant vision, but the "near point" of the eye (the closest position at which objects can be seen distinctly) gets increasingly far away. From perhaps 7 cm at age 10, the near point recedes until it is often 2 m or more at age 60. The correction for presbyopia is a converging lens; if the range of accommodation is severely limited, more than one set of corrective lenses may be required.

The closer an object is to the eye, the larger it seems and the more the detail that can be made out. Since the eye cannot focus on objects closer than the near point, this sets a limit to the magnification of which the eye is capable. By convention, the distance of most distinct vision is taken as 25 cm (about 10 in.), which is a comfortable object distance for most people. For purposes of calculation, an optical instrument such as a microscope or a telescope is therefore assumed to form a virtual image 25 cm behind the lens (or lens system) nearest the eye.

Astigmatism can be corrected by a cylindrical lens

Astigmatism is a defect of vision caused by the cornea (or sometimes the lens) having different curvatures in different planes. When light rays that lie in one plane are in focus on the retina of an astigmatic eye, those in other planes will be in focus either in front or in back of the retina. As a result only one of the bars of a cross can be in focus at the same time (Fig. 24–17). Astigmatism is a source of eyestrain because the mechanism of accommodation continually varies the focus of the lens in an effort to produce a completely sharp image. The remedy is a corrective lens that has a cylindrical curvature, as in Fig. 24–18.

24–8   THE MICROSCOPE

A microscope produces enlarged images of nearby objects

In many applications a system of two or more lenses is superior to a simple lens. An ordinary magnifying glass, for instance, cannot produce images enlarged beyond 3 × or so without severe distortion. Greater magnifications of nearby objects can be satisfactorily obtained by using instead two converging lenses arranged as a *microscope*. The optical system of a microscope is shown in Fig. 24–19. The *objective* is a lens of short focal length that forms an enlarged real image of the object. This image is further enlarged by the *eyepiece,* which acts as a simple magnifier to form a virtual final image.

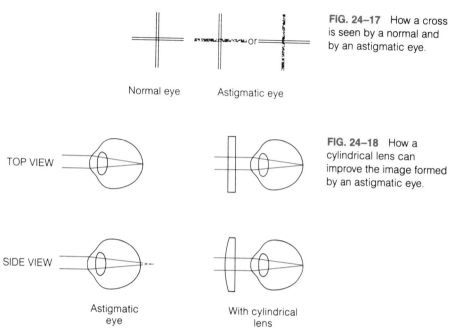

FIG. 24–17 How a cross is seen by a normal and by an astigmatic eye.

Normal eye   Astigmatic eye

TOP VIEW

SIDE VIEW

Astigmatic eye

With cylindrical lens

FIG. 24–18 How a cylindrical lens can improve the image formed by an astigmatic eye.

The image produced by the objective is thus the object of the eyepiece, and the final image has been magnified twice.

The total magnification produced by a two-lens system such as a microscope is the product of the magnification m_1 of the first lens (the objective) and the magnification m_2 of the second lens (the eyepiece):

$$m = m_1 m_2 \qquad (24\text{--}10)$$

Typical objectives yield magnifications of $10\times$ to $100\times$, and standard eyepiece magnifications are $5\times$, $10\times$, and $15\times$. Total magnifications of up to $1500\times$ are therefore possible with a good laboratory microscope. However, as discussed in the next chapter, the wave nature of light limits the useful magnification of a microscope to a maximum of perhaps $500\times$; higher magnifications give larger images but do not show finer details. Actual microscopes use compound lenses that consist of two to six elements in place of the single lenses shown in Fig. 24–19, but their optical behavior is the same.

The limit of useful magnification is about $500\times$ with a microscope

Example A microscope has an objective of focal length 4 mm and an eyepiece of focal length 20 mm. If the image distance of the objective is 160 mm and that of the

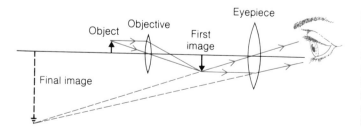

Object   Objective   First image   Eyepiece

Final image

FIG. 24–19 In a microscope, the image formed by the objective is further magnified by the eyepiece. In this diagram the rays used to locate the final image are not the same as those used to locate the first image.

eyepiece is 250 mm (which are the usual figures for these quantities), find the magnification produced by each lens and by the entire microscope. What is the distance between the objective and the eyepiece?

Solution   (a) The object distance of the objective is

$$p_1 = \frac{q_1 f}{q_1 - f_1} = \frac{(160 \, \text{mm})(4 \, \text{mm})}{160 \, \text{mm} - 4 \, \text{mm}} = 4.10 \, \text{mm}$$

and so its magnification is

$$m_1 = -\frac{q_1}{p_1} = -\frac{160 \, \text{mm}}{4.10 \, \text{mm}} = -39$$

The minus sign means the image is inverted.

(b) With $q_2 = -250$ mm and $f_2 = 20$ mm the same procedure yields for the eyepiece

$$p_2 = \frac{q_2 f_2}{q_2 - f_2} = \frac{(-250 \, \text{mm})(20 \, \text{mm})}{-250 \, \text{mm} - 20 \, \text{mm}} = 18.5 \, \text{mm}$$

$$m_2 = -\frac{q_2}{p_2} = -\frac{-250 \, \text{mm}}{18.5 \, \text{mm}} = 13.5$$

(c) The magnification of the microscope is

$$m = m_1 \times m_2 = -39 \times 13.5 = -527$$

and the distance between the objective and the eyepiece (see Fig. 24–19) is

$$L = q_1 + p_2 = 160 \, \text{mm} + 18.5 \, \text{mm} = 178.5 \, \text{mm} \qquad \blacksquare$$

24–9   THE TELESCOPE

The objective of a telescope has a long focal length

A telescope is a lens system used to examine distant objects. As in a microscope, two lenses are involved, with an eyepiece to enlarge the image produced by the objective. A telescope objective, however, has a long focal length whereas that of a microscope is very short.

Figure 24–20 shows a simple telescope. The image produced by the objective is real, inverted, and smaller than the object. The eyepiece then acts as a simple magnifier to form an enlarged virtual image whose object is the initial image. In practice, the object distance is usually very long relative to the focal length of the objective, and the final image is smaller than the object itself. However, the image seen by the eye is larger than it would be without the telescope, so the effect is the same as if the object were closer to the eye than it actually is.

Angular magnification equals ratio of focal lengths of objective and eyepiece

Telescopes are described in terms of the *angular magnification* they produce. This quantity is the ratio between the angle β subtended at the eye by the image and the angle α subtended at the eye by the object seen directly. For distant objects ($p \gg f$) the angular magnification of a telescope is simply the ratio between the focal lengths of its objective and eyepiece:

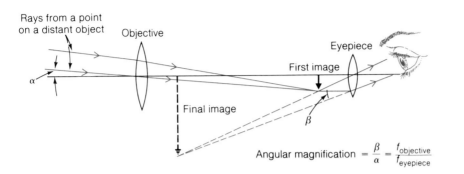

FIG. 24–20 Ray diagram of a simple telescope. The rays used to locate the final image are not the same as those used to locate the first image.

$$m_{\text{ang}} = \frac{f_{\text{objective}}}{f_{\text{eyepiece}}} = \frac{\beta}{\alpha}$$

Angular magnification (24–11)

The higher the magnification, the greater in diameter the objective must be in order to gather in enough light for the image to be visible. This sets a limit to the size of a refracting telescope, since a large glass lens tends to distort under its own weight. The largest refracting telescope in the world is the 1.02-m-diameter instrument at Yerkes Observatory in Wisconsin, whose objective has a focal length of nearly 20 m.

Example A telescope with an objective of focal length 60 cm and an eyepiece of focal length 1.5 cm is used to examine a hummingbird 5 cm long that is 20 m away. If the image distance of the eyepiece is 25 cm (the distance of most distinct vision), what is the apparent length of the hummingbird?

Solution The angular magnification of the telescope is

$$m_{\text{ang}} = \frac{f_{\text{objective}}}{f_{\text{eyepiece}}} = \frac{60\,\text{cm}}{1.5\,\text{cm}} = 40$$

The angle α subtended by the bird from the location of the telescope is

$$\alpha = \frac{\text{object length}}{\text{object distance}} = \frac{0.05\,\text{m}}{20\,\text{m}} = 0.0025\,\text{radian}$$

If L is the length of the bird's image as seen through the telescope, the angle β this image subtends is

$$\beta = \frac{\text{image length}}{\text{image distance}} = \frac{L}{25\,\text{cm}}$$

Since the angular magnification of the telescope is 40, from Eq. (24–11) we have

$$\beta = m_{\text{ang}}\alpha$$

$$\frac{L}{25\,\text{cm}} = 40 \times 0.0025\,\text{radian}$$

$$L = 2.5\,\text{cm}$$

The hummingbird seems to be 2.5 cm long, half its actual length, and to be located 25 cm from the viewer's eye. ∎

Large astronomical telescopes use concave mirrors as objectives

Modern astronomical telescopes invariably use concave parabolic mirrors as their objectives, since such a mirror produces a real image of a distant object and can be adequately supported from behind. A parabolic mirror is not subject to the aberrations mentioned in the next section—a further advantage. A small secondary mirror reflects the image outside the telescope tube for viewing or, more often, for photographing (Fig. 24–21). When used as a camera, a telescope needs no eyepiece: its objective lens or mirror simply acts as a giant telephoto lens. One of the largest reflecting telescopes in the world is at Mount Palomar in California and has a mirror 5.08 m in diameter.

Example   The diameter of the planet Mars is 6.8×10^6 m. What focal length must a telescope objective have in order to produce a photographic image of Mars 1 mm in diameter at a time when Mars is 8×10^{10}m from the earth?

Solution   Here $h = 6.8 \times 10^6$ m and $h' = 1$ mm $= 10^{-3}$ m, so the required magnification is

$$m = \frac{h'}{h} = -\frac{10^{-2}\text{m}}{6.8 \times 10^6\text{m}} = -1.5 \times 10^{-10}$$

A minus sign is used because the image will be inverted. The object distance p is the Mars-earth distance of 8×10^{10} m. Since the focal length f is going to be much smaller than p, $p - f \approx p$ and the image distance q is

$$q = \frac{pf}{p-f} \approx f$$

The formula $m = -q/p$ therefore becomes $m = -f/p$ here, and

$$f = -mp = -(-1.5 \times 10^{-10})(8 \times 10^{10}\text{ m}) = 12\text{ m}$$

The telescope objective should have a focal length of 12 m. Of course, further enlargement of the image can be made from the negative.   ■

A telescope can be constructed that produces an erect image if a third lens is introduced between the objective and the eyepiece (Fig. 24–22). The additional lens merely inverts the initial image, but it has the disadvantage of lengthening the telescope tube. A better scheme employs a pair of prisms which serve both to invert the image so that it is erect and to shorten the instrument length. The latter method is employed in prism binoculars (Fig. 24–23).

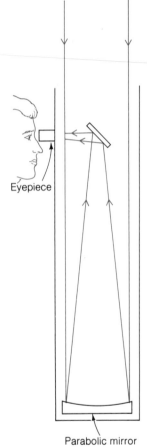

Eyepiece

Parabolic mirror

FIG. 24–21 One type of reflecting telescope.

FIG. 24–22 In a terrestrial telescope, an intermediate lens is used to produce an erect image.

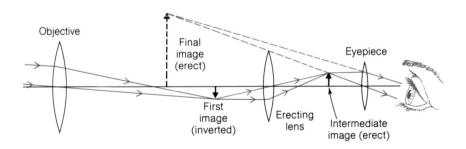

Objective

Final image (erect)

Eyepiece

First image (inverted)

Erecting lens

Intermediate image (erect)

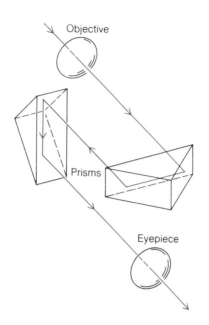

FIG. 24–23 A pair of prisms is used to erect the image in each half of a prism binocular.

24–10 LENS ABERRATIONS

The image formed by a single lens is never a perfect replica of its object. Of the variety of aberrations such an image is subject to, perhaps the most familiar is the presence of fringes of color around whatever is being viewed. This *chromatic aberration* is a consequence of the variation with wavelength of the index of refraction of glass (Fig. 24–24). Because of this variation, the focal length of a lens is slightly different for light of different colors, and the fringes are the result. The remedy for chromatic aberration is to combine a converging and a diverging lens made of different glass so that the dispersion produced by one is canceled by the other while leaving a net converging or diverging power. Such an *achromatic lens* is illustrated in Fig. 24–24.

Chromatic aberration

The lens equation was derived on the basis of light rays that made only small angles with the axis. When a simple lens is used to form images of objects some distance from the axis, a variety of aberrations arise even if the light used is mono-chromatic. One of them is *spherical aberration,* in which rays passing near the lens rim come to a focus closer to the lens than rays near the axis (Fig. 24–25). Another is

Spherical aberration

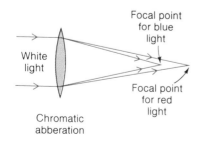

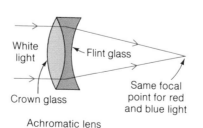

FIG. 24–24 A compound lens made of different types of glass can correct for chromatic aberration.

FIG. 24–25 Spherical aberration.

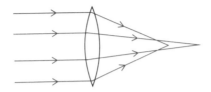

distortion of the image because the magnification of a simple lens varies with the distance of an object from the axis. By using several lenses of different types of glass and different curvatures to replace a single lens, the most conspicuous of the aberrations can be minimized. Compound lenses that consist of two or more elements are invariably used in high-performance optical systems such as those in microscopes, prism binoculars, and quality cameras.

24–11   SPHERICAL MIRRORS

Concave mirrors are converging, convex mirrors are diverging

Mirrors with spherical reflecting surfaces form images in much the same way that lenses do. Figure 24–26(a) shows how a concave mirror converges a parallel beam of light to a real focal point, and Fig. 24–26(b) shows how a convex mirror diverges such a beam so that the reflected rays seem to originate in a virtual focal point behind the mirror. If R is the radius of the reflecting surface in each case, the focal lengths of these mirrors are

$$f = \frac{R}{2} \qquad\qquad\qquad\qquad \textit{Concave mirror} \quad (24\text{–}12)$$

$$f = -\frac{R}{2} \qquad\qquad\qquad\qquad \textit{Convex mirror} \quad (24\text{–}13)$$

The position, size, and nature of the image produced by a spherical mirror of an object in front of it can be determined with the help of a scale drawing. As in the case

FIG. 24–26 (a) A concave spherical mirror converges a parallel beam of incident light. (b) A convex spherical mirror diverges a parallel beam of incident light.

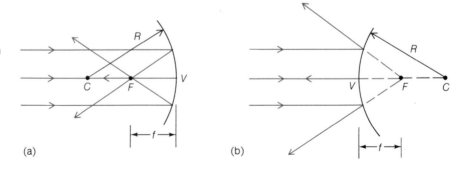

(a)   (b)

R = radius of curvature
V = vertex
C = center of curvature
F = focal point
f = focal length

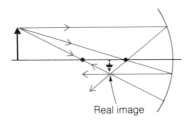

Real image

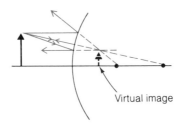

Virtual image

FIG. 24–27 The position and size of an image produced by a spherical mirror can be determined by tracing any two of the rays shown.

of a lens, the procedure is to consider two different light rays that originate at a certain point on the object and to trace their paths until they (or their backward extensions) come together again after reflection. There are three rays that are especially useful because they are so easily traced (Fig. 24–27):

Rays for finding the image produced by a mirror

1. A ray that leaves the object parallel to the mirror axis. When this ray is reflected, it passes through the focal point of a concave mirror or seems to come from the focal point of a convex mirror.

2. A ray that leaves the object and passes through the focal point of a concave mirror, or is directed toward the focal point of a convex mirror. When this ray is reflected, it proceeds parallel to the axis.

3. A ray that leaves the object along a radius of the mirror. When this ray is reflected, it returns along its original path.

In any given situation only two of these rays are needed to locate the image of a reflected object. Figure 24–28 shows how the properties of the image produced by a concave mirror vary with the position of the object. Solid lines represent the actual paths taken by light rays, and dashed lines represent virtual paths.

When the object is closer to the mirror than the focal point *F*, the image is erect (right side up), enlarged, and virtual. The image *seems* to be behind the mirror because the reflected light rays from the mirror diverge as though coming from a point behind the mirror.

Images produced by a concave mirror

When the object is precisely at the focal point, no image is formed because the reflected rays are all parallel and so do not intersect; the image in this case is sometimes said to be at infinity.

An object between the focal point *F* and the center of curvature *C* has an inverted, enlarged image that is real; the image would appear on a screen placed at its position.

When the object is at *C* its image is at the same place and is the same size but is inverted, while an object past *C* has a real image that is reduced in size and is inverted.

The image formed by a convex mirror of a real object is always erect, smaller than the object, and virtual (Fig. 24–29). The field of view of a convex mirror is wider than that of a plane mirror, which accounts for a number of its applications, such as at blind corners in roads.

Images produced by a convex mirror

The object and image distances *p* and *q* of a spherical mirror are related to its focal length by the same formula that holds for a thin lens:

Formulas for spherical mirrors

$$\frac{1}{p} + \frac{1}{q} = \frac{1}{f} \qquad\qquad \textit{Mirror equation} \quad (24\text{–}14)$$

Equations (24–5), (24–6), and (24–7) hold here as well as for lenses. The linear

FIG. 24–28 Image formation by a concave mirror.

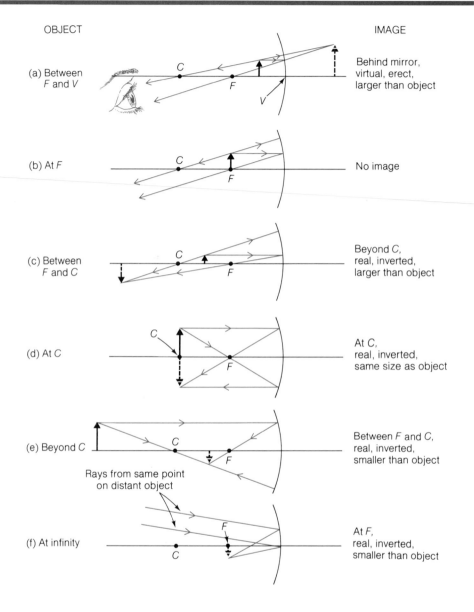

OBJECT

(a) Between F and V

(b) At F

(c) Between F and C

(d) At C

(e) Beyond C

Rays from same point on distant object

(f) At infinity

IMAGE

Behind mirror, virtual, erect, larger than object

No image

Beyond C, real, inverted, larger than object

At C, real, inverted, same size as object

Between F and C, real, inverted, smaller than object

At F, real, inverted, smaller than object

magnification of a mirror also follows the same formula as for a lens:

$$m = \frac{h'}{h} = -\frac{q}{p}$$ *Magnification of a mirror*   (24–15)

The sign conventions for a mirror are given in Table 24–2.

Example   A candle 5 cm high is placed 40 cm from a concave mirror whose radius of curvature is 60 cm. Find the position, size, and nature of the image.

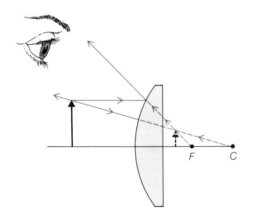

FIG. 24–29 The image of a real object formed by a convex mirror is always virtual, erect, and smaller than the object.

Solution The focal length of the mirror is

$$f = \frac{R}{2} = \frac{60\,\text{cm}}{2} = 30\,\text{cm}$$

The situation therefore corresponds to that shown in Fig. 24–28(c). The image distance q is

$$q = \frac{pf}{p - f} = \frac{(40\,\text{cm})(30\,\text{cm})}{40\,\text{cm} - 30\,\text{cm}} = 120\,\text{cm}$$

The image distance is positive, so the image is a real one. The size of the image is, from Eq. (24–15),

$$h' = -h\frac{q}{p} = (-5\,\text{cm})\left(\frac{120\,\text{cm}}{40\,\text{cm}}\right) = -15\,\text{cm}$$

The image of the candle is 3 times as large as the candle itself, and is inverted. ∎

Example A concave shaving mirror has a focal length of 40 cm. How far away from it should one's face be for the reflected image to be erect and twice its actual size?

Solution Here the magnification is $+2$ since the image is erect. Hence

$$m = -\frac{q}{p} = 2 \qquad q = -2p$$

The negative image distance signifies that the image is virtual; see Fig. 24–28(a). Now

Quantity	Positive	Negative
Focal length f	Concave mirror	Convex mirror
Object distance p	Real object	Virtual object
Image distance q	Real image	Virtual image
Magnification m	Erect image	Inverted image

TABLE 24–2
Sign conventions for spherical mirrors

FIG. 24–30 Spherical aberration. The more highly curved the mirror, the greater the aberration.

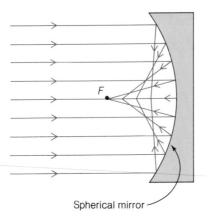

Spherical mirror

we substitute $f = 40$ cm and $q = -2p$ into the mirror equation and solve for the object distance p:

$$\frac{1}{p} + \frac{1}{q} = \frac{1}{f}$$

$$\frac{1}{p} - \frac{1}{2p} = \frac{1}{40\,\text{cm}}$$

$$\frac{1}{2p} = \frac{1}{40\,\text{cm}}$$

$$p = 20\,\text{cm}$$

When one's face is 20 cm from the mirror, the image one sees is magnified twice and is erect. ■

Mirrors do not suffer from chromatic aberration but do exhibit spherical aberration. In the case of a concave miror, rays reflected from the outer part of the mirror cross the axis closer to the vertex than rays reflected from the central part (Fig. 24–30). When a high-quality image is required, as in the case of an astronomical telescope (Fig. 24–21), a concave mirror whose cross-sectional shape is a parabola is the answer. However, despite the presence of spherical aberration, spherical mirrors are common because they are the easiest to manufacture.

IMPORTANT TERMS

A **lens** is a transparent object of regular form that can produce an image of an object placed before it. A **converging lens** brings parallel light to a single **real focal point,** while a **diverging lens** deviates parallel light outward as though it originated at a single **virtual focal point.** The distance from a lens to its focal point is its **focal length.**

The **magnification** of an optical system is the ratio between the size of the image and that of the object.

A **microscope** is a lens system used to produce enlarged images of nearby objects. A **telescope** is a lens system used to produce larger images of distant objects than would be seen by the unaided eye, although the image itself may be smaller than the actual object.

A **concave mirror** curves inward toward its center and converges parallel light to a single real focal point.

A **convex mirror** curves outward toward its center and diverges parallel light as though the reflected light came from a single virtual focal point behind the mirror. The distance from a mirror to its focal point is the focal length of the mirror.

IMPORTANT FORMULAS

Lensmaker's equation:

$$\frac{1}{f} = (n - 1)\left(\frac{1}{R_1} + \frac{1}{R_2}\right) \qquad (R = + \text{ for convex,} \\ - \text{ for concave surface})$$

Focal length of concave mirror: $f = \dfrac{R}{2}$

Focal length of convex mirror: $f = -\dfrac{R}{2}$

Lens and mirror equation: $\dfrac{1}{p} + \dfrac{1}{q} = \dfrac{1}{f}$

Alternate forms of lens and mirror equation:

$$p = \frac{qf}{q - f}$$

$$q = \frac{pf}{p - f}$$

$$f = \frac{pq}{p + q}$$

Linear magnification: $m = -\dfrac{q}{p}$

$$m = m_1 \times m_2$$

Angular magnification of telescope:

$$m_{\text{ang}} = \frac{f_{\text{objective}}}{f_{\text{eyepiece}}}$$

MULTIPLE CHOICE

1. In order to calculate the focal length of a glass lens, it is not necessary to know
(a) the index of refraction of the glass.
(b the index of refraction of the medium in which the lens is located.
(c) the radii of curvature of the lens surfaces.
(d) the diameter of the lens.

2. A converging lens may not have
(a) a positive focal length.
(b) a negative focal length.
(c) one plane surface.
(d) one concave surface.

3. An object infinitely far from a converging lens has an image that is
(a) real.
(b) virtual.
(c) erect.
(d) larger than the object.

4. An object farther from a converging lens than its focal point always has an image that is
(a) inverted. (b) virtual.
(c) the same in size. (d smaller in size.

5. An object closer to a converging lens than its focal point always has an image that is
(a) inverted. (b) virtual.
(c) the same in size. (d) smaller in size.

6. The image of a real object formed by a diverging lens is always
(a) real.
(b) virtual.
(c) inverted.
(d) larger than the object.

7. A positive magnification signifies an image that is
(a) erect.
(b) inverted.
(c) smaller than the object.
(d) larger than the object.

8. A negative image distance signifies an image that is
(a) real. (b) virtual.
(c) erect. (d) inverted.

9. The pupil of the eye controls
(a) the focal length of the eye.
(b) the range of accommodation of the eye.
(c) the distance of most distinct vision.
(d) the amount of light reaching the eye.

10. The human eye has a focal length of about 5 cm. Its power in diopters is
(a) 0.05. (b) 0.2.
(c) 5. (d) 20.

11. Which of the following combinations of shutterspeed and lens opening will admit the most light to the film of a camera?
(a) 1/125 s at $f/8$
(b) 1/125 s at $f/16$
(c) 1/250 s at $f/4$
(d) 1/250 s at $f/5.6$

12. An object is located 10 cm from a converging lens of focal length 12 cm. The image distance is
(a) $+5.45$ cm. (b) -5.45 cm.
(c) $+60$ cm. (d) -60 cm.

13. An object is located 12 cm from a converging lens of focal length 10 cm. The image distance is
  (a) $+5.45$ cm.
  (b) -5.45 cm.
  (c) $+60$ cm.
  (d) -60 cm.

14. The image of an object 10 cm from a lens is located 10 cm behind the object. The focal length of the lens is
  (a) $+6.7$ cm.
  (b) -6.7 cm.
  (c) $+20$ cm.
  (d) -20 cm.

15. A pencil 10 cm long is placed 70 cm in front of a lens of focal length $+50$ cm. The image is
  (a) 4 cm long and erect.
  (b) 4 cm long and inverted.
  (c) 25 cm long and erect.
  (d) 25 cm long and inverted.

16. A pencil 10 cm long is placed 100 cm in front of a lens of focal length $+50$ cm. The image is
  (a) 5 cm long and erect.
  (b) 5 cm long and inverted.
  (c) 10 cm long and erect.
  (d) 10 cm long and inverted.

17 A pencil 10 cm long is placed 175 cm in front of a lens of focal length $+50$ cm. The image is
  (a) 4 cm long and erect.
  (b) 4 cm long and inverted.
  (c) 25 cm long and erect.
  (d) 25 cm long and inverted.

18. A magnifying glass is to be used at the fixed object distance of 1 cm. If it is to produce an erect image magnified 5 times, its focal length should be
  (a) $+0.2$ cm.
  (b) $+0.8$ cm.
  (c) $+1.25$ cm.
  (d) $+5$ cm.

19. Four lenses with the listed focal lengths are being considered for use as a microscope objective. The one that will produce the greatest magnification with a given eyepiece has the focal length
  (a) -5 mm.
  (b) $+5$ mm.
  (c) -5 cm.
  (d) $+5$ cm.

20. Four lenses with the listed focal lengths are being considered for use as a telescope objective. The one that will produce the greatest magnification with a given eyepiece has the focal length
  (a) -1 m.
  (b) $+1$ m.
  (c) -2 m.
  (d) $+2$ m.

21. A concave mirror produces an erect image when the object distance is
  (a) less than f.
  (b) equal to f.
  (c) between f and $2f$.
  (d) greater than $2f$.

22. The image formed by a concave mirror is larger than the object
  (a) when p is less than $2f$.
  (b) when p is more than $2f$.
  (c) for no values of p.
  (d) for all values of p.

23. The image formed by a convex mirror is larger than the object
  (a) when p is less than $2f$.
  (b) when p is more than $2f$.
  (c) for no values of p.
  (d) for all values of p.

24. A pencil 10 cm long is placed 30 cm in front of a mirror of focal length $+50$ cm. The image is
  (a) 2.5 cm long and erect.
  (b) 25 cm long and erect.
  (c) 250 cm long and erect.
  (d) 25 cm long and inverted.

25. A pencil 10 cm long is placed 1 m in front of a mirror of focal length $+50$ cm. The image is
  (a) 3 cm long and erect.
  (b) 10 cm long and erect.
  (c) 3 cm long and inverted.
  (d) 10 cm long and inverted.

2. A pencil 10 cm long is placed 30 cm in front of a mirror of focal length -50 cm. The image is
  (a) 25 cm long and erect.
  (b) 6.25 cm long and erect.
  (c) 25 cm long and inverted.
  (d) 6.25 cm long and inverted.

EXERCISES

1. Is there any way in which a diverging lens, used by itself, can form a real image of a real object? Is there any way in which a converging lens, used by itself, can form a virtual image of a real object?

2. Under what circumstances, if any, is a light ray that passes through a converging lens not deflected? Under what circumstances, if any, is a light ray that passes through a diverging lens not deflected?

3. Under what circumstances, if any, will a diverging lens form an inverted image of a real object? Under what circumstances, if any, will a converging lens form an erect image?

4. Is the mercury column in a thermometer wider or narrower than it appears?

5. A slide is in sharp focus on a screen. The screen is then moved farther away from the projector. Should the projector's lens be moved closer to or farther from the slide to bring it back into focus on the screen?

6. What are the characteristics of the image formed on the retina by the lens of the eye?

7. A fortune-teller's crystal ball is 15 cm in diameter. (a) Would you expect the lensmaker's equation to hold for the focal length of this ball? (b) Would you expect the ball to form undistorted images?

8. Does a spherical bubble of air in a volume of water act to converge or diverge light passing through it?

9. A double-convex lens made of crown glass is placed in a tank of benzene. Will it act as a converging or as a diverging lens there?

10. What is the *f*-number of the Yerkes Observatory telescope described in Sec. 24–9?

11. What can you say about the properties of an image when the magnification is (a) less than -1? (b) between -1 and 0? (c) between 0 and $+1$? (d) greater than $+1$?

12. A double-concave lens has surfaces whose radii of curvature are both 40 cm. The lens is made from flint glass whose index of refraction is 1.55. Find its focal length.

13. A converging meniscus lens has surfaces whose radii of curvature are 20 cm and 30 cm. The lens is made from crown glass whose index of refraction is 1.50. Find its focal length.

14. A diverging meniscus lens has surfaces whose radii of curvature are 20 cm and 30 cm. The lens is made from crown glass whose index of refraction is 1.50. Find its focal length.

15. A plano-concave lens of focal length 9 cm is to be ground from quartz of index of refraction 1.55. Find the required radius of curvature.

16. A double-convex lens whose focal length is 35 cm has surfaces whose radii of curvature are 25 m and 50 cm respectively. Find the index of refraction of the glass.

17. A lens is to be used to focus sunlight on a piece of paper to ignite it. What kind of lens should be used? If the focal length of the lens is 8 cm, how far should it be held from the paper?

18. A candle is placed with its flame 10 cm from a lens whose focal length is $+15$ cm. Find the location of the image of the flame. Is the image larger or smaller than the actual flame? What is the character of the image; that is, is it erect or inverted, real or virtual?

19. The flame of the candle of exercise 18 is 15 cm from the lens. Answer the same questions for this case.

20. The flame of the candle of exercise 18 is 25 cm from the len. Answer the same questions for this case.

21. The flame of the candle of exercise 18 is 30 cm from the lens. Answer the same questions for this case.

22. The flame of the candle of exercise 18 is 50 cm from the lens. Answer the same questions for this case.

23. A lens held 20 cm from a sardine produces a real, inverted image of it 30 cm on the other side. What is the focal length of the lens? Is it converging or diverging?

24. A lens held 20 cm from a sardine produces a virtual, erect image of it that appears to originate 10 cm in front of the sardine. What is the focal length of the lens? Is it converging or diverging?

25. Is there any way in which a convex mirror, used by itself, can form a real image of a real object? Is there any way in which a concave mirror, used by itself, can form a virtual image of a real object?

26. A convex mirror has a radius of curvature of 40 cm. What is its focal length? Where is its focal point?

27. A butterfly is 20 cm in front of a concave mirror whose focal length is 40 cm. Find the location of the image. Is the image larger or smaller than the butterfly? What is the character of the image—that is, is it erect or inverted, real or virtual?

28. A dime is 40 cm in front of the mirror of exercise 27. Answer the same questions for the image of the dime.

29. A peanut is 50 cm in front of the mirror of exercise 27. Answer the same questions for the image of the peanut.

30. A caterpillar is 80 cm in front of the mirror of exercise 27. Answer the same questions for the image of the caterpillar.

31. A button is 100 cm in front of the mirror of exercise 27. Answer the same questions for the image of the button.

32. An object 6 cm high is 30 cm in front of a convex mirror whose focal length is 50 cm. What is the height and character of the image?

33. A dentist's concave mirror has a diameter of 1 cm and a focal length of 2.5 cm. What magnification does it produce when held 1.8 cm from a tooth?

34. A man stands 6 m from a concave mirror, and an inverted image of himself of the same height is formed on a screen beside him. What is the radius of curvature of the mirror?

PROBLEMS

1. A converging lens made from glass of $n = 1.60$ has a focal length in air of 15 cm. (a) Is it converging or diverging when immersed in water? (b) What is its focal length in water?

2. A diverging lens made from glass of $n = 1.55$ has a focal length of -8 cm in air. (a) Is it converging or diverging when immersed in carbon disulfide ($n = 1.63$)? (b) What is its focal length in carbon disulfide?

3. A motion picture director holds a diverging lens 5 m from an actress and sees her one-tenth of her normal size. What is the focal length of the lens?

4. A camera whose lens has a focal length of 90 mm is used to photograph a person 4 m away. (a) How far in front of the film should the lens be placed? (b) If the film is 40 mm square, what is the area of the subject that it covers?

5. An aerial camera whose lens has a focal length of 1 m is used to photograph a military base from an altitude of 7 km. How long is the image on the film of a tank 9 m long?

6. The moon's diameter is 3476 km and its average distance from the earth is 3.8×10^5 km. (a) What is the diameter of its image when a 35-mm camera with a 300-mm telephoto lens is used to photograph it? (b) What would the image diameter be if a camera using 9×12-cm film were used with the same lens?

7. A photographic enlarger is being designed to produce prints 40 cm $\times$ 50 cm from negatives 10 cm $\times$ 12.5 cm. (a) If the maximum distance from negative to print paper is to be 60 cm, what should the focal length of the lens be? (b) How far would such a negative be from the print paper using the same lens if the enlargement size were 20 cm $\times$ 25 cm?

8. The actual size of each frame of 16-mm motion picture film is 7.5 mm $\times$ 10.5 mm. A projector whose lens has a focal length of 25 mm is to be used with a screen 1.5 m wide. How far from the projector should the screen be located?

9. A magnifying glass of 10-cm focal length is held 8 cm from a stamp. What is the actual length of a feature of the stamp that appears to be 1 cm long?

10. A magnifying glass of 50-mm focal length is held 35 mm from a spider egg 0.5 mm long. What is the apparent size of the egg?

11. A farsighted eye has a near point of 60 cm. What is its near point when a correcting lens of $+3.33$ diopters is used?

12. A nearsighted person whose eyes have far points of 60 cm is lent a pair of glasses whose power is -1.5 diopters. How far can he see clearly with these glasses?

13. A certain presbyopic eye has a near point 1.0 m away. What type of lens is needed to permit the eye to see clearly objects 25 cm away, and what focal length should such a lens have?

14. A certain myopic eye cannot bring to a focus objects farther than 15 cm away. What type of lens is needed to permit this eye to see clearly objects at infinity, and what focal length should such a lens have?

15. In a certain microscope an objective of 4 mm focal length is used with a $10 \times$ eyepiece. (a) What is the magnification of the microscope? (b) How far should the objective be from the specimen being examined?

16. In a certain microscope an objective of 10 mm focal length is used with a $5 \times$ eyepiece. (a) What is the magnification of the microscope? (b) How far should the objective be from the specimen being examined?

17. A telescope with an objective of focal length 1.0 m and an eyepiece of focal length 5.0 cm is used to examine a penguin 40 cm high. Find the apparent height of the penguin when it is (a) 50 m, and (b) 5 m away from the telescope. Assume the image to be 25 cm in front of the eyepiece.

18. A telescope has an objective of 750-mm focal length and an eyepiece of 25-mm focal length. The telescope is focussed on a distant albatross. (a) What is the angular magnification? (b) How far apart are the lenses? Assume that the final image is located 25 cm in front of the eyepiece.

19. Verify that the effective focal length f of two thin lenses of focal lengths f_1 and f_2 that are in contact is given by
$$\frac{1}{f} = \frac{1}{f_1} + \frac{1}{f_2}$$
(To do this, let the image produced by the first lens be the object of the second.)

20. An object should be about 25 cm from a normal eye for maximum distinctness of vision. (a) Find a formula for the magnification of a converging lens when it is used as a magnifying glass with an image distance of -25 cm; see Fig. 24–6(a). (b) Find the magnification of a lens whose focal length is 5 cm.

21. A worm crawls toward a polished metal ball 60 cm in diameter lying on a lawn. How far from the surface of the ball is the worm when its image appears to be 10 cm behind the surface?

22. A mirror in an amusement park produces an erect image four times enlarged of anyone standing 3 m away. (a) Is the mirror concave or convex? (b) What is its radius of curvature?

23. A virtual image 6 cm long is formed of a paper clip 2 cm long placed 10 cm in front of a concave mirror of unknown curvature. Where else can the paper clip be placed for an image 6 cm long to be formed? What is the nature of the image in the latter case?

24. A pawn 5 cm high is placed in front of a concave mirror whose radius of curvature is 1 m. What are the two object distances that will lead to images 20 cm high? What is the character of the image in each case?

25. What should the radius of curvature of a convex mirror be if it is to produce an image one-fifth the size of an object 150 cm away?

26. The moon is 3476 km in diameter. What radius of curvature should a concave mirror have if it is to produce a lunar image 1 cm in diameter when the moon is 3.84×10^5 km away?

ANSWERS TO MULTIPLE CHOICE

1. d		**10.** d		**19.** b	
2. b		**11.** c		**20.** d	
3. a		**12.** d		**21.** a	
4. a		**13.** c		**22.** a	
5. b		**14.** c		**23.** c	
6. b		**15.** d		**24.** b	
7. a		**16.** d		**25.** d	
8. b		**17.** b		**26.** b	
9. d		**18.** c			

PHYSICAL OPTICS

25

In the previous chapter we had little need to invoke the wave character of light. But, although the properties of lenses and mirrors are more readily analyzed in terms of rays than in terms of waves, there are other optical phenomena in which the wave nature of light is directly involved. The study of the latter phenomena is called physical optics, whereas the study of those aspects of light behavior that can be understood using a ray treatment is geometrical optics. Since the basic laws of refraction and reflection follow from Huygens' principle, geometrical optics is evidently an approximation of physical optics whose usefulness comes from its simplified view of light propagation. In this chapter we shall find the wave approach of physical optics necessary for explaining interference, diffraction, and polarization.

25–1 INTERFERENCE OF LIGHT

Light was recognized as a wave phenomenon well before its electromagnetic character became known a century ago. The problem of the nature of light was an old one: Newton felt sure light consists of a stream of tiny particles, whereas his contemporary Christian Huygens (1629–1695) thought it to be a succession of waves. Neither man offered any hypothesis as to what kind of particle or wave is involved. Eventually interference, diffraction, and polarization were discovered in light, and these phenomena could only be explained on the basis that light consists of transverse waves. But Newton was not completely wrong. As we shall learn in Chapter 27, light has certain distinctly particle properties in addition to its wave properties, and one of the central problems of contemporary physics has been the resolution of this apparent paradox.

When light waves from one source are mixed with those from another source, the two wave trains are said to *interfere*. We recall from Chapter 12 the principle of superposition, which governs interference: When two or more waves of the same nature travel past a point at the same time, the amplitude at that point is the sum of the instantaneous amplitudes of the individual waves. Constructive interference refers to the reinforcement of waves in phase (in step) with one another, and destructive interference refers to the partial or complete cancellation of waves out of phase with one another (Fig. 25–1).

Constructive and destructive interference

Anyone with a pan of water can see how interference between water waves can lead to a water surface disturbed in a variety of characteristic patterns. Two people who hum fairly pure tones slightly different in frequency will hear beats as the result of interference in the sound waves. But if we shine light from two flashlights at the same place on a screen, there is no evidence of interference: the region of overlap is merely uniformly bright.

There are two reasons for the difficulty of observing interference in light. First, light waves have extremely short wavelengths—the visible part of the spectrum extends only from 400 nm for violet light to 700 nm for red light. Second, every natural source of light emits light waves only as short trains of random phase, so that any interference that occurs is usually averaged out during even the briefest period of observation by the eye or photographic film unless special procedures are used. Interference in light is nevertheless just as real a phenomenon as interference in water or sound waves, and there is one example of it familiar to everybody—the bright colors of a thin film of oil spread out on a water surface.

Why interference is hard to demonstrate in light

Two sources of waves are said to be *coherent* if there is a fixed phase relationship between the waves they emit during the time the waves are being observed. It does not matter whether the waves are exactly in step when they leave the sources, or exactly out of step, or anything in between; the important thing is that the phase relationship stay the same. If the souces shift back and forth in relative phase while the obsevation

Coherent and incoherent light sources

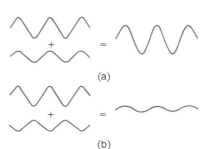

FIG. 25–1 (a) Constructive interference. (b) Destructive interference.

(a)

(b)

FIG. 25–2 (a) Radio waves from an antenna are coherent. (b) Light waves from a gas discharge tube are incoherent.

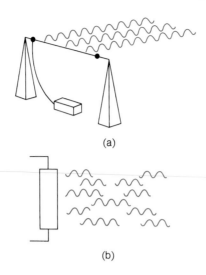

(a)

(b)

is made, the phase differences average out, and there will be no interference pattern. The latter sources are *incoherent*.

Ordinary light sources do not emit continuous wave trains

The question of coherence is especially significant for light waves because an excited atom radiates for 10^{-8} s or less, depending upon its environment. Therefore a monochromatic light source such as a gas discharge tube (a neon sign is an example) does not emit a continuous wave train as a radio antenna does but instead a series of individual wave trains whose phases are random. The light from such a tube actually comes from a great many individual, uncoordinated sources, namely the gas atoms, and these individual sources are in effect being switched on and off rapidly and irregularly (Fig. 25–2).

Coherence depends on the time scale

Suppose we have two point sources of monochromatic light, for instance a discharge tube with a shield that has two pinholes close together. Different atoms are behind each pinhole, so they are independent sources. Therefore we have at most 10^{-8} s to observe the interference of waves from the two sources. If our detecting instruments are fast enough, as some modern electronic devices are, interference can be demonstrated and the soures can be considered coherent. If we are limited to the eye and to photographic film, which average arriving light signals over times far greater than 10^{-8} s, no interference can be observed in the light from the two sources, and they must then be considered incoherent. Like beauty, coherence lies in the eye of the beholder.

Does the brief lifetime of an excited atom mean that interference patterns can never be literally seen but can only be recorded by instruments? Hardly. There are three ways to construct separate sources of light coherent for long enough periods of time to produce visible interference patterns. These are:

Coherent light sources

1. Illuminate two (or more) slits with light from one slit behind them. Then the light waves from the secondary slits are automatically coordinated.
2. Obtain coherent virtual sources from a single source by reflection or refraction. This is how interference is produced by thin films of oil.
3. Coordinate the radiating atoms in each separate source so that they always have the

same phase even though different atoms are radiating at successive instants. This is done in the *laser*.

We shall examine the first two of these methods here, and we shall discuss the laser in Chapter 28.

25-2 DOUBLE SLIT

The interference of light waves was demonstrated in 1801 by Thomas Young, who used an arrangement similar to that shown in Fig. 25–3. A source of monochromatic light (that is, light consisting of only a single wavelength) is placed behind a narrow slit S in an opaque screen, and another screen with two similar slits A and B is placed on the other side. Light from S passes through both A and B and then to the viewing screen. If light were not a wave phenomenon, we would expect to find the viewing screen completely dark, since no light ray can reach it from the source along a straight path. What actually happens is that each slit acts as a source of secondary wavelets—we recall Huygens' principle from Sec. 23–4—so that the entire screen is illuminated. Even the point S', separated from S by the opaque barrier between the slits A and B, turns out to be bright rather than dark (Fig. 25–4).

> **If light were not a wave, the screen in Young's experiment would be totally dark**

Owing to interference the screen is not evenly illuminated but shows a pattern of alternate bright and dark lines. Light waves from slits A and B are exactly in phase, since A and B are the same distance from S. The centerline S' of the screen is equally distant from A and B, so light waves from these slits interfere constructively there to produce a bright line.

> **An interference pattern consists of bright and dark lines**

Let us next see what happens at the position C on the screen located to one side of S'. The distance BC is longer than the distance AC by the amount BD, which is equal to exactly half a wavelength of the light being used. That is,

$$BD = \tfrac{1}{2}\lambda$$

> **Condition for the first dark line**

When a crest from A reaches C, this difference in path length means that a trough from B arrives there at the same time, since $\tfrac{1}{2}\lambda$ separates a crest and a trough in the same

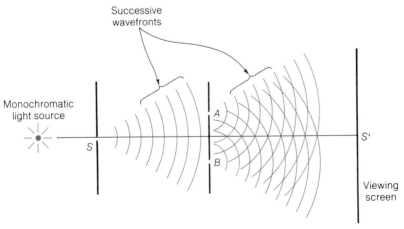

Monochromatic light source

Successive wavefronts

S

A

B

S'

Viewing screen

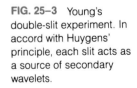

FIG. 25–3 Young's double-slit experiment. In accord with Huygens' principle, each slit acts as a source of secondary wavelets.

FIG. 25–4   The appearance of the screen in Young's experiment.

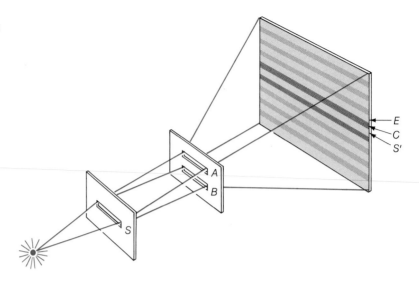

wave. The two cancel each other out, the light intensity at C is zero and a dark line results on the screen there. At S' the equality of path length gives rise to constructive interference; at C the difference of $\frac{1}{2}\lambda$ in path length gives rise to destructive interference (Fig. 25–5).

If we go past C on the screen we will come to a point E such that the distance BE is greater by exactly one wavelength than the distance AE. That is, the difference BF between BE and AE is

Condition for the next bright line

$$BF = \lambda$$

Consequently, when a crest from A reaches E, a crest from B also arrives there, although the latter crest left B earlier than that from A owing to the longer path it had to cover. Because $BF = \lambda$, waves arriving at E from both slits are always in the same part of their cycles, and they constructively interfere to produce a bright line at E.

By continuing the same analysis, we find that the alternate bright and dark lines actually observed on the screen correspond respectively to locations where constructive and destructive interference occurs. Waves reaching the screen from A and B along

Origin of pattern of bright and dark lines

FIG. 25–5   Origin of the double-slit interference pattern.

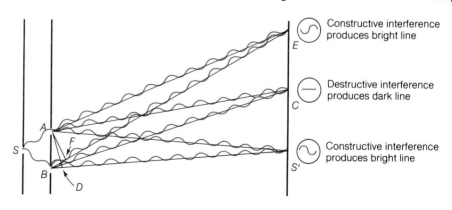

Constructive interference produces bright line

Destructive interference produces dark line

Constructive interference produces bright line

paths that are equal or differ by a whole number of wavelengths (λ, 2λ, 3λ, and so on) reinforce, while those whose paths differ by an odd number of half wavelengths ($\frac{1}{2}\lambda$, $\frac{3}{2}\lambda$, $\frac{5}{2}\lambda$, and so on) cancel. At intermediate locations on the screen the interference is only partial, so that the light intensity on the screen varies gradually between the bright and dark lines.

25–3 WAVELENGTH OF LIGHT

Interference patterns like those produced by the double slit permit us to determine the wavelength of the light used. To illustrate the procedure, we shall consider the double-slit experiment in detail. Figure 25–6 is a diagram of the experiment: d is the separation between the slits, L is the distance from the slits to the screen, and y is the distance from the central point S' on the screen to the point Q whose illumination we are observing.

The waves that travel from slit B to Q must travel s farther than those from slit A. As we have seen, when the path difference s is

Conditions for constructive and destructive interference

$$s = 0, \lambda, 2\lambda, 3\lambda, \ldots \qquad \textit{Constructive interference} \quad (25\text{–}1)$$

where λ is the wavelength of the light from the source, waves from A and B arrive at Q in the same stage of their cycles and reinforce one another to produce a bright line there. On the other hand, when the path difference s is

$$s = \tfrac{1}{2}\lambda, \tfrac{3}{2}\lambda, \tfrac{5}{2}\lambda, \ldots \qquad \textit{Destructive interference} \quad (25\text{–}2)$$

waves from A and B arrive at Q in the opposite stages of their cycles and cancel one another out to produce a dark line there. The triangles ABD and EQS' are similar, since each is a right triangle and two sides of each are perpendicular to two sides of the other. Corresponding sides of similar triangles are proportional, and so

$$\frac{s}{y} = \frac{d}{EQ} \qquad (25\text{–}3)$$

In an actual experiment, y is much smaller than L (their ratio is usually more than 1:1000; in Fig. 25–6, y is exaggerated for clarity), which means that EQ is very nearly equal to L. At the expense of introducing a negligible error, we may substitute L for EQ in Eq. (25–3), with the result that

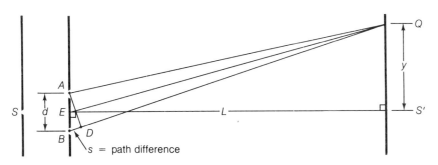

FIG. 25–6 A diagram of the double-slit experiment.

FIG. 25–7

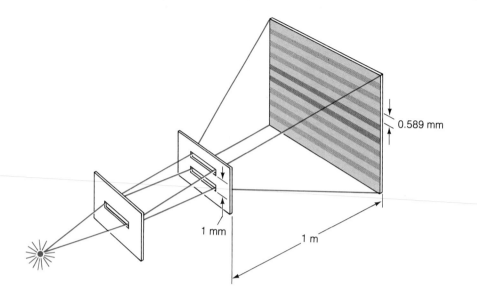

0.589 mm

1 mm

1 m

$$\frac{s}{y} = \frac{d}{L} \qquad s = \frac{dy}{L} \tag{25–4}$$

Combining Eqs. (25–1) and (25–2) with Eq. (25–4) yields the conditions for bright and dark lines to occur at Q:

$$y = 0, \frac{L\lambda}{d}, \frac{2L\lambda}{d}, \frac{3L\lambda}{d}, \dots \qquad\qquad\qquad \textit{Bright lines} \tag{25–5}$$

$$y = \frac{L\lambda}{2d}, \frac{3L\lambda}{2d}, \frac{5L\lambda}{2d}, \dots \qquad\qquad\qquad \textit{Dark lines} \tag{25–6}$$

The central line is always bright

According to Eq. (25–5), there is a bright line when $y = 0$, corresponding to the center of the screen S', a bright line on either side of S' a distance $L\lambda/d$ from it, another bright line on either side a distance $2L\lambda/d$ from S', and so on (Fig. 25–4). Similarly, Eq. (25–6) states that there is a dark line on either side of S' a distance $L\lambda/2d$ from it, another dark line on either side a distance $3L\lambda/2d$ from it, and so on.

In an actual experiment, L and d are known initially, and the distance y from the center of the screen to any particular light or dark line can be measured. The wavelength of the light may then be found by using the equation corresponding to the particular line.

Example   Monochromatic yellow light from a sodium-vapor lamp illuminates two narrow slits 1 mm apart. The viewing screen is 1 m from the slits, and the distance from the central bright line to the bright line nearest it is found to be 0.589 mm (Fig. 25–7). Find the wavelength of the light.

Solution   For the second bright line, $y = L\lambda/d$. Hence

$$\lambda = \frac{yd}{L} = \frac{(5.89 \times 10^{-4}\,\text{m})(10^{-3}\,\text{m})}{1\,\text{m}} = 5.89 \times 10^{-7}\,\text{m} = 589\,\text{nm}$$

25–4 DIFFRACTION GRATING

There are two difficulties in using a double slit for measuring wavelengths. First, the "bright" lines on the screen are actually extremely faint and an intense light source is therefore required; second, the lines are relatively broad and it is hard to locate their centers accurately. A *diffraction grating* that consists, in essence, of a large number of parallel slits overcomes both of these difficulties. Gratings are made by ruling grooves on a glass or metal plate with a diamond; the clear bands between the grooves are the "slits." Replica gratings, made by allowing a transparent liquid plastic to harden in contact with an original grating, are ordinarily used in practice. Replica gratings are often given a thin coating of silver or aluminum and produce their characteristic diffraction patterns by the interference of reflected rather than transmitted light. A phonograph record held at a glancing angle acts as a reflecting grating by virtue of its closely spaced grooves.

A diffraction grating uses interference to disperse light

Gratings are ruled with from 2,000 to 10,000 lines/cm, and a lens is used to focus the light from the slits between them on a screen. The effect of phase differences among the rays from the various slits is accentuated by their great number, and as a result the intensity of light on the screen falls rapidly on either side of the center of each bright line. The bright lines are therefore sharp, and, because there are so many slits, they are also bright in a literal sense. Very accurate wavelength determinations can be made with the help of a grating, and wavelengths that are close together can be resolved. The analysis of a light beam in terms of the particular wavelengths it contains is today almost invariably carried out by using a grating.

A grating produces sharper and brighter interference maxima than a double slit

Figure 25–8 shows a diffraction grating that forms a bright line on a screen at a deviation angle of θ from the original beam of light. The condition for a bright line is that $s = n\lambda$, where $n = 1, 2, 3$, and so on. Since $s = d \sin \theta$, where d is the spacing of the slits, bright lines occur at those angles for which

$$\sin \theta = n\frac{\lambda}{d} \qquad n = 1, 2, 3, \ldots \qquad\qquad \textit{Bright lines} \quad (25\text{–}7)$$

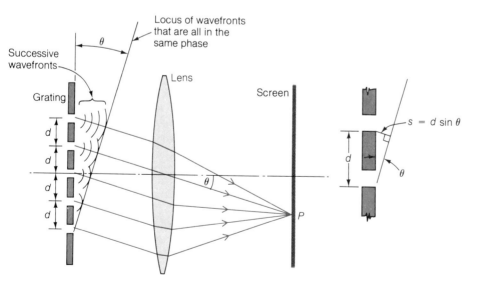

FIG. 25–8 The plane diffraction grating.

Spectral orders

When polychromatic light is incident on a grating, a series of spectra is formed on each side of the original beam corresponding to $n = 1$, $n = 2$, and so on. The *first-order spectrum* contains bright lines for which $n = 1$, the *second-order spectrum* contains bright lines for which $n = 2$, and so on. In some gratings the higher-order spectra overlap, so that, for example, the blue end of the third-order spectrum may be deviated by less than the red end of the second-order spectrum. According to Eq. (25–7), the angle θ increases with wavelength, so that blue light is deviated least and red light most, which is the reverse of what occurs when a prism is used to form a spectrum.

Width of grating spectrum

Example Visible light includes wavelengths from approximately 4×10^{-7} m (violet light) to 7×10^{-7} m (red light). Find the angular width of the first-order spectrum produced by a grating ruled with 8000 lines/cm.

Solution The slit spacing d that corresponding to 8000 lines/cm is

$$d = \frac{10^{-2}\,\text{m/cm}}{8 \times 10^3\,\text{lines/cm}} = 1.25 \times 10^{-6}\,\text{m}$$

Since $n = 1$ for a first-order spectrum, the angular deviations of blue and red light respectively are given by

$$\sin \theta_b = \frac{\lambda_b}{d} = \frac{4 \times 10^{-7}\,\text{m}}{1.25 \times 10^{-6}\,\text{m}} = 0.32 \qquad \theta_b = 19°$$

and

$$\sin \theta_r = \frac{\lambda_r}{d} = \frac{7 \times 10^{-7}\,\text{m}}{1.25 \times 10^{-6}\,\text{m}} = 0.56 \qquad \theta_r = 34°$$

The total width of the spectrum is therefore $34° - 19° = 15°$. ∎

Why a mirror does not produce a spectrum

A legitimate question is why spectra are not formed whenever specular reflection (see Sec. 23–5) occurs, because the atoms of a solid are arranged in regular patterns and so its surface ought to act as a reflection grating. Reflection of light waves takes place when atomic electrons at the surface of the reflecting material are caused to oscillate by the incident electromagnetic waves, and the reradiated waves have their central maximum ($n = 0$) at an angle of reflection equal to the angle of incidence, as Fig. 23–12 shows. However, because atoms in a solid are only about 10^{-10} m apart, λ/d is equal to several thousand for visible light, which means that spectra corresponding to $n = 1$ or more do not occur since $\sin \theta$ cannot exceed 1. But wavelengths comparable with the atomic spacings in solids are found in X rays, and indeed X-ray diffraction is the means whereby the structures of many solids have been determined. The double helix form of the DNA molecule was discovered with the help of X-ray interference patterns.

25–5 THIN FILMS

Thin films produce interference patterns

We have all seen the marvelous rainbow colors that appear in soap bubbles and thin oil films. Some of us may also have observed the patterns of light and dark bands that occur when two glass plates are almost (but not quite) in perfect contact. Both phenomena owe their origins to a combination of reflection and interference.

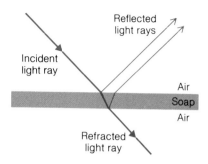

FIG. 25–9 Reflection occurs at both surfaces of a soap film.

Let us consider a beam of monochromatic light that strikes a thin film of soapy water. Figure 25–9 shows a ray picture of what happens. We notice that some reflection takes place at both the air-soap and soap-air interfaces. This is a general result: waves are always partially reflected when they go from one medium to another in which their speed is different. (See the discussion in Sec. 12–2 of the reflection of pulses in a string under similar circumstances.)

A light ray actually consists of a succession of wavefronts. Figure 25–10 is the same diagram with the wavefronts drawn in. In (a) the two reflected wave trains are out of phase and they interfere destructively to partially or completely cancel out. Most or all of the light reaching this part of the soap bubble therefore passes right through.

Another part of the soap film may have a different thickness. When the film is a little thinner than in (a), the waves in the two reflected trains are exactly in phase, and they interfere constructively to reinforce one another, as in Fig. 25–10(b). Light reaching this part of the soap bubble is strongly reflected. Shining monochromatic light on a soap bubble therefore yields a pattern of light and dark that results from the varying thickness of the bubble.

When white light is directed at a soap bubble, light waves of each wavelength present pass through the soap film without reflection at those places where the film is exactly the right thickness for the two reflected rays to destructively interfere. Light waves of the other wavelengths are reflected to at least some extent, and give rise to the vivid colors seen. The varying thickness of the bubble means that the color of the light reflected from the bubble changes from place to place. Exactly the same effect is responsible for the coloration of thin oil films. Generally speaking, soap or oil films whose thickness is comparable with the wavelengths in visible light give rise to the most striking color effects.

Why thin films appear colored in white light

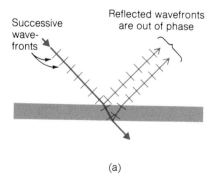

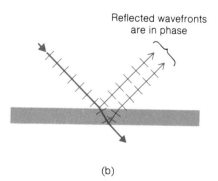

FIG. 25–10 (a) Destructive and (b) constructive interference in a thin film.

(a)   (b)

FIG. 25–11   Newton's rings.

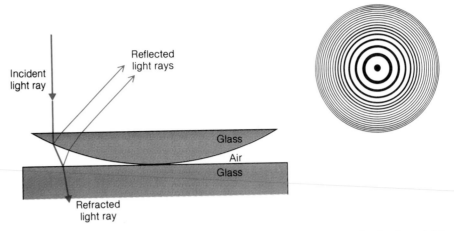

Newton's rings

A thin film of air between two sheets of glass or transparent plastic also yields a pattern of colored bands when illuminated with white light. A notable example is *Newton's rings,* which occur when a slightly curved lens is placed on a flat glass plate (the curvature is exaggerated in Fig. 25–11). Again reflection takes place at both the top and bottom of the film, and again the result is constructive or destructive inteference, depending upon the film thickness. Because the thickness of the air film increases with distance from the central point of contact, the pattern of light and dark bands consists of concentric circles.

Why the center of Newton's rings is a dark spot

The dark spot at the center of a set of Newton's rings is not what we might expect to find (Fig. 25–12). At the center, where the air film between the pieces of glass is minute, the path difference between the waves reflected from the upper and lower surfaces of the film is negligible. Hence there ought to be constructive interference and reinforcement of the light to yield a bright spot. What this analysis overlooks is the fact

FIG. 25–12   Origin of dark spot at center of Newton's rings.

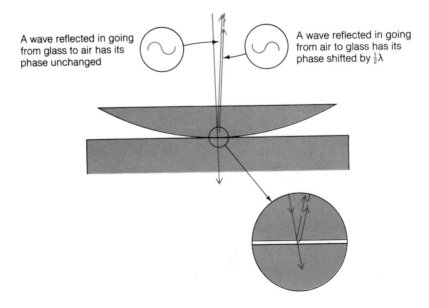

A wave reflected in going from glass to air has its phase unchanged

A wave reflected in going from air to glass has its phase shifted by $\frac{1}{2}\lambda$

FIG. 25–13

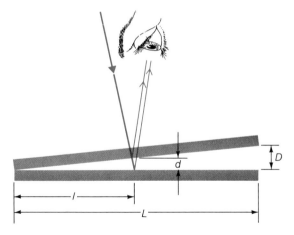

that a wave reflected at the surface of a new medium in which its speed is less (in optical terms, a medium of higher index of refraction) is shifted by half a wavelength. That is, a positive displacement of the wave variable is reflected as a negative one, and vice versa. The same effect was noted in the discussion of pulses in a stretched string and is shown in Fig. 12–6. Thus a $\frac{1}{2}\lambda$ shift occurs when light waves are reflected in going from air to glass, but not in going from glass to air. In consequence the two wave trains reflected at the center of a Newton's ring pattern exactly cancel each other out to yield the dark spot actually observed.

Example   Two flat glass plates 12 cm long are separated at one edge by a piece of foil 0.02 mm thick, as in Fig. 25–13. (a) How far apart are the interference bands when the arrangement is perpendicularly illuminated by red light of $\lambda = 680$ nm? (b) How far apart are the bands if the space between the plates is filled with water?

Interference in a wedge-shaped thin film

Solution   (a) Light reflected from the upper surface of the lower plate is shifted in phase by $\frac{1}{2}\lambda$ whereas light reflected by the lower surface of the upper plate is not shifted in phase. As a result a dark band will occur when the path difference between light rays reflected by the two surfaces is 0, λ, 2λ, 3λ, and so on. At a point where the plates are d apart, the path difference is $2d$, so dark bands occur when $2d = m\lambda$,

$$d = m\frac{\lambda}{2} \qquad m = 0, 1, 2, 3, \ldots$$

Since corresponding sides of similar triangles are proportional to each other,

$$\frac{d}{D} = \frac{l}{L} \qquad l = \frac{dL}{D}$$

Substituting $d = m\lambda/2$, $\lambda = 680$ nm, $L = 12$ cm, and $D = 0.02$ mm yields

$$l = m\frac{\lambda L}{2D} = m\frac{(680 \times 10^{-9}\,\text{m})(0.12\,\text{m})}{2(0.02 \times 10^{-3}\,\text{m})} = m\,(0.002\,\text{m})$$

$$= m\,(2.0\,\text{mm}) \qquad m = 0, 1, 2, 3 \ldots$$

This result means that the first dark line is at $l = 0$, the second at $l = 2.0$ mm, the third at $l = 4.0$ mm, and so on, so that the bands are 2.0 mm apart. The bright bands are similarly spaced, of course.

(b) From Table 23–1 we see that the index of refraction of water is 1.33, which is less than that of glass. Hence the phase shift of $\frac{1}{2}\lambda$ at the upper surface of the lower glass plate still takes place, and the only difference is that the wavelength in the water film is reduced to λ/n. Substituting λ/n for λ in the formula for l in part (a) gives

$$l = m\left(\frac{2.0\,\text{mm}}{1.33}\right) = m\,(1.5\,\text{mm})$$

The dark bands are now 1.5 mm apart.   ∎

Lens reflection is a problem in optical instruments

About 4% of the light striking a glass-air interface is reflected. This is not a lot, but there may be many glass-air interfaces in an optical instrument and the total amount of light lost through reflection may be considerable. There are ten such interfaces in each of the optical systems in a pair of binoculars, for instance, so only about two-thirds of the incoming light actually gets through to the observer's eyes. Even in a camera, where there are fewer glass-air interfaces, reflections are a nuisance because they may lead to secondary images that blur the picture.

Coated lenses reduce reflection

To reduce reflections at a glass-air interface, the glass can be coated with a very thin layer of a transparent substance (usually magnesium fluoride) whose index of refraction is intermediate between those of glass and of air. If the layer is exactly $\frac{1}{4}\lambda$ thick, light reflected at its bottom will have traveled $\frac{1}{2}\lambda$ farther when it rejoins light reflected at the top of the layer, and the two will cancel out exactly, as in Fig. 25–10(a). (Light waves reflected at both the air-fluoride and fluoride-glass surfaces are shifted in phase by $\frac{1}{2}\lambda$, so these shifts have no effect on the cancellation.)

Why coated lenses appear colored

But the cancellation described above is exact only for a particular wavelength λ, whereas white light contains a range of wavelengths. What is therefore done is to choose a wavelength in the middle of the visible spectrum, which corresponds to green light, so that at least partial cancellation occurs over a wide range. The red and violet ends of the spectrum are acordingly least affected and the light reflected from a coated lens is a mixture of these colors, a purplish hue. The average reflectivity of a glass surface coated in this way is only about 1%. Multiple coatings are sometimes used to reduce reflection even further; a triple coating brings the average reflectivity below 0.5%. Despite their lack of perfection at suppressing reflections, coated lenses transmit appreciably more light than uncoated ones, and they are universally used in fine optical instruments.

25–6   DIFFRACTION

Waves diffract around the edge of an obstacle

Waves are able to bend around the edge of an obstacle in their path, a property called *diffraction*. We all have heard sound that originated around the corner of a building from where we were standing, for example. These sound waves cannot have traveled in a straight line from their source to our ears, and refraction cannot account for their behavior. Water waves, too, diffract, as the simple experiment illustrated in Fig. 25–14 shows. The waves on the far side of the gap spread out into the geometrical "shadow"

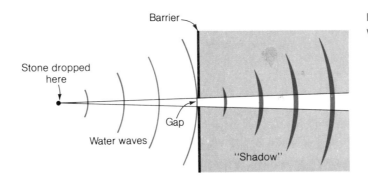

FIG. 25–14 Diffraction in water waves.

of the gap's edges, though with reduced amplitude. The diffracted waves spread out as though they originated at the gap, in accord with Huygens' principle.

If we look very carefully at the edge of the shadow cast by an obstruction in the path of the light spreading out from a pinhole or other point source, we will see that it is not sharp but smeared out (Fig. 25–15). The fuzzy edges of shadows are not easy to observe because the wavelengths in visible light are so short, less than 10^{-6} m, and the extent of diffraction into the shadow zone is correspondingly small. (In contrast, a typical audible sound wave might have a wavelength of 1 m and a typical wave in a pan of water might have a wavelength of 10 cm, and it is easy to observe diffraction effects with such waves.) In fact, because he was not able to perceive any diffraction with his relatively crude apparatus, Newton felt sure that light must be corpuscular in nature and not consist of waves at all.

Shadows do not have sharp edges

A broad light source such as a light bulb or the sun does not produce sharp shadows for another reason. In this case light from different parts of the source passes the edge of the obstacle at different angles, which is not true of light from a point source.

We can refine our observation of diffraction further by using a monochromatic light source and a sheet of photographic film instead of a screen. When we enlarge the developed image on the film, we find a pattern of light and dark fringes at the edge of the shadow. Figure 25–16 shows what the shadow caused by a razor blade looks like. Patterns like this are the result of interference between secondary wavelets from different parts of the same wavefront, not from different sources as in Young's double-slit experiment. The wavefronts in a beam of unobstructed light produce secondary wavelets that

Origin of diffraction fringes

FIG. 25–15 Even under ideal conditions, the edge of a shadow is never completely sharp.

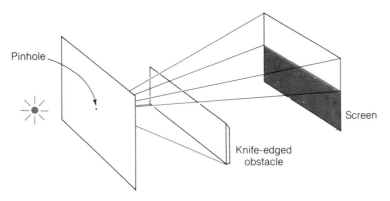

FIG. 25–16 The shadow of a razor blade.

interfere in such a way as to produce new wavefronts exactly like the old ones. By obstructing part of the wavefronts, points in the shadow region are not reached by secondary wavelets from the entire initial wavefronts but only from part of them, and the result is an interference pattern.

Diffraction limits the useful magnification of an optical system

Diffraction sets a limit to the useful magnification of an optical system such as that of a telescope or microscope. Diffraction occurs whenever wavefronts of light are obstructed, and the light that enters a lens (or mirror) is affected by the finite opening which admits only part of each incident wavefront. No matter how perfect a lens is, the image of a point source of light it produces is always a tiny disk of light with bright and dark fringes around it (Fig. 25–17); only if the lens has an infinite diameter can a point source give rise to a point image. The smaller the lens, the larger the image of a point source. The angular width of the radius of this disc of light is about

$$\theta_0 = 1.22 \frac{\lambda}{D} \tag{25–8}$$

in radians, where λ is the wavelength of the light used and D is the lens diameter.

Criterion for resolving nearby objects

Two objects separated by less than θ_0 cannot be *resolved*, that is, distinguished apart, no matter how high the magnification employed, because their images will overlap (Fig. 25–18). Hence there is no advantage in using a higher magnification than will just reveal features that subtend the angle θ_0 at the position of the lens. Although Eq. (25–8) was derived for the image of a point source of light an infinite distance from a lens, it is a reasonable approximation of the resolving power of a telescope or microscope when D is taken as the diameter of the objective lens.

FIG. 25–17 The image a lens produces of a point source of light is always a tiny disk with bright and dark fringes around it. The smaller the lens, the larger the image.

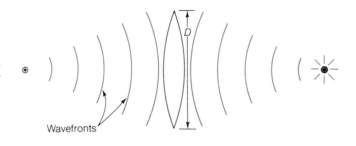

Wavefronts

FIG. 25–18 A large lens or mirror is better able to resolve nearby objects than a small one.

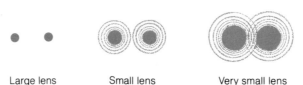

Large lens   Small lens   Very small lens

If two objects d apart are the distance L from an observer, the angle between them, in radians, is

$$\theta = \frac{d}{L}$$

Hence Eq. (25–8) can be rewritten

$$d_0 = 1.22 \frac{\lambda L}{D} \qquad\qquad\qquad \textit{Resolving power} \quad (25\text{–}9)$$

In this formula

d_0 = minimum separation of objects that can be resolved
λ = wavelength of the light used
L = object distance
D = diameter of objective lens or mirror

Example The pupils of a person's eyes under ordinary conditions of illumination are about 3 mm in diameter, and the distance of most distinct vision is 25 cm for most people. What is the resolving power of the eye at this distance under the assumption that it is limited only by diffraction? *Resolving power of the eye*

Solution Using $\lambda = 550$ nm, which is in the middle of the visible spectrum,

$$d_0 = 1.22 \frac{\lambda L}{D} = \frac{(1.22)(5.5 \times 10^{-7}\,\text{m})(0.25\,\text{m})}{0.003\,\text{m}} = 5.6 \times 10^{-5}\,\text{m} = 0.056\,\text{mm}$$

The photoreceptors in the retina are not quite close enough together to permit this degree of resolution, and 0.1 mm is a more realistic figure under ideal conditions. In terms of angular resolving power, $\theta_0 \approx 5 \times 10^{-4}$ radians for the human eye. ■

25-7 NUMERICAL APERTURE

The *numerical aperture* NA of a microscope objective is a measure of its resolving power. If α is the half-angle of the cone of light rays that reach an objective from an object, as in Fig. 25–19, and n is the index of refraction of the medium between the *Numerical aperture and resolving power*

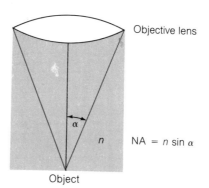

Objective lens

n NA = $n \sin \alpha$

Object

FIG. 25–19 The numerical aperture of a microscope objective depends upon the angular width of the cone of light that reaches it from the object and also upon the index of refraction of the medium between the objective and the object.

objective and the object, then by definition

$$NA = n \sin \alpha \qquad \textit{Numerical aperture} \quad (25\text{–}10)$$

The resolving power of the objective is

$$d_0 = 1.22 \frac{\lambda}{2NA} \qquad (25\text{–}11)$$

The greater the diameter of an objective relative to its focal length and the higher the index of refraction n, the greater the NA and the finer the detail that can be resolved.

Immersion oil for microscope objective

For an objective used in air, the angle α has a maximum possible value of 90°, which would mean NA = 1 and $d_0 = 0.61 \lambda$. In practice, α seldom exceeds 75°, which corresponds to an NA of 0.97. However, if a liquid of high index of refraction is used between the objective and the object, the NA can be significantly increased. An immersion oil of $n = 1.55$ together with a lens for which $\alpha = 75°$ gives an NA of 1.5. The higher the magnification of an objective, the higher the NA it requires in order that the enlarged details be sharp enough to be distinguished. An NA of 0.25 is sufficient for a $10\times$ objective, for instance, whereas an NA of at least 1.25 is usual for a $100\times$ objective.

Maximum useful magnification

The eyepiece of a microscope provides the user with an image 25 cm in front of the eye, and so the limit to the detail one can see clearly is about 0.1 mm, as mentioned at the end of the previous section. The finest detail the microscope itself can resolve is d_0, which for $\lambda = 550$ nm and NA = 1.5 is

$$d_0 = 1.22 \frac{\lambda}{2NA} = \frac{(1.22)(5.5 \times 10^{-7}\,\text{m})}{(2)(1.5)} = 2.2 \times 10^{-7}\,\text{m}$$

This is about half a wavelength of the light being used! The ratio between 0.1 mm and d_0 is

$$M = \frac{10^{-4}\,\text{m}}{2.2 \times 10^{-7}\,\text{m}} = 455$$

which represents the maximum useful magnification of the microscope in the sense that higher magnifications will give larger images but will not reveal any more detail. A larger image is nevertheless useful when examining small objects in order to reduce eyestrain, but magnifications much beyond $1000\times$ serve no purpose. For really high magnifications an electron microscope must be used; the electrons in such a device behave in certain respects like waves with very short wavelengths (see Sec. 27–4) and so greater resolving powers can be attained than are possible with an optical microscope.

25–8   POLARIZATION

Polarized transverse waves lie in a single plane of polarization

A *polarized* beam of transverse waves is one whose vibrations occur in only a single direction perpendicular to the direction in which the beam travels, so that the entire wave motion is confined to a plane called the *plane of polarization* (Fig. 25–20). When many different directions of polarization are present in a beam of transverse waves, vibrations occur equally often in all directions perpendicular to the direction of motion,

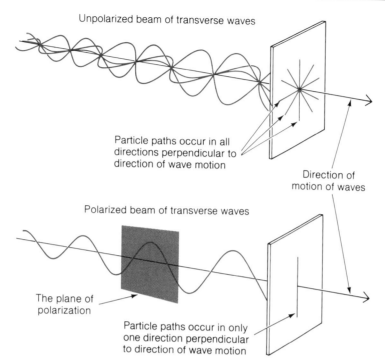

Unpolarized beam of transverse waves

FIG. 25–20 An unpolarized and a polarized beam of transverse waves.

Particle paths occur in all directions perpendicular to direction of wave motion

Direction of motion of waves

Polarized beam of transverse waves

The plane of polarization

Particle paths occur in only one direction perpendicular to direction of wave motion

and the beam is then said to be *unpolarized*. Since the vibrations that constitute longitudinal waves can take place in only one direction, namely that in which the waves travel, longitudinal waves cannot be polarized.

Light waves are transverse, and it is possible to produce and detect polarized light. To clarify the ideas involved, let us first consider the behavior of transverse waves in a stretched string. If the string passes through a tiny hole in a fence, as in Fig. 25–21(a), waves traveling down the string are stopped since the string cannot vibrate there. When the hole is replaced by a vertical slot, waves whose vibrations are vertical can get through the fence, but waves with vibrations in other directions cannot; see Fig. 25–21(b) and (c). In a situation in which several waves vibrating in different directions move down the string, the slot stops all but vertical vibrations (d): an initially unpolarized series of waves has become polarized.

The above approach can be used to determine whether a particular wave phenomenon can be polarized or not. In the case of a stretched string, what we do is erect another fence a short distance from the first, as in Fig. 25–21(e). If the slot in the new fence is also vertical, those waves that can get through the first fence can also get through the second. If the slot in the new fence is horizontal, however, it will stop all waves that reach it from the first fence (f).

Should longitudinal waves (say in a spring) go through the fence, it is possible that their amplitudes might decrease in passing through the slots, but the relative alignments of the slots would not matter; see Fig. 25–21(g) and (h). On the other hand, the alignment of the slots is the critical factor in the case of transverse waves.

Longitudinal waves cannot be polarized

The preceding chain of reasoning made it possible for the polarization of light waves to be demonstrated in the last century. A number of substances—for instance,

FIG. 25–21 Mechanical analogies of fundamental polarization phenomena.

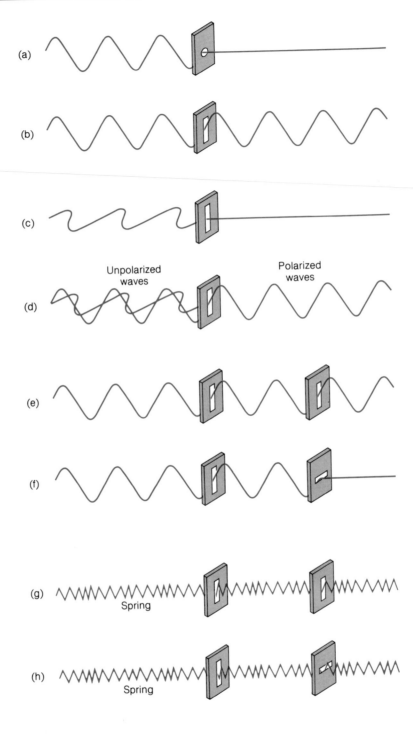

(a)

(b)

(c)

(d)

Unpolarized waves

Polarized waves

(e)

(f)

(g)

Spring

(h)

Spring

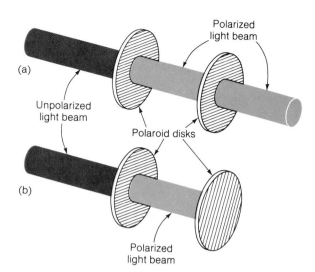

FIG. 25–22 Experiment showing the transverse nature of light waves.

quartz, calcite, and tourmaline—have different indexes of refraction for light with different planes of polarization relative to their crystal structures. and prisms can be made from them that transmit light in only a single plane of polarization. When a beam of unpolarized light is incident upon such a prism, only those of its waves whose planes of polarization are parallel to a particular plane in the prism emerge from the other side. The remainder of the waves are absorbed or deflected.

Polaroid is an artifically made polarizing material in wide use that only transmits light with a single plane of polarization. To exhibit the transverse nature of light waves, we first place two Polaroid disks in line so that their axes of polarization are parallel, Fig. 25–22(a), and note that all light passing through one disk also passes through the other. Then we turn one disk until its axis of polarization is perpendicular to that of the other, Fig. 25–22(b), and note that all light passing through one disk is now *stopped* by the other.

How polarization in light can be demonstrated

Just what is it whose vibrations are aligned in a beam of polarized light? As discussed in Chapter 23, light waves actually consist of oscillating electric and magnetic fields perpendicular to each other. Because it is the electric fields of light waves whose interactions with matter produce nearly all common optical effects, the plane of polarization of a light wave is considered to be that in which both the direction of its electric field and the direction of the wave lie (Fig. 25–23). Even though nothing material moves during the passage of a light wave, it is possible to establish its transverse nature and identify its plane of polarization.

Plane of polarization of a light wave

25–9 SCATTERING

When light waves encounter an obstacle of some sort, they are diffracted around its edges. If the obstacle is comparable in size with the wavelength of the light, the diffracted wavefronts are more or less spherical, and they spread out as though they originated in the obstacle. The incoming light is said to be *scattered* by the obstacle.

FIG. 25–23 The electric field of a light wave defines its plane of polarization.

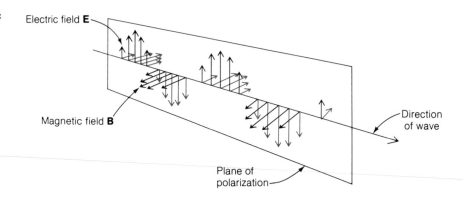

Electric field **E**

Magnetic field **B**

Direction of wave

Plane of polarization

Why the sky is blue

In general, the intensity of light of wavelength λ scattered by an object small compared with λ is proportional to λ^4. The shorter the wavelength, the greater the proportion of the incoming light that is scattered. This is the reason that the sky is blue. When we look at the sky, what we see is light from the sun that has been scattered by molecules in the upper atmosphere. Blue light, which consists of the shortest wavelengths, is scattered about ten times more readily than red light, so the scattered light is chiefly blue in color (Fig. 25–24(a)). At sunrise or sunset, when sunlight must make

FIG. 25–24 (a) Blue light is scattered the most by the earth's atmosphere, red light the least. Hence the sky appears blue. (b) Why the sun appears red at sunrise and sunset.

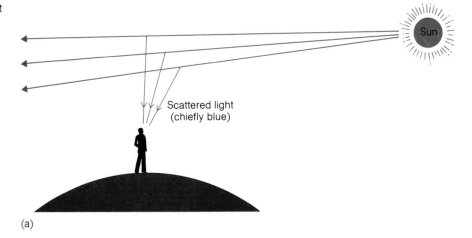

Sun

Scattered light (chiefly blue)

(a)

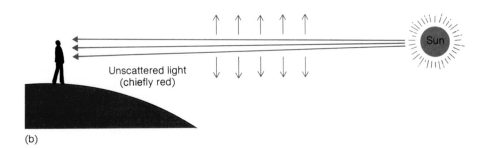

Unscattered light (chiefly red)

Sun

(b)

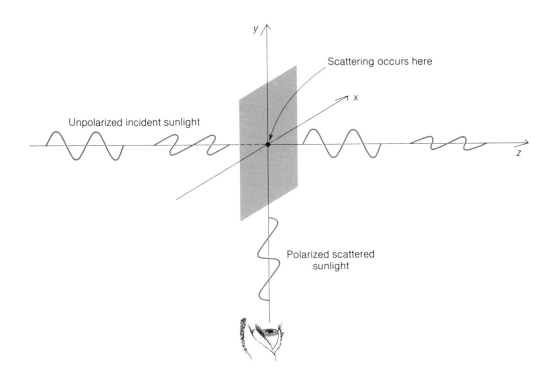

a very long passage through the atmosphere to reach an observer, much of its short-wavelength content is scattered out along the way, and the sun accordingly appears red in color (Fig. 25–24(b)). The water droplets and ice crystals in clouds are larger than λ and the scattering they produce is independent of λ; hence clouds do not appear colored. Above the atmosphere the sky appears black, and the moon, stars, and planets are visible to astronauts in the daytime.

Skylight is not only blue but is partly polarized as well. To verify this statement, all one has to do is hold up a piece of Polaroid against the sky and rotate it. Sunglasses are often made with Polaroid lenses so oriented as to discriminate against the polarized sunlight, which reduces glare while affecting other light to a lesser extent. The eye itself responds equally to all states of polarization.

Figure 25–25 shows why scattered light is polarized. A beam of unpolarized light heading in the $+z$-direction strikes some air molecules and is scattered. In this process electrons in the molecules are set in vibration by the electric fields of the light waves, and the vibrating electrons then reradiate. Because the electric field of an electromagnetic wave is perpendicular to its direction of motion (Fig. 25–23), the initial beam contains electric fields that lie in the xy-plane only. A scattered wave that proceeds downward, which is the $-y$-direction in the figure, can have its electric field in the x-direction only, so it is polarized. Only two planes of polarization for the incident light are shown for clarity; in the case of intermediate planes of polarization, the components of **E** in the x-direction can also lead to scattered waves traveling downward that are similarly polarized. Because skylight arrives at our eyes from a variety of directions, the polarization is not complete, but enough occurs to be easily demonstrated.

FIG. 25–25 Why skylight is polarized. Only incident waves whose electric fields have components in the x-direction can produce scattered waves that move in the y-direction, and such scattered waves must have their electric fields in the x-direction.

Skylight is polarized

IMPORTANT TERMS

Two sources of waves are **coherent** if there is a fixed phase relationship between the waves they emit during the time the waves are being observed. Interference can be observed only in waves from coherent sources.

The ability of waves to bend around the edges of obstacles in their paths is called **diffraction**. A **diffraction grating** is a series of parallel slits that produces a spectrum through the interference of light that is diffracted by them.

The **resolving power** of an optical system refers to its ability to produce separate images of nearby objects; resolving power is limited by diffraction, and the larger the objective lens of an optical system, the greater its resolving power.

A **polarized** beam of transverse waves is one whose vibrations occur in only a single direction perpendicular to the direction in which the beam travels, so that the entire wave motion is confined to a plane called the **plane of polarization**. An **unpolarized** beam of transverse waves is one whose vibrations occur equally often in all directions perpendicular to the direction of motion.

IMPORTANT FORMULAS

Double slit:

$$y = 0, \ \frac{L\lambda}{d}, \ \frac{2L\lambda}{d}, \ \frac{3L\lambda}{d}, \ \ldots \quad \text{(bright lines)}$$

Diffraction grating: $\quad \sin\theta = n\dfrac{\lambda}{d} \quad$ (bright lines)

Resolving power: $\quad \theta_0 = 1.22\dfrac{\lambda}{D}$

$$d_0 = 1.22\frac{\lambda L}{D}$$

MULTIPLE CHOICE

1. For two light beams to interfere, their sources must be
   (a) coherent.   (b) incoherent.
   (c) lasers.   (d) slits.

2. Coherent electromagnetic waves are not emitted by
   (a) two antennas connected to the same radio transmitter.
   (b) two pinholes in an opaque shield over a sodium-vapor lamp.
   (c) a pinhole in an opaque shield over a sodium-vapor lamp and its reflection in a mirror.
   (d) two lasers.

3. In a double-slit experiment, the maximum intensity of the first bright line on either side of the central one occurs on the screen at locations where the arriving waves differ in path length by
   (a) $\lambda/4$.   (b) $\lambda/2$.
   (c) λ.   (d) 2λ.

4. Wavelength determinations cannot be made with the help of
   (a) a glass prism.
   (b) a pair of narrow parallel slits.
   (c) a diffraction grating.
   (d) a pair of Polaroid disks.

5. An interference pattern is produced whenever
   (a) reflection occurs.
   (b) refraction occurs.
   (c) diffraction occurs.
   (d) polarization occurs.

6. Thin films of oil and soapy water owe their brilliant colors to a combination of reflection and
   (a) refraction.   (b) interference.
   (c) diffraction.   (d) polarization.

7. A characteristic property of the spectra produced by a diffraction grating is the
   (a) sharpness of the bright lines.
   (b) diffuseness of the bright lies.
   (c) absence of bright lines.
   (d) absence of dark lines.

8. The minimum separation of two features of a distant object that can be discerned by a telescope does *not* depend on
   (a) the diameter of the objective lens.
   (b) the focal length of the objective lens.
   (c) the wavelength of the light being used.
   (d) the distance to the object.

9. The wavelength of light plays no role in
   (a) interference.   (b) diffraction.
   (c) resolving power.   (d) polarization.

10. An unpolarized beam of transverse waves is one whose vibrations
   (a) are confined to a single plane.
   (b) occur in all directions.
   (c) occur in all directions perpendicular to their direction of motion.
   (d) have not passed through a Polaroid disk.

11. Longitudinal waves do not exhibit
   (a) refraction.   (b) reflection.
   (c) diffraction.   (d) polarization.

12. The sky is blue because
(a) air molecules are blue.
(b) the lens of the eye is blue.
(c) the scattering of light is more efficient the shorter its wavelength.
(d) the scattering of light is more efficient the longer its wavelength.

13. The greater the number of lines that are ruled on a grating of given width,
(a) the shorter the wavelengths that can be diffacted.
(b) the longer the wavelengths that can be diffracted.
(c) the narrower the spectrum that is produced.
(d) the broader the spectrum that is produced.

14. Monochromatic green light of wavelength 5×10^{-7} m illuminates a pair of narrow slits 1 mm apart. The separation of bright lines on the interference pattern formed on a screen 2 m away is
(a) 0.1 mm. (b) 0.25 mm.
(c) 0.4 mm. (d) 1.0 mm.

15. Monochromatic light is used to illuminate a pair of narrow slits 0.3 mm apart, and the interference pattern is observed on a screen 0.9 m away. The second dark band appears 3 mm from the center of the pattern. The wavelength of the light is
(a) 2.2×10^{-7} m. (b) 3.3×10^{-7} m.
(c) 6.7×10^{-7} m. (d) 1.3×10^{-7} m.

16. A pair of binoculars designated "7 × 50" has a magnification of 7 and an objective lens diameter of 50 mm. The smallest detail that in principle can be perceived with such an instrument when viewing an object 1 km away in light of wavelength 5×10^{-7} m has a linear dimension of approximately
(a) 1 mm. (b) 1 cm.
(c) 10 cm. (d) 1 m.

EXERCISES

1. Can light from incoherent sources interfere? If so, then why is a distinction made between coherent and incoherent sources?

2. The waves used to carry television signals cannot reach receivers beyond the visual horizon of their transmitting antennas, whereas ordinary radio waves readily travel beyond the visual horizon of their transmitting antennas. Can you think of a reason for this contradictory behavior?

3. Radio waves diffract pronouncedly around buildings, whereas light waves, which are also electromagnetic waves, do not. Why?

4. In Young's double-slit experiment, which effects are due to diffraction and which to interference?

5. What do diffraction and interference have in common? How do they differ?

6. Which of the following can occur in (a) transverse waves and (b) longitudinal waves: refraction, dispersion, interference, diffraction, polarization?

7. Explain the peculiar appearance of a distant light source when seen through a piece of finely woven cloth.

8. What advantages has a diffraction grating over a double slit for determining wavelengths of light?

9. What is the difference between the first-order and the second-order spectra produced by a grating? Which is wider? Does a prism produce spectra of different orders?

10. What governs the angular width of the first-order spectrum of white light produced by a grating?

11. What becomes of the energy of the light waves whose destructive interference leads to dark lines in an interference pattern?

12. As a soap bubble is blown up, its wall becomes thinner and thinner. Just before the bubble breaks, the thinnest part of its wall turns black. Why?

13. Give two advantages a large-diameter telescope objective has over a small-diameter one.

14. A camera can be made by using a pinhole instead of a lens. What happens to the sharpness of the picture if the hole is too large? If it is too small?

15. Since light consists of transverse waves, why is not every light beam polarized?

16. What is the relationship between the plane of polarization of a transverse wave and its direction of propagation?

17. Light of unknown wavelength is used to illuminate two parallel slits 1 mm apart. Adjacent bright lines on the interference pattern that results on a screen 1.5 m away are 0.65 mm apart. What is the wavelength of the light?

18. Two parallel slits 0.12 mm apart are illuminated by light of wavelength 500 nm. A viewing screen is 1.5 m from the slits. (a) How far from the central bright line is the next bright line? (b) How far is the first dark line? (c) How far is the fifth dark line?

19. Two parallel slits 0.1 mm apart are illuminated by light of wavelength 546 nm. A viewing screen is 0.8 m from the slits. (a) How far from the central bright line is the second bright line? (b) How far is the third bright line? (c) How far is the third dark line?

20. A 5500-line/cm diffraction grating produces an image deviated by 27° in the second order. Find the wavelength of the light.

21. Light of wavelength 750 nm is directed on a grating ruled with 4000 lines/cm. What is the angular deviation of this light in (a) the first order, and (b) the third order?

22. The index of refraction of a certain soap bubble illuminated by white light is $n = 1.35$. If the bubble appears orange ($\lambda = 630$ nm) at a point nearest the viewer, what is the minimum thickness of the bubble at that point? Be sure to take into account any phase shifts upon reflection.

23. The index of refraction of magnesium fluoride is 1.38. How thick should an antireflection coating of this material be for maximum cancellation at 550 nm, which is the middle of the visible spectrum?

24. A radar has a resolving power of 30 m at a range of 1 km. What is the minimum width of its antenna if its operating frequency is 9500 MHz?

25. The largest telescope in the world, the Hale telescope at Mount Palomar in California, has a concave mirror 5 m in diameter. How many meters apart must two features of the moon's surface be in order to be resolved by this telescope? Take the distance from the earth to the moon as 386,000 km and use 500 nm for the wavelength of the light.

26. An astronaut circles the earth in a satellite at an altitude of 150 km. If the diameter of his pupils is 2 mm and the average wavelength of the light reaching his eyes is 550 nm, is it conceivable that he can distinguish sports stadiums on the earth's surface? Private houses? Cars?

PROBLEMS

1. Two parallel slits are illuminated by light of two wavelengths, one of which is 5.8×10^{-7} m. On a viewing screen an unknown distance from the slits the fourth dark line of the light of the known wavelength coincides with the fifth bright line of the light of the unknown wavelength. Find the unknown wavelength.

2. Two parallel slits 0.25 mm apart are illuminated by light of two wavelengths, 500 and 600 nm. A viewing screen is 2 m away from the slits. How far are the bright lines of one wavelength from the bright lines of the other?

3. How many diffracted images are formed on either side of the central image when radiation of wavelength 600 nm falls on a 4000-line/cm grating?

4. Light containing wavelengths of 500 and 550 nm is directed at a 2000-line/cm grating. How far apart are the

lines formed by these wavelengths on a screen 4 m away in the second order?

5. White light that contains wavelengths from 400 to 700 nm is directed at a 3000-line/cm grating. How wide is the first-order spectrum on a screen 2 m away?

6. White light that contains wavelengths from 400 to 700 nm is directed at a 5000-line/cm grating. Do the first- and second-order spectra overlap? The second- and third-order spectra? If there is an overlap, will the use of a grating with a different number of lines/cm change the situation?

7. A radio transmitter operating at 15 MHz has two vertical antennas 15 m apart located on an east-west line. How many intensity maxima in the horizontal plane are there? In what directions?

8. The antennas of problem 7 are moved so that they are 50 m apart on the same east-west line. How many intensity maxima in the horizontal plane are there now? In what directions?

9. The Jodrell Bank radiotelescope has a parabolic reflecting "dish" 76 m in diameter. (a) What is the angular diameter in degrees of a point source of radio waves of wavelength 21 cm as seen by this telescope? (b) How does the above figure compare with the angular width of the moon, whose diameter is 3476 km and whose average distance from the earth is 3.8×10^5 km?

10. According to a famous battle command, "Don't fire until you see the whites of their eyes." (a) If the diameter of the white of an eye is 20 mm, the diameter of the pupil in bright sunlight is 2 mm, and the light has a wavelength of 500 nm, find this distance on the basis of the theoretical resolution formula. (b) The actual resolving power of the eye, which is limited by the structure of the retina, is about 5×10^{-4} rad. Find the maximum distance at which the white of an eye can be distinguished on the basis of this figure.

11. At night, the pupils of a person's eyes are 8 mm in diameter. (a) How many km away from a car facing a woman will she be able to distinguish its headlights from each other? (b) If her pupils were 4 mm in diameter (say at twilight), how far away from the car could she distinguish its headlights from each other? Assume the headlights are 1.5 m apart, that the average wavelength of their light is 600 nm, and that her eyes are capable of attaining their theoretical resolving power.

12. A certain laser produces a beam of monochromatic light whose wavelength is 550 nm and whose initial diameter is 1 mm. (a) In what distance will diffraction have caused the beam to double its diameter? (b) If the initial diameter of the beam were 1 cm, would the doubling distance be different? If so, what would it be?

13. The smaller the aperture of a camera lens, the greater the depth of field. However, a small aperture means reduced resolution. The criterion for an enlarged print to show sharp detail from a small negative is that the image of a point object on it be not more than about 0.01 mm across. Find the maximum f-number of a camera lens in order that this criterion be met for a distant object in 550-nm light.

ANSWERS TO MULTIPLE CHOICE

1. a	**7.** a	**13.** d
2. b	**8.** b	**14.** d
3. c	**9.** d	**15.** c
4. d	**10.** c	**16.** b
5. c	**11.** d	
6. b	**12.** c	

RELATIVITY

Few physical theories represent so drastic an assault on traditional habits of thought as the theory of relativity. Relativity links time and space, matter and energy, electricity and magnetism—and for all the seeming magic of its conclusions, most of them can be reached with the simplest of mathematics. The theory of relativity was proposed in 1905 by Albert Einstein, and little of physical science since then has remained unaffected by his ideas.

26–1 SPECIAL RELATIVITY

Thus far no special point was raised about how measurements of such physical quantities as length, time, and mass are carried out. It was simply assumed that these quantities could be determined in some way, and that, since standard units have been established for each of them, it doesn't matter who makes a

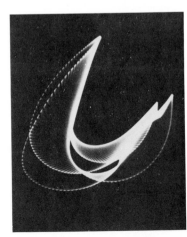

particular determination—everybody ought to get the same figure. There is certainly no question of principle associated with, say, finding the length of an airplane on the ground: all we need do is place one end of a tape measure at the airplane's nose and note the number on the tape at the airplane's tail.

But what if the airplane is in flight and we are on the ground? It is not hard to find the length of a distant object with the help of a surveyor's transit to measure angles, a tape measure to establish a baseline, and a knowledge of trigonometry. Because the airplane is moving, however, things become more complicated because now we must take into account the fact that light does not travel instantaneously from one place to another but does so at a definite, fixed speed—and light is the means by which information is carried from a distant object to our instruments. When a careful analysis is made of the problem of measuring physical quantities when there is relative motion between the instruments and whatever is being observed, many surprising results emerge.

In Chapter 2 the notion of *frame of reference* was introduced. When we observe something moving, what we actually detect is that its position relative to something else is changing (Fig. 26–1). A passenger moves relative to a train; the train moves relative to the earth; the earth moves relative to the sun; the sun moves relative to the galaxy of stars (the "Milky Way") of which it is a member, and so on. In each case a frame of reference is part of the description of the motion; it is meaningless to say that something is moving unless it is understood with respect to what the motion occurs.

There is no universal frame of reference that can be used everywhere. If we see something changing its position with respect to us at constant velocity, we have no way of knowing whether *it* is moving or *we* are moving. If we were isolated from the rest of the universe, we would be unable to find out if we were moving at constant velocity or not—indeed, the question would make no sense. All motion is relative to the observer, and there is no such thing as "absolute motion."

All motion is relative to the observer; "absolute motion" does not exist

The theory of relativity is concerned with the physical consequences of the absence of a universal frame of reference. The special theory of relativity, published in 1905 by Albert Einstein, is confined to problems involving the motion of frames of reference at constant velocity (that is, both constant speed and constant direction) with respect to one another; the general theory of relativity, published 10 years later by Einstein, deals with problems involving frames of reference accelerated with respect to one another. The special theory has had an enormous impact on all of physics, and its chief conclusions will be examined here.

Two principles are fundamental to the special theory of relativity. The principle of relativity states that

The laws of physics are the same in all frames of reference moving at constant velocity with respect to one another.

Principle of relativity

This principle follows directly from the absence of a universal frame of reference. If the laws of physics were different for different observers in relative motion, they could infer from these differences which of them were "stationary" in space and which were "moving." But such a distinction does not exist in nature, and the principle of relativity is an expression of this fact.

FIG. 26–1 Some frames of reference.

Frame of reference: the Milky Way

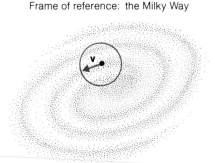

Frame of reference: the sun

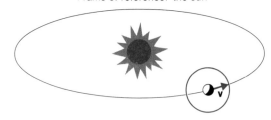

Frame of reference: the earth

Frame of reference: a railway car

Thus experiments of any kind performed, for instance, in an elevator that is ascending at a constant velocity yield exactly the same results as the same experiments performed when the elevator is at rest or is descending at a constant velocity. On the other hand, an isolated observer *can* detect accelerations, as any elevator passenger can verify.

The second principle, which is based on the results of a great many experiments, states that

The speed of light is the same to all observers

The speed of light in free space has the same value for all observers, regardless of their state of motion or the state of motion of the source.

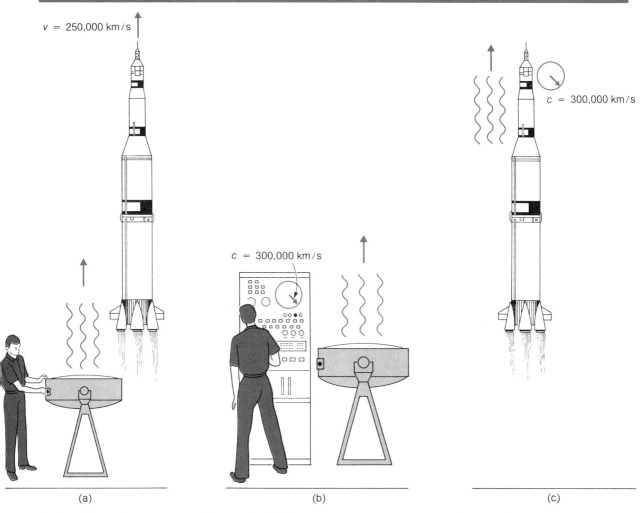

$v = 250,000$ km/s

$c = 300,000$ km/s

$c = 300,000$ km/s

(a) (b) (c)

At first glance the constancy of the speed of light may not seem so very extraordinary, but this is a misleading impression. Let us examine a hypothetical experiment in essence no different from actual experiments that have been performed in a number of ways.

Suppose I turn on a searchlight at the same moment you take off in a spacecraft at a speed of 250,000 km/s (Fig. 26–2). We both measure the speed of the light waves from the searchlight using identical instruments. From the ground I find their speed to be 300,000 km/s, as usual. "Common sense" tells me you ought to find a speed of (300,000 − 250,000) km/s or only 50,000 km/s for the same light waves. But you also find their speed to be 300,000 km/s, even though to me you seem to be moving parallel to the waves at 250,000 km/s. As so often, common sense is wrong.

There is only one way to account for the apparent discrepancy between the above results without violating the principle of relativity, and that is to conclude that measurements of space and time are not absolute but depend upon the relative motion of the observer and that which is observed. If I were to make measurements from the ground of the rate at which your clock ticks and of the length of your meter stick, I

FIG. 26–2 The speed of light is the same to all observers.

would find that the clock ticks more slowly than it did on the ground and that the meter stick is shorter in the direction of motion of the spacecraft. To you, your clock and meter stick are the same as they were on the ground before you took off; to me they are different because of the relative motion, but in such a way that the speed of light you measure is the same 300,000 km/s that I measure. Time intervals and lengths are relative quantities, but the speed of light in free space is the same to all observers.

Electromagnetic theory is relativistically correct, but Newtonian mechanics is not

Before Einstein's work, a conflict had existed between the principles of mechanics, which were then based on Newton's laws of motion, and those of electricity and magnetism, which had been developed into a unified theory by Maxwell. Newtonian mechanics had proved quite satisfactory for more than two centuries, and Maxwell's theory not only incorporated all that was known at the time about electrical and magnetic phenomena but had also been able to predict the existence of electromagnetic waves and to identify light as an example of them. But the equations of mechanics and those of electromagnetism differed in the way they related measurements made in one frame of reference with those made in another frame in relative motion. Einstein showed that Maxwell's theory is consistent with special relativity whereas Newtonian mechanics is not, and his modification of mechanics brought these branches of physics into accord. As we shall find, relativistic and Newtonian mechanics agree for relative speeds much smaller than the speed of light, which is why the latter theory seemed correct for so long, but at higher speeds Newtonian mechanics fails and must be replaced by a relativistic formulation.

26–2 THE RELATIVITY OF TIME

Measurements of time intervals are affected by relative motion between an observer and what is observed. As a result, a clock moving with respect to an observer ticks more slowly than it does without such motion, and all processes (including those of life) occur more slowly to an observer when they take place in a frame of reference in relative motion.

We begin by considering the operation of the particularly simple clock shown in Fig. 26–3. In this clock a pulse of light is reflected back and forth between two mirrors. Whenever the light strikes the lower mirror, an electrical signal is produced that is registered as a mark on the recording tape. Each mark corresponds to the tick of an ordinary clock.

Let us consider two of these clocks, one of them at rest in a laboratory and another in a spaceship moving at the speed v relative to the laboratory. An observer in the laboratory watches both clocks: does he find that they tick at the same rate?

Figure 26–4 shows the laboratory clock in operation. The mirrors are L apart, and the time interval between ticks is t_0. Hence the time needed for the light pulse to travel the distance L between the mirrors at the speed c is $t_0/2$, and so

$$L = c\left(\frac{t_0}{2}\right) \qquad t_0 = \frac{2L}{c}$$

Figure 26–5 shows the moving clock with its mirrors parallel to the direction of the relative velocity as seen from the laboratory. The time interval between ticks is t.

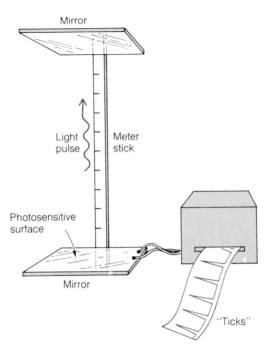

FIG. 26–3 A light-pulse clock.

Because the clock is moving, the pulse of light follows a zigzag path in which it travels the distance $ct/2$ in going from one mirror to the other in the time $t/2$. From the Pythagorean theorem,

Light in the moving clock follows a zigzag path

$$\left(\frac{ct}{2}\right)^2 = L^2 + \left(\frac{vt}{2}\right)^2$$

How is t related to t_0? To find out, we first solve the preceding equation for t:

$$\left(\frac{ct}{2}\right)^2 = L^2 + \left(\frac{vt}{2}\right)^2 \qquad \text{so} \qquad \frac{t^2}{4}\left(c^2 - v^2\right) = L^2$$

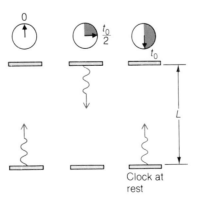

FIG. 26–4 Light-pulse clock in the laboratory as seen by an observer in the laboratory.

FIG. 26–5　Light-pulse clock in the spaceship as seen by an observer in the laboratory.

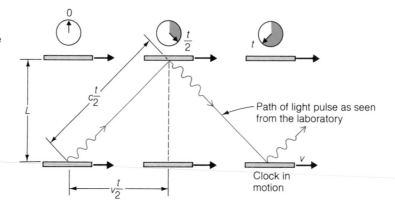

Hence

$$t^2 = \frac{4L^2}{(c^2 - v^2)} = \frac{(2L)^2}{c^2(1 - v^2/c^2)}$$

$$t = \frac{2L/c}{\sqrt{1 - v^2/c^2}}$$

But the quantity $2L/c$ is the time interval t_0 between ticks in the laboratory clock, as we saw. Hence

$$t = \frac{t_0}{\sqrt{1 - v^2/c^2}} \qquad\qquad \textit{Time dilation} \quad (26\text{–}1)$$

Here is a reminder of what the symbols in this important formula represent:

t_0 = time interval on clock at rest
t = time interval on clock in relative motion as determined by outside observer
v = speed of relative motion
c = speed of light

A moving clock ticks more slowly than a clock at rest

Because the quantity $\sqrt{1 - v^2/c^2}$ is always smaller than 1 for a moving object, t is always greater than t_0. *A clock moving with respect to an observer ticks more slowly than a clock that is stationary with respect to the same observer.* This effect is referred to as *time dilation* (to dilate is to become larger).

Now let us turn the situation around and ask what an observer in a spacecraft finds when he compares his clock with one on the ground. The only change needed in the preceding derivation is the direction of motion: if the person on the ground sees the spacecraft moving to the east, the person in the spacecraft sees the laboratory on the ground moving to the west. To the person in the spacecraft the light pulse of the ground clock follows a zigzag path that requires a total time per round trip of

$$t = \frac{t_0}{\sqrt{1 - v^2/c^2}}$$

whereas the light pulse in his own clock takes t_0 for the round trip. Thus to the man in the spacecraft the clock on the ground ticks at a slower rate than his own clock does. A clock moving relative to an observer *always* is slower than a clock at rest relative to him, regardless of where the observer is located.

A light clock is a rather more exotic timepiece than most of us are accustomed to. What if a stationary cuckoo clock and a moving one are compared: do we again find that the moving clock runs more slowly?

The principle of relativity makes it easy for us to predict the outcome of the experiment. Suppose cuckoo clocks tick at exactly the same rate to all observers, whether there is relative motion or not. We put a cuckoo clock and a light-pulse clock (which *does* tick more slowly when in motion) on a spacecraft. On the ground they show the same time.

In flight, the two clocks show different times to an observer on the ground, since the light-pulse clock ticks slower whereas the cuckoo clock (by hypothesis) does not. To an observer in the spacecraft, however, the two clocks agree, since to him the clocks are stationary and it is the ground which is moving away from him. Therefore the laws of physics that govern the operation of the clocks must be different on the spacecraft from what they are on the ground—which contradicts the principle of relativity. *All moving clocks tick more slowly than clocks at rest.*

It is important to keep in mind that the slowing down of a moving clock is significant only at relative speeds not far from the speed of light. Such speeds are readily attained by elementary particles, and most of the experiments that have confirmed time dilation have employed such particles. Today's spacecraft are far too slow to exhibit time dilation. For instance, the highest speed reached by the Apollo 11 spacecraft on its way to the moon was only about 10.8 km/s, or 0.0036% of the speed of light. At this speed, clocks on the spacecraft differ from those on the earth by less than 1 part in 10^9.

Time dilation is significant only at speeds near that of light

Example Find the speed relative to the earth of a spacecraft whose clock runs 1 s slow per day compared with a terrestrial clock.

Solution Here $t_0 = (24\ \text{h})(60\ \text{min/h})(60\ \text{s/min}) = 86{,}400$ s is the time interval on the earth and $t = 86{,}401$ s is the time interval on the spacecraft. We begin by solving Eq. (26–1) for v and then substitute the values of t_0, t, and c:

$$t = \frac{t_0}{\sqrt{1 - v^2/c^2}}$$

$$\sqrt{1 - v^2/c^2} = t_0/t$$

$$1 - v^2/c^2 = (t_0/t)^2$$

Thus

$$v = c\sqrt{1 - (t_0/t)^2} = 3 \times 10^8\ \text{m/s} \times \sqrt{1 - \left(\frac{86{,}400\ \text{s}}{86{,}401\ \text{s}}\right)^2}$$

$$= 1.44 \times 10^6\ \text{m/s}$$

This is more than a thousand times faster than existing spacecraft. ■

26–3   MUON DECAY

A good illustration of time dilation occurs in the decay of unstable elementary particles called *muons,* which are further described in Chapter 32.

A muon at rest decays into a positron or an electron in 2 microseconds

Muons have masses 207 times that of the electron and may have positive or negative electric charges. A muon decays into a positron (positively charged electron) or an electron an average of 2.0×10^{-6} s (2.0 μs) after it comes into being. Muons are created at high altitudes as an ultimate result of collisions between the nuclei of atoms in the earth's atmosphere and fast cosmic-ray particles, which are largely protons, that reach the earth from space. A muon passes through each square centimeter of the earth's surface a little more often than once a minute. The muon speeds are observed to be about 2.994×10^8 m/s, or $0.998c$, where c is the speed of light. But in $t_0 = 2.0 \times 10^{-6}$ s, the average muon lifetime, they can travel a distance of only

$$vt_0 = (2.994 \times 10^8 \text{ m/s})(2.0 \times 10^{-6} \text{ s}) = 600 \text{ m}$$

whereas they actually come into being at elevations ten or more times greater than this.

The key to resolving this paradox is to note that the average muon lifetime of 2 μs is what an observer at rest with respect to a muon would find. If we could collect some muons at the instant of their creation and time their decays when they are at rest, we would find an average of $t_0 = 2$ μs.

However, when we are on the ground and the muons are hurtling toward us at the considerable speed of $0.998c$, we find instead that their lifetimes have been extended by time dilation to

$$t = \frac{t_0}{\sqrt{1 - v^2/c^2}} = \frac{2 \times 10^{-6} \text{ s}}{\sqrt{1 - \dfrac{(0.998c)^2}{c^2}}} = 31.6 \times 10^{-6} \text{ s}$$

Moving muons have longer lifetimes

The fast muons have lifetimes almost 16 times longer than those at rest. In a time interval of $t = 31.6 \times 10^{-6}$ s, a muon whose speed is $0.998c$ can cover the distance

$$vt = (2.994 \times 10^8 \text{ m/s})(31.6 \times 10^{-6} \text{ s}) = 9500 \text{ m}$$

Despite its brief life span of $t_0 = 2$ μs in its own frame of reference, a muon is able to reach the ground from high altitudes because in the frame of reference in which these altitudes are measured, the muon lifetime is $t = 31.6$ μs.

26–4   THE LORENTZ CONTRACTION

As we saw, the arrival of cosmic-ray muons at sea level from high altitudes is not in conflict with the brevity of their lives (in their frames of reference) since these lives are increased 16-fold (in our frame of reference) by their relative motion. But what if somebody could accompany the muons downward at the same speed of $0.998c$, so that to him the muons are at rest? Both the muons and the observer are now in the same frame of reference, and the muon lifetime is only 2 μs in this frame. The question is, does the moving observer find that the muons reach the ground, or does he find that they decay beforehand?

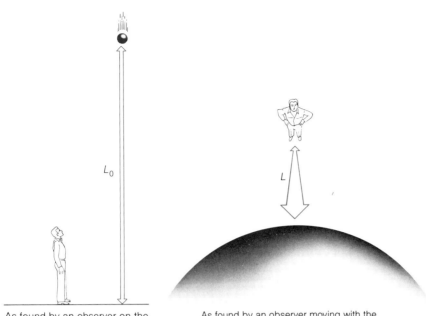

FIG. 26–6 The muon size is greatly exaggerated here; in fact, the muon may actually be a point particle with no extension in space.

As found by an observer on the ground, the muon altitude is L_0.

As found by an observer moving with the muon, the ground is L below it, which is a shorter distance than L_0.

The principle of relativity states that the laws of physics are the same in all frames of reference moving at constant velocity with respect to one another. An observer on the ground and an observer moving with the muons are in relative motion at a constant velocity, namely $0.998c$, and if we on the ground find that the muons reach our apparatus before they decay, then the moving observer must also find the same thing. Though the appearance of an event may be different to different observers, the fact of the event's occurrence is not subject to dispute.

The only way an observer in a muon's frame of reference can reconcile its arrival at sea level with the lifetime of 2 μs he finds is if the distance the muon travels is shortened by virtue of its motion (Fig. 26–6). The principle of relativity enables us to infer at once the extent of the shortening—it must be by the same factor of $\sqrt{1 - v^2/c^2}$ that the muon lifetime is extended from the point of view of a stationary observer. Thus a distance we on the ground measure to be L_0 will appear to the muon as the abbreviated distance

Distances are shorter to a moving observer

$$L = L_0 \sqrt{1 - v^2/c^2}$$

In our frame of reference, the average distance a muon can go at the speed $v = 0.998c$ before it decays is $L_0 = 9500$ m. The corresponding distance in the muon's frame of reference is

$$L = L_0 \sqrt{1 - v^2/c^2} = 9500 \text{ m} \times \sqrt{1 - \frac{(0.998c)^2}{c^2}} = 600 \text{ m}$$

This is precisely how far a muon traveling at $0.998c$ can go in 2 μs. Both points of view—from the frame of reference of someone on the ground, to whom the muon lifetime is dilated, and from the frame of reference of the muon itself, to which the distance to the ground is contracted—give the same result.

The relativistic shortening of distances is an example of the general *Lorentz contraction* of lengths:

$$L = L_0 \sqrt{1 - v^2/c^2}$$ *Lorentz contraction*   (26–2)

The symbols in this formula have these meanings:

L_0 = length measured when the object is at rest
L = length measured when the object is in relative motion
v = speed of relative motion
c = speed of light

Moving objects are shorter than when at rest

The length of an object in motion with respect to an observer is measured by the observer to be shorter than when it is at rest with respect to him. This shortening works both ways; to a person in a spacecraft, measurements indicate that objects on the earth are shorter than they were when he was on the ground, and someone on the ground finds that the spacecraft is shorter in flight than when it was at rest (Fig. 26–7). (To the person in the spacecraft, its length is the same whether on the ground or in flight, since it is always at rest with respect to him.) The length of an object is a maximum when determined in a reference frame in which it is stationary, and its length is less when determined in a reference frame with respect to which it is moving.

The Lorentz contraction occurs only in the direction of motion

Only lengths in the direction of motion undergo contractions. Thus to the outside observer a spacecraft is shorter in flight than on the ground, but it is not narrower.

The relativistic length contraction is negligible for ordinary speeds, but is an important effect at speeds close to the speed of light. A speed of 1000 km/s seems enormous to us, yet it results in a shortening in the direction of motion to only

$$\frac{L}{L_0} = \sqrt{1 - (v^2/c^2)} = \sqrt{1 - \frac{(1000\,\text{km/s})^2}{(300,000\,\text{km/s})}} = 0.999994 = 99.9994\%$$

of the length at rest. On the other hand, something traveling at nine-tenths the speed of light is shortened to

$$\frac{L}{L_0} = \sqrt{1 - \frac{(0.9c)^2}{c^2}} = 0.436 = 43.6\%$$

of its length at rest, a significant change.

26–5   VELOCITY ADDITION

One of the principles that form the basis of the special theory of relativity states that all observers, regardless of their relative motion, will find the same value for c, the speed of light in free space. But if we throw a ball forward at 10 m/s from a car moving at 20 m/s relative to a road, everyday experience tells us that the ball's speed relative to the road will be 30 m/s. Hence we would expect that a pulse of light emitted by a spacecraft moving at a relative speed v toward the earth ought to have a speed of $c + v$ relative to the earth, which contradicts the above principle. However, everyday experience is not an adequate guide to events outside the limits of that experience.

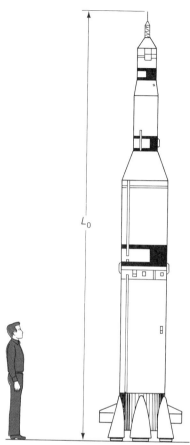

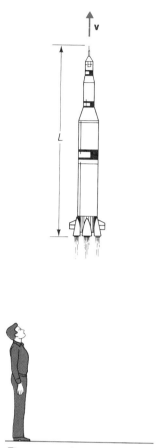

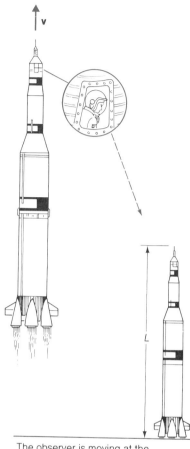

An observer and a spacecraft are at rest on the ground. The observer finds the spacecraft's length to be L_0.

The spacecraft is moving at the speed v. The observer on the ground finds its length to be $L = L_0 \sqrt{1 - v^2/c^2}$.

The observer is moving at the speed v and the spacecraft is on the ground, The observer finds its length to be $L = L_0 \sqrt{1 - v^2/c^2}$.

Because measurements of length and time are different in frames of reference in relative motion, it follows that measurements of speeds, too, will be different in such frames. The correct formula for velocity addition when the velocities are along the same straight line turns out to be

FIG. 26–7

$$V = \frac{V' + v}{1 + vV'/c^2} \qquad \qquad \textit{Velocity addition} \quad (26\text{–}3)$$

In this formula, V' is the speed of something with respect to a frame of reference that is moving at the speed v relative to an observer, and V is the speed the observer measures. (If $\mathbf{V'}$ is opposite to $\mathbf{v}$, then $-V'$ is used in place of V'.)

When V' and v are small compared with the speed of light c, $V \approx V' + v$, which is the case for a ball thrown from a moving car. For a pulse of light emitted from a spacecraft, $V' = c$, and an observer on the earth would find the pulse to have the speed

$$V = \frac{V' + v}{1 + vV'/c^2} = \frac{c + v}{1 + vc/c^2} = \frac{c + v}{1 + v/c} = \frac{c(c + v)}{c + v} = c$$

All observers, regardless of their relative motion, find c for the speed of light in free space.

　　　　Spacecraft Alfa has a speed of $0.9c$ with respect to the earth. If spacecraft Bravo is to pass Alfa in the same direction at a relative speed of $0.5c$, what speed must Bravo have with respect to the earth?

　　　　According to classical mechanics, Bravo would have to travel at $0.9c + 0.5c = 1.4c$ with respect to the earth. By using Eq. (26–3) with $V' = 0.5c$ and $v = 0.9c$ we find instead that the required speed is only

$$V = \frac{V' + v}{1 + vV'/c^2} = \frac{0.5c + 0.9c}{1 + (0.9c)(0.5c)/c^2} = 0.966c$$

which is less than c. Spacecraft Bravo must go only 6.6% faster than a spacecraft traveling at $0.9c$ (both speeds measured from the earth) in order to pass it at a relative speed of $0.5c$ (as measured from either spacecraft).　　　　■

26–6　THE TWIN PARADOX

A person is a biological clock

Since life processes occur with regular rhythms, a person constitutes a biological clock and must behave in the same way as any other clock when in motion relative to an observer. There is no difference in principle between heartbeats and the ticks of a clock. The slowing down of a moving clock thus means that the life processes of a person in a moving spacecraft occur at a slower rate than they do on the ground, so he or she ages more slowly than somebody on the ground does.

　　　　The celebrated case of the twins Dick and Jane illustrates the consequences of time dilation in space travel. Dick is 20 years old when he embarks on a space voyage at a speed of 297,000 km/s, which is 99% of the speed of light. To Jane, who has stayed behind, the pace of Dick's life is slower than her own by a factor of

$$\sqrt{1 - v^2/c^2} = \sqrt{1 - (0.99c)^2/c^2} = 0.141 \approx \tfrac{1}{7}$$

The traveling twin is younger on his return than the twin who remained at home

Dick's heart beats only once for every seven beats of Jane's heart; Dick takes only one breath for every seven of Jane's; Dick thinks only one thought for every seven of Jane's. Eventually Dick returns after 70 years have elapsed by Jane's calendar—but Dick is only 30 years old, whereas Jane, the stay-at-home twin, is 90 years old.

　　　　Jane is baffled by the youth of her astronaut brother. "After all," she argues, "according to the principle of relativity, *my* life processes should have appeared 7 times slower to Dick, so by the same reasoning I ought to be 30 and Dick ought to be 90."

The formulas of relativity can be applied to the journey only by the twin who was not accelerated

　　　　But advanced age has dulled Jane's powers of reasoning. The two situations are not at all interchangeable. Jane, on the earth, has stayed in the same unaccelerated frame of reference at all times (the centripetal acceleration of the earth is negligible here), so she is entitled to apply the time dilation formula to Dick's entire voyage. Dick, however, has had to change from one frame of reference moving at constant velocity relative to the earth to another frame in order to reverse his direction, so his use of the formula is valid only for the outbound trip. The correct conclusion is that Dick will be younger on his return.

If we want to look at Dick's journey from his own point of view, we must take into account that the distance L he covers is Lorentz-contracted by the fraction

The traveling twin covers a Lorentz-contracted distance

$$\frac{L}{L_0} = \sqrt{1 - v^2/c^2} = 0.141 \approx \tfrac{1}{7}$$

relative to the distance L_0 measured by Jane from the earth. To Dick, time passes at the usual rate, and his voyage has taken 10 years because he has traveled the shorter distance L. Thus Dick's life span has not been extended *to him,* since regardless of his sister's 70-year wait, he has spent only 10 years on the trip by his own reckoning.

What has actually happened to make Dick younger than Jane? The only answer is: that is the way the universe works. The asymmetric aging of the twins is a consequence of the laws of nature, just as the more familiar world of our everyday experience is.

26–7 ORIGIN OF MAGNETIC FORCES

The theory of relativity provides the connection between electrical and magnetic phenomena. It is hardly obvious that, when we pick up a nail with a magnet, we are witnessing a consequence of relative motion. Most relativistic effects are imperceptible in everyday life because the speeds of the objects around us are so small compared with the speed of light. Even though experiments show that there are moving electrons in the atoms of the nail and the magnet, their speeds are nowhere near that of light. The puzzle is underscored when we consider that the effective speeds of the electrons that carry a current in a wire are less than 1 mm/s—slower than a caterpillar. Yet current-carrying wires do give rise to appreciable magnetic effects, as anyone who has seen an electric motor in operation can testify.

Magnetism is a relativistic effect

If we think about the matter for a moment, though, the idea that electricity and magnetism are connected via relativity becomes less implausible. For one thing, electric forces are extremely strong, so even a small alteration in their character due to relative motion (which is what magnetic forces represent) may have large consequences. As we saw in Chapter 16, the electric force between the electron and proton in a hydrogen atom is more than 10^{39} times greater than the gravitational force between them. Second, although the individual charges involved in magnetic forces usually do move slowly, there may be such enormous numbers of them that the total effect is not negligible; for example, even a modest current in a wire involves the motion of 10^{20} electrons in each centimeter of the wire.

Magnetic forces are modifications of electric forces that result from relative motion between charges

As an illustration of how relativity accounts for the magnetic forces between moving charges, let us look into how the forces between two parallel currents come into being. In doing so, we must keep in mind that, like the speed of light, **electric charge is relativistically invariant,** so a charge whose magnitude is found to be Q in one frame of reference will be found to be Q in all other frames of reference regardless of their relative velocities.

Electric charge is relativistically invariant

Figure 26–8(a) shows two parallel conductors when no current is present. We imagine them to contain equally-spaced positive and negative charges that are at rest. The conductors are electrically neutral.

In Fig. 26–8(b) we see the same conductors when they carry currents in the same direction. The positive charges move to the right at the speed u and the negative charges move to the left at the same speed u, as seen from the laboratory frame of reference.

FIG. 26–8 (a) Two idealized conductors that contain equal numbers of positive and negative charges. (b) The conductors attract each other when they carry currents in the same direction. (c) Conductor 2 as seen by a negative charge in conductor 1 has a net positive charge. (d) Conductor 2 as seen by a positive charge in conductor 1 has a net negative charge. The various length contractions are greatly exaggerated here.

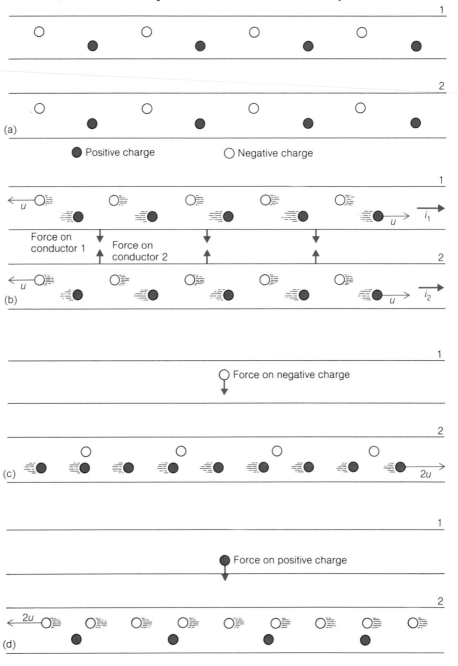

The spacing of the charges is smaller than before by the factor $\sqrt{1 - u^2/c^2}$ because of the relativistic length contraction. Since the charges of both signs have the same speed in the idealized situation we are considering, the contractions in their spacings are the same, and the conductors are still neutral to an observer in the laboratory frame of reference. There is an attractive force between the conductors: How does it arise?

We begin by looking at conductor 2 from the frame of reference of one of the negative charges in conductor 1, as in Fig. 26–8(c). To this charge, the negative charges in conductor 2 are at rest, since they are (as we see the situation from the outside) all moving at the same speed u as it is. The spacing of the negative charges is not contracted, as it is to an observer in the laboratory, so they are farther apart than in the previous diagram. However, in this frame the positive charges in conductor 2 are moving at the speed $2u$, and their spacing accordingly exhibits a greater contraction. Conductor 2 therefore appears positively charged to the negative charge in conductor 1, and there is an attractive electric force on this charge in its own frame of reference.

From the frame of reference of one of the positive charges in conductor 1, the positive charges in conductor 2 are at rest and their spacing, in the absence of any relativistic length contraction, is greater than we find in the laboratory; see Fig. 26–8(d). The negative charges in conductor 2 have the speed $2u$ and they are accordingly closer together than in the laboratory frame of reference. There is a net negative charge on conductor 2 as seen by a positive charge in conductor 1, and it is attracted electrically to conductor 2.

An identical argument shows that both the negative and positive charges in conductor 2 are attracted to conductor 1. To any of the charges in either conductor, the force on it is an "ordinary" electric force that occurs because the charges of opposite sign in the other conductor are closer together than the charges of the same sign, yielding a net attractive force. To an observer in the laboratory both conductors are electrically neutral, and he or she therefore finds it natural to ascribe the force to a special "magnetic" interaction between the currents in the conductors.

As in everything else where there is relative motion, the frame of reference from which a phenomenon is viewed is an essential part of the description of the phenomenon. Although for many purposes it is convenient to think of magnetic forces as something different from electric ones, it is worth keeping in mind that both are manifestations of a single electromagnetic interaction that occurs between charges.

A similar approach accounts for the repulsive force between parallel currents in opposite directions. Again, the "magnetic force" turns out to be an inevitable consequence of Coulomb's law, charge invariance, and the principles of special relativity.

Actual currents in metal wires consist of flows of electrons only, with the positive ions remaining in place. The advantage of considering the idealized currents above, which are electrically equivalent to actual currents, is that they are easier to analyze; the results are exactly the same in both cases.

As we have seen, a current-carrying conductor that is electrically neutral in one frame of reference might not be neutral in another frame. But this observation does not apply to the *entire* circuit of which the conductor is a part. Every electric circuit in which a current exists more than momentarily is a closed circuit, so for every current element in one direction that a moving observer finds to have a positive charge, there must be another current element in the opposite direction, which the same observer finds to have a negative charge. Hence we would expect magnetic forces to occur

Origin of forces on negative charges in conductor 1

Origin of forces on positive charges in conductor 1

Electricity and magnetism are manifestations of a single electromagnetic interaction between charges

between different parts of a circuit, which is experimentally observed, even though all observers agree on the electrical neutrality of the circuit as a whole. The latter agreement is required by charge invariance.

26–8 RELATIVITY OF MASS

Another important finding of the theory of relativity is that the mass of a body is not the same to all observers but depends upon the body's speed with respect to each observer who measures its mass. The variation of mass with speed obeys the formula

$$m = \frac{m_0}{\sqrt{1 - v^2/c^2}} \qquad\qquad \textit{Mass increase} \quad (26\text{–}4)$$

whose symbols have these meanings:

m_0 = mass measured when object is at rest ("rest mass")
m = mass measured when object is in relative motion
v = speed of relative motion
c = speed of light

An obect is more massive when moving than when at rest

Since the denominator of Eq. (26–4) is always less than one, an object will always appear more massive when in relative motion than when at rest. This mass increase is reciprocal; to a measuring device on the rocket ship in flight, its twin ship on the ground also appears to have a mass m greater than its own mass m_0.

Relativistic mass increases are significant only at speeds approaching that of light (Fig. 26–9). The rest mass of the Apollo 11 spacecraft (apart from its launch vehicle, which dropped away after accelerating the spacecraft) was about 63,070 kg. On its way

FIG. 26–9 The relativity of mass.

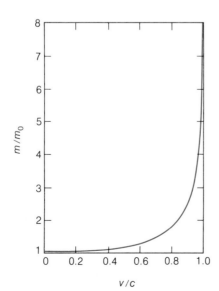

v / c

to the moon, the spacecraft's speed was about 10,840 m/s and so its mass in flight, as measured by an observer on the earth, increased to

$$m = \frac{m_0}{\sqrt{1 - \dfrac{v^2}{c^2}}} = \frac{63,070\,\text{kg}}{\sqrt{1 - \dfrac{(1.084 \times 10^4\,\text{m/s})^2}{(3 \times 10^8\,\text{m/s})^2}}} = 63,070.000041\,\text{kg}$$

This is not much of a change.

Smaller objects can be given much higher speeds. It is not very difficult to accelerate electrons (rest mass 9.109×10^{-31} kg) to speeds of, say, $0.9999c$. The mass of such an electron is

$$m = \frac{m_0}{\sqrt{1 - \dfrac{v^2}{c^2}}} = \frac{9.109 \times 10^{-31}\,\text{kg}}{\sqrt{1 - \dfrac{(0.9999c)^2}{c^2}}} = 644 \times 10^{-31}\,\text{kg}$$

The electron's mass is 71 times its rest mass! The mass increases predicted by the relativistic mass formula, even such remarkable ones as this, have been experimentally verified without exception.

Equation (26–4) has something interesting to say about the greatest speed an object can have. The closer v approaches c, the closer v^2/c^2 approaches one, and the closer $\sqrt{1 - (v^2/c^2)}$ approaches zero. As the denominator of Eq. (26–4) becomes smaller, the mass m becomes larger, so that if the relative speed v actually were equal to the speed of light, the object's mass would be infinite. The concept of an infinite mass anywhere in the universe is, of course, nonsense on many counts; it would have required an infinite force to have accelerated it to the speed of light, its length in the direction of motion would be zero by Eq. (26–2) so that its volume would be zero, and it would exert an infinite gravitational force on all other bodies in the universe. Hence we interpret Eq. (26–4) to mean that no material body can ever equal or exceed the speed of light.

Provided that linear momentum is defined as

$$m\mathbf{v} = \frac{m_0\mathbf{v}}{\sqrt{1 - v^2/c^2}} \qquad \textit{Relativistic Momentum} \quad (26\text{–}5)$$

conservation of momentum is just as valid in special relativity as it is in classical physics. However, Newton's second law of motion is correct only in the form

Force = rate of change of momentum

$$\mathbf{F} = \frac{\Delta\,(m\mathbf{v})}{\Delta\,t} \qquad \textit{Relativistic second law} \quad (26\text{–}6)$$

where $m\mathbf{v}$ is given by Eq. (26–5). This is *not* the same as

Force = mass × rate of change of velocity

$$\mathbf{F} = m\frac{\Delta\,\mathbf{v}}{\Delta\,t} = m\mathbf{a}$$

because m as well as $\mathbf{v}$ changes when an object is accelerated. Relativistic mechanics is therefore more complicated than classical mechanics.

26–9   MASS AND ENERGY

The most famous conclusion of the theory of relativity is the equivalence of mass and energy according to the formula $E = mc^2$. In this formula m is the measured mass of an object, which is greater than its rest mass m_0 if the object is moving. Thus the total energy E of the object is given by

$$E = mc^2 = \frac{m_0 c^2}{\sqrt{1 - v^2/c^2}} \qquad \text{Total energy} \quad (26\text{–}7)$$

If the object is at rest, the quantity

$$E_0 = m_0 c^2 \qquad \text{Rest energy} \quad (26\text{–}8)$$

is its *rest energy*, the energy equivalent of its rest mass; this formula was introduced in Sec. 6–7. If the object is moving, its total energy is the sum of its rest energy and its kinetic energy KE:

$$E = E_0 + \text{KE} = m_0 c^2 + \text{KE} \qquad \text{Total energy} \quad (26\text{–}9)$$

Since the zero level of potential energy is arbitrary, PE is not included here. If the PE of the object changes, its total energy will change by the same amount.

From Eqs. (26–7) and (26–9) we see that the kinetic energy of a moving object is given by

$$\text{KE} = mc^2 - m_0 c^2 = \frac{m_0 c^2}{\sqrt{1 - v^2/c^2}} - m_0 c^2 \qquad \text{Kinetic energy} \quad (26\text{–}10)$$

This formula is rather different from the kinetic energy formula $\frac{1}{2}m_0 v^2$ we have been using thus far. However, it is not hard to verify that the relativistic formula for KE reduces to $\frac{1}{2}m_0 v^2$ when v is small compared with c.

We start by noting that, when x is small, the binomial expansion of algebra shows that $1/\sqrt{1 - x}$ can be approximated by

$$\frac{1}{\sqrt{1 - x}} \approx 1 + \frac{x}{2} \qquad x \ll 1$$

The symbol $\ll$ means "much smaller than." Here we let $x = v^2/c^2$, so that

$$\frac{1}{\sqrt{1 - v^2/c^2}} \approx 1 + \frac{1}{2}\frac{v^2}{c^2} \qquad v \ll c$$

and

$$\text{KE} = \frac{m_0 c^2}{\sqrt{1 - v^2/c^2}} - m_0 c^2 \approx \left(1 + \frac{1}{2}\frac{v^2}{c^2}\right) m_0 c^2 - m_0 c^2 \approx \frac{1}{2}m_0 v^2 \qquad v \ll c$$

$\frac{1}{2}m_0 v^2$ **is the low-speed approximation to the KE of a moving object**

The relativistic formula is correct for all speeds, and the formula $\text{KE} = \frac{1}{2}m_0 v^2$ is the low-speed approximation to it. The greater the speed v, the more the formula $\frac{1}{2}m_0 v^2$ understates the true KE of an object.

The degree of accuracy required is what determines whether the approximation $\frac{1}{2}m_0 v^2$ is appropriate. For instance, when $v = 10^7$ m/s $= 0.033c$, the approximation understates the true KE by only 0.08%; when $v = 3 \times 10^7$ m/s $= 0.1c$, it understates the true KE by 0.8%; but when $v = 1.5 \times 10^8$ m/s $= 0.5c$, the understatement is a significant 19%, and when $v = 0.999c$, the approximation gives too small a KE by a factor of 43. According to KE $= \frac{1}{2}m_0v^2$, an object would need a kinetic energy of $\frac{1}{2}m_0c^2$, half its rest energy, to move at the speed of light; but according to Eq. (26–10) it would need an infinite kinetic energy, and so such a speed is unattainable.

Example How many times greater than its rest mass is the mass of an electron whose kinetic energy is 1 GeV? The rest energy of the electron is 0.511 MeV.

Solution Since 1 GeV $= 10^9$ eV $= 10^3$ MeV,

$$\frac{m}{m_0} = \frac{mc^2}{m_0c^2} = \frac{m_0c^2 + KE}{m_0c^2} = \frac{1000\,\text{MeV} + 0.511\,\text{MeV}}{0.511\,\text{MeV}} = 1958$$

Electrons with energies well in excess of 1 GeV occur naturally in the cosmic radiation in space and have been produced in the laboratory as well. ■

Example An object at rest explodes into two fragments whose rest masses are both 1 g. The fragments move apart at speeds of $0.6c$ relative to the original object. What was the rest mass of the original object?

Solution Let us call the rest mass of the original object m_0 and those of the fragments m_{01} and m_{02}, with the speeds of the latter being respectively v_1 and v_2. Since the total energy m_0c^2 of the original object must equal the sum of the total energies m_1c^2 and m_2c^2 of the fragments,

$$m_0c^2 = m_1c^2 + m_2c^2 = \frac{m_{01}c^2}{\sqrt{1 - v_1^2/c^2}} + \frac{m_{02}c^2}{\sqrt{1 - v_2^2/c^2}}$$

Since $m_{01} = m_{02} = 1$ g and $v_1 = v_2 = 0.6c$, dividing through by c^2 and substituting these values gives

$$m_0 = \frac{2 \times m_{01}}{\sqrt{1 - v_1^2/c^2}} = \frac{2 \times 1\,\text{g}}{\sqrt{1 - (0.6)^2}} = 2.5\,\text{g}$$

Thus 20% of the original object's mass became kinetic energy of the fragments in the explosion. ■

26–10 GENERAL RELATIVITY

The special theory of relativity shows us how to interpret what we observe in frames of reference that move at constant velocity with respect to us. The laws of physics are valid in all such frames of reference, but measurements of some quantities—notably time intervals, lengths, and masses—are affected by relative motion, whereas measurements of others—notably the speed of light and electric charge—are not. The *general theory of relativity* explores the effects of accelerated motion on what we observe. Published by Einstein in 1915, it provides insights into gravitational phenomena as

General relativity concerns accelerated motion and gravitation

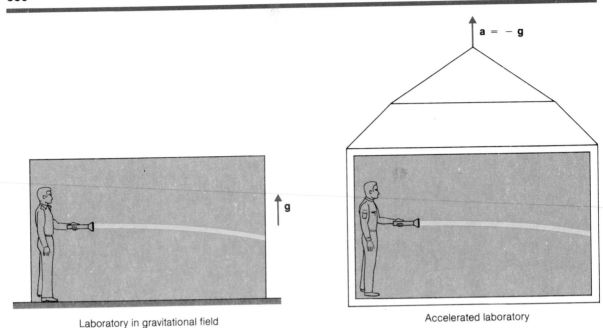

Laboratory in gravitational field   Accelerated laboratory

FIG. 26–10 According to the principle of equivalence, events that take place in an accelerated laboratory cannot be distinguished from those that take place in a gravitational field. Hence the deflection of a light beam relative to an observer in an accelerated laboratory means that light must be similarly deflected in a gravitational field.

profound and far-reaching as did those of the special theory into the relationships between mass and energy and between electricity and magnetism.

One of the basic ideas of general relativity is the *principle of equivalence*: An observer in a closed laboratory cannot distinguish between the effects produced by a gravitational field and those produced by an acceleration of the laboratory. This principle is another way to express the experimental finding that the inertial mass of an object, which determines its acceleration when a force acts on it, is always equal to its gravitational mass, which determines the gravitational pull another object exerts on it. (The two masses are actually proportional; the constant of proportionality is set equal to 1 by an appropriate choice of the gravitational constant G.)

An immediate consequence of the principle of equivalence is that light should be subject to gravity. We can come to this conclusion by considering the passage of a light beam across an accelerated laboratory (Fig. 26–10). The light beam will pursue a curved path relative to the laboratory, which to an observer there is exactly the path that would be taken if it were subject to the same gravitational field the acceleration is equivalent to.

The first confirmation of the gravitational deflection of light came from a comparison of the positions of stars in the sky that appeared near the sun during an eclipse, when they could be seen because the sun's disk was obscured by the moon, with their positions at other times when the light from them did not have to pass close to the sun (Fig. 26–11). The observed deviation of about 0.0005° for light grazing the sun was in agreement with the prediction of general relativity. More recently laboratory experiments based on the quantum theory of light (Chapter 27) have independently verified that light is indeed subject to gravity.

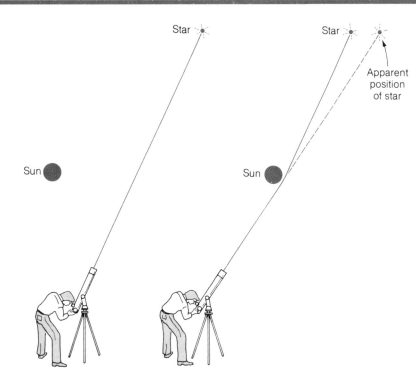

FIG. 26–11 Starlight passing near the sun is deflected by its strong gravitational field. The deflection can be measured during a solar eclipse when the sun's disk is obscured by the moon.

Gas molecules are held in the atmospheres of the earth and other planets by gravity despite their considerable average speeds. Is it possible that an astronomical object both massive enough and small enough to have an enormously strong gravitational field could similarly prevent light from escaping from it? The answers seems to be yes. Such an object is called a "black hole" because, besides emitting no light of its own, it captures light from other sources that happen to pass nearby. A black hole that is a member of a double-star system will manifest its presence by its gravitational effect on the other member, which will rotate about the center of mass of the system. In addition, the intense gravitational field of a black hole will attract matter from space; as it is drawn in, this matter will be compressed and heated to such high temperatures that X rays will be emitted in profusion just outside the region of permanent capture. An invisible object that many astronomers believe to be a black hole is known as Cygnus X-1, whose mass is about eight times that of the sun but whose radius is thought to be only about 10 km.

Black holes are collapsed stars whose gravitational fields prevent light from escaping

IMPORTANT TERMS

The **special theory of relativity** relates measurements made on an object or phenomenon from frames of reference moving at constant velocity with respect to one another.

The relativistic **time dilation** refers to the fact that a clock moving with respect to an observer appears to tick less rapidly than it does to an observer traveling with the clock.

The **Lorentz contraction** refers to the decrease in the measured length of an object when it is moving relative to an observer.

The **relativity of mass** refers to the increase in the measured mass of an object when it is moving relative to an observer. The object's **rest mass** is its mass measured when it is at rest relative to the observer.

The **general theory of relativity** concerns frames of

reference accelerated with respect to one another. One of the conclusions of general relativity is that light is affected by gravitational fields.

IMPORTANT FORMULAS

Lorentz contraction: $L = L_0 \sqrt{1 - v^2/c^2}$

Time dilation: $t = \dfrac{t_0}{\sqrt{1 - v^2/c^2}}$

Mass increase $m = \dfrac{m_0}{\sqrt{1 - v^2/c^2}}$

Kinetic energy:

$$KE = mc^2 - m_0c^2 = \dfrac{m_0c^2}{\sqrt{1 - v^2/c^2}} - m_0c^2$$

MULTIPLE CHOICE

1. The theory of relativity is in conflict with
   (a) experiment.
   (b) Newtonian mechanics.
   (c) electromagnetic theory.
   (d) ordinary mathematics.

2. Which, if any, of the following quantities have the same value to all observers?
   (a) The speed of light
   (b) The charge of the muon
   (c) The mass of the muon
   (d) The average lifetime of the muon

3. According to the principle of relativity, the laws of physics are the same in all frames of reference
   (a) at rest with respect to one another.
   (b) moving toward or away from one another at constant velocity.
   (c) moving parallel to one another at constant velocity.
   (d) all of the above.

4. The lifetime of a muon in motion relative to an observer appears to him to be
   (a) shorter than its lifetime at rest.
   (b) the same as its lifetime at rest.
   (c) longer than its lifetime at rest.
   (d) any of the above, depending upon the relative velocity.

5. An observer in a closed laboratory wishes to determine whether the laboratory is at rest or in motion at constant velocity. The observer
   (a) can find out by measuring the apparent velocity of light in the laboratory.
   (b) can find out by measuring his mass.
   (c) can find out by comparing two different clocks in the laboratory over a period of time.
   (d) cannot find out.

6. A spacecraft has left the earth and is moving toward Mars. An observer on the earth finds that, relative to measurements made when it was at rest, the spacecraft's
   (a) length is greater.
   (b) mass is smaller.
   (c) clocks tick faster.
   (d) momentum is greater.

7. The formula $KE = \frac{1}{2}m_0v^2$ for kinetic energy
   (a) is the correct formula if v is properly interpreted.
   (b) always gives too high a value.
   (c) is the low-speed approximation to the correct formula.
   (d) is the high-speed approximation to the correct formula.

8. An electron is moving through a body of water. The greatest speed it can possibly have is
   (a) the speed of water waves.
   (b) the speed of sound waves in water.
   (c) the speed of light waves in water.
   (d) the speed of light waves in vacuum.

9. When an object whose length is 1 m when at rest approaches the speed of light, its length approaches
   (a) 0.   (b) 0.5 m.
   (c) 2 m.   (d) ∞.

10. When an object whose rest mass is 1 kg approaches the speed of light, its mass approaches
   (a) 0.   (b) 0.5 kg.
   (c) 2 kg.   (d) ∞.

11. Which of the following speeds must an object have if its mass is to be double its rest mass?
   (a) $c/2$ (b) $\sqrt{3}c/2$
   (c) c (d) $2c$

12. A man 6 ft tall lies along the axis of a space vehicle traveling at $0.9c$. His height as measured by a stationary observer is
   (a) 1.9 ft.   (b) 2.6 ft.
   (c) 6.0 ft.   (d) 14 ft.

13. The speed of an electron whose mass is ten times its rest mass is

(a) 2×10^8 m/s. (b) 2.98×10^8 m/s.

(c) 4×10^8 m/s. (d) 3×10^9 m/s.

EXERCISES

1. Can an observer in a windowless laboratory in principle determine whether the earth is (a) moving through space with a uniform velocity; (b) moving through space with a uniform linear acceleration; (c) rotating on its axis?

2. The electron beam in a television picture tube can move across the screen faster than the speed of light. Why does this not violate the special theory of relativity?

3. If the speed of light were smaller than it is, would relativistic phenomena be more or less conspicuous than they are now?

4. Is the mass of an object the same whether it is moving toward an observer or away from him at the same speed? Is it the same whether it moves toward the observer or the observer moves toward it at the same speed? What is the connection between the answers to these questions and the principle of relativity?

5. The length of a rod is measured by several observers, one of whom is stationary with respect to the rod. What must be true of the figure obtained by the latter observer?

6. An object moving in a circle has an average velocity **v** of zero. Would you expect it to have a relativistic mass increase?

7. The hydrogen molecule contains two protons and two electrons. The electrons are moving considerably faster than the protons. The most accurate measurements to date indicate that the hydrogen molecule is electrically neutral to at least one part in 10^{10}. What significance has this result with respect to the question of whether electric charge is an invariant quantity, like the speed of light in free space, or depends upon relative motion, like mass, length, and the duration of a time interval?

8. Does a laboratory at rest on the earth's surface constitute a nonaccelerated frame of reference? If not, where on the surface is the acceleration greatest? Where is it least? Is it zero anywhere in the earth?

9. Is the density of a moving object less than, the same as, or more than it appears to an observer when the object is at rest?

10. A meter stick appears only 30 cm long to an observer.

(a) What is its relative speed? (b) An observer times the passage of the meter stick past a point in her own frame of reference. What does she find?

11. An atomic nucleus 5×10^{-15} m in diameter is moving at a speed of 10^8 m/s. What is its thickness in its direction of motion as measured by a laboratory observer?

12. A certain process requires 10^{-6} s to occur in an atom at rest in the laboratory. How much time will this process require to an observer in the laboratory when the atom is moving at a speed of 5×10^7 m/s?

13. How fast would a spacecraft have to go for each year on the spacecraft to correspond to two years on the earth?

14. A woman leaves the earth in a spacecraft that makes a round trip to the nearest star, 4 light-years distance, at a speed of $0.9c$. How many days younger is she upon her return than her twin sister who remained behind? (A light-year is the distance light travels in free space in a year. It is equal to 9.46×10^{15} m.)

15. A man has a mass of 100 kg on the ground. When he is in a spacecraft in flight, an observer on the earth measures his mass to be 101 kg. How fast is the spacecraft moving?

16. Find the mass of an object whose rest mass is 1000 g when it is traveling at 10%, 90%, and 99% of the speed of light.

17. What is the kinetic energy in MeV of a neutron whose mass is double its rest mass? The rest energy of the neutron is 940 MeV.

18. How many joules of energy per kilogram of rest mass are required to bring a spacecraft from rest to a speed of $0.9c$?

PROBLEMS

1. An airplane is flying at 300 m/s (672 mi/h). How much time must elapse before a clock in the airplane and one on the ground differ by 1 s?

2. Two observers, A on earth and B in a rocket ship whose velocity is 2×10^8 m/s, both set their watches to the same time when the ship is abreast of the earth. (a) How much time must elapse by A's reckoning before the watches differ by 1 s? (b) To A, B's watch seems to run slow. To B, does A's watch seem to run fast, run slow, or keep the same time as his own watch?

3. How much mass does a proton gain when it is accelerated to a kinetic energy of 500 MeV? The rest mass of the proton is 1.67×10^{-27} kg and its rest energy is 938 MeV.

4. An electron has a kinetic energy of 0.10 MeV. Find its speed according to classical and relativistic mechanics. The rest energy of the electron is 0.51 MeV.

5. An object moving at $0.5c$ with respect to an observer disintegrates into two fragments that move in opposite directions relative to their center of mass along the same line of motion as the original object. One fragment has a speed of $0.6c$ in the backward direction relative to the center of mass and the other has a speed of $0.5c$ in the forward direction. What speeds will the observer find?

6. A woman on the moon sees two spacecraft, A and B coming toward her from opposite directions at the respective speeds of $0.8c$ and $0.9c$. (a) What does a man on A measure for the speed with which he is approaching the moon? For the speed with which he is approaching B? (b) What does a man on B measure for the speed with which he is approaching the moon? For the speed with which he is approaching A?

7. An electron whose speed relative to an observer in a laboratory is $0.8c$ is also being studied by an observer moving in the same direction as the electron at a speed of $0.5c$ relative to the laboratory. What is the kinetic energy, in MeV, of the electron to each observer? Note that $m_0c^2 = 0.51$ MeV for an electron.

ANSWERS TO MULTIPLE CHOICE

1. b	**6.** d	**11.** b
2. a, b	**7.** c	**12.** b
3. d	**8.** d	**13.** b
4. c	**9.** a	
5. d	**10.** d	

PARTICLES AND WAVES

27

Particles and waves are entirely separate concepts in everyday life: it is impossible to confuse a stone, for instance, with the waves it produces when it is dropped into a lake. But in the realm of the atom, the situation is very different. Electromagnetic waves—which, as we know, exhibit such characteristic wave phenomena as interference—in many ways behave exactly as particles do. And electrons—whose particle nature is amply shown in the operation of a television picture tube—nevertheless behave in many ways exactly as waves do. On the microscopic level, then, a wave-particle duality replaces the distinction between waves and particles so evident on a macroscopic level. This duality turns out to be the key to understanding the structure of atoms and why they behave as they do.

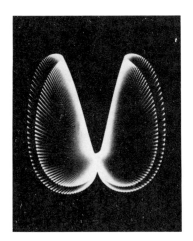

27–1   PHOTOELECTRIC EFFECT

The formulation of the theory of relativity and that of the quantum theory of light, both of which took place early in this century, profoundly altered our

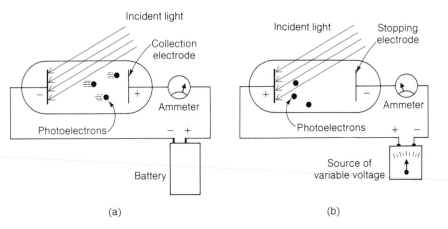

FIG. 27-1   (a) A method of detecting the photoelectric effect. The photoelectrons ejected from the irradiated metal plate are attracted to the positive collection electrode at the other end of the tube, and the current that results is measured with an ammeter. (b) A method of detecting the maximum energy of the photoelectrons. Note polarity opposite to that in (a). As the stopping electrode is made more negative, the slower photoelectrons are repelled before they can reach it. Finally a voltage will be reached at which no photoelectrons whatever are received at the stopping electrode, as indicated by the current dropping to zero, and this voltage corresponds to the maximum photoelectron energy.

Relativity and quantum theory revolutionized physics

ways of thinking about the physical world. We have already examined some of the remarkable consequences of relativity, and now we come to the realm of quanta, which will be no less remarkable. But such a subjective term is really not justified, because it is merely based upon the limits to our imagination that are imposed by our experience; if we were in a world where we were about the same size as an electron, relativistic and quantum phenomena would be familiar (though then most macroscopic phenomena would not).

Photoelectric effect

Toward the end of the nineteenth century a number of experiments were performed that revealed the emission of electrons from a metal surface when light (particularly ultraviolet light) falls on it (Fig. 27–1). This phenomenon is known as the *photoelectric effect*. It is not, at first glance, anything to surprise us, for light waves carry energy, and some of the energy absorbed by the metal may somehow concentrate on individual electrons and reappear as kinetic energy. Upon closer inspection of the data, however, we find that the photoelectric effect can hardly be explained in so straightforward a manner.

Photoelectrons are emitted immediately

The first peculiarity of the photoelectric effect is that, even when the metal surface is only faintly illuminated, the emitted electrons (which are called *photoelectrons*) leave the surface immediately. But according to the electromagnetic theory of light, the energy content of light waves is spread out across the width of the wavefronts of the light beam involved. Calculations show that a definite period of time—several months in the case of a beam of very low intensity—must elapse before any individual electrons accumulate enough energy to leave the metal. Instead, the electrons are found to be emitted as soon as the light is turned on.

Another unexpected discovery is that the energy of the photoelectrons does not depend upon the intensity of the light. A bright light yields more electrons than a dim

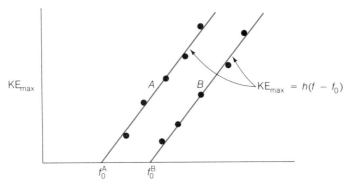

$$KE_{max} = h(f - f_0)$$

Frequency of incident light, f

FIG. 27–2 The variation of maximum photoelectron energy with the frequency of the incident light for two target metals. No photoelectrons are emitted for frequencies below f_0^A in the case of metal A and below f_0^B in the case of metal B. In both cases, however, the angle between the experimental line and either axis is the same. Hence we may write the equation of the lines as $KE_{max} = h(f - f_0)$, where h has the same value in all cases but where f_0, the minimum frequency required for photoelectric emission to occur, depends upon the nature of the target metal.

one, but their average energy remains the same. This behavior contradicts the electromagnetic theory of light, which predicts that the energy of photoelectrons should depend upon the intensity of the light beam responsible for them.

The energies of the photoelectrons emitted from a given metal surface turn out, most surprisingly of all, to depend upon the *frequency* of the light employed. At frequencies below a certain critical one (which is characteristic of the particular metal), no electrons whatever are given off. Above this threshold frequency the photoelectrons have a range of energies from zero to a certain maximum value, and *this maximum energy increases with increasing frequency* (Fig. 27–2). High-frequency light yields high maximum photoelectron energies; low-frequency light yields low maximum photoelectron energies. Thus dim blue light produces electrons with more energy than those produced by intense red light, although the latter results in a greater number of them (Fig. 27–3).

Photoelectron energy increases with frequency of incident light

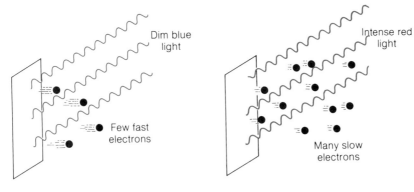

Dim blue light

Few fast electrons

Intense red light

Many slow electrons

FIG. 27–3

27–2   QUANTUM THEORY OF LIGHT

The energy of light waves travels in bursts called quanta

Aware that the electromagnetic theory of light, despite its notable success in accounting for other optical phenomena, failed to explain the photoelectric effect, Albert Einstein in 1905 sought some other basis for interpreting it. He found what he needed in a novel assumption that Max Planck, a German physicist, had had to make a few years earlier in order to understand the origin of blackbody radiation (see Sec. 14–8). Planck found that the accepted physical laws of the time predicted the observed characteristics of this radiation *provided* that the radiation is considered as though emitted in little bursts of energy rather than continuously. These bursts of energy are called *quanta*.

Planck showed that the energy E of each quantum had to be related to the light frequency f by the formula

$$E = hf \hspace{5cm} Quantum\ energy \quad (27\text{--}1)$$

Planck's constant

where h is a constant, known today as Planck's constant, whose value is

$$h = 6.626 \times 10^{-34}\ \text{joule-second} \hspace{3cm} Planck's\ constant$$

Although the energy radiated by a heated object must be regarded as coming out intermittently, in order for theory and experiment to agree, Planch held to the conventional view that it nevertheless travels through space as continuous waves.

Einstein saw that Planck's idea could be used to interpret the photoelectric effect if light not only is emitted a quantum at a time but also propagates as separate quanta. Then the h of the photoelectric effect equation (Fig. 27–2) is the same as the h of the formula $E = hf$, and the significance of the former equation becomes clear when it is rewritten

$$hf = \text{KE}_{\text{max}} + hf_0 \hspace{4cm} Photoelectric\ effect \quad (27\text{--}2)$$

What this equation states is that

$$\text{Quantum energy} = \text{maximum electron energy} + \text{energy required to eject an electron}$$

The work function is the minimum energy needed to remove an electron from a surface

The reason for a threshold energy hf_0 (often called *work function*) is clear: it is the minimum energy required to dislodge an electron from the metal surface, as in Fig. 27–4. (There must be such a minimum energy, or electrons would leave metals all the time.) There are several plausible reasons why not all photoelectrons have the same energy even though a single frequency of light is used. For instance, not all the quantum energy hf may be transferred to a single electron, and an electron may lose some of its initial energy in collisions with other electrons within the metal before it actually emerges from the surface.

Example   The photoelectric work function for copper is 4.5 eV. Find the maximum energy of the photoelectrons when ultraviolet light of 1.5×10^{15} Hz falls on a copper surface.

Solution   The quantum energy of the incident photons in eV is

$$hf = \frac{(6.63 \times 10^{-34}\ \text{J·s})(1.5 \times 10^{15}\ \text{Hz})}{1.6 \times 10^{-19}\ \text{J/eV}} = 6.2\ \text{eV}$$

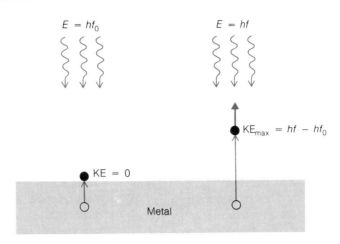

FIG. 27-4 If the energy hf_0 is required to remove an electron from a metal surface, the maximum electron kinetic energy will be $hf - hf_0$ when light of frequency f is directed at the surface.

Hence the maximum photoelectron energy is

$$\text{KE}_{max} = hf - hf_0 = (6.2 - 4.5)\,\text{eV} = 1.7\,\text{eV}$$

Einstein's notion that light travels as a series of little packets of energy (sometimes referred to as quanta, sometimes as *photons*) is in complete contradiction with the wave theory of light (Fig. 27–5). And the wave theory, as we know, has some powerful observational evidence on its side. There is no other way to explain interference effects, for example. According to the wave theory, light spreads out from a source in a manner analogous to the spreading out of ripples on the surface of a lake when a stone is dropped into it, with the energy of the light distributed continuously throughout the wave pattern. According to the quantum theory, light spreads out from a source as a succession of localized packets of energy, each sufficiently small to permit its being absorbed by a single electron. Yet, despite the particle picture of light that it presents, the quantum theory requires a knowledge of the light frequency f, a wave quantity, in order to determine the energy of each quantum.

Photons are quanta of light

Light as a wave phenomenon

On the other hand, the quantum theory of light is able to explain the photoelectric effect. It predicts that the maximum photoelectron energy should depend upon the frequency of the incident light and not upon its intensity, precisely the opposite of what the wave theory suggests, and it is able to explain why even the feeblest light can lead to the immediate emission of photoelectrons. The wave theory can give no reason why there should be a threshold frequency below which no photoelectrons are observed, no

Light as a quantum phenomenon

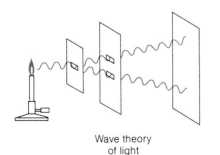

Wave theory of light

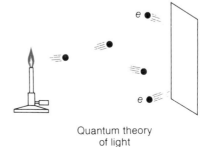

Quantum theory of light

FIG. 27-5 The wave theory of light is needed to explain diffraction and interference phenomena, which the quantum theory cannot explain. The quantum theory of light is needed to explain the photoelectric effect, which the wave theory cannot explain.

matter how strong the light beam, something that follows naturally from the quantum theory.

Which theory is correct? The history of physics is filled with examples of physical ideas that required revision or even replacement when new empirical data conflicted with them, but this is the first occasion in which two completely different theories are both required to explain a single physical phenomenon. In thinking about this, it is important for us to note that, in a particular situation, light behaves *either* as though it has a wave nature *or* a particle nature. While light sometimes assumes one guise and sometimes the other, there is no physical process in which both are simultaneously exhibited. The same light beam can diffract around an obstacle and then impinge on a metal surface to eject photoelectrons, but these two processes occur separately.

The wave and quantum theories of light are complementary

The electromagnetic theory of light and the quantum theory of light complement each other; by itself, each theory is "correct" in certain experiments, and there are no relevant experiments that neither can account for. Light must be thought of as a phenomenon that incorporates both particle and wave characters. Although we cannot visualize its "true nature," these complementary theories of light are able to account for its behavior, and we have no choice but to accept them both.

27–3   X RAYS

If photons of light can give up their energy to electrons, can the kinetic energy of moving electrons be converted into photons? The answer is that such a transformation is not only possible, but had in fact been discovered (though not understood) prior to the work of Planck and Einstein. In 1895 Roentgen found that a mysterious, highly penetrating radiation is emitted when high-speed electrons impinge on matter. The X rays (so called because their nature was then unknown) caused phosphorescent substances to glow, exposed photographic plates, traveled in straight lines and were not affected by electric or magnetic fields. The more energetic the electrons, the more penetrating the X rays, and the greater the number of electrons, the greater the density of the resulting X-ray beam.

X-rays are short-wavelength em waves

After over 10 years of study, it was finally established that X rays exhibit, under certain circumstances, both interference and polarization effects, leading to the conclusion that they are electromagnetic waves. From the interference experiments their wavelengths were found to be very short, shorter than those in ultraviolet light. Electromagnetic radiation in the approximate wavelength interval from 0.01 to 10 nm is today classed as X-radiation.

Figure 27–6 is a diagram of an X-ray tube. Battery A sends a current through the filament, heating it until it emits electrons. These electrons are then accelerated toward a metallic target by the potential difference V provided by battery B. The tube is evacuated to permit the electrons to reach the target unimpeded. The impact of the electrons causes the evolution of X rays from the target.

How x-rays are produced experimentally

What is the physical process involved in the production of X rays? It is known that charged particles emit electromagnetic waves whenever they are accelerated, and so we may reasonably identify X rays as the radiation accompanying the slowing down of fast electrons when they strike matter. The great majority of the incident electrons, to be sure, lose their kinetic energy too gradually for X rays to be evolved, and merely

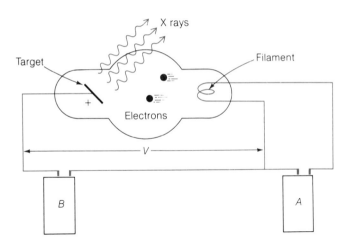

FIG. 27–6 An X-ray tube.

act to heat the target. (Consequently the targets in X-ray tubes are made of metals with high melting points, and a means for cooling the target is often provided.) A few electrons, however, lose much or all their energy in single collisions with target atoms, and this is the energy that appears as X rays. In other words, we may regard X-ray production as an inverse photoelectric effect.

Since the threshold energy hf_0 needed to remove an electron from a metal is only a few electron volts, whereas the accelerating potential V in an X-ray tube usually exceeds 10,000 volts, we can neglect the work function hf_0 here. The highest frequency f_{max} found in the X rays emitted from a particular tube should therefore correspond to a quantum energy of hf_{max}, where hf_{max} equals the kinetic energy

$$KE = Ve$$

of an electron that has been accelerated through a potential difference of V. We conclude that

$$hf_{max} = Ve \qquad\qquad \textit{X-ray energy} \quad (27\text{–}3)$$

in the operation of an X-ray tube.

Example   Find the highest frequency present in the radiation from an X-ray machine whose operating potential is 50,000 volts.

Solution   From the formula $hf_{max} = Ve$ we find that

$$f_{max} = \frac{Ve}{h} = \frac{(5.0 \times 10^4\,\text{V})(1.6 \times 10^{-19}\,\text{C})}{6.6 \times 10^{-34}\,\text{J}\cdot\text{s}} = 1.2 \times 10^{19}\,\text{Hz}$$

The corresponding wavelength is $\lambda = c/f = 2.5 \times 10^{-11}$ m $= 0.025$ nm.   ■

27–4   PHOTONS AS PARTICLES

According to the quantum theory of light, photons behave like particles except for their lack of rest mass. How far can this analogy be carried? For instance, can we consider a collision between a photon and an electron just as if both were billiard balls?

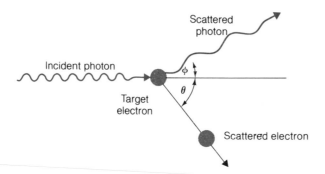

FIG. 27–7 The scattering of a photon by an electron is called the Compton effect. Energy and momentum are conserved in such an event, with the result that the scattered photon has a lower frequency (longer wavelength) than the incident photon.

Figure 27–7 shows how such a collision can be represented, with a photon of frequency f striking an electron, initially at rest, and being scattered away from its original direction of motion while the electron receives an impulse and begins to move. As a result of the impact, the photon loses an amount of energy equal to the kinetic energy KE gained by the electron, so the scattered photon has a lower frequency f' specified by the formula

Change in photon energy = electron kinetic energy

$$hf - hf' = \text{KE}$$

Photons have momentum as well as energy

A moving electron has the momentum $p = mv$ as well as energy, and a photon must also have momentum in order to transfer any to an electron. Although a photon has no rest mass, we can take its total energy hf as equivalent to the total energy mc^2 of a particle whose mass is m. On this basis $hf = mc^2$ and the photon "mass" is

$$m_p = \frac{hf}{c^2} \qquad\qquad\qquad \textit{Photon "mass"}$$

Since the photon travels at the speed of light, its momentum ought to be $p = m_p c$, which is

$$p = \frac{hf}{c} \qquad\qquad\qquad \textit{Photon momentum} \quad (27\text{–}4)$$

When a collision between a photon and an electron is analyzed with the help of special relativity, starting from conservation of energy and conservation of linear momentum, the result is a formula that relates the change in photon frequency to the angle ϕ between the original photon direction and the scattered photon direction:

$$\frac{1}{f'} - \frac{1}{f} = \frac{h}{m_0 c^2}(1 - \cos \phi)$$

The Compton effect

This formula can be readily checked by experiment: a beam of X rays of a single, known frequency f is directed at a target, and the frequencies of the scattered X rays are measured at various angles ϕ. The greater the scattering angle, the greater should be the change in frequency. Theory and experiment agree, which signifies that photons do indeed possess the momentum $p = hf/c$ and do indeed behave like particles in collisions. Photon scattering is called the *Compton effect* in honor of its discoverer,

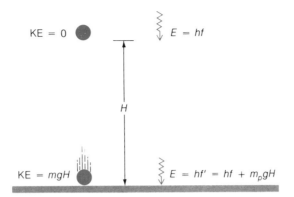

FIG. 27-8 A photon falling in a gravitational field gains energy, just as a stone does. This gain of energy is manifested as an increase in frequency.

A. H. Compton. The Compton effect is the chief means by which X rays lose energy in passing through matter.

As we have seen, a photon behaves in collisions as if it has the inertial mass hf/c^2 even though it lacks rest mass. According to the principle of equivalence discussed in Sec. 26–10, the gravitational and inertial masses of a material object are always the same. If this principle applies to photons—a reasonable supposition, since light is subject to gravity—then photons ought to exhibit a gravitational mass of hf/c^2.

When we drop a stone of mass m from a height H, the gravitational pull of the earth accelerates it as it falls and the stone gains the energy mgH. The stone's final kinetic energy $\frac{1}{2}mv^2$ is equal to mgH, so its final speed is $\sqrt{2gH}$ (Fig. 27–8) which we can measure. But a photon has the speed of light and cannot go any faster. However, its initial quantum energy hf will increase by $m_p gH$ if it is affected by gravity, which means that a photon's frequency ought to increase after it has "fallen" in a gravitational field. In a laboratory experiment, the frequency change is so small that we can neglect the corresponding change in the photon's "mass" m_p, and so, if f' is the new frequency,

Photons and gravity

$$\text{Final photon energy} = \text{initial photon energy} + \text{increase in energy}$$

$$hf' = hf + m_p gH$$

Since $m_p = hf/c^2$,

$$hf' = hf + \frac{hfgH}{c^2} = hf\left(1 + \frac{gH}{c^2}\right)$$

For $H = 20$ m, a reasonable height for a laboratory experiment, the relative change in frequency is

$$\frac{\Delta f}{f} = \frac{f' - f}{f} = \frac{gH}{c^2} = \frac{(9.8\text{ m/s}^2)(20\text{ m})}{(3 \times 10^8\text{ m/s})^2} = 2.2 \times 10^{-15}$$

A frequency change of this magnitude is detectable, and in experiments using gamma rays, which are electromagnetic waves of higher frequency than X rays, the effect has been confirmed.

27–5   MATTER WAVES

Quantum mechanics is a probabilistic description of nature

As we have seen, electromagnetic waves under certain circumstances have properties indistinguishable from those of particles. It requires no greater stretch of the imagination to speculate whether what we normally think of as particles might not have wave properties, too. This speculation was first made by Louis de Broglie in 1924. Soon afterward de Broglie's idea was taken up and developed by a number of other physicists (notably Heisenberg, Schrödinger, Born, Pauli, and Dirac) into the elaborate, mathematically difficult—but very beautiful—theory called *quantum mechanics*.

The advent of quantum mechanics did more than provide a supremely accurate and complete description of atomic phenomena; it also altered the way in which physicists approach nature, so that they now think in terms of probabilities instead of in terms of certainties. The universe is closer in many respects to a roulette wheel than to a clock—but it is a roulette wheel that obeys certain rules, and the laws of physics are these rules.

De Broglie started with Eq. (27–4) for the linear momentum of a photon. Since $\lambda f = c$, this momentum can be expressed in terms of wavelength as $p = h/\lambda$. Hence for a photon

$$\lambda = \frac{h}{p} \qquad\qquad \textit{Photon wavelength} \quad (27\text{–}5)$$

De Broglie proposed that moving objects have wave properties

De Broglie suggested that this equation for wavelength is a perfectly general one, applying to material objects as well as to photons. In the case of an object $p = mv$, and so its *de Broglie wavelength* is

$$\lambda = \frac{h}{mv} \qquad\qquad \textit{De Broglie wavelength} \quad (27\text{–}6)$$

The more momentum an object has, the shorter its wavelength. The relativistic formula $m = m_0/\sqrt{1 - v^2/c^2}$ must usually be used for m in computing de Broglie wavelengths.

Matter waves exhibit diffraction and interference

How can de Broglie's hypothesis be verified? Perhaps the most striking example of wave behavior is a diffraction pattern, which depends upon the ability of waves both to bend around obstacles and to interfere constructively and destructively with one another. Several years after de Broglie's work, Davisson and Germer, in the United States, and G. P. Thomson in England independently demonstrated that streams of electrons are diffracted when they are scattered from crystals. The diffraction patterns they observed were in complete accord with the electron wavelengths predicted by de Broglie's formula.

Criterion for type of behavior

In certain aspects of its behavior, a moving object resembles a wave, and in other aspects it resembles a particle. Which type of behavior is most conspicuous depends upon how the object's de Broglie wavelength compares with its dimensions and with the dimensions of whatever it interacts with. Two examples will help us appreciate this statement.

Example   In one of their experiments, Davisson and Germer aimed a beam of 54-eV electrons at a nickel crystal (Fig. 27–9). Compare the de Broglie wavelength of these electrons with the spacing of the atomic planes in the crystal, which is 0.91 $\times$ 10^{-10} m.

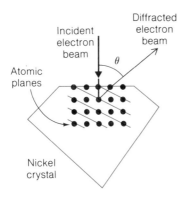

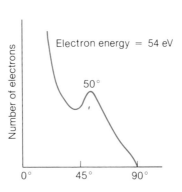

FIG. 27–9 The Davisson-Germer experiment. The peak in the number of diffracted 54-eV electrons at $\theta = 50°$ is in agreement with de Broglie's formula for the wavelength of a moving particle.

Solution The kinetic energy of a 54-eV electron is

$$KE = (54 \text{ eV})(1.6 \times 10^{-19} \text{ J/eV}) = 8.6 \times 10^{-18} \text{ J}$$

The momentum of such an electron can be calculated nonrelativistically. Since $KE = \frac{1}{2}mv^2$, $mv = \sqrt{2mKE}$, and

$$\lambda = \frac{h}{mv} = \frac{h}{\sqrt{2mKE}} = \frac{6.63 \times 10^{-34} \text{ J·s}}{\sqrt{2(9.1 \times 10^{-31} \text{ kg})(8.6 \times 10^{-18} \text{ J})}} = 1.7 \times 10^{-10} \text{ m}$$

This is the same order of magnitude as the spacing of the atoms in the nickel crystal, and we recall that diffraction is prominent only when the wavelength of the waves involved is comparable with the spacing of the scattering centers. Davisson and Germer found that the scattered 54-eV electrons were concentrated in just the direction ($\theta = 50°$) predicted by the theory of diffraction for waves of $\lambda = 1.7 \times 10^{-10}$ m, instead of the more even distribution expected for purely billiard-ball scattering. They interpreted the result as support for de Broglie's hypothesis. ■

Example Find the de Broglie wavelength of a 1500-kg car whose speed is 30 m/s.

Solution The car's wavelength is

$$\lambda = \frac{h}{mv} = \frac{6.63 \times 10^{-34} \text{ J·s}}{(1.5 \times 10^3 \text{ kg})(30 \text{ m/s})} = 1.5 \times 10^{-38} \text{ m}$$

The wavelength is so small compared with the car's dimensions that no wave behavior is to be expected. ■

The above examples are extreme ones. More ambiguous is the case of a moving atom or molecule. In Chapter 13 we saw how a model of a gas in which molecules are considered as particles was very successful in explaining the properties of gases. But at 0°C the average wavelength of helium atoms, to give a specific illustration, is 7.6×10^{-11} m, which is of the same order of magnitude as atomic dimensions. Such atoms may exhibit either particle or wave characteristics, depending upon the situation.

Not long after their discovery, matter waves were used for a very practical purpose, namely as replacements for light waves in microscopes. The resolving power of a microscope is proportional to the wavelength of whatever is used to illuminate the

FIG. 27–10 An electron microscope uses electrons instead of light waves to produce an enlarged image. Because electron de Broglie wavelengths are shorter than optical wavelengths, an electron microscope has a higher resolving power and hence can be used to produce higher magnifications. The magnetic "lenses" are current-carrying coils whose magnetic fields focus an electron beam just as the glass lenses in an ordinary microscope focus a light beam.

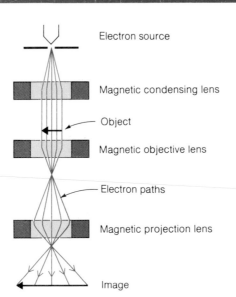

Electron source

Magnetic condensing lens

Object

Magnetic objective lens

Electron paths

Magnetic projection lens

Image

The electron microscope achieves high magnifications because of the short wavelength of electron waves

specimen, which limits optical microscopes to useful magnifications of less than $500 \times$ (see Sec. 25–7). Moving electrons, however, may have wavelengths much shorter than those of light waves. X rays also have short wavelengths, but it has proved difficult to focus them, whereas electrons, by virtue of their charge, are easily controlled by electric and magnetic fields. In an electron microscope, a beam of electrons passes through a very thin specimen in a vacuum chamber, to prevent scattering of the beam and hence blurring of the image, and is brought to a focus on a fluorescent screen or a photographic plate by a system of magnetic fields that act as lenses (Fig. 27–10). Because the numerical aperture of a magnetic "lens" cannot be made as high as that of an optical lens, the full theoretical resolution of electron waves cannot be realized in practice. For example, a 100-keV electron has a wavelength of 3.7×10^{-12} m, which is 0.0037 nm, but the actual resolution available ordinarily would not exceed perhaps 0.1 nm. However, this is still a great improvement on the 200 nm or more resolution of an optical microscope, and magnifications of $100,000 \times$ or more are commonly achieved with electron microscopes.

27–6 WAVE FUNCTION

Wave function

In water waves, the physical quantity that varies periodically is the height of the water surface. In sound waves, the variable quantity is pressure in the medium the waves travel through. In light waves, the variable quantities are the electric and magnetic fields. What is it that varies in the case of matter waves?

The quantity whose variations constitute the matter waves of a moving object is known as its *wave function*. The symbol for wave function is ψ, the Greek letter *psi*.

The value of ψ^2 for a particular object at a certain place and time is proportional to the probability of finding the object at that place at that time.

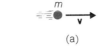

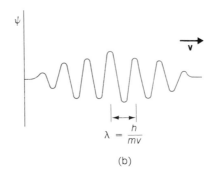

(a)

(b)

$$\lambda = \frac{h}{mv}$$

FIG. 27–11 (a) Particle description of moving object. (b) Wave description of moving object.

For this reason the quantity ψ^2 is called the *probability density* of the object. A large value of ψ^2 signifies that the body is likely to be found at the specified place and time; a small value of ψ^2 signifies that the body is unlikely to be found at that place and time. Thus matter waves may be regarded as waves of probability.

Why does the probability of finding an object depend upon ψ^2 and not upon the wave function ψ itself? The answer is subtle. The amplitude of every wave varies from $-A$ to $+A$ to $-A$ to $+A$ and so on, where A is the maximum absolute value of whatever the wave variable is. But a negative probability is meaningless: the probability that an object be in a given place at a given time must lie between 0 (the object is definitely not there) and 1 (the object is definitely there). An intermediate probability, say 0.4, means that there is a 40% chance of finding the object there at that time. A probability of -0.4, however, makes no sense at all. Procedures are known for calculating the wave function ψ for moving objects in a great many situations, and each value of ψ must be squared to obtain a positive quantity that can be compared with experiment. Actually, ψ sometimes turns out to have an imaginary component that contains the factor $\sqrt{-1}$, and in such cases ψ^2 is obtained by another method.

A group or packet of matter waves is associated with every moving object. The packet travels with the same velocity as the object does. The waves in the packet have the average wavelength $\lambda = h/mv$ given by de Broglie's formula (Fig. 27–11). Even though we cannot visualize what is meant by ψ and so cannot form a mental image of matter waves, the agreement between theory and experiment signifies that the notion of matter waves is a meaningful way to describe moving objects.

Why probability density depends upon the square of the wave function

27-7 PARTICLE IN A BOX

When an object is confined to a certain region of space instead of being able to move freely, its wave properties lead to certain remarkable consequences. Let us see what these consequences are in the simplest possible case—that of a particle trapped in a box L wide whose walls are infinitely hard, so that the particle does not lose energy as

FIG. 27–12 A moving particle in a box L wide.

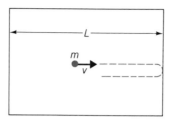

it bounces back and forth (Fig. 27–12). We shall assume that the particle's speed v is sufficiently small so that relativistic considerations can be ignored.

A trapped particle is equivalent to a standing de Broglie wave

The wave equivalent of a particle in a box is a standing de Broglie wave. The reason for this is the same as in the analogous phenomenon of standing waves in a stretched string (Sec. 12–6): the wave variable—transverse displacement in the case of a string, wave function ψ in the case of a moving particle—must be 0 at the walls of the box in order that the waves be reflected there. Hence the possible de Broglie wavelengths depend upon the width L of the box (Fig. 27–13). The longest possible wavelength is $\lambda = 2L$, next is $\lambda = L$, then $\lambda = 2L/3$, and so on; in general, the permitted wavelengths are given by

$$\lambda_n = \frac{2L}{n} \qquad n = 1, 2, 3, \ldots \qquad \qquad \textit{Permitted wavelengths} \quad (27\text{–}7)$$

The wavelength of a particle of mass m and velocity v is

$$\lambda = \frac{h}{mv}$$

Wavelength restrictions lead to energy restrictions

so the restrictions on λ imposed by the size of the box amount to restrictions on the speed v the particle can have and, in turn, to restrictions on its energy. The kinetic energy of a moving particle of wavelength λ is

FIG. 27–13 Wave functions ψ and probability densities ψ^2 of a particle in a box.

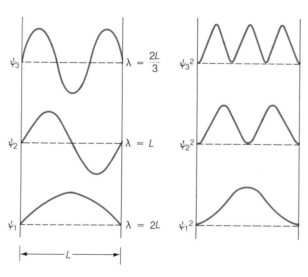

$$KE = \tfrac{1}{2}mv^2 = \frac{1}{2m}(mv)^2 = \frac{h^2}{2m\lambda^2}$$

Our trapped particle has only kinetic energy, and because the only wavelengths it can have are $\lambda = 2L/n$, the only energies it can have are

$$E_n = \frac{n^2h^2}{8mL^2} \qquad n = 1, 2, 3, \ldots \qquad \text{Energy levels} \quad (27\text{–}8)$$

Each permitted energy is called an *energy level,* and the integer n that corresponds to a given energy level is called its *quantum number.*

    We can draw three important general conclusions from the preceding analysis. These conclusions apply to *any* particle confined to a certain region of space, such as an atomic electron held captive by the attraction of the positively charged nucleus, and not just to the artificial situation of a particle in a rigid-walled box.

    1. A trapped particle can possess only certain specific energies and no others. The energy of such a particle is said to be *quantized,* and the magnitudes of the energy levels depend upon the manner in which the particle's motion is restricted.

    2. A trapped particle cannot have zero energy; it must have a certain minimum amount E_1. Since the de Broglie wavelength of a particle is $\lambda = h/mv$, a speed of $v = 0$ means an infinite wavelength. But there is no way to reconcile an infinite wavelength with a trapped particle, so such a particle must possess at least some kinetic energy. This is the origin of the "zero-point" energy mentioned in Sec. 13–7.

    3. Because Planck's constant is so small—only 6.63×10^{-34} J·s—quantization of energy is conspicuous only when m and L are also small. Two examples will make this clear.

A quantum number characterizes every energy level

Energy quantization

Minimum energy

When energy quantization is significant

Example An electron is confined to a box 10^{-10} m across, which is the order of magnitude of atomic dimensions. Find its permitted energies.

Solution From Eq. (27–8),

$$E_n = \frac{n^2h^2}{8mL^2} = \frac{n^2 \times (6.63 \times 10^{-34}\,\text{J·s})^2}{8(9.1 \times 10^{-31}\,\text{kg})(10^{-10}\,\text{m})^2}$$

$$= 6.0 \times 10^{-18}\,n^2\,\text{J} = 38n^2\,\text{eV} \qquad n = 1, 2, 3, \ldots$$

The energies the electron can have are therefore 38 eV, 152 eV, 342 eV, 608 eV, and so on (Fig. 27–14). If such a box existed, the quantization of a trapped electron's energy would be a prominent feature of the system. (And, indeed, energy quantization *is* prominent in the case of an atomic electron, as we shall find in Chapter 28.) ■

Example A 10-g marble is confined to a box 10 cm across. Find its permitted energies.

Solution Here

$$E_n = \frac{n^2h^2}{8mL^2} = \frac{(6.63 \times 10^{-34}\,\text{J·s})^2 n^2}{8(10^{-2}\,\text{kg})(0.1\,\text{m})^2} = 5.5 \times 10^{-64}\,n^2\,\text{J} \qquad n = 1, 2, 3, \ldots$$

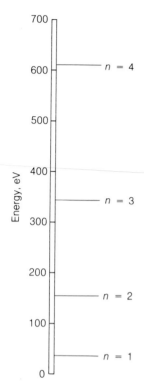

FIG. 27–14 Energy levels of an electron confined to a box 10^{-10} m wide. A trapped particle cannot have zero energy.

Here energy quantization is hardly likely to be detected. When $n = 1$, the marble's speed is 3.3×10^{-31} m/s, a speed no experiment could distinguish from zero, and the spacing between the higher energy levels is too small for measurement. ∎

On a macroscopic scale, then, quantum effects are unobservable, which is why classical physics is adequate on this scale. But on the scale of the atom, quantum effects are dominant, and the concepts and principles of classical physics must be replaced by others of a more sophisticated character. In fact, classical physics turns out to be just an approximation of quantum physics, which is perfectly general in its range of application.

27–8   UNCERTAINTY PRINCIPLE

To regard a moving object as a wave packet raises the problem of the ultimate accuracy with which such "particle" properties as its position and momentum can be determined. In the macroscopic world, where the wave aspects of matter are insignificant, the limit to how accurately position and momentum can be measured depends solely upon our instruments, and there is, in principle, no absolute limit at all. But in the microscopic world, where the wave aspects of matter are very significant indeed, these wave aspects set a fundamental limit to the accuracy of measurements of position and momentum. Of course, poor instruments will give poor results, but even if perfect instruments were available, they would not yield exact results. The *uncertainty principle* is the physical law which expresses quantitatively the basic indeterminacy which the wave nature of matter imposes on measurements of moving objects.

Figure 27–15 shows how the wave nature of matter leads to the uncertainty principle. A narrow wave packet means a small uncertainty in the position of the corresponding object, and a wide wave packet means a large uncertainty in position. On the other hand, the wavelength of the waves in a wide packet is fairly well defined, and the momentum of the object has only a small uncertainty Δmv. The wavelength of a narrow wave packet is imprecise, and the momentum of the object therefore has a large uncertainty Δmv. We conclude that

The uncertainty principle follows from the wave nature of matter

1. If an object has a well-defined position at a certain time, its momentum must have a large uncertainty.
2. If an object has a well-defined momentum at a certain time, its position must have a large uncertainty.

Evidently a reciprocal relationship of some kind exists between the uncertainty Δx in an object's position and the uncertainty Δmv in its momentum. The details of this relationship can be found from the mathematical nature of the wave packet that corresponds to a moving object. Such a packet can be regarded as being formed by the superposition of many trains of sinusoidal waves, as discussed in Chapter 12, with each train having a different wavelength (Fig. 27–16). The narrower the wave packet, the greater the spread of wavelengths involved and hence the greater the uncertainty Δmv. A smaller range of wavelengths is needed to construct a wide wave packet, but now Δx is necessarily large. The advantage of this approach is that it permits a precise analysis of the uncertainty principle, with the result that

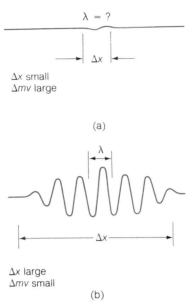

FIG. 27–15 (a) A narrow wave packet means a small uncertainty Δx in the position of a moving object but a large uncertainty in its wavelength and hence momentum. (b) A wide wave packet means a large uncertainty in the position of a moving object but a small uncertainty in its wavelength and hence momentum. It is impossible to have a narrow wave packet with a well-defined wavelength.

$$\Delta x \, \Delta mv \geq \frac{h}{2\pi} \qquad\qquad \textit{Uncertainty principle} \quad (27\text{–}9)$$

In words, the uncertainty principle states that

$$\left(\begin{array}{c} \text{Uncertainty} \\ \text{in position} \end{array} \right) \times \left(\begin{array}{c} \text{uncertainty} \\ \text{in momentum} \end{array} \right) \quad \begin{array}{c} \textit{is equal to or} \\ \textit{greater than} \end{array} \quad \frac{\text{Planck's constant}}{2\pi}$$

Example The radius of the hydrogen atom is 5.3×10^{-11} m. Assuming that the uncertainty Δx in the location of its single electron has the same value, estimate the minimum energy of the electron.

Solution From the uncertainty principle,

$$\Delta mv \geq \frac{h}{2\pi \, \Delta x} \geq \frac{6.63 \times 10^{-34}\, \text{J·s}}{(2\pi)(5.3 \times 10^{-11}\,\text{m})} \geq 2.0 \times 10^{-24}\, \text{kg·m/s}$$

This momentum uncertainty corresponds to a kinetic energy uncertainty of

$$\Delta \text{KE} = \frac{(\Delta mv)^2}{2m} = \frac{(2.0 \times 10^{-24}\,\text{kg·m/s})^2}{2(9.1 \times 10^{-31}\,\text{kg})} = 2.2 \times 10^{-18}\, \text{J}$$

which is

$$\Delta \text{KE} = \frac{2.2 \times 10^{-18}\,\text{J}}{1.6 \times 10^{-19}\,\text{J/eV}} = 14\,\text{eV}$$

This is the lower limit to the kinetic energy of the electron, and in fact is equal to the kinetic energy of an electron in the lowest energy level of a hydrogen atom.

When a similar calculation is made of the minimum energy of an electron in an atomic nucleus, whose radius is typically 5×10^{-15} m, the result turns out to be over

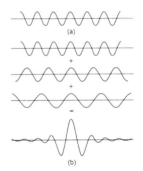

FIG. 27–16 (a) If a moving object can be represented by a wave with a single wavelength, its position cannot be established at all. (b) A wave packet is the result of superposing waves of different wavelengths; the greater the range of wavelengths, the narrower the packet.

20 MeV. (Such a calculation must be made relativistically.) However, the electrons associated with nuclei, even unstable nuclei, never have more than a small fraction of this energy, and we conclude that electrons cannot be present in atomic nuclei. ■

On a macroscopic scale, since Planck's constant h is such a very small quantity, the limitation imposed upon measurement by the uncertainty principle is negligible, but on a microscopic scale the uncertainty principle dominates many phenomena and probability considerations are correspondingly important.

27–9 UNCERTAINTY PRINCIPLE: ALTERNATIVE APPROACH

Particle approach to uncertainty principle

The uncertainty principle makes sense from the point of view of the particle properties of waves as well as from the point of view of the wave properties of particles. Suppose we wish to measure the position and momentum of something at a certain moment. To do so, we must touch it with something else that will carry the required information back to us; that is, we must poke it with a stick, shine light on it, or perform some similar act. The measurement process itself thus requires that the body be interfered with in some way, and if we consider this interference in detail, we are led to the same uncertainty principle as before even without taking into account the wave nature of moving objects.

Detecting an electron changes its momentum

Let us imagine we are looking at an electron with the help of light whose wavelength is λ, as in Fig. 27–17. Each photon of this light has the momentum h/λ. When one of these photons bounces off the electron (which must occur if we are to "see" it), the electron's original momentum will be changed. The exact amount of the change Δmv cannot be predicted, but it will be of the same order of magnitude as the photon momentum h/λ. Hence

$$\Delta mv \approx \frac{h}{\lambda}$$

The *larger* the wavelength of the observing photon, the smaller the uncertainty in momentum.

Because light is a wave phenomenon as well as a particle phenomenon, we cannot expect to determine the electron's location with perfect accuracy even with the best of instruments. A reasonable estimate of the irreducible uncertainty Δx in the measurement might be one photon wavelength. That is,

$$\Delta x \geq \lambda$$

Light of short wavelength is needed to measure position accurately

The *smaller* the wavelength, the smaller the uncertainty in location. Hence if we use light of short wavelength to increase the accuracy of our position measurement, there will be a corresponding decrease in the accuracy of our momentum measurement because the higher photon energy will disturb the electron's motion to a greater extent, whereas light of long wavelength will yield an accurate momentum but an inaccurate position.

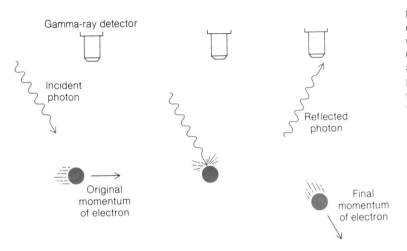

FIG. 27–17 An electron cannot be observed without affecting its momentum. The more accurately the electron's position is determined, the greater the change in momentum.

By combining the two formulas above, we find that

$$\Delta x \, \Delta mv \geqslant h$$

A more detailed calculation shows that somewhat better accuracy is in principle possible, so that the limit of the product $\Delta x \, \Delta mv$ is $h/2\pi$ instead of just h.

A form of the uncertainty principle that involves energy and time is easy to obtain. Suppose we are measuring the energy E emitted in some process during the period of time Δt. If the energy is in the form of electromagnetic waves, the time Δt limits the accuracy with which we can measure the wave frequency f. Because the minimum uncertainty in the number of waves we count in a given wave train is of the order of one wave, and

Uncertainty principle for energy and time

$$\text{Frequency} = \frac{\text{number of waves}}{\text{time interval}}$$

the uncertainty in frequency Δf of our determination is

$$\Delta f \geqslant \frac{1}{\Delta t}$$

The corresponding uncertainty in energy ΔE is

$$\Delta E = h \, \Delta f$$

and so

$$\Delta E \geqslant \frac{h}{\Delta t} \quad \text{or} \quad \Delta E \, \Delta t \geqslant h$$

A more detailed calculation changes this result to

$$\Delta E \, \Delta t \geqslant \frac{h}{2\pi} \qquad \textit{Uncertainties in energy and time} \quad (27\text{–}10)$$

Equation (27–10) states that the product of the uncertainty in an energy measurement and the time available for the measurement is at best equal to $h/2\pi$. This form of the uncertainty principle can be derived in other ways as well and is a general conclusion that is not limited to electromagnetic waves. Equation (27–10) will be used in Chapter 32 to estimate the mass of a certain type of elementary particle.

IMPORTANT TERMS

The **photoelectric effect** is the emission of electrons from a metal surface when light shines on it.

The **quantum theory of light** states that light travels in tiny bursts of energy called **quanta** or **photons.** The quantum theory of light is required to account for the photoelectric effect.

X rays are high-frequency electromagnetic waves emitted when fast electrons impinge on matter.

A moving body behaves as though it has a wave character. The waves representing such a body are **matter waves,** also called **de Broglie waves.** The wave variable in a matter wave is its **wave function,** whose square is the **probability density** of the body. The value of the probability density of a particular body at a certain place and time is proportional to the probability of finding the body at that place at that time. Matter waves may thus be regarded as waves of probability.

Because of its wave nature, a particle restricted to a definite region of space can have only certain specific energies, each of which is called an **energy level** and corresponds to a **quantum number.**

The **uncertainty principle** is an expression of the limit set by the wave nature of matter on finding both the position and state of motion of a moving body.

IMPORTANT FORMULAS

Quantum energy: $E = hf$

Photoelectric effect: $hf = KE_{max} + hf_0$

X ray energy: $hf_{max} = eV$

De Broglie wavelength: $\lambda = \dfrac{h}{mv}$

Uncertainty principle: $\Delta x\, \Delta mv \geq \dfrac{h}{2\pi}$

$\Delta E\, \Delta t\ \geq \dfrac{h}{2\pi}$

MULTIPLE CHOICE

1. Photoelectrons are emitted by a metal surface only when the light directed at it exceeds a certain minimum
   (a) wavelength.
   (b) frequency.
   (c) speed.
   (d) charge.

2. When light is directed on a metal surface, the energies of the emitted electrons
   (a) vary with the intensity of the light.
   (b) vary with the frequency of the light.
   (c) vary with the speed of the light.
   (d) are random.

3. The photoelectric effect can be understood on the basis of
   (a) the electromagnetic theory of light.
   (b) the special theory of relativity.
   (c) the principle of superposition.
   (d) none of the above.

4. When the speed of the electrons striking a metal surface is increased, the result is an increase in
   (a) the number of X rays emitted.
   (b) the frequency of the X rays emitted.
   (c) the speed of the X rays emitted.
   (d) the size of the X rays emitted.

5. Modern physical theories indicate that
   (a) all particles exhibit wave behavior.
   (b) only moving particles exhibit wave behavior.
   (c) only charged particles exhibit wave behavior.
   (d) only uncharged particles exhibit wave behavior.

6. The description of a moving body in terms of matter waves is legitimate because
   (a) it is based upon common sense.
   (b) matter waves have been actually seen.
   (c) the analogy with electromagnetic waves is plausible.
   (d) theory and experiment agree.

7. The wave packet that corresponds to a moving particle
  (a) has the same size as the particle.
  (b) has the same speed as the particle.
  (c) has the speed of light.
  (d) consists of X rays.

8. A moving body is described by the wave function ψ at a certain time and place. The value of ψ^2 is proportional to the body's
  (a) electric field.
  (b) speed.
  (c) energy.
  (d) probability of being found.

9. The dimensions of the region in which a particle is trapped govern
  (a) its minimum mass.
  (b) its minimum wavelength.
  (c) its minimum energy.
  (d) the number of possible energies it can have.

10. The narrower the wave packet of a particle is,
  (a) the shorter its wavelength.
  (b) the more precisely its position can be established.
  (c) the more precisely its momentum can be established.
  (d) the more precisely its energy can be established.

11. If Planck's constant were larger than it is,
  (a) moving bodies would have shorter wavelengths.
  (b) moving bodies would have higher energies.
  (c) moving bodies would have higher momenta.
  (d) the uncertainty principle would be significant on a larger scale of size.

12. Light of wavelength 5×10^{-7} m consists of photons whose energy is
  (a) 1.1×10^{-48} J.
  (b) 1.3×10^{-27} J.
  (c) 4×10^{-19} J.
  (d) 1.7×10^{-15} J.

13. The de Broglie wavelength of an electron whose speed is half that of light is
  (a) 3.6×10^{-12} m.
  (b) 4.2×10^{-12} m.
  (c) 4.9×10^{-12} m.
  (d) 1.2×10^{-11} m.

14. The lowest energy possible for a certain particle trapped in a certain box is 2 eV. The next highest energy the particle can have is
  (a) 3 eV.   (b) 4 eV.
  (c) 6 eV.   (d) 8 eV.

EXERCISES

1. Why do you think the wave aspect of light was discovered earlier than its particle aspect?

2. If Planck's constant were smaller than it is, would quantum phenomena be more or less conspicuous than they are now in everyday life?

3. The atoms in a solid possess a certain minimum zero-point energy even at 0 K, whereas no such restriction holds for the molecules of an ideal gas. Explain this observation with the help of the uncertainty principle.

4. A photon and a proton have the same wavelength. What can be said about how their linear momenta compare? About how the photon's energy compares with the particle's kinetic energy?

5. Must a particle have an electric charge in order for matter waves to be associated with its motion?

6. Can the rest mass of a moving particle be determined by measuring its de Broglie wavelength?

7. What is the simplest experimental procedure that can distinguish between a gamma ray whose wavelength is 10^{-11} m and an electron whose de Broglie wavelength is also 10^{-11} m?

8. The uncertainty principle applies to all objects, yet its consequences are only significant for such tiny particles as electrons, protons, and neutrons. Why?

9. The eye can detect as little as 10^{-18} J of electromagnetic energy. How many photons of $\lambda = 6 \times 10^{-7}$ m does this energy represent?

10. How many photons per second are emitted by a 150-W amateur radio transmitter operating at a wavelength of 20 m?

11. Find the frequency and wavelength corresponding to a 5-MeV photon.

12. The threshold frequency for photoelectric emission in calcium is 7.7×10^{14} Hz. Find the maximum energy in electron volts of the photoelectrons when light of frequency 1.20×10^{15} Hz is directed on a calcium surface.

13. What is the maximum wavelength of light that will lead to photoelectric emission from platinum, whose work function is 5.6 eV? In what part of the spectrum is such light?

14. A silver ball is suspended by a string in a vacuum chamber and ultraviolet light of wavelength 200 nm is directed at it. What electric potential will be the ball acquire as a result? The work function of silver is 4.7 eV.

15. Electrons are accelerated in television tubes through potential differences of about 10,000 V. Find the highest frequency of the electromagnetic waves that are emitted when these electrons strike the screen of the tube. What type of waves are these?

16. What voltage must be applied to an X-ray tube for it to emit X rays with a minimum wavelength of 3×10^{-11} m?

17. What is the de Broglie wavelength of a 1-mg grain of sand blown by the wind at a velocity of 20 m/s?

18. Find the de Broglie wavelength of a 1-MeV proton. Since the rest energy of the proton is 938 MeV, the calculation may be made nonrelativistically.

19. (a) An electron is confined in a box 10^{-9} m in length. What is the uncertainty in its speed? (b) A proton is confined in the same box. What is the uncertainty in its speed?

20. Derive a formula for the energy levels (in MeV) of a neutron confined to a one-dimensional box 10^{-14} m wide. What is its minimum energy? (The diameter of an atomic nucleus is of this order of magnitude.)

PROBLEMS

1. Light from the sun reaches the earth at the rate of about 1.4×10^3 W/m^2 of area perpendicular to the direction of the light. Assume sunlight is monochromatic with a frequency of 5×10^{14} Hz. (a) How many photons fall per second on each square meter of the earth's surface directly facing the sun? (b) How many photons are present in each m^3 near the earth on the sunlit side?

2. Light of wavelength 420 nm falls on the cesium surface of a photoelectric cell at the rate of 5 mW. If one photoelectron is emitted for every 10^4 incident photons, find the current produced by the cell. The work function of cesium is 1.9 eV.

3. Calculate the de Broglie wavelength of (a) an electron whose speed is 1×10^8 m/s, and (b) an electron whose speed is 2×10^8 m/s. Use relativistic formulas.

4. Show that the de Broglie wavelength of an oxygen molecule in thermal equilibrium in the atmosphere at 20°C is smaller than its diameter of about 4×10^{-10} m.

5. The position and momentum of a 1000-eV electron are determined at the same time. If the position is found to within 10^{-10} m, what is the percentage of uncertainty in the momentum?

6. A proton's position is to be determined without changing its KE by more than 1 keV. Find the maximum accuracy with which the position determination can be made.

7. At a certain time a measurement establishes the position of an electron with an accuracy of $\pm 10^{-11}$ m. Calculate the uncertainty in the electron's momentum and, from this, the uncertainty in its position 1 s later.

8. From the formula for the possible energies of a particle trapped in a box, find the minimum momentum the particle can have and compare it with the prediction of the uncertainty principle.

9. Verify that the uncertainty principle can be expressed in the form $\Delta L \, \Delta \theta \geq h/2\pi$, where ΔL is the uncertainty in the angular momentum of a body and $\Delta \theta$ is the uncertainty in its angular position. To do this, consider a particle of mass m moving in a circle.

10. (a) How much time is needed to measure the KE of an electron whose speed is 10 m/s with an uncertainty of no more than 0.1%? How far will the electron have traveled in this time interval? (b) Make the same calculations for a 1-g insect with the same speed. What do these sets of figures indicate?

11. An atom in an energy level above its lowest one gives up its excess energy by emitting one or more photons, each with a characteristic frequency, as described in Chapter 28. Usually about 10^{-8} s elapses between the excitation of an atom and the time it emits a photon. Find the uncertainty in the photon frequency in hertz.

ANSWERS TO MULTIPLE CHOICE

1. b	**6.** d	**11.** d
2. b	**7.** b	**12.** c
3. d	**8.** d	**13.** b
4. b	**9.** c	**14.** d
5. b	**10.** b	

THE HYDROGEN ATOM

28

A hydrogen atom consists of a proton with an electron moving around it. In the Bohr model of this atom, the particle and wave properties of the electron are combined: its orbital motion is such that the centripetal force needed is provided by the electric force exerted by the proton, and the circumference of its orbit is exactly one de Broglie wavelength. This model predicts that the hydrogen atom can have certain specific energies only, and no others. This prediction agrees remarkably well with experiment. Despite this success, however, the Bohr theory has certain deficiencies, and has been superseded by the more general (and more complicated) quantum theory of the atom which is the subject of Chapter 29.

28–1 PARADOX OF ATOMIC STABILITY

The picture of the atom that emerged from Rutherford's work, as we learned in Sec. 16–5, consists of a tiny, massive, positively charged nucleus sur-

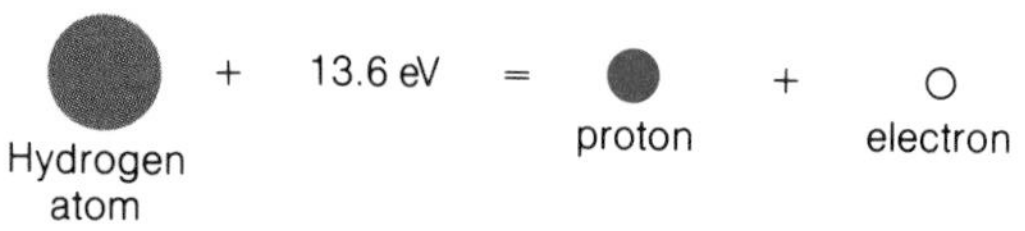

FIG. 28–1 To break a hydrogen atom apart requires 13.6 eV.

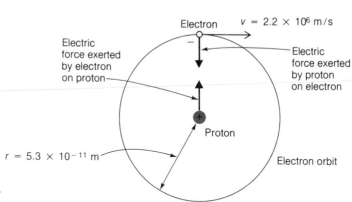

FIG. 28–2 The hydrogen atom consists of an electron circling a proton. The electric force exerted by the proton on the electron provides the centripetal force required to hold it in a circular path. The proton is nearly 2000 times as heavy as the electron, so its motion under the influence of the electric force that the electron exerts is insignificant.

rounded by enough negatively charged electrons to leave the atom as a whole electrically neutral. These electrons cannot be stationary, since the electric attraction of the nucleus would pull them in at once. If the electrons are in motion around the nucleus, however, stable orbits like those of the planets about the sun would seem to be possible.

Mechanical model of hydrogen atom

A hydrogen atom, the simplest of all, has a single electron and a nucleus that consists of a single proton. Experiments indicate that 13.6 eV of work must be performed to break apart a hydrogen atom into a proton and electron that go their separate ways (Fig. 28–1). With the help of calculations based upon what we already know about mechanics and electricity (see Sec. 28–5) this figure leads to an orbital radius of $r = 5.3 \times 10^{-11}$ m and an orbital speed of $v = 2.2 \times 10^6$ m/s for the electron in the hydrogen atom. The proton mass is 1836 times the electron mass, so we can consider the proton as stationary with the electron revolving around it in such a way that the required centripetal force is provided by the electric force exerted by the proton (Fig. 28–2).

Mechanical model is inconsistent with electromagnetic theory

What the preceding analysis overlooks is that, according to electromagnetic theory, all accelerated electric charges radiate electromagnetic waves—and an electron moving in a circular path is certainly accelerated. Thus an atomic electron circling its nucleus to keep from falling into it by electrical attraction cannot help radiating away energy, which means that it must spiral inward until it is swallowed up by the nucleus (Fig. 28–3). Clearly the ordinary laws of physics cannot account for the stability of the hydrogen atom, the simplest atom of all, whose electron must be whirling around the nucleus to keep from being pulled into it and yet must be radiating electromagnetic energy continuously. However, since other phenomena similarly impossible to understand—the photoelectric effect, for instance—find explanation in terms of quantum concepts, it is appropriate for us to inquire whether this might also be true for the atom.

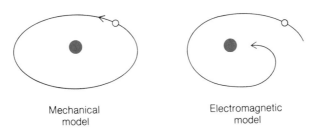

FIG. 28–3 The mechanical and electromagnetic models of the hydrogen atom are in conflict. Only quantum theory can account for the stability of atoms.

Mechanical model

Electromagnetic model

In the discussion that follows of the Bohr model of the atom, the initial argument is somewhat different from that of Bohr, who in 1913 did not have the notion of matter waves to guide his thinking. The results are exactly the same as Bohr obtained, however.

28–2 ELECTRON WAVES IN THE ATOM

Let us begin by looking into the wave properties of the electron in the hydrogen atom. The de Broglie wavelength λ of an object of mass m and speed v is $\lambda = h/mv$ where h is Planck's constant. The speed of the electron in a hydrogen atom is $v = 2.2 \times 10^6$ m/s, and so the wavelength of its matter waves is

Wavelength of electron in hydrogen atom

$$\lambda = \frac{h}{mv} = \frac{6.63 \times 10^{-34}\,\text{J·s}}{(9.1 \times 10^{-31}\,\text{kg})(2.2 \times 10^6\,\text{m/s})} = 3.3 \times 10^{-10}\,\text{m}$$

This is a most exciting result, because the electron's orbit has a circumference of exactly

$$2\pi r = 2\pi \times 5.3 \times 10^{-11}\,\text{m} = 3.3 \times 10^{-10}\,\text{m}$$

We therefore conclude that *the orbit of the electron in a hydrogen atom corresponds to one complete electron wave joined on itself* (Fig. 28–4).

The fact that the electron orbit in a hydrogen atom is one electron wavelength in circumference is just the clue we need to construct a theory of the atom. If we examine

The ground-state electron orbit is one wavelength in circumference

FIG. 28–4 The electron orbit in a ground-state hydrogen atom is exactly one de Broglie wavelength in circumference.

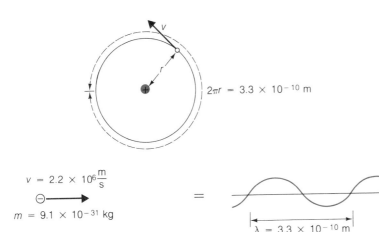

$$2\pi r = 3.3 \times 10^{-10}\,\text{m}$$

$$v = 2.2 \times 10^6 \frac{\text{m}}{\text{s}}$$

$$m = 9.1 \times 10^{-31}\,\text{kg}$$

$$\lambda = 3.3 \times 10^{-10}\,\text{m}$$

FIG. 28–5 Three possible modes of vibration of a wire loop.

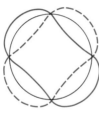

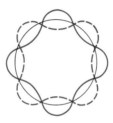

Circumference = 2 wavelengths

Circumference = 4 wavelengths

Circumference = 9 wavelengths

the vibrations of a wire loop, as in Fig. 28–5, we see that their wavelengths always fit a whole number of times into the loop's circumference, each wave joining smoothly with the next. In the absence of dissipative effects, such vibrations would persist indefinitely. Why are these the only vibrations possible in a wire loop? A fractional number of wavelengths cannot be fitted into the loop and still allow each wave to join smoothly with the next (Fig. 28–6); the result would be destructive interference as the waves travel around the loop, and the vibrations would die out rapidly.

By considering the behavior of electron waves in the hydrogen atom as analogous to the vibrations of a wire loop, then, we may postulate that

Condition for orbit

An electron can circle an atomic nucleus only if its orbit is a whole number of electron wavelengths in circumference.

The above postulate is the decisive one in our understanding of the atom. We note that it combines both the particle and wave characters of the electron into a single statement; although we can never observe these antithetical characters at the same time in an experiment, they are inseparable in nature.

It is easy to express in a formula the condition that a whole number of electron wavelengths fit into the electron's "orbit." The circumference of a circular orbit of radius r is $2\pi r$, and so the condition for orbit stability is

Each orbit is characterized by a quantum number n

$$n\lambda = 2\pi r_n \qquad n = 1, 2, 3, \ldots \qquad \textit{Condition for orbit stability} \quad (28\text{–}1)$$

where r_n designates the radius of the orbit that contains n wavelengths. The quantity n is called the *quantum number* of the orbit.

A straightforward calculation (given in Sec. 28–5) shows that the stable electron orbits are those whose radii are given by the formula

FIG. 28–6 Unless a whole number of wavelengths fits into the wire loop, destructive interference causes the vibrations to die out rapidly.

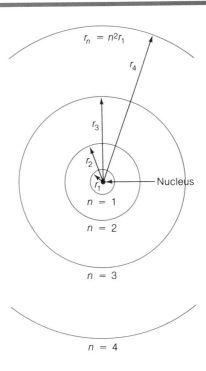

FIG. 28–7 Electron orbits in the hydrogen atom according to the Bohr model. The orbit radii are proportional to n^2.

$$r_n = n^2 r_1 \qquad n = 1, 2, 3, \ldots \qquad \textit{Orbital radii in Bohr atom} \quad (28\text{–}2)$$

where

$$r_1 = 5.3 \times 10^{-11} \text{ m}$$

is the radius of the innermost orbit (Fig. 28–7).

28–3 ENERGY LEVELS

The total energy of a hydrogen atom is not the same in the various permitted orbits. The energy E_n of a hydrogen atom whose electron is in the nth orbit is given by

Each orbit has a different energy

$$E_n = \frac{E_1}{n^2} \qquad n = 1, 2, 3, \ldots \qquad \textit{Energy levels of hydrogen atom} \quad (28\text{–}3)$$

where $E_1 = -13.6$ eV $= -2.18 \times 10^{-18}$ J is the energy corresponding to the innermost orbit. The energies specified by the above formula are called the *energy levels* of the hydrogen atom.

The energy levels of the hydrogen atom are all less than zero, which signifies that the electron does not have enough energy to escape from the atom. The lowest energy level E_1, corresponding to the quantum number $n = 1$, is called the *ground state* of the atom; the highest levels E_2, E_3, E_4, and so on are called *excited states* (Fig. 28–8).

Ground and excited states

As the quantum number n increases, the energy E_n approaches closer and closer to zero. In the limit of $n = \infty$, $E_\infty = 0$ and the electron is no longer bound to the

FIG. 28–8   Energy levels of the hydrogen atom.

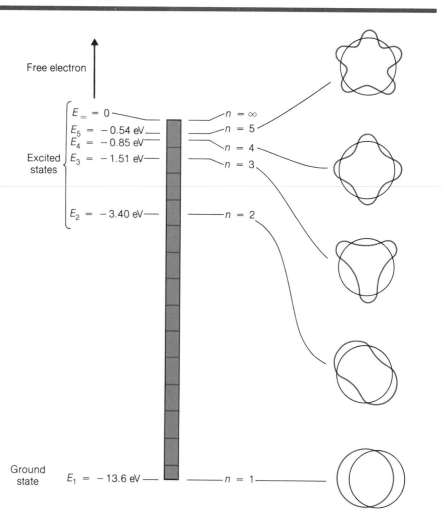

Free electron

$E_\infty = 0$ ⸺ $n = \infty$
$E_5 = -0.54$ eV⸺ $n = 5$
$E_4 = -0.85$ eV⸺ $n = 4$
Excited $E_3 = -1.51$ eV⸺ $n = 3$
states

$E_2 = -3.40$ eV⸺ $n = 2$

Ground
state $E_1 = -13.6$ eV⸺ $n = 1$

The higher the quantum number, the less tightly bound the electron

Ionization energy

proton to form an atom. An energy greater than zero signifies an unbound electron which, since it has no closed orbit that must satisfy quantum conditions, may have any positive energy whatever.

The work needed to remove an electron from an atom in its ground state is called its *ionization energy*. The ionization energy is therefore equal to $-E_1$, the amount of energy that must be provided to raise an electron from its ground state to an energy of $E = 0$, when it is free. In the case of hydrogen, the ionization energy is 13.6 eV, since the ground-state energy of the hydrogen atom is -13.6 eV.

Example   An electron collides with a hydrogen atom initially in its ground state and excites it to a state of $n = 3$. How much energy was transferred to the hydrogen atom in this inelastic (KE not conserved) collision?

Solution   From Eq. (28–3) the energy change of a hydrogen atom that goes from an initial state of quantum number n_i to a final state of quantum number n_f is

$$\Delta E = E_f - E_i = \frac{E_1}{n_f^2} - \frac{E_1}{n_i^2} = E_1 \left(\frac{1}{n_f^2} - \frac{1}{n_i^2} \right)$$

Here $n_i = 1$ and $n_f = 3$, and so, since $E_1 = -13.6$ eV,

$$\Delta E = E_1 \left(\frac{1}{3^2} - \frac{1}{1^2} \right) = -13.6 \left(\frac{1}{9} - 1 \right) \text{eV} = -13.6 \left(-\frac{8}{9} \right) \text{eV}$$

$$= 12.1 \text{ eV}$$

The presence of definite energy levels in an atom—which is true for all atoms, not just the hydrogen atom—is another example of the fundamental graininess of physical quantities on a microscopic scale. In the everyday world, matter, electric charge, energy, and so on seem continuous and capable of being cut up, so to speak, into parcels of any size we like. In the world of the atom, however, matter consists of elementary particles of fixed rest masses which join together to form atoms of fixed rest masses; electric charge always comes in multiples of $+e$ and $-e$; energy in the form of electromagnetic waves of frequency f always comes in separate photons of energy hf; and stable systems of particles, such as atoms, can have only certain energies and no others.

Quantization is characteristic of many quantities in nature

Other quantities in nature are also grainy, or *quantized,* and it has turned out that this graininess is the key to understanding how the properties of matter we are familiar with in everyday life originate in the interactions of elementary particles. In the case of the atom, the quantization of energy is a consequence of the wave nature of moving bodies: in an atom the electron wave functions can only occur in the form of standing waves, much as a violin string can vibrate only at those frequencies that give rise to standing waves.

28–4 ATOMIC SPECTRA

The Bohr theory is strikingly confirmed in its successful explanation of atomic spectra. When a gas or vapor at somewhat less than atmospheric pressure is "excited" by the passage of an electric current through it, light whose spectrum consists of a limited number of individual wavelengths is emitted (Fig. 28–9). The characteristic red-orange color of a neon sign is an example of this phenomenon.

Atomic spectra are called *line spectra* from their appearance. Every element exhibits a unique line spectrum when a sample of it is suitably excited, and the presence of any element in a substance of unknown composition can be ascertained by the appearance of its characteristic wavelengths in the spectrum of the substance. Spectral lines are found in the infrared and ultraviolet as well as in the visible region.

Every element has a unique line spectrum

It is worth noting that whereas unexcited gases and vapors do not radiate their characteristic spectral lines, they do *absorb* light of certain of those wavelengths when white light is passed through samples of them. In other words, the *absorption spectrum* of an element is closely related to its *emission spectrum.* Emission spectra consist of bright lines on a dark background; absorption spectra consist of dark lines on a bright background. Figure 28–10 shows the absorption and emission spectra of sodium vapor.

Absorption and emission spectra

FIG. 28–9 (a) The spectrum of a substance may be obtained by "exciting" a sample of its vapor in a tube by means of an electrical discharge and then directing the light it gives off through a prism or grating. (b) The most prominent lines in the spectra of hydrogen, helium, and mercury in the visible region.

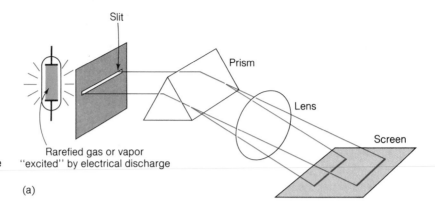

(a)

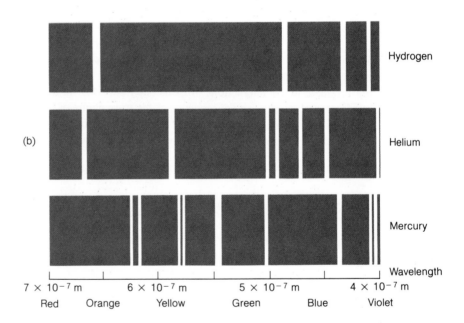

A photon is emitted when an atom in an excited state falls to a lower state

The wavelengths present in atomic spectra fall into definite series. The spectral series of hydrogen are shown in Fig. 28–11. The presence of a sequence of definite, discrete energy levels in the hydrogen atom suggests a connection with line spectra. Let us assert that when an electron in an excited state drops to a lower state, the difference in energy between the states is emitted as a single photon of light. Because electrons cannot, according to our model, exist in an atom except in certain specific energy levels, a rapid "jump" from one level to the other, with the energy difference being given off all at once in a photon rather than in some gradual manner, fits in well with this model.

If the quantum number of the initial (higher energy) state is n_i and the quantum number of the final (lower energy) state is n_f, what we assert is that

FIG. 28–10 (a) Absorption spectrum of sodium vapor. (b) Emission spectrum of sodium vapor. Each dark line in the absorption spectrum corresponds to a bright line in the emission spectrum.

$$E_i - E_f = hf \qquad \text{\textit{Origin of spectral lines}} \quad (28\text{–}4)$$

Initial energy $-$ final energy $=$ quantum energy

where f is the frequency of the emitted photon and h is Planck's constant. This formula agrees with the experimental data on the spectrum of hydrogen.

Figure 28–12 is an energy-level diagram of the hydrogen atom showing the possible transitions from initial quantum states to final ones. Each transition—or "jump"—involves a characteristic amount of energy, and hence a photon of a certain characteristic frequency. The larger the energy difference between initial and final energy levels, as

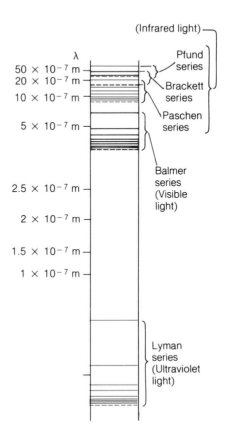

FIG. 28–11 The line spectrum of hydrogen with the various series of spectral lines indicated. The wavelength scale is not linear in order to cover the entire spectrum.

FIG. 28–12 Energy-level diagram of the hydrogen atom with some of the transitions that give rise to spectral lines indicated.

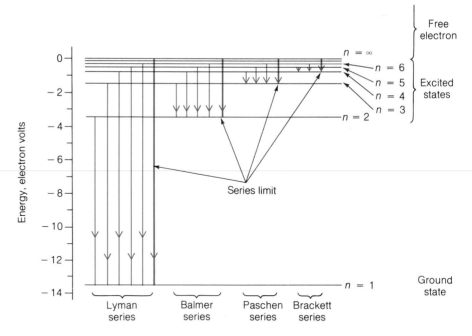

indicated by the lengths of the arrows, the higher the frequency of the emitted photon. The origins of the various series of spectral lines are indicated on the diagram.

Example   Obtain a formula for the wavelength of the photon emitted when a hydrogen atom in an excited state of quantum number n_i drops to a state of lower quantum number n_f.

Solution   Since $hf = E_i - E_f$ and $E_n = E_1/n^2$,

$$hf = E_1\left(\frac{1}{n_i^2} - \frac{1}{n_f^2}\right)$$

$$f = \frac{E_1}{h}\left(\frac{1}{n_i^2} - \frac{1}{n_f^2}\right) = -\frac{E_1}{h}\left(\frac{1}{n_f^2} - \frac{1}{n_i^2}\right)$$

Because $\lambda = c/f$, $1/\lambda = f/c$ and

$$\frac{1}{\lambda} = -\frac{E_1}{ch}\left(\frac{1}{n_f^2} - \frac{1}{n_i^2}\right) = R\left(\frac{1}{n_f^2} - \frac{1}{n_i^2}\right)$$

where $R = -E_1/ch = 1.097 \times 10^7 \text{ m}^{-1}$. We recall that E_1 is a negative quantity (-13.6 eV, in fact), so R is a positive quantity. The latter formula was originally obtained from studies of the spectra themselves, and its derivation was a triumph for Bohr's theory of the hydrogen atom. The various spectral series correspond to different values of n_f, as is clear from Fig. 28–12, with the Lyman series being specified by $n_f = 1$, the Balmer series by $n_f = 2$, and so on. ∎

Example   Find the longest wavelength present in the Balmer series of hydrogen.

Solution   In the Balmer series the quantum number of the final state is $n_f = 2$. The longest wavelength in this series corresponds to the smallest energy difference between energy levels, and hence to an initial state of $n_i = 3$:

$$\frac{1}{\lambda} = R\left(\frac{1}{n_f^2} - \frac{1}{n_i^2}\right) = R\left(\frac{1}{2^2} - \frac{1}{3^2}\right) = 0.139R$$

$$\lambda = \frac{1}{0.139R} = \frac{1}{0.139(1.097 \times 10^7 \, \text{m}^{-1})} = 6.56 \times 10^{-7} \, \text{m}$$

$$= 656 \, \text{nm}$$

This wavelength is near the red end of the visible spectrum.   ■

28–5   THE BOHR MODEL

Let us see how the various results of the Bohr theory of the hydrogen atom quoted earlier can be obtained.

We start with the classical picture of the hydrogen atom in which the electron is assumed to circle the proton in a stable orbit. If the orbit radius is r, the centripetal force $F_c = mv^2/r$ on the electron is provided by the electric force $F_e = ke^2/r^2$ between the electron and the proton. Hence

$$\frac{mv^2}{r} = k\frac{e^2}{r^2}$$

so the electron speed is related to its orbit radius r by the formula

$$v = e\sqrt{\frac{k}{mr}} \qquad (28\text{–}5)$$

Because the proton is nearly 2000 times heavier than the electron, we are justified in ignoring the proton's motion under the influence of the electric force the electron exerts on it.

The total energy E of a hydrogen atom is the sum of the electron's kinetic energy $\text{KE} = \frac{1}{2}mv^2$ and the electric potential energy of the system of electron and proton, which is $\text{PE} = -ke^2/r$. The latter formula was discussed in Sec. 17–5. Hence

$$E = \text{KE} + \text{PE} = \frac{mv^2}{2} - k\frac{e^2}{r}$$

and so, in view of Eq. (28–5),

$$E = k\left(\frac{e^2}{2r} - \frac{e^2}{r}\right) = -\frac{ke^2}{2r} \qquad (28\text{–}6)$$

The total energy of the hydrogen atom is negative, which expresses the fact that the electron is bound to the proton. If E were greater than zero, the electron would not follow a closed orbit about the proton.

Experiments indicate that 13.6 eV of work must be performed to break a hydrogen atom apart into a proton and an electron. In other words, the total energy of the hydrogen atom in its ground state is $E = -13.6$ eV and is made up of 13.6 eV of KE and -27.2 eV of PE. We can find the orbital radius of the electron from Eq. (28–6) for E. Since 13.6 eV $= 2.18 \times 10^{-18}$ J and $k = 9.0 \times 10^9$ N·m^2/C^2, we have

$$r = -\frac{k}{2}\frac{e^2}{E} = -\frac{(9.0 \times 10^9 \, \text{N·m}^2/\text{C}^2)(1.6 \times 10^{-19}\,\text{C})^2}{2(-2.18 \times 10^{-18}\,\text{J})}$$

$$= 5.3 \times 10^{-11}\,\text{m}$$

The electron's speed is, from Eq. (28–5),

$$v = e\sqrt{\frac{k}{mr}} = 1.6 \times 10^{-19}\,\text{C}\sqrt{\frac{9.0 \times 10^9 \, \text{N·m}^2/\text{C}^2}{(9.1 \times 10^{-31}\,\text{kg})(5.3 \times 10^{-11}\,\text{m})}}$$

$$= 2.2 \times 10^6\,\text{m/s}$$

This speed is well below that of light ($c = 3 \times 10^8$ m/s), and so the above nonrelativistic calculation is justified.

To derive a formula for the radii r_n of the possible electron orbits in the hydrogen atom, we start from the quantum condition of Eq. (28–1),

$$n\lambda = 2\pi r_n \qquad n = 1, 2, 3, \ldots$$

The formula for the de Broglie wavelength of the electron is $\lambda = h/mv$. Using Eq. (28–5) for v with r_n in place of r yields

$$\lambda = \frac{h}{me}\sqrt{\frac{mr_n}{k}} = \frac{h}{e}\sqrt{\frac{r_n}{mk}}$$

and so the quantum condition becomes

$$n\lambda = \frac{nh}{e}\sqrt{\frac{r_n}{mk}} = 2\pi r_n$$

Finally we square both sides and solve for r_n to obtain

$$\frac{n^2h^2}{e^2}\frac{r_n}{mk} = 4\pi^2 r_n^2$$

$$r_n = \frac{h^2 n^2}{4\pi^2 kme^2} \qquad n = 1, 2, 3, \ldots \tag{28–7}$$

which we can write as

$$r_n = r_1 n^2 \qquad n = 1, 2, 3, \ldots \tag{28–8}$$

where r_1 is the radius of the first (innermost) orbit. This result was given as Eq. (28–2).

The total energy E_n of a hydrogen atom whose electron is in the nth orbit depends only upon the radius r_n of that orbit, since

$$E = -\frac{k}{2}\frac{e^2}{r}$$

for all orbits. Hence the possible energies of a hydrogen atom are limited to

$$E_n = -\frac{ke^2}{2h^2n^2/4\pi^2kme^2} = -\frac{2\pi^2k^2me^4}{h^2n^2} \qquad n = 1, 2, 3, \ldots \qquad (28\text{–}9)$$

which we can write as

$$E_n = \frac{E_1}{n^2} \qquad n = 1, 2, 3, \ldots \qquad (28\text{–}10)$$

where E_1 is the energy of the innermost orbit. This result was given as Eq. (28–3).

It is worth noting that the formula for E_n contains only the directly measurable quantities k, m, e, and h, all of which are constants of nature. The formula is in complete agreement with the experimentally determined values of E_n.

28–6 ATOMIC EXCITATION

There are two principal mechanisms by which an atom may be excited to an energy level above that of its ground state and thereby become capable of radiating. One of these mechanisms is a collision with another atom during which part of their kinetic energy is transformed into electron energy within either or both of the participating atoms. An atom excited in this way will then lose its excitation energy by emitting one or more photons in the course of returning to its ground state (Fig. 28–13). In an electric discharge in a rarefied gas, an electric field accelerates electrons and charged atoms and molecules (whose charge arises from either an excess or a deficiency in the electrons required to neutralize the positive charge of their nuclei) until their kinetic energies are sufficient to excite atoms with which they happen to collide. A neon sign is a familiar example of how applying a strong electric field between electrodes in a gas-filled tube leads to the emission of the characteristic spectral radiation of that gas, which happens to be orange light in the case of neon.

Excitation by collision

Another excitation mechanism is the absorption by an atom of a photon of light whose energy is just the right amount to raise it to a higher energy level. A photon of wavelength 121.7 nm is emitted when a hydrogen atom in the $n = 2$ state drops to the $n = 1$ state; hence the absorption of a photon of wavelength 121.7 nm by a hydrogen atom initially in the $n = 1$ state will bring it up to the $n = 2$ state. This process explains the origin of absorption spectra (Fig. 28–14).

Excitation by absorption of radiation

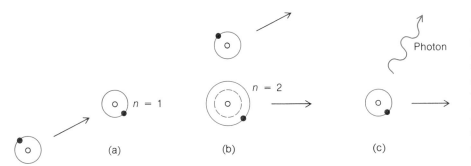

FIG. 28–13 In (a) both atoms are in their ground states. During the collision some kinetic energy is transformed into excitation energy, and in (b), the target atom is in an excited state. In (c) the target atom has returned to its ground state by emitting a photon.

FIG. 28–14 The origins of emission and absorption spectra.

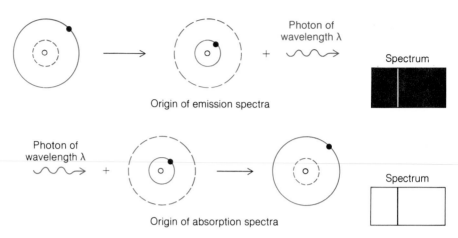

When white light (in which all wavelengths are present) is passed through hydrogen gas, photons of those wavelengths that correspond to transitions between hydrogen energy levels are absorbed. The resulting excited hydrogen atoms reradiate their excitation energy almost at once, but these photons come off in random directions, not all in the same direction as in the original beam of white light (Fig. 28–15). The dark lines in an absorption spectrum are therefore never totally dark, but only appear so by contrast with the bright background of transmitted light. We would expect the lines in the absorption spectrum of a particular substance to be the same as lines in its emission spectrum, which is what is found. As Fig. 28–10 shows, only a few of the emission lines appear in absorption. The reason is that reradiation is so rapid that essentially all the atoms in an absorbing gas or vapor are initially in their lowest ($n = 1$) energy states and therefore take up and reemit only photons that represent transitions to and from the $n = 1$ state.

28–7   THE LASER

The *laser* is a device for producing a light beam with a number of remarkable properties:

FIG. 28–15 The dark lines in an absorption spectrum are never totally dark.

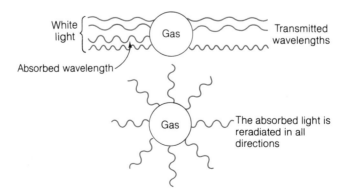

1. The beam is extremely intense, more intense by far than the light from any other source. To achieve an energy density equal to that in some laser beams, a hot object would have to be at a temperature of 10^{30} K.

2. There is very little divergence, so that a laser beam from the earth that was reflected by a mirror left on the moon during the Apollo 11 expedition remained collimated enough to be detected upon its return to the earth—a total journey of more than 750,000 km.

3. The light is essentially monochromatic.

4. The light is coherent, with the waves all exactly in step with one another (Fig. 28–16). It is possible to obtain interference patterns not only by merely placing two slits in a laser beam but also by using beams from two separate lasers.

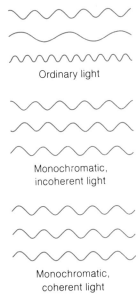

The term "laser" stands for *light amplifaction by stimulated emission of radiation.* Let us look into how a laser functions.

Left to itself, an atom always remains in its ground state, the quantum state in which it has the least possible energy. The atom can be raised to an excited state by acquiring energy in a collision with an electron or another atom or by absorbing a photon of exactly the right frequency. An excited atom normally falls to its ground state almost at once, sometimes first dropping to an intermediate excited state, with the emission of a photon during each transition. The lifetime of most excited states is only about 10^{-8} s, but certain states are *metastable* (temporarily stable). An atom may remain in a metastable state for 10^{-3} second or more before radiating, provided it does not undergo a collision in the meantime in which its excitation energy would be lost (Fig. 28–17). The operation of lasers depends upon the existence of metastable states in atoms.

Three kinds of transition involving electromagnetic radiation are possible between two energy levels in an atom whose energy difference is $E_2 - E_1 = hf$. In *induced absorption* the atom when in the lower level absorbs a photon of energy hf and is elevated to the upper level (Fig. 28–18). In *spontaneous emission* the atom when in the upper level emits a photon of energy hf and falls to the lower one. There is also a third possibility, *induced emission,* in which the presence of radiation of frequency f causes a transition from the upper level to the lower one. In induced emission, the emitted light waves are exactly in step with the incident ones, so that the result is an enhanced beam of coherent light.

FIG. 28–16 A laser produces a beam of light whose waves all have the same frequency (monochromatic) and are in phase with one another (coherent). The beam is also well collimated and spreads out very little even over a long distance.

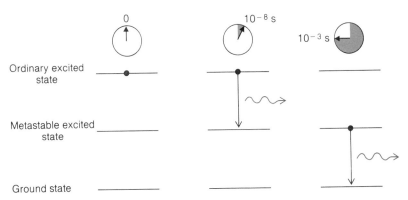

FIG. 28–17 An atom can exist in a metastable energy level for a longer time before radiating than it can in an ordinary energy level.

FIG. 28–18 The three types of transition between energy levels in an atom.

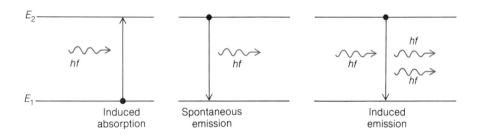

Induced absorption   Spontaneous emission   Induced emission

Now let us suppose we have an assembly of atoms of some kind which have metastable states of excitation energy hf. If we can somehow raise a majority of the atoms to the metastable level and then shine light of frequency f on the assembly, there will be more induced emission than induced absorption and the result will be an amplification of the original light. This is the concept that underlies the operation of the laser. The term *population inversion* is given to an assembly of atoms in which a majority are in excited states, because under normal circumstances the ground states are the more highly populated.

Optical pumping

There are a number of ways in which population inversions can be produced. One of them, called "optical pumping," employs an external light source some of whose photons have the right frequency to raise the atoms to excited states that decay into the desired metastable ones. This is the method used in the ruby laser, in which the chromium ions Cr^{+++} in a ruby crystal are excited by light from a xenon-filled flash lamp (Fig. 28–19). (A ruby is a crystal of Al_2O_3 a small number of whose Al^{+++} ions are replaced by Cr^{+++} ions, which are responsible for the reddish color. Such ions are Cr atoms that have lost three electrons each.)

The Cr^{+++} ions are raised in the pumping process to a level E_3 from which they decay to the metastable level E_2 by losing energy to other atoms in the crystal. This

FIG. 28–19 The ruby laser produces pulses of light.

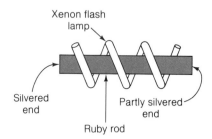

Xenon flash lamp

Silvered end   Partly silvered end

Ruby rod

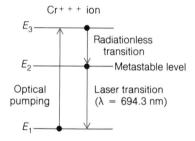

Cr^{+++} ion

E_3 ——————— Radiationless transition

E_2 ——————— Metastable level

Optical pumping   Laser transition (λ = 694.3 nm)

E_1 ———————

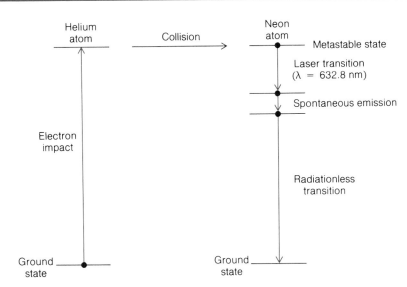

FIG. 28–20 Energy levels involved in the helium-neon laser, which operates continuously.

Helium atom

Collision

Neon atom

Metastable state

Laser transition ($\lambda = 632.8$ nm)

Spontaneous emission

Electron impact

Radiationless transition

Ground state

Ground state

metastable level has a lifetime of about 0.003 s. Because there are comparatively few Cr^{+++} ions in the ruby rod and a xenon flash lamp emits a great deal of light, most of the Cr^{+++} ions can be pumped to the E_2 level in this way. A few Cr^{+++} ions in the E_2 level then spontaneously fall to the E_1 level, and some of the resulting photons bounce back and forth between the reflecting ends of the ruby rod. The presence of light of exactly the right frequency now stimulates the other Cr^{+++} ions in the E_2 level to radiate, and the result is an avalanche process that produces a large pulse of red light in a few microseconds that emerges from the partly silvered end of the rod. The length of the rod is made to be exactly an integral number of half-wavelengths long, so the radiation trapped in it forms an optical standing wave. (The rod constitutes a "resonant cavity.") Since the induced emissions are stimulated by the standing wave, their waves are all in step with it, which is why the final pulse of light is both coherent and collimated.

The helium-neon gas laser achieves a population inversion in a different way. A mixture of about 7 parts of helium and 1 part of neon at a low pressure (1 mm of mercury) is placed in a glass tube that has parallel mirrors at both ends. An electric discharge is then produced in the gas by electrodes connected to a source of high-frequency alternating current, and collisions with electrons from the discharge excite He and Ne atoms to metastable states respectively 20.61 and 20.66 eV above their ground states (Fig. 28–20). Some of the excited He atoms transfer their energy to ground-state Ne atoms in collisions, with the 0.05 eV of additional energy being provided by the kinetic energy of the atoms. The purpose of the He atoms is thus to assist in producing a population inversion in the Ne atoms.

An excited Ne atom emits a photon of wavelength $\lambda = 632.8$ nm in the transition to another excited state 18.70 eV above the ground state; this is one of the transitions that leads to laser action. Then another photon is spontaneously emitted in a transition to a lower metastable state (this transition produces only incoherent radiation because it is spontaneous rather than induced), and the remaining excitation energy is lost in

Operation of ruby laser

Operation of helium-neon laser

The helium-neon laser operates continuously

collisions with the tube walls. Because the electron impacts that produce a population inversion in the Ne atoms are not intermittent but occur all the time, a He-Ne laser operates continuously.

In the ruby laser, at least half the Cr^{+++} ions must be in the metastable state or absorption rather than induced emission will predominate and laser amplification will not occur. In the He-Ne laser, the final state of the laser transition is 18.70 eV above the ground state and hence almost entirely unpopulated at ordinary temperatures, which means that very few Ne atoms will be able to absorb photons with the laser wavelength of 632.8 nm. For this reason the fraction of Ne atoms that must be excited in order for laser amplification to occur can be much less than half the total, which is a definite advantage.

Many other types of laser have been constructed, including solid-state and chemical lasers. Solid-state lasers are small and have limited power outputs, but are ideal for use in fiber-optic transmission lines in which signals are carried in the form of modulated light beams instead of as modulated electric currents. Such transmission lines are already being used between telephone exchanges on an experimental basis. In a chemical laser, molecules in metastable excited states are produced by chemical reactions, and the overall efficiency of the system is thereby improved.

IMPORTANT TERMS

An **emission spectrum** consists of the various wavelengths of light emitted by an excited substance. An **absorption spectrum** consists of the various wavelengths of light absorbed by a substance when white light is passed through it.

According to the **Bohr theory of the atom,** an electron can circle an atomic nucleus indefinitely without radiating energy if its orbit is a whole number of electron wavelengths in circumference. The number of wavelengths that fit into a particular permitted orbit is called the **quantum number** of that orbit. The electron energies corresponding to the various quantum numbers constitute the **energy levels** of the atom, of which the lowest is the **ground state** and the rest are **excited states.** When an electron in an excited state drops to a lower state, the difference in energy between the states is emitted as a single photon of light; when a photon of the same wavelength is absorbed, the electron goes from the lower to the higher state. The above two processes account for the properties of **emission** and **absorption spectra** respectively.

A **laser** is a device for producing a narrow, monochromatic, coherent beam of light. The terms stands for light amplification by stimulated emission of radiation. Laser operation depends upon the existence of **metastable states,** which are excited atomic states that can persist for unusually long periods of time.

IMPORTANT FORMULAS

Condition for orbit stability: $n\lambda = 2\pi r_n$

Energy levels of hydrogen atom:

$$E_n = \frac{E_1}{n^2} \qquad n = 1, 2, 3, \ldots$$

Origin of spectral lines: $E_i - E_f = hf$

MULTIPLE CHOICE

1. The classical model of the hydrogen atom fails because
   (a) a moving electron has more mass than an electron at rest.
   (b) a moving electron has more charge than an electron at rest.
   (c) the attractive force of the nucleus is not enough to keep an electron in orbit around it.
   (d) an accelerated electron radiates electromagnetic waves.

2. According to the Bohr model, an electron can revolve around the nucleus of a hydrogen atom indefinitely if its orbit is
  (a) a perfect circle.
  (b) sufficiently far from the nucleus to avoid capture.
  (c) less than one de Broglie wavelength in circumference.
  (d) exactly one de Broglie wavelength in circumference.

3. An atom emits a photon when one of its electrons
  (a) collides with another of its electrons.
  (b) is removed from the atom.
  (c) undergoes a transition to a quantum state of lower energy.
  (d) undergoes a transition to a quantum state of higher energy.

4. The bright-line spectrum produced by the excited atoms of an element contains wavelengths that
  (a) are the same for all elements.
  (b) are characteristic of the particular element.
  (c) are evenly distributed throughout the entire visible spectrum.
  (d) are different from the wavelengths in its dark-line spectrum.

5. A neon sign does not produce
  (a) a line spectrum.
  (b) an emission spectrum.
  (c) an absorption spectrum.
  (d) photons.

6. Which of the following transitions in a hydrogen atom emits the photon of highest frequency?
  (a) $n = 1$ to $n = 2$ (b) $n = 2$ to $n = 1$
  (c) $n = 2$ to $n = 6$ (d) $n = 6$ to $n = 2$

7. Which of the following transitions in a hydrogen atom absorbs the photon of highest frequency?
  (a) $n = 1$ to $n = 2$ (b) $n = 2$ to $n = 1$
  (c) $n = 2$ to $n = 6$ (d) $n = 6$ to $n = 2$

8. Which of the following transitions in a hydrogen atom emits the photon of lowest frequency?
  (a) $n = 1$ to $n = 2$ (b) $n = 2$ to $n = 1$
  (c) $n = 2$ to $n = 6$ (d) $n = 6$ to $n = 2$

9. A hydrogen atom is in its ground state when its orbital electron
  (a) is within the nucleus.
  (b) has escaped from the atom.
  (c) is in its lowest energy level.
  (d) is stationary.

10. With increasing quantum number, the energy difference between adjacent energy levels
  (a) decreases.
  (b) remains the same.
  (c) increases.
  (d) sometimes decreases and sometimes increases.

11. Most excited states of an atom have lifetimes of about
  (a) 10^{-8} s.   (b) 10^{-3} s.
  (c) 1 s.   (d) 10 s.

12. The operation of the laser is based upon
  (a) the uncertainty principle.
  (b) the interference of de Broglie waves.
  (c) induced emission of radiation.
  (d) spontaneous emission of radiation.

13. The light produced by a laser is not
  (a) incoherent.
  (b) monochromatic.
  (c) in the form of a narrow beam.
  (d) electromagnetic.

EXERCISES

1. How are the Bohr and Rutherford models of the hydrogen atom related? In the Bohr model, why is the electron pictured as revolving around the nucleus?

2. In the Bohr theory of the hydrogen atom, the electron is in constant motion. How can such an electron have a negative amount of energy?

3. Explain why the spectrum of hydrogen has many lines, although a hydrogen atom contains only one electron.

4. Would you expect the fact that the atoms of an excited gas are in rapid random motion to have any effect on the sharpness of the spectral lines they produce?

5. What kind of spectrum is observed in (a) light from the hot filament of a light bulb; (b) light from a sodium-vapor highway lamp; (c) light from an electric light bulb that has passed through cool sodium vapor?

6. When radiation with a continuous spectrum is passed through a volume of hydrogen gas whose atoms are all in the ground state, which spectral series will be present in the resulting absorption spectrum?

7. The three kinds of transition involving electromagnetic radiation that can occur between two energy levels in an atom are induced absorption, spontaneous emission, and induced emission. Why is spontaneous absorption impossible?

8. Why is the length of the optical cavity of a laser so important?

9. A proton and an electron, both at rest initially, combine to form a hydrogen atom in the ground state. A single photon is emitted in this process. What is its wavelength?

10. Find the wavelength of the spectral line that corresponds to a transition in hydrogen from the $n = 6$ state to the $n = 3$ state. In what part of the spectrum is this?

11. Find the wavelength of the spectral line that corresponds to a transition in hydrogen from the $n = 10$ state to the ground state. In what part of the spectrum is this?

12. How much energy is needed to ionize a hydrogen atom when it is in the $n = 4$ state?

13. What is the shortest wavelength present in the Brackett series of spectral lines?

14. What is the shortest wavelength present in the Paschen series of spectral lines?

15. A beam of electrons is used to bombard gaseous hydrogen. What is the minimum energy in electron volts the electrons must have if the second line of the Paschen series, corresponding to a transition from the $n = 5$ state to the $n = 3$ state, is to be emitted?

PROBLEMS

1. (a) Calculate the de Broglie wavelength of the earth. (b) What is the quantum number that characterizes the earth's orbit about the sun? (The earth's mass is 6.0×10^{24} kg, its orbital radius is 1.5×10^{11} m, and its orbital speed is 3×10^4 m/s.)

2. Calculate the average kinetic energy per molecule in a gas at room temperature (20°C), and show that this is much less than the energy required to raise a hydrogen atom from its ground state ($n = 1$) to its first excited state ($n = 2$).

3. To what temperature must a hydrogen gas be heated if the average molecular kinetic energy is to equal the binding energy of the hydrogen atom?

4. (a) Derive a formula for the frequency of revolution of an electron in the nth orbit of the Bohr atom. (b) Find the frequencies of revolution of the electron when it is in the $n = 1$ and $n = 2$ orbits. (c) An electron spends about 10^{-8} s

in an excited state before it drops to a lower state by giving up energy in the form of a photon. How many revolutions does an electron in the $n = 2$ state of the Bohr atom make before dropping to the $n = 1$ state? How does this compare with the number of revolutions the earth has made around the sun in the 4.5×10^9 years of its existence?

5. A beam of electrons whose energy is 13 eV is used to bombard gaseous hydrogen. What series of wavelengths will be emitted?

6. An excited hydrogen atom emits a photon of wavelength λ in returning to the ground state. (a) Derive a formula that gives the quantum number of the initial excited state in terms of λ and R. (b) Use this formula to find n_i for a 102.55-nm photon.

7. Repeat the derivation of the Bohr theory for a one-electron ion whose nuclear charge is Ze and show that the energy of the electron is proportional to Z^2.

8. Calculate the radius and speed of an electron in the ground state of doubly-ionized lithium and compare them with the radius and speed of the electron in the ground state of the hydrogen atom. (Li^{++} has a nuclear charge of $3e$.)

9. A negative muon ($m = 207\ m_e$, $Q = -e$) can be captured by a proton to form a *muonic atom*. What is the radius of the first Bohr orbit of such an atom? What is the ionization energy of such an atom? Assume that the proton remains stationary as the muon revolves around it; in reality, both revolve around a common center of mass.

10. Show that the angular momentum of an electron in the nth orbit of a Bohr atom is equal to $nh/2\pi$. (In fact, this was the starting point of Bohr's original formulation of his theory of the hydrogen atom, which was carried out before de Broglie waves had been discovered.)

ANSWERS TO MULTIPLE CHOICE

1. d	**6.** b	**10.** a
2. d	**7.** a	**11.** a
3. c	**8.** d	**12.** c
4. b	**9.** c	**13.** a
5. c		

QUANTUM THEORY OF THE ATOM

29

The Bohr theory of the atom is indeed impressive in its agreement with experiment, but it has certain serious limitations. These limitations are absent from the quantum theory of the atom, which was developed a decade after Bohr's work, and whose refinement and application to new problems continues to the present day. Not only has the quantum theory of the atom provided the theoretical framework for understanding the structure of the atom itself, but it has also furnished key insights that explain how and why atoms join together to form molecules, solids, and liquids.

29–1   BOHR THEORY VERSUS QUANTUM THEORY

The Bohr theory is unable to account for many important atomic phenomena. While correctly predicting the wavelengths of the spectral lines in hydrogen,

Deficiencies of Bohr theory

which has but a single atomic electron, the Bohr theory fails when attempts are made to apply it to more complex atoms. Even with hydrogen, it is not possible to calculate from the Bohr theory the relative probabilities of the various transitions between energy levels, for instance, whether it is more likely that an atom in the $n = 3$ quantum state will go directly to the $n = 1$ state or instead first drop to the $n = 2$ state. (In other words, we cannot find from the Bohr theory which of the spectral lines of hydrogen will show up brightest in an emission spectrum and which will be fainter). For another thing, the careful study of spectral lines show that many of them actually consist of two or more separate lines that are close together, something that the Bohr theory cannot account for.

In cataloging these objections to the Bohr theory, the intent is not to detract from its eminence in the history of science, which is certainly secure, but instead to emphasize that a more general approach capable of wider application is necessary. Such an approach was developed in the 1920s by Schrödinger, Heisenberg, and others, under the name of the quantum theory of the atom. Instead of trying to visualize an atomic electron as a kind of hybrid of a particle and a wave and thinking of it as occupying one of various possible orbits, the quantum theory of the atom avoids all reference to anything not capable of direct measurement and restricts itself only to such observable quantities as photon energies, the mass and charge of the electron, and so on.

The quantum theory of the atom is consistent with the uncertainty principle

In the Bohr theory we compute the radius of an electron orbit from a knowledge of its de Broglie wavelength, which depends upon the electron's momentum; however, according to the uncertainty principle, the position (and hence orbital radius) and momentum of an electron can never simultaneously have well-defined values, and so even in principle the Bohr theory is unsatisfactory. The quantum theory sacrifices such easily pictured notions as that of electrons circling a nucleus like planets around the sun in favor of an abstract mathematical formulation dealing with probabilities, and it is able to tackle successfully a broad range of atomic problems.

Classical physics is an approximation of quantum physics

The traditional laws of mechanics and electromagnetism do not "work" when applied to the atomic world, as we have seen—these laws lead to predictions that do not agree with experimental findings. But quantum theory works on all scales of size, and it turns out that the traditional laws of physics are merely approximate versions of quantum theory that are successful for large-scale phenomena because on a large scale all that is apparent is the average behavior of a great many separate particles. In dealing with individual particles, however, no averaging is possible, and only quantum theory can be applied.

Unfortunately the quantum theory of the hydrogen atom involves advanced mathematics and so cannot be developed here. The conclusions of the theory, however, are fairly straightforward.

In the Bohr model of the hydrogen atom, the electron is regarded as being confined to a definite circular orbit, and the only quantity that changes as it revolves around its nucleus is its position on this fixed circle. A single quantum number is all that is required to specify the physical state of such an electron.

Quantum numbers of an atomic electron

In the more general quantum theory of the atom, the electron is not restricted to a specific orbit, and three quantum numbers turn out to be necessary instead of the single one of the Bohr theory. These are:

1. The principal quantum number n.
2. The orbital quantum number l.
3. The magnetic quantum number m_l.

The value of n is the chief factor that governs the total energy of an electron bound to a nucleus. The energy levels of a hydrogen atom are the same in the Bohr and quantum theories and are specified by the formula

The principal quantum number n governs E

$$E = \frac{E_1}{n^2} \qquad n = 1, 2, 3, \ldots \qquad \textit{Energy of atomic electron} \quad (29\text{–}1)$$

where $E_1 = -13.6$ eV is the energy of the ground state.

29–2 ANGULAR MOMENTUM QUANTIZATION

The orbital quantum number l governs the magnitude L of the electron's angular momentum **L** about the nucleus. Even though the quantum theory discards the picture of an atomic electron as circling the nucleus in favor of a certain distribution of probability density ψ^2 (see Sec. 27–6), angular momentum can be a property of the atom. The possible angular momenta are restricted to

The orbital quantum number l governs the magnitude of L

$$L = \sqrt{l(l+1)}\,\frac{h}{2\pi} \qquad l = 0, 1, 2, \ldots, (n-1) \qquad \begin{matrix}\textit{Angular momentum}\\\textit{of atomic electron}\end{matrix} \quad (29\text{–}2)$$

There are several interesting things about this result. The most conspicuous is the quantization of angular momentum itself, which is a universal phenomenon: like mass, electric charge, and the energy of a trapped particle, the angular momentum of a particle or system of particles can have only certain specific values. The quantity $h/2\pi$ is the natural unit of angular momentum. Because h is so small, the quantization of angular momentum is only perceptible in the physics of atoms and molecules, where it plays an important role.

The orbital quantum number l can be 0 or any integer up to $n-1$, where n is the principal quantum number of the electron's quantum state. If $n = 3$, for instance, l can be 0, 1, or 2. When $l = 0$, the angular momentum $L = 0$ also. Here is a significant difference from the Bohr theory, since an electron revolving in a circle *must* possess angular momentum. No such requirement emerges from the quantum theory of the atom.

The magnetic quantum number m_l governs the *direction* of the electron's angular momentum **L**. Angular momentum is a vector quantity with both direction and magnitude, so it is natural (though perhaps surprising) that the direction of **L** should also be restricted in some way. Figure 29–1 shows the right-hand rule for the direction of **L**.

The magnetic quantum number m_l governs the direction of L

What can be the significance of the direction of **L**? And with respect to what is this direction reckoned? The answers to these questions become clear when we consider that an electron with angular momentum behaves like a loop of electric current and interacts with an external magnetic field in a manner similar to that of a bar magnet. The greater the angular momentum **L**, the stronger the equivalent bar magnet. The

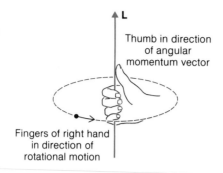

FIG. 29–1 The right-hand rule for the direction of angular momentum **L**.

Thumb in direction of angular momentum vector

Fingers of right hand in direction of rotational motion

potential energy of a bar magnet in a magnetic field **B** varies with the strength of the magnet and with its direction relative to **B**; the energy is least when the magnet is parallel to the field, and is most when it is antiparallel (Fig. 29–2).

The magnetic quantum number m_l determines the angle between **L** and **B** and hence governs the extent of the magnetic contribution to the total energy of the atom when the atom is in a magnetic field. An atomic electron characterized by a certain value of m_l will assume a certain corresponding orientation of its angular momentum **L** relative to a magnetic field when placed in such a field.

How the direction of L is specified

The direction of **L** relative to **B** is specified in terms of the component of **L** parallel to the magnetic field **B**. By convention this component is called L_z, and its possible values are also in units of $h/2\pi$:

$$L_z = m_l \frac{h}{2\pi}$$

Component of angular momentum of atomic electron (29–3)

The magnetic quantum number m_l can be any integer from $-l$ through 0 to $+l$. An atomic electron for which $l = 2$, for instance, could have a magnetic quantum number of $-2, -1, 0, +1,$ or $+2$. The angular momentum vector **L** of the atom can never be exactly parallel to **B**, even when $m_l = l$ since L_z is always smaller than L (Fig. 29–3).

The uncertainty principle enables us to understand why the angular momentum **L** of an atomic electron can never be exactly parallel to **B**. If **L** were parallel to **B**, or if

FIG. 29–2 The potential energy of a bar magnet in a magnetic field depends upon the orientation of the magnet relative to **B**.

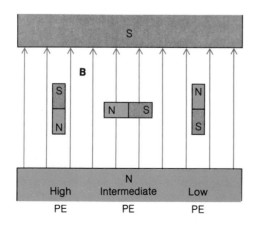

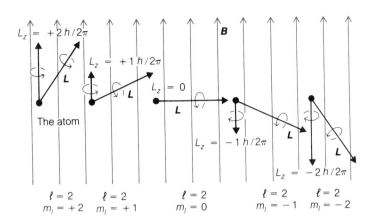

FIG. 29–3 Possible orientations of the angular momentum vector **L** of an atomic electron in a magnetic field when the orbital quantum number of the electron is $l = 2$.

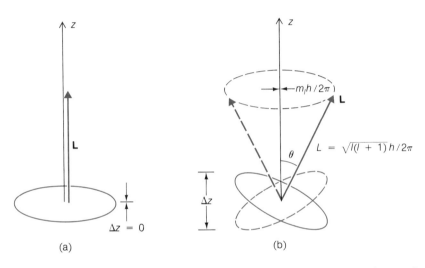

(a)   (b)

FIG. 29–4 (a) If **L** were parallel to a given direction in space (for instance, that of a magnetic field), then the electron would be confined to a definite plane. (b) In reality, L is always greater than L_z, and so there is an intrinsic uncertainty in the electron's position in three dimensions.

L pointed in any other definite direction in space, it would mean that the electron is always present in a specific plane perpendicular to **B**. But if there is such precision in the z-coordinate of the electron, then the electron's momentum component in the z-direction must be infinitely uncertain. An infinite uncertainty in the momenta of atomic electrons cannot be reconciled with the existence of stable atoms. However, the problem does not in fact arise because $L > L_z$ for all values of m_l. As Fig. 29–4 shows, **L** may point anywhere along the circle at its tip and still have the component $m_l h/2\pi$ in the direction of **B**. Thus there is a built-in uncertainty in the electron's position, and its momentum uncertainty in consequence does not exceed an amount compatible with its presence in an atom.

Although the interpretation of the magnetic quantum number m_l was discussed in terms of an external magnetic field, actually the component of **L** in *any* specific direction must be $m_l h/2\pi$. The point is that, for an isolated atom, an external magnetic field provides a definite, experimentally meaningful reference direction. But when the atom is part of a molecule, for instance, there are then other experimentally meaningful reference directions, namely those defined by the relative positions of the various atoms

Why L cannot have a fixed direction

Reference direction for L_z

in the molecule. Thus a line joining the two H atoms in the hydrogen molecule H_2 is a quite specific reference direction, and along this line the components of the angular momenta of the two H atoms are fixed by their m_l values.

The Zeeman effect is the splitting of spectral lines due to a magnetic field

Because the magnetic quantum number m_l usually has several possible values, the presence of an external magnetic field splits the energy levels of a particular atom into two or more sublevels. The emission spectrum of an element in a magnetic field accordingly differs from its ordinary spectrum in that the spectral lines of the latter now have several components whose spacing varies with B (Fig. 29–5). This phenomenon is known as the *Zeeman effect*. For example, a field of 1 T leads to an energy difference of 5.8×10^{-4} eV between adjacent sublevels of different m_l, which is observed as a wavelength difference between the split components of a given spectral line of order of magnitude 0.1%. The Zeeman effect is especially valuable in astronomy, where it was responsible, among other findings, for the discovery of the magnetic field of the sun and the magnetic nature of sunspots.

29–3 PROBABILITY CLOUDS

In place of the picture of a hydrogen atom as an electron circling a proton in a definite orbit, the quantum theory of the atom provides a description in terms of the probability density of the electron. We recall from Sec. 27–6 what is meant by the probability density ψ^2: The value of ψ^2 for a particular object at a certain place and time is proportional to the probability of finding the object at that place at that time. A large value of ψ^2 signifies that the object is likely to be found; a small value of ψ^2 signifies that the object is unlikely to be found.

Each quantum state has a different probability cloud

The calculation of probability densities starts from *Schrödinger's equation*, an equation as central to atomic physics as Newton's laws of motion are to classical mechanics, but far more complicated. The results show that the electron probability density distribution is different for each quantum state of a hydrogen atom, that is, for each set of quantum numbers n, l, and m_l. We might call the ψ^2 distribution that corresponds to a particular quantum state a *probability cloud*. In the Bohr model, an atomic electron travels in a specific orbit; in the quantum theory, an atomic electron moves about within a probability cloud that forms a certain pattern in space.

Probability cloud for $n = 1$ electron

There is only a single probability cloud possible for a ground-state hydrogen atom, since $l = 0$ and $m_l = 0$ when $n = 1$. This cloud is spherically symmetric—that is, the likelihood of finding the electron at a given distance from the nucleus is the same in all directions. We can picture ψ^2 for an $n = 1$ electron by a shaded drawing in which the darker the shading, the greater the value of ψ^2. As we can see in Fig. 29–6, the probability cloud does not have a sharp boundary. The electron spends most of the time near the nucleus, but there is a certain probability, which becomes vanishingly small at large distances, that it may be found anywhere.

An electron with the principal quantum number $n = 2$ may exist in one of four quantum states:

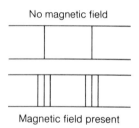

No magnetic field

Magnetic field present

FIG. 29–5 An example of the Zeeman effect.

$$n = 2 \begin{cases} l = 0 & \{\, m_l = 0 \\ l = 1 & \begin{cases} m_l = +1 \\ m_l = 0 \\ m_l = -1 \end{cases} \end{cases}$$

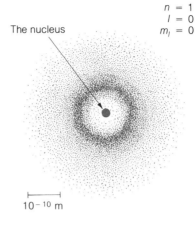

$n = 1$
$l = 0$
$m_l = 0$

FIG. 29–6 Probability cloud for the ground state of the hydrogen atom.

The nucleus

10⁻¹⁰ m

$n = 2$
$l = 0$
$m_l = 0$

FIG. 29–7 The probability cloud for the $n = 2$, $l = 0$ state of the hydrogen atom.

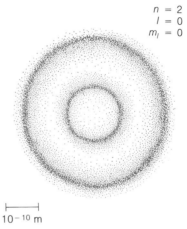

10⁻¹⁰ m

The $n = 2$, $l = 0$ probability cloud is spherically symmetric (indeed, *all* $l = 0$ clouds are spherically symmetric), but it does not fade out uniformly with distance as does the $n = 1$, $l = 0$ cloud (Fig. 29–7). Instead there is a gap at a radius of 1.06×10^{-10} m; the electron is *never* exactly this far from the nucleus. This gap corresponds to a node in a standing-wave pattern—in fact, we can think of the entire probability cloud as a kind of standing-wave pattern in three dimensions. The dense part of the cloud, which covers the region in which the electron is nearly always to be found, is four times as far across as the $n = 1$ cloud. This is in accord with the Bohr theory, in which an electron's orbital radius is proportional to n^2. We can think of the cloud in the drawing as representing a fuzzy sphere inside a fuzzy spherical shell.

    The sizes of the $n = 2$, $l = 1$ probability clouds are roughly the same as that of the $n = 2$, $l = 0$ cloud, but they have different shapes (Fig. 29–8). The $m_l = +1$ cloud has the form of a doughnut centered on the z-axis. Because $m_l = +1$, the cloud possesses angular momentum whose component in the z-direction is up (that is, in the $+z$-direction). This is indicated in the figure by the arrow, which is in accord with the right-hand rule of Fig. 29–1. If we like, we can think of the cloud as a rotating distribution of charge, with the rotation furnishing both the observed angular momentum and the observed magnetic behavior. In probability terms, if we were to determine

Probability cloud for $n = 2$, $l = 0$ electron

A probability cloud is a standing wave in three dimensions

An $n = 2$, $l = 1$ electron has three possible probability clouds

FIG. 29–8 Probability clouds for the three possible $n = 2, l = 1$ states of the hydrogen atom.

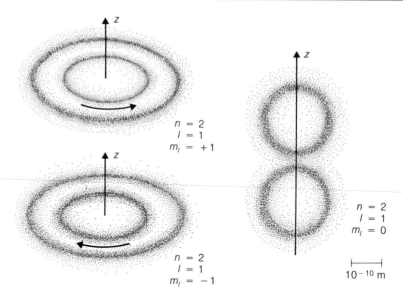

$$n = 2$$
$$l = 1$$
$$m_l = +1$$

$$n = 2$$
$$l = 1$$
$$m_l = -1$$

$$n = 2$$
$$l = 1$$
$$m_l = 0$$

10^{-10} m

the electron's direction of motion within an atom in this state in a series of experiments, more often than not we would find it to be moving counterclockwise around the z-axis.

The $m_l = -1$ probability cloud is exactly the same as the $m_l = +1$ cloud except that the angular momentum is in the opposite direction. The rotational motion associated with the cloud is now in the clockwise sense around the z-axis.

The $m_l = 0$ probability cloud has two concentrations around the z-axis on either side of the nucleus. These concentrations are symmetric about the z-axis—the pattern is the same on all sides. Because $m_l = 0$, there is no angular momentum about the z-axis. A series of measurements of the electron's direction of motion would show it going as often clockwise as counterclockwise. But since $l = 1$, the atom nevertheless has angular momentum. Evidently the rotational motion that corresponds to this angular momentum occurs in such a way that **L** is always perpendicular to the z-axis; then **L** has no component along this axis.

The probability clouds for quantum states whose principal quantum number n is greater than 2 are more complicated than those for $n = 1$ and $n = 2$. However, when $l = 0$, the clouds are always spherically symmetric, and when $l = 1$, the $m_l = 0 \pm 1$ clouds are similar in form to the corresponding ones for $n = 2, l = 1$ depicted above.

29–4   ELECTRON SPIN

We mentioned earlier that one of the many problems facing the atomic theorist is the observed splitting of many spectral lines into several components close together. For example, the first line of the Balmer series of hydrogen, which both theory and experiment place at a wavelength of 656.280 nm, actually consists of a pair of lines 0.014 nm apart.

An electron behaves much like a spinning charged sphere

In 1925 Goudsmit and Uhlenbeck pointed out that this fine structure of spectral lines would occur if the electron behaves like a charged sphere spinning on its axis rather than like a single point charge; several years later the British physicist P. A. M.

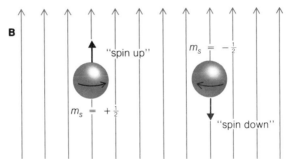

FIG. 29–9 The spin magnetic quantum number m_s of an atomic electron has two possible values, $+\frac{1}{2}$ and $-\frac{1}{2}$. Electrons behave in certain respects like spinning charged spheres, but experiments suggest they are more likely to be point particles with no extension in space.

Dirac was able to show on the basis of a relativistic version of quantum theory that the electron *must* have the spin attributed to it. A spinning electron is in effect a tiny bar magnet, and it interacts with the magnetic field produced by its own motion in an atom as well as with any magnetic fields originating outside the atom.

Electron spin is described by the *spin magnetic quantum number, m_s*, whose values are $+\frac{1}{2}$ and $-\frac{1}{2}$. The magnitude of the angular momentum associated with its spin is the same for every electron; the quantum number m_s refers to the *direction* of the spin. Figure 29–9 shows how electrons with $m_s = +\frac{1}{2}$ and $m_s = -\frac{1}{2}$ align themselves in an external magnetic field. The component L_{sz} of spin angular momentum in the direction of **B** is $m_s h/2\pi$.

The spin magnetic quantum number has two values

The notion of an electron as a spinning charged sphere is simply a model to make it easier to understand what is going on. No one knows what an electron really looks like, and in fact it is unlikely that any picture based on our perceptions of the macroscopic world can be relevant to elementary particles; indeed, observations of the scattering of electrons by other electrons at high energies suggest that the electron must be less than 10^{-16} m across, and quite possibly is a point particle. But we *can* say that the possession of angular momentum and the magnetic behavior that goes with it are properties as fundamental to an electron as its rest mass and charge.

In its own frame of reference, an atomic electron "sees" the positively charged nucleus moving around *it*, and the electron accordingly experiences a magnetic field. Because the energy of an electron is different in each orientation of its spin relative to the magnetic field it experiences, the various energy levels of an atom are split into sublevels. The presence of these sublevels of energy is responsible for the observed fine structure of spectral lines. The introduction of electron spin into the theory of the atom means that a total of four quantum numbers, n, l, m_l, and m_s, is required to describe each possible state of an atomic electron. These are listed in Table 29–1.

Origin of fine structure in atomic spectra

There are $2n^2$ possible quantum states for each value of the principal quantum number n. For example, when $n = 3$ there are 18 states, as shown in Table 29–2.

Name	Symbol	Possible Values	Quantity Determined
Principal	n	$1, 2, 3, \ldots$	Electron energy
Orbital	l	$0, 1, 2, \ldots, n-1$	Magnitude of angular momentum
Magnetic	m_l	$-l, \ldots, 0, \ldots, +l$	Direction of angular momentum
Spin magnetic	m_s	$-\frac{1}{2}, +\frac{1}{2}$	Direction of electron spin

TABLE 29–1
Quantum numbers of an atomic electron

TABLE 29–2

Quantum states of an electron of principal quantum number $n = 3$

	$m_l = 0$	$m_l = -1$	$m_l = +1$	$m_l = -2$	$m_l = +2$	
$l = 0$:	⇅					↑ $m_s = +\frac{1}{2}$
$l = 1$:	⇅	⇅	⇅			↓ $m_s = -\frac{1}{2}$
$l = 2$:	⇅	⇅	⇅	⇅	⇅	

29–5   THE EXCLUSION PRINCIPLE

The probability clouds of atoms that contain more than one electron are similar in shape to the corresponding ones in the hydrogen atom. Normally the electron in a hydrogen atom is in its ground state of $n = 1$. What about the electrons in a more complex atom? Are they all in the same $n = 1$ state, jammed together in a single probability cloud, or are they distributed in some special way among the various other quantum states? The answer to this question is fundamental to chemistry because it shows how the properties of the various elements arise, and explains the mechanisms by which elements combine to form compounds.

The arrangement of the electrons in a complex atom is governed by the *exclusion principle* discovered by Wolfgang Pauli. This principle states that

Exclusion principle

No two electrons in an atom can exist in the same quantum state.

That is, each electron in a complex atom must have a different set of the quantum numbers n, l, m_l, and m_s. The exclusion principle can be generalized to refer to the electrons in *any* small region of space, regardless of whether or not they constitute an atom, as we shall find in Chapter 30.

Pauli was led to the exclusion principle by a study of atomic spectra. It is possible to determine the quantum states of an atom empirically by analyzing its spectrum, since states with different quantum numbers differ in energy (even if only slightly) and the various wavelengths present correspond to transitions between these states. Pauli found several lines missing from the spectra of every element except hydrogen; the missing lines correspond to transitions to and from atomic states having certain sets of quantum numbers. Every one of these absent states has two or more electrons with identical sets of quantum numbers, a result that is expressed in the exclusion principle. Hydrogen, with a single electron, naturally has no absent states.

29–6   FAMILIES OF ELEMENTS

Atomic number is the number of electrons in an atom

Certain elements resemble one another so closely in their physical and chemical properties that it is natural to think of them as forming a "family." Three particularly striking examples of such families are the *halogens*, the *inert gases*, and the *alkali metals*. The members of these groups of elements are listed in Table 29–3 together with the number of electrons the respective atoms contain. This number is called the *atomic number* (symbol Z) of the element.

Halogens	Inert Gases	Alkali Metals
	(2) Helium	(3) Lithium
(9) Fluorine	(10) Neon	(11) Sodium
(17) Chlorine	(18) Argon	(19) Potassium
(35) Bromine	(36) Krypton	(37) Rubidium
(53) Iodine	(54) Xenon	(55) Cesium
(85) Astatine	(86) Radon	(87) Francium

TABLE 29–3
Three families of elements. The atomic numbers of the elements are in parentheses.

The halogens are nonmetallic elements with a high degree of chemical activity. At room temperature fluorine and chlorine are gases, bromine is a liquid, and iodine and astatine are solids. The halogens have chemical valences of -1 and form diatomic molecules in the vapor state. The inert gases, as their name suggests, are inactive chemically: they form virtually no compounds with other elements, and their atoms do not join together into molecules. The inert gases have valences of 0. The alkali metals, like the halogens, are chemically very active, but they are active as reducing agents rather than as oxidizing agents. They are soft, not very dense, and have low melting points (only lithium is solid above 100°C). The alkali metals have valences of $+1$.

A curious feature of the three groups listed in Table 29–3 is that, while the atomic numbers of the member elements in each group bear no obvious relation to one another, each inert gas is preceded in atomic number by a halogen (except in the case of helium) and followed by an alkali metal. Thus fluorine, neon, and sodium have the atomic numbers 9, 10, and 11, respectively, a sequence that persists through astatine (85), radon (86), and francium (87).

When all the elements are listed in order of atomic number, elements with similar chemical and physical properties recur at regular intervals.

This observation, first formulated in detail by Dmitri Mendeleev about 1869, is known as the *periodic law.* As Mendeleev presciently remarked, "The periodic law, together with the revelations of spectrum analysis, have contributed to again revive an old but remarkably long-lived hope—that of discovering, if not by experiment, at least by mental effort, the *primary matter."*

A periodic table is a listing of the elements according to atomic number in a series of rows such that elements with similar properties form vertical columns. Table 29–4 is a simple form of periodic table; the number above the symbol of each element is its atomic mass, and the number below the symbol is its atomic number. The elements whose atomic masses appear in parentheses are radioactive and are not found in nature, but have been prepared in nuclear reactions. The atomic mass in each such case is the mass number of the longest-lived isotope of the element.

The columns in the periodic table are called *groups* (Fig. 29–10). We recognize group I as the alkali metals plus hydrogen, group VII as the halogens, and group VIII as the inert gases. In addition to the elements forming the eight principal groups, there are a number of *transition elements* falling between groups II and III. The transition elements are metals which share certain general properties: most are hard and brittle, have high melting points, and exhibit several different valences.

Group

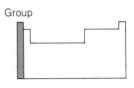

Period

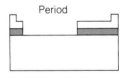

Nonmetals

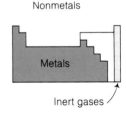

Metals

Inert gases

FIG. 29–10 (a) The elements in a group of the periodic table have similar properties. (b) The elements in a period have progressively different properties, ranging from an alkali metal at the left to an inert gas at the right. (c) Most elements are metals.

Elements in a period have different properties

The rows in the periodic table are called *periods*. Each period starts with an active alkali metal and proceeds through less active metals to weakly active nonmetals to an active halogen and an inactive inert gas. The transition elements in each period may be very much alike; the rare earths and actinides are so much alike that they are usually considered as separate categories.

For nearly a century the periodic law has been a mainstay of the chemist by permitting him to predict the behavior of undiscovered elements and by providing a framework for organizing his knowledge. It is one of the triumphs of the quantum theory of the atom that it enables us to account for the periodic law in complete detail without invoking any new assumptions or postulates.

TABLE 29–4
Periodic classification of the elements. The number above the symbol of each element is its atomic mass (see Sec. 13–6), and the number below the symbol is its atomic number.

Group Period	I	II						
1	1.008 H 1							
2	6.94 Li 3	9.01 Be 4						
3	22.99 Na 11	24.31 Mg 12						
4	39.10 K 19	40.08 Ca 20	44.96 Sc 21	47.90 Ti 22	50.94 V 23	52.00 Cr 24	54.94 Mn 25	55.85 Fe 26
5	85.47 Rb 37	87.62 Sr 38	88.91 Y 39	91.22 Zr 40	92.91 Nb 41	95.94 Mo 42	(97) Tc 43	101.1 Ru 44
6	132.91 Cs 55	137.34 Ba 56	* 57–71	178.49 Hf 72	180.95 Ta 73	183.85 W 74	186.2 Re 75	190.2 Os 76
7	(223) Fr 87 *Alkali metals*	226.05 Ra 88	† 89–104					

	138.91 La 57	140.12 Ce 58	140.91 Pr 59	144.27 Nd 60
* *Rare earths*				

	227.03 Ac 89	232.04 Th 90	231.04 Pa 91	238.03 U 92
† *Actinides*				

29–7 ATOMIC STRUCTURES

Two basic principles determine the structures of atoms with more than one electron:

1. An atom is stable when its total energy is a minimum. This means that in the normal configuration of an atom its various electrons are present in the lowest energy states available to them.

2. Only one electron can exist in each quantum state of an atom. This is the exclusion principle.

The notion of shells and subshells of electrons in an atom is a useful one. The

				III	*IV*	*V*	*VI*	*VII*	*VIII*
									4.00
									He
									2
				10.81	12.01	14.01	16.00	19.00	20.18
				B	C	N	O	F	Ne
				5	6	7	8	9	10
				26.98	28.09	30.98	32.07	35.46	39.94
				Al	Si	P	S	Cl	Ar
				13	14	15	16	17	18
58.93	58.71	63.54	65.37	69.72	72.59	74.92	78.96	79.91	83.80
Co	Ni	Cu	Zn	Ga	Ge	As	Se	Br	Kr
27	28	29	30	31	32	33	34	35	36
102.91	106.4	107.87	112.40	114.82	118.69	121.76	127.61	126.90	131.30
Rh	Pd	Ag	Cd	In	Sn	Sb	Te	I	Xe
45	46	47	48	49	50	51	52	53	54
192.2	195.09	196.97	200.59	204.37	207.19	208.98	(209)	(210)	222
Ir	Pt	Au	Hg	Tl	Pb	Bi	Po	At	Rn
77	78	79	80	81	82	83	84	85	86
								Halogens	*Inert gases*

(145)	150.35	151.96	157.25	158.92	162.50	164.93	167.26	168.93	173.04	174.97
Pm	Sm	Eu	Gd	Tb	Dy	Ho	Er	Tm	Yb	Lu
61	62	63	64	65	66	67	68	69	70	71
237.05	(244)	(243)	(247)	(247)	(251)	(254)	(257)	(258)	(259)	(260)
Np	Pu	Am	Cm	Bk	Cf	Es	Fm	Mv	No	Lr
93	94	95	96	97	98	99	100	101	102	103

Elements created in the laboratory

Electrons in a shell have the same quantum number n

Electrons in a subshell have the same quantum numbers n and l

electrons in an atom that share the same principal quantum number n are said to occupy the same *shell*. These electrons average about the same distance from the nucleus and have comparable, though not identical, energies as a rule. Electrons in a given shell that also have the same orbital quantum number l are said to occupy the same *subshell*. In complex atoms the various subshells of the same shell vary in energy because electrons with different angular momenta have different probability density distributions and hence different average distances from the nucleus. In the same shell, the higher the value of l, the higher the energy.

We can account for the periodic table of the elements with the help of the above considerations. Our procedure will be to investigate the status of a new electron added to an existing electronic structure. (Of course, the nuclear charge must also increase by

Helium

$+e$ each time this is done.) In the simplest case we add an electron to a hydrogen atom ($Z = 1$) to give a helium atom ($Z = 2$). Both electrons in helium fall into the same $n = 1$ shell. Since $l = 0$ is the only value l can have when $n = 1$, both electrons have $l = m_l = 0$. The exclusion principle is not violated here since one electron can have the spin magnetic quantum number $m_s = +\frac{1}{2}$ while the other has $m_s = -\frac{1}{2}$. It is customary to describe this situation by saying that the electrons have *opposite spins*, that is, that they behave as though they rotate in opposite directions.

Closed shells and subshells

Because no more than two electrons can occupy the $n = 1$ shell, helium atoms have *closed shells*. The characteristic properties of closed shells and subshells are that the orbital and spin angular momenta of their constituent electrons cancel out independently and that their effective electric charge distributions are perfectly symmetrical. As a result, atoms with closed shells do not tend to interact with other atoms, which we know to be true of helium.

Lithium

Lithium, with $Z = 3$, has one more electron than helium; it is the lightest of the alkali metals. There is no room left in the $n = 1$ shell, and so the additional electron goes into the $l = 0$ subshell of the $n = 2$ shell. The outer electron is relatively far from the nucleus in this shell and is much less tightly bound. The chemical activity of lithium is a consequence of the low binding energy of this electron, which is readily lost to other atoms.

Beryllium

The next element, beryllium, has two electrons of opposite spin in the $l = 0$ subshell of the $n = 2$ shell. The nuclear charge is $+4e$, and so these outer electrons are more tightly held than the single outer electron in lithium; beryllium is accordingly less reactive than lithium.

Boron

Boron, with $Z = 5$, has an electron in the $l = 1$ subshell as well as two in the $l = 0$ one. The $l = 1$ subshell can contain a total of six electrons, corresponding to two electrons of opposite spin in the $m_l = +1$, $m_l = 0$, and $m_l = -1$ states. This subshell is closed (and the $n = 2$ shell is also closed) in neon, whose atomic number is 10. We therefore expect neon to be chemically inert, as indeed it is.

Fluorine

Fluorine, the element just before neon in the periodic table and the lightest of the halogens, lacks but one electron for having a closed outer shell. Just as lithium tends to lose its single outermost electron in interacting with other elements, thereby leaving it with a closed shell configuration, fluorine tends to gain a single electron to close its outer shell. The very different behaviors of the alkali metals, the inert gases, and the halogens thus find explanation in terms of their respective atomic structures. Further analysis of the preceding kind is able to account for all the other relationships revealed in the periodic table.

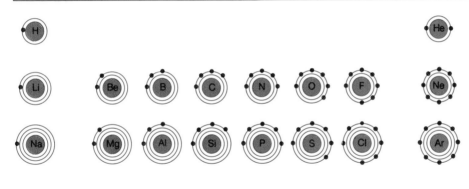

FIG. 29–11 Schematic representation of the electron structures of the elements in the first three periods of the periodic table. The electrons in the filled inner shells of the second and third period elements are not shown.

Figure 29–11 is a highly schematic representation of the electron structures of the elements in the first three periods of the periodic table. A more realistic picture of each atom would show the probability clouds associated with each electron having a particular set of quantum numbers n, l, and m_l (two electrons of opposite spin can have the same probability cloud in an atom).

Table 29–5 shows the electron configurations of the 36 lightest elements. We note that the sequence of electron addition becomes irregular with potassium, which has an electron in its $n = 4$ shell even though its $n = 3$ shell is incomplete. The origin of this apparent anomaly is that electrons in the $n = 4$, $l = 0$ subshell have less energy, and therefore are more tightly bound, than those in the $n = 3$, $l = 2$ subshell, since the probability clouds of the former electrons have concentrations that are closer to the nucleus than those of the latter electrons. The energy difference between the $n = 3$, $l = 2$ and $n = 4$, $l = 0$ subshells is actually quite small, as we can see from the configurations of chromium ($Z = 24$) and copper ($Z = 29$). In both of these elements an additional electron is present in the $n = 3$, $l = 2$ subshell at the expense of the $n = 4$, $l = 0$ subshell, leaving a vacancy in the latter that is filled in the succeeding element.

The ferromagnetic properties of iron, cobalt, and nickel are a consequence of their unfilled $n = 3$, $l = 2$ subshells; without violating the exclusion principle, the electrons in these subshells do *not* pair off, and their spins do not cancel out. In the $n = 3$, $l = 2$ subshell of iron, for instance, five of the six electrons have parallel spins, leaving each iron atom with a strongly magnetic character due to its four unpaired electrons. The reason atomic electrons in the same subshell have parallel spins whenever possible can be traced to their mutual electric repulsion. A consequence of this repulsion is that the more widely separated electrons are, the lower the energy of the system to which they belong (everything else being equal). Electrons in the same subshell which have parallel spins must have different values of m_l and therefore different probability clouds. Hence such electrons are farther apart on the average than paired electrons with the same m_l values, and this arrangement is the more stable one since it represents the lowest possible energy.

Ferromagnetism is due to unpaired electrons

29–8 ATOMIC SIZES

Strictly speaking, an atom of a certain element cannot be said to have a definite size because the probability clouds of its outer electrons do not have a sharp boundary.

TABLE 29–5
Electron configurations of the 36 lightest elements

Atomic number	Symbol	Element	$n = 1$ $l = 0$	$n = 2$ $l = 0, l = 1$		$n = 3$ $l = 0, l = 1, l = 2$			$n = 4$ $l = 0, l = 1, \ldots$	
1	H	Hydrogen	1							
2	He	Helium	2 *Inert gas*							
3	Li	Lithium	2	1 *Alkali metal*						
4	Be	Beryllium	2	2						
5	B	Boron	2	2	1					
6	C	Carbon	2	2	2					
7	N	Nitrogen	2	2	3					
8	O	Oxygen	2	2	4					
9	F	Fluorine	2	2	5 *Halogen*					
10	Ne	Neon	2	2	6 *Inert gas*					
11	Na	Sodium	2	2	6	1 *Alkali metal*				
12	Mg	Magnesium	2	2	6	2				
13	Al	Aluminum	2	2	6	2	1			
14	Si	Silicon	2	2	6	2	2			
15	P	Phosphorus	2	2	6	2	3			
16	S	Sulfur	2	2	6	2	4			
17	Cl	Chlorine	2	2	6	2	5 *Halogen*			
18	Ar	Argon	2	2	6	2	6 *Inert gas*			
19	K	Potassium	2	2	6	2	6		1 *Alkali metal*	
20	Ca	Calcium	2	2	6	2	6		2	
21	Sc	Scandium	2	2	6	2	6	1	2	
22	Ti	Titanium	2	2	6	2	6	2	2	
23	V	Vanadium	2	2	6	2	6	3	2	
24	Cr	Chromium	2	2	6	2	6	5	1	
25	Mn	Manganese	2	2	6	2	6	5	2	
26	Fe	Iron	2	2	6	2	6	6	2	⎫ *Ferro-*
27	Co	Cobalt	2	2	6	2	6	7	2	⎬ *magnetic*
28	Ni	Nickel	2	2	6	2	6	8	2	⎭ *metals*
29	Cu	Copper	2	2	6	2	6	10	1	
30	Zn	Zinc	2	2	6	2	6	10	2	
31	Ga	Gallium	2	2	6	2	6	10	2	1
32	Ge	Germanium	2	2	6	2	6	10	2	2
33	As	Arsenic	2	2	6	2	6	10	2	3
34	Se	Selenium	2	2	6	2	6	10	2	4
35	Br	Bromine	2	2	6	2	6	10	2	5 *Halogen*
36	Kr	Krypton	2	2	6	2	6	10	2	6 *Inert gas*

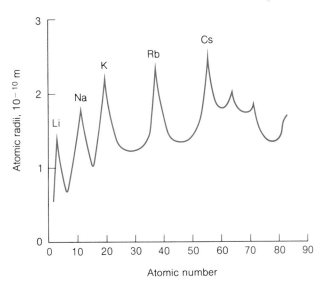

FIG. 29–12 Atomic radii of the elements.

However, from a practical point of view a fairly definite size can usually be attributed to such an atom on the basis of interatomic spacings in molecules and solids that contain that element. Figure 29–12 shows how atomic radii obtained in this way vary with atomic number. The periodicity is quite conspicuous, with the alkali metals of Group I having the largest radii.

An alkali metal atom, such as lithium or sodium, has a single electron in its outer shell, which is shielded by the inner electrons from all but an effective nuclear charge of $+e$ rather than the full $+Ze$. Because the attractive force on the outer electron is relatively weak, the probability cloud of the electron is large in extent. A Group II atom, such as beryllium or magnesium, has two outer electrons, each acted upon by an effective nuclear charge of $+2e$. This double charge produces a greater attractive force on the outer electrons that has the effect of contracting their probability clouds to make the atom smaller than a Group I atom of the same period. A Group III atom, such as boron or aluminum, has three outer electrons, each acted upon by an effective nuclear charge of $+3e$, and is accordingly still smaller. Atomic radii therefore tend to decrease going from left to right across each period. Near the middle of a period a contrary effect becomes significant—namely, the mutual repulsion of the electrons in the outer shell, which results in a gradual increase in size toward the end of the period.

Although atomic radii in general increase in going from period to period as successive shells become occupied, the increase is not very great because the inner electrons, too, have their probability clouds progressively contracted by the higher and higher nuclear charge $+Ze$. Hence the range of atomic sizes is surprisingly small: The heaviest atoms, with over 90 electrons, have radii only about three times that of the hydrogen atom, which has only one electron. Even the cesium atom, the largest of all, has a radius only 4.4 times that of the hydrogen atom.

As we can see, the quantum theory of the atom is the most powerful and fruitful approach yet devised for understanding the properties of matter.

Why sizes vary

The range of atomic sizes is small

IMPORTANT TERMS

In the **quantum theory of the atom** only experimentally measurable quantities are considered, and no use is made of models inconsistent with the uncertainty principle. According to this theory, each electron in an atom is described by its **principal quantum number** n, which governs its energy, its **orbital quantum number** l, which governs the magnitude of its angular momentum, and its **magnetic quantum number** m_l, which governs the orientation of its angular momentum. To each set of quantum numbers there is a corresponding **probability cloud** that governs the likelihood the electron thus described will be found in any particular location in an atom.

Every electron has a certain intrinsic amount of angular momentum called its **spin**. The **spin magnetic quantum number** m_s of an atomic electron has two possible values, $+\frac{1}{2}$ and $-\frac{1}{2}$. Owing to its spin, every electron acts like a tiny bar magnet.

According to the Pauli **exclusion principle,** no two electrons in an atom can exist in the same quantum state.

The **periodic law** of chemistry states that if we list the elements in order of atomic number, elements with similar properties recur at regular intervals. The quantum theory of the atom together with the exclusion principle is able to explain the origin of the periodic law.

The electrons in an atom that have the same total quantum number n are said to occupy the same **shell.** Electrons in a given shell which have the same orbital quantum number l are said to occupy the same **subshell.** Shells and subshells containing the maximum number of electrons permitted by the exclusion principle are **closed.** Atoms whose subshells are all closed possess unusual stability.

MULTIPLE CHOICE

1. The quantum theory of the atom
   (a) is based upon the Bohr theory.
   (b) is more comprehensive but less accurate than the Bohr theory.
   (c) cannot be reconciled with Newton's laws of motion.
   (d) is not based upon a mechanical model and considers only observable quantities.

2. The energy of a quantum state of an atom depends chiefly upon its
   (a) principal quantum number n.
   (b) orbital quantum number l.
   (c) magnetic quantum number m_l.
   (d) spin magnetic quantum number m_s.

3. The angular momentum of an atomic electron is
   (a) not quantized.
   (b) quantized in magnitude only.
   (c) quantized in direction only.
   (d) quantized in both magnitude and direction.

4. The splitting of the spectral lines of an element when it radiates in a magnetic field is known as
   (a) the Schrödinger effect.
   (b) the Zeeman effect.
   (c) induced emission.
   (d) population inversion.

5. A large value of the probability density ψ^2 of an atomic electron at a certain place and time signifies that the electron
   (a) is likely to be found there.
   (b) is certain to be found there.
   (c) has a great deal of energy there.
   (d) has a great deal of charge there.

6. The probability cloud of an atomic electron is spherically symmetric when
   (a) $n = 0$.   (b) $l = 0$.
   (c) $m_i = 0$.   (d) $m_s = 0$.

7. The natural unit of angular momentum in atomic physics is the quantity
   (a) $2h$.   (b) $2\pi h$.
   (c) $h/2$.   (d) $h/2\pi$.

8. The number of possible orientations of an atomic electron in a magnetic field is
   (a) 1.   (b) 2.
   (c) 4.   (d) 2π.

9. Pauli's exclusion principle states that no two electrons in an atom can
   (a) be present in the same probability cloud.
   (b) have the same spin.
   (c) have the same quantum numbers.
   (d) interact with each other.

10. The two electrons in a helium atom
   (a) occupy different shells.
   (b) occupy different subshells.
   (c) have opposite spins.
   (d) have parallel spins.

11. Each vertical column of the periodic table contains elements whose atoms have
   (a) similar chemical properties.
   (b) different chemical properties.
   (c) the same number of electrons.
   (d) the same atomic mass.

12. Each horizontal row of the periodic table contains elements whose atoms have
(a) similar chemical properties.
(b) different chemical properties.
(c) the same number of electrons.
(d) the same atomic mass.

13. In the periodic table, elements are listed in order of
(a) atomic number.   (b) atomic mass.
(c) atomic size.   (d) density.

14. At room temperature and atmospheric pressure, most elements are
(a) metals.   (b) nonmetals.
(c) gases.   (d) liquids.

15. An alkali metal atom
(a) has a closed outer shell.
(b) is one electron short of having a closed outer shell.
(c) has one electron in its outer shell.
(d) has two electrons in its outer shell.

16. A halogen atom
(a) has a closed outer shell.
(b) is one electron short of having a closed outer shell.
(c) has one electron in its outer shell.
(d) has two electrons in its outer shell.

17. Atoms of the inert gases
(a) form diatomic molecules.
(b) combine chemically only with the alkali metals and halogens.
(c) exhibit chemical behavior like that of hydrogen.
(d) are almost never found in molecules.

EXERCISES

1. The quantum theory of the atom is consistent with what fundamental principle that is violated by the Bohr theory?

2. What quantity is governed by each of the quantum numbers n, l, m_l, and m_s?

3. What are the possible values of the magnetic quantum number of an atomic electron with the orbital quantum number $l = 4$?

4. What are the possible values for the orbital and magnetic quantum numbers of an atomic electron with the principal quantum number $n = 3$?

5. Why is the probability cloud for an atomic electron with no orbital angular momentum the same in all directions?

6. Which quantum number is not involved in describing the probability cloud of an atomic electron?

7. According to the Bohr model, the radius of an atomic electron's orbit is proportional to n^2. Is there any aspect of the probability-cloud model that might correspond to this relationship?

8. Under what circumstances can two electrons share the same probability cloud in an atom?

9. Under what circumstances do electrons exhibit spin? Do electrons in inner shells spin faster than those in outer shells?

10. What significance does electron spin have in atomic structure?

11. How many elements would there be if atoms with occupied electron shells up through $n = 6$ could exist?

12. What is the minimum orbital angular momentum of an atomic electron according to the Bohr theory? According to the quantum theory?

13. How many electrons do the elements in group II of the periodic table have in their outermost shells?

14. How can you account for the fact that fluorine and chlorine exhibit similar chemical behavior?

15. How can you account for the fact that lithium and sodium exhibit similar chemical behavior?

ANSWERS TO MULTIPLE CHOICE

1. d	**7.** d	**13.** a
2. a	**8.** b	**14.** a
3. d	**9.** c	**15.** c
4. b	**10.** c	**16.** b
5. a	**11.** a	**17.** d
6. b	**12.** b	

ATOMS IN COMBINATION **30**

Individual atoms are rare on the earth and in its lower atmosphere; only the inert gas atoms occur by themselves. All other atoms are found linked in small groups called molecules or in larger ones as liquids or solids. Some of these groups consist exclusively of atoms of the same element, others consist of atoms of different elements, but in every case the arrangement is favored over separate atoms by virtue of interactions between the atoms that reduce the total energy of the system. The quantum theory of the atom is able to account for these interactions in a natural way, with no special assumptions, which is further testimony to the power of this approach.

30–1   THE HYDROGEN MOLECULE

A molecule is a group of atoms that stick together strongly enough to act as a single particle. A molecule always has a certain definite structure; hydrogen

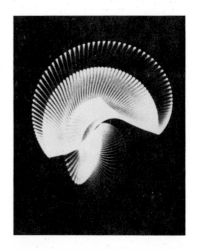

molecules always consist of two hydrogen atoms each, for instance, and water molecules always consist of one oxygen atom and two hydrogen atoms each. A piece of iron is not a molecule because, even though its atoms stay together, any number of them do so to form an object of any size or shape.

What is a molecule?

A molecule of a given kind is complete in itself with little tendency to gain or lose atoms. If one of its atoms is somehow removed or another atom is somehow attached, the result is a molecule of a different kind with different properties. A liquid or a solid, on the other hand, can gain or lose additional atoms of the kinds already present without changing its character.

The chief mechanism that bonds atoms together to form molecules involves the sharing of electrons by the atoms involved. As the shared electrons circulate around the atoms, they are more often between the atoms than they are on the outside, which produces an attractive force. To see why this force should occur, we may think of the atoms that are sharing the electrons as positive ions, so the presence of the shared electrons between them means a negative charge that holds the positive ions together. This is especially easy to see in the case of the hydrogen molecule, H_2, which consists of two hydrogen atoms.

The atoms in a molecule are held together by electron sharing

In the hydrogen molecule, the two protons are 7.42×10^{-11} m apart and the two electrons, one contributed by each atom, belong to the entire molecule rather than to their parent nuclei. Figure 30–1(a) is a picture of the H_2 molecule in terms of electron orbits. A more realistic picture of this molecule is provided by the quantum theory, in which the notion of electrons with definite, predictable positions and speeds at every moment is replaced by the notion of probability clouds. Figure 30–1(b) shows how the electron probability clouds of two hydrogen atoms join to form the probability cloud of a hydrogen molecule.

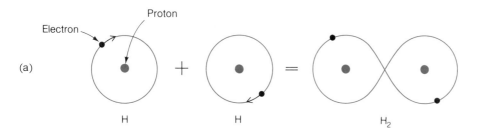

(a)

FIG. 30–1 (a) Orbit model of the hydrogen molecule. (b) Probability cloud model of the hydrogen molecule.

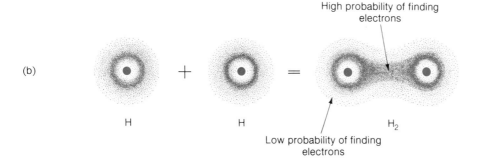

(b)

Origin of bonding force in H_2

Because the electrons spend more time on the average between the protons than they do on the outside, there is effectively a net negative charge between the protons. The attractive force this charge exerts on the protons is more than enough to counterbalance the direct repulsion between them. If the protons are too close together, however, their repulsion becomes dominant and the molecule is not stable. The balance between attractive and repulsive forces occurs at a separation of 7.42×10^{-11} m, where the total energy of the H_2 molecule is -4.5 eV. Hence 4.5 eV of work must be done to break a H_2 molecule into two H atoms:

$$H_2 + 4.5 \text{ eV} \rightarrow H + H$$

By comparison, the binding energy of the hydrogen atom is 13.6 eV:

$$H + 13.6 \text{ eV} \rightarrow p^+ + e^-$$

Atoms are more stable than molecules

This is an example of the general rule that it is easier to break up a molecule than to break up an atom.

An argument based on the uncertainty principle makes it possible, even without a detailed calculation, to understand why a hydrogen molecule should have less energy (and hence more stability) than two separate hydrogen atoms. According to the uncertainty principle, the smaller the uncertainty in a particle's position, the greater the uncertainty in its momentum. Therefore the smaller the region of space in which an electron is confined, the greater its momentum and hence energy must be. An electron shared by two protons has more room in which to move about than an electron bound to a single proton to form a hydrogen atom, so its energy is less.

Why do only two hydrogen atoms join together to form a molecule? Why not three, or four, or a hundred?

Molecular size and the exclusion principle

The basic reason for the limited size of molecules is the exclusion principle, which prohibits more than one electron in an atomic system from having the same set of quantum numbers. If a third H atom were to be brought up to an H_2 molecule, its electron would have to leave the $n = 1$ shell and go to the $n = 2$ shell in order for an H_3 molecule to be formed, since only two electrons can occupy the $n = 1$ shell. But an $n = 2$ electron in the hydrogen atom has over 10 eV more energy than an $n = 1$ electron, and the binding energy in H_2 is only 4.5 eV. Hence an H_3 molecule, if it could somehow be put together, would immediately break apart into H_2 + H with the release of energy. Similar considerations hold for other molecules.

The ability of atoms to join with only a limited number of others to form molecules is referred to as *saturation*.

Why the inert gases do not form molecules

The exclusion principle is also responsible for the fact that the inert gases—helium, neon, argon, and so on—do not occur as molecules. The electron shells of the inert gas atoms are all filled to capacity, so in order for two of them to share an electron pair, one of the electrons would have to go into an empty shell of higher energy. The increase in energy involved would be more than the energy decrease produced by sharing the electrons, and therefore no such molecules as He_2 or Ne_2 occur.

30–2 COVALENT BOND

The mechanism by which electron sharing holds atoms together to form molecules is known as *covalent bonding*. Often it is convenient to think of the atoms as being held together by *covalent bonds*, with each shared pair of electrons constituting a bond.

A shared pair of electrons constitutes a covalent bond

More complex atoms than hydrogen also join together to form molecules by sharing electrons. Depending upon the electron structures of the atoms involved, there may be one, two, or three covalent bonds—shared electron pairs—between the atoms. For example, in the oxygen molecule O_2 there are two bonds between the O atoms, and in the nitrogen molecule N_2 there are three bonds between the N atoms. Covalent bonds are represented either by a pair of dots or a single dash for each shared pair of electrons. Thus the H_2, O_2, and N_2 molecules can be represented as follows:

More than one pair of electrons may be shared

$$
\begin{array}{lll}
\text{H:H} & \text{or} & \text{H} - \text{H} \\
\text{O::O} & \text{or} & \text{O} = \text{O} \\
\text{N:::N} & \text{or} & \text{N} \equiv \text{N}
\end{array}
$$

Covalent bonds are not limited to atoms of the same element nor to only two atoms per molecule. Here are two examples of more complicated molecules, with a line representing each covalent bond:

$$
\begin{array}{ll}
\text{Water, } H_2O &
\begin{array}{c}
\text{H} \\
| \\
\text{O} - \text{H}
\end{array}
\qquad
\text{Ammonia, } NH_3 \qquad
\begin{array}{c}
\text{H} \\
| \\
\text{N} - \text{H} \\
| \\
\text{H}
\end{array}
\end{array}
$$

We notice that oxygen participates in two bonds in H_2O and nitrogen in three bonds in NH_3, whereas H participates in only one bond in both cases. This behavior is consistent with the fact that the hydrogen, oxygen, and nitrogen molecules respectively have one, two, and three bonds between their atoms.

Carbon atoms tend to form four covalent bonds at the same time, since they have four electrons in their outer shells and these shells lack four electrons to be complete. Various distributions of these bonds are possible, including bonds between adjacent carbon atoms in a complex molecule. The structures of the common covalent molecules methane, carbon dioxide, and acetylene illustrate the different bonds in which carbon atoms can participate to form molecules.

A carbon atom participates in four covalent bonds

$$
\begin{array}{ccc}
\begin{array}{c}
\text{H} \\
| \\
\text{H} - \text{C} - \text{H} \\
| \\
\text{H}
\end{array}
& \qquad \text{O} = \text{C} = \text{O} \qquad &
\text{H} - \text{C} \equiv \text{C} - \text{H} \\
\\
\text{methane} & \text{carbon dioxide} & \text{acetylene}
\end{array}
$$

Carbon atoms are so versatile in forming covalent bonds with each other as well as with other atoms that literally millions of carbon compounds are known, some whose molecules contain tens of thousands of atoms. Such compounds were once thought to originate only in living things, and their study is accordingly known even today as organic chemistry.

Organic chemistry is the study of carbon compounds

When atoms join together to form a molecule, their inner, complete electron shells undergo little change. In molecules whose atoms are the same, such as H_2, O_2, and

Polar and nonpolar molecules

N_2, the atoms share their outer electrons evenly, and on the average have uniform distributions of electric charge.

In a molecule composed of different atoms, the shared electrons favor one or another of the atoms, depending upon which elements are involved. The resulting molecule has a nonuniform distribution of electric charge, with some parts of the molecule being positively charged and others parts being negatively charged. A molecule of this kind, as we learned earlier, is called a *polar molecule*, whereas a molecule whose charge distribution is symmetric is called a *nonpolar molecule*. Thus the covalent molecule HCl is polar because a chlorine atom has greater attraction for an electron than a hydrogen atom, a situation we can represent by placing the pair of dots that indicate a covalent bond closer to the Cl atom:

Hydrochloric acid molecule   H   :Cl

Water molecules are highly polar

The water molecule is highly polar because the two O—H bonds are 104.5° apart (not 180° apart), and the two shared electron pairs spend more time near the O atom than near the H atoms:

Water molecule   H   :O
.．
H

30–3   SOLIDS AND LIQUIDS

The same covalent bonds that can tie several atoms together into a molecule can in certain circumstances also tie an unlimited number of them together into a solid or a liquid. Other bonding mechanisms are found in solids and liquids as well, one of which is responsible for the ability of metals to conduct electric current. Although only a minute proportion of the universe is in the solid state and even less is in the liquid state, matter in these forms constitutes much of the physical world of our experience, and modern technology is to a large extent based upon the exploitation of the unique characteristics of various solid materials.

Crystals have long-range order

Most solids are crystalline in nature, with their constitutent atoms or molecules arranged in regular, repeated patterns. A crystal is thus characterized by the presence of *long-range order* in its structure.

Amorphous solids have short-range order only

Other solids lack the definite arrangements of atoms and molecules so conspicuous in crystals. They can be thought of as liquids whose stiffness is due to an exaggerated viscosity. Examples of such *amorphous* ("without form") solids are pitch, glass, and many plastics. The structures of amorphous solids exhibit *short-range order* only.

Some substances, for instance B_2O_3, can exist in either crystalline or amorphous forms. In both cases each boron atom is surrounded by three larger oxygen atoms, which represents a short-range order. In a B_2O_3 crystal a long-range order is also present, as shown in a two-dimensional representation in Fig. 30–2, whereas amorphous B_2O_3, a glassy material, lacks this additional regularity.

The lack of long-range order in amorphous solids means that the various bonds vary in strength. When an amorphous solid is heated, the weakest bonds are ruptured

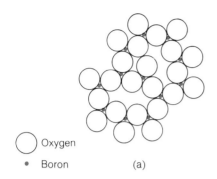

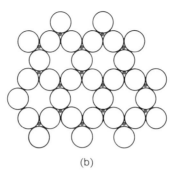

FIG. 30–2 Structure of B_2O_3. (a) Amorphous B_2O_3, a glass, has short-range order only in its structure. (b) Crystalline B_2O_3 has long-range order as well.

○ Oxygen

• Boron   (a)   (b)

at lower temperatures than the others and the solid softens gradually, whereas in a crystalline solid the bonds break simultaneously and melting is a sudden process.

Liquids have more in common with solids than with gases, even though liquids **Nature of the liquid** share with gases the ability to flow from place to place. Because the density of a given **state** liquid is usually approximately the same as that of the same substance in solid form, we infer that the bonding mechanism is similar in both cases. When a solid is heated to its melting point, its atoms or molecules acquire enough energy to shift the bonds holding them together so that they form into separate clusters, but not until the liquid is heated to its vaporization point are the atoms or molecules able to break loose completely and form a gas. This interpretation of the liquid state is confirmed by X-ray studies that reveal definite clusters of atoms or molecules in a liquid (that is, short-range order like that in amorphous solids), but the clusters are constantly shifting their arrangements unlike the permanent arrangements in a solid.

The unusual behavior of water near the freezing point that was mentioned in Sec. **Why ice floats** 13–2 can be traced to the above effect. Ice crystals have very open structures because each H_2O molecule can participate in only 4 bonds with other H_2O molecules; in other solids, each atom or molecule may have as many as 12 nearest neighbors, which allows the assemblies to be more compact. Because clusters of molecules are smaller and less stable in the liquid state, water molecules are on the average packed more closely together than are ice molecules, and water has the higher density: hence ice floats. The density of water increases from 0°C to a maximum at 4°C as large clusters of H_2O molecules break up into smaller ones that occupy less space in the aggregate, and only above 4°C does the normal thermal expansion of a liquid show up as a decreasing density with increasing temperature.

Few crystals are perfect in structure, with completely regular arrangements of atoms. The defects in the structure of a crystal—atoms in the "wrong" place, missing atoms, extra atoms, impurity atoms, and so on—are extremely important in determining its mechanical strength.

A common type of crystal defect is a missing line of atoms. Such a defect is called **A dislocation is a** a *dislocation*. The existence of dislocations makes it possible to understand why a solid **missing line of atoms** can be deformed permanently by bending, squeezing, or stretching without breaking. We might think that such plastic deformations occur when a force is applied to a solid by the sliding of layers of atoms over one another, but calculations show that such sliding—which involves the breaking of millions of bonds between atoms simultane-

FIG. 30–3 The motion of a dislocation in crystal under stress results in a permanent deformation.

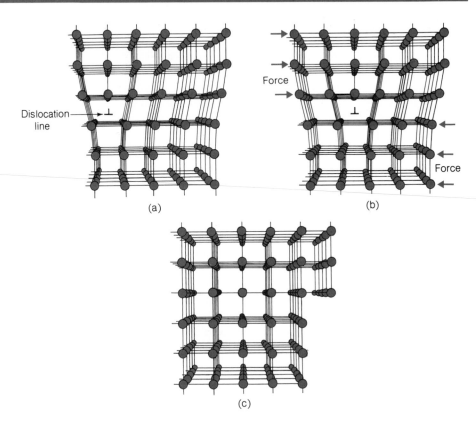

(a)

(b)

(c)

ously—would require forces about a thousand times stronger than those actually able to produce the deformations experimentally.

How dislocations contribute to plastic deformation

Figure 30–3(a) shows a crystal with a missing row of atoms. When equal and opposite forces are applied to the crystal along different lines of action, as in Fig. 30–3(b), the dislocation shifts to the right as the atoms in the layer containing the missing row shift their bonds, one row at a time, with the atoms in the layer above. In Fig. 30–3(c) the dislocation has reached the end of the crystal, which is now permanently deformed. Because the bonds were shifted one row at a time, instead of all at once, much less force was needed than would have been required to deform a perfect solid.

Hardening a metal

To increase the hardness of a solid, it is necessary to hinder the motion of dislocations in its structure. There are two chief ways to do this in a metal. One is to increase the number of dislocations by hammering the metal or squeezing it between rollers. The dislocations then become so numerous and tangled together that they interfere with each other's motion. This effect is called *work hardening*. Another approach is to add foreign atoms to the metal that act as roadblocks to the progress of dislocations. Thus the addition of small amounts of carbon, chromium, manganese, and other elements to iron turns it into the much stronger steel by impeding the ability of dislocations to move through the material.

30-4 COVALENT AND IONIC SOLIDS

Molecular bonds are all essentially covalent in character, even though in a few highly polar molecules the shared electrons stay so close to one of the partner atoms that there is effectively a transfer rather than a sharing of them. In solids there are four bonding mechanisms that occur, depending upon the nature of the solid: covalent, ionic, van der Waals, and metallic.

Four bonding mechanisms in solids

The covalent bonds between atoms that are responsible for the formation of molecules also act to hold certain crystalline solids together. Figure 30-4 shows the array of carbon atoms in a diamond crystal, with each carbon atom sharing electron pairs with the four other carbon atoms adjacent to it. All the electrons in the outer shells of the carbon atoms participate in the bonding, and it is therefore not surprising that diamond is extremely hard and must be heated to over 3500°C before its crystal structure is disrupted and it melts. Purely covalent crystals are relatively few in number. In addition to diamond, some examples are silicon, germanium, and silicon carbide ("Carborundum"). Like diamond, all are hard and have high melting points.

Covalent bonding in solids

In covalent bonding, two atoms share one or more pairs of electrons. In *ionic bonding,* one or more electrons from one atom transfer to another atom, producing a positive ion and a negative ion that attract each other. Let us see how ionic bonding functions in the case of NaCl, whose crystals constitute ordinary table salt. We shall consider a hypothetical NaCl "molecule" for convenience; salt crystals are aggregates of Na and Cl atoms which, although they do not pair off into individual molecules, do interact through ionic bonds whose nature can be most easily examined in terms of a single Na-Cl unit.

Ionic bonding

Figure 30-5 shows schematically an electron shifted from a sodium atom, which becomes an Na^+ ion, to a chlorine atom, which becomes a Cl^- ion. The ions then attract each other electrically. The combination is stable because more energy is needed to pull the Na^+ and Cl^- ions apart than will be supplied by the return of the shifted electron from the Cl^- ion to the Na^+ ion. Not all pairs of atoms can interact by electron transfer to form stable combinations; let us see why Na and Cl have this ability.

The single outermost electron of the Na atom is relatively easy to detach because of the presence of the ten inner electrons, which shield the outer electron from all but

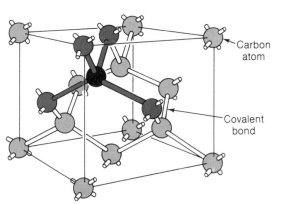

FIG. 30-4 Covalent structure of diamond. Each carbon atom shares electron pairs with four other carbon atoms.

Carbon atom

Covalent bond

FIG. 30–5 Electron transfer in NaCl. The resulting ions attract each other electrically.

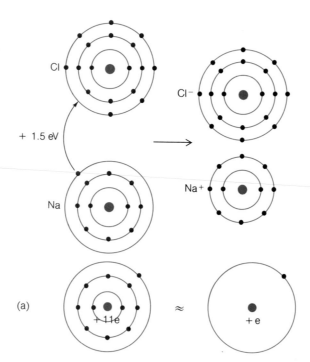

FIG. 30–6 (a) A sodium atom. The presence of 10 electrons in the $n = 1$ and $n = 2$ shells effectively shields the outer $n = 3$ electron from all but $+e$ of the nuclear charge. In consequence an Na atom tends to lose its outer electron to become an Na^+ atom. (b) A chlorine atom. The inner electrons leave unshielded $+7e$ of the nuclear charge. In consequence a Cl atom tends to pick up another electron to become a Cl^- ion.

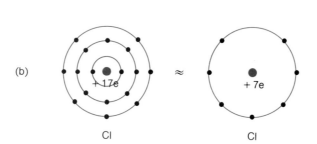

Why an Na atom tends to lose an electron and a Cl atom tends to gain an electron

$+e$ of the actual nuclear charge of $+11e$ (Fig. 30–6(a)). In the Cl atom, on the other hand, the ten inner electrons leave $+7e$ of the nuclear charge unshielded, so the attractive force on the outer electrons is much greater (Fig. 30–6(b)). Accordingly, a Cl atom tends to pick up an additional electron to become a Cl^- ion. Thus Na and Cl are an ideal match: one readily loses an electron, the other readily gains an electron. The same is true for the partners in other ionic solids, which owe these tendencies to their electron structures. In general, a compound of an element from Group I or II of the periodic table with one from Group VI or VII exhibits ionic bonding in the solid state.

What keeps ions apart

If the Na^+ and Cl^- ions were to come very close to each other, their electron structures would mesh together (Fig. 30–7). Two different effects prevent this from occurring. First, when such meshing takes place, the Na^+ and Cl^- nuclei cease being

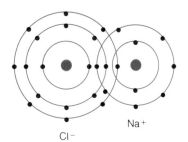

FIG. 30–7 Strong repulsive forces whose origin is associated with the exclusion principle prevent atoms from coming so close together that their electron structures mesh.

shielded by their electrons and repel electrically. Second but more important, such meshing means that the inner electrons of the Na^+ and Cl^- ions no longer constitute separate atomic systems but instead constitute a single one. According to the Pauli exclusion principle, no two electrons in the same system can be in the same quantum state, and so some of the electrons must go into unoccupied states of higher energy. The result of both effects is that work must be done to push the ions together to make their electron structures overlap. The force between the ions is now repulsive.

The structure of an NaCl crystal is shown in Fig. 30–8. Each ion behaves essentially like a point charge and thus tends to attract to itself as many ions of opposite sign as can fit around it. In a NaCl crystal each Na^+ ion is surrounded by six Cl^- ions and vice versa. In crystals having different structures the number of "nearest neighbors" around each ion may be 3, 4, 6, 8, or 12. Ionic bonds are usually fairly strong, and consequently ionic crystals are strong, hard, and have high melting points.

Many crystalline bonds are partly ionic and partly covalent in origin. An example of such mixed bonding is quartz, SiO_2.

30–5   VAN DER WAALS BONDS

A number of molecules and nonmetallic atoms exist whose electronic structures do not lend themselves to either of the above kinds of binding. The inert gas atoms, which

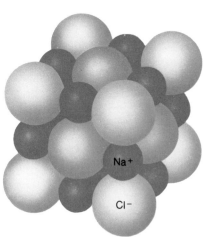

FIG. 30–8 A NaCl crystal is composed of Na^+ and Cl^- ions in an array such that each ion is surrounded by six ions of the other kind. Na^+ ions are small because they have lost their outer electrons.

FIG. 30–9 Polar molecules attract each other electrically.

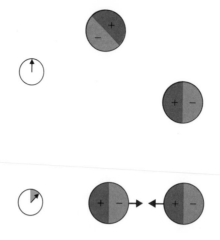

have filled outer electron shells, and molecules such as methane,

$$H - \underset{\displaystyle \overset{\displaystyle H}{|}}{\underset{\displaystyle \underset{\displaystyle H}{|}}{C}} - H$$

whose valence electrons are fully involved in the molecular bond itself, fall into this category. However, even these virtually noninteracting substances condense into solids and liquids at low enough temperatures through the action of what are known collectively as *van der Waals forces*.

Polar-polar attraction

The electric attraction between polar molecules that was discussed in Sec. 16–7 is an example of a van der Waals force. These molecules have asymmetric distributions of charge, with one end positive and the other negative. When one such molecule is near another, the ends of opposite polarity attract each other to hold the molecules together (Fig. 30–9).

Polar-nonpolar attraction

A somewhat similar effect occurs when a polar molecule is near a nonpolar one. The electric field of the polar molecule distorts the initially symmetric charge distri-

FIG. 30–10 A polar molecule attracts a nonpolar one by first distorting the latter's originally symmetric charge distribution.

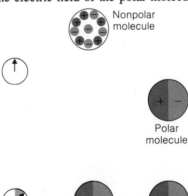

Nonpolar molecule

Polar molecule

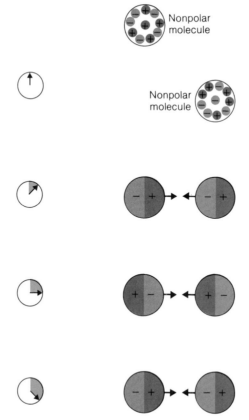

FIG. 30–11 Nonpolar molecules have, on the average, symmetric charge distributions, but at any instant the distributions are not necessarily symmetric. The fluctuations in the charge distributions of adjacent nonpolar molecules keep in step, which leads to an attractive force between them.

bution of the nonpolar one, as in Fig. 30–10, and the two then attract each other in the same way as any other pair of polar molecules. This phenomenon is similar to that involved in the attraction of bits of paper by a charged rubber comb shown in Fig. 16–8.

Nonpolar molecules attract one another in much the same way that polar molecules attract nonpolar ones. In a nonpolar molecule, the electrons are distributed symmetrically *on the average,* but *at any moment* one part of the molecule contains more electrons than usual and the rest of the molecule contains fewer. Thus *every* molecule (and atom) behaves as though it is polar, though the direction and magnitude of the polarization vary constantly. The fluctuations in the charge distributions of nearby nonpolar molecules keep in step through the action of electric forces, and these forces also hold the molecules together (Fig. 30–11).

Nonpolar-nonpolar attraction

In general, van der Waals bonds are considerably weaker than ionic, covalent, and metallic bonds; usually less than 1% as much energy is needed to remove an atom or molecule from a van der Waals solid as is required in the case of ionic or covalent crystals. As a result the inert gases and compounds with symmetric molecules liquify and vaporize at rather low temperatures. Thus the boiling point of argon is $-186°C$ and the boiling point of methane is $-161°C$. Molecular crystals, whose lattices consist of individual molecules held together by van der Waals forces, generally lack the mechanical strength of other kinds of crystals.

Van der Waals bonds are weak

30–6   METALLIC BOND

An electron "gas" bonds metals

A fourth important type of cohesive force in crystalline solids is the *metallic bond,* which has no molecular counterpart.

A characteristic property of all metal atoms is the presence of only a few electrons in their outer shells, and these electrons can be detached relatively easily to leave behind positive ions. According to the theory of the metallic bond, a metal in the solid state consists of an assembly of atoms that have given up their outermost electrons to a common "gas" of freely moving electrons that pervades the entire metal. The electric interaction between the positive ions and the negative electron gas holds the metal together.

This theory has much in its favor. The high electrical and thermal conductivity of metals follows from the ability of the free electrons to migrate through their crystal structures, while all the electrons in ionic and covalent crystals are bound to particular atoms or pairs of atoms. Also, since the atoms in a metal interact through the medium of a common electron gas, the properties of mixtures of different metal atoms should not depend critically on the relative proportions of each kind of atom, provided that their sizes are similar. This prediction is fulfilled in the observed behavior of alloys, in contrast to the specific atomic proportions characteristic of ionic and covalent solids.

Table 30–1 summarizes the properties of the four types of crystalline solids that have been discussed here.

TABLE 30–1   Types of crystalline solids

Type	Covalent	Ionic	Molecular	Metallic
Lattice	Shared electrons	Negative ion Positive ion	Instantaneous charge separation in molecule	Metal ion Electron gas
Bond	Shared electrons	Electric attraction	Van der Waals forces	Electron gas
Properties	Very hard; high melting point; soluble in very few liquids	Hard; high melting point; may be soluble in polar liquid such as water	Soft; low melting and boiling points; soluble in covalent liquids	Ductile; metallic luster; ability to conduct heat and electric current readily

30–7 ENERGY BANDS

The notion of energy bands provides a useful framework for understanding the electrical behavior of solids. The nature and properties of semiconductors in particular are clarified with the help of an energy-band analysis.

    When atoms are brought as close together as those in a crystal, they interact with one another to such an extent that their outer electron shells constitute a single system of electrons common to the entire array of atoms. The Pauli exclusion principle prohibits more than two electrons (one with each spin) in any energy level of a system. This principle is obeyed in a crystal because the energy levels of the outer electron shells of the various atoms are all slightly altered by their mutual interaction. (The inner shells do not interact and therefore do not undergo a change.) As a result of the shifts in the energy levels, an *energy band* exists in a crystal in place of each sharply defined energy level of its component atoms (Fig. 30–12). While these bands are actually composed of a multitude of individual energy levels, as many as there are atoms in the crystal, the levels are so near one another as to form a continuous distribution.

    The energy bands in a crystal correspond to energy levels in an atom, and an electron in a crystal can only have an energy that falls within one of these bands. The various energy bands in a crystal may or may not overlap, depending upon the composition of the crystal (Fig. 30–13). If they do not overlap, the gaps between them

Origin of energy bands in a crystal

A forbidden band may separate energy bands

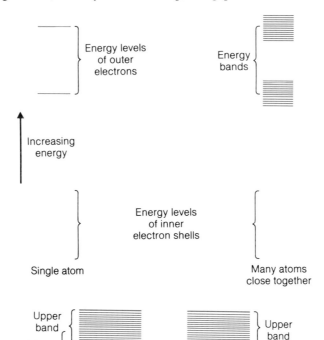

FIG. 30–12 Energy bands replace energy levels of outer electrons in an assembly of atoms that are close together.

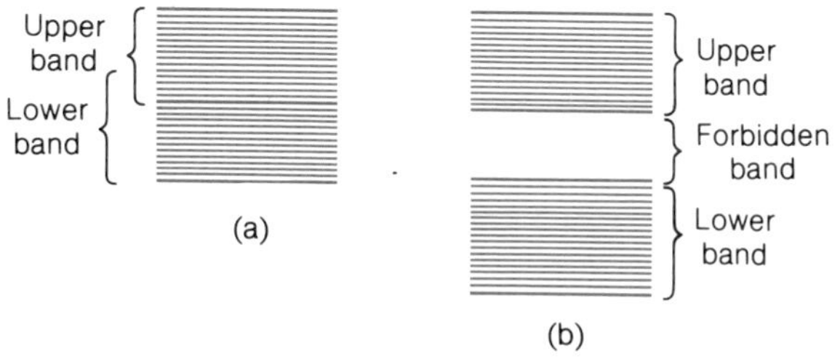

FIG. 30–13 (a) Overlapping energy bands. (b) The gap between energy bands that do not overlap is called a forbidden band.

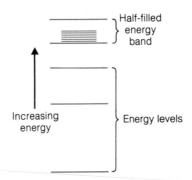

FIG. 30–14 Energy levels and bands in solid sodium, which is a metal. (Not to scale.)

represent energy values which electrons in the crystal cannot have. The gaps are accordingly known as *forbidden bands*.

The energy bands we have been speaking of contain all the possible energies that can be possessed by electrons. The electrical properties of a crystalline solid depend upon both its energy-band structure and the way in which the bands are normally occupied by electrons. We shall examine a few specific cases to see how the energy-band approach accounts for the observed electrical behavior of such solids.

Energy bands of a metal

Figure 30–14 shows the energy bands of solid sodium. A sodium atom has only one electron in its outer shell. This means that the upper energy band in a sodium crystal is only half filled with electrons, since each level within the band, like each level in the atom, is capable of containing *two* electrons. When an electric field is established in a sodium crystal, electrons readily acquire the small additional energy they need to move up in their energy band. The additional energy is in the form of kinetic energy, and the moving electrons constitute an electric current. Sodium is therefore a good conductor, as are other crystalline solids with energy bands that are only partially filled. Such solids are metals.

Energy bands of an insulator

Figure 30–15 shows the energy-band structure of diamond. The two lower energy bands are completely filled, and there is a gap of 6 eV between the top of the higher of these bands and the empty band above it. Hence a minimum of 6 eV of additional energy must be given to an electron in a diamond crystal if it is to be capable of free motion, since it cannot have an energy lying in the forbidden band. Such an energy increment cannot readily be imparted to an electron in a crystal by an electric field.

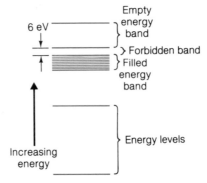

FIG. 30–15 Energy levels and bands in diamond, which is an electrical insulator. (Not to scale.)

Diamond, like other solids with similar energy-band structures, is therefore an electrical insulator.

Silicon has a crystal structure similar to that of diamond, and, as in diamond, a gap separates the top of a filled energy band from an empty higher band. However, whereas the gap is 6 eV wide in diamond, it is only 1.1 eV wide in silicon. At very low temperatures silicon is hardly better than diamond as a conductor, but at room temperature a small proportion of its electrons can possess enough kinetic energy of thermal origin to exist in the higher band. These few electrons are sufficient to permit a limited amount of current to flow when an electric field is applied. Thus silicon has a resistivity intermediate between those of conductors (such as sodium) and those of insulators (such as diamond), and is classified as a *semiconductor*.

Energy bands of a semiconductor

The optical properties of solids and their energy-level structures are closely related. Photons of visible light have energies from about 1 to 3 eV. A free electron in a metal can readily absorb such a photon since its allowed energy band is only partly filled, and metals are accordingly opaque. The characteristic luster of a metal is due to the reradiation of light absorbed by its free electrons. If the metal surface is smooth, the reradiated light appears as a reflection of the original incident light.

Optical properties of solids

For an electron in an insulator to absorb a photon, on the other hand, the photon energy must be more than 3 eV in order for the electron to jump across the forbidden band to the next allowed band. Insulators are therefore unable to absorb photons of visible light and are transparent. To be sure, most samples of insulating materials do not appear transparent, but this is due to the scattering of light by irregularities in their structures. Insulators are opaque to ultraviolet light, whose higher frequencies mean high enough photon energies to enable electrons to cross the forbidden band. Because the forbidden bands in semiconductors are comparable in width to the photon energies of visible light, they are usually opaque to visible light but are transparent to infrared light whose lower frequencies mean photon energies too low to be absorbed. Thus infrared lenses can be made from the semiconductor germanium, whose appearance is that of a solid metal.

30–8 IMPURITY SEMICONDUCTORS

The conductivity of semiconductors is markedly affected by slight amounts of impurity. Suppose that we add several arsenic atoms to a silicon crystal. Arsenic atoms have five electrons in their outermost shells, while silicon atoms have four. When an arsenic atom replaces a silicon atom in a silicon crystal, four of its electrons participate in covalent bonds with its nearest neighbors. The remaining electron needs very little energy to be detached and move about freely in the crystal. Such a solid is called an *n-type* semiconductor because electric current in it is carried by the motion of negative charges toward the positive end of a sample of. it (Fig. 30–16). In an energy-band diagram, as in Fig. 30–17, the effect of arsenic as an impurity in silicon is to supply occupied energy levels just below an empty energy band. These levels are called *donor levels*.

Electrons carry current in an *n*-type semiconductor

If we add gallium atoms to a silicon crystal, a different effect occurs. Gallium atoms have only three electrons in their outer shells, and their presence leaves vacancies

FIG. 30–16 Current in an *n*-type semiconductor is carried by excess electrons that do not fit into the electron-bond structure of the crystal.

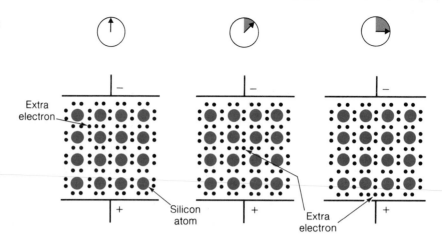

Extra electron

Silicon atom

Extra electron

FIG. 30–17 Donor levels due to arsenic atoms in a silicon crystal.

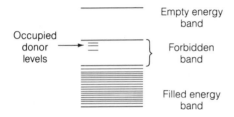

Empty energy band

Occupied donor levels

Forbidden band

Filled energy band

A "hole" is a missing electron

Holes carry current in a *p*-type semiconductor

called *holes* in the electron structure of the crystal. An electron requires little energy to move into a hole, but as it does so it leaves a new hole in its previous location. When an electric field is applied to a silicon crystal in which gallium is present as an impurity, electrons move toward the positive electrode by successively filling holes. The flow of current here is best described in terms of the motion of the holes, which behave as though they are positive charges since they move toward the negative electrode (Fig. 30–18). A material of this kind is therefore called a *p-type* semiconductor. In the energy-band diagram of Fig. 30–19 we see that the effect of gallium as an impurity is to provide energy levels, called *acceptor levels,* just above the highest filled band; electrons that enter these levels leave behind unoccupied levels in the formerly filled band which make possible the conduction of current.

Adding an impurity to a semiconductor is called *doping*. Phosphorus, antimony, and bismuth as well as arsenic have atoms with five outer electrons and so can be used as donor impurities in doping silicon and germanium to yield an *n*-type semiconductor. Similarly indium and tellurium as well as gallium have atoms with three outer electrons and so can be used as acceptor impurities. A minute amount of impurity can lead to a dramatic change in the conductivity of a semiconductor; for instance, 1 part of a donor impurity per 10^8 parts of germanium increases its conductivity by a factor of 12. Silicon and germanium are not the only semiconducting materials with practical applications: another important class of semiconductors consists of such compounds as GaAs, GaP, InSb, InAs, and InP.

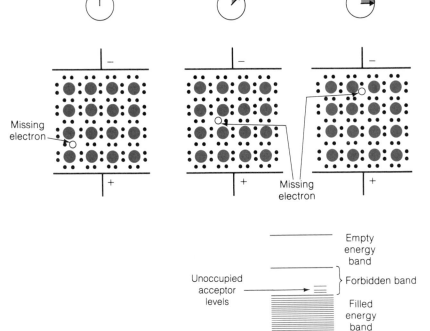

FIG. 30–18 Current in a
p-type semiconductor is
carried by the motion of
"holes," which are sites of
missing electrons. Holes
move toward the negative
electrode as a succession
of electrons move into
them.

Missing
electron

Missing
electron

Empty
energy
band

Unoccupied
acceptor
levels

Forbidden band

Filled
energy
band

FIG. 30–19 Acceptor
levels due to gallium
atoms in a silicon crystal.

30–9   SEMICONDUCTOR DEVICES

The significance of semiconductors in technology arises from the degree of control of **Semiconductor**
electric current that can be accomplished by combining *n*- and *p*-type semiconductors **junctions**
in various ways. Most semiconductor devices depend for their action on the properties
of junctions between *n*- and *p*-type materials. Such junctions can be produced in several
ways. One of them is to gradually pull out a crystal that is forming in molten silicon
that contains, say, a donor impurity, and quickly add an acceptor impurity to the melt
during the process. The first part of the resulting crystal to solidify will be *n*-type and
the rest will be *p*-type if the proportion of acceptor impurity exceeds that of donor
impurity. Another technique, which is especially adapted to the manufacture of integrated circuits, involves diffusing impurities in gas form into a semiconductor wafer
in regions defined by masks. A series of diffusion steps using donor and acceptor
impurities and other materials can produce circuits that contain as many as 50,000
resistors, capacitors, diodes, and transistors on a silicon chip 5 mm square.

An important property of a *p-n* semiconductor junction is that electric current can **Reverse bias**
pass through it much more readily in one direction than in the other. In the crystal
shown in Fig. 30–20, the left-hand end is a *p*-type region (current carried by holes)
and the right-hand end is an *n*-type region (current carried by electrons). When a voltage
is applied across the crystal so that the *p*-end is negative and the *n*-end positive, the
holes in the *p*-region migrate to the left and the electrons in the *n*-region migrate to the

FIG. 30–20 (a) Current is carried in a *p*-type semiconductor by the motion of holes and in an *n*-type semiconductor by the motion of electrons. (b) Reverse bias, little current flows. (c) Forward bias, much current flows.

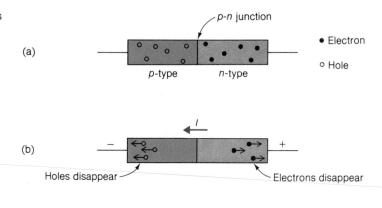

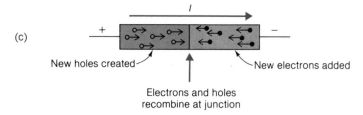

right. Only a limited number of holes and electrons are in the respective regions and new ones appear spontaneously at only a very slow rate; the current through the entire crystal is therefore negligible. This situation is called *reverse bias*.

Figure 30–20(c) shows the same crystal with the connections changed so that the *p*-end is positive and the *n*-end negative. Now new holes are created continuously by the removal of electrons at the *p*-end while new electrons are fed into the *n*-end of the crystal. The holes migrate to the right and the electrons to the left to produce a net positive current flowing from + to −. The electrons and holes meet at the junction between the *p*- and *n*- regions and recombine there: a hole is a missing electron, and when electrons and holes come together they disappear into the regular structure of the crystal and can no longer act as current carriers. Thus current can flow readily through a semiconductor junction from the *p*- to the *n*-region, but hardly at all in the opposite direction. Semiconductor rectifiers ("diodes") are widely used today.

Light-emitting diodes

Energy is needed to create an electron-hole pair, and this energy is released when an electron and a hole recombine. In silicon and germanium the recombination energy is absorbed by the crystal as heat, but in certain other semiconductors, notably gallium arsenide, a photon is emitted when recombination occurs. This is the basis of the *light-emitting diode*. Solid-state lasers have also been made that make use of this phenomenon.

Junction transistors

Transistors are semiconductor devices that are able to amplify weak signals into strong ones. Figure 30–21 shows an *n-p-n* junction transistor, one of a variety of such devices, which consists of a *p*-type material sandwiched between *n*-type materials. The *p*-type region is called the *base,* and the two *n*-type regions are the *emitter* and the *collector.* In the figure the transistor is connected as a simple amplifier. There is a forward bias across the emitter-base junction in this circuit, so electrons pass readily from the emitter to the base. Depending upon the rate at which holes are produced in

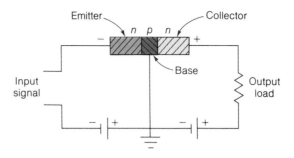

FIG. 30–21 A simple transistor amplifier.

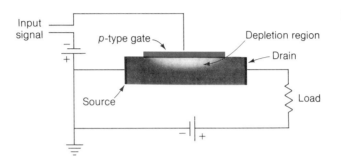

FIG. 30–22 A field-effect

the base, which in turn depends upon its potential relative to the emitter and hence upon the signal input, a certain proportion of the electrons from the emitter will recombine there. The rest of the electrons migrate across the base to the collector to complete the emitter-collector circuit. If the base is at a high positive potential relative to the emitter, many holes are produced there to recombine with electrons from the emitter, and little current can flow through the transistor. If the relative positive potential of the base is low, the number of electrons arriving from the emitter will exceed the number of holes being formed in the base, and the surplus electrons will continue across the base to the collector. Changes in the input-circuit current are thus mirrored by changes in the output-circuit current, which is only a few percent smaller. The ability of the transistor of Fig. 30–21 to produce amplification comes about because the reverse bias across the base-collector junction permits a much higher voltage in the output circuit than that in the input circuit. Since electric power = current × voltage, the power of the output signal can greatly exceed the power of the input signal.

One problem with the junction transistor is that it is difficult to incorporate large numbers of them in an integrated circuit. The *field-effect transistor* (FET) lacks this disadvantage and is widely used today. As in Fig. 30–22, an *n*-channel FET consists of a strip of *n*-type material with contacts at each end together with a strip of *p*-type material, called the *gate,* on one side. When connected as shown, electrons move from the *source* terminal to the *drain* terminal through the *n*-type channel. The *p-n* junction is given a reverse bias, and as a result both the *n*- and *p*-materials near the junction are depleted of charge carriers (see Fig. 30–19(b)). The higher the reverse potential on the gate, the larger the depleted region in the channel, and the fewer the electrons available to carry the current. Thus the gate voltage controls the channel current. In a metal oxide semiconductor FET (MOSFET), the semiconductor gate is replaced by a metal film

Field-effect transistors

separated from the channel by an insulating layer of silicon dioxide. The metal film is thus capacitively coupled to the channel, and its potential controls the drain current through the number of induced charges in the channel. A MOSFET is easier to manufacture than a FET and occupies only a few percent of the area needed for a junction transistor.

IMPORTANT TERMS

A **molecule** is a group of atoms that stick together strongly enough to act as a single particle. A molecule of a given compound always has a certain definite structure and is complete in itself with little tendency to gain or lose atoms.

In a **covalent bond** between atoms in a molecule or a solid, the atoms share one or more electron pairs.

Solids whose constituent atoms or molecules are arranged in regular, repeated patterns are called **crystalline**. When only short-range order is present, the solid is **amorphous**.

In an **ionic bond**, electrons are transferred from one atom to another, and the two then attract each other electrically. The **ionization energy** of an atom is the amount of work that must be done to remove an electron from it. The **electron affinity** of an atom is the amount of energy released when it acquires an additional electron.

Van der Waals forces arise from the electric attraction between asymmetrical charge distributions in atoms and molecules.

The **metallic bond** that holds metal atoms together in the solid state arises from a "gas" of freely moving electrons that pervades the entire metal.

Because the atoms in a crystal are so close together, the energy levels of their outer electron shells are altered slightly to produce **energy bands** characteristic of the entire crystal, in place of the individual sharply defined energy levels of the separate atoms. Gaps between energy bands represent energies forbidden to electrons in the crystal and are called **forbidden bands.**

Semiconductors are intermediate in their ability to carry electric current between conductors and insulators. An **n-type semiconductor** is one in which electric current is carried by the motion of electrons. A **p-type semiconductor** is one in which electric current is carried by the motion of **holes**, which are vacancies in the electron structure that behave like positive charges.

MULTIPLE CHOICE

1. Relative to the energy needed to separate the electrons of an atom from its nucleus, the energy needed to separate the atoms of a molecule is
(a) smaller.
(b) about the same.
(c) larger.
(d) much larger.

2. The reason only two H atoms join to form a hydrogen molecule is a consequence of
(a) the uncertainty principle.
(b) the exclusion principle.
(c) the size of the H atom.
(d) the mass of the H atom.

3. In a diatomic molecule,
(a) only a single pair of electrons can be shared.
(b) more than one pair of electrons can be shared.
(c) one of the atoms must be a hydrogen atom.
(d) one of the atoms must be a carbon atom.

4. The number of covalent bonds a hydrogen atom forms when it combines chemically is
(a) 1.
(b) 2.
(c) 3.
(d) 4.

5. The number of covalent bonds a carbon atom usually forms when it combines chemically is
(a) 1.
(b) 2.
(c) 3.
(d) 4.

6. When two or more atoms join to form a stable molecule,
(a) energy is absorbed.
(b) energy is given off.
(c) there is no energy change.
(d) any of the above, depending upon the circumstances.

7. Short-range order is never found in
(a) crystalline solids.
(b) amorphous solids.
(c) liquids.
(d) gases.

8. A crystalline solid always
(a) is transparent.
(b) is held together by covalent bonds.
(c) is held together by ionic bonds.
(d) has long-range order in its structure.

9. An amorphous solid is closest in structure to
  (a) a covalent crystal.
  (b) an ionic crystal.
  (c) a semiconductor.
  (d) a liquid.

10. The lowest melting points are usually found in solids held together by
  (a) covalent bonds.
  (b) ionic bonds.
  (c) metallic bonds.
  (d) van der Waals bonds.

11. Van der Waals forces arise from
  (a) electron transfer.
  (b) electron sharing.
  (c) symmetrical charge distributions.
  (d) asymmetrical charge distributions.

12. The particles that make up the lattice of the van der Waals crystal of a compound are
  (a) electrons.　　(b) atoms.
  (c) ions.　　(d) molecules.

13. The particles that make up the lattice of an ionic crystal are
  (a) electrons.　　(b) atoms.
  (c) ions.　　(d) molecules.

14. The particles that make up the lattice of a covalent crystal are
  (a) electrons.　　(b) atoms.
  (c) ions.　　(d) molecules.

15. A property of metals that is not due to the electron "gas" that pervades them is their unusual ability to
  (a) conduct electricity.　　(b) conduct heat.
  (c) reflect light.　　(d) form oxides.

16. Defects of crystal structure are not involved in
  (a) the flow of electrons in a metal.
  (b) the flow of electrons in a semiconductor.
  (c) the flow of holes in a semiconductor.
  (d) the mechanical strength of a solid.

17. A crystal whose upper energy band is partly occupied by electrons is a
  (a) conductor.
  (b) insulator.
  (c) n-type semiconductor.
  (d) p-type semiconductor.

18. A hole in a p-type semiconductor is
  (a) an excess electron.
  (b) a missing electron.
  (c) a missing atom.
  (d) a donor level.

19. Current in an n-type semiconductor is carried by
  (a) electrons.
  (b) holes.
  (c) positive ions.
  (d) electrolytes.

EXERCISES

1. What is wrong with the model of a hydrogen molecule in which the two electrons are supposed to follow figure-eight orbits that encircle the two protons?

2. What must be true of the spins of the two electrons in the H_2 molecule?

3. What property of carbon atoms enables them to form so many varied and complex molecules?

4. Why do the inert gas atoms almost never participate in covalent bonds?

5. The atoms in a molecule are said to share electrons, yet some molecules are polar. Explain.

6. Small "perfect crystals" have been prepared with very few structural defects. How would you expect their strength to compare with that of ordinary crystals of the same kind?

7. (a) An iron bar is hammered flat on an anvil, and becomes harder as a result. What change occurred in the structure of the iron? (b) The piece of iron is then strongly heated and allowed to cool slowly, and it becomes softer. Why do you think this happened?

8. The energy needed to detach the electron from a hydrogen atom is 13.6 eV, but the energy needed to detach an electron from a hydrogen molecule is 15.7 eV. Why do you think the latter energy is greater?

9. Why are electrons much more readily liberated from lithium when it is irradiated with ultraviolet light than from fluorine?

10. Why are Cl atoms more active chemically than Cl^- ions?

11. Why are Na atoms more active chemically than Na^+ ions?

12. The ionic crystals of NaF, NaCl, NaBr, and NaI have the respective melting points 988°C, 801°C, 740°C, and 660°C. Can you account for this regular decrease in melting point?

13. The ionization energies of Li, Na, K, Rb, and Cs are, respectively, 5.4, 5.1, 4.3, 4.2, and 3.9 eV. All are in Group I of the periodic table. Account for the decrease in ionization energy with increasing atomic number.

14. Does the "gas" of freely moving electrons in a metal include all the electrons present? If not, which electrons are members of the "gas"?

15. Van der Waals forces can hold inert gas atoms together to form solids, but they cannot hold such atoms together to form molecules in the gaseous state. Why not?

16. The temperature of a gas falls when it passes slowly from a full container to an empty one through a porous plug. Since the expansion is into a rigid container, no mechanical work is done. What is the origin of the fall in temperature?

17. The separation between Na^+ and Cl^- ions in an NaCl crystal is 2.8×10^{-10} m, whereas it is 2.4×10^{-10} m in an NaCl molecule such as might exist in the gaseous state. Why is it reasonable that these separations be different?

18. How does the energy-band structure of a solid determine whether it is a conductor, a semiconductor, or an insulator of electricity?

19. What is the basic reason that energy bands rather than specific energy levels exist in a solid?

20. Does the addition of a small amount of indium to germanium result in the formation of an *n*- or a *p*-type semi-conductor? An indium atom has three electrons in its outermost shell; a germanium atom has four.

21. What is the connection between the abiltiy of a metal to conduct electricity and its ability to conduct heat?

22. The forbidden energy band in germanium that lies between the highest filled band and the empty band above it has a width of 0.7 eV. Compare the conductivity of germanium with that of silicon at (a) very low temperatures and (b) room temperature.

ANSWERS TO MULTIPLE CHOICE

1. a	**6.** b	**11.** d	**16.** a
2. b	**7.** d	**12.** d	**17.** a
3. b	**8.** d	**13.** c	**18.** b
4. a	**9.** d	**14.** b	**19.** a
5. d	**10.** d	**15.** d	

THE NUCLEUS

Until now we have not had to regard the nucleus of an atom as anything but a tiny positively charged object whose sole functions are to provide the atom with most of its mass and to hold its electrons in place. Since the behavior of atomic electrons is responsible for the behavior of matter in bulk, the properties of matter we have been exploring, save for mass, have nothing directly to do with atomic nuclei. Nevertheless, the nucleus turns out to be of supreme importance in the universe: the elements exist by virtue of the ability of nuclei to hold multiple electric charges, and the energy involved in nearly all natural processes has its ultimate origin in nuclear reactions and transformations.

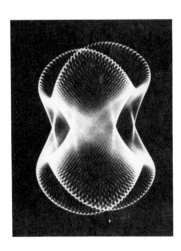

31–1   NUCLEAR STRUCTURE

The nature and behavior of the electron structure of the atom was understood before even the composition of its nucleus was known. The reason is that the nucleus is held together as a unit by forces vastly stronger than the electric forces that hold the electrons to the nucleus, and it is correspondingly harder to break apart a nucleus to find out what is inside. Changes in the electron structure of an atom, such as those that occur in the emission of photons or in the formation of chemical bonds, involve energies of only several eV, whereas changes in nuclear structure involve energies of several MeV, a million times more. Let us first inquire into the composition of the nucleus: What is it that gives a nucleus its characteristic mass and charge?

The nucleus of the hydrogen atom consists of a single proton, whose charge is $+e$ and whose mass is

$$m_{\text{proton}} = 1.673 \times 10^{-27} \text{ kg}$$

Atomic nuclei consist of protons and neutrons, jointly called nucleons

The proton mass is 1836 times that of the electron, so by far the major part of the hydrogen atom's mass resides in its nucleus. This is true of all other atoms as well.

Elements more complex than hydrogen have nuclei that contain *neutrons* as well as protons. The neutron, as its name suggests, is uncharged. The neutron mass is slightly more than that of the proton:

$$m_{\text{neutron}} = 1.675 \times 10^{-27} \text{ kg}$$

Neutrons and protons are jointly called *nucleons*.

Significance of atomic number

Every neutral atom contains the same number of protons and electrons; this number is the *atomic number Z* of the element involved. Except in the case of ordinary hydrogen atoms, the number of neutrons in a nucleus equals or, more often, exceeds the number of protons. The compositions of atoms of the four lightest elements are illustrated in Fig. 31–1.

The mass of an atom can be determined with the help of a mass spectrometer, such as the one described in Sec. 19–7. As mentioned earlier, atomic masses are conventionally expressed in terms of the atomic mass unit, abbreviated u, whose value is

Atomic mass unit

$$1 \text{ u} = 1.66 \times 10^{-27} \text{ kg}$$

The electron, proton, neutron, and hydrogen atom masses in u are respectively

FIG. 31–1 The electronic and nuclear compositions of hydrogen, helium, lithium, and beryllium atoms.

$$m_e = 0.000549 \text{ u} \qquad m_p = 1.007277 \text{ u}$$
$$m_n = 1.008665 \text{ u} \qquad m_{\text{H}} = 1.007825 \text{ u}$$

Because m_{H} and m_n are so close to 1 u, we would expect atomic masses expressed in u to be very nearly whole numbers. This is often true. For example, the atomic mass

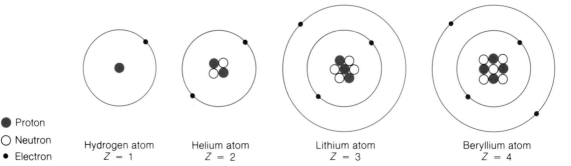

○ Proton
○ Neutron
• Electron

Hydrogen atom
$Z = 1$

Helium atom
$Z = 2$

Lithium atom
$Z = 3$

Beryllium atom
$Z = 4$

TABLE 31-1
The isotopes of hydrogen and chlorine found in nature

| Element | Properties of Element | | Properties of Isotope | | | |
	Atomic Number	Average Atomic Mass, u	Protons in Nucleus	Neutrons in Nucleus	Atomic Mass, u	Relative Abundance
Hydrogen	1	1.008	1	0	1.008	99.985%
			1	1	2.014	0.015%
			1	2	3.016	Very small
Chlorine	17	35.46	17	18	34.97	75.53%
			17	20	36.97	24.47%

of helium is 4.003 u, that of lithium is 6.939 u, and that of beryllium is 9.012 u. (Mass values such as these always include the masses of the surrounding electrons in the neutral atom.) However, the chlorine found in nature has an atomic mass of 35.46 u, which does not fit in with this picture.

Chlorine is an example of an element whose nuclei do not all have the same composition. The several varieties of an element are called its *isotopes*. The number of protons is always equal to the atomic number Z of the element, of course, but the number of neutrons may be different. Thus chlorine consists of two isotopes, one whose nuclei contain 17 protons and 18 neutrons and another whose nuclei contain 17 protons and 20 neutrons. There are about three times as many nuclei of the former type as there are of the latter, which yields an average atomic mass of 35.46 u (Table 31–1).

The isotopes of an element differ in number of nuclear neutrons

All elements have isotopes, even hydrogen. The most abundant hydrogen isotope has nuclei that each consist of a single proton. Less common is the isotope *deuterium*, whose nuclei each consist of a proton and a neutron, and the isotope *tritium*, whose nuclei each consist of a proton and two neutrons (Fig. 31–2). Deuterium is stable, but tritium is radioactive and a sample of it gradually changes to an isotope of helium. The flux of cosmic rays from space continually replenishes the earth's tritium by nuclear reactions in the atmosphere; only about 2 kg of tritium of natural origin is present on the earth's surface, nearly all of it in the oceans.

Deuterium and tritium are hydrogen isotopes

Because the chemical properties of an element depend upon the distribution of the electrons in its atoms, which in turn depends upon the nuclear charge, nuclear structure beyond the number of protons present has little significance for the chemist. The physical properties of an element, however, depend strongly on the nuclear structures of its isotopes, whose behavior may be very different from one another although chemically they are indistinguishable.

Isotopes of an element have almost identical chemical behavior

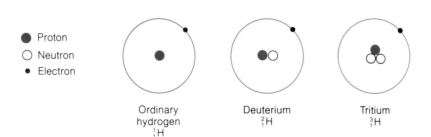

FIG. 31-2 The three isotopes of hydrogen.

● Proton
○ Neutron
• Electron

Ordinary hydrogen
$^1_1$H

Deuterium
$^2_1$H

Tritium
$^3_1$H

The conventional symbols for nuclear species, or *nuclides*, all follow the pattern $^A_Z X$, where

X = chemical symbol of the element
Z = atomic number of the element
 = number of protons in the nucleus
A = mass number of the nuclide
 = number of protons and neutrons in the nucleus

Hence ordinary hydrogen is designated $^1_1 H$, since its atomic number and mass number are both 1, while tritium is designated $^3_1 H$. The two isotopes of chlorine mentioned above are designated $^{35}_{17} Cl$ and $^{37}_{17} Cl$ respectively.

31–2   NUCLEAR SIZE AND COMPOSITION

The Rutherford scattering experiment (Sec. 16–5) provides information on nuclear dimensions as well as on atomic structure. The observed distribution of scattering angles is consistent with a nucleus of infinitely small size provided the alpha particles are not too energetic. That is, below a certain particle energy the size of the nucleus is small compared with the minimum distance to which the incident alpha particles approach it. At higher energies, discrepancies occur between theory and data suggesting that the particles have come so close to the nucleus that it no longer can be thought of as a point charge; from the energy at which these discrepancies appear, an estimate can be made of nuclear dimensions.

Example   The scattering of 7.7-MeV alpha particles (which are $^4_2 He$ nuclei) by a gold foil follows the predictions of Coulomb's law. Use this observation to set an upper limit to the size of the gold nucleus.

Solution   An alpha particle comes closest to a nucleus when it is headed directly toward it. At the distance of closest approach r_0 the initial KE of the particle is entirely converted to electric potential energy PE, and thus

$$KE = PE = \frac{kq_A q_B}{r_0}$$

$$r_0 = \frac{kq_A q_b}{KE}$$

Here $q_A = 2e$ is the charge of an alpha particle, $q_B = 79e$ is the charge of a gold nucleus, and

$$KE = (7.7 \times 10^6 \text{ eV})(1.6 \times 10^{-19} \text{ J/eV}) = 1.2 \times 10^{-12} \text{ J}$$

is the alpha-particle energy. Since $k = 9 \times 10^9 \text{ N·m}^2/\text{C}^2$ we have for the upper limit to the radius of the gold nucleus

$$r_0 = \frac{(9 \times 10^9 \text{ N·m}^2/\text{s}^2)(2)(79)(1.6 \times 10^{-19} \text{ C})^2}{1.2 \times 10^{-12} \text{ J}} = 3.0 \times 10^{-14} \text{ m}$$

This is less than 1/10,000 of the radius of the gold atom. The actual radius of the gold nucleus is about 7.0×10^{-15} m. ■

More recent experiments that employ high-energy electrons, protons, and neutrons yield more precise figures for nuclear sizes. Nuclear radii are found to range from about 1.1×10^{-15} m for the proton to about 7.4×10^{-15} m for the $^{235}_{92}$U nucleus. These experiments are in essence observations of the diffraction of the matter waves of the fast particles by the nuclei of the target atoms; the result is a diffraction pattern formed by the scattered particles, from which the size of the "obstacle"—the nucleus—can be inferred. The observations show that the diffraction patterns are almost, but not quite, the same as those that would be produced by a black disk. The nucleus does not have a sharp boundary, but, like the atom itself, a fuzzy one.

Scattering experiments reveal sizes of nuclei

The density of nuclear matter is about 2.4×10^{17} kg/m^3, which is equivalent to 4 billion tons per cubic inch. Certain stars, called *neutron stars,* are made up of atoms that have been so compressed that their protons and electrons have fused into neutrons, which constitute the most stable form of matter under enormous pressures. The densities of neutron stars are comparable with those of nuclei: a neutron star packs the mass of one or two suns into a sphere only about 20 km in radius. Neutron stars are the remnants of supernova explosions. Pulsars, which emit radio waves in regular bursts typically a second apart, are believed to be rotating neutron stars.

Neutron stars have the same density as nuclei

Only certain combinations of protons and neutrons form stable nuclei. In the lightest nuclei there are about as many neutrons as protons, while in heavier ones there are more neutrons than protons. Figure 31–3 is a plot of neutron number N versus Z for stable nuclei. Evidently the number of neutrons in the nuclei of a given element is a fairly critical quantity. Let us see why.

Most nuclei contain more neutrons than protons

There are two opposing tendencies in a nucleus. The first is a tendency for N to equal Z, which arises from the existence of quantum states of different energy in a nucleus. Just like an electron in an atom, a nucleon in a nucleus can possess only certain specific energies, and, also like an electron, it possesses spin. Because neutrons and protons obey the exclusion principle, at most two of each kind of nucleon (one whose spin is "up" and one whose spin is "down") can occupy each quantum state. Again as in the case of atomic energy levels, nuclear energy levels are filled in sequence to achieve nuclei of minimum energy and hence maximum stability. Thus the boron isotope $^{12}_{5}$B has more energy than the carbon isotope $^{12}_{6}$C because one of its neutrons is in a higher energy level, and $^{12}_{5}$B is accordingly unstable (Fig. 31–4). If created in a nuclear reaction, a $^{12}_{5}$B nucleus changes by beta decay into a $^{12}_{6}$C nucleus in a fraction of a second (see Sec. 31–5).

Nucleons occupy energy levels and possess spin

The other tendency in a nucleus is for the number of neutrons to exceed the number of protons. This is a consequence of the strong electric repulsion exerted by the protons upon one another, which must be balanced by the attractive nuclear forces that act between nucleons. The repulsive electric forces increase more rapidly with Z than the attractive nuclear forces increase with A, the total number of nucleons. Hence a greater proportion of neutrons, which produce only attractive forces owing to their electrical neutrality, is necessary for stability in large nuclei.

Why N exceeds Z

FIG. 31–3 The number of neutrons versus the number of protons in stable nuclei. The larger the nucleus, the greater the proportion of neutrons.

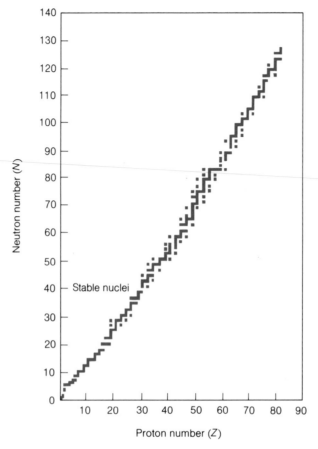

FIG. 31–4 Each nuclear energy level can contain two protons of opposite spins and two neutrons of opposite spins. In the light nuclei, energy levels are filled in sequence so the resulting configuration is one of minimum energy.

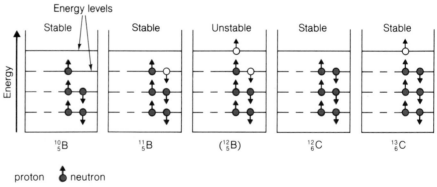

31–3 BINDING ENERGY

As mentioned earlier, the nucleus of a deuterium atom consists of a proton and a neutron. Thus we would expect that the mass of a deuterium atom, ${}^2_1\text{H}$, should be equal to the mass of an ordinary hydrogen atom, ${}^1_1\text{H}$, plus the mass of a neutron. However, it turns out that the mass of ${}^2_1\text{H}$ is 0.002388 u *less* than the combined masses of a ${}^1_1\text{H}$ atom and a neutron (Fig. 31–5).

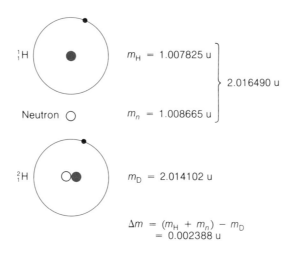

FIG. 31–5 The mass of every atom is less than the total of the masses of its constituent neutrons, protons, and electrons. This phenomenon is illustrated here for the deuterium atom.

$$\Delta m = (m_H + m_n) - m_D$$
$$= 0.002388 \text{ u}$$

The case of deuterium is an example of a general observation: *stable atoms always have less mass than the combined masses of their constituent particles.* The energy equivalent of the "missing" mass of a nucleus is called its *binding energy.* In order to break a nucleus apart into its constituent nucleons, an amount of energy equal to its binding energy must be supplied either in a collision with another particle or by the absorption of a sufficiently energetic photon. Binding energies are due to the action of the nuclear forces that hold nuclei together, just as ionization energies of atoms, which must be supplied to remove electrons from them, are due to the action of the electric forces that hold them together.

The binding energy of a nucleus is the energy needed to break it up into separate neutrons and protons

The difference between the ${}^{2}_{1}H$ atomic mass and the combined masses of ${}^{1}_{1}H$ and a neutron is 0.002388 u. Since the energy equivalent of 1 u is 931 MeV, the binding energy of the deuteron (as the deuterium nucleus is called) is therefore

$$0.002388 \text{ u} \times 931 \frac{\text{MeV}}{\text{u}} = 2.22 \text{ MeV}$$

This figure is confirmed by experiments that show that the minimum energy a photon must have in order to disrupt a deuteron is 2.22 MeV (Fig. 31–6).

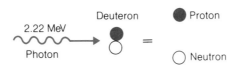

FIG. 31–6 The binding energy of the deuteron is 2.22 MeV, which means that this much energy is required to make up the difference between the deuteron's mass and the combined mass of its constituent neutron and proton. Absorbing 2.22 MeV, for instance by being struck by a 2.22-MeV gamma-ray photon, furnishes a deuteron with enough additional mass to split into a neutron and a proton.

FIG. 31–7 The binding energy per nucleon versus mass number. The higher the binding energy per nucleon, the more stable the nucleus. When a heavy nucleus is split into two lighter ones, a process called *fission*, the greater binding energy of the product nuclei causes the liberation of energy. When two very light nuclei join to form a heavier one, a process called *fusion*, the greater binding energy of the product nucleus again causes the liberation of energy.

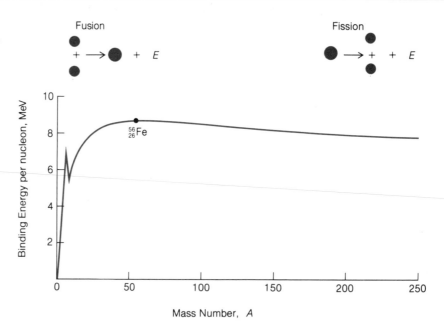

The binding-energy curve is perhaps the most significant in all of science because it is the key to energy production in the universe

The binding energy per nucleon in a nucleus is equal to the total binding energy (calculated from the mass deficiency of the nucleus) divided by the number of neutrons and protons it contains. A very interesting and important curve results when we plot binding energy per nucleon versus mass number, as in Fig. 31–7. Except for an anomalously high peak for $_2^4$He, the curve is a quite regular one. Nuclei of intermediate size have the highest binding energies per nucleon, which means that their nucleons are held together more securely than the nucleons in both heavier and lighter nuclei. The maximum in the curve is 8.8 MeV/nucleon at $A = 56$, which corresponds to the iron nucleus $_{26}^{56}$Fe.

Nuclei of intermediate size are the most stable

A remarkable feature of nuclear structure is illustrated by this curve. Suppose that we split the nucleus $_{92}^{235}$U, whose binding energy is 7.6 MeV/nucleon, into two fragments. Each fragment will be the nucleus of a much lighter element, and therefore will have a higher binding energy per nucleon than the uranium nucleus. The difference is about 0.8 MeV/nucleon, and so, if such *nuclear fission* were to take place, an energy of

Nuclear fission

$$0.8\,\frac{\text{MeV}}{\text{nucleon}} \times 235\ \text{nucleons} = 188\ \text{MeV}$$

would be given off per splitting. This is a truly immense amount of energy to be produced in a single atomic event. As a comparison, chemical processes involve energies of the order of magnitude of 1 electron volt per reacting atom, 10^{-8} the energy involved in fission.

Figure 31–7 also shows that if two of the extremely light nuclei are combined to form a heavier one, the higher binding energy of the latter will also result in the evolution of energy. For instance, if two deuterons were to join to make a $_2^4$He nucleus, more than 23 MeV would be released. This process is known as *fusion*, and, together with fission, it promises to be the source of more and more of the world's energy as

Nuclear fusion

reserves of fossil fuels are depleted. Nuclear fusion is the means by which the sun and stars obtain their energy.

31–4 STRONG NUCLEAR INTERACTION

The existence of stable nuclei can only be explained on the basis of a special interaction between nucleons. This interaction cannot be electrical, because neutrons are uncharged and the positive charges of protons lead to repulsive forces only. It cannot be gravitational, because gravitational forces are far too weak to be able to counterbalance the repulsive electric forces between protons. Thus we have a third fundamental interaction, the *strong nuclear interaction*, which is responsible for the existence of atomic nuclei more complex than that of $_1^1H$, which is a single proton.

The strong interaction is what holds nuclei together

Two properties of the strong interaction stand out. First, it is by far the strongest of all the fundamental interactions, as we can tell from the magnitude of nuclear binding energies. To pull apart the neutron and proton in a deuterium nucleus takes 2.22 MeV, but to pull the electron in a deuterium atom away from the nucleus takes only 13.6 eV, or more than 100,000 times less work (Fig. 31–8).

The second noteworthy aspect of the strong interaction is its short range. Electric and gravitational forces fall off with distance as $1/r^2$, and are effective at considerable separations between the interacting objects; for example, although the planet Pluto is 6×10^{12} m from the sun, it is kept in orbit by the gravitational attraction of the sun. But nuclear forces are effective only over a range of a few nucleon diameters. Up to a separation of about 3×10^{-15} m, the nuclear attraction between two protons is about 100 times stronger than the electric repulsion between them, but beyond this distance the nuclear force dies out rapidly. The nuclear interactions between protons and protons, between protons and neutrons, and between neutrons and neutrons appear to be identical.

The strong interaction has a very short range

The short range of nuclear forces is responsible for the restricted number of stable elements. The larger a nucleus, the stronger the electric repulsive forces that act on

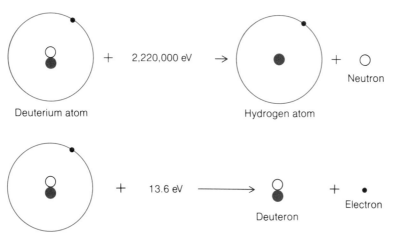

FIG. 31–8 Nuclear binding energies are much greater than atomic binding energies.

each of its protons, but the attractive nuclear forces on each nucleon cannot increase indefinitely because only a limited number of other nucleons are close enough to interact with it. The largest stable nucleus is the bismuth isotope $^{209}_{83}\text{Bi}$, and nuclei larger than the uranium isotope $^{238}_{92}\text{U}$ are too unstable to have survived on earth since its formation.

31–5   RADIOACTIVITY

Not all atomic nuclei are stable. At the beginning of the twentieth century it became known, as the result of research by Becquerel, the Curies, and others, that some nuclei exist that spontaneously transform themselves into other nuclear species with the emission of radiation. Such nuclei are said to be *radioactive*.

Unstable nuclei emit three kinds of radiation:

1. *Alpha particles,* which are the nuclei of ^4_2He atoms.
2. *Beta particles,* which are electrons or positrons (positively charged electrons).
3. *Gamma rays,* which are photons of high energy.

The early experimenters identified these radiations with the help of a magnetic field. Figure 31–9 shows a radium sample in a magnetic field directed into the paper: the positively charged alpha particles are deflected to the left and the negatively charged beta particles are deflected to the right. Gamma rays carry no charge and are not affected by the magnetic field.

To understand why alpha, beta, and gamma decays take place, let us return to Fig. 31–3, which is a plot showing the number of neutrons versus the number of protons in stable nuclei. For light, stable nuclei the number of neutrons and protons are approximately the same, while for heavier nuclei somewhat more neutrons than protons are

FIG. 31–9 The radiations from a radium sample may be analyzed with the help of a magnetic field. In the figure the direction of the field is into the paper; hence the positively charged alpha particles (which are helium nuclei) are deflected to the left and the negatively charged beta particles (which are electrons), to the right. Gamma rays (which are energetic photons) carry no charge and are not affected by the magnetic field.

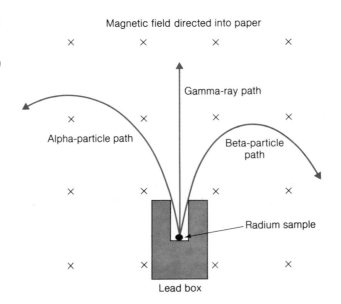

Magnetic field directed into paper

Gamma-ray path

Alpha-particle path

Beta-particle path

Radium sample

Lead box

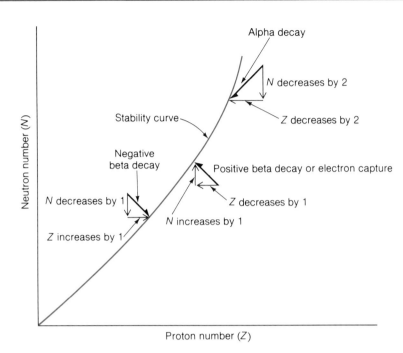

FIG. 31–10 How alpha and beta decays tend to bring an unstable nucleus to a stable configuration.

required for stability. It is evident that an element of a given atomic number has only a very narrow range of possible numbers of neutrons if it is to be stable.

Suppose now that a nucleus exists that has too many neutrons for stability relative to the number of protons present. If one of the excess neutrons transforms itself into a proton, this will simultaneously reduce the number of neutrons while increasing the number of protons (Fig. 31–10). To conserve electric charge, such a transformation requires the emission of a negative electron, and we may write it in equation form as

Negative beta decay occurs in nuclei with too many neutrons

$$n^0 \rightarrow p^+ + e^- \qquad\qquad \textit{Electron emission} \quad (31\text{–}1)$$

The electron leaves the nucleus, and is detectable as a "beta particle." The residual nucleus may be left with some extra energy as a consequence of its shifted binding energy, and this energy is given off in the form of gamma rays. Sometimes more than one such *beta decay* is required for a particular unstable nucleus to reach a stable configuration.

Should the nucleus have too few neutrons, the inverse reaction

Positron emission occurs in nuclei with too few neutrons

$$p^+ \rightarrow n^0 + e^+ \qquad\qquad \textit{Positron emission} \quad (31\text{–}2)$$

in which a proton becomes a neutron with the emission of a positron, may take place. This is also called beta decay, since it resembles the emission of negative electrons from an unstable nucleus in every way save for the difference in charge.

A process that competes with positron emission is the capture by a nucleus with too small a neutron/proton ratio of one of the electrons in its innermost atomic shell. The electron is absorbed by a nuclear proton which becomes a neutron in so doing. This process can be expressed as

Electron capture is an alternative to positron emission

$$p^+ + e^- \rightarrow n^0 \qquad\qquad \textit{Electron capture} \quad (31\text{–}3)$$

TABLE 31–2
Radioactive decay. The asterisk (*) denotes an excited nuclear state and γ denotes a gamma-ray photon.

Decay	Nuclear Transformation	Example
Alpha	$^A_Z X \rightarrow ^{A-4}_{Z-2} Y + ^4_2 He$	$^{238}_{92} U \rightarrow ^{234}_{90} Th + ^4_2 He$
Electron emission	$^A_Z X \rightarrow _{Z+1}^A Y + e^-$	$^{14}_6 C \rightarrow ^{14}_7 N + e^-$
Positron emission	$^A_Z X \rightarrow _{Z-1}^A Y + e^+$	$^{64}_{29} Cu \rightarrow ^{64}_{28} Ni + e^+$
Electron capture	$^A_Z X + e^- \rightarrow _{Z-1}^A Y$	$^{64}_{29} Cu + e^- \rightarrow ^{64}_{28} Ni$
Gamma	$^A_Z X^* \rightarrow ^A_Z X + \gamma$	$^{87}_{38} Sr^* \rightarrow ^{87}_{38} Sr + \gamma$

Electron capture does not lead to the emission of a particle, but can be detected by the X-ray photon that is produced when one of the atom's outer electrons falls into the vacancy left by the absorbed electron.

Alpha decay

Another way of altering its structure to achieve stability may involve a nucleus in *alpha decay,* in which an alpha particle consisting of two neutrons and two protons is emitted. Thus negative beta decay increases the number of protons by one and decreases the number of neutrons by one; positive beta decay and electron capture decrease the number of protons by one and increase the number of neutrons by one; and alpha decay decreases both the number of protons and the number of neutrons by two. These processes are summarized in Table 31–2 and shown schematically in Fig. 31–11. Very often a succession of alpha and beta decays, with accompanying gamma decays to carry off excess energy, is required before a nucleus reaches stability.

Example   The nucleus $^6_2 He$ is unstable. What kind of decay would you expect it to undergo?

Solution   The most stable helium nucleus is $^4_2 He$, as shown by its position on the binding-energy curve of Fig. 31–7 where it is responsible for the peak at the left of the curve. Since $^6_2 He$ has four neutrons whereas $^4_2 He$ has only two, the instability of $^6_2 He$ must arise from an excess of neutrons. Accordingly it would seem likely that $^6_2 He$ undergoes negative beta decay to become $^6_3 Li$, whose neutron/proton ratio is more consistent with stability:

$$^6_2 He \rightarrow ^6_3 Li + e^-$$

This is in fact the manner in which $^6_2 He$ decays.   ∎

Example   The polonium isotope $^{210}_{84} Po$ is unstable and emits a 5.30-MeV alpha particle. The atomic mass of $^{210}_{84} Po$ is 209.9829 u and that of $^4_2 He$ is 4.0026 u. Identify the daughter nuclide and find its atomic mass.

Solution   (a) The daughter nuclide will have an atomic number of $84 - 2 = 82$ and a mass number of $210 - 4 = 206$. From Table 29–4 we see that $Z = 82$ corresponds to lead, so the symbol of the daughter nuclide is $^{206}_{82} Pb$.

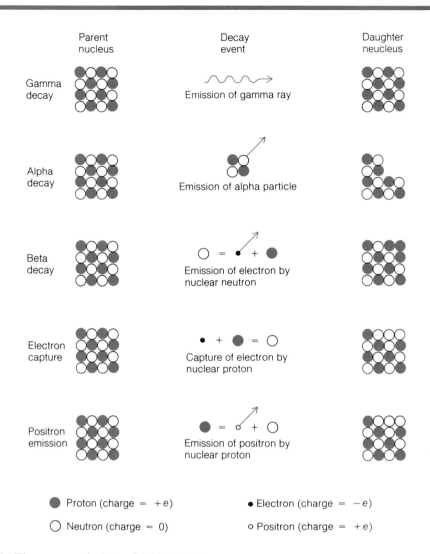

Parent nucleus	Decay event	Daughter neucleus

Gamma decay — Emission of gamma ray

Alpha decay — Emission of alpha particle

Beta decay — Emission of electron by nuclear neutron

Electron capture — Capture of electron by nuclear proton

Positron emission — Emission of positron by nuclear proton

FIG. 31–11 Five kinds of radioactive decay.

● Proton (charge $= +e$)   • Electron (charge $= -e$)

○ Neutron (charge $= 0$)   ○ Positron (charge $= +e$)

(b) The mass equivalent of 5.30 MeV is

$$m_E = \frac{5.30\,\text{MeV}}{931\,\text{MeV/u}} = 0.0057\,\text{u}$$

The mass lost by $^{210}_{84}\text{Po}$ in its decay equals the atomic mass of the alpha particle $^{4}_{2}\text{He}$ plus m_E, which is $(4.0026 + 0.0057)\,\text{u} = 4.0083\,\text{u}$, so that

Mass of $^{210}_{84}\text{Po}$ − (mass of $^{4}_{2}\text{He} + m_E)$ = mass of $^{206}_{82}\text{Pb}$

$$209.9829\,\text{u} - 4.0083\,\text{u} = 205.9746\,\text{u} \qquad ■$$

The strong interaction that holds nucleons together to form nuclei cannot account for the occurrence of beta decay, positron emission, and electron capture. Another

The weak interaction is responsible for beta decay

fundamental interaction turns out to be responsible: the *weak interaction*. The range of the latter is so short ($\sim 10^{-17}$ m) that it operates only *within* certain elementary particles and leads to their transformation into other particles. The four fundamental interactions—gravitational, electromagnetic, strong, and weak—are discussed further in the next chapter.

Insofar as the structure of matter is concerned, the role of the weak interaction seems to be confined to causing beta decays in nuclei whose neutron/proton ratios are not appropriate for stability. This interaction also affects elementary particles that are not part of a nucleus. The name "weak interaction" arose because the other short-range force acting upon nucleons is extremely strong, as the high binding energies of nuclei attests. Actually the gravitational interaction is weaker than the weak interaction at distances where the latter is a factor.

31–6   ALPHA DECAY

The heaviest nuclei undergo alpha decay

Nuclei that contain more than about 210 nucleons are so large that the short-range forces holding them together are barely able to counterbalance the long-range electric repulsive forces of their protons. Such a nucleus can reduce its bulk and thereby achieve greater stability by emitting an alpha particle, which decreases its mass number A by 4.

Why alpha particles are emitted

It is appropriate to ask why it is that only alpha particles are given off by excessively heavy nuclei, and not, for example, individual protons or $_2^3$He nuclei. The reason is a consequence of the high binding energy of the alpha particle, which means that it has significantly less mass than four individual nucleons. Because of this small mass, an alpha particle can be ejected by a heavy nucleus with energy to spare. Thus the alpha particle released in the decay of $_{92}^{232}$U has a kinetic energy of 5.4 MeV, while 6.1 MeV would have to be supplied from the outside to this nucleus if it is to release a proton, and 9.6 MeV supplied if it is to release a $_2^3$He nucleus.

A PE barrier surrounds every nucleus

Even though alpha decay may be energetically possible in a particular nucleus, it is not obvious just how the alpha particle is able to break away from the nuclear forces that bind it to the rest of the nucleus. Typically, an alpha particle has available about 5 MeV of energy with which to escape. However, an alpha particle located at a point near the nucleus but just outside the range of its nuclear forces has an electric potential energy of perhaps 25 MeV; that is, if released from this position it will have a kinetic energy of 25 MeV when it is an infinite distance away as a result of electric repulsion (Fig. 31–12). An alpha particle inside the nucleus therefore should require a minimum of 25 MeV in energy, five times more than is available, in order to break loose.

The alpha particle, then, is located in a box whose walls are of such a height that an energy of 25 MeV is needed to surmount them, while the particle itself has only 5 MeV for the purpose.

Quantum mechanics provides the answer to the paradox of alpha decay. Two assumptions are needed: (1) an alpha particle can exist as an individual entity within a nucleus, and (2) it is in constant motion there.

According to quantum theory, a moving particle has a wave character, so the proper

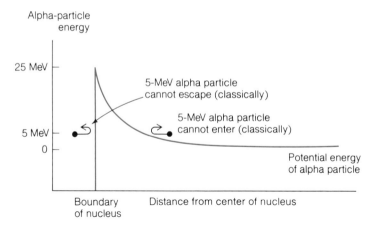

FIG. 31–12 The variation in alpha-particle potential energy near a typical heavy nucleus. The alpha particle in this nucleus has a kinetic energy of 5 MeV.

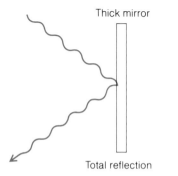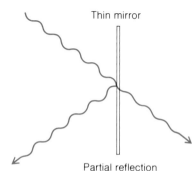

FIG. 31–13 In reflection, a wave penetrates the reflecting surface for a short distance and may pass through it if the mirror is sufficiently thin.

classical analog of an alpha particle in a nucleus is a light wave trapped between mirrors and not a particle bouncing back and forth between solid walls. Now in order for a light wave to be reflected from a mirror, it must actually penetrate the reflecting surface for a short distance (Fig. 31–13). The intensity of the wave drops off quite rapidly inside the reflecting surface, to be sure, but it *must* penetrate to some extent. If the mirror is thick, all of the incident light is reflected. However, if the mirror is very thin, some of the incident light can pass right through the mirror, as shown. The formal theory of this partial transmission, with only minor changes, is able to account quantitatively for alpha decay. The very existence of alpha decay, in fact, is further confirmation of the validity of quantum ideas, since the principles of physics that follow from Newton's laws of motion prohibit such decay.

Quantum theory of alpha decay

Of course, a 25-MeV energy barrier is not very "transparent" to a 5-MeV alpha particle. An alpha-radioactive nucleus is usually about 1.5×10^{-14} m in diameter, and an alpha particle within it might oscillate back and forth with a speed of 2×10^7 m/s. Hence the alpha particle strikes the confining potential-energy wall nearly 10^{21} times per second, but may nevertheless have to wait as much as 10^{10} years to escape from certain nuclei.

The quantum-mechanical phenomenon of barrier penetration is sometimes called the *tunnel effect*, because the particle escapes *through* the barrier and not over it.

Tunnel effect

31–7   HALF-LIFE

The rate at which a sample of radioactive material decays is called its *activity*. The SI unit of activity is the *becquerel*, where 1 Bq = 1 event per second. The activities encountered in practice are usually so high that the MBq (10^6 Bq) and GBq (10^9 Bq) are more suitable. The traditional unit of activity, which is still in common use, is the *curie*. Originally the curie was defined as the activity of 1 g of radium ($^{226}_{88}$Ra), and its precise value accordingly changed as measuring techniques improved. For this reason the curie is now defined arbitrarily as

$$1 \text{ curie} = 3.70 \times 10^{10} \text{ events/s} = 37 \text{ GBq}$$

The activity of 1 g of radium is a few percent smaller. The *millicurie* (10^{-3} curie) and *microcurie* (10^{-6} curie) are frequently employed to supplement the curie. A luminous watch dial contains several microcuries of $^{226}_{88}$Ra; ordinary potassium has an activity of about 1 millicurie/kg owing to the presence of the radioactive isotope $^{40}_{19}$K; "cobalt-60" sources of 1 or more curies are widely used in medicine for radiation therapy and industrially for the inspection of metal castings and welded joints.

One of the characteristics of all types of radioactivity is that the rate at which the nuclei in a given sample decay always follows a curve whose shape is like that shown in Fig. 31–14. If we start with a sample whose rate of decay is, say, 100 events/s, it will not continue to decay at that rate but instead fewer and fewer disintegrations will occur in each successive second.

Some radioactive nuclides decay faster than others, but in each case a certain definite time is required for half of an original sample to decay. This time is called the *half-life* of the nuclide. For instance, the radon isotope $^{222}_{86}$Rn undergoes alpha decay to the polonium isotope $^{218}_{84}$Po with a half-life of 3.8 days. Should we start with 1 mg of radon in a closed container (since it is a gas), $\frac{1}{2}$ mg will remain undecayed after 3.8 days; $\frac{1}{4}$ mg will remain undecayed after 7.6 days; $\frac{1}{8}$ mg will remain undecayed after 11.4 days; and so on (Fig. 31–15).

If $t_{1/2}$ is the half-life of a nuclide and N_0 is the original amount of it at the time $t = 0$, then at a time t later the amount remaining undecayed is given by

$$N = \frac{N_0}{2^{t/t_{1/2}}} \qquad\qquad\qquad \textit{Radioactive decay} \quad (31\text{–}4)$$

Thus after t = 1 half-life, $t/t_{1/2}$ = 1 and $N = N_0/2$; after 2 half-lives, $t/t_{1/2}$ = 2 and $N = N_0/2^2 = N_0/4$; after 3 half-lives, $t/t_{1/2}$ = 3 and $N = N_0/2^3 = N_0/8$; and so on, which is just what was given above for the decay of radon.

FIG. 31–14 The rate at which a sample of radioactive substance decays is not constant but varies with time in the manner shown in the curve.

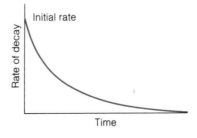

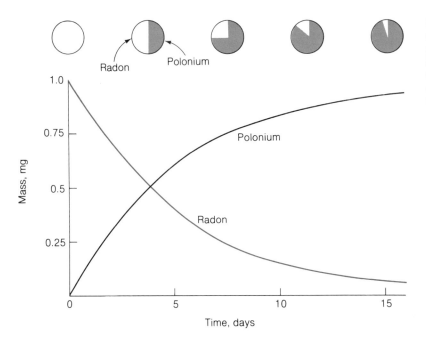

FIG. 31–15 The alpha decay of radon ($^{222}_{86}$Rn) to polonium ($^{218}_{84}$Po) has a half-life of 3.8 days. The sample of radon whose decay is graphed here had an initial mass of 1.0 mg.

For times that are not multiples of $t_{1/2}$, the calculation of $2^{t/t_{1/2}}$ can be made using logarithms (see Appendix B-5). Since $\log a^b = b \log a$ in general,

$$\log 2^{t/t_{1/2}} = \frac{t}{t_{1/2}} \log 2 = 0.301 \frac{t}{t_{1/2}}$$

and we have

$$2^{t/t_{1/2}} = \log^{-1}\left(0.301 \frac{t}{t_{1/2}}\right)$$

where $\log^{-1}$ signifies the antilogarithm. To find the antilogarithm of a number x on an electronic calculator, simply enter the number and then press the $[10^x]$ or [INV LOG] key.

Example The hydrogen isotope tritium, ^3_1H, is radioactive and emits an electron with a half-life of 12.5 years. (a) What does ^3_1H become after beta decay? (b) What percentage of an original sample of tritium will remain undecayed 5 years after its preparation?

Solution (a) When a nucleus emits an electron, its atomic number increases by 1 unit (corresponding to an increase in nuclear charge of $+e$) and its mass number is unchanged. Helium has the atomic number 2, and so

$$^3_1\text{H} \rightarrow {}^3_2\text{He} + e^-$$

(b) Here $t/t_{1/2} = (5 \text{ years}/12.5 \text{ years}) = 0.4$ and

$$2^{t/t_{1/2}} = 2^{0.4} = \log^{-1}(0.301 \times 0.4) = \log^{-1} 0.120 = 1.32$$

Hence from Eq. (31–4)

$$\frac{N}{N_0} = \frac{1}{2^{t/t_{1/2}}} = \frac{1}{1.32} = 0.758 = 75.8\%$$

of the original amount of tritium is left after 5 years.   ■

Half-lives vary widely

Half-lives range from billionths of a second to billions of years. Samples of radioactive isotopes decay in the manner illustrated because a great many individual nuclei are involved, each having a certain probability of decaying. The fact that radon has a half-life of 3.8 days signifies that every radon nucleus has a 50% chance of decaying in any 3.8-day period. Because a nucleus does not have a memory, this does *not* mean that a radon nucleus has a 100% chance of decaying in 7.6 days: the likelihood of decay of a given nucleus stays the same until it actually does decay. Thus a half-life of 3.8 days means a 75% probability of decay in 7.6 days, an 87.5% probability of decay in 11.4 days, and so on, because in each interval of 3.8 days the probability is 50%.

Half-lives are unaffected by external conditions

The above discussion suggests that radioactive decay involves individual events that take place within individual nuclei, rather than collective processes that involve more than one nucleus in interaction. This idea is confirmed by experiments which show that the half-life of a particular isotope is invariant under changes of pressure, temperature, electric and magnetic fields, and so on, which might, if strong enough, influence internuclear phenomena.

Classical physics cannot explain radioactivity

Three aspects of radioactivity are wholly remarkable from the point of view of classical (pre-relativity and pre-quantum theory) physics:

1. The atomic number of a nucleus that undergoes alpha or beta decay changes, so that the nucleus becomes one characteristic of a different element. Elements *can* be transmuted into other elements, though hardly in a manner anticipated by alchemy.

2. Radioactive decay liberates energy that can only come from *within* individual atoms. One gram of radium in a sealed container (to prevent the escape of radon, a gaseous product of its decay that is itself radioactive) evolves energy at the rate of 0.16 W; this rate decreases so slowly that after 1600 years it has dropped by only 50%. Where does all this energy come from? Not until 1905, when Einstein proposed the equivalence of mass and energy, was this puzzle understood.

3. Radioactivity is a statistical process. Every nucleus of a radioisotope has a certain likelihood of decaying, but, because the decay obeys the laws of chance, we have no way of predicting *which* nuclei will actually decay at a particular time. There is no cause-effect relationship in radioactivity as there is in all classical physics.

31–8   RADIATION DOSAGE

Ionizing radiation is harmful to living things

Radiation energetic enough to ionize matter in its path—which means X rays as well as the various radiations from radioactive nuclei—is harmful to living tissue, and exposure to it should be kept to a minimum. In conflict with this aim is the desirability of many processes that involve ionizing radiation, from the use of X rays in diagnosis and therapy to the operation of nuclear reactors. The most insidious aspect of the problem is the delay between an exposure and certain of the consequences it may have, among them leukemia in the case of the person irradiated and genetic defects in his or her descendants.

Particularly unfortunate is the widespread abuse of diagnostic X rays. For example, careful studies have shown that children whose mothers were X-rayed while pregnant develop cancer more often than those whose mothers were not, yet some pregnant women are still X-rayed as a matter of routine rather than solely for specific cause. Another example is the mass screening by X ray of symptomless young women for breast cancer, which many experts feel has increased the overall rate of cancer mortality rather than decreased it. The only dosage of ionizing radiation that is unquestionably harmless is zero, and the benefits claimed for any irradiation, whether of a person or of the public in general, must be weighed against the hazard that accompanies it.

The SI unit of radiation dosage is the *gray,* where 1 Gy is equal to 1 J of energy absorbed per kilogram of target material. The gray is a large unit and the *rad,* equal to 0.01 Gy, is more widely used. The dosage in such units is insufficient as an index of the biological effectiveness of a certain irradiation, however: radiations of different kinds and different energies do not have the same effects on a given tissue, and different tissues respond differently to the same radiation. The relative biological effectiveness (RBE) of a particular radiation is therefore an important quantity, even though hard to specify with precision. The RBE of 250-KeV X rays is taken as 1, and a dosage of 1 rad of such X rays to a person is considered 1 *rem* (rad equivalent man). The RBEs of other X rays and of gamma and beta rays are close to 1, but the RBE of fast neutrons is about 10 and that of 1-MeV alpha particles is about 25; 1 rad absorbed from such alpha particles therefore means a dosage of about 25 rem.

How radiation dosage is measured

A radiation dosage of 400 rem is considered lethal in the sense that a person receiving such a dosage in a short time has only a 50% likelihood of survival. A fatal cancer is estimated to occur once per 5000–6000 rem of radiation exposure of a population, a genetic defect slightly less often. The lifetime dosage per person from such unavoidable natural sources as cosmic rays and radioactive nuclides in the earth and in the body itself (radon liberated in the burning of coal is significant here) is perhaps 10 rem. Averaged over the U.S. population, diagnostic X rays add somewhat over half that amount, nuclear power stations 0.003% as much.

Consequences of radiation dosage

The recommended maximum dosage rate for people who are exposed to radiation in the course of their work is 5 rem/year, 40 times the background rate, and for the general public it is 0.5 rem/year, four times greater. These figures, in particular the latter one, are not free from dispute; even if such dosage rates cannot be unambiguously associated with specific maladies, their genetic consequences are without question, and an acceleration of the mutation rate is not something to be embarked upon without considerable thought on the part of all sections of the community.

31–9 RADIOMETRIC DATING

Radioactive decay is the basis of a valuable method for establishing the ages of a variety of geological and biological specimens. Because the decay of a given radioactive nuclide proceeds at a constant rate regardless of its environment, the ratio between the amounts of that nuclide and of its stable daughter in a specimen depends upon the specimen's age. The greater the proportion of the daughter nuclide, the older the specimen. Let us see how this procedure is used to date objects of biological origin using *radiocarbon,* the beta-active carbon isotope $^{14}_{6}C$.

Radiocarbon is a byproduct of cosmic-ray bombardment

Cosmic rays are high-energy atomic nuclei, mainly protons, that circulate through the Milky Way galaxy of which the sun is a member. About 10^{18} of them arrive at the earth each second and disrupt the nuclei of atoms they encounter in the atmosphere to produce showers of secondary particles. Among these secondaries are neutrons that can react with nitrogen nuclei in the atmosphere to form radiocarbon with the emission of a proton:

$$^{14}_{7}N + {}^{1}_{0}n \rightarrow {}^{14}_{6}C + {}^{1}_{1}H$$

The proton picks up an electron and becomes a hydrogen atom. The ^{14}C isotope has too many neutrons for stability and ultimately undergoes beta decay to become ^{14}N with a half-life of 5760 years. At the present time a total of about 90 tons of radiocarbon is distributed around the world, which is replenished by the cosmic-ray bombardment as it decays so that the amount remains constant.

Shortly after their formation, radiocarbon atoms combine with oxygen molecules to form carbon dioxide (CO_2) molecules. Green plants convert water and carbon dioxide into carbohydrates in the process of photosynthesis, which means that every plant contains some radiocarbon. Animals eat plants, and so they too contain radiocarbon. Thus every living thing on earth has a very small proportion of radiocarbon in its tissues. Since the mixing of radiocarbon is relatively efficient, living plants and animals all have the same ratio of radiocarbon to ordinary carbon, which is ^{12}C.

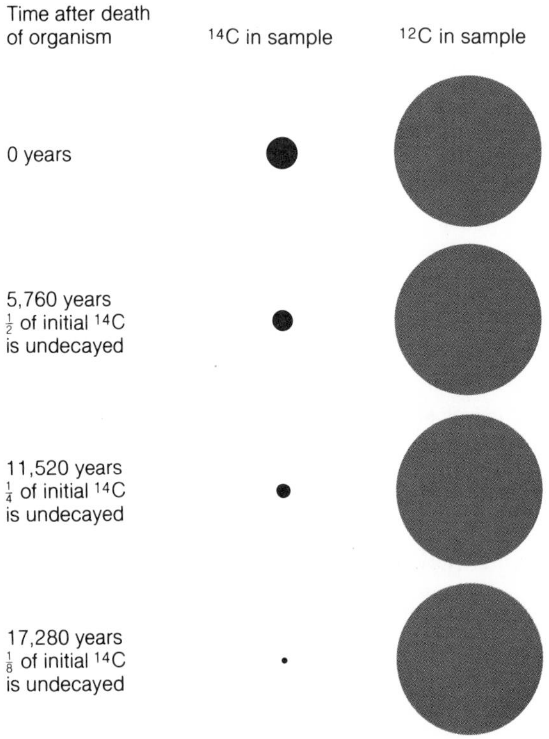

FIG. 31–16 The radioactive ^{14}C content of a specimen of plant or animal tissue decreases steadily, whereas its ^{12}C content does not change. Hence the $^{14}C/^{12}C$ ratio indicates the time that has passed since the organism died. The half-life of ^{14}C is 5760 years.

Time after death of organism

^{14}C in sample

^{12}C in sample

0 years

5,760 years
$\frac{1}{2}$ of initial ^{14}C is undecayed

11,520 years
$\frac{1}{4}$ of initial ^{14}C is undecayed

17,280 years
$\frac{1}{8}$ of initial ^{14}C is undecayed

Method	Parent Nuclide	Daughter Nuclide	Half-Life	
Potassium-argon	^{40}K	^{40}Ar	1.3×10^9 years	**TABLE 31–3**
Rubidium-strontium	^{87}Rb	^{87}Sr	4.7×10^{10} years	Geological dating methods
Uranium-lead	^{238}U	^{206}Pb	4.5×10^9 years	

When a plant or animal dies, however, it no longer takes in radiocarbon atoms, but the radiocarbon atoms it already contains continually decay to nitrogen. After 5760 years a dead organism has left only half as much radiocarbon as it had while alive, after 11,520 years only a quarter as much, and so on. By determining the ratio of radiocarbon to ordinary carbon it is therefore possible to evaluate the ages of ancient objects and remains of organic origin (Fig. 31–16). This technique permits the dating of mummies, wooden implements, cloth, leather, charcoal from campfires, and similar artifacts from ancient civilizations as much as 50,000 years old, about nine half-lives of ^{14}C.

Principle of radiocarbon dating

Because the earth's history goes back 4.5 or so billion years, geologists use radioactive nuclides of much longer half-lives than that of radiocarbon to date rocks. The most common geological dating methods use the nuclides ^{40}K, ^{87}Rb, and ^{238}U (Table 31–3). In each case the assumption is made that all of the stable daughter nuclide found in a particular rock sample originated from the decay of the parent nuclide. Although ^{40}K and ^{87}Rb decay into stable nuclides in a single step, ^{238}U requires no less than 14 successive alpha and beta decays to become ^{206}Pb. However, the half-lives of the intermediate products are so short compared with the 4.5-billion-year half-life of ^{238}U itself that only the ^{238}U and ^{206}Pb contents of a particular sample need be considered. What is meant by the age of a rock depends upon the nature of the rock; it may refer to the time at which the minerals of the rock crystallized, for example. Meteorites and lunar rocks as well as terrestrial rocks have been dated by the methods of Table 31–3.

Geological dating methods

31–10   FISSION

When two nuclei approach close enough together, it is possible for a rearrangement of their constituent nucleons to occur with one or more new nuclei formed. Such a process is called a nuclear reaction, by analogy with chemical reactions in which two or more compounds may combine to form new ones.

Atoms and molecules are neutral, so it is easy for them to come together and react. Nuclei all have positive charges, as much as $+92e$ in nuclei found in nature, and the electric repulsion between them is sufficient to keep them beyond the range where they can interact unless they are moving very fast to begin with. In the sun and other stars, whose interior are at temperatures of many millions of K, the nuclei present are moving fast enough on the average for nuclear reactions to be frequent, and indeed nuclear reactions provide the energy that maintains these temperatures.

Mutual repulsion hinders nuclear reactions

In the laboratory it is easy enough to produce nuclear reactions on a very small scale, either with alpha particles from radioactive substances or with protons, deuterons, or even heavier nuclei accelerated in cyclotrons and similar devices. But only one type of nuclear reaction has as yet proved to be a practical source of energy on

FIG. 31–17 In nuclear fission an absorbed neutron causes a heavy nucleus to split in two parts, with the emission of several neutrons and gamma rays.

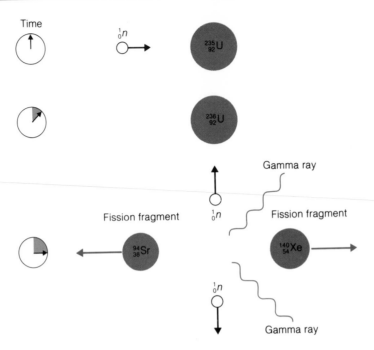

Fission products are highly radioactive and may remain so for long periods

the earth, namely the fission that occurs when neutrons strike the nuclei of certain very heavy nuclei.

In nuclear fission, which can take place only in certain very heavy nuclei such as $^{235}_{92}\text{U}$, the absorption of an incoming neutron causes the target nucleus to split into two smaller nuclei called *fission fragments* (Fig. 31–17). Because stable light nuclei have proportionately fewer neutrons than do heavy nuclei, the fragments are unbalanced when they are formed and at once release one or two neutrons each. Usually the fragments are still unstable and undergo radioactive decay to achieve appropriate neutron-proton ratios. The products of fission, such as the wastes from a nuclear reactor and the fallout from a nuclear bomb burst, are accordingly highly radioactive. Since some of these products have long half-lives, their radioactivity will persist for many generations and their disposal remains a problem without a really satisfactory solution as yet.

Although a variety of nuclear species may appear as fission fragments, we might cite as a typical fission reaction

$$^{235}_{92}\text{U} + {}^{1}_{0}n \longrightarrow {}^{236}_{92}\text{U} \longrightarrow {}^{140}_{54}\text{Xe} + {}^{94}_{38}\text{Sr} + {}^{1}_{0}n + {}^{1}_{0}n + 200\,\text{MeV} \qquad (31\text{–}5)$$

The $^{236}_{92}\text{U}$ that is first formed lasts for only a small fraction of a second before it splits into two parts. About 84% of the total energy liberated during fission appears as kinetic energy of the fission fragments, about 2.5% as kinetic energy of the neutrons, and about 2.5% in the form of instantaneous emitted gamma rays, with the remaining 11% being given off in the decay of the fission fragments.

Chain reaction

Because each fission event liberates two or three neutrons while only one neutron is required to initiate it, a rapidly multiplying sequence of fissions can occur in a lump of suitable material (Fig. 31–18). When uncontrolled, such a *chain reaction* evolves

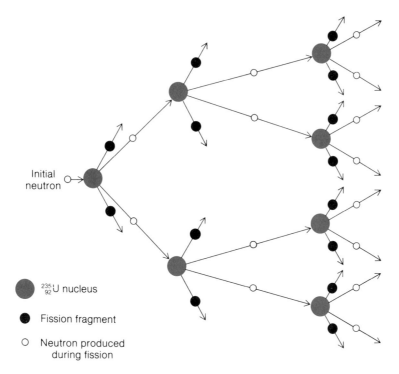

FIG. 31–18 Simplified sketch of a chain reaction in $^{235}_{92}$U.

Initial neutron

$^{235}_{92}$U nucleus

Fission fragment

○ Neutron produced during fission

an immense amount of energy in a short time. If we assume that two neutrons emitted in each fission are able to induce further fissions and that 10^{-8} s elapses between the emission of a neutron and its subsequent absorption, a chain reaction starting with a single fission will release 2×10^{13} J of energy in less than 10^{-6} s! An uncontrolled chain reaction evidently can cause an explosion of exceptional magnitude and in fact is the basis of the "atomic" bomb.

31–11 NUCLEAR REACTORS

When controlled so that exactly one neutron per fission causes another fission, a chain reaction occurs at a constant power output. Such a reaction makes a very efficient source of energy: an output of about 1 MW (1000 kW) is produced by the fission of 1 g of a suitable nuclide per day, as compared with the consumption of 2.6 tons of coal per day in a conventional power plant. A device in which a chain reaction can be initiated and controlled is called a *nuclear reactor*.

Fission occurs at a steady rate in a nuclear reactor

The energy liberated in a nuclear reactor appears as heat, and it is extracted by circulating a coolant liquid or gas. The hot coolant can then be used as the heat source of a conventional steam turbine, which in turn may power an electric generator, a ship, or a submarine. In 1978, 236 reactors in 22 countries (72 of them in the United States) rated at a total of 125,000 MW were responsible for producing about 8% of the world's electrical energy. A somewhat larger number of reactors were in use in various naval vessels in the same year.

Each fission in ^{235}U liberates an average of 2.5 neutrons, hence no more than 1.5 neutrons per fission can be lost if a self-sustaining chain reaction is to take place. Neutrons can be lost in two ways: through the reactor surface and by absorption within the reactor without inducing fission. Since a large object has less surface area relative to its volume than a small object of the same shape, neutron escape can be minimized simply by increasing the reactor size. It is also helpful to surround the reactor with a material that scatters neutrons without much tendency to absorb them so that some of the emerging neutrons will be reflected back.

Neutron absorption inside the reactor is a more difficult problem because natural uranium contains only 0.7% of the fissionable isotope ^{235}U. The more abundant ^{238}U readily captures the fast neutrons liberated in fission, but usually does not undergo fission itself. The neutrons absorbed by ^{238}U are therefore wasted. However, ^{238}U has little ability to capture *slow* neutrons, whereas slow neutrons are much more apt to induce fission in ^{235}U than fast ones. Slowing down the fast neutrons produced by fission will thus prevent them from being taken up unproductively by ^{238}U and at the same time promote further fissions in ^{235}U.

Neutrons are slowed down by the moderator

In order to slow down fission neutrons, the uranium fuel in a reactor is embedded in a suitable substance called a *moderator*. When a fast neutron collides with a nucleus, some of its initial kinetic energy is imparted to the target nucleus. Just how much energy is lost by the neutron depends on the details of the interaction, but in general the more nearly equal the masses of the particles involved are, the more the energy transferred (see Sec. 7–6). Hence hydrogen would seem the best moderator, since hydrogen nuclei are simply protons whose mass is nearly identical to the neutron mass, and indeed ordinary ("light") water is the most common moderator in nuclear reactors since each water molecule contains two hydrogen atoms. The water also serves as the coolant in such reactors. The slowed-down neutrons are in thermal equilibrium with the moderator, which means energies of less than 1 eV at typical reactor operating temperatures.

Enriched uranium contains more ^{235}U than natural uranium does

Unfortunately, protons have a marked affinity for neutrons, so a neutron colliding with a proton may well become attached to it to form a deuteron, $^{2}_{1}$H. The neutron losses in a light-water reactor make it necessary to use *enriched* uranium whose ^{235}U content has been increased fourfold to about 3%. Most enriched uranium is produced by the gaseous diffusion process in which uranium hexafluoride (UF_6) gas is exposed to a succession of semipermeable barriers. Because of their smaller mass, molecules of $^{235}UF_6$ are slightly more likely to diffuse through each barrier than molecules of $^{238}UF_6$, and any desired degree of enrichment can be achieved in this way. Gaseous diffusion requires a large plant and much energy, so it is an expensive procedure. Other approaches, such as the use of gas centrifuges, are under development.

The actual operation of a reactor begins when enough fissionable material is brought together with a moderator. A stray neutron—from a spontaneous fission, perhaps, or a cosmic-ray secondary—causes a ^{235}U nucleus to split, which releases two or three other neutrons. These neutrons are slowed down from energies of several MeV to thermal energies in 0.001 s or so by collisions with moderator nuclei and then are absorbed by other ^{235}U nuclei to induce further fissions. To control the rate of the resulting chain reaction, the reactor has movable rods of cadmium or boron, which are good absorbers of slow neutrons. As these rods are inserted further and further into the reactor, the reaction rate is progressively reduced.

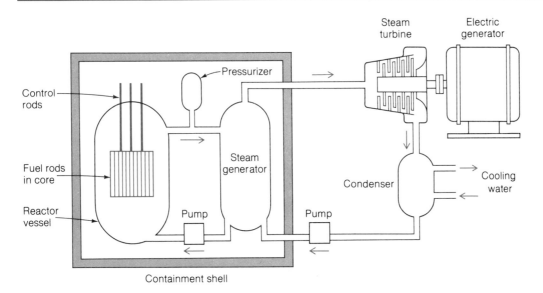

The fuel for a light-water reactor consists of uranium oxide (UO_2) pellets sealed in long, thin zirconium alloy tubes that are assembled along with control rods into a core that is enclosed in a steel pressure vessel. A typical pressure vessel is 10 m high, has an inside diameter of 3 m, and has walls 20 cm thick. In a *pressurized-water reactor* (PWR), the most common type, the water that circulates through the core is kept at a sufficiently high pressure, about 155 atm, to prevent boiling. The water enters the pressure vessel at perhaps 280°C and leaves at 320°C, passing through a heat exchanger to produce steam that drives a turbine (Fig. 31–19). Such a PWR used to produce electricity might contain 70 tons of UO_2 and operate at 2700 MW to yield 900 MW of electric power.

In a *boiling-water reactor* (BWR), a lower pressure of about 70 atm inside the pressure vessel allows steam to form, and this steam is sent directly to a turbine without using a separate steam generator. The BWR has the advantage over the PWR of greater simplicity, but it has the disadvantage that, if the coolant becomes contaminated by fission products from a leaking fuel rod or by corrosion products that have become radioactive due to neutron bombardment, there is one less barrier to the possible contamination of the outside world. Several hundred PWRs and BWRs are today in use for generating electricity and for warship propulsion; although their capital costs exceed those of fossil-fuel plants, their fuel costs are lower.

"Heavy" water, whose molecules contain deuterium ($_1^2H$) atoms instead of ordinary hydrogen ($_1^1H$) atoms, is a better moderator than light water because deuterons are less likely to capture neutrons. Although heavy water is more expensive than light water, natural uranium can be used as the fuel instead of enriched uranium. Reactors of this kind are the most efficient of nonbreeder reactors.

Early reactors, including the very first one (which was built in 1942 at the University of Chicago by Enrico Fermi, Walter Zinn, and their collaborators) used natural uranium with graphite as the moderator. Graphite, a form of pure carbon, has little tendency to pick up neutrons, has good mechanical properties, and is readily available, all of which help compensate for its relative inefficiency at slowing down neutrons.

FIG. 31–19 The basic design of a pressurized-water nuclear reactor. Water serves both as moderator and as coolant for the core. In a boiling-water reactor, steam is allowed to form inside the reactor vessel and then is used directly to power the turbine.

Boiling-water reactor

Heavy water contains deuterium instead of ordinary hydrogen

**High-temperature gas-
cooled reactor**

(A neutron loses 28% of its kinetic energy in a head-on collision with a ^{12}C nucleus, compared with 89% when the collision is with a deuteron and nearly 100% when it is with a proton.) More recently graphite has been used as the moderator in high-temperature gas-cooled reactors (HTGRs). An HTGR uses a gas such as helium or carbon dioxide for heat transfer and operates at a much higher temperature than a water-cooled reactor, which means a higher thermal efficiency (see Sec. 15–4). In a typical HTGR, helium enters at 319°C and emerges at 756°C for an overall efficiency of 38%, as compared with 32% to 33% for a PWR or BWR. HTGRs are more difficult to build than light-water reactors, but a number of them are operating in various countries.

**Neptunium and
plutonium are
transuranic elements
not found in nature**

Certain nonfissionable nuclides can be transmuted into fissionable ones by absorbing neutrons. A notable example is ^{238}U, which becomes ^{239}U when it captures a fast neutron. This uranium isotope beta decays with a half-life of 24 min into $^{239}_{93}$Np, an isotope of the element *neptunium,* which is also beta-radioactive. The decay of ^{239}Np has a half-life of 2.3 days and yields $^{239}_{94}$Pu, a fissionable isotope of *plutonium* whose half-life against alpha decay is 24,000 years. The entire sequence is thus

$$^{238}_{92}\text{U} + {}^{1}_{0}n \rightarrow {}^{239}_{92}\text{U} \xrightarrow{\hspace{2cm}} {}^{239}_{93}\text{Np} + e^{-}$$
$$\text{24 min}$$
$$\xrightarrow{\hspace{2cm}} {}^{239}_{94}\text{Pu} + e^{-}$$
$$\text{2.3 d}$$

Because plutonium is chemically different from uranium, the separation of ^{239}Pu from the ^{238}U remaining after neutron irradiation is easier to carry out than the separation of ^{235}U from the more abundant ^{238}U in natural uranium. Both neptunium and plutonium are transuranic elements, none of which are found on the earth because their half-lives are too short for them to have survived even if they had been present when the earth came into being 4.5 billion years ago. Transuranic elements up to $Z = 105$ have been produced by the neutron bombardment of lighter nuclides.

Another "fertile" nuclide besides ^{238}U is the thorium isotope ^{232}Th, whose transmutation into the fissionable ^{233}U proceeds via the protactinium isotope ^{233}Pa:

$$^{232}_{90}\text{Th} + {}^{1}_{0}n \rightarrow {}^{233}_{90}\text{Th} \xrightarrow{\hspace{2cm}} {}^{233}_{91}\text{Pa} + e^{-}$$
$$\text{22 min}$$
$$\xrightarrow{\hspace{2cm}} {}^{233}_{92}\text{U} + e^{-}$$
$$\text{27 d}$$

Breeder reactors

A *breeder reactor* is one designed to produce as much or more fissionable material, in the form of ^{239}Pu or ^{233}U, than the ^{235}U it consumes. Because the otherwise useless ^{238}U comprises 99.3% of natural uranium, and ^{232}Th is plentiful, the significance of breeder reactors in a world of dwindling energy resources is obvious. The widespread use of breeder reactors would mean that known reserves of nuclear fuel would be adequate for centuries to come, rather than decades, which would give enough time for alternative energy technologies to be perfected and implemented. On the other hand, the problems of reactor safety, nuclear waste disposal, and the proliferation of nuclear weapons would be more severe.

The thorium breeding cycle is most effective with the slow neutrons of the reactors described above. To convert ^{238}U to ^{239}Pu efficiently, on the other hand, requires fast neutrons, and a *fast breeder reactor* functions with fissions induced by fast rather than slow neutrons. Present breeder reactors are all of the fast type and use liquid sodium as the coolant; because they operate at high temperatures, their overall efficiencies are 40% or more. Such reactors yield as much as 60 to 70 times more energy from a given amount of natural uranium than conventional reactors do.

31–12 FUSION ENERGY

The basic energy-producing process in stars—and hence the source, direct or indirect, of nearly all the energy in the universe—is the fusion of hydrogen nuclei into helium nuclei. This can take place under stellar conditions in two different series of nuclear reactions. In stars whose internal temperatures are below 1.6×10^7 K, such as the sun, the *proton-proton cycle* predominates:

Stars obtain their energy from the proton-proton and carbon cycles of fusion reactions

$$^1_1H + {}^1_1H \rightarrow {}^2_1H + e^+ + 0.4\,MeV$$
$$^1_1H + {}^2_1H \rightarrow {}^3_2He + 5.5\,MeV \qquad \text{\textit{Proton-proton cycle}}$$
$$^3_2He + {}^3_2He \rightarrow {}^4_2He + 2\,{}^1_1H + 12.9\,MeV$$

The first two of these reactions must each occur twice for every synthesis of ^4_2He, so the total energy liberated is 24.7 MeV. In the other reaction sequence, the *carbon cycle,* which predominates in stars hotter than the sun, a $^{12}_6C$ nucleus absorbs three protons, with two positive beta decays along the way, until it becomes the nitrogen nucleus $^{15}_7N$. When a $^{15}_7N$ nucleus reacts with a fourth proton, the result is the formation of a ^4_2He nucleus and the reappearance of a $^{12}_6C$ nucleus: the original $^{12}_6C$ nucleus acts as a kind of catalyst for the process, since it emerges at the end ready to start another cycle.

The energy liberated by nuclear fusion is often called *thermonuclear energy.* High temperatures and densities are necessary for fusion reactions to occur in such quantity that a substantial amount of thermonuclear energy is produced. The high temperature assures that the initial light nuclei have enough thermal energy to overcome their mutual electric repulsion and come close enough together to react, and the high density assures that such collisions are frequent. A further condition for the proton-proton and carbon cycles is a large reacting mass, such as that of a star, since a number of separate steps is involved in each cycle and much time may elapse between the initial fusion of a particular proton and its ultimate incorporation in an alpha particle.

Fusion reactions that produce helium are not the only ones that occur in stars. In fact, nuclei up to $^{56}_{26}Fe$ are produced, this being the nucleus with the highest binding energy per nucleon (Sec. 31–3). The formation of nuclei heavier than $^{56}_{26}Fe$ by fusion requires a net energy input, whereas the formation of lighter nuclei evolves energy and so is favored under ordinary stellar conditions. Then where do the heavier elements in the universe come from? In a star, the inward pull of gravity is balanced by the pressure due to heat given off in fusion reactions. As a star ages, its fuel supply becomes exhausted, and it contracts. A star whose mass is comparable with that of the sun gradually shrinks down to about the size of the earth to become a *white dwarf.* A star much larger than the sun has a more violent fate, with a sudden collapse followed by a huge explosion. The exploding star is called a *supernova,* and when one occurs (every few hundred years on the average) it outshines other stars, galaxies, and planets for

Origin of the elements

some days or weeks. Most nuclei heavier than $^{56}_{26}$Fe are believed to have been created in supernova explosions, where sufficient energy is available to synthesize them from lighter nuclei. The heavy nuclei become dispersed in interstellar matter and are incorporated in new stars and planets that come into being from this matter.

Practical fusion reactions for the earth

On the earth, where any reacting mass must be very limited in size, an efficient fusion process cannot involve more than a single step. Two reactions that appear promising as sources of commercial power involve the combination of two deuterons to form a triton and a proton,

$$\,^2_1H + \,^2_1H \rightarrow \,^3_1H + \,^1_1H + 4.0\,\text{MeV}$$

or their combination to form a $^3_2$H nucleus and a neutron,

$$\,^2_1H + \,^2_1H \rightarrow \,^3_2He + \,^1_0n + 3.3\,\text{MeV}$$

Both reactions have about equal probabilities. A major advantage of these reactions is that deuterium is present in seawater; although its concentration is only 0.015%, this adds up to a total of about 10^{15} tons in the world's oceans.

The first fusion reactors are more likely to employ a deuterium-tritium mixture because the reaction

$$\,^3_1H + \,^2_1H \rightarrow \,^4_2He + \,^1_0n + 17.6\,\text{MeV}$$

has a higher yield than the others at relatively low temperatures. Seawater contains too little tritium for its economical extraction, but it can be produced by the neutron bombardment of the two isotopes of natural lithium:

$$\,^6_3Li + \,^1_0n \rightarrow \,^3_1H + \,^4_2He$$
$$\,^7_3Li + \,^1_0n \rightarrow \,^3_1H + \,^4_2He + \,^1_0n$$

In fact, tritium can be made by the fusion reactor itself by providing a lithium shield to absorb the neutrons liberated during the reactor's operation. Liquid lithium, heated by energetic neutrons, could also be used to extract the evolved energy by being circulated between the reactor and a steam generator. The steam then would power a turbine connected to an electric generator, as in the case of fossil-fuel and fission electric plants.

Possible methods for producing fusion energy

The big problem in exploiting fusion energy is to achieve the required combination of temperature, density, and containment time for a deuterium or deuterium-tritium mixture to react sufficiently to produce a net energy yield. Such a combination occurs in the explosion of a fission ("atomic") bomb, and incorporating the ingredients for fusion in such a bomb leads to a much more destructive weapon, the "hydrogen" bomb. One approach to the more controlled release of fusion energy involves using energetic beams to both heat and compress tiny deuterium-tritium pellets to produce what are, in effect, miniature hydrogen bomb explosions, of which a succession could furnish a steady supply of energy. Laser beams have received the most attention for this purpose, but electron and proton beams also show promise. However, the greatest amount of development has been concentrated on containing an extremely hot, dense plasma (fully ionized gas) of the reactants by means of a strong magnetic field, and considerable progress has been made in this direction. Although practical fusion reactors are still in the future (how far, nobody knows), the advantages of thermonuclear energy—cheap, essentially limitless fuel; much less hazard than for fission energy; relatively minor production of radioactive wastes; no possibility of military use—are such that eventually it will almost certainly supply a large part of the world's energy needs.

IMPORTANT TERMS

The **atomic number** of an element is the number of electrons in each of its atoms or, equivalently, the number of protons in each of its atomic nuclei.

The **neutron** is an electrically neutral particle, slightly heavier than the proton, that is present in nuclei together with protons. Neutrons and protons are jointly called **nucleons.** The **mass number** of a nucleus is the number of nucleons it contains.

Isotopes of an element have the same atomic number but different mass numbers. Symbols of **nuclides** (nuclear species) follow the pattern

$$_Z^A X$$

where X is the chemical symbol of the element, Z its atomic number, and A the mass number of the particular nuclide.

The **binding energy** of a nucleus is the energy equivalent of the difference between its mass and the sum of the masses of its individual constituent nucleons. This amount of energy must be supplied to the nucleus if it is to be completely disintegrated.

Radioactive nuclei spontaneously transform themselves into other nuclear species with the emission of radiation. The radiation may consist of **alpha particles,** which are the nuclei of helium atoms, or **beta particles,** which are positive or negative electrons. Positive electrons are known as **positrons.** Electron capture by a nucleus is an alternative to positron emission. The emission of **gamma rays,** which are energetic photons, enables an excited nucleus to lose its excess energy.

The time required for half of a given sample of a radioactive nuclide to decay is called its **half-life.**

In **nuclear fission,** the absorption of neutrons by certain heavy nuclei causes them to split into smaller **fission fragments.** Because each fission also liberates several neutrons, a rapidly multiplying sequence of fissions called a **chain reaction** can occur if a sufficient amount of the proper material is assembled. A **nuclear reactor** is a device in which a chain reaction can be initiated and controlled.

In **nuclear fusion,** two light nuclei combine to form a heavier one with the emission of energy. The energy liber-

ated in the process when it takes place on a large scale is called **thermonuclear energy.**

MULTIPLE CHOICE

1. The chemical behavior of an atom is determined by its
   (a) atomic number.
   (b) mass number.
   (c) binding energy.
   (d) number of isotopes.

2. The mass number of a nucleus is equal to the number of
   (a) electrons it contains.
   (b) protons it contains.
   (c) neutrons it contains.
   (d) nucleons it contains.

3. Atoms whose atomic numbers are the same but whose mass numbers are different are called
   (a) alpha particles.   (b) beta particles.
   (c) nuclides.   (d) isotopes.

4. The number of neutrons in the nucleus of the aluminum isotope $_{13}^{27}$Al is
   (a) 13.   (b) 14.
   (c) 27.   (d) 40.

5. Relative to the sum of the masses of its constituent nucleons, the mass of a nucleus is
   (a) greater.
   (b) the same.
   (c) smaller.
   (d) sometimes greater and sometimes smaller.

6. The ionization energy of an atom relative to the binding energy of its nucleus is
   (a) greater.
   (b) the same.
   (c) smaller.
   (d) sometimes greater and sometimes smaller.

7. The mass of a $_3^7$Li nucleus is 0.042 u less than the sum of the masses of 3 protons and 4 neutrons. The binding energy per nucleon in $_3^7$Li is
   (a) 5.6 MeV   (b) 10 MeV.
   (c) 13 MeV.   (d) 39 MeV.

8. The binding energy per nucleon is
(a) the same for all nuclei.
(b) greatest for very small nuclei.
(c) greatest for nuclei of intermediate size.
(d) greatest for very large nuclei.

9. The element whose nuclei contain the most tightly-bound nucleons is
(a) helium.
(b) carbon.
(c) iron.
(d) uranium.

10. A consequence of the limited range of the strong nuclear interaction is
(a) the existence of isotopes.
(b) the magnitude of nuclear binding energies.
(c) the instability of large nuclei.
(d) the ratio of atomic size to nuclear size.

11. In a stable nucleus the number of neutrons is always
(a) less than the number of protons.
(b) less than or equal to the number of protons.
(c) more than the number of protons.
(d) more than or equal to the number of protons.

12. As a sample of a radioactive nuclide decays, its half-life
(a) decreases.
(b) remains the same.
(c) increases.
(d) any of the above, depending upon the isotope.

13. After 2 hours, $\frac{1}{16}$ of the initial amount of a certain radioactive isotope remains undecayed. The half-life of the isotope is
(a) 15 min.
(b) 30 min.
(c) 45 min.
(d) 60 min.

14. When a nucleus undergoes radioactive decay, its new mass number is
(a) always less than its original mass number.
(b) always more than its original mass number.
(c) never less than its original mass number.
(d) never more than its original mass number.

15. A nucleus with an excess of neutrons may decay radio-actively with the emission of
(a) a neutron.
(b) a proton.
(c) an electron.
(d) a positron.

16. The first nuclear reaction ever observed occurred when $^{14}_{7}$N was bombarded with alpha particles and protons were ejected. This reaction produces
(a) $^{17}_{7}$N.
(b) $^{17}_{8}$O.
(c) $^{17}_{9}$F.
(d) $^{17}_{10}$Ne.

17. A reaction between a proton and $^{11}_{5}$B that produces $^{11}_{6}$C must also liberate
(a) a proton.
(b) a neutron.
(c) an electron.
(d) an alpha particle.

18. The process by which a heavy nucleus splits into two lighter nuclei is known as
(a) fission.
(b) fusion.
(c) alpha decay.
(d) a chain reaction.

19. By "chain reaction" is meant
(a) the joining together of protons and neutrons to form atomic nuclei.
(b) the joining together of light nuclei to form heavy ones.
(c) the successive fissions of heavy nuclei induced by neutrons emitted in the fissions of other heavy nuclei.
(d) the burning of uranium in a special type of furnace called a nuclear reactor.

20. Enriched uranium is a better fuel for nuclear reactors than natural uranium because it has a greater proportion of
(a) slow neutrons.
(b) deuterium.
(c) $^{235}_{92}$U.
(d) $^{238}_{92}$U.

21. From the point of view of a power plant engineer, a nuclear reactor is a source of
(a) heat.
(b) electric current.
(c) slow neutrons.
(d) gamma rays.

22. The sun's energy comes from
(a) nuclear fission.
(b) radioactivity.
(c) the conversion of hydrogen to helium.
(d) the conversion of helium to hydrogen.

23. Fusion reactions on the earth are likely to use as fuel
(a) ordinary hydrogen
(b) deuterium
(c) plutonium
(d) uranium

EXERCISES

1. In what ways are the isotopes of an element similar to one another? In what ways are they different?

2. State the number of neutrons and protons in each of the following nuclei: $^{6}_{3}$Li; $^{13}_{6}$C; $^{31}_{15}$P; $^{94}_{40}$Zr; $^{137}_{56}$Ba.

3. State the number of neutrons and protons in each of the following nuclei: $^{10}_{5}$Be; $^{22}_{10}$Ne; $^{36}_{16}$S; $^{88}_{38}$Sr; $^{180}_{72}$Hf.

4. Why do stable nuclei never have more protons than neutrons?

5. What limits the size of a stable nucleus?

6. What happens to the atomic number and mass number of a nucleus that emits a gamma ray? What happens to the actual mass of the nucleus?

7. Radium undergoes spontaneous decay into helium and radon. Why is radium regarded as an element rather than as a chemical compound of helium and radon?

8. What is the limitation on the fuel that can be used in a reactor whose moderator is ordinary water? Why is the situation different if the moderator is heavy water?

9. What are the differences and similarities between nuclear fission and nuclear fusion? Where does the evolved energy come from in each case?

10. Why are the conditions in the interior of a star favorable for nuclear fusion reactions?

11. The nuclide $^{238}_{92}$U decays into a lead isotope through the successive emissions of eight alpha particles and six electrons. What is the symbol of the lead isotope?

12. A $^{80}_{35}$Br nucleus can decay by beta decay, positron emission, or electron capture. What is the daughter nucleus in each case?

13. The nucleus $^{233}_{90}$Th undergoes two negative beta decays in becoming an isotope of uranium. What is the symbol of the isotope?

14. (a) Under what circumstances does a nucleus emit an electron? A positron? (b) The nuclei $^{14}_{8}$O and $^{19}_{8}$O both undergo beta decay in order to become stable nuclei. Which would you expect to emit a positron and which an electron?

15. How many disintegrations per second occur in a 25 millicurie sample of thorium?

16. The activity of a sample of $^{227}_{90}$Th is observed to be 12.5% of its original amount 54 days after the sample was prepared. What is the half-life of $^{227}_{90}$Th?

17. The half-life of radium is 1600 years. How long will it take for $\frac{15}{16}$ of a given sample of radium to decay?

18. A nucleus of $^{15}_{7}$N is struck by a proton. A nuclear reaction can take place with the emission of either a neutron or an alpha particle. Give the atomic number, mass number, and chemical name of the remaining nucleus in each of the above cases.

19. Complete the following nuclear reactions:

$$^{6}_{3}\text{Li} + ? \rightarrow {}^{7}_{4}\text{Be} + {}^{1}_{0}n$$
$$^{10}_{5}\text{B} + ? \rightarrow {}^{7}_{3}\text{Li} + {}^{4}_{2}\text{He}$$
$$^{35}_{17}\text{Cl} + ? \rightarrow {}^{32}_{16}\text{S} + {}^{4}_{2}\text{He}$$

20. A reaction often used to detect neutrons occurs when a neutron strikes a $^{10}_{5}$B nucleus, with the subsequent emission of an alpha particle. What is the atomic number, mass number, and chemical name of the remaining nucleus?

PROBLEMS

1. Using the figures in Table 31–1, verify that the average atomic mass of chlorine is 35.46 u.

2. Ordinary boron is a mixture of the $^{10}_{5}$B and $^{11}_{5}$B isotopes and has a composite atomic mass of 10.82 u. Find the percentage of each isotope present in ordinary boron. The atomic masses of $^{10}_{5}$B and $^{11}_{5}$B are, respectively, 10.01 u and 11.01 u.

3. The mass of $^{4}_{2}$He is 4.002603 u. Find its binding energy and binding energy per nucleon.

4. The binding energy of $^{20}_{10}$Ne is 160.64 MeV. Find its atomic mass.

5. The neutron decays in free space into a proton and an electron. What must be the minimum binding energy contributed by a neutron to a nucleus in order that the neutron not decay inside the nucleus? How does this figure compare with the observed binding energies per nucleon in stable nuclei?

6. The atomic masses of $^{15}_{7}$N, $^{15}_{8}$O, and $^{16}_{8}$O are, respectively, 15.0001, 15.0030, and 15.9949 u. (a) Find the average binding energy per nucleon in $^{16}_{8}$O. (b) How much energy is needed to remove one proton from $^{16}_{8}$O? (c) How much energy is needed to remove one neutron from $^{16}_{8}$O? (d) Why are these figures different from one another?

7. The distance between the two protons in a $^{3}_{2}$He nucleus is roughly 1.7×10^{-5} m. (a) Calculate the electric potential energy of these protons. (b) Show that this energy is of the right order of magnitude to account for the difference in binding energy between $^{3}_{1}$H and $^{3}_{2}$He. What conclusion can be drawn from this result about the dependence of nuclear forces upon electric charge? The atomic mass of $^{3}_{1}$H is 3.016050 u and that of $^{3}_{2}$He is 3.016030 u.

8. The electric potential energy of two protons a distance r apart is ke^2/r. Compare the electric potential energy of two protons 5×10^{-15} m apart with the binding energy per nucleon of nuclei with $A \approx 60$. Such nuclei are about 5×10^{-15} m in radius.

9. The nuclear reaction

$$^{6}_{3}\text{Li} + {}^{2}_{1}\text{H} \rightarrow 2\,{}^{4}_{2}\text{He}$$

evolves 22.4 MeV. Calculate the mass of $^{6}_{3}$Li in u.

10. The atomic mass of $^{226}_{88}$Ra is 226.0254 u and the energy liberated in its alpha decay is 4.87 MeV. (a) Identify the daughter nucleus and find its atomic mass. (b) The alpha particle emitted in the decay is observed to have a KE of 4.78 MeV. Where do you think the other 0.09 MeV goes?

11. The nuclide $^{232}_{92}$U (mass 232.0372 u) alpha decays into $^{228}_{90}$Th (mass 228.0287 u). (a) Find the amount of energy released in the decay. (b) Is it possible for $^{232}_{92}$U to decay into $^{231}_{92}$U (mass 231.0364 u) by emitting a neutron? Why? (c) Is it possible for $^{232}_{92}$U to decay into $^{231}_{91}$Pa (mass 231.0359 u) by emitting a proton? Why?

12. The half-life of the alpha-emitter $^{210}_{84}$Po is 138 days. This nuclide is used to power a thermonuclear cell whose initial output is 1 W. What will the output be after 1 year?

13. In free space the neutron is unstable and beta-decays into a proton and an electron with a half-life of 10.8 minutes. What proportion of the original number of neutrons in a beam will remain undecayed after 1 minute? After 1 hour?

14. (a) How much mass is lost per day by a nuclear reactor operated at a 1 GW (10^9 W) power level? (b) If each fission in $^{235}_{92}$U releases 200 MeV, how many fissions must occur per second to yield a power level of 1 GW?

15. $^{235}_{92}$U loses about 0.1% of its mass when it undergoes fission. (a) How much energy is released when 1 kg of $^{235}_{92}$U undergoes fission? (b) One ton of TNT releases about 4×10^9 J when it is detonated. How many tons of TNT are equivalent in destructive power to a bomb that contains 1 kg of $^{235}_{92}$U?

16. Fusion reactions involving deuterons can take place when the deuterons have energies of about 10 keV or more. Find the temperature of a deuterium plasma in which the average KE of the deuterons is 10 keV.

17. Old stars obtain part of their energy by the fusion of three alpha particles to form a $^{12}_{6}$C nucleus, whose mass is 12.0000 u. How much energy is evolved in each such reaction?

ANSWERS TO MULTIPLE CHOICE

1. a	7. a	13. b	19. c
2. d	8. c	14. d	20. c
3. d	9. c	15. c	21. a
4. b	10. c	16. b	22. c
5. c	11. d	17. b	23. b
6. c	12. b	18. a	

ELEMENTARY PARTICLES 32

Ordinary matter is composed of neutrons, protons, and electrons. Another significant elementary particle is the neutrino, which is emitted along with the electron in beta decay. All these particles have antiparticle counterparts whose mass is the same but whose charge (and certain other properties) differ. When a particle and its antiparticle come together, they both disappear as their mass is turned into energy. A great many other elementary particles have been discovered in recent years, all of which are unstable and decay rapidly after being created in high-energy collisions between other elementary particles. It seems likely that most of these short-lived particles, along with the neutron and the proton, should not be considered as true "elementary particles" but instead should be considered as composites of a small number of rather unusual particles called quarks that have not as yet been isolated experimentally—and may never be.

32–1   PARTICLE DETECTION

A variety of methods have been devised for detecting photons and subatomic particles. Nearly all of them are based on the ionization that results, directly or indirectly, from the passage of a photon or particle through matter.

Heavy charged particles

Heavy charged particles, such as protons and alpha particles, lose energy chiefly by electric interactions with atomic electrons. These electrons are either raised to excited states or, more often, are pulled away from their parent atoms entirely. The ejected electrons may have energies of several hundred electron volts and so are able to ionize atoms themselves. The incident particle slows down gradually until it comes to a stop or reacts with a nucleus in its path.

Electrons

Electrons, too, lose energy by ionization, but for them another mechanism may also be important. As mentioned in Sec. 23–2, electromagnetic radiation is given off whenever an electric charge is accelerated. Energy loss by radiation is more significant for electrons than for heavier particles because they are more violently accelerated when passing near nuclei in their paths. The more energetic the electron and the greater the nuclear charge of the atoms it encounters, the more it radiates. In lead the rate of energy loss by radiation becomes equal to that by ionization for an electron energy of about 10 MeV, whereas in air ionization remains dominant below an electron energy of about 100 MeV.

Neutrons

Because neutrons are uncharged, they do not interact with atomic electrons in their paths but only with nuclei. Most such interactions are "billiard ball" collisions in which part of the neutron's kinetic energy is given to the target nucleus. As discussed in Sec. 7–6, the closer the mass of the target nucleus is to the neutron mass, the more the energy that will be transferred. Thus a 2-MeV neutron on the average needs 18 collisions with hydrogen nuclei to reach equilibrium at room temperature (when its most probable energy is 0.025 eV), 25 with deuterium nuclei, 114 with carbon nuclei, and 2150 with uranium nuclei.

The three principal ways in which the energetic photons of X and gamma rays lose energy when they pass through matter are the photoelectric effect, Compton scattering, and pair production (Fig. 32–1). In each case photon energy is conveyed to electrons, which in turn lose energy mainly by exciting or ionizing atoms in the absorbing material. At low photon energies the photoelectric effect is the principal mechanism of energy loss, to be replaced by Compton scattering at higher energies. As the photon energy increases beyond the minimum energy of 1.02 MeV required, pair production

FIG. 32–1   X and gamma rays interact with matter chiefly through the photoelectric effect, Compton scattering, and pair production. Pair production requires a photon energy of at least 1.02 MeV.

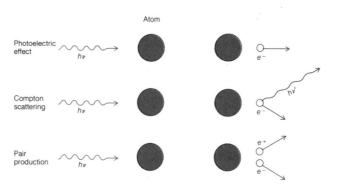

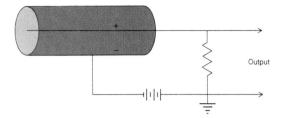

FIG. 32–2 A gas-filled detector of this kind can be used either as an ionization chamber or as a Geiger counter. When it is to be an ionization chamber, the applied voltage is low, and the output pulse that occurs when a photon or charged particle passes through the tube is a measure of the amount of ionization produced in the gas. When it is to be a Geiger counter, the applied voltage is high enough to give the initial ions enough energy of their own to cause further ionization, and the result is a large output pulse that is the same for all incident photons or particles.

becomes more and more important. Gamma rays from radioactive decay and X rays interact with matter largely through Compton scattering.

A number of devices can be used to detect the ionization produced by an energetic photon or charged particle. Some, such as the *ionization chamber* and the *Geiger counter,* are gas-filled tubes with electrodes that attract the electrons and ions to produce voltage pulses in external circuits (Fig. 32–2). Another device, the *scintillation counter,* is based on the emission of a flash of light by certain substances (such as those used to make the screen of a television picture tube) when struck by ionizing radiation. The light is then picked up by a sensitive photomultiplier tube (Fig. 32–3). *Semiconductor*

Ionization detectors

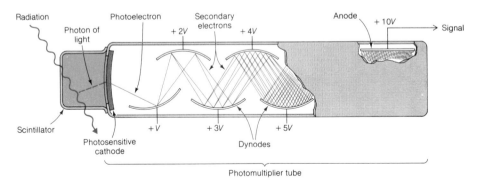

FIG. 32–3 Essential features of a scintillation counter. The scintillator may be a crystalline or plastic solid or a liquid. The number of dynodes in the photomultiplier tube may be ten or more, with each dynode several hundred volts higher in potential than the one before it. Each dynode produces 2 to 5 secondary electrons per incident electron for a total amplification of as much as 10^6 or 10^7. A scintillation counter is more rapid in its operation than an ionization chamber or Geiger counter.

detectors make use of the fact that a thin layer on both sides of a *p-n* junction is normally depleted in charge carriers when the crystal has an external voltage across it; electrons and holes produced there by an ionizing particle lead to a measurable pulse.

Neutron detectors

Although neutrons do not themselves produce ionization and so cannot be detected directly, indirect techniques have been devised. For example, a counter whose sensitive volume is rich in hydrogen will respond to fast neutrons through the ionization produced by protons recoiling after collisions with the neutrons. In the case of slow neutrons, one method is to employ the nuclear reaction

$$\_0^1 n + \_5^{10}B \rightarrow \_2^4 He + \_3^7 Li$$

in which a neutron reacts with a nucleus of the boron isotope $\_5^{10}B$ to produce an alpha particle and a nucleus of the lithium isotope $\_3^7 Li$. If $\_5^{10}B$ is incorporated in the sensitive volume of a suitable counter (the gas BF_3 is often used), the appearance of ionization that corresponds to an alpha particle signifies the capture of a slow neutron.

Bubble chamber

Instruments of the above kinds are useful when photons or particles of a certain kind that arrive at a certain place are to be counted or have their energies measured. If the interactions and decays of elementary particles are being studied, however, a means of displaying the paths of the various particles for visual inspection is needed. The most widely used device for this purpose is the *bubble chamber*. In a bubble chamber a suitable liquid is heated to a temperature above its normal boiling point while under sufficient pressure to keep it from boiling. The pressure is then reduced suddenly, which leaves the liquid in a superheated condition. A superheated liquid is unstable, and bubbles of vapor form around any ions present. A charged particle moving through the liquid at just this time will leave a track of bubbles that can be photographed. A magnetic field is usually applied to a bubble chamber to permit determining the charges and momenta of the various particles involved in an event that takes place inside it (Fig. 32–4). Bubble chambers for elementary particle research are usually filled with hydrogen, whose nuclei are protons, at a temperature of −246°C and a pressure of 5 bars.

In some experiments a *spark chamber* is preferred to a bubble chamber. A spark chamber consists of a series of parallel metal plates in a gas-filled container. A high

FIG. 32–4 How pair production appears in a bubble chamber placed in a magnetic field. Bubbles form along the ion trails of the electron and positron in the superheated liquid. As the particles slow down, the radii of their paths decrease in accord with Eq. (19–9), and the result is a pair of spiral tracks.

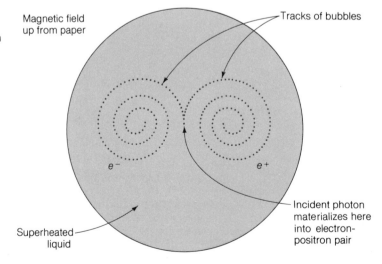

Magnetic field up from paper

Tracks of bubbles

e^- $e+$

Superheated liquid

Incident photon materializes here into electron-positron pair

voltage is applied between each pair of plates. The passage of a charged particle through the chamber leaves behind a trail of ions that increases the conductivity of the gas, and sparks occur along this trail. The string of sparks can be photographed, and, as with a bubble chamber, a magnetic field enables the charge and momentum of the responsible particle to be established from the curvature of the track.

Spark chamber

32–2   THE NEUTRINO

In radioactive decay, as in all other natural processes, energy (including mass energy) is conserved. For this reason the total mass of the products of a particular decay must be less than the mass of the initial nucleus, with the missing mass appearing as photon energy in the case of gamma decay and as kinetic energy in the cases of alpha and beta decay. In gamma and alpha decay the liberated energy is indeed precisely equal to the energy equivalent of the lost mass, but in beta decay a strange effect occurs: instead of all having the same energy, the emitted electrons from a particular nuclide exhibit a variety of energies. These energies range from zero up to a maximum figure equal to the energy equivalent of the missing mass in the transformation. This effect is illustrated in Fig. 32–5, which shows the spread in electron energy in the decay of $^{210}_{83}\text{Bi}$.

Momentum as well as energy is apparently not conserved in beta decay. When an object at rest disintegrates into two parts, they must move apart in opposite directions in order that the total momentum of the system remain zero. Experiments show, however, that the emitted electron and the daughter nucleus do *not* in general travel in opposite directions after beta decay occurs, so their momenta cannot cancel out to equal the initial momentum of zero.

Energy, momentum, and angular momentum are apparently not conserved in beta decay

A third difficulty concerns angular momentum. The spin angular momentum component in a given direction of the neutron, the proton, the electron, and the positron is in every case $\pm \frac{1}{2}h/2\pi$. The conversion of a neutron into a proton and an electron or of a proton into a neutron and a positron therefore leaves an angular momentum discrepancy of $\frac{1}{2}h/2\pi$.

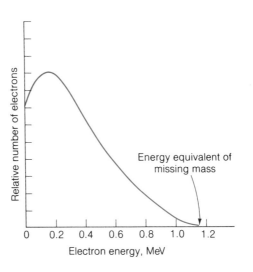

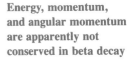
Energy equivalent of missing mass

FIG. 32–5 The spread of electron energies found in the beta decay of $^{210}_{83}\text{Bi}$. The maximum electron energy is equal to the energy equivalent of the mass lost by the decaying nucleus minus the electron mass.

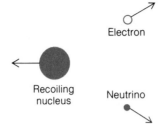

FIG. 32–6 An electron and a neutrino are simultaneously emitted in the beta decay of a nucleus, which makes possible the conservation of energy, momentum, and angular momentum in the process.

The neutrino was postulated to avoid discarding conservation principles

The process of beta decay is the first one we have encountered in which the conservation laws of energy, linear momentum, and angular momentum do not seem to hold. To account for the above discrepancies without abandoning three of the most fundamental and otherwise well-established physical principles, the existence of a new particle was postulated, the *neutrino,* symbol ν_e (Greek letter "nu"). The neutrino has an electric charge and either no mass or very little, but is able to possess both energy and momentum and has an intrinsic spin. (Lest this seem unlikely, we might reflect that the photon, also massless, has energy, momentum, and angular momentum. The neutrino is *not* a photon, however, but an entirely different entity.) According to the neutrino theory, an electron and neutrino are simultaneously emitted in beta decay, which permits energy and momentum to be conserved (Fig. 32–6).

For a quarter of a century the neutrino hypothesis was accepted despite the absence of any direct evidence in its support. This was a very unusual situation: here was a particle of a rather odd kind, which nobody had ever detected experimentally, yet practically no physicists doubted its existence. It was far from being a matter of blind faith, however. There were no theoretical objections to the neutrino hypothesis, whereas there were strong theoretical and experimental objections to dropping the principles of conservation of energy, linear momentum, and angular momentum. Furthermore, the neutrino, which lacks charge, has at most about 10^{-5} the mass of the electron, and is not electromagnetic in nature as is the photon, interacts only feebly with matter, so it was not easy to think of a way to detect it.

Experimental confirmation of the neutrino

Finally, in 1956, an experiment was performed in which a nuclear reaction that, in theory, could only be caused by a neutrino, was actually found to take place. In the operation of a nuclear reactor, a great many beta decays take place, and as a result more than 10^{16} neutrinos per second may emerge from each square meter of the shielding around a reactor. A neutrino striking a proton has a small probability of inducing the reaction

$$\nu_e + p \rightarrow n + e^+$$

in which a neutron and positron are created. By placing a sensitive detecting chamber containing hydrogen near a nuclear reactor, the simultaneous appearance of a neutron and a positron could be registered each time the above reaction occurred. Calculations were made initially of how many such reactions per second should occur based on the known properties of the detecting apparatus and on the theoretical properties of the neutrino. When this reaction rate was actually found, there was no doubt that neutrinos indeed exist.

Neutrinos are able to travel unimpeded through vast amounts of matter because they are limited to the weak interaction. On the average, a neutrino must traverse over 100 light-years of solid iron before being absorbed—and a light-year, the distance light travels in a year, is 9.5×10^{15} m.

In the sequence of nuclear reactions by which hydrogen is converted to helium in the sun and other stars, two beta decays occur for each helium nucleus formed. A vast number of neutrinos is therefore produced in the sun at all times. Because neutrinos travel freely through matter, almost all of these neutrinos escape into space and take with them 6 to 8% of the total energy generated by the sun. The flux of neutrinos from the sun is such that every cubic centimeter on the earth contains several neutrinos at any moment. The considerable energy carried by the neutrinos created in the sun and the other stars is apparently lost forever from the universe in the sense that it is no longer available for conversion into other forms of energy, such as matter.

Only the weak interaction affects neutrinos

The universe is flooded with neutrinos

32–3 ANTIPARTICLES

Another experimental discovery of a particle whose existence had been predicted theoretically decades earlier is that of the negative proton, or *antiproton,* whose symbol is $\bar{p}$. This is a particle with the same properties as the proton except that it has a negative electric charge. The existence of antiprotons was predicted largely on the basis of symmetry arguments: since the electron has a positive counterpart in the positron, why should the proton not have a negative counterpart as well? Actually, as sophisticated theories show, this is an excellent argument, and few physicists were surprised when the antiproton was actually found.

The antiproton has a negative charge

The reason positrons and antiprotons are so difficult to find is that they are readily *annihilated* upon contact with ordinary matter (Fig. 32–7). When a positron is in the vicinity of an electron, they attract one another electrically, come together, and then both vanish simultaneously, with their missing mass appearing in the form of two gamma-ray photons:

Particles and their antiparticles annihilate each other

$$e^+ + e^- \rightarrow \gamma + \gamma$$

The total mass of the two particles is the equivalent of 1.02 MeV, and so each photon has an energy of 0.51 MeV. (Their energies must be equal and they must be emitted in opposite directions in order that momentum be conserved.)

While the similar reaction

$$p + \bar{p} \rightarrow \gamma + \gamma$$

can occur when a proton and antiproton undergo annihilation, it is more usual for the

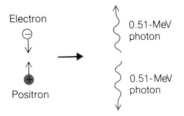

FIG. 32–7 The mutual annihilation of an electron and a positron.

FIG. 32–8 The production of (a) an electron-positron pair and (b) a proton-antiproton pair by the materialization of sufficiently energetic photons. Pair production can occur only in the presence of a nucleus.

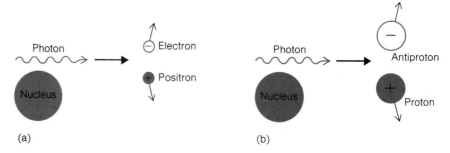

(a)   (b)

vanished mass to reappear in the form of several mesons, particles which we shall consider in the next section.

In pair production, a particle and its antiparticle are created

The reverse of annihilation can also take place, with the electromagnetic energy of a photon materializing into a positron and an electron or, if it is energetic enough, into a proton and an antiproton (Fig. 32–8). This phenomenon is known as *pair production,* and requires the presence of a nucleus in order that momentum as well as energy be conserved. Any photon energy in excess of the amount required to provide the mass of the created particles (1.02 MeV for a positron-electron pair, 1872 MeV for a proton-antiproton pair) appears as kinetic energy.

Antineutrons and antineutrinos

Antineutrons (symbol $\bar{n}$) and antineutrinos (symbol $\bar{\nu}_e$) have also been identified. Antineutrons can be detected through their mutual annihilation with neutrons, while more indirect, though equally definite, evidence supports the existence of antineutrinos. The antineutrino differs from the neutrino in that, while the spin axes of both are parallel to their directions of motion, the spin of the former is clockwise and that of the latter is counterclockwise when viewed from behind. A moving neutrino may be thought of as resembling a left-handed screw, and a moving antineutrino as resembling a right-handed screw (Fig. 32–9). An antineutrino is released during a beta decay in which an electron is emitted, and a neutrino is released during a beta decay in which a positron is emitted or an electron is captured. Thus the correct equations of beta decay are

$$n \rightarrow p + e^- + \bar{\nu}_e \qquad \text{Electron emission} \quad (32\text{–}1)$$
$$p \rightarrow n + e^+ + \nu_e \qquad \text{Positron emission} \quad (32\text{–}2)$$
$$p + e^- \rightarrow n + \nu_e \qquad \text{Electron capture} \quad (32\text{–}3)$$

Neutrino

Antineutrino

FIG. 32–9 The spins of the neutrino and antineutrino are in opposite senses with respect to their directions of motion.

Annihilation and pair production are consequences of the facts that matter is a form of energy. Conversions from matter to energy and from energy to matter are therefore no more improbable than conversions from, say, gravitational potential energy to kinetic energy.

Ordinary atoms are composed of neutrons, protons, and electrons. There is apparently no reason why atoms composed of antineutrons, antiprotons, and positrons should not be stable and behave in every way like ordinary atoms. It would seem to be an attractive notion that equal amounts of matter and *antimatter* came into being at the origin of the universe which became segregated into separate galaxies. The spectra of the light emitted by the members of antimatter galaxies would be exactly the same as the spectra of the light emitted by the members of galaxies of ordinary matter, which gives us no way to distinguish between the two—except when antimatter comes in contact with ordinary matter, whereupon their mutual annihilation would occur with the release of a great deal of energy. But in fact such events have never been observed, so it seems that the universe consists entirely of ordinary matter. Current theories of the fundamental interactions seem able to account for the absence of antimatter in terms of the character of these interactions in the early moments of the universe, which is likely to have been different from what it is now.

Matter and antimatter

32–4   MESON THEORY OF NUCLEAR FORCES

In 1935 the Japanese physicist Yukawa suggested that the strong nuclear interaction could be regarded as the result of an interchange of certain particles between nucleons. Today these particles are called *pions*. Pions may be charged or neutral; those with charges of $+e$ or $-e$ have rest masses of 273 times the electron mass, while neutral pions have rest masses of 264 times the electron mass. Pions are members of a class of elementary particles collectively called *mesons*—the word pion is a contraction of the original name π-meson.

The strong interaction is mediated by pion exchange

The crude analogy illustrated in Fig. 32–10 may help in understanding how meson exchange can lead to both attractive and repulsive forces between nuclei. Each child in the figure has a pillow. When the children exchange pillows by snatching them from each other's grasp, the effect is like that of a mutually attractive force. On the other hand, the children may also exchange pillows by throwing them at each other. Here conservation of momentum requires that the children move apart, just as if a repulsive force were present between them.

According to Yukawa's theory, nearby nucleons constantly exchange mesons without themselves being altered. We note that the emission of a meson by a nucleon at rest

FIG. 32–10 Particle exchange can lead to attractive or repulsive forces.

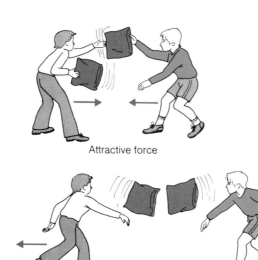

Attractive force

Repulsive force

FIG. 32–11 A meson can come into being, travel to another nucleon, and disappear if the entire process occurs rapidly enough.

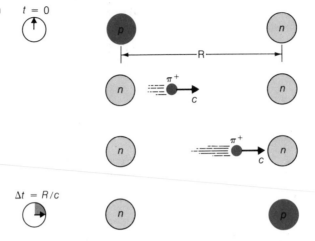

which does not lose a corresponding amount of mass violates the law of conservation of energy. However, the law of conservation of energy, like all physical laws, deals only with measurable quantities. Because the uncertainty principle restricts the accuracy with which we can perform certain measurements, it limits the range of application of physical laws such as that of energy conservation.

In Sec. 27–9 we saw that the uncertainty principle could also be expressed in the form

$$\Delta E \, \Delta t \geqslant \frac{h}{2\pi} \tag{32–4}$$

The uncertainty principle permits the temporary nonconservation of energy

On the basis of this formula, we can infer that a process can take place in which energy is *not* conserved by an amount ΔE *provided* that the time interval Δt in which the process occurs is not more than $h/2\pi\Delta E$. Thus the creation, transfer, and disappearance of a meson do not conflict with the conservation of energy if the sequence takes place fast enough. The latter condition provides a way to estimate the mass of the pion.

Let us assume that the temporary energy discrepancy ΔE is of the same magnitude as the rest energy mc^2 of the pion, and that the pion travels at very nearly the speed of light c as it goes from one nucleon to another. (These assumptions are crude because the kinetic energy of the pion is ignored, but all we are after is an approximate figure for m.) The time Δt the pion spends between its creation in one nucleon and its absorption in another cannot be greater than R/c, where R is the maximum distance that can separate interacting nucleons (Fig. 32–11).

Estimate of pion mass

We therefore have, using the symbol $\approx$ to indicate that the result is only a rough approximation,

$$\Delta E \, \Delta t \approx \frac{h}{2\pi}$$

$$mc^2 \times \frac{R}{c} \approx \frac{h}{2\pi}$$

$$m \approx \frac{h}{2\pi Rc} \tag{32–5}$$

The strong nuclear interaction responsible for the attractive forces between nucleons has a range of about 1.7×10^{-15} m. When we substitute $R = 1.7 \times 10^{-15}$ m in the above formula, we obtain $m \approx 2.1 \times 10^{-28}$ kg for the pion mass, which is about 230 electron masses. Of course, the preceding calculation is hardly a rigorous one, but if Yukawa's theory has any validity, the pions he postulated should have masses some-where in this vicinity—as they do. Heavier mesons than the pion have also been dis-covered, some over a thousand times more massive than the electron. The contribution of these mesons to nuclear forces is, from Eq. (32–5), limited to shorter distances than those characteristic of pions.

Some years before Yukawa's work, particle exchange had been suggested as the mechanism of a different kind of interaction, that responsible for electromagnetic forces. In this case the particles are photons which, being massless, are not limited in range by Eq. (32–5). However, the greater the distance between two charges, the smaller must be the energies of the photons that pass between them (and hence the less the momenta of the photons and the weaker the resulting force) in order that the uncertainty principle not be violated. For this reason electric forces decrease with distance. Because the photons exchanged in the interactions of electric charges cannot be detected, they are called *virtual photons*; as in the case of pions, they can become actual photons if enough energy is somehow supplied to liberate them from the energy conservation constraint. The idea of photons as carriers of electromagnetic forces is attractive on many counts, an obvious one being that it explains why such forces are transmitted with the speed of light and not, say, instantaneously. As subsequently developed, the full theory is called *quantum electrodynamics,* and its conclusions have turned out to be in extraordinarily precise agreement with the data on such phenomena as the pho-toelectric effect, pair production and annihilation, and the emission of photons by excited atoms and by accelerated charges.

The em interaction is mediated by virtual photon exchange

32–5 PIONS AND MUONS

Not long after Yukawa's work, charged particles of about the right mass were experi-mentally discovered in the cosmic radiation. Their discovery was not unexpected because a sufficiently energetic nuclear collision should be able to liberate mesons by providing enough energy to create them without violating conservation of energy, and nuclear collisions between fast cosmic-ray protons from space and oxygen and nitrogen nuclei occur constantly in the atmosphere. However, these particles did not behave at all in the way they were expected to behave. Far from strongly interacting with nuclei, as Yukawa's mesons were supposed to do, they barely interacted at all. Instead of being absorbed in at most a meter of earth, they penetrated thousands of meters into the ground.

Finally, in 1947, the explanation for their peculiar behavior was found. The weakly interacting mesons, known as *muons* (a constraction of μ-meson, where μ is the Greek letter "mu") and the chief constituent of cosmic rays at sea level, are not the direct products of nuclear collisions, but are secondary particles that result from the decay of pions.

At sea level, most cosmic rays are muons

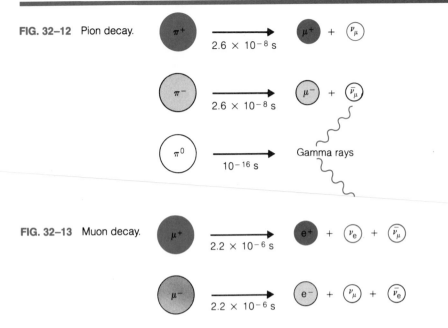

FIG. 32–12   Pion decay.

FIG. 32–13   Muon decay.

Pion and muon decay schemes

Outside a nucleus a charged pion decays in an average of 2.6×10^{-8} s into a muon and a neutrino (or antineutrino) plus kinetic energy (Fig. 32–12). A small proportion of charged pions also decay directly into an electron or positron plus a neutrino or antineutrino. The positive and negative muons decay into positrons and electrons respectively together with a neutrino-antineutrino pair in each case (Fig. 32–13). Neutral pions, whose masses of 264 m_e are a little less than the charged pion mass of 273 m_e, decay in about 10^{-16} s into a pair of gamma-ray photons: $\pi^0 \rightarrow \gamma + \gamma$.

A different kind of neutrino is associated with the muon

The neutrinos involved in pion decay have been denoted ν_μ, whereas those involved in beta decay have been denoted ν_e. Until 1962 it had been thought that there is only a single kind of neutrino. In that year an experiment was performed in which pions were produced by bombarding a metal target with high-energy protons from an accelerator. The pion decays liberated neutrinos, and the interactions of these neutrinos with matter was studied. The only inverse reactions found led to the production of muons; no electrons whatever were created. Hence the neutrinos set free in pion decay are different from those set free in beta decay.

32–6   CATEGORIES OF ELEMENTARY PARTICLES

Several hundred elementary particles have been discovered in recent years in addition to the ones discussed above. Such abundance where scarcity had been expected stimulated an intense research effort, with the result that a number of suggestive regularities have been found in the properties of the various particles. The search for a comprehensive theory of elementary particles—perhaps more difficult, frustrating, and exciting than any other in the history of science—seems at last to be on the road to success, as we shall see later in this chapter.

Table 32–1 is a list of the 37 longest-lived elementary particles, together with some of their properties. The other known particles have much briefer lifetimes, in most cases so short ($10^{-22} - 10^{-23}$ s) that their existence is inferred on the basis of indirect evidence. The particles in the table, together with the unlisted ones, fall into four categories: *photons, leptons, mesons,* and *baryons.* The photon, π^0 meson, and η^0 mesons are their own antiparticles. Only the principal mode or modes of decay are given; some particles may decay in a variety of other ways.

TABLE 32–1
The 37 longest-lived elementary particles. Particles subject to the strong nuclear interaction are called hadrons; the various mesons and baryons are hadrons.

Name	Particle	Anti-particle	Mass, in Electron Masses	Stability and Chief Decay Mode	Average Lifetime, s	Category
Photon	γ	(γ)	0	Stable		Photon
Neutrino	ν_e	$\bar{\nu}_e$	0	Stable		Leptons
	ν_μ	$\bar{\nu}_\mu$	0	Stable		
Electron	e^-	e^+	1	Stable		
Muon	μ^-	μ^+	207	Unstable; decays into electron plus two neutrinos	2.2×10^{-6}	
Pion	π^+	π^-	273	Unstable; decays into muon plus neutrino	2.6×10^{-8}	Mesons
	π^0	(π^0)	264	Unstable; decays into two gamma rays	8.9×10^{-17}	
Kaon	K^+	K^-	966	Unstable; decays into muon plus neutrino or into two or three pions	1.2×10^{-8}	
	K_1^0	$\overline{K_1^0}$	974	Unstable; decays into two pions	8.7×10^{-11}	
	K_2^0	$\overline{K_2^0}$	974	Unstable; decays into three pions or into pion and neutrino plus muon or electron	5.3×10^{-8}	
Eta meson	η^0	(η^0)	1073	Unstable; decays into three pions or two gamma rays	2.5×10^{-19}	
Proton	p^+	p^-	1836	Stable		Baryons
Neutron	n^0	$\overline{n^0}$	1839	Unstable in free space; decays into electron, proton, and neutrino	932	
Lambda hyperon	Λ^0	$\overline{\Lambda^0}$	2182	Unstable; decays into pion plus neutron or proton	2.5×10^{-10}	
Sigma hyperon	Σ^+	$\overline{\Sigma^+}$	2328	Unstable; decays into pion plus neutron or proton	8×10^{-11}	
	Σ^-	$\overline{\Sigma^-}$	2341	Unstable; decays into neutron plus pion	1.5×10^{-10}	
	Σ^0	$\overline{\Sigma^0}$	2332	Unstable; decays into lambda hyperon plus gamma ray	10^{-14}	
Xi hyperon	Ξ^-	$\overline{\Xi^-}$	2583	Unstable; decays into lambda hyperon plus pion	1.7×10^{-10}	
	Ξ^0	$\overline{\Xi^0}$	2571	Unstable; decays into lambda hyperon plus pion	3.0×10^{-10}	
Omega hyperon	Ω^-	$\overline{\Omega^-}$	3290	Unstable; decays into lambda hyperon plus K meson or into xi hyperon plus pion	1.3×10^{-10}	

Photons. As we know, photons are quanta of electromagnetic energy that are stable in space and have zero rest mass. It is possible to think of the electromagnetic interaction as being "carried" from one charge to another by the circulation of photons between them, just as the strong nuclear interaction is carried from one nucleon to another by the circulation of mesons between them.

The graviton is the hypothetical quantum of the gravitational interaction

Since both mesons and photons actually exist, it is tempting to ask whether the gravitational and weak interactions also might have force-carrying particles associated with them. Since gravitational forces are unlimited in range, Eq. 32–5 requires that the hypothetical *graviton* have zero rest mass. (The same argument accounts for the zero rest mass of the photon.) Also like the photon, the graviton should be stable, electrically neutral, and travel with the speed of light. Unlike the photon, the graviton should interact only feebly with matter, and so would be extremely hard to detect. There is no definite experimental evidence either for or against the existence of the graviton.

The intermediate boson is the hypothetical quantum of the weak interaction

The quanta of the weak interaction are called *intermediate bosons,* of which there should be more than one kind. Because the weak interaction has a shorter range than the strong interaction, the rest masses of the intermediate bosons should be very large, more than 30 proton masses. They should be charged and decay rapidly. A number of careful searches has turned up no traces of intermediate bosons, but indirect evidence for them remains strong.

Leptons. Leptons include electrons, muons, and neutrinos. One family of leptons consists of the electron, the positron, the ν_e neutrino, and the $\bar{\nu}_e$ antineutrino; another consists of the positive and negative muon, the ν_μ neutrino, and the $\bar{\nu}_\mu$ antineutrino; a third family has recently been discovered as well. A conservation law applies to each lepton family: in every known process, the number of leptons of each kind remains constant, reckoning a lepton as $+$ and an antilepton as $-$.

Lepton conservation

An easy way to apply these conservation laws is to assign the quantum numbers L and M to the various leptons as follows:

$L = +1$	$L = -1$	$M = +1$	$M = -1$
e^-	e^+	μ^-	μ^+
ν_e	$\bar{\nu}_e$	ν_μ	$\bar{\nu}_\mu$

All other particles have $L = 0$ and $M = 0$. Here are a few processes that involve leptons, showing that L and M have the same values before and after:

$$\text{Pion decay} \quad \pi^- \rightarrow \mu^- + \bar{\nu}_\mu$$
$$M = 0 \qquad +1 \quad -1$$

$$\text{Muon decay} \quad \mu^- \rightarrow e^- + \nu_\mu + \nu_e$$
$$L = 0 \qquad +1 \quad 0 \quad -1$$
$$M = +1 \qquad 0 \quad +1 \quad 0$$

$$\text{Pair production} \quad \gamma \rightarrow e^- + e^+$$
$$L = 0 \qquad +1 \quad -1$$

Other conservation principles, such as those of energy, linear momentum, angular momentum, and electric charge, can be traced to certain symmetries in the natural world. What symmetries the conservation of L and M are associated with, if any, remain unknown.

Mesons. The role of mesons as carriers of the strong interaction that binds nucleons together into nuclei has been examined earlier. There seems to be no conservation principle associated specifically with the number of mesons and antimesons involved in a process, although a number of other conservation principles must be obeyed.

Baryons. Nucleons and heavier particles subject to the strong nuclear interaction belong to the category of baryons. Baryons heavier than the neutron are called *hyperons* and are all unstable with mean lifetimes of less than a billionth of a second. Hyperons decay in a variety of ways, but the end result is always a proton or a neutron. As an example, here is one sequence which the omega hyperon can follow in its decay:

$$\Omega^- \rightarrow \Xi^0 + \pi^-$$
$$\hookrightarrow \Lambda^0 + \pi^0$$
$$\hookrightarrow p^+ + \pi^-$$

As in the case of the lepton families, a quantum number B can be assigned that is $+1$ for a baryon and -1 for an antibaryon. Thus far every particle interaction and decay studied in the laboratory is consistent with the conservation of B. An example is the decay of the neutron, which also illustrates the conservation of L:

Neutron decay $n^0 \rightarrow p^+ + e^- + \bar{\nu}_e$

$$L = 0 \quad 0 \quad +1 \quad -1$$
$$B = +1 \quad +1 \quad 0 \quad 0$$

This is the only way in which the neutron can decay and still conserve both energy (since the proton is the only lighter baryon) and quantum number B. The apparent stability of the proton fits into the same picture: there is no way in which it can decay without violating either conservation of energy or conservation of B. However, as discussed in Sec. 32–8, current theories of the fundamental interactions suggest that the proton may not be immune to decay into a lepton after all. Experimental confirmation is being sought, but the expected mean proton lifetime of 10^{30} years will make it hard to find. The conservation of L, M, and B still seems to hold for processes with more modest time scales.

Baryon conservation

32–7 QUARKS

Leptons seem to be truly elementary in nature, with no hint of internal structures or even of extensions in space. They are as close to being point particles as present measurements can establish. On the other hand, mesons and baryons, which are jointly called *hadrons* and are subject to the strong nuclear interaction, apparently not only have definite sizes (they are about 10^{-15} m across) but also have internal structures.

Leptons are point particles but hadrons (mesons and baryons) have definite sizes and internal structures

What is known about the various hadrons suggests that they are probably composite objects whose constituent particles are known as *quarks*. Quarks are thought to be elementary in the same sense as leptons, but unlike leptons they have never been isolated experimentally. The present status of quarks is much like that of neutrinos for 25 years

Hadrons are composed of quarks

after they were proposed: a wealth of indirect evidence supports their existence, but something in their basic character makes them hard to detect. The parallel may not be too precise, however, because the elusiveness of the neutrino arises merely from its minimal interaction with matter, whereas there may be a basic reason why quarks cannot occur independently outside of hadrons. (Such a reason would be an attractive force between quarks that increases in strength with distance.)

Quarks have fractional electric charges

Quarks have a number of remarkable properties. One is their possession of fractional electric charges, something unknown elsewhere in nature. As originally formulated in 1963 by Murray Gell-Mann and, independently, by George Zweig, the quark model of hadrons involved three such particles, labeled u (for "up"), d (for "down"), and s (for "strange"). The up and down quarks were so named to distinguish their spin directions; the significance of "strange" is given below. These quarks and their antiparticles $\bar{u}$, $\bar{d}$, and $\bar{s}$ were assigned charges as follows:

$$u: +\tfrac{2}{3}e \qquad \bar{u}: -\tfrac{2}{3}e$$
$$d: -\tfrac{1}{3}e \qquad \bar{d}: +\tfrac{1}{3}e$$
$$s: -\tfrac{1}{3}e \qquad \bar{s}: +\tfrac{1}{3}e$$

A baryon consists of three quarks, a meson of two quarks

Quarks have the baryon number $B = +\tfrac{1}{3}$ and antiquarks have the baryon number $B = -\tfrac{1}{3}$. Each baryon is made up of three quarks, so $B = 1$; each antibaryon of three antiquarks so $B = -1$; and each meson of one quark and one antiquark, so $B = 0$. All the hadrons known until recently can be accounted for on this basis, and every permitted combination of quarks is associated with a known hadron. Thus the proton is believed to consist of two u and one d quarks, which gives a total charge of $(+\tfrac{2}{3} + \tfrac{2}{3} - \tfrac{1}{3})e = +e$, and the neutron of one u and two d quarks, which gives a total charge of $(+\tfrac{2}{3} - \tfrac{1}{3} - \tfrac{1}{3})e = 0$. The π^{+} pion, a meson, consists of a u quark and a $\bar{d}$ antiquark, which gives a total charge of $(+\tfrac{2}{3} + \tfrac{1}{3})e = +e$ (see Fig. 32–14).

Strangeness

The s quark is associated with unstable particles whose production and decay had earlier led to the introduction of a quantum number S, called *strangeness*, which is conserved under certain circumstances. The s quark is assigned $S = -1$ and the $\bar{s}$

FIG. 32–14 Compositions of the proton, neutron, and negative pion, together with their antiparticles, according to the quark model of hadrons. Electric charges are in units of e.

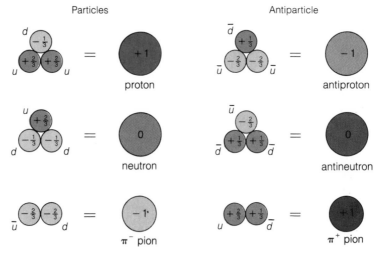

antiquark is assigned $S = +1$. Thus the K^+ hadron ought to be composed of a u quark ($S = 0$) and an $\bar{s}$ antiquark ($S = +1$), which yields the observed strangeness of $S = +1$ and charge of $(+\frac{2}{3} + \frac{1}{3})e = +e$. For another example, the Ω^- hyperon of $S = -3$ can be accounted for on the basis of three s quarks, with a resulting charge of $(-\frac{1}{3} - \frac{1}{3} - \frac{1}{3})e = -e$.

A serious objection to the idea that hadrons are composed of quarks was that the presence of more than one quark of the same kind in a single particle (for instance, two u quarks in a proton, three s quarks in a Ω^- hyperon) violates the exclusion principle, to which quarks ought to be subject. To take care of this objection, it was suggested that quarks and antiquarks have an additional property of some kind that can be manifested in different ways. An analogy is electric charge, a property that can be manifested in the two ways that have come to be called positive and negative. In the case of quarks, the new property was whimsically called *color,* and its modes of expression were termed red, green, and blue; the antiquark colors are antired, antigreen, and antiblue. According to this scheme, all three quarks in a baryon have different colors, which satisfies the exclusion principle. Such a combination can be thought of as white by analogy with the way red, green, and blue light combine to make white light. (Of course, there is no connection between quark colors and actual visual colors.) A meson is supposed to consist of a quark of one color and an antiquark of the corresponding anticolor, thereby canceling out the color. As a result both baryons and mesons are colorless: quark color is a property that is significant within hadrons but cannot be observed in the outside world.

The notion of quark color has turned out to be more than merely a device for getting around the exclusion principle. In particular, it seems possible to consider the strong interaction as being based on quark color, in the same sense as the electromagnetic interaction is based on electric charge. The resulting theory, called quantum chromodynamics, has had a number of successes in explaining elementary-particle phenomena.

In time the original u, d, and s quarks were supplemented by three more, called "charmed," "top," and "bottom," each in three colors like the others. (The top and bottom quarks have the alternative names "truth" and "beauty." Actually, of course, charm, truth, and beauty are nothing new in physics.) Nevertheless, all the properties of ordinary matter can be understood in terms of just two leptons, the electron and its associated neutrino, and just two quarks, u and d. This is an astonishing achievement. The other leptons, and hadrons that incorporate the other quarks, are all unstable particles created in high-energy collisions with no obvious role in ordinary matter.

Quarks may be "red", "green", or "blue"; antiquarks may be "antired", "antigreen", or "antiblue"

Mesons and baryons are colorless, hence color cannot be observed

The strong interaction is based on color just as the em interaction is based on charge

Two leptons and two quarks can account for ordinary matter

32–8 THE FUNDAMENTAL INTERACTIONS

Only four fundamental interactions between elementary particles apparently govern the structure and behavior of the entire physical universe on all scales of size from atomic nuclei to galaxies of stars. These interactions have been described earlier; their basic properties are summarized in Table 32–2. Although the relative strengths of the interactions vary enormously, the distances through which they are effective are very different. Thus the strong force between nearby nucleons completely dominates the grav-

The fundamental interactions have different ranges and different strengths

Interaction	Particles Affected	Range	Relative Strength	Particles Exchanged	Role in Universe
Strong	Hadrons	$\sim 10^{-15}$ m	1	Mesons	Holds protons and neutrons together to form atomic nuclei
Electromagnetic	Charged particles	∞	$\sim 10^{-2}$	Photons	Determines structures of atoms, molecules, solids, and liquids; is important factor in astronomical universe
Weak	Hadrons and leptons	$\sim 10^{-13}$ m	$\sim 10^{-13}$	Intermediate bosons	Helps determine compositions of atomic nuclei
Gravitational	All	∞	$\sim 10^{-40}$	Gravitons	Assembles matter into planets, stars, and galaxies

TABLE 32–2
The four fundamental interactions. The graviton and the intermediate bosons have not been experimentally observed.

itational force between them, but when the nucleons are even a millimeter apart the reverse is true. The existence of nuclei is a consequence of the strong interaction, the existence of atoms is a consequence of the electromagnetic interaction. Because matter in bulk is electrically neutral and the strong and weak interactions are severely limited in range, the gravitational interaction, which is completely negligible on a small scale, becomes the dominant one on a large scale. The function of the weak interaction in the structure of matter seems to be limited to enabling nuclei with inappropriate neutron/proton ratios to undergo corrective beta decays.

Gluons mediate the interaction between quarks

If hadrons are indeed composed of quarks, as appears likely, then the strong interaction between hadrons ought to have its origin in an interaction between quarks. The particles that quarks exchange to produce their interaction are called *gluons,* of which eight have been postulated. Gluons are supposed to be massless and to move at the speed of light, and each one carries a color and an anticolor. When a quark emits or absorbs a gluon, its color changes accordingly, just as the emission of a charged meson by a nucleon changes its electric charge. For instance, a blue quark that emits a blue-antired gluon becomes a red quark, and a red quark that absorbs this gluon becomes a blue quark. Quantum chromodynamics, the theory of how quarks interact with each other, not only is able to explain the behavior of quarks within hadrons but also has predicted certain effects that were afterward observed in high-energy particle experiments. However, this theory has not yet been developed to the point where it can account in more than a general way for the longer-range interaction between quarks that is observed as the strong interaction between hadrons.

The weak and em interactions are parts of the same basic phenomenon, and the strong interaction may be as well

Studies independently carried out in the 1960s by Steven Weinberg and Abdus Salam showed that the weak and electromagnetic interactions are in reality different manifestations of the same essential phenomenon. Supported by experiment, this conclusion is now widely accepted. More recently it has become apparent that the strong interaction can be linked to the unified weak and electromagnetic ones as well. Leptons and quarks find natural places in such a grand unified theory, and it is possible in this way to explain, among other things, why the electron (a lepton) and the proton (a composite of quarks) have electric charges of exactly the same magnitude.

Part of the new theory is an interaction between leptons and quarks that enables a member of one class to be transformed into a member of the other. This has as a consequence the instability of protons, hitherto regarded as completely stable, which ought eventually to decay into leptons. The lepton-quark interaction is extremely feeble, to be sure, so the estimated mean life of the proton is around 10^{30} years; by comparison, the universe is less than 10^{11} years old. Experiments are planned to search for proton decays, but the task will not be an easy one.

Proton decay

Although much remains to be done, and no doubt further surprises are in store, at least the outline of a unified picture of the strong, weak, and electromagnetic interactions is in existence—a remarkable achievement. Still more remarkable would be a way to weave the gravitational interaction into a single skein with the others, and there are hints that this goal is not beyond reach.

IMPORTANT TERMS

The **neutrino** is an uncharged particle of small or zero mass that is emitted in the decay of certain elementary particles. It can possess energy and both linear and angular momentum.

Nearly every elementary particle has an **antiparticle** counterpart whose electric charge and certain other properties have the opposite sign. Thus the antiparticle of the electron is the positron (charge $+e$) and that of the proton is the antiproton (charge $-e$). When a particle and an antiparticle of the same kind come together, they **annihilate** each other, with the vanished mass reappearing as photons or mesons.

The four categories of elementary particles are the photons, the leptons, the mesons, and the baryons. **Leptons** are point particles with no internal structures; examples are the electron and the neutrino. **Mesons** are particles that can be regarded as "carriers" of the strong interaction; an example is the pion. **Baryons** include nucleons and heavier unstable particles. Mesons and baryons, jointly known as **hadrons,** are subject to the strong interaction and are thought to be composed of **quarks,** which are particles with fractional electric charges that have not as yet been experimentally isolated.

The four **fundamental interactions,** in order of strength, are the strong, electromagnetic, weak, and gravitational. The electromagnetic and weak interactions, and probably the strong as well, are closely related.

MULTIPLE CHOICE

1. The neutrino does not possess
(a) charge.
(b) linear momentum.
(c) angular momentum.
(d) energy.

2. The mass of the neutrino is
(a) less than that of the electron.
(b) equal to that of the electron.
(c) between that of the electron and that of the muon.
(d) greater than that of the muon.

3. The strong interaction between nucleons is believed to arise through the exchange of
(a) neutrinos.   (b) mesons.
(c) leptons.   (d) baryons.

4. The particle whose properties are most nearly the same as those of the proton is the
(a) positron.   (b) neutron.
(c) antiproton.   (d) positive pion.

5. Of the following particles, the one that is its own antiparticle is the
(a) photon.   (b) neutrino.
(c) neutron.   (d) proton.

6. The heaviest class of elementary particles consists of the
(a) photons.   (b) leptons.
(c) mesons.   (d) baryons.

7. The only one of the following particles that is not an elementary particle is the
(a) neutron.   (b) pion.
(c) muon.   (d) alpha particle.

8. The muon most closely resembles the
(a) pion.   (b) electron.
(c) proton.   (d) neutrino.

9. The only one of the following particles that does not decay in free space is the
(a) neutrino.   (b) neutron.
(c) muon.   (d) pion.

10. A quantity not necessarily conserved in elementary-particle decays is
(a) charge.   (b) mass.
(c) lepton number.   (d) baryon number.

11. All leptons are
    (a) stable.
    (b) composed of quarks.
    (c) subject to the weak interaction.
    (d) subject to the strong interaction.

12. Which of the following would be violated if the quarks in a hadron had the same color?
    (a) Uncertainty principle
    (b) Exclusion principle
    (c) Conservation of energy
    (d) Conservation of charge

13. A particle unlikely to be composed of quarks is the
    (a) muon.   (b) pion.
    (c) proton.   (d) neutron.

14. The weakest of the four fundamental interactions is the
    (a) gravitational interaction.
    (b) electromagnetic interaction.
    (c) strong nuclear interaction.
    (d) weak interaction.

EXERCISES

1. Discuss the similarities and difference between the photon and the neutrino.

2. Why does a free neutron not decay into an electron and a positron? Into a proton and an antiproton?

3. A negative muon collides with a proton, and a neutron plus another particle are created. What is the other particle?

4. Why must a neutrino be emitted in the process of electron capture by a nucleus? Could an antineutrino be emitted instead?

5. According to the theory of the continuous creation of matter (which has turned out to be inconsistent with astronomical observations), the evolution of the universe can be traced to the spontaneous appearance of neutrons in free space. Which conservation laws would this process violate?

6. What enables neutrinos to travel immense distances through matter without interacting?

7. No particle of fractional charge has yet been observed. If none is found in the future either, does this necessarily mean that the quark hypothesis is wrong?

8. The gravitational interaction is the weakest of all, yet it alone governs the motions of the planets around the sun. Why?

PROBLEMS

1. One proton strikes another, and the reaction

$$p + p \rightarrow n + p + \pi^+$$

takes place. What is the minimum energy the incident proton must have had? (Ignore momentum conservation.)

2. Find the energy of each of the gamma-ray photons produced in the decay of a neutral pion at rest. Why must their energies be the same?

3. One way to account for the decay of the π^0 meson into two photons, which are electromagnetic quanta, despite the absence of either charge or magnetic moment is to assume that the π^0 first becomes a nucleon-antinucleon pair that can interact electromagnetically to produce two photons whose energies are consistent with the π^0 mass. (a) How long does the uncertainty principle permit such a nucleon-antinucleon pair to exist? (b) Do any conservation principles other than conservation of energy prohibit the transformation of a π^0 meson into a nucleon-antinucleon pair?

4. (a) Find the energy of the photon emitted in the decay $\Sigma^0 \rightarrow \Lambda^0 + \gamma$. Ignore the recoil energy of the Λ^0. (b) What difference will the recoil of the Λ^0 make with respect to the photon energy?

5. How much energy must a gamma-ray photon have if it is to materialize into a neutron-antineutron pair? Is this more or less than that required to form a proton-antiproton pair?

6. A 2-MeV positron collides head on with a 2-MeV electron and the two are annihilated. Find the energy of each of the resulting photons. How would these photons be classified?

ANSWERS TO MULTIPLE CHOICE

1. a	6. d	11. c
2. a	7. d	12. b
3. b	8. b	13. a
4. c	9. a	14. a
5. a	10. b	

APPENDIX A
USEFUL MATHEMATICS

A–1 ALGEBRA

Algebra is a generalized arithmetic in which symbols are used in place of numbers. Algebra thus provides a language in which general relationships can be expressed among quantities whose numerical values need not be known in advance.

The arithmetical operations of addition, subtraction, multiplication, and division have the same meanings in algebra. The symbols of algebra are normally letters of the alphabet. If we have two quantities a and b and add them to give the sum c, we would write

$$a + b = c$$

If we subtract b from a to give the difference d, we would write

$$a - b = d$$

Multiplying a and b together to give e may be written in any of these ways:

$$a \times b = e \qquad ab = e \qquad (a)(b) = e$$

Whenever two algebraic quantities are written together with nothing between them, it is understood that they are to be multiplied.

Dividing a by b to give the quotient f is usually written

$$\frac{a}{b} = f$$

but it may sometimes be more convenient to write

$$a/b = f$$

which has the same meaning.

Parentheses and brackets are used to show the order in which various operations are to be performed. Thus

$$\frac{(a + b)c}{d} - e = f$$

means that, in order to find f, we are first to add a and b together, then multiply their sum by c and divide by d, and finally subtract e.

A–2 EQUATIONS

An *equation* is simply a statement that a certain quantity is equal to another one. Thus

$$7 + 2 = 9$$

which contains only numbers, is an arithmetical equation, and

$$3x + 12 = 27$$

which contains a symbol as well, is an algebraic equation. The symbols in an algebraic equation usually cannot have any arbitrary values if the equality is to hold. Finding the possible values of these symbols is called *solving* the equation. The *solution* of the latter equation above is

$$x = 5$$

since only when x is 5 is it true that $3x + 12 = 27$.

In order to solve an equation, a basic principle must be kept in mind.

Any operation performed on one side of an equation must be performed on the other.

An equation therfore remains valid when the same quantity, numerical or otherwise, is added to or subtracted from both sides, or when the same quantity is used to multiply or divide both sides. Other operations, for instance squaring or taking the square root, also do not alter the equality if the same thing is done to both sides. As a simple example, to solve $3x + 12 = 27$, we first subtract 12 from both sides:

$$3x + 12 - 12 = 27 - 12$$
$$3x = 15$$

To complete the solution we divide both sides by 3:

$$\frac{3x}{3} = \frac{15}{3}$$
$$x = 5$$

To check a solution, we substitute it back in the original equation and see whether the equality is still true. Thus we can check that $x = 5$ by reducing the original algebraic equation to an arithmetical one:

$$3x + 12 = 27$$
$$(3)(5) + 12 = 27$$
$$15 + 12 = 27$$
$$27 = 27$$

Two helpful rules follow directly from the principle stated above. The first is,

Any term on one side of an equation may be transposed to the other side by changing its sign.

To verify this rule, we subtract b from each side of the equation

$$a + b = c$$

to obtain

$$a + b - b = c - b$$
$$a = c - b$$

We see that b has disappeared from the left-hand side and $-b$ is now on the right-hand side.

The second rule is,

A quantity which multiplies one side of an equation may be transposed in order to divide the other side, and vice versa.

To verify this rule, we divide both sides of the equation

$$ab = c$$

by b. The result is

$$\frac{ab}{b} = \frac{c}{b}$$
$$a = \frac{c}{b}$$

We see that b, a multiplier on the left-hand side, is now a divisor on the right-hand side.

EXAMPLE Solve the following equation for x:

$$4(x - 3) = 7$$

SOLUTION The above rules are easy to apply here:

$$4(x - 3) = 7$$
$$x - 3 = \frac{7}{4}$$
$$x = \frac{7}{4} + 3 = 1.75 + 3 = 4.75$$

When each side of an equation consists of a fraction, all we need to do to remove the fractions is to *cross multiply*:

$$\frac{a}{b} = \frac{c}{d}$$
$$ad = bc$$

What was originally the denominator (lower part) of each fraction now multiplies the numerator (upper part) of the other side of the equation.

EXAMPLE Solve the following equation for y:

$$\frac{5}{y + 2} = \frac{3}{y - 2}$$

SOLUTION First we cross multiply to get rid of the fractions, and then solve in the usual way:

$$5(y - 2) = 3(y + 2)$$
$$5y - 10 = 3y + 6$$
$$5y - 3y = 6 + 10$$
$$2y = 16$$
$$y = 8$$

EXERCISES

1. Solve each of the following equations for x. The answers are given at the end of Appendix A.

(a) $\dfrac{x + 7}{6} = x + 2$

(b) $\dfrac{3x - 42}{9} = 2(7 - x)$

(c) $\dfrac{1}{x + 1} = \dfrac{1}{2x - 1}$

(d) $\dfrac{8}{x} = \dfrac{1}{4 - x}$

(e) $\dfrac{x}{2x - 1} = \dfrac{5}{7}$

A–3 SIGNED NUMBERS

The rules for multiplying and dividing positive and negative quantities are straightforward. First, perform the indicated operation on the absolute value of the quantity (the absolute value of -7 is 7, for instance). If the quantities are both positive or both negative, the result is positive:

$$(+a)(+b) = (-a)(-b) = +ab$$
$$\frac{+a}{+b} = \frac{-a}{-b} = +\frac{a}{b}$$

If one quantity is positive and the other negative the result is negative:

$$(-a)(+b) = (+a)(-b) = -ab$$
$$\frac{-a}{+b} = \frac{+a}{-b} = -\frac{a}{b}$$

A few examples might be helpful:

$$(-6)(-3) = 18 \qquad \frac{-20}{-5} = 4$$

$$(4)(-5) = -20$$

$$(-10)(7) = -70 \qquad \frac{6}{-2} = -3$$

$$\frac{-30}{15} = -2$$

EXAMPLE Find the value of

$$z = \frac{xy}{x - y}$$

when $x = -12$ and $y = 4$.

SOLUTION We begin by evaluating xy and $x - y$, which are

$$xy = (-12) \times 4 = -48$$
$$x - y = (-12) - 4 = -16$$

Hence

$$z = \frac{xy}{x - y} = \frac{-48}{-16} = 3$$

EXERCISES

2. Evaluate the following:

(a) $\dfrac{3(x + y)}{2}$ when $x = 5, y = -2$

(b) $\dfrac{1}{x - y} - \dfrac{1}{x + y}$ when $x = 3, y = -5$

(c) $\dfrac{3(x + 7)}{y + 2}$ when $x = 3, y = -6$

(d) $\dfrac{x + y}{2z} + \dfrac{z}{x - y}$ when $x = -2, y = 2, z = 4$

(e) $\dfrac{x + z}{y} - \dfrac{xy}{2}$ when $x = 2, y = -8, z = 10$

A–4 EXPONENTS

It is often necessary to multiply a quantity by itself a number of times. This process is indicated by a superscript number called the exponent, according to the following scheme:

$$A = A^1$$
$$A \times A = A^2$$
$$A \times A \times A = A^3$$
$$A \times A \times A \times A = A^4$$
$$A \times A \times A \times A \times A = A^5$$

We read A^2 as "A squared" because it is the area of a square of length A on a side; similarly A^3 is called "A cubed" because it is the volume of a cube each of whose sides is A long. More generally we speak of A^n as "A to the nth power." Thus A^5 is read as "A to the fifth power."

When we multiply a quantity raised to some particular power (say A^n) by the same quantity raised to another power (say A^m), the result is that quantity raised to a power equal to the sum of the original exponents. That is,

$$A^n A^m = A^{(n + m)}$$

For example,

$$A^2 A^5 = A^7$$

which we can verify directly by writing out the terms:

$$(A \times A)(A \times A \times A \times A \times A)$$

$$= A \times A \times A \times A \times A \times A \times A = A^7$$

From the above result we see that when a quantity raised to a particular power (say A^n) is to be multiplied by itself a total of m times, we have

$$(A^n)^m = A^{nm}$$

For example,

$$(A^2)^3 = A^6$$

since

$$(A^2)^3 = A^2 \times A^2 \times A^2 = A^{(2 + 2 + 2)} = A^6$$

Reciprocal quantities are expressed in a similar way with the addition of a minus sign in the exponent, as follows:

$$\frac{1}{A} = A^{-1} \qquad \frac{1}{A^2} = A^{-2} \qquad \frac{1}{A^3} = A^{-3} \qquad \frac{1}{A^4} = A^{-4}$$

Exactly the same rules as before are used in combining quantities raised to negative powers with one another and with some quantity raised to a positive power. Thus

$$A^5 A^{-2} = A^{(5-2)} = A^3$$
$$(A^{-1})^{-2} = A^{-1(-2)} = A^2$$
$$(A^3)^{-4} = A^{-4 \times 3} = A^{-12}$$
$$A A^{-7} = A^{(1-7)} = A^{-6}$$

It is important to remember that any quantity raised to the zeroth power, say A^0, is equal to 1. Hence

$$A^2 A^{-2} = A^{(2-2)} = A^0 = 1$$

This is more easily seen if we write A^{-2} as $1/A^2$:

$$A^2 A^{-2} = A^2 \times \frac{1}{A^2} = \frac{A^2}{A^2} = 1$$

A–5 ROOTS

The *square root* of A, $\sqrt{A}$, is that quantity which, when multiplied by itself, is equal to A:

$$\sqrt{A} \times \sqrt{A} = A$$

The square root has that name because the length of each side of a square of area A is given by $\sqrt{A}$. Some examples of square roots are as follows:

$$\sqrt{1} = 1 \qquad \text{because } 1 \times 1 = 1$$
$$\sqrt{16} = 4 \qquad \text{because } 4 \times 4 = 16$$
$$\sqrt{100} = 10 \qquad \text{because } 10 \times 10 = 100$$
$$\sqrt{42.25} = 6.5 \qquad \text{because } 6.5 \times 6.5 = 42.25$$
$$\sqrt{9A^2} = 3A \qquad \text{because } 3A \times 3A = 9A^2$$

The square root of a number less than 1 is larger than the number itself:

$$\sqrt{0.01} = 0.1 \qquad \text{because } 0.1 \times 0.1 = 0.01$$
$$\sqrt{0.49} = 0.7 \qquad \text{because } 0.7 \times 0.7 = 0.49$$

The square root of a quantity may be either positive or negative because $(+A)(+A) = (-A)(-A) = A^2$. Whether the positive or negative root (or even both) is correct in a given problem depends upon the details of the problem and is usually obvious. Using exponents we see that, because

$$(A^{1/2})^2 = A^{2 \times (1/2)} = A^1 = A$$

we can express square roots by the exponent $\frac{1}{2}$:

$$\sqrt{A} = A^{1/2}$$

Other roots may be expressed similarly. The *cube root* of a quantity A, written $\sqrt[3]{A}$, when multiplied by itself twice equals A. That is,

$$\sqrt[3]{A} \times \sqrt[3]{A} \times \sqrt[3]{A} = A$$

which may be more conveniently written

$$(A^{1/3})^3 = A$$

where $\sqrt[3]{A} = A^{1/3}$.

In general the nth root of a quantity, $\sqrt[n]{A}$, may be written $A^{1/n}$, which is a more convenient form for most purposes. Some examples may be helpful:

$$\sqrt{A^4} = (A^4)^{1/2} = A^{1/2 \times 4} = A^2$$
$$(\sqrt{A})^4 = (A^{1/2})^4 = A^{4 \times 1/2} = A^2$$
$$\sqrt{\sqrt{A}} = (A^{1/2})^{1/2} = A^{1/2 \times 1/2} = A^{1/4}$$
$$(A^{1/4})^{-7} = A^{-7 \times (1/4)} = A^{-7/4}$$
$$(A^3)^{-1/3} = A^{-(1/3) \times 3} = A^{-1}$$
$$(A^{1/4})^{1/4} = A^{(1/4) \times (1/4)} = A^{1/16}$$

Although procedures exist for finding roots arithmetically with pencil and paper, in practice it is much easier to use an electronic calculator. Calculators for science and engineering usually have a square root [$\sqrt{\ }$] key. For other roots, such as $A^{1/3}$, and for nonintegral exponents in general, such as $A^{3.27}$, logarithms must be used, as described in Appendix B-5. A calculator with [LOG] and [10^x] or [INV LOG] keys makes such quantities easy to evaluate.

A–6 EQUATIONS WITH POWERS AND ROOTS

An equation that involves powers or roots or both is subject to the same basic principle that governs the manipulation of simpler equations: Whatever is done to one side must be done to the other. Hence the following rules:

An equation remains valid when both sides are raised to the same power, that is, when each side is multiplied by itself the same number of times as the other side.

An equation remains valid when the same root is taken of both sides.

EXAMPLE Newton's law of gravitation states that the force F between two bodies the distance r apart whose masses are m_A and m_B is given by the formula

$$F = G\frac{m_A m_B}{r^2}$$

where G is a universal constant. Solve this formula for r.

SOLUTION We begin by transposing the r^2 to give

$$Fr^2 = Gm_A m_B$$

Next we transpose the F to give

$$r^2 = \frac{Gm_A m_B}{F}$$

and finally take the square root of both sides:

$$r = \sqrt{r^2} = \sqrt{\frac{Gm_A m_B}{F}}$$

The positive value of the root is correct here since a negative separation r between the bodies is meaningless.

Quadratic equations, which have the general form

$$ax^2 + bx + c = 0$$

are often encountered in algebraic calculations. In such an equation, a, b, and c are constants. A quadratic equation is satisfied by the values of x given by the formulas

$$x_+ = \frac{-b + \sqrt{b^2 - 4ac}}{2a}$$

$$x_- = \frac{-b - \sqrt{b^2 - 4ac}}{2a}$$

By looking at these formulas, we can see that the nature of the solution depends upon the value of the quantity $b^2 - 4ac$. When $b^2 = 4ac$, $\sqrt{b^2 - 4ac} = 0$ and the two solutions are equal to just $x = -b/2a$. When $b^2 > 4ac$, $\sqrt{b^2 - 4ac}$ is a real number and the solutions x_+ and x_- are different. When $b^2 < 4ac$, $\sqrt{b^2 - 4ac}$ is the square root of a negative number, and the solutions are different. The square root of a negative number is called an *imaginary number* because squaring a real number, whether positive or negative, always

gives a positive number: $(+2)^2 = (-2)^2 = +4$, hence $\sqrt{-4}$ cannot be either $+2$ or -2. Imaginary numbers do not occur in the sort of physical problems treated in this book.

EXAMPLE Solve the quadratic equation

$$2x^2 - 3x - 9 = 0$$

SOLUTION Here $a = 2$, $b = -3$, and $c = -9$, so that

$$b^2 - 4ac = (-3)^2 - 4(2)(-9) = 9 + 72 = 81$$

The two solutions are

$$x_+ = \frac{-b + \sqrt{b^2 - 4ac}}{2a}$$

$$= \frac{-(-3) + \sqrt{81}}{2(2)} = \frac{3 + 9}{4} = \frac{12}{4} = 3$$

$$x_- = \frac{-b - \sqrt{b^2 - 4ac}}{2a}$$

$$= \frac{-(-3) - \sqrt{81}}{2(2)} = \frac{3 - 9}{4} = \frac{-6}{4} = -\frac{3}{2}$$

EXAMPLE Solve the quadratic equation

$$x^2 - 4x + 4 = 0$$

SOLUTION Here $a = 1$, $b = -4$, and $c = 4$, so that

$$b^2 - 4ac = (-4)^2 - 4(1)(4) = 16 - 16 = 0$$

Hence the only solution is

$$x = \frac{-b}{2a} = \frac{-(-4)}{2} = 2$$

EXERCISES

3. Solve each of the following equations for x.

(a) $9x^2 + 12x + 4 = 0$

(b) $x^2 + x - 2 = 0$

(c) $2x^2 - 5x + 2 = 0$

(d) $x^2 + x - 20 = 0$

(e) $4x^2 - 4x - 11 = 0$

A–7 TRIGONOMETRIC FUNCTIONS

A *right triangle* is a triangle two sides of which are perpendicular. Such triangles are frequently encountered in physics, and it is necessary to know how their sides and angles

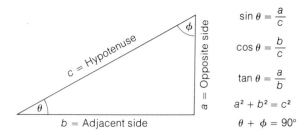

$$\sin \theta = \frac{a}{c}$$

$$\cos \theta = \frac{b}{c}$$

$$\tan \theta = \frac{a}{b}$$

$$a^2 + b^2 = c^2$$

$$\theta + \phi = 90°$$

are related. The *hypotenuse* of a right triangle is the side opposite the right angle, as in Fig. A–1; it is always the longest side. The three basic trigonometric functions, the *sine, cosine,* and *tangent* of an angle, are defined as follows:

$$\sin \theta = \frac{a}{c} = \frac{\text{opposite side}}{\text{hypotenuse}}$$

$$\cos \theta = \frac{b}{c} = \frac{\text{adjacent side}}{\text{hypotenuse}}$$

$$\tan \theta = \frac{a}{b} = \frac{\text{opposite side}}{\text{adjacent side}}$$

From these definitions we see that

$$\frac{\sin \theta}{\cos \theta} = \frac{a/c}{b/c} = \frac{a}{b} = \tan \theta$$

Numerical values of $\sin \theta$, $\cos \theta$, and $\tan \theta$ for angles form 0° to 90° are given in Appendix C. These figures may be used for angles from 90° to 360° with the help of this table:

	θ	$90° + \theta$	$180° + \theta$	$270° + \theta$
sin	$\sin \theta$	$\cos \theta$	$-\sin \theta$	$-\cos \theta$
cos	$\cos \theta$	$-\sin \theta$	$-\cos \theta$	$\sin \theta$
tan	$\tan \theta$	$-\dfrac{1}{\tan \theta}$	$\tan \theta$	$-\dfrac{1}{\tan \theta}$

For example, if we require the value of sin 120°, we first note that 120° = 90° + 30°. Then, since

$$\sin (90° + \theta) = \cos \theta$$

we have

$$\sin 120° = \sin (90° + 30°) = \cos 30° = 0.866$$

Similarly, to find tan 342°, we begin by noting that 342° = 270° + 72°. Then, since

$$\tan (270° + \theta) = -\frac{1}{\tan \theta}$$

we have

$$\tan 342° = \tan (270° + 72°) = -\frac{1}{\tan 72°} = -\frac{1}{3.078}$$

$$= -0.325$$

The *inverse* of a trigonometric function is the angle whose function is given. For instance, the inverse of $\sin \theta$ is the angle θ. If $\sin \theta = x$, then the angle θ may be designated either as $\theta = \arcsin x$ or as $\theta = \sin^{-1} x$. It is important to keep in mind that an expression such as $\sin^{-1} x$ does *not* mean $1/\sin x$. The inverse trigonometric functions are as follows:

$$\sin \theta = x$$
$$\theta = \arcsin x = \sin^{-1} x = \text{angle whose sine is } x$$
$$\cos \theta = y$$
$$\theta = \arccos y = \cos^{-1} y = \text{angle whose cosine is } y$$
$$\tan \theta = z$$
$$\theta = \arctan z = \tan^{-1} z = \text{angle whose tangent is } z$$

To find an inverse function, say the angle whose cosine is 0.907, the procedure with some calculators is to enter 0.907 and then press in succession the [ARC] and [COS] keys. The result is $\theta = 25°$ to the nearest degree. With other calculators, a shift key must first be pressed that converts the [SIN], [COS], and [TAN] keys, respectively, to $\sin^{-1}$, $\cos^{-1}$, and $\tan^{-1}$, and then the [COS] key is pressed. With a table of cosines, we look for the value nearest to 0.907 in the body of the table, which is 0.906, and then read across to find the corresponding angle.

EXERCISES

4. Express the following functions in terms of angles between 0 and 90°.

    (a)   sin 100°
    (b)   sin 300°
    (c)   cos 150°
    (d)   cos 350°
    (e)   tan 180°
    (f)   tan 220°

A–8   SOLVING A RIGHT TRIANGLE

To solve a given triangle means to find the values of any unknown sides or angles in terms of the values of the known sides and angles. A triangle has three sides and three angles, and we must know the values of at least three of these six quantities, including one of the sides, to solve the triangle for the others. In a right triangle, one of the angles is always 90°, and so all we need here are the lengths of any two of its sides or the length of one side and the value of one of the other angles to find the remaining sides and angles.

Suppose we know the length of the side b and the angle θ in the right triangle of Fig. A–1. From the definitions of sine and tangent we see that

$$\tan \theta = \frac{a}{b}$$

$$a = b \tan \theta$$

$$\sin \theta = \frac{a}{c}$$

$$c = \frac{a}{\sin \theta}$$

This gives us the two unknown sides a and c. To find the unknown angle ϕ, we can use any of these formulas:

$$\phi = \sin^{-1}\frac{b}{c} \qquad \phi = \cos^{-1}\frac{a}{c} \qquad \phi = \tan^{-1}\frac{b}{a}$$

Alternatively, we can use the fact that the sum of the angles in any triangle is 180°. Because one of the angles in a right triangle is 90°, the sum of the other two must be 90°:

$$\theta + \phi = 90°$$

Hence $\phi = 90° - \theta$ here.

Another useful relationship in a right triangle is the Pythagorean theorem, which states that the sum of the squares of the sides of such a triangle adjacent to the right angle is equal to the square of its hypotenuse. For the triangle of Fig. A–1,

$$a^2 + b^2 = c^2$$

Thus we can always express the length of any of the sides of a right triangle in terms of the other sides:

$$a = \sqrt{c^2 - b^2}$$
$$b = \sqrt{c^2 - a^2}$$
$$c = \sqrt{a^2 + b^2}$$

EXAMPLE In the triangle of Fig. A–1, $a = 7$ cm and $b = 10$ cm. Find c, θ, and ϕ.

SOLUTION   From the Pythagorean theorem,

$$c = \sqrt{a^2 + b^2} = \sqrt{7^2 + 10^2} \text{ cm}$$

$$= \sqrt{149} \text{ cm} = 12.2 \text{ cm}$$

Since $\tan \theta = a/b$,

$$\theta = \tan^{-1}\frac{a}{b} = \tan^{-1}\frac{7 \text{ cm}}{10 \text{ cm}}$$

$$= \tan^{-1} 0.7 = 35°$$

The value of the other angle ϕ is given by

$$\phi = 90° - \theta = 90° - 35° = 55°$$

EXERCISES

5. Find the values of the unknown sides and angles in the right triangles for which the following data are known:

    (a) $\theta = 45°, a = 10$

    (b) $\theta = 15°, b = 4$

    (c) $\theta = 25°, c = 5$

    (d) $a = 3, b = 4$

    (e) $a = 5, c = 13$

ANSWERS

1.   (a) -1
      (b)   8
      (c)   2
      (d)   32/9
      (e)   5/3

2.   (a)   4.5
      (b)   0.625
      (c) -3
      (d) -1
      (e)   6.5

3.   (a) $-2/3$
      (b) $1, -2$
      (c) $2, 0.5$
      (d) $4, -5$
      (e) $0.5 \pm \sqrt{3} = 2.232, -1.232$

4.   (a) $\cos 10°$
      (b) $-\cos 30°$
      (c) $-\sin 60°$
      (d) $\sin 80°$
      (e) $-1/\tan 90°$ or $\tan 0$. Since $\tan 90° = \infty$ and $\tan 0 = 0$, both expressions give the same value.
      (f) $\tan 40°$

5.   (a) $b = 10, c = 14.1, \phi = 45°$
      (b) $a = 1.07, c = 4.14, \phi = 75°$
      (c) $a = 2.11, b = 4.53, \phi = 65°$
      (d) $c = 5, \theta = 37°, \phi = 53°$
      (e) $b = 12, \theta = 23°, \phi = 67°$

APPENDIX B
POWERS OF TEN
AND LOGARITHMS

B–1 POWERS OF TEN

Very small and very large numbers are common in physics. For example, the mass of an electron is 0.000, 000,000,000,000,000,000,000,000,910,9 kilogram, and the mass of the earth is 5,983,000,000,000, 000,000,000,000,000 kilograms. Such numbers in ordinary decimal form are clumsy to write and to make calculations with, and it is hard to appreciate their precise magnitudes because of the sea of zeros.

A better method for expressing numbers makes use of powers-of-ten notation. This method is based on the fact that all numbers may be represented by a number between 1 and 10 multiplied by a power of 10. In powers-of-ten notation the mass of an electron is written simply as 9.109×10^{-31} kilogram and the mass of the earth is written as 5.983×10^{24} kilogram.

Table B-1 contains powers of 10 from 10^{-10} to 10^{10}. Evidently positive powers of 10 (which cover numbers greater than 1) follow this pattern:

TABLE B-1

$10^{-10} = 0.000,000,000,1$	$10^0 = 1$
$10^{-9} = 0.000,000,001$	$10^1 = 10$
$10^{-8} = 0.000,000,01$	$10^2 = 100$
$10^{-7} = 0.000,000,1$	$10^3 = 1000$
$10^{-6} = 0.000,001$	$10^4 = 10,000$
$10^{-5} = 0.000,01$	$10^5 = 100,000$
$10^{-4} = 0.000,1$	$10^6 = 1,000,000$
$10^{-3} = 0.001$	$10^7 = 10,000,000$
$10^{-2} = 0.01$	$10^8 = 100,000,000$
$10^{-1} = 0.1$	$10^9 = 1,000,000,000$
$10^0 = 1$	$10^{10} = 10,000,000,000$

$10^0 = 1.$ $= 1$ $= 1$ with decimal point moved 0 places,

$10^1 = 1.0$ $= 10$ $= 1$ with decimal point moved 1 place to the right,

$10^2 = 1.00$ $= 100$ $= 1$ with decimal point moved 2 places to the right,

$10^3 = 1.000$ $= 1,000$ $= 1$ with decimal point moved 3 places to the right,

$10^4 = 1.0000$ $= 10,000$ $= 1$ with decimal point moved 4 places to the right,

$10^5 = 1.00000$ $= 100,000$ $= 1$ with decimal point moved 5 places to the right,

$10^6 = 1.000000$ $= 1,000,000$ $= 1$ with decimal point moved 6 places to the right, and so on.

The exponent of the 10 indicates how many places the decimal point is moved *to the right* from $1.000 \cdots$

A similar pattern is followed by negative powers of 10, whose values always lie between 0 and 1:

$$10^0 = 1. \qquad = 1 \qquad = 1 \text{ with decimal point moved 0 places,}$$

$$10^{-1} = 01. \qquad = 0.1 \qquad = 1 \text{ with decimal point moved 1 place to the left,}$$

$$10^{-2} = 001. \qquad = 0.01 \qquad = 1 \text{ with decimal point moved 2 places to the left,}$$

$$10^{-3} = 0001. \qquad = 0.001 \qquad = 1 \text{ with decimal point moved 3 places to the left,}$$

$$10^{-4} = 00001. \qquad = 0.000,1 \qquad = 1 \text{ with decimal point moved 4 places to the left,}$$

$$10^{-5} = 000001. \qquad = 0.000,01 \qquad = 1 \text{ with decimal point moved 5 places to the left,}$$

$$10^{-6} = 0000001. = 0.000,001 = 1 \text{ with decimal point moved 6 places to the left, and so on.}$$

The exponent of the 10 now indicates how many places the decimal point is moved *to the left* from 1.

Here are a few examples of powers-of-ten notation:

$$600 = 6 \times 100 = 6 \times 10^2$$
$$7940 = 7.94 \times 1000 = 7.94 \times 10^3$$
$$93,000,000 = 9.3 \times 10,000,000 = 9.3 \times 10^7$$
$$0.023 = 2.3 \times 0.01 = 2.3 \times 10^{-2}$$
$$0.000,035 = 3.5 \times 0.000,01 = 3.5 \times 10^{-5}$$

EXERCISES

1. Express the following numbers in decimal notation.

(a) 2×10^5

(b) 8×10^{-2}

(c) 7.819×10^2

(d) 4.51×10^8

(e) 1.003×10^{-6}

(f) 10^{-10}

(g) 9.56×10^{-5}

2. Express the following numbers in powers-of-ten notation.

(a) 70

(b) 0.14

(c) 3.81

(d) 8400

(e) 1,000,000

(f) 0.007,890

(g) 351,600

B–2 USING POWERS OF TEN

Let us see how to make calculations using numbers written in powers-of-ten notation. To add or subtract numbers written in powers-of-ten notation, they must be expressed in terms of the *same* power of ten.

$$7 \times 10^4 + 2 \times 10^5 = 0.7 \times 10^5 + 2 \times 10^5$$
$$= 2.7 \times 10^5$$
$$5 \times 10^{-2} + 3 \times 10^{-4} = 5 \times 10^{-2} + 0.03 \times 10^{-2}$$
$$= 5.03 \times 10^{-2}$$
$$8 \times 10^{-3} - 7 \times 10^{-4} = 8 \times 10^{-3} - 0.7 \times 10^{-3}$$
$$= 7.3 \times 10^{-3}$$
$$4 \times 10^5 - 1 \times 10^6 = 4 \times 10^5 - 10 \times 10^5$$
$$= -6 \times 10^5$$

To multiply powers of ten together, add their exponents:

$$(10^n)(10^m) = 10^{n+m}$$

Be sure to take the sign of each exponent into account.

$$(10^2)(10^3) = 10^{2+3} = 10^5$$
$$(10^7)(10^{-3}) = 10^{7-3} = 10^4$$
$$(10^{-2})(10^{-4}) = 10^{-2-4} = 10^{-6}$$

To multiply numbers written in powers-of-ten notation, multiply the decimal parts of the numbers together and add the exponents to find the power of ten of the product:

$$(A \times 10^n)(B \times 10^m) = AB \times 10^{n+m}$$

If necessary, rewrite the result so the decimal part is a number between 1 and 10.

$$(3 \times 10^2)(2 \times 10^5) = (3 \times 2) \times 10^{2+5}$$
$$= 6 \times 10^7$$
$$(8 \times 10^{-5})(3 \times 10^7) = (8 \times 3) \times 10^{-5+7}$$
$$= 24 \times 10^2 = 2.4 \times 10^3$$
$$(1.3 \times 10^{-3})(4 \times 10^{-5}) = (1.3 \times 4) \times 10^{-3-5}$$
$$= 5.2 \times 10^{-8}$$
$$(-9 \times 10^{17})(6 \times 10^{-18}) = (-9 \times 6) \times 10^{17-18}$$
$$= -54 \times 10^{-1} = -5.4$$

To divide one power of ten by another, subtract the exponent of the denominator from the exponent of the numerator:

$$\frac{10^n}{10^m} = 10^{n-m}$$

Be sure to take the sign of each exponent into account.

$$\frac{10^5}{10^3} = 10^{5-3} = 10^2$$

$$\frac{10^{-2}}{10^4} = 10^{-2-4} = 10^{-6}$$

$$\frac{10^{-3}}{10^{-7}} = 10^{-3-(-7)} = 10^{-3+7} = 10^4$$

To divide a number written in powers-of-ten notation by another number written that way, divide the decimal parts of the numbers in the usual way and use the above rule to find the exponent of the power of ten of the quotient:

$$\frac{A \times 10^n}{B \times 10^m} = \frac{A}{B} \times 10^{n-m}$$

If necessary, rewrite the result so the decimal part is a number between 1 and 10:

$$\frac{6 \times 10^5}{3 \times 10^2} = \frac{6}{3} \times 10^{5-2} = 2 \times 10^3$$

$$\frac{2 \times 10^{-7}}{8 \times 10^4} = \frac{2}{8} \times 10^{-7-4} = \frac{1}{4} \times 10^{-11}$$

$$= 0.25 \times 10^{-11} = 2.5 \times 10^{-12}$$

$$\frac{-7 \times 10^5}{10^{-2}} = -7 \times 10^{5-(-2)} = -7 \times 10^{5+2}$$

$$= -7 \times 10^7$$

$$\frac{5 \times 10^{-2}}{-2 \times 10^{-9}} = -\frac{5}{2} \times 10^{-2-(-9)} = -2.5 \times 10^{-2+9}$$

$$= -2.5 \times 10^7$$

To find the reciprocal of a power of ten, change the sign of the exponent:

$$\frac{1}{10^n} = 10^{-n}$$

$$\frac{1}{10^{-m}} = 10^m$$

$$\frac{1}{10^5} = 10^{-5}$$

$$\frac{1}{10^{-3}} = 10^3$$

Hence the prescription for finding the reciprocal of a number written in powers-of-ten notation is

$$\frac{1}{A \times 10^n} = \frac{1}{A} \times 10^{-n}$$

For example,

$$\frac{1}{2 \times 10^3} = \frac{1}{2} \times 10^{-3} = 0.5 \times 10^{-3} = 5 \times 10^{-4}$$

$$\frac{1}{4 \times 10^{-8}} = \frac{1}{4} \times 10^8 = 0.25 \times 10^8 = 2.5 \times 10^7$$

The powers-of-ten method of writing large and small numbers makes arithmetic involving such numbers relatively easy to carry out. Here is a calculation that would be very tedious if each number were kept in decimal form.

$$\frac{(3800)(0.0054)(0.000,001)}{(430,000,000)(73)}$$

$$= \frac{(3.8 \times 10^3)(5.4 \times 10^{-3})(10^{-6})}{(4.3 \times 10^8)(7.3 \times 10^1)}$$

$$= \frac{(3.8)(5.4)}{(4.3)(7.3)} \times \frac{(10^3)(10^{-3})(10^{-6})}{(10^8)(10^1)}$$

$$= 0.65 \times 10^{(3-3-6-8-1)} = 0.65 \times 10^{-15}$$

$$= 6.5 \times 10^{-16}$$

EXERCISES

3. Perform the following additions and subtractions.

   (a) $3 \times 10^2 + 4 \times 10^3$
   (b) $7 \times 10^{-2} + 2 \times 10^{-3}$
   (c) $4 \times 10^{-5} + 5 \times 10^{-3}$
   (d) $6.32 \times 10^2 + 5$
   (e) $4 \times 10^3 - 3 \times 10^2$
   (f) $3.2 \times 10^{-4} - 5 \times 10^{-5}$
   (g) $7 \times 10^4 - 2 \times 10^{-5}$
   (h) $4.76 \times 10^{-3} - 4.81 \times 10^{-3}$

4. Evaluate the following reciprocals.

   (a) $\dfrac{1}{10^2}$ (c) $\dfrac{1}{10^{-2}}$

   (b) $\dfrac{1}{2 \times 10^2}$ (d) $\dfrac{1}{4 \times 10^{-4}}$

5. Perform the following calculations.

   (a) $\dfrac{(500,000)(18,000)}{9,000,000}$

   (b) $\dfrac{(30)(80,000,000,000)}{0.0004}$

(c) $\dfrac{(30,000)(0.000,000,6)}{(1000)(0.02)}$

(d) $\dfrac{(0.002)(0.000,000,05)}{0.000,004}$

(e) $\dfrac{(0.06)(0.0001)}{(0.000,03)(40,000)}$

(f) $\dfrac{(3 \times 10^4)(5 \times 10^{-12})}{10^3}$

(g) $\dfrac{9 \times 10^{12}}{9 \times 10^{-12}}$

(h) $\dfrac{(8 \times 10^{10})(3)}{6 \times 10^{-4}}$

(i) $\dfrac{10^{-3}}{(5 \times 10^4)(2 \times 10^2)}$

(j) $\dfrac{(5 \times 10^5)(2 \times 10^{-18})}{4 \times 10^4}$

B–3　POWERS AND ROOTS

To square a power of ten, multiply the exponent by 2; to cube a power of ten, multiply the exponent by 3:

$$(10^n)^2 = 10^{2n}$$
$$(10^n)^3 = 10^{3n}$$

In general, to raise a power of ten to the mth power, multiply the exponent by m:

$$(10^n)^m = 10^{m \times n}$$

Be sure to take the sign of each exponent into account, as in these examples:

$$(10^3)^2 = 10^{2 \times 3} = 10^6$$
$$(10^{-2})^5 = 10^{5 \times -2} = 10^{-10}$$
$$(10^{-4})^{-2} = 10^{-2 \times -4} = 10^8$$

To raise a number written in powers-of-ten notation to the mth power, multiply the decimal part of the number by itself m times and multiply the exponent of the power of ten by m:

$$(A \times 10^n)^m = A^m \times 10^{m \times n}$$

If necessary, rewrite the result so the decimal part is a number between 1 and 10:

$$(2 \times 10^5)^2 = 2^2 \times 10^{2 \times 5} = 4 \times 10^{10}$$

$$(3 \times 10^{-3})^3 = 3^3 \times 10^{3 \times -3} = 27 \times 10^{-9}$$
$$= 2.7 \times 10^{-8}$$
$$(5 \times 10^{-2})^{-4} = 5^{-4} \times 10^{-4 \times -2} = \dfrac{1}{5^4} \times 10^8$$
$$= \dfrac{1}{625} \times 10^8 = 0.0016 \times 10^8$$
$$= 1.6 \times 10^5$$

To take the mth root of a power of ten, divide the exponent by m:

$$\sqrt[m]{10^n} = (10^n)^{1/m} = 10^{n/m}$$

Thus

$$\sqrt{10^4} \quad (10^4)^{1/2} = 10^{4/2} = 10^2$$
$$\sqrt[3]{10^9} = (10^9)^{1/3} = 10^{9/3} = 10^3$$
$$\sqrt[3]{10^{-9}} = (10^{-9})^{1/3} = 10^{-9/3} = 10^{-3}$$

In powers-of-ten notation, the exponent of the 10 must be an integer. Hence in taking the mth root of a power of ten, the exponent should be an integral multiple of m. Instead of, for example,

$$\sqrt{10^5} = (10^5)^{1/2} = 10^{2.5}$$

which, while correct, is hardly useful, we would write

$$\sqrt{10^5} = \sqrt{10^1 \times 10^4}$$
$$= \sqrt{10} \times \sqrt{10^4}$$
$$= 3.16 \times 10^2$$

Here are two other examples:

$$\sqrt{10^{-3}} = \sqrt{10^1 \times 10^{-4}} = \sqrt{10} \times \sqrt{10^{-4}}$$
$$= 3.16 \times 10^{-2}$$
$$\sqrt[3]{10^8} = \sqrt[3]{10^2 \times 10^6} = \sqrt[3]{100} \times \sqrt[3]{10^6}$$
$$= 4.64 \times 10^2$$

To take the mth root of a number expressed in powers-of-ten notation, first write the number so the exponent of the 10 is an integral multiple of m. Then take the mth root of the decimal part of the number and divide the exponent by m to find the power of ten of the result:

$$(A \times 10^n)^{1/m} = \sqrt[m]{A} \times 10^{n/m}$$

Thus

$$\sqrt{9 \times 10^4} = \sqrt{9} \times 10^{4/2} = 3 \times 10^2$$
$$\sqrt{9 \times 10^{-6}} = \sqrt{9} \times 10^{-6/2} = 3 \times 10^{-3}$$

$$\sqrt{4 \times 10^7} = \sqrt{40 \times 10^6} = \sqrt{40} \times 10^{6/2}$$
$$= 6.32 \times 10^3$$
$$\sqrt[3]{3 \times 10^{-5}} = \sqrt[3]{30 \times 10^{-6}} \quad \sqrt[3]{30} \times 10^{-6/3}$$
$$= 3.11 \times 10^{-2}$$

EXERCISES

6. Evaluate the following powers.

(a) $(2 \times 10^7)^2$ (c) $(3 \times 10^{-8})^2$

(b) $(2 \times 10^7)^{-2}$ (d) $(5 \times 10^{-4})^{-3}$

7. Evaluate the following roots. Assume $\sqrt{4} = 2$, $\sqrt{40} = 6.3$, $\sqrt[3]{4} = 1.6$, $\sqrt[3]{40} = 3.4$, and $\sqrt[3]{400} = 7.4$.

(a) $(4 \times 10^6)^{1/2}$ (f) $(4 \times 10^{13})^{1/3}$

(b) $(4 \times 10^7)^{1/2}$ (g) $(4 \times 10^{14})^{1/3}$

(c) $(4 \times 10^{-4})^{1/2}$ (h) $(4 \times 10^{-6})^{1/3}$

(d) $(4 \times 10^{-5})^{1/2}$ (i) $(4 \times 10^{-7})^{1/3}$

(e) $(4 \times 10^{-12})^{1/3}$ (j) $(4 \times 10^{-8})^{1/3}$

B–4 SIGNIFICANT FIGURES

An advantage of powers-of-ten notation is that it gives no false impression of the degree of accuracy with which a number is stated. For instance, the equatorial radius of the earth is 6378 km, but it is often taken as 6400 km for convenience in making rough calculations. To indicate the approximate character of the latter figure, all we have to do is write

$$r = 6.4 \times 10^3 \text{ km}$$

whose meaning is

$$r = (6.4 \pm 0.05) \times 10^3 \text{ km}$$

With this method, how large the number is and how accurate it is are both clear. The accurately known digits, plus one uncertain digit, are called *significant figures;* in the above case, r has two significant figures, 6 and 4. If we require greater accuracy, we would write

$$r = 6.38 \times 10^3 \text{ km}$$

which contains three significant figures, or

$$r = 6.378 \times 10^3 \text{ km}$$

which contains four significant figures.

Sometimes one or more zeros in a number are significant figures, and it is proper to retain them when expressing the number in powers-of-ten notation. There is quite a difference between 3×10^5 and 3.00×10^5:

$$3 \times 10^5 = (3 \pm 0.5) \times 10^5$$
$$3.00 \times 10^5 = (3 \pm 0.005) \times 10^5$$

When quantities are combined arithmetically, the result is no more accurate than the quantity with the largest uncertainty. Suppose a 75-kg person picks up a 0.23-kg apple. The total mass of person plus apple is still 75 kg because all we know of the person's mass is that it is somewhere between 74.5 and 75.5 kg, which means an uncertainty greater than the apple's mass. If the person's mass is instead quoted as 75.0 kg, the mass of person plus apple is 75.2 kg; if it is quoted as 75.00 kg, the mass of person plus apple is 75.23 kg. Thus

$$75 \text{ kg} + 0.23 \text{ kg} = 75 \text{ kg}$$
$$75.0 \text{ kg} + 0.23 \text{ kg} = 75.2 \text{ kg}$$
$$75.00 \text{ kg} + 0.23 \text{ kg} = 75.23 \text{ kg}$$

Significant figures must be taken into account in multiplication and division also. For example, if we divide 1.4×10^5 by 6.70×10^3, we are not justified in writing

$$\frac{1.4 \times 10^5}{6.70 \times 10^3} = 20.89552 \cdots$$

We may properly retain only two significant figures, corresponding to the two significant figures in the numerator, and so the correct answer is just 21.

In a calculation with several steps, however, it is a good idea to keep an extra digit in the intermediate steps and to round off the result only at the end. As an example,

$$\frac{5.7 \times 10^4}{3.3 \times 10^{-2}} + \sqrt{1.8 \times 10^{12}} = 1.73 \times 10^6 + 1.34 \times 10^6$$
$$= 3.07 \times 10^6 = 3.1 \times 10^6$$

If the intermediate results had been rounded off to two digits, however, the result would have been the incorrect

$$1.7 \times 10^6 + 1.3 \times 10^6 = 3.0 \times 10^6$$

B–5 LOGARITHMS

The logarithm of a number N is the exponent n to which a given base number a must be raised in order that $a^n = N$. That is, if

$$N = a^n, \quad \text{then} \quad n = \log_a N$$

Here are some examples using the base $a = 10$:

$1000 = 10^3$,   therefore $\log_{10} 1000 = 3$;

$5 = 10^{0.699}$,   therefore $\log_{10} 5 = 0.699$;

$0.001 = 10^{-3}$, therefore $\log_{10} 0.001 = -3$.

Since the decimal system of numbers has the base 10, this is a convenient number to use as the base of a system of logarithms. Logarithms to the base 10 are called *common logarithms* and are denoted simply as "log N." Another widely used system of logarithms uses $e = 2.718 \cdots$ as the base. Such logarithms are called *natural logarithms* (because they arise in a natural way in calculus) and are denoted "ln N." To go from one system to the other these formulas are needed:

$\log N = 0.43429 \ln N$

$\ln N = 2.3026 \log N$

Logarithms are defined only for positive numbers because the quantity a^n is positive whether n is positive, negative, or zero. Since n is the logarithm of a^n, n can describe only a positive number.

Let us consider a number N that is the product of two numbers x and y, so that $N = xy$. If $x = 10^n$ and $y = 10^m$, then

$N = xy = 10^n \times 10^m = 10^{n+m}$

$\log N = n + m$

Since $n = \log x$ and $m = \log y$, $n + m = \log x + \log y$, and so

$\log N = \log x + \log y$

Thus we have the general rule for the logarithm of a product:

$\log xy = \log x + \log y$

Similar reasoning gives the additional rules

$\log \dfrac{x}{y} = \log x - \log y$

$\log x^n = n \log x$

Before the days of electronic calculators, logarithms were widely used to simplify arithmetical work because they permit replacing multiplication and division by addition and subtraction, which are easier to do and less prone to error. Today, of course, calculators are used for such routine arithmetic. However, logarithms are still needed for finding powers and roots, with calculators replacing tables of logarithms. For example, to find $(2.13)^4$ we proceed as follows:

$\log (2.13)^4 = 4 \log 2.13 = (4)(0.3284) = 1.314$

$(2.13)^4 = \log^{-1} 1.314 = 20.6$

To obtain $\log 2.13$ with a calculator, enter 2.13 and press the [LOG] key. To obtain $\log^{-1} 1.314$ (the *antilogarithm* of 1.314, which is the number whose logarithm is 1.314), the 1.314 is entered and the $[10^x]$ key pressed, since 10^x is the number whose logarithm is x. This key is alternatively designated [INV LOG].

The same procedure is followed for negative exponents, for instance $1/\sqrt[4]{7}$:

$$\frac{1}{\sqrt[4]{7}} = \frac{1}{7^{1/4}} = 7^{-1/4}$$

$$\log 7^{-1/4} = -\tfrac{1}{4} \log 7 = -\frac{1}{4} \times 0.8451 = -0.2113$$

$$7^{-1/4} = \log^{-1} (-0.2113) = 0.615$$

The logarithms of numbers written in powers-of-ten notation can be found with a calculator without first converting them to decimal notation. Let us consider 6.04×10^9. Since

$\log xy = \log x + \log y$

we have

$\log (6.04 \times 10^9) = \log 6.04 + \log 10^9$

But from the definition of logarithms to the base 10, $\log 10^9 = 9$. We therefore have

$\log (6.04 \times 10^9) = 0.7810 + 9 = 9.7810$

The same procedure holds for a number smaller than 1, for instance 2.4×10^{-5}:

$\log (2.4 \times 10^{-5}) = \log 2.4 + \log 10^{-5}$
$= 0.3802 - 5 = -4.6198$

EXERCISES

7. Evaluate the following with the help of logarithms.

    (a) $0.0181^{1.5}$ (d) $\sqrt[4]{156}$

    (b) $62.2^{7.13}$ (e) $(6.24 \times 10^{-4})^{1/3}$

    (c) $(8.15 \times 10^{14})^6$ (f) $(2.71 \times 10^5)^{1/8}$

ANSWERS

1.   (a)   200,000   (e)   0.000,001,003

     (b)   0.08   (f)   0.000,000,000,1

     (c)   781.9   (g)   0.000,0956

     (d)   451,000,000

2. (a) 7×10^1
(b) 1.4×10^{-1}
(c) 3.81
(d) 8.4×10^3
(e) 1×10^6
(f) 7.890×10^{-3}
(g) 3.516×10^5

3. (a) 4.3×10^3
(b) 7.2×10^{-2}
(c) 5.04×10^{-3}
(d) 6.37×10^2
(e) 3.7×10^3
(f) 2.7×10^{-4}
(g) -1.3×10^5
(h) -5×10^{-5}

4. (a) $10^{-2} = 0.01$
(b) $5 \times 10^{-3} = 0.005$
(c) $10^2 = 100$
(d) $2.5 \times 10^3 = 2500$

5. (a) 1×10^3
(b) 6×10^{15}
(c) 9×10^{-4}
(d) 2.5×10^{-5}
(e) 5×10^{-6}
(f) 1.5×10^{-10}
(g) 1×10^{24}
(h) 4×10^{14}
(i) 1×10^{-10}
(j) 2.5×10^{-17}

6. (a) 2×10^3
(b) 6.3×10^3
(c) 2×10^{-2}
(d) 6.3×10^{-3}
(e) 1.6×10^{-4}
(f) 3.4×10^4
(g) 7.4×10^4
(h) 1.6×10^{-2}
(i) 7.4×10^{-3}
(j) 3.4×10^{-3}

7. (a) 2.44×10^{-3}
(b) 6.16×10^{12}
(c) 2.93×10^{89}
(d) 3.53
(e) 8.55×10^{-2}
(f) 4.78

APPENDIX C
TABLES

TABLE C–1
The Elements

Atomic number	Element	Symbol	Atomic mass*	Atomic number	Element	Symbol	Atomic mass*
1	Hydrogen	H	1.008	36	Krypton	Kr	83.80
2	Helium	He	4.003	37	Rubidium	Rb	85.47
3	Lithium	Li	6.939	38	Strontium	Sr	87.62
4	Beryllium	Be	9.012	39	Yttrium	Y	88.91
5	Boron	B	10.81	40	Zirconium	Zr	91.22
6	Carbon	C	12.01	41	Niobium	Nb	92.91
7	Nitrogen	N	14.01	42	Molybdenum	Mo	95.94
8	Oxygen	O	16.00	43	Technetium	Tc	(97)
9	Fluorine	F	19.00	44	Ruthenium	Ru	101.1
10	Neon	Ne	20.18	45	Rhodium	Rh	102.9
11	Sodium	Na	22.99	46	Palladium	Pd	106.4
12	Magnesium	Mg	24.31	47	Silver	Ag	107.9
13	Aluminum	Al	26.98	48	Cadmium	Cd	112.4
14	Silicon	Si	28.09	49	Indium	In	114.8
15	Phosphorus	P	30.98	50	Tin	Sn	118.7
16	Sulfur	S	32.07	51	Antimony	Sb	121.8
17	Chlorine	Cl	35.46	52	Tellurium	Te	127.6
18	Argon	Ar	39.94	53	Iodine	I	126.9
19	Potassium	K	39.10	54	Xenon	Xe	131.3
20	Calcium	Ca	40.08	55	Cesium	Cs	132.9
21	Scandium	Sc	44.96	56	Barium	Ba	137.3
22	Titanium	Ti	47.90	57	Lanthanum	La	138.9
23	Vanadium	V	50.94	58	Cerium	Ce	140.1
24	Chromium	Cr	52.00	59	Praseodymium	Pr	140.9
25	Manganese	Mn	54.94	60	Neodymium	Nd	144.2
26	Iron	Fe	55.85	61	Promethium	Pm	(145)
27	Cobalt	Co	58.93	62	Samarium	Sm	150.4
28	Nickel	Ni	58.71	63	Europium	Eu	152.0
29	Copper	Cu	63.54	64	Gadolinium	Gd	157.3
30	Zinc	Zn	65.37	65	Terbium	Tb	158.9
31	Gallium	Ga	69.72	66	Dysprosium	Dy	162.5
32	Germanium	Ge	72.59	67	Holmium	Ho	164.9
33	Arsenic	As	74.92	68	Erbium	Er	167.3
34	Selenium	Se	78.96	69	Thulium	Tm	168.9
35	Bromine	Br	79.91	70	Ytterbium	Yb	173.0

*The unit of mass is the u. Elements whose atomic masses are given in parentheses have not been found in nature but have been produced by nuclear reactions in the laboratory. The atomic mass in such a case is the mass number of the longest-lived radioactive isotope of the element.

TABLE C-1
The Elements (continued)

Atomic number	Element	Symbol	Atomic mass*	Atomic number	Element	Symbol	Atomic mass*
71	Lutetium	Lu	175.0	88	Radium	Ra	226.0
72	Hafnium	Hf	178.5	89	Actinium	Ac	227.0
73	Tantalum	Ta	181.0	90	Thorium	Th	232.0
74	Tungsten	W	183.9	91	Protactinium	Pa	231.0
75	Rhenium	Re	186.2	92	Uranium	U	238.0
76	Osmium	Os	190.2	93	Neptunium	Np	(237)
77	Iridium	Ir	192.2	94	Plutonium	Pu	(244)
78	Platinum	Pt	195.1	95	Americium	Am	(243)
79	Gold	Au	197.0	96	Curium	Cm	(247)
80	Mercury	Hg	200.6	97	Berkelium	Bk	(247)
81	Thallium	Tl	204.4	98	Californium	Cf	(251)
82	Lead	Pb	207.2	99	Einsteinium	Es	(254)
83	Bismuth	Bi	209.0	100	Fermium	Fm	(257)
84	Polonium	Po	(209)	101	Mendelevium	Md	(258)
85	Astatine	At	(210)	102	Nobelium	No	(259)
86	Radon	Rn	222	103	Lawrencium	Lr	(260)
87	Francium	Fr	(223)				

Natural Trigonometric Functions

Degree	Radian	Sine	Cosine	Tangent	Degree	Radian	Sine	Cosine	Tangent
0°	.000	0.000	1.000	0.000					
1°	.017	.018	1.000	.018	16°	.279	.276	.961	.287
2°	.035	.035	0.999	.035	17°	.297	.292	.956	.306
3°	.052	.052	.999	.052	18°	.314	.309	.951	.325
4°	.070	.070	.998	.070	19°	.332	.326	.946	.344
5°	.087	.087	.996	.088	20°	.349	.342	.940	.364
6°	.105	.105	.995	.105	21°	.367	.358	.934	.384
7°	.122	.122	.993	.123	22°	.384	.375	.927	.404
8°	.140	.139	.990	.141	23°	.401	.391	.921	.425
9°	.157	.156	.988	.158	24°	.419	.407	.914	.445
10°	.175	.174	.985	.176	25°	.436	.423	.906	.466
11°	.192	.191	.982	.194	26°	.454	.438	.899	.488
12°	.209	.208	.978	.213	27°	.471	.454	.891	.510
13°	.227	.225	.974	.231	28°	.489	.470	.883	.532
14°	.244	.242	.970	.249	29°	.506	.485	.875	.554
15°	.262	.259	.966	.268	30°	.524	.500	.866	.577

TABLE C-2
Natural Trigonometric Functions (continued)

Angle					Angle				
Degree	*Radian*	*Sine*	*Cosine*	*Tangent*	*Degree*	*Radian*	*Sine*	*Cosine*	*Tangent*
31°	.541	.515	.857	.601	61°	1.065	.875	.485	1.804
32°	.559	.530	.848	.625	62°	1.082	.883	.470	1.881
33°	.576	.545	.839	.649	63°	1.100	.891	.454	1.963
34°	.593	.559	.829	.675	64°	1.117	.899	.438	2.050
35°	.611	.574	.819	.700	65°	1.134	.906	.423	2.145
36°	.628	.588	.809	.727	66°	1.152	.914	.407	2.246
37°	.646	.602	.799	.754	67°	1.169	.921	.391	2.356
38°	.663	.616	.788	.781	68°	1.187	.927	.375	2.475
39°	.681	.629	.777	.810	69°	1.204	.934	.358	2.605
40°	.698	.643	.766	.839	70°	1.222	.940	.342	2.747
41°	.716	.658	.755	.869	71°	1.239	.946	.326	2.904
42°	.733	.669	.743	.900	72°	1.257	.951	.309	3.078
43°	.751	.682	.731	.933	73°	1.274	.956	.292	3.271
44°	.768	.695	.719	.966	74°	1.292	.961	.276	3.487
45°	.785	.707	.707	1.000	75°	1.309	.966	.259	3.732
46°	.803	.719	.695	1.036	76°	1.326	.970	.242	4.011
47°	.820	.731	.682	1.072	77°	1.344	.974	.225	4.331
48°	.838	.743	.669	1.111	78°	1.361	.978	.208	4.705
49°	.855	.755	.656	1.150	79°	1.379	.982	.191	5.145
50°	.873	.766	.643	1.192	80°	1.396	.985	.174	5.671
51°	.890	.777	.629	1.235	81°	1.414	.988	.156	6.314
52°	.908	.788	.616	1.280	82°	1.431	.990	.139	7.115
53°	.925	.799	.602	1.327	83°	1.449	.993	.122	8.144
54°	.942	.809	.588	1.376	84°	1.466	.995	.105	9.514
55°	.960	.819	.574	1.428	85°	1.484	.996	.087	11.43
56°	.977	.829	.559	1.483	86°	1.501	.998	.070	14.30
57°	.995	.839	.545	1.540	87°	1.518	.999	.052	19.08
58°	1.012	.848	.530	1.600	88°	1.536	.999	.035	28.64
59°	1.030	.857	.515	1.664	89°	1.553	1.000	.018	57.29
60°	1.047	.866	.500	1.732	90°	1.571	1.000	.000	∞

TABLE C-3
The Greek Alphabet

A	α	Alpha
B	β	Beta
Γ	γ	Gamma
Δ	δ	Delta
E	ε	Epsilon
Z	ζ	Zeta
H	η	Eta
Θ	θ	Theta
I	ι	Iota
K	κ	Kappa
Λ	λ	Lambda
M	μ	Mu
N	ν	Nu
Ξ	ξ	Xi
O	o	Omicron
Π	π	Pi
P	ρ	Rho
Σ	σ	Sigma
T	τ	Tau
Υ	υ	Upsilon
Φ	φ	Phi
X	χ	Chi
Ψ	ψ	Psi
Ω	ω	Omega

Glossary

Absolute temperature scale *Absolute zero* is the lowest temperature possible; in the elementary kinetic theory of matter, it is that temperature at which random molecular movement would cease, although in reality a small amount of movement would still persist. Absolute zero corresponds to $-273°C$. The *absolute temperature scale* expresses temperature in °C above absolute zero; its unit is the *Kelvin*, denoted K.

Acceleration The *acceleration a* of a body is the rate at which its velocity changes with time; the change in velocity may be in magnitude or direction or both.

Acceleration of gravity The *acceleration of gravity g* is the acceleration of a freely falling body near the earth's surface. Its value is 32 ft/s^2 in British units and 9.8 m/s^2 in metric units.

Alpha particle An *alpha particle* is the nucleus of a helium atom. It consists of two neutrons and two protons, and it is emitted in the radioactive decay of certain isotopes.

Alternating current The direction of an *alternating electric current* reverses itself periodically.

Ampere The *ampere* (A) is the unit of electric current. It is equal to a flow of charge at the rate of 1 C/s.

Amplitude The *amplitude* of a body undergoing simple harmonic motion is its maximum displacement on either side of its equilibrium position. The amplitude of a wave is the maximum value of the wave variable (for instance, displacement, pressure, electric field) regardless of sign.

Angular momentum The *angular momentum L* of a rotating body is the product $I\omega$ of its moment of inertia and angular velocity. The principle of *conservation of angular momentum* states that the total angular momentum of a system of particles remains constant when no net external torque acts upon the system.

Angular velocity The *angular velocity* ω of a rotating body is the angle through which it turns per unit time. The *angular acceleration* $\propto$ of a rotating body is the rate of change of its angular velocity with respect to time.

Antiparticle Nearly every elementary particle has an *antiparticle* counterpart whose electric charge and certain other properties have the opposite sign. Thus the antiparticle

of the electron is the *positron* (charge $+e$) and that of the proton is the *antiproton* (charge $-e$). When a particle and an antiparticle of the same kind come together, they *annihilate* each other, with the vanished mass reappearing as photons or mesons.

Archimedes' principle *Archimedes' principle* states that the buoyant force on a submerged object is equal to the weight of fluid it displaces.

Atom An *atom* is the ultimate particle of an element. Every atom consists of a very small positively charged nucleus and a number of electrons at some distance from it. The nucleus contains nearly all the mass of the atom.

Atomic mass unit Atomic and nuclear masses are expressed in *atomic mass units* (u), equal to 1.66×10^{-27} kg.

Atomic number The *atomic number* of an element is the number of electrons in each of its atoms or, equivalently, the number of protons in each of its atomic nuclei.

Back emf A *back* (or *counter*) *emf* is induced in the rotating coils of an electric motor and is opposite in direction to the external voltage applied to them.

Beta decay *Beta decay* is a type of radioactive decay in which a nucleus emits an electron or a positron.

Binding energy The *binding energy* of a nucleus is the energy equivalent of the difference between its mass and the sum of the masses of its individual constituent nucleons. This amount of energy must be supplied to the nucleus if it is to be completely disintegrated. The binding energy per nucleon is least for very light and very heavy nuclei; hence the *fusion* of very light nuclei to form heavier ones and the *fission* of very heavy nuclei to form lighter ones are both processes that liberate energy.

Bohr theory of the atom According to the *Bohr theory of the atom,* an electron can circle an atomic nucleus indefinitely without radiating energy if its orbit is an integral number of electron wavelengths in circumference. The number of wavelengths that fit into a particular permitted orbit is called the *quantum number* of that orbit. The electron energies corresponding to the various quantum numbers constitute the *energy levels* of the atom, of which the lowest is the *ground state* and the rest are *excited states.*

Boyle's law *Boyle's law* states that, at constant temperature, the absolute pressure of a sample of a gas is inversely proportional to its volume, so that $pV =$ constant at that temperature regardless of changes in either p or V individually.

Bulk modulus The *bulk modulus B* of a material is equal to the pressure on a sample of it divided by the fractional decrease in its volume.

Capacitor A *capacitor* is a device that stores electrical energy in the form of an electric field. The ratio between the charge on either plate of a capacitor and the potential difference between the plates is called its *capacitance C*. The unit of capacitance is the *farad* (F) which is equal to 1 C/V.

Capacitive reactance The *capacitive reactance X_C* of a capacitor is a measure of its effect on an alternating current; it is equal to $1/2\pi fC$, where C is the capacitance. If V_C is the effective voltage across a capacitor, the effective current that flows into and out of it is $I = V_C/X_C$. The unit of reactance is the ohm.

Carnot engine A *Carnot engine* is an idealized engine which is not subject to such practical difficulties as friction or heat losses by conduction or radiation but which obeys all physical laws. The efficiency of a Carnot engine which absorbs heat at the absolute temperature T_1 and exhausts heat at the absolute temperature T_2 is equal to $1 - T_2/T_1$; no engine operating between the same two temperatures can be more efficient than a Carnot engine operating between them.

Cathode ray tube In a *cathode ray tube*, an electron beam controlled by electric or magnetic fields traces out an image on a fluoresent screen.

Celsius temperature scale In the *celsius* (centigrade) *temperature scale*, the freezing point of water is assigned the value 0°C and the boiling point of water the value of 100°C.

Center of gravity The *center of gravity* of an object is that point from which it can be suspended in any orientation without tending to rotate. The weight of an object can be considered as a downward force acting on its center of gravity.

Centripetal acceleration The velocity of an object in uniform circular motion continually changes in direction although its magnitude remains constant. The object's acceleration is called *centripetal acceleration*, and it points toward the center of the circular path. Centripetal acceleration is proportional to the square of the object's speed and inversely proportional to the radius of its path.

Centripetal force The inward force that provides an object in uniform circular motion with its centripetal acceleration is called *centripetal force*.

Charles's law *Charles's law* states that, at constant pressure, the volume of a sample of a gas is directly proportional to its absolute temperature, so that $V/T =$ constant at that pressure regardless of changes in either V or T individually.

Coherence Two sources of waves are *coherent* if there is a fixed phase relationship between the waves they emit during the time the waves are being observed. Interference can be observed only in waves from coherent sources.

Component of a vector A *component of a vector* is its projection in a specified direction.

Compound Two or more elements may combine chemically to form a *compound*, a new substance whose properties are different from those of the elements that compose it.

Compression When equal and opposite forces that act toward each other are applied to a body, it is said to be in *compression*.

Concave mirror A *concave mirror* curves inward toward its center and converges parallel light to a single *real focal point*. The distance from the mirror to the focal point is the *focal length* of the mirror.

Concurrent forces When the lines of action of the various forces that act on an object intersect at a common point, the forces are said to be *concurrent;* when their lines of action do not intersect, the forces are *nonconcurrent*.

Conduction In *conduction*, heat is transferred from one place to another by successive molecular collisions.

Convection In *convection*, heat is transferred from one place to another by the motion of a volume of hot fluid from one place to another.

Convex mirror A *convex mirror* curves outward toward its center and diverges parallel light as though the reflected light came from a single *virtual focal point* behind the mirror. The distance from the mirror to the focal point is the *focal length* of the mirror.

Coulomb The *coulomb* (C) is the unit of electric charge.

Coulomb's law *Coulomb's law* states that the force one charge exerts upon another is directly proportional to the magnitudes of the charges and inversely proportional to the square of the distance between them. The force between like charges is repulsive, and that between unlike charges is attractive.

Covalent bond In a *covalent bond* between adjacent atoms of a molecule or solid, the atoms share one or more electron pairs.

Critical point The *critical point* is the upper limit of the vaporization curve of a substance, which cannot exist in the liquid state at a temperature above that of its critical point.

Crystalline solid Solids whose constituent atoms or molecules are arranged in regular, repeated patterns are called *crystalline*. When only short-range order is present, the solid is *amorphous*.

De Broglie waves A moving object behaves as though it has a wave nature. The waves representing such a particle are called *de Broglie waves*.

Density The *density d* of a substance is its mass per unit volume.

Dielectric constant The *dielectric constant K* of a particular material is a measure of how effective it is in reducing an electric field set up across a sample of the material.

Diffraction The ability of waves to bend around the edges of obstacles in their paths is called *diffraction*. A *diffraction grating* is a series of parallel slits that produces a spectrum through the interference of light that is diffracted by them.

Dispersion *Dispersion* refers to the splitting up of a beam of light containing different frequencies by passage through a substance whose index of refraction varies with frequency.

Domain An assembly of atoms in a ferromagnetic material whose atomic magnetic moments are aligned is called a *domain*.

Doppler effect The *Doppler effect* refers to the change in frequency of a wave when there is relative motion between its source and an observer.

Effective value The *effective value* of an alternating current is such that a direct current of this magnitude produces heat in a resistor at the same rate as the alternating current. The relationship between I_{eff} and I_{max} is $I_{eff} = 0.707I_{max}$. The similar relationship $V_{eff} = 0.707V_{max}$ holds for the effective and maximum voltages in an ac-circuit.

Elastic collision A *completely elastic collision* is one in which kinetic energy is conserved. A *completely inelastic collision* is one in which the bodies stick together upon impact, which results in the maximum possible kinetic energy loss.

Elastic limit The *elastic limit* is the maximum stress a solid can experience without being permanently altered. Hooke's law is only valid when the elastic limit is not exceeded.

Elastic potential energy A body under stress possesses *elastic potential energy* which is equal to the work done in deforming it.

Electric charge *Electric charge, Q*, like rest mass, is a fundamental property of certain of the elementary particles of which all matter is composed. There are two kinds of electric charge, *positive charge* and *negative charge;* charges of like sign repel, unlike charges attract. The unit of charge is the *coulomb* (C). All charges, of either sign, occur in multiples of the fundamental *electron charge* of 1.6×10^{-19} C. The principle of *conservation of charge* states that the net electric charge in an isolated system remains constant. Electric charge is relativistically invariant.

Electric current A flow of electric charge from one place to another is called an *electric current*. The unit of electric current is the *ampere* (A), which is equal to a flow of 1 C/s.

Electric field An *electric field* E exists wherever an electric force acts on a charged particle. The magnitude E of an electric field at a point is defined as the force that would act on a charge of $+1$ C placed there. The unit of electric field is the volt/m, which is equal to 1 N/C.

Electrolysis *Electrolysis* is the process by which free elements are liberated from a liquid by the passage of an electric current.

Electrolyte A substance that separates into free ions when dissolved in water is called an *electrolyte* since the resulting solution is able to conduct electric current.

Electromagnetic induction *Electromagnetic induction* refers to the production of an electric field wherever magnetic lines of force are in motion.

Electromagnetic waves *Electromagnetic waves* consist of coupled electric and magnetic oscillations. The electric field of such a wave is perpendicular to its magnetic field, and both fields are perpendicular to the direction in which the wave travels. Radio waves, light waves, x-rays, and gamma rays are all electromagnetic waves differing only in their wavelength. Electromagnetic waves are produced by accelerated electric charges, and in free space have the speed $c = 3.00 \times 10^8$ m/s.

Electromotive force The *electromotive force* (emf) of a battery, generator, or other source of electrical energy is the potential difference across its terminals when no current flows. When a current is flowing, the terminal voltage is less than the emf owing to the potential drop in the *internal resistance* of the source.

Electron The *electron* is the least massive elementary particle found in matter. The electron has a charge of $-e$, where $e = 1.60 \times 10^{-19}$ coulomb.

Electron volt The *electron volt* is the energy acquired by an electron that has been accelerated by a potential difference of 1 volt. It is equal to 1.6×10^{-19} J.

Elementary particles The four categories of *elementary particles* are the photons, the leptons, the mesons, and the baryons. *Leptons* are point particles with no internal structures; examples are the electron and the neutrino. *Mesons* are particles that can be regarded as "carriers" of the strong interaction; an example is the pion. *Baryons* include nucleons and heavier unstable particles. Mesons and baryons, jointly known as *hadrons,* are subject to the strong interaction and are thought to be composed of *quarks,* which are particles with fractional electric charges that have not as yet been experimentally isolated.

Elements *Elements* are the simplest substances encountered in bulk. They cannot be decomposed or transformed into one another by ordinary chemical or physical means.

Energy *Energy E* is that which may be converted into work. When something possesses energy, it is capable of performing work or, in a general sense, of accomplishing a change in some aspect of the physical world. The unit of energy is the *joule* (J). The three broad categories of energy are *kinetic energy,* which is the energy something possesses by virtue of its motion; *potential energy,* which is the energy something possesses by virtue of its position in a force field; and *rest energy,* which is the energy something possesses by virtue of its mass. The principle of *conservation of energy* states that the total amount of energy in a system isolated from the rest of the universe always remains constant, although energy transformations from one form to another, including rest energy, may occur within the system.

Energy band The *energy bands* in a crystal are ranges of energy that correspond to energy levels in an atom. An electron in a crystal can only have an amount of energy that falls within one of its energy bands.

Equilibrium An object not acted upon by a net force is in *translational equilibrium* and may be at rest or have a constant linear velocity. An object not acted upon by a net torque is in *rotational equilibrium* and may be at rest or have a constant angular velocity.

Equivalent resistance The *equivalent resistance* of a set of interconnected resistors is the value of the single resistor that can be substituted for the entire set without affecting the current that flows in the rest of any circuit of which it is a part.

Escape speed The minimum speed needed by an object to permanently escape from the gravitational attraction of an astronomical body such as a planet or star is called the *escape speed* of the body.

Exclusion principle According to the Pauli *exclusion principle,* no two electrons in an atom can exist in the same quantum state.

Farad The unit of capacitance is the *farad* (F), which is equal to 1 C/V.

Force A *force F* is any influence that can cause a body to be accelerated. The unit of force is the *newton* (N); in the British system it is the *pound* (lb).

Force field A *force field* is a region of space at every point in which an appropriate test object would experience a force. Thus a *gravitational field* is a region of space in which an object by virtue of its mass is acted on by a force, and an *electric field* is a region of space in which an object by virtue of its electric charge is acted upon by a force.

Frame of reference A *frame of reference* is something with respect to which observations are made on something else. When a person standing beside a road sees a car moving along the road, the road is the frame of reference; when a person in the car sees the roadside moving backward past him, the car is the frame of reference. All measurements, including those of time, have meaning only when the frame of reference from which they are made is specified. All frames of reference are equally valid — there is no universal frame of reference. The *special theory of relativity* relates measurements made on an object or phenomenon from frames of reference moving at constant velocity with respect to one another.

Frequency The *frequency f* of something undergoing harmonic motion is the number of oscillations it makes per unit time. The frequency of a train of waves is the number of waves that pass a particular point per unit time. The unit of frequency is the *hertz* (Hz), which is equal to 1 cycle/s.

Frequency modulation In *frequency modulation,* information is contained in variations in the frequency of the carrier wave.

Friction The term *friction* refers to the resistive forces that arise to oppose the motion of a body past another with which it is in contact. *Sliding friction* is the frictional resistance a body in motion experiences, while *static friction* is the frictional resistance a stationary body must overcome in order to be set in motion. The *coefficient of friction* is the constant of proportionality for a given pair of contacting surfaces that relates the frictional force between them to the normal force with which one presses against the other. Usually the coefficient of static friction is greater than that of sliding friction. *Rolling friction* refers to the resistance a circular object experiences as it rolls over a smooth, flat surface; coefficients of rolling friction are much smaller than those of sliding friction.

Fusion, heat of The *heat of fusion* of a substance is the amount of heat that must be supplied to change a unit quantity of it at its melting point from the solid to the liquid state; the same amount of heat must be removed from a unit quantity of the substance in the liquid state at its melting point to change it to a solid.

Gamma ray A *gamma ray* is an energetic photon emitted by certain radioactive nuclei.

Gauge pressure *Gauge pressure* is the difference between true pressure and atmospheric pressure.

Generator A *generator* is a device that converts mechanical energy into electrical energy.

Gravitation Newton's *law of universal gravitation* states that every body in the universe attracts every other body with a force directly proportional to both their masses and inversely proportional to the square of the distance separating them.

Half-life The time required for half of a given sample of a radioactive substance to decay is called its *half-life*.

Harmonic motion *Simple harmonic motion* is an oscillatory motion that occurs whenever a force acts on a body in the opposite direction to its displacement from its normal position, with the magnitude of the force proportional to the magnitude of the displacement. The period of a simple harmonic oscillator is independent of its amplitude. In a *damped* harmonic oscillator, friction progressively reduces the amplitude of the vibrations.

Heat *Heat* is internal energy in transit from one body of matter to another. If a body of matter does not change state during the addition or removal of heat and neither does work nor has work done on it, the change in its internal energy results in a corresponding change in its temperature. The SI unit of heat is the joule. Another unit in common use is the *kilocalorie* (kcal), which is the amount of heat required to change the temperature of 1 kg of water by 1°C.

Henry The unit of inductance is the *henry* (H), which is equal to 1 V·s/A.

Hertz The unit of frequency is the *hertz* (Hz), which is equal to 1 cycle/s.

Hooke's law *Hooke's law* states that the strain (relative amount of deformation) experienced by a body under stress is proportional to the magnitude of the stress. Thus the elongation of a wire is proportional to the tension applied to it. Hooke's law is only valid when the elastic limit of the body is not exceeded.

Hydraulic press The *hydraulic press* is a machine consisting of two fluid-filled cylinders of different diameters connected by a tube. The input force is applied to a piston in one of the cylinders and the output force is exerted by a piston in the other cylinder. The TMA of a hydraulic press is equal to the inverse ratio of the cylinder diameters.

Hysteresis The permeability of a ferromagnetic material in a given magnetizing field B_0 depends upon its past history as well as upon B_0, a phenomenon called *hysteresis*.

Ideal gas law The equation $pV/T =$ constant, a combination of Boyle's and Charles's laws, is called the *ideal gas law* and is obeyed approximately, though not exactly, by all gases.

Image A *real image* of an object is formed by light rays that pass through the image; the image would therefore appear on a properly placed screen. A *virtual image* can only be seen by the eye because the light rays that seem to come from the image actually do not pass through it.

Impedance The *impedance, Z,* of an ac-circuit is analogous to the resistance of a dc-circuit. When the effective alternating potential difference V is applied to a circuit of impedance Z, the effective current that flows is $I = V/Z$. The unit of impedance is the ohm.

Impulse The *impulse* of a force is the product of the force and the time during which it acts. Impulse is a vector quantity having the direction of the force. When a force acts on a body that is free to move, its change in momentum equals the impulse given it by the force.

Inductance The *inductance L* of a circuit is the ratio between the magnitude of the self-induced emf $\mathscr{E}$ due to a changing current in it and the rate of change $\Delta I/\Delta t$ of the current. The unit of inductance is the *henry* (H), which is equal to 1 V·s/A.

Inductive reactance The *inductive reactance* X_L of an inductor is a measure of its effect on an alternating current; it is equal to $2\pi f L$, where f is the frequency of the current and L the inductance. If V_L is the effective voltage across an inductor, the effective current that flows through it is $I = V_L/X_L$. The unit of reactance is the ohm.

Inertia The term *inertia* refers to the apparent resistance a body offers to changes in its state of motion.

Interactions, fundamental There are only four *fundamental interactions* that are responsible for all the forces in the universe. In order of decreasing strength, these are the *strong nuclear, electromagnetic, weak nuclear,* and *gravitational* interactions. The two nuclear interactions are effective only over short distances. The electromagnetic and gravitational interactions are unlimited in range, but become weaker with increasing distance. The electromagnetic and weak interactions, and probably the strong as well, are closely related.

Interference The interaction of different waves of the same nature is called *interference: constructive interference* occurs when the resulting composite wave has an amplitude greater than that of either of the original waves, and *destructive interference* when the resulting composite wave has an amplitude less than that of either of the original waves.

Ion An *ion* is an atom or group of atoms that carries a net electric charge. An atom or group of atoms becomes a negative ion when it picks up one or more electrons in addition to its normal number, and becomes a positive ion when it loses one or more of its usual number.

Ionic bond Electrons are transferred between the atoms of certain solids so that the resulting crystal consists of positive and negative ions rather than of neutral atoms. Such a solid is said to be held together by *ionic bonds*. The attractive forces between ions of opposite charge balance the repulsive forces between ions of like charge in an ionic solid.

Isotope The *isotopes* of an element have the same atomic number but different mass numbers. Thus the nuclei of the isotopes of an element all contain the same number of protons but have different numbers of neutrons.

Joule The *joule* (J) is the unit of work and energy. It is equal to $1 \text{ kg-m}^2/\text{s}^2$.

Kilocalorie The *kilocalorie* (kcal) is a unit of heat equal to that amount of heat required to change the temperature of 1 kg of water by 1°C. 1 kcal = 4185 J.

Kilogram The *kilogram* (kg) is the unit of mass. One kilogram weighs 2.21 lb at the earth's surface.

Kinetic energy *Kinetic energy* is the energy a moving object possesses by virtue of its motion. If the object has the mass *m* and speed *v*, its kinetic energy is $\frac{1}{2}mv^2$.

Kinetic theory According to the *kinetic theory of gases*, a gas consists of a great many tiny individual molecules that do not interact with one another except when collisions occur. The molecules are far apart compared with their dimensions and are in constant random motion. The ideal gas law may be derived from the kinetic theory of gases.

Kirchhoff's rules *Kirchhoff's rules* for network analysis are: (1) The sum of the currents flowing into a junction of three or more wires is equal to the sum of the currents flowing out of the junction; (2) The sum of the emf's around a closed conducting loop is equal to the sum of the *IR* potential drops around the loop.

Laminar flow In *laminar* (or *streamline*) *flow* every particle of fluid passing a particular point follows the same path, whereas in *turbulent flow* irregular whirls and eddies occur.

Laser A *laser* is a device for producing a narrow, monochromatic, coherent beam of light. The term stands for *l*ight *a*mplification by *s*timulated *e*mission of *r*adiation. Laser operation depends upon the existence of *metastable states,* which are excited atomic states that can persist for unusually long periods of time.

Lens A *lens* is a transparent object of regular form that can produce an image of an object placed before it. A *converging lens* brings parallel light to a single real focal point, while a *diverging lens* deviates parallel light outward as though it originated at a single virtual focal point.

Lenz's law *Lenz's law* states that the direction of an induced current must be such that its own magnetic field opposes the changes in flux that are inducing it.

Lines of force *Lines of force* are means for visualizing a force field. Their direction at any point is that in which a test body would move if released there, and their concentration in the neighborhood of a point is proportional to the magnitude of the force on a test particle at that point. (In the case of a magnetic field, the direction of a line of force at a point is that in which a moving charge would experience no force.)

Longitudinal waves *Longitudinal waves* occur when the individual particles of a medium vibrate back and forth in the direction in which the waves travel. Sound consists of longitudinal waves.

Magnetic field A *magnetic field* **B** exists wherever a magnetic force acts on a moving charged particle. The magnitude *B* of a magnetic field at a point is defined as the force that would act on a charge of +1 C moving at a speed of 1 m/s past that point, when the direction of the motion is such as to result in the maximum force. The unit of magnetic field is the *tesla* (T), equal to 1 N/A·m.

Magnetic force   Electric charges in motion relative to an observer appear to exert forces upon one another that are different from the forces they exert when at rest. These differences are by custom attributed to *magnetic forces*. In reality, magnetic forces represent relativistic corrections to electric forces due to the motion of the charges involved.

Mass   The property of matter that manifests itself as inertia is called *mass m*. The *rest mass* of a body is its mass when stationary with respect to an observer. The unit of mass is the *kilogram* (kg).

Mass number   The *mass number* of a nucleus is the number of nucleons (protons and neutrons) it contains.

Matter waves   A moving body behaves as though it has a wave character. The waves representing such a body are *matter waves*, also called *de Broglie waves*. The wave variable in a matter wave is its *wave function*, whose square is the *probability density* of the body. The value of the probability density of a particular body at a certain place and time is proportional to the probability of finding the body at that place at that time. Matter waves may thus be regarded as waves of probability.

Mechanical advantage   The *mechanical advantage* of a machine is the ratio between the output force it exerts and the input force that is furnished to it. The *theoretical mechanical advantage* (TMA) is its value under ideal circumstances, while the *actual mechanical advantage* (AMA) is its value when friction is taken into account.

Metallic bond   The *metallic bond* which holds metal atoms together in the solid state arises from a "gas" of freely moving electrons pervading the entire metal.

Meter   The *meter* (m) is the unit of length. One meter is equal to 3.28 ft.

Mirror   A *mirror* is a specular reflecting surface of regular form that can produce an image of an object placed before it.

Molecule   A *molecule* is a group of atoms that stick together strongly enough to act as a single particle. A molecule of a given compound always has a certain definite structure and is complete in itself with little tendency to gain or lose atoms.

Moment arm of a force   The *moment arm* of a force is the perpendicular distance from its line of action to a pivot point.

Moment of inertia   The *moment of inerita I* of a body about a given axis is the rotational analog of mass in linear motion. Its value depends upon the way in which the mass of the body is distributed about the axis.

Momentum, linear   The *linear momentum* **p** of a body is the product of its mass and velocity. Linear momentum is a vector quantity whose direction is that of the body's velocity. The principle of *conservation of linear momentum* states that the total linear momentum of a system of particles isolated from the rest of the universe remains constant regardless of what events occur within the system.

Motion, laws of   *Newton's first law of motion* states that an object at rest will remain at rest and an object in motion will continue in motion in a straight line at constant velocity in the absence of any interaction with the rest of the universe. *Newton's second law of motion* states that the net force acting on an object is equal to the rate of change of the object's linear momentum. *Newton's third law of motion* states that, when an object exerts a force on another object, the second object exerts a force on the first object of the same magnitude but in the opposite direction.

Neutrino   The *neutrino* is an uncharged particle of little or no mass that is emitted in the decay of certain elementary particles. It can possess energy and both linear and angular momentum.

Neutron   The *neutron* is an electrically neutral elementary particle, slightly heavier than the proton, which is present together with protons in atomic nuclei.

Newton   The *newton* (N) is the SI unit of force. It is equal to 1 kg-m/s^2. One newton is equal to 0.225 pound.

Nuclear fission   In *nuclear fission*, the absorption of neutrons by certain heavy nuclei causes them to split into smaller *fission fragments* with the release of energy. Because each fission also liberates several neutrons, a rapidly multiplying sequence of fissions called a *chain reaction* can occur if a sufficient amount of the proper material is assembled. A *nuclear reactor* is a device in which a chain reaction can be intitiated and controlled.

Nuclear fusion.   In *nuclear fusion*, two light nuclei combine to form a heavier one with the evolution of energy. The sun and stars obtain their energy from fusion reactions.

Nucleon   Neutrons and protons, the constituents of atomic nuclei, are jointly called *nucleons*.

Nucleus   The *nucleus* of an atom is located at its center and contains all of the positive charge and most of the mass of the atom. The nucleus consists of protons and neutrons.

Nuclide   A *nuclide* is a nuclear species characterized by a certain atomic number Z and mass number A. The symbol of a nuclide is $_Z^A X$, where X is the chemical symbol of the element of atomic number Z.

Ohm   The *ohm* (Ω) is the unit of electrical resistance. It is equal to 1 V/A.

Ohm's law   *Ohm's law* states that the current in a metallic conductor is proportional to the potential difference between its ends. Thus in such a conductor $I = V/R$.

Pascal's principle   *Pascal's principle* states that an external pressure exerted on a fluid is transmitted uniformly throughout its volume.

Period   *The period T* of a body undergoing simple harmonic motion is the time required for it to make one complete oscillation. The period of a wave is the time required for one complete wave to pass a particular point.

Periodic law   The *periodic law* of chemistry states that if the elements are listed in order of atomic number, elements with similar properties recur at regular intervals. The quantum theory of the atom together with the exclusion principle is able to explain the origin of the periodic law.

Permanent magnet   A *permanent magnet* is an object composed of a ferromagnetic material whose atomic current loops have been aligned by an external magnetic field and which remain aligned after the external field is removed.

Permeability   The *permeability* of a medium is a measure of its magnetic properties. *Paramagnetic* and *ferromagnetic* substances have higher permeabilities than free space and *diamagnetic* ones have lower permeabilities.

Phase relationships   The *phase relationships* between the instantaneous voltage and instantaneous current in ac-circuit components are as follows: the voltage across a pure resistor is in phase with the current; the voltage across a pure inductor leads the current by ¼ cycle; the voltage across a pure capacitor lags behind the current by ¼ cycle.

Phasor   A *phasor* is a vector rotating f times per second whose projection on a tangent line can represent either current or voltage in an ac-circuit whose frequency is f.

Photoelectric effect   The *photoelectric effect* refers to the emission of electrons from a metal surface when light shines on it.

Photon   Electromagnetic waves transport energy in tiny bursts called *photons* that resemble particles in a number of respects. The energy of a photon is related to the frequency f of the corresponding wave by $E = hf$, where h is Planck's constant.

Plasma   A *plasma* is a gas composed of electrically charged particles, and its behavior depends strongly upon electromagnetic forces. Most of the matter in the universe is in the plasma state.

Polar molecule   A *polar molecule* is one whose charge distribution is asymmetrical, so that one end is positive and the other negative even though the molecule as a whole is electrically neutral.

Polarization   A *polarized* beam of transverse waves is one whose vibrations occur in only a single direction perpendicular to the direction in which the beam travels, so that the entire wave motion is confined to a plane called the *plane of polarization*. An *unpolarized* beam of transverse waves is one whose vibrations occur equally often in all directions perpendicular to the direction in which the beam travels.

Pole, magnetic   The ends of a permanent magnet are called its *poles*. Magnetic lines of force leave the *north pole* of a magnet and enter its *south pole*.

Positron   A *positron* is a positively charged electron.

Potential difference   The electrical *potential difference V* between two points is the work that must be done to take a charge of 1 C from one point to the other. The unit of potential difference is the *volt* (V), which is equal to 1 J/C.

Potential energy   *Potential energy* is the energy an object has by virtue of its position. The gravitational potential energy of an object of mass m at a height h above a particular reference point is mgh; if its weight w is specified, its potential energy is wh. Other examples of potential energy are that of a planet with respect to the sun, that of a piece of iron with respect to a magnet, and that of a body at the end of a stretched spring with respect to its equilibrium position.

Power   The rate at which work is done is called *power*. The unit of power in the metric system is the *watt* (W), which is equal to 1 J/s, and in the British system it is the *ft·lb/s*. One *horsepower* is equal to 746 W or 550 ft·lb/s.

Power factor   The *power factor* of an ac-circuit is the ratio between the power consumed in the circuit and the product of the effective current and voltage there; this ratio is equal to that between the resistance and the impedance of the circuit, and is less than 1, except at resonance. The unit of apparent power $V_{eff} I_{eff}$ is the *volt-ampere*, as distinct from the watt, which is the unit of consumed power.

Pressure The *pressure* on a surface is the perpendicular force per unit area that acts upon it. *Gauge pressure* is the difference between true pressure and atmospheric pressure.

Proton The *proton* is an elementary particle found in all atomic nuclei; its charge is $+e$ and its mass is 1836 times that of the electron.

Quantum theory of the atom In the *quantum theory of the atom* only experimentally measurable quantities are considered, and no use is made of mechanical models inconsistent with the uncertainty principle. According to this theory, four quantum numbers are required to describe each electron in an atom. These are the *principal quantum number n*, which governs the electron's energy, the *orbital quantum number l*, which governs the magnitude of its angular momentum, the *magnetic quantum number m_l*, which governs the orientation of its angular momentum, and the *spin magnetic quantum number m_s*, which governs the orientation of its spin.

Quantum theory of light The *quantum theory of light* states that light travels in tiny bursts of energy called *quanta* or *photons*. If the frequency of the light is *f*, each burst has the energy *hf*, where *h* is known as *Planck's constant*. The quantum theory of light complements the wave theory of light.

Quarks Elementary particles subject to the strong nuclear interaction are believed to be composed of *quarks*, which are entities with fractional electric charges that have not as yet been experimentally isolated but whose existence is supported by indirect evidence.

Radian The *radian* is a unit of angular measure equal to $57.30°$. If a circle is drawn whose center is at the vertex of an angle, the angle in radian measure is equal to the ratio between the arc of the circle cut by the angle and the radius of the circle. A full circle contains 2π radians.

Radiation In *radiation*, energy is transferred from one place to another in the form of electromagnetic waves, which require no material medium for their passage.

Radioactivity *Radioactive nuclei* spontaneously transform themselves into other nuclear species by the emission of *alpha particles*, which are the nuclei of helium atoms, or *beta particles*, which are positive or negative electrons. *Electron capture* is an alternative to positron emission. The emission of *gamma rays*, which are energetic photons, enables an excited nucleus to lose its excess energy.

Reflection In *diffuse reflection* an incident beam of parallel light is spread out in many directions, while in *specular reflection* the angle of reflection is equal to the angle of incidence. In *total internal reflection,* light arriving at a medium of lower index of refraction at an angle greater than the *critical angle* of incidence is reflected back into the medium it came from.

Refraction The bending of a light beam when passing from one medium to another is called *refraction*. The quantity that governs the degree to which a light beam will be deflected in entering a medium is its *index of refraction,* defined as the ratio between the speed of light in free space and its value in the medium.

Refrigerator A *refrigerator* is a device that transfers heat from a cold reservoir to a hot one, and it must expend energy in order to do this. In essence, it is a heat engine operating in reverse.

Relative humidity The *relative humidity* of a volume of air is the ratio between the amount of water vapor it contains and the amount that would be present at saturation.

Relativity The *special theory of relativity* relates measurements made on an object or phenomenon from frames of reference moving at constant velocity with respect to one another. The *Lorentz contraction* refers to the decrease in the measured length of an object when it is moving relative to an observer. The relativistic *time dilation* refers to the fact that a clock moving with respect to an observer appears to tick less rapidly than it does to an observer traveling with the clock. The *relativity of mass* refers to the increase in the measured mass of an object when it is moving relative to an observer. The *general theory of relativity* concerns frames of reference accelerated with respect to one another. One of the conclusions of general relativity is that light is affected by gravitational fields.

Resistance The *resistance R* of a body of matter is a measure of the extent to which it impedes the passage of electric current. It is defined as the ratio between the potential difference applied across the ends of the body and the resulting current. The unit of resistance is the *ohm (Ω)*, which is equal to 1 V/A. The resistance of a conductor is proportional to its length and to the *resisitivity ρ* of the material of which it is made, and inversely proportional to its cross-sectional area.

Resolution of vectors A vector can be *resolved* into two or more other vectors whose sum is equal to the original vector. The new vectors are called the *components* of the original vector, and are normally chosen to be perpendicular to one another.

Resolving power The *resolving power* of an optical system refers to its ability to produce separate images of nearby objects. Resolving power is limited by diffraction; the larger the objective lens of an optical system, the greater its resolving power.

Resonance *Resonance* occurs when periodic impulses are given to an object at a frequency equal to one of its natural frequencies of oscillation.

Resonance frequency The *resonance frequency* of a series ac-circuit is that frequency for which the impedance is a minimum; its value is $f_0 = 1/2\pi \sqrt{LC}$.

Right-hand rule for magnetic field According to the *right-hand rule for magnetic field*, when a current-carrying wire is grasped with the right hand so that the thumb points in the direction of the current, the curled fingers of that hand then point in the direction of the magnetic field.

Right-hand rule for magnetic force Open the right hand so the fingers are together and the thumb sticks out. According to the *right-hand rule for magnetic force*, when the thumb is in the direction of motion of a positive charge and the fingers are in the direction of a magnetic field, the palm faces in the direction of the force acting on the charge. When a negative charge is involved, the force is in the opposite direction.

Scalar quantity A *scalar quantity* is one that has magnitude only.

Semiconductors *Semiconductors* are intermediate between conductors and insulators in their ability to carry electric current. An *n-type semiconductor* is one in which electric current is carried by the motion of electrons. A *p-type semiconductor* is one in which electric current is carried by the motion of *holes*, which are vacancies in the electron structure of the material that behave like positive charges.

Shear When equal and opposite forces that do not act along the same line are applied to a body, it is said to be in *shear*. The *shear modulus S* of a material is equal to the shear force per unit cross-sectional area applied to a sample of it divided by the relative distortion of the sample.

Shells, atomic The electrons in an atom that have the same total quantum number *n* are said to occupy the same *shell*. Electrons in a given shell which have the same orbital quantum number *l* are said to occupy the same *subshell*. Shells and subshells containing the maximum number of electrons permitted by the exclusion principle are *closed*. Atoms whose subshells are all closed possess unusual stability.

Shock wave A *shock wave* is a shell of high pressure produced by the motion of an object whose speed is greater than that of sound.

Snell's law *Snell's law* states that the ratio between the sine of the angle of incidence of a light ray upon a boundary between two media and the sine of the angle of refraction is equal to the ratio of the speeds of light in the two media.

Sound *Sound* is a longitudinal wave phenomenon that consists of successive compressions and rarefactions of the medium through which it travels. *Infrasound* and *ultrasound* respectively refer to sounds whose frequencies are too low and too high to be audible. Sound intensity level in air is measured in *decibels*.

Specific heat capacity The *specific heat capacity c* of a substance is the amount of heat required to change the temperature of a unit quantity of it by 1°.

Spectrum An *absorption spectrum* results when white light is passed through a cool gas; it is a *dark line spectrum* because it appears as a series of dark lines on a bright background, with the lines representing characteristic wavelengths absorbed by the gas. An *emission spectrum* consists of the various wavelengths of light emitted by an excited substance; it may be a *continuous spectrum*, in which all wavelengths are present, or a *bright line spectrum*, in which only a few wavelengths characteristic of the individual atoms of the substance are present.

Speed The *average speed* of a moving object is the distance it covers in a time interval divided by the time interval. The object's *instantaneous speed* at a certain moment is the rate at which it is covering distance at that moment. Speed is a scalar quantity.

Spin Every electron has a certain intrinsic amount of angular momentum called its *spin*. The spin of an electron is as fundamental a property as its mass or electric charge. Owing to its spin, every electron acts like a tiny bar magnet. Most other elementary particles (such as the proton and neutron) also have spin associated with them.

Standing waves *Standing waves* in a stretched string are transverse harmonic oscillations that result in a wave pattern whose amplitude varies in a sinusoidal manner along the string. Points of zero amplitude are called *nodes*. Other types of waves can also occur as standing waves.

Stress and strain The *stress* on an object is the force (tension, compression, or shear) applied to it per unit area; the *strain* is the resulting change in a dimension of the object relative to its original value. A *modulus of elasticity* of a material is the ratio between a particular kind of applied stress and the resulting strain, providing the elastic limit is not exceeded.

Sublimation   *Sublimation* is the direct conversion of a substance from the solid to the vapor state, or vice versa, without it first becoming a liquid.

Superposition, principle of   The *principle of superposition* states that when two or more waves of the same nature travel past a given point at the same time, the amplitude at the point is the sum of the amplitudes of the individual waves.

Surface tension   The *surface tension* of a liquid refers to the tendency of its surface to contract to the minimum possible area in any situation.

Système International   The modern version of the metric system of units is the *Système International* (SI). SI units are almost universally used by scientists and are in everyday use in most of the world as well.

Temperature   The *temperature, T,* of a body of matter is a measure of the average kinetic energy of random translational motion of its constituent particles. When two bodies are in contact, heat flows from the one at the higher temperature to the one at the lower temperature. In the Celsius temperature scale, the unit is the °C; in the absolute temperature scale, the unit is the *kelvin* (K). A *thermometer* is a device for measuring temperature.

Tension   When equal and opposite forces that act away from each other are applied to a body, it is said to be in *tension*.

Terminal speed   Because air resistance varies with velocity, a falling body eventually reaches a *terminal speed* after which it ceases to be accelerated downward.

Tesla   The *tesla* (T) is the unit of magnetic field. It is equal to 1 N/A·m.

Thermal expansion   The *coefficient of linear expansion* is the ratio between the change in length of a solid rod of a particular material and its original length per 1° change in termperature. The *coefficient of volume expansion* is the ratio between the change in volume of a sample of a particular solid or liquid and its original volume per 1° change in temperature.

Thermodynamics   The *first law of thermodynamics* states that the work output of any engine is equal to its net energy input plus any decrease in its stored energy. This law is thus a restatement of the principle of conservation of energy. The *second law of thermodynamics* states that no engine can be completely efficient in converting energy to work — some of the input energy must be wasted as heat. This law is a consequence of the tendency of all physical systems to become more and more disordered as time goes on.

Thermonuclear energy   The energy liberated by nuclear fusion is called *thermonuclear energy*.

Torque   The *torque* τ of a force about a particular axis is the product of the magnitude of the force and the perpendicular distance from the line of action of the force to the axis. The latter distance is called the *moment arm* of the force. Torque plays the same role in rotational motion that force does in linear motion.

Transformer   An alternating current flowing in the primary coil of a *transformer* induces another alternating current in the secondary coil. The ratio of the emfs is proportional to the ratio of turns in the coils.

Transverse waves   *Transverse waves* occur when the individual particles of a medium vibrate from side to side perpendicular to the direction in which the waves travel. The vibrations of a stretched string are transverse waves. Electromagnetic waves are transverse because the varying electric and magnetic fields of which they consist are perpendicular to the wave direction, even though nothing material is in motion.

Triple point   The *triple point* of a substance refers to the temperature and pressure at which its solid, liquid, and vapor states can exist in equilibrium with one another.

Uncertainty principle   The *uncertainty principle* is an expression of the limits set by the wave nature of matter on finding both the position and state of motion of a moving object. According to this principle, the product of the uncertainties in simultaneous measurements of the position and momentum of an object cannot be less than $h/2\pi$.

Uniform circular motion   An object traveling in a circle at constant speed is said to be undergoing *uniform circular motion*.

Van der Waals force   *Van der Waals forces* originate in the electric attraction between asymmetrical charge distributions in atoms and molecules. Molecular solids and liquids are held together by van der Waals forces.

Vaporization, heat of   The *heat of vaporization* of a substance is the amount of heat that must be supplied to change 1 kg of it at its boiling point from the liquid to the gaseous (or vapor) state; the same amount of heat must be removed from 1 kg of the substance at its boiling point to change it into a liquid.

Vector   A *vector* is an arrowed line whose length is proportional to the magnitude of some vector quantity and whose direction is that of the quantity. A *vector diagram* is a scale drawing of the various forces, velocities, or other vector quantities involved in the motion of a body. In *vector addition,* the tail of each successive vector is placed at the head of the previous one, with their lengths and original directions kept unchanged. The *resultant* is a vector drawn from the tail of the first vector to the head of the last. A vector can be *resolved* into two or more other vectors called the *components* of the original vector. Usually the components of a vector are chosen to be in mutually perpendicular directions.

Vector quantity   A *vector quantity* is one that has both magnitude and direction. The symbol of a vector quantity is printed in bold-face type, for instance **A.**

Velocity   The *velocity* **v** of an object is a specification of both its speed and the direction in which it is moving. Velocity is a vector quantity. The *instantaneous velocity* of an object is its velocity at a specific instant of time. The *average velocity* of an object is the total displacement through which it has moved in a time interval divided by the interval.

Viscosity   The *viscosity* of a fluid is a measure of its internal friction.

Volt   The unit of electrical potential difference is the *volt* (*V*). It is equal to 1 J/C.

Watt   The unit of power is the *watt* (**W**), which is equal to 1 J/s.

Wave motion   *Wave motion* is characterized by the propagation of a change in a medium, rather than by the net motion of the medium itself. The passage of a wave across the surface of a body of water, for instance, involves the motion of a pattern of alternate crests and troughs, with the individual water molecules themselves ideally executing uniform circular motion.

Wavefront   A *wavefront* is an imaginary surface that joins points where all the waves from a source are in the same phase of oscillation. According to *Huygens' principle,* every point on a wavefront can be considered as a point source of secondary wavelets which spread out in all directions with the wave speed of the medium. The wavefront at any time is the envelope of these wavelets.

Weight   The *weight w* of an object is the gravitational force exerted on it by the earth. The weight of an object is proportional to its mass.

Work   Whenever a force affects the motion of a body, the body undergoes a displacement while the force acts on it. The product of the force and the component of the displacement of the body in the direction of the force is called the *work W* done by the force on the body. Work is a measure of the change (in a general sense) a force gives rise to when it acts upon something. The unit of work is the same as that of energy, namely the *joule (J).*

X rays   *X rays* are high-frequency electromagnetic waves emitted when fast electrons impinge on matter.

Young's modulus   *Young's modulus Y* of a particular material is equal to the tension or compression force per unit cross-sectional area applied to a sample of that material divided by the fractional change in the length of the sample.

SOLUTIONS TO ODD-NUMBERED EXERCISES AND PROBLEMS

CHAPTER 1

EXERCISES

1. No. The distinction between vector and scalar quantities is simply that vector quantities have directions associated with them, and both kinds of quantity are found in the physical world.

3. Yes.

5. $(368 \text{ ft})(0.305 \text{ m/ft}) = 112 \text{ m} = 0.112 \text{ km}$.

7. (a) $(4840 \text{ yd}^2)(3 \text{ ft/yd})^2(0.3048 \text{ m/ft})^2 = 4047 \text{ m}^2$. (b) $(1 \text{ km}^2)(10^3 \text{ m/km})^2/(4047 \text{ m}^2/\text{acre}) = 247 \text{ acres/km}^2$.

9. (a) $1 \text{ bd ft} = (12 \text{ in})^2(1 \text{ in}) = 144 \text{ in}^3$. (b) Since $1 \text{ ft}^3 = (12 \text{ in})^3 = 1728 \text{ in}^3$, $1 \text{ bd ft} = 144 \text{ in}^3/(1728 \text{ in}^3/\text{ft}^3) = 0.0833 \text{ ft}^3$. (c) Since $1 \text{ in}^3 = (2.54 \text{ cm})^3 = 16.4 \text{ cm}^3$, $1 \text{ bd ft} = (144 \text{ in}^3)(16.4 \text{ cm}^3/\text{in}^3) = 2.36 \times 10^3 \text{ cm}^3$.

11. $v = (50 \text{ km/h})(0.621 \text{ mi/km}) = 31 \text{ mi/h}$.

13. $s = \sqrt{(70 \text{ m})^2 + (40 \text{ m})^2} = 80.6 \text{ m}$; $\tan \theta = 40 \text{ m}/70 \text{ m} = 0.571$, $\theta = \tan^{-1} 0.571 = 30°$ above the horizontal.

15. $F_y = 50 \text{ N} + (25 \text{ N})(\sin 45°) = 67.7 \text{ N}$.

17. Alpha: $v_w = v_\alpha \cos 40° = 3.83 \text{ km/h}$. Beta: $v_w = v_\beta \cos 50° = 3.86 \text{ km/h}$. Beta has the higher windward component of velocity.

19. $F = \sqrt{(100 \text{ N})^2 + (40 \text{ N})^2} = 108 \text{ N}$.

PROBLEMS

1. The angle between the path of ball 1 and the $+x$ direction is specified by $\tan \theta_1 = v_y/v_x = 2$, $\theta_1 = 63°$. Similarly $\tan \theta_2 = v_y/v_x = 1.5$, $\theta_2 = 56°$. The difference in angle is 7°.

3. $v = v_x/\cos 37° = 87.6 \text{ km/h}$.

5. (a) $v_y = v \sin 25° = 84.6 \text{ km/h}$. (b) $v_x = v \cos 25° = 181 \text{ km/h}$. (c) $v_{\text{south}} = v_x \sin 45° = 128 \text{ km/h}$.

7. $F_x = F \cos 40° = 7.66 \text{ N}$; $F_y = F \sin 40° = 6.43 \text{ N}$.

9. One way to solve this problem is to note that the angle between each towrope and the direction in which the ship moves is 15°. The force each tugboat exerts has a component in this direction of $(5.0 \text{ tons})(\cos 15°) = 4.8 \text{ tons}$. The total force on the ship is 9.6 tons since the force components perpendicular to this direction are equal and opposite and so cancel out.

11. Calling E the $+x$ direction and N the $+y$ direction, $s_x = 200 \text{ km} - (100 \text{ km})(\sin 45°) = 129 \text{ km}$ and $s_y = -200 \text{ km} + (100 \text{ km})(\cos 45°) = -129 \text{ km}$. Hence $s = \sqrt{s_x^2 + s_y^2} = 182 \text{ km}$. The direction is NW since s_x and s_y are equal in length.

13. The two velocities are perpendicular, so $v = \sqrt{(7 \text{ kn})^2 + (3 \text{ kn})^2} = 7.6 \text{ kn}$. If θ is the angle between v and NW, then $\tan \theta = 3 \text{ kn}/7 \text{ kn} = 0.429$, $\theta = 23°$. Since NW is itself 45° W of N, the direction of v is $45° + 23° = 68°$ W of N.

15. (a) Relative to L: $v_{Qx} = -v_L$; $v_{Qy} = v_Q$; $v_Q = \sqrt{v_{Qx}^2 + v_{Qy}^2} = 10.6 \text{ kn}$. If ϕ is angle between v_Q (relative to L) and N, $\phi = \tan^{-1} (v_{Qx}/v_{Qy}) = -49°$, so ϕ is 49° W of N (counterclockwise from N). (b) v_L (relative to Q) is equal in magnitude but opposite in direction to v_Q (relative to L). Hence v_L is 10.6 kn at 229° counterclockwise from N, which is 131° clockwise from N or 41° S of E.

17. Here $A_x = A \sin 37° = 6 \text{ cm}$; $A_y = A \cos 37° = 8 \text{ cm}$; $B_x = B \cos 37° = 8 \text{ cm}$; $B_y = -B \sin 37° = -6 \text{ cm}$. (a) $R_x = A_x + B_x = 14 \text{ cm}$, $R_y = A_y + B_y = 2 \text{ cm}$; $R = \sqrt{R_x^2 + R_y^2} = 14.1 \text{ cm}$. R points θ clockwise from the $+y$ direction where $\tan \theta = R_x/R_y = 7$, $\theta = 82°$. (b) $R_x = A_x - B_x = -2 \text{ cm}$, $R_y = A_y - B_y = 14 \text{ cm}$; $R = \sqrt{R_x^2 + R_y^2} = 14.1 \text{ cm}$. R points θ counterclockwise from the $+y$ direction where $\tan \theta = R_x/R_y = 0.143$, $\theta = 8°$. (c) $R_x = B_x - A_x = 2 \text{ cm}$, $R_y = B_y - A_y = -14 \text{ cm}$; $R = \sqrt{R_x^2 + R_y^2} = 14.1 \text{ cm}$. R points θ clockwise from the $+x$ direction where $\tan \theta = R_y/R_x = 0.143$, $\theta = 8°$.

CHAPTER 2

EXERCISES

1. Yes.

3. The squirrel should stay where it is, since if it lets go, it will fall with the same acceleration as the bullet and so will be struck.

5. Yes.

7. $s_1 = v_1 t_1 = 1200$ km, $s_2 = v_2 t_2 = 1000$ km; $\bar{v} = (s_1 + s_2)/(t_1 + t_2) = 338$ km/h.

9. (a) $a = (v_f - v_0)/t = 1.2$ m/s². (b) -2 m/s².

11. (a) C, D, G, H, I. (b) A, B, E. (c) E. (d) H. (e) E. (f) G, I. (g) F. (h) D. (i) C. (j) B.

13. (a) $a = (v_f - v_0)/t = 1.1$ m/s². (b) $t = (v_f - v_0)/a = 7.27$ s.

15. $a = v_f/t$; $s = \frac{1}{2}at^2 = v_f t/2 = 90.5$ m.

17. (a) $a = (80 \text{ m/s})/35$ s; $t = (v_f - v_0)/a = 8.75$ s; $s = \frac{1}{2}at^2 = 87.5$ m. (b) $t = (v_f - v_0)/a = 8.75$ s; $s = v_0 t + \frac{1}{2}at^2 = 612.5$ m. (c) $s = \frac{1}{2}at^2 = 1400$ m.

19. $v = \sqrt{2gh} = 26.6$ m/s.

21. $h = \frac{1}{2}gt^2 = 78.4$ m.

23. (a) $v_t = v_0 + gt = 19.8$ m/s. (b) $v_t = 29.6$ m/s.

25. $h = v^2/2g = 13.1$ m.

27. The time of rise and the time of fall are both $t = 1.5$ s. Hence $v = gt = 14.7$ m/s is the value of both the initial and final speeds.

29. Yes; the speed is greatest at both ends of the path and is least at the highest point.

31. At maximum range, $\theta = 45°$ and $v_0 = \sqrt{Rg} = 10.4$ m/s.

33. $R = (v_0^2 \sin 2\theta)/g$, $v_0 = \sqrt{Rg/\sin 2\theta} = 20$ m/s.

PROBLEMS

1. The airplane's speed is 550 km/h from 4:00 P.M. to 5:41 P.M., 620 km/h from 5:41 P.M. to 7:54 P.M., and 520 km/h from 7:54 P.M. to 10:00 P.M.

3. The car had an initial speed of 6 m/s and accelerated at 0.1 m/s² for 3 min to 24 m/s. It remained at 24 m/s for 4 min and then slowed down to a stop at 0.2 m/s² in 2 min.

5. If s = total distance, $t_1 = s/2v_1$ and $t_2 = s/2v_2$. Hence $\bar{v} = s/(t_1 + t_2) = 2/[(1/v_1) + (1/v_2)] = 2v_1 v_2/(v_1 + v_2) = 6.86$ km/h.

7. $s = v_0 t + \frac{1}{2}at^2$, $a = 2s/t^2 - 2v_0/t = -0.238$ m/s²; $v = v_0 + at = 4.42$ m/s = 16 km/h.

9. $s_1 = v_0 t = 20$ m, $s_2 = v_0^2/2a = 100$ m; $s = s_1 + s_2 = 120$ m.

11. $s_1 = \frac{1}{2}a_1 t_1^2 = 135$ m; $v = a_1 t_1 = 9$ m/s, $s_2 = vt_2 = 1080$ m; $a_2 = -v/t_3 = -0.6$ m/s², $s_3 = vt_3 + \frac{1}{2}a_2 t_3^2 = 67.5$ m; $s = s_1 + s_2 + s_3 = 1282.5$ m.

13. $t_1 = v/a = 4$ s; $s_1 = \frac{1}{2}at^2 = 12$ m; $t_3 = t_1 = 4$ s; $s_3 = s_1 = 12$ m. Hence constant-speed distance is $s_2 = 50$ m $- 24$ m $= 26$ m and $t_2 = s_2/v = 4.33$ s; $t = t_1 + t_2 + t_3 = 12.33$ s.

15. Since the coconuts met halfway down the cliff, the one thrown upward must have had the same initial speed as the speed of the dropped coconut at the place of meeting, which is $v = \sqrt{2gh} = 19.8$ m/s.

17. (a) $h_1 = v_0^2/2g = 5.10$ m, $t_1 = \sqrt{2h_1/g} = 1.02$ s; $h_2 = 20$ m $+ 5.1$ m $= 25.1$ m, $t_2 = \sqrt{2h_2/g} = 2.26$ s; $t = t_1 + t_2 = 3.28$ s. (b) $v = gt_2 = 22$ m/s.

19. Man: $t = h/v = 25$ s; monocle: $t = \sqrt{2h/g} = 5.5$ s; $\Delta t = 19.5$ s.

21. (a) $v_m = at = 800$ m/s. (b) $h_1 = \frac{1}{2}at^2 = 3.20$ km; $h_2 = v_m^2/2g = 32.65$ km; $h = h_1 + h_2 = 35.85$ km. (c) $t_1 = 8$ s, $t_2 = v_m/g = 81.6$ s; $t = t_1 + t_2 + t_3 = 175$ s. (d) $v = gt = 878$ m/s.

23. $t = s/v$, $h = \frac{1}{2}gt^2 = \frac{1}{2}gs^2/v^2 = 0.10$ m $= 10$ cm.

25. (a) $v_y = \sqrt{2gh} = 19.8$ m/s; $v_x = 30$ m/s; $v = \sqrt{v_x^2 + v_y^2} = 35.9$ m/s. (b) $\theta = \tan^{-1} v_y/v_x = 33°$ below the horizontal.

27. (a) $v_x = \sqrt{v_N^2 + v_W^2} = 12.8$ m/s, $v_y = gt = 19.6$ m/s, $v = \sqrt{v_x^2 + v_y^2} = 23.4$ m/s. (b) $\tan \theta_1 = v_x/v_y = 0.653$, $\theta_1 = 33°$. (c) $\tan \theta_2 = v_W/v_N = 1.25$, $\theta_2 = 51°$.

29. (a) $v_0 = \sqrt{2gh}$, $R_{max} = v_0^2/g = 2h = 200$ m. (b) $v = \sqrt{2gh} = 44.3$ m/s; 44.3 m/s.

31. (a) $R = (v_0^2/g)\sin 2\theta_1 = 7953$ m; $T_1 = (2v_0/g)\sin \theta_1 = 30.6$ s; $h_1 = v_y t - \frac{1}{2}gt^2 = (v_0 \sin \theta_1)(T_1/2) - \frac{1}{2}g(T_1/2)^2 = 1148$ m. (b) $\theta_2 = 90° - \theta_1 = 60°$; $T_2 = (2v_0/g)\sin \theta_2 = 53$ s; $h_2 = (v_0 \sin \theta_2)(T_2/2) - \frac{1}{2}g(T_2/2)^2 = 3444$ m.

CHAPTER 3

EXERCISES

1. No. Only a net force produces an acceleration, and when a force is applied to an object, other forces may come into being that cancel it out. Thus pushing down on a book lying on a table does not accelerate the book because the table pushes back with an equal and opposite force.

3. The deceleration of a person falling onto loose earth is more gradual than if he falls onto concrete, hence the force acting on him is less.

5. The net forces are, in the order given: zero, constant, and increasing so as to be proportional to t.

7. Here $F = w = mg$, hence $a = F/m = g$.

9. No. Action and reaction forces act upon different objects, and so a single force can certainly act upon an object.

11. A propeller works by pushing backward on the air, whose reaction force in turn pushes the propeller itself forward. No air, no reaction force, so the idea is no good.

13. $m = w/g = 0.816$ kg; $a = F/m = 24.5$ m/s^2.

15. $F_{max} = m_0 a_0 = 2000$ N. Hence $a = F_{max}/m = 0.67$ m/s^2.

17. $a = F/m$, $v = at = Ft/m = 28.6$ m/s.

19. (a) $v_f = 55.6$ m/s; $a = (v_f - v_0)/t = 18.5$ m/s^2 = $1.89g$. (b) $F = ma = 2.22 \times 10^5$ N.

21. $a = (m_1 - m_2)g/(m_1 + m_2) = 1.96$ m/s^2.

23. $N = mg$, $F = \mu N = \mu mg = 294$ N.

25. $F = \mu N = 2$ N.

27. $a = v_0^2/2s$; $\mu = F/mg = ma/mg = a/g = v_0^2/2sg = 0.041$.

29. (a) $a = v_0^2/2s$; $\mu = F/mg = ma/mg = a/g = v_0^2/2sg = 0.23$. (b) $F = \mu mg = 4.5$ N.

PROBLEMS

1. A force equal to ma is needed in addition to the force $w = mg$ required just to lift the box without accelerating it. Hence $F = mg + ma = 59$ N.

3. $F = w - T = ma$, $T = w - ma = mg - ma = 8800$ N.

5. (a) Net upward force $= F = T - mg$, $a = F/m = (T - mg)/m = 2.7$ m/s^2. (b) 9.8 m/s^2.

7. $v_f^2 = v_0^2 + 2as$, $a = (v_f^2 - v_0^2)/2s = 30.6$ m/s^2; $F = ma + mg = 4.04 \times 10^3$ N; $F/mg = 4.1$.

9. Here $s = 0.6$ m, $h_1 = 2.0$ m, $h_2 = 1.8$ m, $m_1 = 40$ kg, $m_2 = ?$ Kangaroo by herself: $v_1^2 = 2gh_1$, $a_1 = v_1^2/2s = gh_1/s$, $F = m_1 g + m_1 a_1 = m_1 g(1 + h_1/s) = 1699$ N. With baby: $v_2^2 = 2gh_2$, $a_2 = v_2^2/2s = gh_2/s$; $F = (m_1 + m_2)(g + a_2)$, $m_1 + m_2 = F/(g + a_2) = F/g(1 + h_2 s) = 43.3$ kg, $m_2 = 43.3$ kg $- 40$ kg $= 3.3$ kg.

11. (a) $F = m_2 g = (m_1 + m_2)a$, $a = m_2 g/(m_1 + m_2) = 4.9$ m/s^2. (b) $F = m_2 g - \mu m_1 g = (m_1 + m_2)a$, $a = (m_2 g - \mu m_1 g)/(m_1 + m_2) = 3.92$ m/s^2.

13. (a) $F_1 = \mu m_1 g = 14{,}700$ N. (b) The maximum force the man can exert without slipping is $F_2 = \mu m_2 g = 784$ N. If he exerts more force than this, he will be accelerated toward the elephant by the reaction force of the rope on him.

His acceleration would then be $a = (F_1 - F_2)/m = 139$ m/s^2 assuming $\mu = \mu_s$; since $\mu < \mu_s$, the acceleration would be even greater.

15. $F = 3w$, $F_x = F \cos 50° = 1.93w$; $m = w/g$; $a = F_x/m = 1.93g = 18.9$ m/s^2.

17. The force $\mathbf{F}$ that accelerates each box is the component of its weight $\mathbf{w}$ parallel to the plane. Hence $F = w \sin \theta = mg \sin \theta = ma$ and $a = g \sin \theta = 3.4$ m/s^2 for each box.

19. $\mu = \tan \theta = 0.176$.

21. $N = mg \cos \theta$; $F = mg \sin \theta - \mu N = mg(\sin \theta - \mu \cos \theta) = ma$; $a = g(\sin \theta - \mu \cos \theta) = 4.11$ m/s^2. Since $s = \frac{1}{2}at^2$, $t = \sqrt{2s/a} = 2.09$ s.

23. (a) From solution to Problem 21, $a = g(\sin \theta - \mu \cos \theta) = 5.55$ m/s^2; $v = \sqrt{2as} = 23.6$ m/s. (b) $F = \mu mg = ma$, $a = \mu g = 0.98$ m/s^2, $s = v^2/2a = 283$ m.

25. (a) $N = w + F_y = w + F\sin \theta$, $F_x = F \cos \theta = \mu N = \mu(w + F \sin \theta)$, $F(\cos \theta - \mu \sin \theta) = \mu w$, $F = \mu w/(\cos \theta - \mu \sin \theta)$. (b) The force becomes infinite when $\cos \theta = \mu \sin \theta$, which means at an angle such that $\tan \theta = 1/\mu$. When $\mu = 0.25$, $\theta = 76°$.

Chapter 4

EXERCISES

1. (a) Perpendicular to the wall. (b) The force on the ground is always greater than the weight of the ladder since the force has a horizontal component equal to the reaction force of the wall on the ladder as well as a vertical component equal to the ladder's weight.

3. The location of the pivot point.

5. About whatever point makes the calculations easiest; the results will be the same regardless of the point chosen.

7. MA greater than 1; MA less than 1.

9. $F = \tau/L = 286$ N.

11. $F_x = 80$ N, $F_y = mg = 49$ N, $F = \sqrt{F_x^2 + F_y^2} = 94$ N. $\theta = \tan^{-1} F_y/F_x = 31°$ above the horizontal.

13. $T_y = T \sin \theta = w$, $T = w/\sin \theta = 156$ N.

15. Let x be the distance of the two men from the far end of the ladder and w be the ladder's weight. Calculating torques about the CG of the ladder yields $(w/3)(2$ m$) = (2w/3)(2$ m $- x)$, $x = 1$ m.

17. Let T_1 and T_2 be the tensions in the two parts of the wire, where $T_1 = T_2 = T$. If θ is the angle either part makes with the horizontal, $\theta = \tan^{-1} (0.5 \text{ m})/(10 \text{ m}) = 2.9°$. For the bird to be in equilibrium, $\Sigma F_y = T_{1y} + T_{2y} - mg = 2T \sin \theta - mg = 0$, $m = 0.93$ kg.

19. Let $x = 0$ be at the right-hand end of the object. We imagine the object to consist of the cross member of mass $M_1 = 10$ whose CG is at $x_1 = -20$ cm and the horizontal member of mass $m_2 = 20$ whose CG is at $x_2 = -10$ cm. From Eq. (4-9) the CG of the object is at $X = (m_1 x_1 + m_2 x_2)/(m_1 + m_2) = -13.3$ cm, which is 6.7 cm to the right of the cross member.

21. Each pulley yields an MA of 2. Hence the total MA is $2 \times 2 \times 2 = 8$.

23. (a) The output RPM does not depend upon the efficiency, so $\text{RPM}_{out}/\text{RPM}_{in} = N_{in}/N_{out}$, $\text{RPM}_{out} = (N_{in}/N_{out})\text{RPM}_{in} = 2$ rev/min. (b) $\tau_{out} = (\text{Eff})(N_{out}/N_{in})(\tau_{in}) = 96$ N·m.

25. $\text{Eff} = F_{out}/(\text{MA})(F_{in}) = F_{out}/(2\pi L/p)(F_{in}) = 0.73 = 73\%$.

PROBLEMS

1. Let $m_1 = 15$ kg, $m_2 = 20$ kg, and $m_3 = 25$ kg. Calculating torques about the pivot with $x_1 = 2$ m and $x_3 = 2$ m gives $m_1 g x_1 + m_2 g x_2 = m_3 g x_3$, $x_2 = 1.00$ m.

3. If w is the weight supported by the front axle, then $(24 \text{ N} - w)$ is the weight supported by the rear axle. Calculating torques about the CG of the truck, $(w)(2.5 \text{ m}) = (24 \text{ kN} - w)(1.5 \text{ m})$, $w = 9$ kN.

5. The bear's weight is $w = mg = 1470$ N, the force shared by both feet is F_1 and that shared by both hands is F_2. Calculating torques about the feet yields $F_2 L_2 = w L_1$, $F_2 = 817$ N. Hence $F_1 = w - F_2 = 653$ N. The force on each hand is $F_2/2 = 408.5$ N and that on each foot is $F_1/2 = 326.5$ N.

7. Calculating torques about the ankle, $F_1 L_1 = mg L_2$, $F_1 = mg L_2/L_1 = 3mg = 2205$ N. If L_1 were greater, F_1 would decrease. $F_2 = F_1 + mg = 2940$ N.

9. At the point of tipping over, the CG is directly above the line of the lower wheels. Hence $h = .09 \text{ m}/\tan 30° = 1.56$ m.

11. The angle between the boom and the rope is $\theta = \tan^{-1} (1 \text{ m})/(3 \text{ m}) = 18.4°$ and the CG of the boom is at its midpoint. Calculating torques about the hinge pin of the boom, $(40 \text{ kg})(g)(1.5 \text{ m}) = (T)(3 \text{ m})(\sin 18.4°)$, $T = 620$ N.

13. Calculating torques about the base of the rod, $(F \sin \theta)(1 \text{ m}) = (mg \cos \theta)(3 \text{ m})$, $F = 3mg/\tan \theta = 247$ N.

15. Here $F_y = F \sin 15°$, $w_1 = m_1 g = 29.4$ N, $w_2 = m_2 g = 9.8$ N, $L_1 = 0.13$ m, $L_2 = 0.28$ m, $L_3 = 0.60$ m. (a) Calculating torques about the shoulder joint yields $F_y L_1 = w_1 L_2$, $F = w_1 L_2/(L_1 \sin 15°) = 244$ N. (b) $F_y' L_1 = w_1 L_2 + w_2 L_2$, $F' = (w_1 L_2 + w_2 L_3)/(L_1 \sin 15°) = 419$ N.

17. (a) Let **F** be the force the upper hinge exerts on the door and $w = 200$ N be the door's weight acting from its CG. Since the door's weight is supported by the upper hinge, $F_y = w = 200$ N. To find F_x, we calculate torques about the lower hinge: $(F_x)(2.4 \text{ m}) = (w)(0.4 \text{ m})$, $F_x = 33.3$ N. Hence $F = \sqrt{F_x^2 + F_y^2} = 203$ N. If θ is the angle between **F** and the vertical, $\theta = \tan^{-1} F_x/F_y = 9.5°$. The force the door exerts on the upper hinge is equal and opposite to **F**, hence 285 N at 9.5° away from vertically downward. (b) Since the lower hinge exerts a horizontal force on the door, it must be equal in magnitude to $F_x = 33.3$ N.

19. Since the wall is frictionless, $F_y = w = mg$ and F_x is equal in magnitude to the force the ladder exerts on the wall. Calculating torques about the foot of the ladder, $(F_x)(2.4 \text{ m}) = (mg)\frac{1}{2}\sqrt{(3 \text{ m})^2 - (2.4 \text{ m})^2} = \frac{1}{2}(mg)(1.8 \text{ m})$, $F_x = 55$ N.

21. (a) If F_1 is the force on the wall, $F_{1y} = 0$. To find F_{1x} it is easiest to calculate torques about the foot of the ladder, whose length is L. Since the ladder's weight of mg acts from its CG and the force the wall exerts on the ladder is equal in magnitude to F_{1x}, $(F_{1x})(L \sin 60°) = (mg)(\frac{1}{2}L \cos 60°)$, $F_1 = F_{1x} = 65.6$ N. (b) If F_2 is the force on the ground, $F_{2y} = mg = 245$ N and $F_{2x} = 65.6$ N. Hence $F_2 = \sqrt{F_{2x}^2 + F_{2y}^2} = 254$ N. If θ is the angle between F_2 and the horizontal, $\theta = \tan^{-1} F_{2y}/F_{2x} = 75°$.

Chapter 5

EXERCISES

1. Under no circumstances.

3. At the bottom of the circle, since here the string must support all the ball's weight of mg as well as provide the centripetal force mv^2/r. At other positions in the circle, only part of the weight is supported by the string.

5. No.

7. Shorter.

9. (a) The record's circumference is $2\pi r$, so $v = (33\frac{1}{3})(2\pi)(0.15 \text{ m})/(60 \text{ s}) = 0.524$ m/s. (b) $a_c = v^2/r = 1.83 \text{ m/s}^2$.

11. Since $a = v^2/r = 4g$, $r = v^2/4g = 2296$ m.

13. $F_c = mv^2/r = 384$ N.

15. (a) The tension in the string at the top of the circle is equal to the centripetal force of mv^2/r minus the weight mg of the stone, so that $T = mv^2/r - mg$. If the string is to be just taut, $T = 0$ and $mv^2 r = mg$, $v = \sqrt{rg} = 2.80$ m/s. (b) Since the mass of the stone does not matter here, the speed is again 2.80 m/s.

17. $mv^2/r = mg$, $v = \sqrt{rg} = 10.8$ m/s.

19. $F = Gm^2/r^2 = 7.4 \times 10^{-6}$ N.

21. From Sec. 5-5, the moon's acceleration toward the earth is 2.7×10^{-3} m/s^2. When $t = 1$ s, $s = \frac{1}{2}at^2 = 1.4 \times 10^{-3}$ m $= 1.4$ mm. The moon never reaches the earth because its "falling" causes it to move in an orbit around the earth.

23. $a = (r_e/r)^2 g = 6.3$ m/s^2.

PROBLEMS

1. $mv_{max}^2/r = \mu mg$, $v_{max} = \sqrt{\mu gr}$. Since $f = v/2\pi r$, $f_{max} = \sqrt{\mu gr}/2\pi r = 0.863$ rev/s $= 52$ rev/min. Hence the dime will stay where it is at $33\frac{1}{3}$ rev/min but will fly off at 78 rev/min.

3. $v = 139$ m/s; $\tan\theta = v^2/gr$, $r = v^2/(g\tan 45°) = 1971$ m $= 1.97$ km.

5. $= \tan^{-1} v^2/gr = 6.75°$; $h = (8$ m$)(\sin 6.75°) = 0.94$ m.

7. The circumference of the circle followed by the particle is $2\pi r$, so $f = v/2\pi r$. Since $r = L\sin\theta$, $\tan\theta = \sin\theta/\cos\theta = v^2/gr = 4\pi^2 f^2 r/g$ and $\cos\theta = g/4\pi^2 f^2 L$.

9. Let R = radius of moon's orbit, r = distance of object from moon, and m the object's mass. Then $Gmm_m/r^2 = Gmm_e/(R - r)^2$, $(R - r)/r = \sqrt{m_e/m_m}$, $r = R/(1 + \sqrt{m_e/m_m}) = 3.8 \times 10^7$ m.

11. (a) From Sec. 5-5, $g = GM/r^2$ at the surface of a body of mass M and radius r. Hence $g_m/g = (M_m/M_e)/(r_m/r_e)^2$, $g_m = 1.6$ m/s^2. (b) $W_m = mg_m = 96$ N.

13. From solution to Problem 11, $g_p = g(M_p/M_e)/(r_p/r_e)^2 = g/2$.

15. (a) $v = \sqrt{r_e^2 g_0/r} = 7.7 \times 10^3$ m/s. (b) $T = 2\pi r/v = 5.5 \times 10^3$ s $= 92$ min.

17. If M = sun's mass and m = earth's mass, $GMm/r^2 = mv^2/r$, $M = v^2 r/G = 2.0 \times 10^{30}$ kg.

19. F_1 = force of sun on $m = 1$ kg at earth's center $= Gm_s m/R_e^2$; F_2 = force of sun on $m = 1$ kg at earth's surface $= Gm_s m/(R_e - r_e)^2 = Gm_s m(1 + 2r_e/R_e)/R_e^2$, where R_e = radius of earth's orbit, r_e = radius of earth. $F_2 - F_1 = 2Gm_s mr_e/R_e^3 = 5.06 \times 10^{-7}$ N. Similarly, if R_m =

radius of moon's orbit and F_3 and F_4 are respectively the force of the moon on $m = 1$ kg at the earth's center and at its surface, $F_4 - F_3 = 2Gm_m mr_e/R_m^3 = 11.5 \times 10^{-7}$ N, which is more than double $F_2 - F_1$.

Chapter 6

EXERCISES

1. No work is done by a net force acting on a moving body when the force is perpendicular to the direction of the body's motion.

3. $h = W/mg = 13.6$ m.

5. (a) $W = Fs = 1040$ J. (b) 1040 J.

7. (a) 1 year $= 3.15 \times 10^7$ s; $P = W/t = 6.3 \times 10^{12}$ W. (b) 1.8×10^3 W $= 2.4$ hp.

9. $P = mgh/t = 4.04$ kW.

11. $P = Fv = mgv = 3.92$ kW.

13. $F = P/v = 23.3$ kN.

15. $v = s/t = 8.89$ m/s, KE $= \frac{1}{2}mv^2 = 2.77$ kJ.

17. (a) $mgh_1 = 1176$ J. (b) $mgh_2 = 3528$ J.

19. $Fs = \frac{1}{2}mv^2$, $v = \sqrt{2Fs/m} = 7.1$ m/s.

21. (a) Here $h = 2$ m $- 0.8$ m $= 1.2$ m and, since $mgh = \frac{1}{2}mv^2$, $v = \sqrt{2gh} = 4.85$ m/s. (b) Since v is independent of m, $v = 4.85$ m/s here also.

23. $\frac{1}{2}mv^2 = mg\Delta h$, $\Delta h = v^2/2g = 3.3$ m; $h = 3.3$ m $+ 1.1$ m $+ 0.6$ m $= 5.0$ m.

25. KE/PE $= v^2/2gh = 0.204$, hence 79.6% of the initial PE is dissipated.

PROBLEMS

1. $Fs = mgh$, $F = mgh/s = 127$ kN.

3. $W = Fs\cos\theta = 376$ kJ.

5. In both cases $h = 3000$ m and $W = mgh = 2.65$ MJ.

7. $P = mgh/t = 191$ W.

9. $P = \frac{1}{2}m(v_2^2 - v_1^2)/t$, $t = m(v_2^2 - v_1^2)/2P = 8.8$ s.

11. $P = (0.2)(60$ kg$)(6$ W.kg$) = 72$ W. Since $P = mgh/t$ where $m = 72$ kg here, $h/t = P/mg = 0.102$ m/s $= 367$ m/h.

13. (a) KE $= W = Fs = 24$ J. (b) KE $= W - PE = Fs - mgh = 9.3$ J.

15. (a) $F_1 = P_1/v = 2698$ N. (b) $F_2 = F_1 + mg\sin\theta = 4335$ N, $P_2 = F_2 v = 48.2$ kW.

17. (a) $P = W/t = mgh/t = 2.94$ MW. (b) 29,400 bulbs.

19. The force needed to raise 200 liters ($= 200$ kg) of water is $F = mg = 1960$ N. Since $t = 60$ s and $s = 1.2$ m, the power output is $P = Fs/t = 39.2$ W. Because the efficiency is 60% $= 0.60$, the power input is $P/0.60 = 65.3$ W.

21. At top of loop, weight $=$ centripetal force, $mg = mv^2/r$, $v^2 = rg$; also, KE $=$ ΔPE $= mg(h - 2r)$, so $\frac{1}{2}mv^2 = mg(h - 2r) = \frac{1}{2}mrg$, $h = 5r/s$.

Chapter 7

EXERCISES

1. Yes; yes.

3. The momentum increases by $\sqrt{2}$ since $2 \times (\frac{1}{2}mv^2) = \frac{1}{2}m(\sqrt{2}v)^2$.

5. The definition of mass and the first law of motion are both included in the principle of conservation of linear momentum. The latter principle goes considerably further as well, since it can be applied to systems that consist of any number of bodies that interact with one another in any manner whatsoever, and it holds in three dimensions, not just in one.

7. (a) Yes. (b) In the opposite direction to that in which the man walks. (c) The car also comes to a stop.

9. $mv = ms/t = 622$ kg·m/s.

11. (a) $v = 242$ m/s, $mv = 3.87 \times 10^7$ kg·m/s. (b) $t = mv/F = 114$ s.

13. $\Delta m/\Delta t = 2$ kg/s, $F = \Delta(mv)/\Delta t = v(\Delta m/\Delta t) = 16$ N.

15. $v_2 = m_1 v_1/m_2 = 2.25$ m/s.

17. Number of bullets $= (mv)_{\text{leopard}}/(mv)_{\text{bullet}} = 7.4$, so 8 bullets are needed.

19. $h' = e^2 h = 1.88$ m.

PROBLEMS

1. From Sec. 2–10 the range of the shell is $R = v_x t$, where T is the time of flight which is unchanged by the explosion since it depends only on the height the shell reaches. The shell explodes at $x = R/2$. The half that continues takes all of the horizontal component of momentum, hence its speed v_x doubles and it continues for an additional distance of $2(R/2)$ for a total range of $R/2 + R = 3R/2 = 3$ km.

3. $F = v(\Delta m/\Delta t)$, $a = (F/m) - g = [v(\Delta m/m)/\Delta t] - g = 30.2$ m/s^2.

5. (a) $(m_1 + m_2)V = m_1 v_1 + m_2 v_2$, $V = (m_1 v_1 + m_2 v_2)/(m_1 + m_2) = 56$ km/h. (b) KE$_{\text{initial}}$ $-$ KE$_{\text{final}}$ $= (\frac{1}{2}m_1 v_1^2 + \frac{1}{2}m_2 v_2^2) - \frac{1}{2}(m_1 + m_2)V^2 = 37.1$ kJ.

7. $m_1 v_1 = (m_1 + m_2)V$, $V = m_1 v_1/(m_1 + m_2) = 3.3 \times 10^4$ km/s.

9. If V is final velocity, $V_N = m_1 v_1/(m_1 + m_2) = 0.444$ m/s; $V_w = m_2 v_2/(m_1 + m_2) = 0.889$ m/s; $V = \sqrt{V_N^2 + V_w^2} = 0.99$ m/s. If θ is angle between V and north, $\tan \theta = V_w/V_N = 2$, $\theta = 63°$ west of north.

11. From conservation of momentum, $v_{1i} = v_{1f}\cos \theta_1 + v_{2f}\cos \theta_2$ and $0 = v_{1f}\sin \theta_1 - v_{2f}\sin \theta_2$, $v_{2f} = v_{1f}(\sin \theta_1/\sin \theta_2)$. Hence $v_{1i} = v_{1f}\cos \theta_1 + v_{1f}(\sin \theta_1/\sin \theta_2)\cos \theta_2$, $v_{1f} = v_{1i}/[\cos \theta_1 + (\sin \theta_1/\sin \theta_2)\cos \theta_2] = 3.83$ m/s and $v_{2f} = v_{1f}(\sin \theta_1/\sin \theta_2) = 3.21$ m/s. Alternatively, conservation of KE, here $v_{1i}^2 = v_{1f}^2 + v_{2f}^2$, could have been used with either of the momentum equations.

13. From $(v_2' - v_1') = -(v_2 - v_1)$ we have $v_1 + v_1' = v_2 + v_2'$. From conservation of momentum, $m_1 v_1 + m_2 v_2 = m_1 v_1' + m_2 v_2'$, we have $m_1(v_1 - v_1') = m_2(v_2' - v_2)$. Now we multiply the left-hand sides of the second and fourth equations together and their right-hand sides together to give $m_1 v_1^2 + m_2 v_2^2 = m_1 v_1'^2 + m_2 v_2'^2$, which agrees with conservation of kinetic energy.

15. If V is initial velocity of wooden block, $mv = (M + m)V$ and $V = mv/(M + m)$. From PE $=$ KE, $(M + m)gh = \frac{1}{2}(M + m)V^2 = m^2 v^2/2(M + m)$ and $v = (1 + M/m)\sqrt{2gh}$.

Chapter 8

EXERCISES

1. The solid cylinder reaches the bottom first because its moment of inertia is smaller and hence less of its initial PE becomes KE of rotation.

3. The car will coast downhill faster with light tires because their moments of inertia are smaller.

5. The length of the day will increase; the earth's moment of inertia will be greater when the water from the icecaps becomes uniformly distributed and hence, according to conservation of angular momentum, the angular velocity must decrease.

7. Force, MLT^{-2}; torque, ML^2T^{-2}; energy, ML^2T^{-2}; power, ML^2T^{-3}; linear momentum, MLT^{-1}; angular momentum, ML^2T^{-1}; impulse, MLT^{-1}.

9. 2π rad/9 $= 0.698$ rad.

11. (a) $1' = 1°/60 = 0.01745$ rad/$60 = 2.91 \times 10^{-4}$ rad. (b) With such a small angle, the chord and the arc are very nearly equal. Hence $s = \theta r = 7.27 \times 10^{-3}$ cm $= 0.073$ mm.

13. (a) $\omega = 2\pi$ rad/12 hr $= 1.45 \times 10^{-4}$ rad/s. (b) $\omega = 2\pi$ rad/60 min $= 1.75 \times 10^{-3}$ rad/s. (c) $\omega = 2\pi$ rad/60 s $= 0.105$ rad/s.

15. $\omega = \theta/t = (2\pi$ rad/turn)(49,299 turns)/20,220 s $= 15.3$ rad/s.

17. (a) $v = \omega r = 94.5$ m/s. (b) $\omega = (315$ rad/s)/[0.1047 (rad/s)/rpm] $= 3009$ rpm.

19. $v = \omega r = 226$ km/h.

21. $\alpha = a_T/r = 0.5$ rad/s^2.

23. (a) $\theta = \frac{1}{2}\alpha t^2$, $\alpha = 2\theta/t^2 = 10$ rad/s^2. (b) $\omega = \alpha t = 100$ rad/s.

25. It is easiest to use the revolution as the angular unit here. Since $\omega_f^2 = \omega_0^2 + 2\alpha\theta$, $\theta = (\omega_f^2 - \omega_0^2)/2\alpha = 60$ revolutions.

27. $I = mL^2/12 = 0.016$ kg·m^2, KE $= \frac{1}{2}I\omega^2 = 0.80$ J.

29. $\tau = P/\omega = 8$ N·m.

31. $\alpha = (\omega_f - \omega_0)/t = 5$ rad/s, $I = \tau/\alpha = 100$ kg·m^2.

PROBLEMS

1. At any instant, the barrel can be regarded as rotating about its point of contact with the ground. Hence in (a) $r = 0.8$ m, $v = \omega r = 4$ m/s; in (b) $r = 0.4$ m, $v = 2$ m/s; and in (c) $r = 0$, $v = 0$. Alternatively we can proceed by noting that the center of the barrel is moving at $+2$ m/s. Hence the top has this velocity plus the velocity of $+2$ m/s corresponding to its motion relative to the center for a total of 4 m/s. The bottom has the velocity of the center plus its own -2 m/s velocity relative to the center for a total of $v = 0$.

3. (a) $I = \frac{1}{2}MR^2$, KE $= \frac{1}{2}I\omega^2 = \frac{1}{4}MR^2\omega^2 = 1.6 \times 10^{10}$ J. (b) $t = $ KE/$P = 1.6 \times 10^4$ s $= 4.4$ h.

5. (a) $I = mr^2$, $v = \omega r$, $h = L \sin \theta = 3.42$ m; $mgh = \frac{1}{2}I\omega^2 + \frac{1}{2}mv^2 = mr^2\omega^2$, $\omega = \sqrt{gh}/r = 11.58$ rad/s. (b) $v = \omega r = 5.8$ m/s. (c) KE(rot) $= \frac{1}{2}I\omega^2 = 50.3$ J. (d) KE $= \frac{1}{2}\omega^2 + \frac{1}{2}mv^2 = 101$ J.

7. (a) $Fs = \frac{1}{2}I\omega^2$, $\omega = \sqrt{2Fs/I} = 15.8$ rad/s. (b) KE $= \frac{1}{2}I\omega^2 = 2.5$ J. (c) $W = Fs = 2.5$ J.

9. When the rod is released, its center of gravity falls by $h = L/2$. Hence its initial potential energy is PE $= mgh = mgL/2$. The rod's KE when it has fallen is $\frac{1}{2}I\omega^2$, where I

$= mL^2/3$. Since $\frac{1}{2}I\omega^2 = mgh$, $mL^2\omega^2/6 = mgL/2$, $\omega = \sqrt{3g/L}$.

11. (a) The torque on the spool is $\tau = FR = MR(g - a)$. Since $a = \alpha R$, $\tau = MR(g - R)$ and $\alpha = \tau/I = (g - \alpha R)/R$, $\alpha R = g - \alpha R$, $\alpha = g/2R$, $a = \alpha R = g/2$. (b) $Mgh = \frac{1}{2}Mv^2 + \frac{1}{2}I\omega^2 = \frac{1}{2}Mv^2 + \frac{1}{2}(MR^2)(v^2/R^2) = Mv^2$, $v^2 = gh$. Since $v^2 = 2ah$ in general, $2ah = gh$, $a = g/2$.

13. (a) $I = Mk^2 = \frac{2}{5}MR^2$. (b) $k = \sqrt{I/M} = \sqrt{2/5}\,R$.

15. $I = \frac{1}{2}mr^2 = 9$ kg·m^2. (a) Since $\alpha = \tau/I$, $t = \omega/\alpha = \omega I/\tau = 4.5$ s. (b) KE $= \frac{1}{2}I\omega^2 = 450$ J.

17. (a) $P = Fv = mgv = 235.2$ kW. (b) $\omega = v/r = 6\frac{2}{3}$ rad/s.

19. The net force on the belt is $F = 180$ N $- 70$ N $= 110$ N and $\omega = (30$ rev/s)(2π rad/rev) $= 188$ rad/s. The torque exerted by the motor is $\tau = Fr$ so $P = \tau\omega = Fr\omega = 2.07$ kW.

21. (a) $I_1 = mR_1^2 = 0.2$ kg-m^2, $I_2 = mR_2^2 = 0.288$ kg-m^2, $\omega_1 = 31.4$ rad/s, $\omega_2 = I_1\omega_1/I_2 = 21.8$ rad/s $= 3.5$ rev/s. (b) KE$_1 = \frac{1}{2} \times \frac{1}{2}I_1\omega_1^2 = 49.3$ J, KE$_2 = \frac{1}{2} \times \frac{1}{2}I_2\omega_2^2 = 34.2$ J; the decrease in KE is due to the work that had to be done to stretch the string.

Chapter 9

EXERCISES

1. $V = $ length $\times$ width $\times$ height $= 60$ m^3, $m = dV = 78$ kg.

3. (a) $d_{water} = 1$ kg/liter, so $V = m/d = 55$ liters. (b) $V = m/d = 140$ m^3.

5. (a) $d = m/V = m/(4\pi r^3/3) = 5.52 \times 10^3$ kg/m^3. (b) The interior must consist of denser materials than those at the surface; no.

7. $\Delta L = L_0 F/YA$, $A = \pi d^2/4$, hence $\Delta L_B = (L_{0B}/L_{0A})(d_A/d_B)^2 \Delta L_A = 2\Delta L_A$.

9. The breaking strength F is proportional to the cross-sectional area $A = \pi d^2/4$ of the rope. Hence (a) $F = (25$ kN)(6 mm/10 mm)$^2 = 9$ kN and (b) $F = (25$ kN)(14 mm/10 mm)$^2 = 49$ kN.

11. $F/A = mg/\pi r^2 = 1.04 \times 10^8$ N/m^2, which is about half the ultimate strength of aluminum.

13. $s = F/k = mg/k = 0.0784$ m.

15. $A = 10^{-6}$ m^2. $\Delta L = L_0 F/YA = L_0 mg/YA = 2.94 \times 10^{-4}$ m.

17. $\Delta L = L_0 F/YA = L_0 mg/YA = 2.0 \times 10^{-5}$ m.

19. $S = Fd/sA = 3333$ N/m^2.

21. $\Delta V = pV_0/B = 4 \times 10^{-8}$ m$^3 = 40$ mm^3.

PROBLEMS

1. Each cylinder has the same cross-sectional area A and volume V. The force corresponding to an ultimate strength of U is $F = UA$ and the mass of a cylinder is $m = dV$. Hence $F/m = UA/dV = U/dL$ where L is the cylinder length. In tension, $(F/m)/L = 8.1 \times 10^4, 7.4 \times 10^4$, and 6.4×10^4 N·m/kg for bone, aluminum, and steel; in compression the figures are the same for aluminum and steel and 10.6×10^4 for bone. Bone is clearly an excellent structural material.

3. If m_0 is the mass per atom, the volume per atom is $V_0 = m_0/d$. We imagine each atom to occupy a cube h long on each edge, so that $V_0 = h^3$ and $h = \sqrt[3]{m_0/d} = 2.57 \times 10^{-10}$ m. The foil thickness is $t = V/A = (m/A)/(m/V) = (10^{-3}$ kg/m$^2)/d = 5.26 \times 10^{-8}$ m, and $t/h = 204$ atoms.

5. (a) $Fs = \frac{1}{2}mv^2$, $F = mv^2/2s = 7.2 \times 10^4$ N. (b) $F/A = 7.2 \times 10^4$ N/0.2 m$^2 = 3.6 \times 10^5$ N/m^2, hence she is likely to survive.

7. $\Delta L = L_0F/YA = L_0F/Y\pi(r_{outside}^2 - r_{inside}^2) = 1.44 \times 10^{-4}$ m $= 0.144$ mm.

9. $F_{out} = (F_{in})(2\pi L/p) = 9425$ N, $\Delta L = L_0F_{out}/YA = 1.9 \times 10^{-5}$ m.

11. $L = 0.01$ m, $A = 4Lh$, $(F/A)_{max} = F/4Lh$, $h = F/4L(F/A)_{max} = 7.1 \times 10^{-3}$ m $= 7.1$ mm.

13. $\theta = 0.436$ rad, $R = 5 \times 10^{-4}$ m, $\tau = Fs = 3.0 \times 10^{-3}$ N·m; $S = 2\tau L/\pi\theta R^4 = 4.91 \times 10^{10}$ N/m^2.

Chapter 10

EXERCISES

1. As the steam inside the can condenses, the internal pressure falls below the external pressure of the atmosphere.

3. The forces are the same because the height of water is the same.

5. The water level is unchanged.

7. Since the heights of liquid above both openings are the same, the rates of flow will be the same.

9. Ice on the wings interferes with the intended flow of air past them and so decreases the lift produced. Ice on the fuselage increases the weight of the airplane and increases

air resistance, but these are usually less significant than the reduction in lift due to ice on the wings.

11. The heavier one, because the buoyant force is the same for both.

13. $p = F/A = F/\pi r^2 = 6.37$ MPa $= 63$ atm.

15. $F = pA = p_{atm}\pi r^2 = 127$ N.

17. $F_{out} = mg$, $m = F_{out}/g = (F_{in})(MA_{lever})(d_{out}^2/d_{in}^2)/g = 4408$ kg.

19. $p = p_{atm} + dgh = 1.10 \times 10^8$ Pa.

21. $\Delta p = 0.1$ atm $= 1.013 \times 10^4$ Pa, $h = \Delta p/dg = 1.03$ m.

23. $V = m/d_{water} = 0.07$ m^3, $F = Vd_{air}g = 0.89$ N.

25. Because the buoyancy of the balloon is not affected by the loss of ballast, the upward force on it is $F = mg = 196$ N. The new mass of the balloon is $M = 780$ kg, so $a = F/M = 0.25$ m/s.

27. $p = 120$ torr $= 1.60 \times 10^4$ Pa. This is gauge pressure, so $h = p/dg = 1.55$ m.

PROBLEMS

1. $\Delta p = dgh$, $F = \Delta pA = dghA = 194$ kN.

3. $\Delta p = 2128$ Pa, $h = \Delta p/dg = 21$ cm.

5. If $\bar{d}$ is the average density of tank and contents, the condition for floating is that $\bar{d} \le d_{water}$. (a) $\bar{d} = m/V = 180$ kg/m^3, so it will float. (b) $m_{water} + m_{tank} = 236$ kg, $\bar{d} = 1180$ kg/m^3, so it will sink. (c) $\bar{d} = 860$ kg/m^3, so it will float.

7. $F = Vg(d_{air} - d_{hydrogen}) - mg = 892$ N.

9. Mass of displaced water $= 100$ g, so volume of statue $= 100$ cm^3. Mass of displaced benzene $= 88$ g, so $d = m/V = 0.88$ g/cm$^3 = 880$ kg/m^3.

11. $A_1 = \pi r_1^2$, $A_2 = 3\pi r_2^2$. Since $v_1A_1 = v_2A_2$, $v_2 = (A_1/A_2)v_1 = (r_1^2/3r_2^2)v_1 = \frac{1}{3}v_1 = 1.33$ m/s.

13. (a) $h = \frac{1}{2}gt^2$, $t = \sqrt{2h/g} = 0.45$ s, $v = x/t = 4.43$ m/s. (b) $p = \frac{1}{2}dv^2 = 9.8$ kPa.

15. $v = \sqrt{2gh}$, $A = 2 \times 10^{-3}$ m^2; $R = vA = \sqrt{2gh} A = 7.14 \times 10^{-3}$ m^3/s $= 7.14$ liters/s.

17. The power output of the left ventricle is $P_L = Fv = (pA)(R/A) = pR = 6.2$ W. Since $P_R = 0.2 P_L = 1.2$ W, the total power output is $P_L + P_R = 7.4$ W.

19. $V = 0.1$ m^3, $W = mgh = dVgh = 1.51$ kJ. Since $t = 60$ s, $P = W/(Eff)(t) = 84$ W.

21. (a) $v_2 = v_1A_1/A_2 = v_1r_1^2/r_2^2 = 1.28$ m/s. (b) Here

$h_1 = h_2$ so $p_2 = p_1 + \frac{1}{2}d(v_1^2 - v_2^2) = 2.31 \times 10^5 \text{ N/m}^2 = 2.31$ bars.

23. (a) $R = A_1\sqrt{2gh/[(A_1/A_2)^2 - 1]} = 2.97 \times 10^{-4} \text{ m}^3/\text{s} = 0.297$ liters/s. (b) 0.297 liter/s.

25. $R = 16$ liters/min $= 2.67 \times 10^{-4} \text{m}^2/\text{s}$, $p/L = 8\eta R/\pi r^4 = 525$ Pa/m.

27. (a) The sphere's weight is $4\pi r^3 dg/3$, the drag force on it at the terminal speed v is $6\pi r\eta v$, and the upward buoyant force on it is $4\pi r_3 d'g/3$. At the terminal speed the sphere is in equilibrium and its weight must be equal to the sum of the other two forces, hence the quoted formula. (b) The quantity $(d - d')\eta$ has the greater value in water, hence the ball dropped in water has the greater terminal speed.

29. Here $\theta = 0$, $\cos\theta = 1$. From Eq. (10-17), $r = 2\gamma\cos\theta/dgh = 0.74$ mm.

Chapter 11

EXAMPLES

1. No, but only if it obeys Hooke's law will the oscillations be simple harmonic in character.

3. The maximum KE and maximum PE of a harmonic oscillator are always equal.

5. When the object has vertical sides, so the restoring force is proportional to the displacement of the object above or below its equilibrium level.

7. Damping permits the cone to follow exactly the electrical oscillations that drive it with no tendency to vibrate at its own natural frequencies of oscillation.

9. $\frac{1}{2}mv^2 = \frac{1}{2}ks^2$, $v = s\sqrt{k/m} = 10$ m/s.

11. $k = F/s = mg/s = 163$ N/m. (a) $T = 2\pi\sqrt{m/k} = 0.49$ s. (b) $f = 1/T = 2.0$ Hz.

13. Let $s = $ maximum sag. Then PE(gravitational) $=$ PE(elastic), $mg(h + s) = \frac{1}{2}ks^2$, $k = 2mg(h + s)/s^2 = 8575$ N/m; $T = 2\pi\sqrt{m/k} = 0.57$ s.

15. $f = 1/T$, $v_{max} = 2\pi fA = 2\pi A/T = 314$ cm/s $= 3.14$ m/s.

17. $T = 2\pi\sqrt{L/g} = 4.9$ s.

19. $I = \frac{2}{5}mr^2$, $T = 2\pi\sqrt{I/K} = 2\pi r\sqrt{2m/5K} = 0.428$ s.

PROBLEMS

1. $T = 2\pi\sqrt{m/k} = 2\pi\sqrt{w/kg}$, $k = 4\pi^2 w/T^2 g = 80.6$ N/m, $s = F/k = 0.621$ m.

3. $f = 1/T = 25$ Hz. (a) $a_{max} = 4\pi^2 f^2 A = 247 \text{ m/s}^2$. (b) $F = ma = 1.23$ N. (c) $a = 4\pi^2 f^2 s = 123 \text{ m/s}^2$. (d) $F = ma = 0.617$ N.

5. (a) $T = 2\pi\sqrt{L/g} = 1.42$ s. (b) The motion of the pendulum bob is unaffected by motion of its support at constant velocity, hence $T = 1.42$ s. (c) The downward force on the bob here is $m(g - a)$, hence $T = 2\pi\sqrt{L/(g - a)} = 1.59$ s. (d) As in part (b), $T = 1.42$ s. (e) The downward force on the bob here is $m(g + a)$, hence $T = 2\pi\sqrt{L/(g + a)} = 1.29$ s.

7. (a) $I = mL^2/3$; $h = L/2$; $T = 2\pi\sqrt{I/mgh} = 2\pi\sqrt{2L/3g} = 1.47$ s. (b) $L' = I/mh = 2L/3 = 53.3$ cm.

9. $T = 2\pi\sqrt{m/2k}$ in each case because the restoring force on the object is twice the magnitude with a single spring.

11. The requirement for simple harmonic motion to occur is that the acceleration of the object be proportional to its displacement from the equilibrium position and in the opposite direction. Here $a = (g/R)r$ and a is opposite to r, which fulfills this condition. From Eq. (11-3) the period is $T = 2\pi\sqrt{r/a} = 2\pi\sqrt{R/g} = 5.08 \times 10^3$ s $= 1$ hr 25 min.

Chapter 12

EXERCISES

1. Sound travels fastest in solids whose constituent particles are more tightly coupled together than those of gases or liquids.

3. The energy becomes dissipated as heat, and the string becomes warmer as a result.

5. 1 atm $= 1.013 \times 10^5 \text{ N/m}^2$, hence the amplitude here is 1.97×10^{-10} atm which is $1.97 \times 10^{-8}\%$ of sea level pressure.

7. The energy absent from locations of destructive interference is added to the wave energy at locations of constructive interference to give a greater amplitude there than if no interference has occurred. The total energy remains the same as it would be without interference, but its distribution is different.

9. If the other fork had a frequency of 246 Hz, 10 beats per second would also be heard.

11. (a) $\lambda = v/f = 0.0429$ m. (b) 0.191 m. (c) 0.625 m.

13. $\lambda = 10^{-4}$ m, $f = v/\lambda = 3000$ Hz.

15. (a) $f = v/\lambda = 0.857$ Hz. (b) $T = 1/f = 1.17$ s.

17. (a) Wavelength of fundamental frequency $= \lambda = 2L$

$= 2m$, hence $v = \lambda f = 600$ m/s. (b) $f_2 = 2f_1 = 600$ Hz; $f_3 = 3f_1 = 900$ Hz; $f_4 = 4f_1 = 1200$ Hz.

19. Air: $t_1 = s/v_{air} = 1.458$ s; rail, $t_2 = s/v_{steel} = 0.100$ s; $t_1 - t_2 = 1.358$ s.

21. (a) $f = 440$ Hz, $\lambda = v_{water}/f = 3.40$ m. (b) $f = 440$ Hz, $\lambda = v_{air}/f = 0.780$ m.

23. Here $v_s = 0$, so $f_L = f_S(v + v_L)/v = f_S + f_Sv_L/v$ and $v_L = v(f_L - f_S)/f_S = 1.34$ m/s.

PROBLEMS

1. From Eq. (12-8), $(m/L)_B = (m/L)_A f_A^2/f_B^2 = 0.6$ g/m.

3. (a) From Eq. (12-8), $f_A/f_B = L_B/L_A$, so with $L_B = 270$ mm $f_B = f_A L_A/L_B = 805$ Hz. $f_2 = 2f_1 = 1610$ Hz; $f_3 = 3f_1 = 2415$ Hz.

5. $\sin\theta = v_s/v = 1/1.3$, $\theta = 50°$.

7. Since $f_1 = \sqrt{T/(m/L)}/2L$, increasing the tension in one of the wires to 1.02 of its former value increases f_1 to $\sqrt{1.02}f_1 = 1.01 f_1 = 404$ Hz. Hence 4 beats/s will occur.

9. The frequency of the waves as received by the reflector is $f_R = f(v + u)/v$. The frequency of the waves as received by the listener is $f_L = f_R v/(v - u) = f[(v + u)/v][v/(v - u)] = f(v + u)/(v - u)$. The number of beats heard by the listener is therefore $f_L - f = f[(v + u)/(v - u) - 1] = f(v + u - v + u)/(v - u) = 2fu/(v - u)$.

11. From Eq. (12-10) $v = c(f^2 - f_s^2)/(f^2 + f_s^2)$. Here $f = c/\lambda = 5.56 \times 10^{14}$ Hz, $f_s = c/\lambda_s = 4.84 \times 10^{14}$ Hz, hence $v = 0.138c = 4.14 \times 10^7$ m/s $= 1.49 \times 10^8$ km/h. The fine is therefore $149 million less $80.

13. Since $\lambda = c/f$ and the source is receding, $\lambda = \lambda_s\sqrt{(1 + v/c/(1 - v/c)} = 5.78 \times 10^{-7}$ m.

15. $I = I_0 \log^{-1}(\beta/10) = 10^{-12}$ W/m^2 [log^{-1} (120/10)] $= 1$ W/m^2.

17. Since $\log x - \log y = \log x/y$, $\beta_2 - \beta_1 = 5$ dB $= 10[\log (I_2/I_0) - \log (I_1/I_0)] = 10 \log (I_2/I_1)$, $\log (I_2/I_1) = 0.5$ and $I_2/I_1 = \log^{-1} 0.5 = 3.16$.

19. $\beta_{20} = 70$ dB $= 10 \log (20I_1/I_0) = 10 \log 20 + 10 \log (I_1/I_0) = 13$ dB $+ \beta_1$, $\beta_1 = (70 - 13)$dB $= 57$ dB.

21. $P_{out}/P_{in} = \log^{-1}[G/10] = \log^{-1} (40/10) = \log^{-1} 4 = 10^4 = 10,000$.

Chapter 13

EXERCISES

1. Glass expands less than copper with a rise in temperature, so the bulb would crack as it heats up. Lead-in wires must have the same coefficient of thermal expansion as glass if they are to be used successfully.

3. At constant volume $p/T = $ constant.

5. No. The only temperature scale in which such comparisons make any sense is the absolute temperature scale.

7. Gas molecules undergo frequent collisions with one another, which considerably increases the time needed for a particular molecule to travel from one place to another.

9. The air in the room was originally cold and therefore had a low moisture content, even though its relative humidity may have been high. When this air is heated, its moisture content remains the same, hence its relative humidity decreases.

11. $T_F = \frac{9}{5}T_C + 32°$, hence 330°C $= 626°$F and 1170°C $= 2138°$F.

13. If $T = T_F = T_C$, $\frac{9}{5}T + 32° = \frac{5}{9}(T - 32°)$, $T = -40°$.

15. $T_C = \frac{5}{9}(T_F - 32°) = -80°$C.

17. $a = \Delta L/L_0\Delta T = 8.06 \times 10^{-6}/°$C.

19. $\Delta L = aL_0\Delta T = 0.015$ m so the true width is $L_0 - L = 39.985$ m.

21. Glycerin: $\Delta V = bV_0\Delta T = 1.275$ cm^3; beaker: $\Delta V = bV_0\Delta T = 0.023$ cm^3; thus $(1.275 - 0.023)$ cm$^3 = 1.252$ cm^3 overflows.

23. (a) $\Delta L = aL_0\Delta T = 0.011$ m. (b) $F = YA\Delta L/L_0 = 1.93 \times 10^5$ N.

25. $T_1 = 293$ K, $T_2 = T_1V_2/V_1 = 586$ K $= 313°$C.

27. $T_1 = 283$ K, $T_2 = 323$ K, $p_1 = 2.8$ bars; $p_2 = p_1T_2/T_1 = 3.2$ bars for a gauge pressure of 2.2 bars.

29. (a) $p_2 = p_1V_1/V_2 = 4 \times 10^5$ Pa. (b) $V_2 = p_1V_1/p_2 = 2.67$ m^3. (c) $V_2 = V_1T_2/T_1 = 2.67$ m^3.

31. 6×12.0 u $+ 12 \times 1.008$ u $+ 6 \times 16.00$ u $= 180.16$ u; m (C$_6$H$_{12}$O$_6$) $= (180.16$ u$)(1.66 \times 10^{-27}$ kg/u$) = 2.99 \times 10^{-25}$ kg.

33. $T_0 = 27°$C $= 300$ K. (a) 2 (KE$_0$) $= \frac{3}{2}k(2T_0)$, so $T = 2T_0 = 600$ K $= 327°$C. (b) $\frac{1}{2}m(2v_0)^2 = \frac{3}{2}k(4T_0)$, so $T = 4T_0 = 1200$ K $= 927°$C.

35. At the same temperature, $\frac{1}{2}m_Hv_H^2 = \frac{1}{2}m_0v_0^2$, $v_0 = v_H\sqrt{m_H/m_0} = 0.25$ m/s.

PROBLEMS

1. $b = 3a = 9 \times 10^{-5}/°$C and $\Delta T = 180°$C. Since $m = d_1V_1 = d_2V_2$ and $V_2 = V_1(1 + b \Delta T)$, $d_2 = d_1/(1 + b \Delta T) = 10.8$ g/cm^3.

3. Alcohol: $V_0 = 0.5000$ liter, $\Delta V_1 = bV_0\Delta T = 0.0094$ liter; water: $V_0 = 0.5000$ liter, $V_2 = bV_0\Delta T = 0.0026$ liter; $\Delta V = \Delta V_1 + \Delta V_2 = 0.012$ liter so profit is ($10)($\Delta V$) $= $0.12.

5. $p_1 = p_{atm} + dgh = 1.993 \times 10^5$ Pa, $p_2 = p_{atm} = 1.013 \times 10^5$ Pa, $T_1 = 278$ K, $T_2 = 293$ K; $V_2/V_1 = p_1T_2/p_2T_1 = 2.07 = D_2^3/D_1^3$, hence $D_2 = \sqrt[3]{2.07}\,D_1 = 1.27$ cm.

7. $v_{rms} = \sqrt{3kT/m} = 3.5$ km/s (H_2), 2.5 km/s (He), 0.94 km/s (N_2), and 0.88 km/s (O_2). These are respectively 32%, 22%, 8.4%, and 7.9% of the escape speed, so a much greater proportion of any H_2 molecules and He atoms originally present in the atmosphere could escape than in the case of N_2 or O_2 molecules. In time, most of the H_2 and He disappeared into space, leaving behind an atmosphere rich in N_2 and O_2.

9. (a) $T = 273$ K, $m = (12.01 + 2 \times 16.00)(1.66 \times 10^{-27}$ kg$) = 7.306 \times 10^{-26}$ kg; $v_{av} = \sqrt{3kT/m} = 393$ m/s. (b) $\frac{1}{2}m(2v_0)^2 = \frac{3}{2}k(4T_0)$, so $T = 4T_0 = 1092$ K $= 819°$C.

11. The average time interval between collisions is $t = s/v = 2.5 \times 10^{-10}$ s. Hence each molecule makes an average of $1/t = 4 \times 10^9$ collisions per second.

13. $V = m/d = 1.2 \times 10^{-9}$ m^3, so the film thickness is $a = V/A = 8.0 \times 10^{-10}$ m. If the film is 1 molecule thick, the volume of each molecule is $a^3 = 5.1 \times 10^{-28}$ m^3/molecule. The volume per mole is V_m = molecular mass/density $= (282.5$ g/mole$)/(8.95 \times 10^5$ g/m$^3) = 3.16 \times 10^{-4}$ m^3/mole. Hence the number of oleic acid molecules per mole on the basis of the information given is $N_0 = (3.16 \times 10^{-4}$ m^3/mole$)/(5.1 \times 10^{-28}$ m^3/molecule$) = 6.2 \times 10^{23}$ molecules/mole, which is not far from the actual value.

15. (a) 1000 liters/(22.4 liters/mole) $= 44.6$ moles. (b) $(44.6$ moles$)(6.023 \times 10^{23}$ molecules/mole$) = 2.69 \times 10^{25}$ molecules.

17. 12.01 u + 2 × 16.00 u = 44.01 u = 44.01 g/mole and 10 g of CO_2 contains 10 g/(44.01 g/mole) $= 0.227$ mole. Hence $V = (0.227$ mole$)(22.4$ liters/mole$) = 5.1$ liters.

19. Since $pV = nRT$ and $T = 473$ K, $n = pV/RT = 0.916$ mole. The molecular mass of N_2 is 2 × 14.01 u = 28.02 u = 28.02 g/mole, so the mass of N_2 is $(0.916$ mole$)(28.02$ g/mole$) = 25.7$ g.

Chapter 14

EXERCISES

1. The work goes into the internal energy of the water, as manifested by an increase in its temperature.

3. The ice, because of its heat of fusion.

5. The highest temperature that water can have while remaining liquid is its boiling point. Increasing the rate of heat supply thus increases the rate at which steam is produced but does not change the temperature of the boiling water.

7. Its ends must be at different temperatures.

9. The vacuum prevents heat transfer by conduction or convection, and the silvered surfaces are poor radiators.

11. Under all circumstances; the higher the temperature, the shorter the predominant wavelength.

13. $Q = mc\Delta T = 17$ kcal.

15. $\Delta T = Q/mc = 91°$C, $T_{final} = 209°$C.

17. $t = 1250$ s, $W = Pt = 1.75 \times 10^6$ J, hence 1.75×10^6 J/(4 × 10^4 J/g) $= 44$ g of fat is metabolized.

19. For each lift, $W = mgh = 117.6$J, so at 10% efficiency the energy used is 1176 J. The barbell must therefore be lifted 4.6×10^5 J/1.176×10^3 J $= 391$ times.

21. Heat gained = heat lost, $m_{cop}c_{cop}(T - 20°$C$) + m_{water}c_{water}(T - 20°C) = m_{iron}c_{iron}(120°C - T)$, $T = 20.7°$C.

23. From Fig. 15-11, at $p = 70$ atm and $T = 20°$C, CO_2 is in the liquid state.

25. For the same insulating ability the ratio k/d must be the same, so here $d_2 = d_1k_2/k_1$. Hence (a) 200 cm, (b) 150 cm, (c) 32.5 cm.

27. $d = 6 \times 10^{-3}$ m, $t = 8.64 \times 10^4$ s; $Q = kAt\Delta T/d = 5.18 \times 10^{11}$ J.

29. $A = \pi r^2 = \pi \times 10^{-2}$ m^2; $Q/t = kA\Delta T/d = 189$ kcal/s. The rate decreases as the milk gets warmer since this reduces ΔT.

31. The average temperature difference in the first case is 42.4°C and in the second case it is 22.5°C, so from Eq. (15-3) $t_2 = t_1\Delta T_1/\Delta T_2 = 9.4$ min.

33. $e = 1$, $T = 973$ K, $P = e\sigma AT^4 = 51$ W.

35. Since $R_1/R_2 = T_1^4/T_2^4$, $T_2 = T_1(R_2/R_1)^{1/4} = 800$ K $= 527°$C. Note that $X^{1/4} = log^{-1}(\frac{1}{4}\log X)$.

PROBLEMS

1. $Q = mc\Delta T = 1.2$ kcal, which is 0.34% of the 350 kcal available from the digestion of the piece of pie.

3. $mL_f = \frac{1}{2}mv^2$, $v = \sqrt{2JL_f} = 818$ m/s.

5. $mc\Delta T + mL_f = \frac{1}{2}mv^2$, $v = \sqrt{2(c\Delta T + L_f)} = 327$ m/s, where ΔT is the difference between the melting point of lead and the initial temperature of 100°C.

7. $mc\Delta T = \frac{1}{2}mv^2$, $\Delta T = v^2/2c = 3.39 \times 10^4$ °C, which is well above the melting point of aluminum. Hence special high-melting-point materials are used for the forward face of a spacecraft, which is sometimes designed to melt or vaporize to dissipate heat rapidly. In addition, the spacecraft may first be slowed by its motors to minimize the energy to be dissipated by friction.

9. If M is the mass of the block of ice and m is the mass that melts and if $M>>m$ (as it is here), then $mL_f = Fs = \mu Mgs$, $\mu = mL_f/Mgs = 0.854$.

11. $\Delta T = 80°C$; $t = Q/P = (m_1c_1 + m_2c_2 + m_3c_3)\Delta T/P = 151$ s $= 2.5$ min.

13. Heat gained by ice = heat lost by alcohol-water mixture, $m_{ice}c_{ice}[0°C - (-10°C)] + m_{ice}L_f + m_{ice}c_{water}(5°C - 0°C) = m_{water}c_{water}(20°C - 5°C) + m_{alcohol}c_{alcohol}(20°C - 5°C)$, $m_{ice} = 0.0263$ kg $= 26.3$ g.

15. The boiling point of nitrogen is $-196°C$, hence $\Delta T = 196°C$; heat lost by water = heat gained by nitrogen, $m_wL_{fw} + m_wc_{ice}\Delta T = m_nL_{fn}$, $m_n = 0.742$ kg.

17. To establish the physical state of the mixture, we note that if all the steam is cooled to 100°C, the heat lost is $Q_1 = m_sc_s\Delta T_1 = 80.4$ kJ; if all the water warms to 100°C, the heat gained is $Q_2 = m_wc_w\Delta T_2 = 293.3$ kJ; and if all the steam condenses to water, $Q_3 = m_sL_v = 22,600$ kJ. We conclude that the final temperature is 100°C and that some of the steam condenses. If m is the mass of steam that condenses, $Q_1 + mL_v = Q_2$, $m = (Q_2 - Q_1)/L_v = 0.094$ kg $= 94$ g.

19. We consider an area of $A = 1$ m^2 of the lake and ignore the increase in ice thickness in calculating the heat flow through the ice during the night, which on this basis is $Q = kAt\Delta T/d = 3456$ kJ. The mass of newly formed ice is $m = Q/L_f = 10.3$ kg. If h is the thickness of the newly formed ice and d is its density, $m = dAh$, $h = m/dA = 0.011$ m $= 1.1$ cm and the total thickness is 6.1 cm.

21. Solar energy absorption rate = $(1.2 \times 0.5 \times 0.7)$kW $= 0.42$ kW, so the heat loss rate is $(0.42 + 0.08)$ kW $= 0.5$ kW $= 1800$ kJ/h. Hence $(1800$ kJ/h$)/(2430$ kJ/kg$) = 0.74$ kg/h of sweat must be evaporated.

5. Heat is absorbed in (2), heat is rejected in (4), and work is done on the outside world in (3).

7. Eff $= 1 - T_2/T_1 = 44\%$.

9. Maximum efficiency is $1 - T_2/T_1 = 65\%$, hence actual efficiency is 62% of maximum.

11. $T_2/T_1 = Q_2/Q_1$, $T_2 = T_1Q_2/Q_1 = 375$ K.

13. Eff $= P_{out}/P_{in} = 746$ W/$(0.25$ kg$)(4.6 \times 10^7$ J/kg$)(1/3600$ s$) = 0.234 = 23.4\%$.

PROBLEMS

1. $T_1 = 773$ K, Eff $= 1 - T_2/T_1$, $T_2 = T_1(1 - $ Eff$) = 502$ K. If T_1' is the intake temperature for an efficiency of 50%, then $T_1' = T_2/(1 - $ Eff$) = 1005$ K $= 732°C$.

3. $mgh = mc\Delta T$, $\Delta T = gh/c = 1.1°C$; $T_1 = 11.1°C = 284.1$ K, $T_2 = 10°C = 283$ K, Eff $= 1 - T_2/T_1 = 0.0039 = 0.39\%$.

5. For simplicity we consider a cylinder of cross-sectional area A which has a piston that the gas pushes out a distance s. The force on the piston is $F = pA$, hence $W = Fs = pAs = p\Delta V$, where $\Delta V = As$ is the increase in volume of the gas due to the piston's motion.

7. The volume of 1 kg of water is $V_1 = m/d = 0.001$ m^3, and the volume of 1 kg of steam is $V_2 = 1.667$ m^3. Hence $\Delta V = V_2 - V_1 = 1.666$ m^3, $p = 1.013 \times 10^5$ N/m^2, and $W = p\Delta V = 1.69 \times 10^5$ J. But to vaporize 1 kg of water requires $Q = 22.6 \times 10^5$ J, so $W/Q = 0.075 = 7.5\%$.

9. In a two-stroke engine, a power stroke occurs in each cylinder once per revolution. Since the engine has 4 cylinders, $p = P/4LAN = 6.51 \times 10^5$ Pa.

11. $T_1 = 268$ K, $T_2 = 298$ K, $Q = 1$ J; $W = Q[(T_2/T_1) - 1] = 0.112$ J.

13. $T_1 = 253$ K, $T_2 = 313$ K, $W = Q[(T_2/T_1) - 1] = 992$ J/kcal for a Carnot refrigerator, which is the most efficient possible. Hence only design C is capable of operating.

Chapter 15

EXERCISES

1. The wound watch evolves more heat as it dissolves since its energy content is greater.

3. The kitchen will become warmer owing to the energy dissipated in the fan's operation.

Chapter 16

EXERCISES

1. The notion of electricity as a fluid can explain phenomena that involve charges whose magnitudes are large compared with the electron charge; it cannot explain the behavior of elementary particles or the structure of matter on an atomic level.

3. There are many reasons, but a convincing one is that all gravitational forces are observed to be attractive, whereas if they were electric in origin some would have to be repulsive.

5. An equal amount of charge of the opposite sign appears on the fur.

7. The positive charge is located in the nucleus, the negative charge in the electrons that surround the nucleus.

9. Most of an atom consists of empty space.

11. An electric current in a metal consists of moving electrons, in an ionized gas it consists of positive and negative ions that move in opposite directions.

13. The need for extremely low temperatures.

15. A molecular ion is a molecule with a net charge. A polar molecule is electrically neutral but has a nonsymmetrical distribution of charge.

17. Water is the most notable polar liquid and dissolves ionic compounds such as NaCl and some non-ionic compounds such as HCl, sugar, and alcohol. Gasoline is a non-polar liquid and dissolves such substances as fats and oils.

19. 8; 8; 10.

21. $F_1/F_2 = (r_2/r_1)^2$, $r_2 = r_1\sqrt{F_1/F_2} = 2$ cm.

23. $F = ke^2/r^2 = 4.18 \times 10^{-8}$ N.

25. $r = \sqrt{kQ_A Q_B/F} = 1$ cm.

PROBLEMS

1. $F = mg = ke^2/r^2$, $r = e\sqrt{k/mg} = 5.08$ m.

3. $mv^2/r = ke^2/r^2$, $v = \sqrt{ke^2/mr} = 2.19 \times 10^6$ m/s.

5. Each outer charge exerts on the middle charge a force of magnitude $F = kQ_A Q_B/r^2 = 0.135$ N. Since both forces are directed toward the negative charge, the total force is 0.27 N.

7. (a) If r_1 is the distance from the test charge Q to $Q_1 = -1 \times 10^{-6}$ C, then $r_2 = (0.4 \text{ m} - r_1)$ is the distance to $Q_2 = -3 \times 10^{-6}$ C. The forces on Q will be equal when $kQQ_1/r_1^2 = kQQ_2/r_2^2$, $r_2/r_1 = \sqrt{Q_2/Q_1} = (0.4 \text{ m} - r_1)/r_1$, $r_1 = 0.4 \text{ m}/(\sqrt{Q_2/Q_1} + 1) = 0.146$ m $= 14.6$ cm. (b) Since r_1 is independent of the magnitude and sign of Q, $r_1 = 14.6$ cm here also.

9. The force each of the other charges exerts has the magnitude $F = kQ^2/a^2$. The lines of action of the forces are 120° apart (since each of the interior angles of the triangle is 60°), so the vector sum of the forces is $F' = 2F \cos 60°$ $= 2F \times 0.5 = kQ^2/a^2$. The force is parallel to the side of the triangle opposite to the $+Q$ charge and points toward the $-Q$ charge.

Chapter 17

EXERCISES

1. Compare the forces each field exerts on a charged and on an uncharged body.

5. No, because at a given point an appropriate test object travels along the line of force passing through that point and it can travel in only one direction at any point.

7. (a) It will not move. (b) It will rotate until aligned with the field.

9. $E = kQ/r^2 = 3.94 \times 10^6$ N/C.

11. $s = V/E = 0.005$ m $= 5$ mm.

13. $E = F/e = V/s$, $V = Fs/e = 2.5 \times 10^3$ V.

15. $W = QV = 3.2 \times 10^8$ J.

17. KE $= \frac{1}{2}mv^2 = 3.25 \times 10^{-14}$ J/1.6×10^{-19} J/eV $= 2.03 \times 10^5$ eV.

19. KE $= 8 \times 10^{-18}$ J, $\frac{1}{2}mv^2 = $ KE, $v = \sqrt{2KE/m} = 4.2 \times 10^6$ m/s.

PROBLEMS

1. $E = 0$ at a point between the charges and the distance s from the charge $Q_1 = 1.5 \times 10^{-6}$ C. At this point $kQ_1/s^2 = kQ_2/(0.2 \text{ m} - s)^2$, $(0.2 \text{ m} - s)/s = \sqrt{Q_2/Q_1}$, $s = 0.2$ m/$(\sqrt{Q_2/Q_1} + 1) = 0.0828$ m.

3. (a) $F = QE = 5 \times 10^{-4}$ N. (b) KE $= Fs = 5 \times 10^{-4}$ J.

5. $E = V/s = 500$ V/m. The acceleration of the electron is $a = F/m = eE/m = 8.78 \times 10^{13}$ m/s². Since $s = 0.01$ m $= \frac{1}{2}at^2$, the electron strikes the plate $t = \sqrt{2s/a} = 1.51 \times 10^{-8}$ s later. In this time it travels $d = vt = 0.151$ m $= 15.1$ cm.

7. (a) $W = QV = 2.4 \times 10^6$ J. (b) $h = W/mg = 1.22 \times 10^4$ m. (c) $v = \sqrt{2W/m} = 490$ m/s.

9. $W = mc^2$, $m = W/c^2 = 1.78 \times 10^{-36}$ kg.

11. The components of $\mathbf{E}_+$ and $\mathbf{E}_-$ along the perpendicular to the dipole axis are equal in magnitude and opposite in direction and so cancel out. The components parallel to the dipole axis are each of magnitude $E_{\parallel} = E \cos \theta = $

$(kQ/r^2) \cos \theta = kQ \cos \theta/[(A^2/4) + R^2]$. Since $\cos \theta = (A/2)/ \sqrt{A^2 + R^2}$ from the geometry, $E_\parallel = (kQA/2)/[(A^2/4) + R^2]^{3/2}$. When $R >> A$, $[(A^2/4) + R^2] \approx R^2$ and $E_\parallel = kQA/2R^3$. The total field of the two charges is $E = 2E_\parallel = kQA/R^3$ and it is parallel to the dipole axis in the direction from the $+Q$ to the $-Q$ charge.

Chapter 18

EXERCISES

1. Energy.

3. In series, because each battery separately does work on a charge that passes through it.

5. Bulb 1 is the brightest because the greatest current flows in it; bulbs 3 and 4 are the dimmest because the least current flows in them.

7. The increase in resistivity cannot be due to the expansion of the metal because its cross-sectional area expands faster than its length and so the resistivity should decrease on this basis. Also, the coefficient of linear expansion is too much smaller than the temperature coefficient of resistivity for the expansion of the metal to have any major effect on resistivity.

9. The rates of flow of charge are the same since the currents are the same. The rate of flow of energy when $V = 120$ V is twice as great as when $V = 60$ V since $P = IV$.

11. $R = V/I = 15 \ \Omega$.

13. $R = \rho L/A = 2.0 \ \Omega$.

15. $R = \rho L/\pi r^2$, $r = \sqrt{\rho L/\pi R} = 1.427 \times 10^{-4}$ m, $d = 2r = 2.85 \times 10^{-4}$ m.

17. (a) $\Delta T = -20°C$, so $\Delta R = \alpha R \Delta T = -8 \ \Omega$ and $R = 100 \ \Omega - 8 \ \Omega = 92 \ \Omega$. (b) $\Delta T = 60°C$, so $\Delta R = 24 \ \Omega$ and $R = 124 \ \Omega$.

19. $\alpha = \Delta R/R\Delta T = 0.0051/°C$.

21. $I = P/V = 1.67$ A.

23. $W = Pt = IVt = 2.7 \ \text{kWh} = 9.72$ MJ.

25. (a) $Q = (1.5 \ \text{C/s})(3600 \ \text{s}) = 5400$ C, $W = QV = 7290$ J. For $P = 0.1 \ \text{mW} = 10^{-4}$ W, $t = W/P = 7.29 \times 10^7$ s $= 844$ days $= 2.3$ years. (b) $I = PV = 7.41 \times 10^{-5}$ A $= 74.1 \ \mu$A.

27. $I = V_1/R_1 = 0.02$ A, $V_2 = IR_2 = 5$ V.

29. A resistor in parallel is needed whose resistance is $R_2 = R_1 R/(R_1 - R) = 6.67 \ \Omega$.

31. (a) $R = 12 \times 5 \ \Omega = 60 \ \Omega$. (b) $P = V^2/R = 240$ W.

PROBLEMS

33. (a) 5-Ω bulb: $I_1 = V/R_1 = 2.4$ A; 10-Ω bulb: $I_2 = V/R_2 = 1.2$ A. (b) 12 V. (c) $P_1 = V^2/R_1 = 28.8$ W; $P_2 = V^2/R_2 = 14.4$ W; $P = P_1 + P_2 = 43.2$ W.

37. (a) From $I = \mathcal{E}/(R + r)$, $r = \mathcal{E}/I - R = 0.9 \ \Omega$. (b) $V = \mathcal{E} - Ir = 22$ V.

39. $I_1 = \mathcal{E}/(R + r) = 5.0$ A, $I_2 = $ A. Since $P = I^2 R$, $P_2/P_1 = I_2^2/I_1^2 = 0.44$.

1. $Q/s = 10^{20}$ e/cm $= 16$ C/cm $= 1.6 \times 10^3$ C/m, $t = s/v$, $I = Q/t = v(Q/s)$, $v = I/(Q/s) = 6.25 \times 10^{-4}$ m/s.

3. (a) $P = IV = 5 \times 10^3$ W. (b) 5×10^6 eV. (c) $(5 \times 10^6 \text{ eV})(1.6 \times 10^{-19} \text{ J/eV}) = 8 \times 10^{-13}$ J.

5. $W = I^2 Rt = 4608$ J; $Q = W = mc\Delta T$, $\Delta T = W/mc = 0.28°C$.

7. If I is the current through each battery, $I = 3i$ is the current in the external resistor. In the loop that contains the external resistor and one of the batteries, $\mathcal{E} = ir + IR = ir + 3iR$, $i = \mathcal{E}/(3R + r) = 0.909$ A; hence $I = 3i = 2.73$ A.

9. (a) $R' = (5 \ \Omega \times 8 \ \Omega)/(5 \ \Omega + 8 \ \Omega) = 3.08 \ \Omega$; $R'' = 3.08 \ \Omega + 5 \ \Omega = 8.08 \ \Omega$; $R = (8.08 \ \Omega \times 3 \ \Omega)/(8.08 \ \Omega + 3 \ \Omega) = 2.19 \ \Omega$. (b) The current through the upper branch is $I'' = V''/R'' = 1.49$ A, hence the potential difference across the 8-Ω resistor is $V = I''r' = 4.59$ V and the current through it is $I = V/R = 0.57$ A.

11. (a) $R_1 = (5 \ \Omega \times 10 \ \Omega)/(5 \ \Omega + 10 \ \Omega) = 3.33 \ \Omega$, $R_2 = (8 \ \Omega \times 8 \ \Omega)/(8 \ \Omega + 8 \ \Omega) = 4 \ \Omega$, $R = R_1 + R_2 = 7.33 \ \Omega$. (b) The resistance of the entire circuit is $R' = 7.33 \ \Omega + 1.5 \ \Omega = 8.83 \ \Omega$, so the current is $I = V/R' = 2.72$ A. The potential difference across the first pair of resistors is $V_1 = IR_1 = 9.06$ V, hence in the 5-Ω resistor $I = V_1/5 \ \Omega = 1.81$ A and in the 10-Ω resistor $I = V_1/10 \ \Omega = 0.906$ A. The potential difference across the second pair of resistors is $V_2 = IR_2 = 10.88$ V, hence in each 8-Ω resistor $I = V_2/8 \ \Omega = 1.36$ A.

13. Let I_1 be the current to the right through the upper battery, I_2 be the current to the left through the lower battery, and I_3 be the current to the right through the resistor R. At the left-hand junction $I_2 = I_1 + I_3$. Proceeding clockwise in the upper loop, $\mathcal{E}_2 - \mathcal{E}_1 = I_1 r_1 + I_2 r_2$. Proceeding counterclockwise in the outermost loop, $\mathcal{E}_1 = I_3 R - I_1 r_1$. Since $I_2 = I_1 + I_3$, the first loop equation becomes $\mathcal{E}_2 - \mathcal{E}_1 = I_1 r_1 + I_1 r_2 + I_3 r_2$. Solving this equation and the second loop equation for I_1 and setting the two expressions for I_1 equal gives $(I_3R - \mathcal{E}_1)/r_1 = (\mathcal{E}_2 - \mathcal{E}_1 - I_3r_2)/(r_1 + r_2)$, from which $I_3 = (\mathcal{E}_2r_1 + \mathcal{E}_1r_2)/(Rr_1 + Rr_2 + r_1r_2) = 0.533$ A.

15. Assume currents downward in each resistor. At the top junction, $I_1 + I_2 + I_3 = 0$. Proceeding counterclockwise in the left-hand loop, 10 V $+ 6$ V $= I_1 \times 10 \ \Omega - I_2 \times 5 \ \Omega$. Proceeding clockwise in the right-hand loop, 10 V $+ 10$ V $= I_3 \times 20 \ \Omega - I_2 \times 5 \ \Omega$. Solving yields $I_1 = 0.857$ A, $I_2 = -1.486$ A, $I_3 = 0.629$ A.

17. (a) To find the potential difference between a and b we can disregard the 8-V battery and the 6-Ω resistor. Applying Kirchhoff's second rule to the outside loop, going clockwise and assuming a clockwise current of I, yields 10 V $- 5$ V $= I(1 \ \Omega + 1 \ \Omega + 3 \ \Omega + 9 \ \Omega)$, $I = 0.357$ A. In the upper part of the loop a potential drop occurs in both the battery and the 3-Ω resistor, hence $V = 5$ V $+ I (1 \ \Omega + 3 \ \Omega) = 6.43$ V. The lower part of the loop gives the same result. Here there is a potential increase of 10 V in the battery and a potential drop in the battery's internal resistance and in the 9-Ω resistor, so $V = 10$ V $- I (1 \ \Omega + 9 \ \Omega) = 6.43$ V. (b) Point b is at a potential of -6.43 V relative to a. No current passes through the 6-Ω resistor or the 8-V battery, so there is no IR drop in either. The 8-V battery's positive terminal is nearest to c, hence the potential of c relative to a is 8 V $- 6.43$ V $= 1.57$ V.

19. When the bridge is balanced, no current passes through the galvanometer. Hence the same current I_1 passes from left to right through R and C, and the same current I_2 passes from left to right through A and B. In order for the junction of R, C, and G to be at the same potential as that of A, B, and G, it must be true that $I_1R = I_2A$ and that $I_1C = I_2B$. Dividing the first of these equations by the second, $R/C = A/B$, $R = AC/B$.

21. (a) $I_{coil} = 2 \times 10^{-5}$ A, $I_{shunt} = (1 \text{ A} - I_{coil}) \approx 1$ A, $R_{shunt} = R_{coil}I_{ccil}/I_{shunt} = 0.0012 \ \Omega$. (b) $R = V/I = 5$ V/2×10^{-5} A $= 2.5 \times 10^5 \ \Omega$, so $R_{series} = 2.5 \times 10^5 \ \Omega - R_{coil} \approx 2.5 \times 10^5 \ \Omega$.

23. The meter's resistance is $R_{coil} = 10$ V/6×10^{-5} A $= 1.67 \times 10^5 \ \Omega$. The required resistance is $R = 100$ V/6×10^{-5} A $= 1.667 \times 10^6 \ \Omega$, so $R_{series} = (1.667 - 0.167) \times 10^6 \ \Omega = 1.5 \times 10^6 \ \Omega$.

25. The equivalent resistance of the resistor and voltmeter is $R_e = V/I = 300 \ \Omega$. Since the unknown resistance R is in parallel with the resistance $R_v = 2000 \ \Omega$ of the voltmeter, $R_e = RR_v/(R + R_v)$, $R = R_vR_e/(R_v - R_e) = 353 \ \Omega$. The voltmeter resistance is significant; it is not correct to assume that the equivalent resistance V/I is equal to the unknown resistance R unless $R_v >> R$.

Chapter 19

EXERCISES

1. East; west.

3. South.

5. $0°$; $90°$.

7. The electric force between the streams is attractive and the magnetic force is repulsive. Which force predominates depends upon the velocities of the particles, since the magnetic force increases with velocity whereas the electric force does not change.

9. There is never a net force on the loop in such a field. There is no torque on the loop when its plane is perpendicular to the field.

11. A magnetic compass is more accurate near the equator, where the horizontal component of the earth's magnetic field is strongest. In the polar regions the field has only a very small horizontal component and so can exert relatively little torque on a compass needle.

13. (a) Cobalt steel, because its retentivity is greater. (b) Alnico 2, because its coercive force is greater.

15. $s = \mu_0I/2\pi B = 0.08$ m $= 8$ cm.

17. $I = 2Br/\mu_0 = 7.96$ A.

19. $N/L = 3000$ turns/m, $B = \mu_0I(N/L) = 0.0143$ T.

21. (a) $\theta = 45°$, $F = Il\Delta/B \sin \theta = 0.283$ N; the force is downward according to the right-hand rule. (b) $F = 0.283$ N; the force is upward.

23. $R = mv/eB = 2.84 \times 10^{-3}$ m.

25. $F_{grav} = 8.9 \times 10^{-30}$ N; If v is perpendicular to $\mathbf{B}$ and $B = 3 \times 10^{-5}$ T (a typical value), $F_{mag} = evB = 1.4 \times 10^{-16}$, which is over 10^{13} times greater than F_{grav}.

27. $P = IV$, $I = P/V = 1.67$ A $= I_1 = I_2$; $F/l = \mu_0I_1I_2/2\pi s = 2.78 \times 10^{-4}$ N/m; the force is repulsive, since the currents are in opposite directions.

PROBLEMS

1. $\mathbf{B} = 0$ midway between the wires because the fields of the currents are opposite in direction and of the same magnitude there.

3. $B_1 = \mu_0 I/2\pi s = 2 \times 10^{-5}$ T, $B_2 = 0.67 \times 10^{-5}$ T. Since the fields are in opposite directions, $B = B_1 - B_2 = 1.33 \times 10^{-5}$ T.

5. The fields of the two layers are in opposite directions, hence the result is the same as if the solenoid had 200 turns. Thus $B = \mu_0 NI/L = 1.26 \times 10^{-4}$ T.

7. $B_{\text{loop}} = B_{\text{solenoid}}$, $\mu_0 I_1/2r = \mu_0 I_2(N/L)$, $I_1 = 2rI_2(N/L) = 20$ A since $(N/L) = 2 \times 10^3$ turns/m.

9. $\mu/\mu_0 \approx 8900$ at $B_0 \approx 9 \times 10^{-5}$ T.

11. (a) The magnetic field 0.1 m from the wire is $B = \mu_0 I/2\pi s = 2 \times 10^{-4}$ T, so the force on the electron is $F = evB = 3.2 \times 10^{-16}$ N since $\mathbf{v} \perp \mathbf{B}$. The force is toward the wire since the moving electron is equivalent to a current in the opposite direction. (b) Here also $\mathbf{v} \perp \mathbf{B}$, so $F = 3.2 \times 10^{-16}$ N. The force is in the same direction as the current.

13. (a) $B = \mu_0 I/2\pi s = 1.6 \times 10^{-4}$ T, $F = evB = 1.54 \times 10^{-15}$ N. The force is directed away from the wire. (b) The force has the same magnitude but is directed toward the wire.

15. (a) The force has the same magnitude as in Problem 13(a), but is in the direction of the current. (b) The force has the same magnitude but is opposite to the direction of the current.

17. $E = V/s = vB$, $V = vBs = 2500$ V.

19. If R is the orbit radius, $T = 2\pi R/v = 2\pi mv/vqB = 2\pi m/QB$, which is independent of R and v.

21. For one segment of the loop, $\Delta B = (\mu I \Delta L \sin \theta)/4\pi r^2 = \mu I \Delta L/4\pi r^2$; for the whole loop, the total value of ΔL is $2\pi r$ and $B = \mu I/2r$.

Chapter 20

EXERCISES

1. Counterclockwise; zero current; clockwise.

3. The left-hand wheels have positive charges, the right-hand wheels have negative charges. See sketch of geomagnetic field in Fig. 19-3.

5. The induced emf is zero when the plane of the coil is perpendicular to the magnetic field, which happens twice per rotation. Hence $T = 1$ s/$(2 \times 100) = 0.005$ s.

7. (a) An emf is induced in the wire loop and a current therefore comes into being. This current interacts with the magnetic field to produce a torque that, by Lenz's law, opposes the rotation. (b) The current in a loop of aluminum wire would be smaller because of the greater electrical resis-

tance of aluminum, and the torque opposing the rotation would be correspondingly smaller.

9. Light bulb and electric heater.

11. Alternating current is necessary in order that there be a changing magnetic field to induce a current in the secondary winding.

13. The changing magnetic field produced by an alternating current in the primary winding.

15. $\mathscr{E} = Blv = 1.8 \times 10^{-3}$ v.

17. (a) $A = 5 \times 10^{-3}$ m^2, $\Phi = BA = 5 \times 10^{-6}$ Wb. (b) $\mathscr{E} = \Delta\Phi/\Delta t = 1.67 \times 10^{-6}$ V.

19. $\mathscr{E} = N\Delta\Phi/\Delta t = 6$ V.

21. $V_2 = V_1 N_2/N_1 = 600$ V; $I_2 = I_1 N_1/N_2 = 0.6$ A.

23. $N_1/N_2 = V_1/V_2 = 20.8$; $I_2 = P/V_2 = 41.7$ A.

PROBLEMS

1. (a) The emf decreases because the greater the current, the greater the IR drop in the armature and the less the potential difference across the generator. Hence the current in the magnet windings drops and, since the emf is proportional to B, it too decreases. (b) A similar argument explains the increase in the emf when the current drops.

3. $V = \mathscr{E} - IR$, $I = (\mathscr{E} - V)/R = 25$ A, $P = IV = 2875$ W.

5. Total current is $I = P/V = 25$ A, $I_f = V/R_f = 1$ A, $I_a = I - I_f = 24$ A. $P_a = I_a^2 R_a = 57.6$ W, $P_f = I_f^2 R_f = 120$ W, total power dissipated is $P' = P_a + P_f = 178$ W. Efficiency $= P(P + P') = 94\%$.

7. (a) Total current is $I = P/V = 3.5$ A, field current is $I_f = V/R_f = 0.6$ A, armature current is $I_a = 3.5$ A $- 0.6$ A $= 2.9$ A. Back emf is $\mathscr{E}_b = V - I_a R_a = 117$ V, power output is $\mathscr{E}_b I_a = 340$ W, efficiency is 340 W/420 W $= 81\%$. (b) $R = R_a R_f/(R_a + R_f) = 1.0$ Ω, I (start) $= V/R = 120$ A. (c) Total resistance needed is 120 V/25 A $= 4.8$ Ω, so series resistor should be 4.8 $\Omega - 1.0$ $\Omega = 3.8$ Ω.

9. 2×10^7 eV/1.5×10^5 rev $= 133$ eV/rev, so $\mathscr{E} = 133$ V and $\Delta B/\Delta t = \mathscr{E}/A = \mathscr{E}/2\pi r = 42.3$ T/s.

Chapter 21

EXERCISES

1. No.

3. Since for a parallel-plate capacitor $C = KA/4\,\pi kd$, C'

$= C/2; W_1 = \frac{1}{2}Q^2/C, W_2 = \frac{1}{2}Q^2/C' = Q^2/C$, hence the work needed is $\Delta W = W_2 - W_1 = \frac{1}{2}Q^2/C$.

5. Since $V = Q/C$, when a certain charge Q is given to a capacitor, the greater C is, the smaller V is and hence the weaker the electric field E in the capacitor is. The inductance L, on the other hand, is a measure of the magnitude of and volume occupied by the magnetic field produced by a current I in an inductor, hence the greater L is, the greater the energy of the inductor.

7. (a) From Lenz's law, the selfinduced emf will be opposite to the direction of the current. (b) For the same reason, the emf here will be in the same direction as that of the current.

9. (a) $Q = CV = 0.025$ C. (b) $W = \frac{1}{2}CV^2 = 12.5$ J.

11. $C = KC_0, K = C/C_0 = 6.25$.

13. (a) Here $K = 1, A = 5 \times 10^{-3}$ m^2, $d = 10^{-3}$ m, so $C = A/4\pi kd = 4.42 \times 10^{-11}$ F $= 44.2$ pF. (b) $Q = CV = 1.99 \times 10^{-9}$ C. (c) $W = \frac{1}{2}CV^2 = 4.48 \times 10^{-8}$ J.

15. $Q_1 = C_0V, Q_2 = KC_0V$, hence if $Q_2 = 2Q_1, K = 2$.

17. $C = C_1C_2/(C_1 + C_2) = 14.3$ μF.

19. $L = \mu N^2A/l = 3.96$ mH.

21. (a) $W = \frac{1}{2}LI^2 = 4 \times 10^{-4}$ J. (b) $I = \sqrt{2W/L} = 10$ A.

23. $\mathscr{E} = L\Delta I/\Delta t = 200$ V.

25. $f = 1/2\pi\sqrt{LC} = 650$ Hz.

27. $C = 1/4\pi^2Lf^2 = 0.317$ μF.

PROBLEMS

1. (a) $W = wV = B^2V/2\mu = 3.98 \times 10^6$ J. (b) $E = \sqrt{8\pi kw} = \sqrt{8\pi kB^2/2\mu} = 3 \times 10^{10}$ V/m.

3. There are several ways to proceed, one of which is this. (a) $C = A/4\pi kd = 2.21 \times 10^{-11}$ F, $W = \frac{1}{2}CV^2 = 1.11 \times 10^{-7}$ J. The stored energy W is equal to the work Fd that would be done if the plates were released and flew together so $F = W/d = 1.11 \times 10^{-4}$ N. (b) The force is attractive since the plates have opposite charges. (c) $w = W/Ad = 0.044$ J/m^3 $= E^2/8\pi k$.

5. $1/C = 1/C_1 + 1/C_2 + 1/C_3, C = 3.125$ μF, $Q = Q_1 = Q_2 = Q_3 = CV = 3.75 \times 10^{-5}$ C; $V_1 = Q_1/C_1 = 7.5$ V, $V_2 = Q_2/C_2 = 3.75$ V, $V_3 = Q_3/C_3 = 0.75$ V.

7. $V_1 = V_2 = V_3 = 12$ V; $Q_1 = C_1 = C_1V_1 = 6 \times 10^{-5}$ C, $Q_2 = C_2V_2 = 1.2 \times 10^{-4}$ C, $Q_3 = C_3V_3 = 6 \times 10^{-4}$ C.

9. At first, $Q_{01} = C_1 V_0 = 1.2 \times 10^{-3}$ C, $Q_{02} = C_2V_0 = 2.4 \times 10^{-3}$ C. Since terminals of opposite sign are

joined, the total charge that remains on the combination is $Q = Q_{02} - Q_{01} = 1.2 \times 10^{-3}$ C. The charge distributes itself on the capacitors so there is the same new potential difference across each. Hence $V_1 = V_2, Q_1/C_1 = Q_2/C_2$, and $Q_2 = Q - Q_1; Q_1 = Q/(1 + C_2/C_1) = Q/3 = 4 \times 10^{-4}$ C, $Q_2 = Q - Q_1 = 8 \times 10^{-4}$ C.

11. (a) $L = \mu N^2A/l = 4.09$ mH. (b) $I = V/(R + r) = 0.286$ A. (c) $W = \frac{1}{2}LI^2 = 1.67 \times 10^{-4}$ J.

13. (a) $I = 0, \Delta I/\Delta t = \mathscr{E}/L = 4.17 \times 10^3$ A/s. (b) $\Delta I/\Delta t = (\mathscr{E} - IR)/L, I = (\mathscr{E} - L\Delta I/\Delta t)R = 3.25$ A. (c) $I = \mathscr{E}/R = 6.25$ A, $\Delta I/\Delta t = 0$.

15. (a) $f = 1/2\pi\sqrt{LC} = 1592$ Hz. (b) $Q = CV = 10^{-5}$ C. (c) $\frac{1}{2}CV^2 = \frac{1}{2}LI^2, I = V\sqrt{C/L} = 0.1$ A.

17. (a) $C = 2W/V^2 = 998$ μF. (b) $t = RC = 0.0148$ s.

19. (a) $t = RC = 0.025$ s. (b) $I = V/R = 0.2$ A. (c) $I = Q/t, t = Q/I = CV/I = 0.025$ s.

21. $t = L/R, L = Rt = 1.4$ H.

Chapter 22

EXERCISES

1. The resistance of an ac-circuit determines how much of the electrical energy that passes through it is dissipated as heat. The capacitive reactance and inductive reactance respectively determine the extent to which the capacitors and inductors in the circuit oppose the flow of current without dissipating energy. X_L and X_C vary with frequency whereas R does not, and their combined effects depend upon $X_L - X_C$, so their analogy with resistance is very limited. The impedance determines the current in the circuit when a potential difference V of a certain frequency is applied according to $I = V/Z$; thus Z has the same role as R does in a dc circuit with respect to I, though the rate of energy dissipation is equal to I^2R in both cases.

3. Under these circumstances $X_L > X_C$, so the voltage leads the current.

5. The power factor of a circuit is the ratio between its resistance and impedance and governs the rate at which power is absorbed by the circuit for a given I and V; it varies with frequency. The power factor can be 0 only when $R = 0$; it is 100% at resonance, when $Z = R$.

7. (a) $I_{max} = \sqrt{2} I_{eff} = 14.1$ A. (b) $V_{max} = \sqrt{2} V_{eff} = 84.9$ V. (c) No; they are coincident only when the frequency is the resonant frequency.

9. $C = 1/2\pi fX_C = 1.59 \times 10^{-5}$ F $= 15.9$ μF.

11. (a) $X_L = 2\pi fL = 0.314$ Ω. (b) 314 Ω.

13. $X_L = 2\pi fL$, $I = V/X_L = V/2\pi fL = 0.159$ A.

15. $R = V/I = 120$ Ω. To halve the current, we require $Z = 240$ Ω. Since $Z = \sqrt{R^2 + X_L^2}$, $X_L = 2\pi fL = \sqrt{Z^2 - R^2}$ and $L = \sqrt{Z^2 - R^2}/2\pi f = 0.662$ H.

17. (a) $X_C = 1/2\pi fC = 637$ Ω; $Z = \sqrt{R^2 + X_C^2} = 704$ Ω; $I = V/Z = 0.170$ A. (b) $P = I^2R = 8.7$ W.

19. Since $V_C = V_L$, the circuit is in resonance, and $V_R = 60$ V.

21. (a) Since $X_L = X_C$, $X_L - X_C = 0$ and $Z = R = 20$ Ω. (b) $\tan \phi = 0$, $\phi = 0$. (c) $V = IZ = 400$ V.

23. (a) The power line must have a rating of at least 10 kW/ (0.70 kW/kVA) = 14.3 kVA. (b) The minimum rating is now 10 kVA.

PROBLEMS

1. (a) $2\pi ft = 2\pi/8 = 45°$, $V = V_{max}\sin 2\pi ft = V_{max}\sin 45° = 70.7$ V. (b) $2\pi ft = 2\pi/4 = 90°$, $V = V_{max}\sin 90° = 100$ V. (c) $2\pi ft = 6\pi/8 = 135°$, $V = V_{max}\sin 135° = V_{max}\sin 45° = 70.7$ V. (d) $2\pi ft = 2\pi/2 = 180°$, $V = V_{max}\sin 180° = 0$.

3. (a) Let Z_1 be the impedance at $f_1 = 100$ Hz and Z_2 be the impedance at $f_2 = 500$ Hz. Then $Z_1^2 = R^2 + X_{L1}^2 = R^2 + 4\pi^2 f_1^2 L^2$, $Z_2^2 = R^2 + 4\pi 2 f_2^2 L^2$, and $L = \sqrt{(Z_2^2 - Z_1^2)/4\pi^2(f_2^2 - f_1^2)} = 28.1$ mH. (b) $R = \sqrt{Z_1^2 - 4\pi^2 f_1^2 L^2} = 46.8$ Ω.

5. At $f = 200$ Hz, $L = X_L/2\pi f = 0.0955$ H. At $f = 60$ Hz, $X_L = 2\pi fL = 36$ Ω, $I = V/X_L = 6.67$ A.

7. (a) The voltage leads the currents when X_C is less than X_L. (b) $I_{max} = \sqrt{2} I_{eff} = 7.07$ A; when $V = 0$, $I = I_{max}\sin (360° - 30°) = -I_{max}\sin 30° = -3.54$ A if V is going from $+$ to $-$, or $+3.54$ A if V is going from $-$ to $+$.

9. (a) $R = V_1/I_1 = 5$ Ω, $Z = V_2/I_2 = 8$ Ω; $Z = \sqrt{R^2 + X_L^2}$, $L = \sqrt{Z^2 - R^2}/2\pi f = 16.6$ mH. (b) $P_1 = I_1^2R = 320$ W; $P_2 = I_2^2R = 125$ W.

11. $L = X_L/2\pi f$, $C = 1/2\pi fX_C$, $f_0 = 1/2\pi \sqrt{LC} = 1/\sqrt{X_L/f^2 X_C} = 632$ Hz.

13. (a) $X_L = 2\pi fL = 113$ Ω, $X_C = 1/2\pi fC = 44$ Ω, $Z = \sqrt{R^2 + (X_L - X_C)^2} = 85$ Ω. (b) $I = V/Z = 1.41$ A. (c) $\tan \phi = (X_L - X_C)/R = 1.38$, $\phi = 54°$, power factor $= \cos \phi = 0.588$. (d) $P = I^2R = 99.4$ W. (e) $IV = P/\cos \phi = 169$ V $\cdot$ A.

15. (a) $f_0 = 1/2\pi \sqrt{LC} = 37.5$ Hz. (b) $I = V/R = 2.4$ A.

17. (a) $X_L = 2\pi fL = 38$ Ω, $X_C = 1/2\pi fC = 265$ Ω, $Z = \sqrt{R^2 + (X_L - X_C)^2} = 235$ Ω, $I = V/Z = 0.51$ A. (b)

$P = I^2R = 15.6$ W. (c) $V_L = IX_L = 19.4$ V; $V_C = 135$ V; $V_R = IR = 30.6$ V.

19. (a) $\cos \phi = P/IV = 0.625$. (b) Since $\cos \phi = 0.625$, $\phi = 51°$, $\tan \phi = 1.235 = X_L/R$; $R = P/I^2 = 75$ Ω, $X_L = 1.235R = 92.6$ Ω; if $X_C = X_L$, $C = 1/2\pi fX_L = 28.6$ μF. (c) $I = V/R = 1.6$ A. (d) $P = I^2R = 192$ W. (e) 192 V $\cdot$ A since V and I are in phase at the resonant frequency.

21. (a) $X_C = 1/2\pi fC = 20$ Ω, $I_C = V/X_C = 0.5$ A; $I_R = V/R = 1.0$ A. (b) $I = \sqrt{I_R^2 + I_C^2} = 1.12$ A. (c) $Z = V/I = 8.9$ Ω. (d) $\phi = \tan^{-1} I_C/I_R = 27°$; the current leads the voltage. (e) $P = I_R^2R = 10$ W. (Alternatively, $P = IV \cos \phi = 10$ W.)

23. (a) $X_C = 1/2\pi fC = 20$ Ω, $X_L = 2\pi fL = 12.6$ Ω; $I_C = V/X_C = 0.5$ A; $I_L = V/X_L = 0.8$ A; $I_R = V/R = 1.0$ A. (b) $I = \sqrt{I_R^2 + (I_C - I_L)^2} = 1.04$ A. (c) $Z = V/I = 9.6$ Ω. (d) $\phi = \tan^{-1} (I_C - I_L)/I_R = -17°$; the current lags behind the voltge. (e) $P = I_R^2R = 10$ W. (Alternatively, $P = IV \cos \phi = 10$ W.)

Chapter 23

EXERCISES

1. The electric fields produced by electromagnetic induction are readily detected through the currents they produce in metal wires, but no such simple means is available for detecting the weak magnetic fields produced by changing electric fields.

3. Light waves consist of coupled electric and magnetic field fluctuations and need no material medium in which to occur. Sound waves consist of pressure fluctuations in a material medium and hence cannot occur in a vacuum.

5. The variations are always in phase.

7. Electromagnetic waves travel more slowly in material media than they do in free space.

9. A real image is formed by light rays that pass through it; a screen placed at the location of the image would show the image. A virtual image is formed by the backward extension of light rays that were diverted by refraction or reflection. The light rays that seem to come from it do not actually pass through a virtual image, and so it cannot appear on a screen although it can be seen by the eye.

11. Since both are transparent with the same index of refraction, light passes through the bottle unaffected by the presence of the rod.

13. The flashes of color are the result of dispersion. In red light the flashes would be red only.

15. (a) $f = 1/T = 10^9$ Hz. (b) $\lambda = c/f = 0.3$ m. (c) Microwaves.

17. The image is always twice as far from the man as the mirror; hence it must be moving twice as fast as the mirror, namely at 60 km/h.

19. (a) $v = c/n = 2.999 \times 10^8$ m/s. (b) 1.24×10^8 m/s. (c) 1.97×10^8 m/s. (d) 2.26×10^8 m/s.

21. $\sin r = (n_1/n_2)\sin i = 0.789$, $r = 52°$.

23. $n_2 = n_1 \sin i/\sin r = 1.51$.

25. (a) $\sin i_c = n_2/n_1 = 0.413$, $i_c = 24°$. (b) $\sin i_c = 0.550$, $i_c = 33°$.

PROBLEMS

1. $\lambda_2/\lambda_1 = n_1/n_2$.

3. (a) By the same procedure as in the second example of Sec. 23-7, the final angle of refraction is 70°. (b) 70°.

5. $n_1 = 1.00$. Red light: $\sin r = (n_1/n_2)\sin i = 0.4263$, $r = 25.23°$. Violet light: $\sin r = (n_1/n_2)\sin i = 0.4201$, $r = 24.84°$; $\Delta r = 0.39°$.

7. $h = h'n_1/n_2 = 5.4$ cm.

9. $h' = hn_2/n_1 = 0.59$ m.

11. $A = 1800$ m^2 so $F_{total} = EA = 5.4 \times 10^5$ lm. The useful flux emitted by each lamp is $F = \frac{2}{3}(4\pi I) = 2.09 \times 10^4$ lm/lamp so $N = F_{total}/F = 26$ lamps.

13. $F = 4\pi I = 2513$ lm, $A = \pi d^4/4 = 0.785$ m^2, $E = F/A = 3200$ luxes.

Chapter 24

EXERCISES

1. No; yes, if the object is placed between the lens and the focal point.

3. (a) Under no circumstances. (b) When the object is between the lens and the focal point.

5. To increase the image distance, the object distance must be reduced, hence the lens should be moved closer to the slide.

7. No; no.

9. Crown glass has a smaller index of refraction than benzene. The immersed lens will act as a diverging lens because wavefronts passing through it are retarded less going through its center where the path length in the medium of higher light speed is greater. Thus the wavefronts are affected as in the right-hand diagram of Fig. 24-3 even though the lens is convex.

11. (a) Inverted and larger than the object. (b) Inverted and smaller than the object. (c) Erect and smaller than the object. (d) Erect and larger than the object.

13. $R_1 = -30$ cm, $R_2 = +20$ cm, $1/f = (n - 1)(1/R_1 + 1/R_2)$, $f = 12$ cm.

15. $1/R_2 = 0$, $R_1 = (n - 1)f = 4.95$ in.

17. (a) Converging. (b) $p = \infty$, $q = f = 8$ cm.

19. No image is formed when the object is at the focal point.

21. $q = pf/(p - f) = 30$ cm, $m = -q/p = -1$. The image is real, inverted, and the same size as the object.

23. $f = pq/(p + q) = 12$ cm; the lens is converging since its focal length is positive.

25. (a) No. (b) A concave mirror forms a virtual image of any object that is between the mirror and its focal point.

27. $q = pf/(p - f) = -40$ cm, $m = -q/p = 2$; the image is therefore erect, virtual, and larger than the object.

29. $q = pf/(p - f) = 200$ cm, $m = -q/p = -4$; the image is therefore inverted, real, and larger than the object.

31. $q = pf/(p - f) = 66.7$ cm, $m = -q/p = -0.667$; the image is therefore inverted, real, and smaller than the object.

33. $q = pf/(p - f) = -6.43$ cm, $m = -q/p = 3.57$.

PROBLEMS

1. (a) Converging. (b) The index of refraction of water is 1.33, so the relative index of refraction is $n' = 1.60/1.33 = 1.20$. From the lensmaker's equation, $f/f' = (n' - 1)/(n - 1) = 1/3$, and $f' = 3f = 45$ cm.

3. $m = 1/10 = -q/p$, $q = -p/10 = -0.5$ m; $f = pq/(p + q) = -0.556$ m.

5. $p = 7000$ m, $f = 1$ m, $q = pf/(p - f) \approx 1$ m; $m = -q/p = -1.43 \times 10^{-4}$; $h' = mh = -1.29 \times 10^{-3}$ m $= -1.29$ m.

7. (a) $p + q = 60$ cm, $p = 60$ cm $- q$; $m = -4 = -q/p$, $p = q/4 = 60$ cm $- q$, $q = 48$ cm, $p = 12$ cm; $f = pq/(p + q) = +9.6$ cm. (b) $m = -2 = -q/p$, $p = q/2$; $1/f = 1/p + 1/q = 3/q$, $q = 3f = +28.8$ cm, $p = q/2 = 14.4$ cm, $p + q = 43.2$ cm.

9. $q = pf/(p - f) = -40$ cm, $m = -q/p = +5$, $h = h'/m = 0.2$ cm $= 2$ mm.

11. $q = -60$ cm and $f = 1$ m/3.33 $= 0.30$ m $= 30$ cm, hence $p = qf/(q - f) = 20$ cm.

13. (a) Converging lens. (b) $p = 25$ cm, $q = -100$ cm, $f = pq/(p + q) = +33$ cm.

15. (a) $q = 160$ mm, $p = q f/(q - f) = 4.1$ mm, $m = -q/p = -39$; the total magnification is $10 \times 39 = 390\times$. (b) 4.1 mm.

17. (a) Objective: $q = pf/(p - f) = 1.02$ m, $m_1 = -q/p = -0.020$. Eyepiece: $q = -25$ cm, $p = q f/(q - f) = 4.17$ m, $m_2 = -q/p = 6$. Hence $m = m_1 m_2 = -0.12$, $h' = mh = -4.8$ cm, where the minus sign signifies an inverted image. (b) Here $m_1 = -0.25$, so $m = -1.5$ and $h' = -60$ cm.

19. The object distance of the first lens is $p_1 = \infty$ and its image distance is $q_1 = f_1$. The object distance of the second lens is therefore $p_2 = -q_1 = -f_1$, and, since $1/p_2 + 1/q_2 = 1/f_2$, we obtain $1/q_2 = 1/f_1 + 1/f_2$. But q_2 is the distance from the lenses at which the original beam is brought to a focus. Hence $q_2 = f$ and we have the desired result.

21. $f = -R/2 = -15$ cm, $q = -10$ cm, $p = qf/(q - f) = 30$ cm.

23. Here $m = +3$, $p = 10$ cm, $q = -mp = -30$ cm, so $f = pq/(p + q) = 15$ cm. When $m = -3$, $q = -mp = 3p$ and $1/p + 1/3p = 1/f$, $p = 4f/3 = 20$ cm. In this case the image is real and inverted.

25. $m = 1/5$, $p = 150$ cm, $q = -mp = -30$ cm, $f = pq/(p + q) = -37.5$ cm; $R = -2f = 75$ cm.

Chapter 25

EXERCISES

1. Light from incoherent sources can interfere, but the resulting interference pattern shifts continually because there is no definite phase relationship between the beams from the sources. When the sources are coherent, there is such a relationship, and the pattern is stable and therefore discernible.

3. The wavelengths in visible light are very small relative to the size of a building, whereas those in radio waves are more nearly comparable.

5. Diffraction and interference are both consequences of the wave nature of light. Interference refers in general to the reinforcement or cancellation of waves of the same kind from different sources that meet at some point; diffraction refers in general to the "bending" of waves around the edge of an obstacle. Interference patterns occur in diffraction, as discussed in the text.

7. This is a diffraction effect.

9. The first-order spectrum contains bright lines for which $\sin \theta = \lambda/d$, where d is the spacing of the slits, and the second-order spectrum contains bright lines for which $\sin \theta = 2\lambda/d$. The second-order spectrum is deviated by more than the first-order spectrum, and is wider as well (twice as wide, in fact). A prism produces only a single spectrum.

11. The "missing" energy is found in the bright lines, which are brighter than would be the case without interference. See also Exercise 7 of Chapter 12.

13. Higher resolving power; greater light-gathering ability, hence ability to form images of faint objects.

15. The individual waves in an ordinary light beam are polarized, but their planes of polarization are random so the beam itself is unpolarized.

17. $\Delta y = L\lambda/d$, $\lambda = d\Delta y/L = 4.33 \times 10^{-7}$ m.

19. (a) $y = 2L\lambda/d = 8.74 \times 10^{-3}$ m. (b) $y = 3L\lambda/d = 1.31 \times 10^{-2}$ m. (c) $y = 5L\lambda/2d = 1.09 \times 10^{-2}$ m.

21. (a) $d = 2.5 \times 10^{-6}$ m, $\sin \theta = \lambda/d = 0.30$, $\theta = 17°$. (b) $\sin \theta = 3\lambda/d = 0.90$, $\theta = 64°$.

23. Since $n_{glass} > 1.38$, light reflected from both top and bottom of the fluoride film undergoes a phase shift of $\lambda/2$, so this factor can be ignored here. For destructive interference, the path length $2d$ in a film d thick must be an odd number of half wavelengths, the smallest of which is $2d = \lambda/2$; the thinnest coating gives maximum cancellation over the greatest angle of incidence, so it is normally used. The wavelength λ in the film is λ_0/n, where λ_0 is the wavelength in air, so $d = \lambda/4 = \lambda_0/4n = 99.6$ nm.

25. $d_0 = 1.22 \lambda L/D = 47.1$ m. In practice, irregularities in the earth's atmosphere prevent this resolution from being attained.

PROBLEMS

1. If the known wavelength is λ_1 and the unknown wavelength is λ_2, $7L\lambda_1/2d = 5L\lambda_2/d$, $\lambda_2 = 0.7\lambda_1 = 4.06 \times 10^{-7}$ m.

3. Here $d = 2.5 \times 10^{-6}$ m. Since $\sin \theta = n\lambda/d$ and the largest value $\sin \theta$ can have is 1, the maximum number of images possible is $n = d/\lambda = 4$.

5. Here $d = 3.33 \times 10^{-6}$ m; $\sin \theta_1 = \lambda_1/d = 0.12$, $\theta_1 = 7°$; $\sin \theta_2 = \lambda_2/d = 0.21$, $\theta_2 = 12°$; $h_1 = L \tan \theta_1 = 0.246$ m, $h_2 = L \tan \theta_2 = 0.426$ m, $h_2 - h_1 = 0.18$ m.

7. $\lambda = c/f = 20$ m and $d = 15$ m, so $\lambda > d$ and the only intensity maxima are in the north and south directions.

9. (a) $\theta_0 = 1.22\ \lambda/D = 3.37 \times 10^{-3}$ rad $= 0.193°$; the angular diameter is $2\theta_0 = 0.386°$. (b) The apparent angular diameter of the moon is 3476 km/3.8×10^5 km $= 0.0091$ rad $= 0.52°$.

11. (a) $L = d_0 D/1.22\lambda = 1.64 \times 10^4$ m $= 16.4$ km. (b) 8.2 km.

13. For a distant object, the image distance L equals the focal length f of the lens. The ratio f/D for the required resolution, with $d_0 = 10^{-5}$ m, is therefore $d_0/1.22\ \lambda = 15$. This is an aperture of $f/15$.

Chapter 26

EXERCISES

1. No; yes; yes.

3. More conspicuous.

5. The rod appears longest to the stationary observer.

7. Because the hydrogen molecule is electrically neutral despite the very different speeds of its electrons and protons, electric charge must be an invariant quantity.

9. The density is greater because the object's mass is greater and its length and hence volume is smaller when it is in motion.

11. $L = L_0 \sqrt{1 - v^2/c^2} = 4.71 \times 10^{-15}$ m.

13. $t = 2t_0$, $\sqrt{1 - v^2/c^2} = \frac{1}{2}$, $v^2/c^2 = \frac{3}{4}$, $v = 2.6 \times 10^8$ m/s.

15. $m = 1.01\ m_0$, $\sqrt{1 - v^2/c^2} = 1/1.01$, $v^2/c^2 = 0.0197$, $v = 4.2 \times 10^7$ m/s.

17. Since $m = 2m_0$, KE $= mc^2 - m_0 c^2 = m_0 c^2 = 940$ MeV.

PROBLEMS

1. $t = t_0/\sqrt{1 - v^2/c^2}$ in general, here $v << c$ so $t = t_0 + \frac{1}{2}t_0 v^2/c^2$, $t - t_0 = \frac{1}{2}t_0 v^2/c^2 = 5 \times 10^{-13}\ t_0 = 1$ s, $t_0 = 2 \times 10^{12}$ s $= 6.35 \times 10^4$ years.

3. $m/m_0 = mc^2/m_0 c^2 = (m_0 c^2 + \text{KE})/m_0 c^2 = 1.533$ so $\Delta m = m - m_0 = 0.533\ m_0 = 8.90 \times 10^{-27}$ kg.

5. (a) Backward particle: $V' = -0.6c$, $v = 0.5c$, $V = (V' + v)/(1 + vV'/c^2) = -0.077c$, which is $0.077c$ in the backward direction. (b) Forward particle: $V' = +0.5c$, $v = 0.5c$, $V = 0.8\ c$.

7. Laboratory observer: KE $= (m_0 c^2/\sqrt{1 - V^2/c^2} - m_0 c^2$ $= 0.34$ MeV. Moving observer: $V' = -0.8c$, $v = 0.5c$, $V = (V' + v)/(1 + vV'/c^2) = 0.5c$, KE $= (m_0 c^2/\sqrt{1 - V^2/c^2}) = m_0 c^2 = 0.079$ MeV.

Chapter 27

EXERCISES

1. Such wave phenomena as diffraction and interference are easier to demonstrate than such quantum phenomena as the photoelectric effect.

3. Each atom in a solid is limited to a certain definite region of space; otherwise the assembly of atoms would not be a solid. The uncertainty in position of each atom is therefore a finite quantity, and the atom's momentum and hence energy cannot be 0. The position of a molecule in a gas is not restricted, so the uncertainty in its position is effectively infinite and its momentum and energy can accordingly be 0.

5. No.

7. Apply an electric or magnetic field and look for a deflection.

9. $f = c/\lambda = 5 \times 10^{14}$ Hz; $n = E/hf = 3$ photons.

11. $f = E/h = 1.21 \times 10^{15}$ Hz, $\lambda = c/f = 2.49 \times 10^{-7}$ m.

13. $\lambda = c/f = ch/E = 2.22 \times 10^{-7}$ m; ultraviolet.

15. $f = eV/h = 2.41 \times 10^{18}$ Hz; X rays.

17. $\lambda = h/mv = 3.3 \times 10^{-29}$ m.

19. (a) $\Delta v = h/2\pi m \Delta x = 1.16 \times 10^5$ m/s. (b) $\Delta v = h/2\pi m \Delta x = 63$ m/s.

PROBLEMS

1. (a) $P/hf = 4.22 \times 10^{21}$ photons/m$^2 \cdot$ s. (b) The photon velocity is c, hence there are $(4.22 \times 10^{21}$ photons/m$^2 \cdot$ s)/$c = 1.41 \times 10^{13}$ photons/m^3.

3. (a) $m = m_0/\sqrt{1 - v^2/c^2}$, $\lambda = h/mv = h\sqrt{1 - v^2/c^2}/m_0 v = 6.86 \times 10^{-12}$ m. (b) $\lambda = h\sqrt{1 - v^2/c^2}/m_0 v = 2.71 \times 10^{-12}$ m.

5. KE $= 10^3$ eV $= 1.6 \times 10^{-16}$ J $= \frac{1}{2}mv^2$, $mv = \sqrt{2m\text{KE}} = 1.71 \times 10^{-23}$ kg$\cdot$m/s; $\Delta mv = h/2\pi\Delta x = 1.06 \times 10^{-24}$ kg$\cdot$m/s; $\Delta mv/mv = 0.062 = 6.2\%$.

7. $\Delta mv = h/2\pi\Delta x = 1.055 \times 10^{-23}$ kg$\cdot$m/s, $\Delta v = \Delta mv/m = 1.16 \times 10^7$ m/s, which is much smaller than c and so this nonrelativistic calculation is satisfactory; $\Delta x' = t\Delta v$ $= 1.16 \times 10^7$ m.

9. If the radius of the circle is r, $L = I\omega = mvr$ and $\Delta L = r\Delta mv$. The uncertainty in the particle's position along the circle is $\Delta x = r\Delta\theta$. Hence $\Delta x \, \Delta mv = \Delta L \, \Delta\theta$ and $\Delta L \, \Delta\theta \geqslant h/2\pi$.

11. $\Delta E \geqslant h/2\pi\Delta t$ so $\Delta f = \Delta E/h = 1/2\pi\Delta t = 1.59 \times 10^7$ Hz.

Chapter 28

EXERCISES

1. (a) The Rutherford model indicates the division of the hydrogen atom into a nucleus and a relatively distant electron, and the Bohr model goes on from there to specify the motion and energy of the electron. (b) In order not to be pulled into the nucleus by the electric force exerted by the nucleus.

3. A hydrogen sample contains a great many atoms, each of which can undergo a variety of possible transitions.

5. (a) Continuous emission spectrum (see Sec. 14-8). (b) Emission line spectrum. (c) Absorption line spectrum.

7. With no incident photons, there is nothing to absorb.

9. $E = E_1 = hf$, $\lambda = c/f = ch/E_1 = 9.12 \times 10^{-8}$ m.

11. $1/\lambda = R(1 - 1/10^2)$, $\lambda = 9.21 \times 10^{-8}$ m $= 92.1$ nm; ultraviolet.

13. The shorter the wavelength, the higher the frequency, which in turn means the greater the energy change. From Fig. 28-12 the maximum energy change in the Bracket series corresponds to a transition from $n = \infty$ to $n = 4$, so that $hf = E_i - E_f = -E_1/4^2 = 1.36 \times 10^{-19}$ J, $f = 2.06 \times 10^{14}$ Hz, and $\lambda = c/f = 1.46 \times 10^{-6}$ m.

15. The H atoms must be raised to the $n = 5$ state from the ground state, which requires $\Delta E = E_1(1/n_i^2 - 1/n_f^2) = E_1(1 - 1/5^2) = 13.06$ eV.

PROBLEMS

1. (a) $\lambda = h/mv = 3.68 \times 10^{-63}$ m. (b) $n\lambda = 2\pi r$, $n = 2\pi r/\lambda = 2.56 \times 10^{74}$.

3. $\frac{3}{2}kT = E_1$, $T = 2E_1/3k = 1.05 \times 10^5$ °K.

5. To find the quantum number n_f of the highest state that can be reached by a hydrogen atom in its ground state ($n_i = 1$) when excited by a maximum of $\Delta E = 13$ eV, we

proceed as follows, with $E_1 = 13.6$ eV: $\Delta E = E_1(1/n_i^2 - 1/n_f^2) = E_1(1 - 1/n_f^2)$, $1/n_f^2 = 1 - \Delta E/E_1 = (E_1 - \Delta E)/E_1$, $n_f = \sqrt{E_1/(E_1 - \Delta E)} = 4.76$. Hence the highest state has $n = 4$, and lines from the following transitions will be present: $4 \rightarrow 3$ (Paschen series); $4 \rightarrow 2$, $3 \rightarrow 2$ (Balmer series); $4 \rightarrow 1$, $3 \rightarrow 1$, $2 \rightarrow 1$ (Lyman series).

9. (a) From Eq. (28-7) with $n = 1$, $r_1(\text{muon}) = h^2/4\pi^2 kme^2 = r_1(\text{electron})/207 = 2.56 \times 10^{-13}$ m. (b) From Eq. (28-9) with $n = 1$, $E_1(\text{muon}) = 2\pi^2 k^2 me^4/h^2 = 207 E_1(\text{electron}) = 2815$ eV.

Chapter 29

EXERCISES

1. The uncertainty principle.

3. $-4, -3, -2, -1, 0, +1, +2, +3, +4$.

5. Such an electron has no vector property associated with it on the average, hence its probability cloud must vary in the same way in all directions.

7. The densest part of a probability cloud, which corresponds to the maximum probability of finding the electron, occurs at a radius proportional to n^2.

9. (a) Spin is an intrinsic property of electrons, which they always exhibit. (b) All electrons have the same spin.

11. 182 elements.

13. Two electrons.

15. Both Li and Na atoms have one electron outside a closed inner shell.

Chapter 30

EXERCISES

1. The existence of such definite orbits is not consistent with the uncertainty principle.

3. Their ability to bond with each other as well as with other atoms.

5. In a polar covalent molecule the shared electrons spend more time on the average near one of the atoms.

7. (a) The number of dislocations in the structure of the iron increased. (b) The strong heating increased the energies of the individual iron atoms, which were then able to shift their positions to produce a more regular crystal structure.

9. Li atoms have a single electron outside a closed shell, and this electron is held relatively weakly since the inner electrons shield most of the full nuclear charge. In fluorine, the outer shells lack an electron of completion, and the electrons in these shells are held very tightly to the nucleus since little of the nuclear charge is shielded from them.

11. Na^+ ions have closed shells, whereas Na atoms each have a single, relatively easily-detached electron.

13. The outermost electrons of these atoms are, in the stated order, farther and farther from their respective nuclei and hence are less and less tightly bound.

15. Van der Waals' forces are too weak to hold inert gas atoms together against the forces exerted during collisions in the gaseous state.

17. In a NaCl molecule, the Na^+ and Cl^- ions interact only with each other. In an NaCl crystal, each ion of a given kind has six nearest neighbors of the other kind of ion with which it interacts strongly, plus twelve neighbors of the same kind farther away, and so on. There is no reason why the equilibrium separation in each case should be the same.

19. The exclusion principle.

21. A good conductor of heat is also a good conductor of electricity, because the mechanism of transport in each case involves the electron gas in the metal.

Chapter 31

EXERCISES

1. The isotopes have the same atomic number and hence the same electron structures, therefore they have the same chemical behavior. They have different numbers of neutrons, hence different atomic masses.

3. $5n$, $5p$; $12n$, $10p$; $20n$, $16p$; $50n$, $38p$; $108n$, $72p$.

5. The limited range of the strong nuclear interaction.

7. Helium and radon cannot be combined chemically to form radium, nor can radium be broken down into helium and radon by chemical means.

9. In fission a large nucleus splits into smaller ones; in fusion small nuclei join to form a large one. In both, the product nucleus or nuclei have less mass than the initial nucleus or nuclei, with the missing mass being evolved as energy.

11. $^{206}_{82}Pb$.

13. $^{233}_{92}U$.

15. Since 1 curie $= 3.7 \times 10^{10}$ disintegrations/s, 25 millicuries $= 0.025$ curie $= 9.25 \times 10^8$ disintegrations/s.

17. One-sixteenth is left, and since $\frac{1}{2} \times \frac{1}{2} \times \frac{1}{2} \times \frac{1}{2} = \frac{1}{16}$, this means four half lives or 6400 years.

19. ^2_1H; ^1_0n; ^1_1H.

PROBLEMS

1. $(0.7553)(34.97\ u) + (0.2447)(36.97\ u) = 35.46\ u$. The discrepancy between this figure and the 35.45 u quoted in the table is due to the rounding-off of the various numbers.

3. $2m_H + 2m_n = 4.032980\ u$, $\Delta m = 0.030377\ u = 28.3$ MeV. There are four nucleons in ^4_2He, hence the binding energy per nucleon is 7.1 MeV.

5. $m_p + m_e = 1.00728\ u + 0.00055\ u = 1.00783\ u$. The difference between this and m_n is $0.00084\ u = 0.78$ MeV, which is less than the observed binding energies per nucleon in stable nuclei; hence neutrons do not decay inside nonradioactive nuclei.

7. (a) $PE = ke^2/r = 1.355 \times 10^{-13}\ J = 0.847$ MeV. (b) $m_H + 2m_n = 3.025155\ u$, $\Delta m = 0.009105\ u$, hence the binding energy of ^3_1H is 8.477 MeV; $2m_H + m_n = 3.024315$ u, $\Delta m = 0.008285\ u$, hence the binding energy of ^3_2He is 7.713 MeV. The difference is 0.764 MeV, almost the same as the electric potential energy, hence nuclear forces must be very nearly independent of charge.

9. $E = 22.4$ MeV $= 0.02406\ u$; $m_{Li} = 2m_{He} + E - m_H = 6.01516\ u$.

11. (a) $m = m_U - m_{Th} - m_{He} = 0.0059\ u = 5.5$ MeV. (b) No, because the products of the decay would total more mass than the original atom. (c) No, for the same reason.

13. (a) $N/N_0 = 1/2^{t/t_{1/2}} = 1/2^{1/10.8} = 1/\log^{-1}[(\log 2)/10.8] = 0.938 = 93.8\%$. (b) $t/t_{1/2} = 60/10.8$ and $N/N_0 = 0.0213 = 2.13\%$.

15. (a) $m = 10^{-3}$ kg, $E = mc^2 = 9 \times 10^{13}$ J. (b) 2.15×10^4 tons of TNT.

17. $\Delta m = 3m_a - m_c = 0.007809\ u = 7.27$ MeV.

Chapter 32

EXERCISES

1. The photon has no rest mass; the neutrino either has no rest mass or very little. The photon is associated with the electromagnetic interaction, whereas the neutrino is associated with the weak nuclear interaction. The neutrino has an antiparticle, whereas the photon is its own antiparticle. The photon can materialize (if it has enough energy) into an electron-positron pair, whereas the neutrino cannot materialize.

3. A μ-neutrino.

5. Energy would not be conserved.

7. No; it is possible there is a reason why quarks cannot exist except in combination with each other as hadrons or mesons.

PROBLEMS

1. $E_{min} = (m_n + m_\pi - m_p)c^2 = 142$ MeV.

3. $\Delta E = (2m_p - m_\pi)c^2 = 2.79 \times 10^{-10}$ J, $\Delta t = h/2\pi\Delta E = 3.78 \times 10^{-25}$ s; no.

5. $2m_nc^2 = 1879$ MeV; more, because the neutron mass exceeds the proton mass.

INDEX

CONVERSION FACTORS

TIME

1 hour = 60 min = 3600 s
1 day = 1440 min = 8.64×10^4 s
1 year = 365.2 days = 3.156×10^7 s

ANGLE

1 radian (rad) = 57.30° = 57°18′
1° = 0.01745 rad (180° = π rad)
1 rev/min (rpm) = 0.1047 rad/s
1 rad/s = 9.549 rpm

LENGTH

1 meter (m) = 100 cm = 39.37 in. = 3.281 ft
1 centimeter (cm) = 10 millimeters (mm)
$\qquad$ = 0.3937 in.
1 kilometer (km) = 1000 m = 0.6214 mi
1 foot (ft) = 12 in. = 0.3048 m = 30.48 cm
1 inch (in.) = 2.540 cm
1 mile (mi) = 5280 ft = 1.609 km
1 nautical mile (nmi) = 6076 ft. = 1.152 mi
$\qquad$ = 1.852 km

AREA

1 m^2 = 10^4 cm^2 = 10.76 ft^2
1 cm^2 = 10^{-4} m^2 = 0.1550 $in.^2$
1 ft^2 = 144 $in.^2$ = 9.290×10^{-2} m^2 = 929.0 cm^2
1 $in.^2$ = 6.452 cm^2
1 hectare (ha) = 10^4 m^2 = 2.471 acres
1 acre = 43,560 ft^2 = 0.4049 ha

VOLUME

1 m^3 = 10^3 liters = 10^6 cm^3 = 35.32 ft^3
1 liter = 10^3 cm^3 = 10^{-3} m^3 = 0.2642 gal
$\qquad$ = 1.056 quart
1 ft^3 = 1728 $in.^3$ = 2.832×10^{-2} m^3 = 28.32 liters
$\qquad$ = 7.481 gal
1 U.S. gallon (gal) = 4 quarts = 0.1337 ft^3
$\qquad$ = 3.785 liters

SPEED

1 m/s = 3.281 ft/s = 2.237 mi/h = 3.600 km/h
1 ft/s = 0.3048 m/s = 0.6818 mi/h = 1.097 km/h
1 km/h = 0.2778 m/s = 0.9113 ft/s
$\qquad$ = 0.6214 mi/h

1 mi/h = 1.467 ft/s = 0.4470 m/s = 1.609 km/h
1 knot = 1 nmi/h = 1.152 mi/h = 1.852 km/h
$\qquad$ = 1.688 ft/s = 0.5144 m/s

MASS

1 kilogram (kg) = 1000 grams (g) = 0.0685 slug
(*Note:* 1 kg corresponds to 2.21 lb in the sense that
the *weight* of 1 kg is 2.21 lb at sea level. Similarly 1
lb corresponds to 453.6 g and 1 oz to 28.35 g.)
1 slug = 14.59 kg
1 atomic mass unit (u) = 1.660×10^{-27} kg
$\qquad$ = 1.492×10^{-10} J = 931.5 MeV

FORCE

1 newton (N) = 0.2248 lb
1 pound (lb) = 4.448 N

PRESSURE

1 pascal (Pa) = 1 N/m^2 = 1.450×10^{-4} $lb/in.^2$
1 bar = 10^5 Pa = 14.50 $lb/in.^2$
1 $lb/in.^2$ = 144 lb/ft^2 = 6.895×10^3 Pa
1 atm = 1.013×10^5 Pa = 1.013 bar
$\qquad$ = 14.70 $lb/in.^2$
1 torr = 133.3 Pa

ENERGY

1 joule (J) = 0.7376 ft · lb = 2.390×10^4 kcal
$\qquad$ = 9.484×10^{-4} Btu
$\qquad$ = 2.778×10^{-7} kWh
1 foot-pound (ft · lb) = 1.356 J = 1.29×10^{-3} Btu
$\qquad$ = 3.25×10^{-4} kcal
1 kilocalorie (kcal) = 4185 J = 3.968 Btu
$\qquad$ = 3077 ft · lb
1 Btu = 0.252 kcal = 778 ft · lb = 1054 J
1 electron volt (eV) = 10^{-6} MeV = 10^{-9} GeV
$\qquad$ = 1.602×10^{-19} J
1 kilowatt-hour (kWh) = 3.600×10^6 J
$\qquad$ = 2.655×10^6 ft · lb